AMP-Activated Protein Kinase Signalling

AMP-Activated Protein Kinase Signalling

Special Issue Editors

Dietbert Neumann
Benoit Viollet

MDPI • Basel • Beijing • Wuhan • Barcelona • Belgrade

MDPI

Special Issue Editors

Dietbert Neumann
CARIM School of Cardiovascular Diseases
Maastricht University
The Netherlands

Benoit Viollet
Institut Cochin INSERM U1016
Department of Endocrinology
Metabolism and Diabetes (EMD)
France

Editorial Office
MDPI
St. Alban-Anlage 66
4052 Basel, Switzerland

This is a reprint of articles from the Special Issue published online in the open access journal *International Journal of Molecular Sciences* (ISSN 1422-0067) from 2018 to 2019 (available at: https://www.mdpi.com/journal/ijms/special_issues/AMPK)

For citation purposes, cite each article independently as indicated on the article page online and as indicated below:

LastName, A.A.; LastName, B.B.; LastName, C.C. Article Title. *Journal Name* **Year**, *Article Number*, Page Range.

ISBN 978-3-03897-662-2 (Pbk)
ISBN 978-3-03897-663-9 (PDF)

Contents

About the Special Issue Editors

Dietbert Neumann studied chemistry, as well as biology at the University of Cologne, Germany. After his double diploma, he worked at the NMI in Reutlingen Germany under the supervision of Dr. T. Joos. Subsequently, he joined the lab of Prof. T. Wallimann at ETH Zurich. Dietbert received his doctorate with distinction and was awarded the ETH medal. After a short postdoctoral period in Oxford, UK, under the guidance of Prof. L. N. Johnson, he returned to ETH Zurich to complete his postdoctoral training, where he became Research Group Leader at the Institute of Cell Biology. At present, he is Associate Professor at Maastricht University. His research focus is on the integration of cellular metabolism into the protein kinase signalling network, in particular AMPK. Among his major achievements is the development of a method for bacterial production of the AMPK heterotrimer, a prerequisite for its structural investigation, which was consequently adopted by virtually all AMPK enthusiasts working in structural biology labs around the globe. Dietbert Neumann is also involved in coordination of various educational activities and teaches in biomedical courses using student-centred approaches. Furthermore, he initiated the 'European Workshop on AMPK' conference series.

Benoit Viollet studied at École Supérieure de Biotechnologie de Strasbourg (ESBS), a state school for biotechnology within the University of Strasbourg. After obtaining his M.Sc. and engineering degree in biotechnology, he undertook his Ph.D. research at the INSERM U129 in Paris, under the supervision of Prof Axel Kahn. He obtained a doctorate in cellular and molecular biology at the University Pierre and Marie Curie. After brief postdoctoral training at IGBMC in Strasbourg in the group of Jean-Marc Egly, he returned to Paris to Institut Pasteur for a post-doctoral fellowship in the group of Moshe Yaniv. Since his appointment at INSERM, he is a group leader at Institut Cochin/INSERM U1016/CNRS UMR 8104/ University Paris Descartes where his research is focused on the pathophysiology and the treatment of obesity and type 2 diabetes. He is currently working to decipher the role of the energy sensor AMPK in the regulation of cellular energy balance and how AMPK integrates stress responses such as exercise, as well as nutrient and hormonal signals, into control energy expenditure and substrate utilization at the whole body level. He edited a number of journal/book series and organized international meetings focusing on AMPK

International Journal of
Molecular Sciences

MDPI

Editorial

AMP-Activated Protein Kinase Signalling

Dietbert Neumann [1,*] and Benoit Viollet [2,3,4,*]

[1] Department of Pathology, CARIM School for Cardiovascular Diseases, Faculty of Health, Medicine and Life Sciences, Maastricht University, 6200 MD Maastricht, The Netherlands

[2] INSERM U1016, Institut Cochin, Department of Endocrinology, Metabolism and Diabetes (EMD), 24 rue du faubourg Saint-Jacques, 75014 Paris, France

[3] CNRS, UMR8104, 75014 Paris, France

[4] Université Paris Descartes, Sorbonne Paris Cité, 75014 Paris, France

* Correspondence: d.neumann@maastrichtuniversity.nl (D.N.); benoit.viollet@inserm.fr (B.V.); Tel.: +31-43-387-7167 (D.N.); +33-1-4441-2401 (B.V.)

Received: 29 January 2019; Accepted: 4 February 2019; Published: 12 February 2019

AMP-activated protein kinase (AMPK) regulates energy homeostasis in eukaryotic cells and organisms. As such, AMPK has attracted enormous interest in various disciplines. Accordingly, the current Special Issue "AMP-Activated Protein Kinase Signalling" is a present-day reflection of the field covering a wide area of research. Although widely conserved throughout evolution and expressed ubiquitously, the functions of AMPK in different tissues and cell types may vary to some extent. Therefore, it is worthwhile to focus on cellular functions of various different origins, or tissues and organs, as well as their interplay in the context of the whole organism. This Special Issue includes research articles and reviews addressing AMPK regulation and function in all biological organization levels in health and disease.

Starting from the AMPK molecule, Yan et al. summarize the knowledge derived from crystal structures and provide expert insight into the molecular mechanisms of kinase activity modulation by adenine nucleotides [1]. As presented, recent research provided a detailed understanding of the molecular mechanisms leading to allosteric activation. The binding of AMP changes the AMPK's conformational landscape, providing direct AMPK activation and protection against dephosphorylation of Thr-172 within the activation loop within the catalytic subunit. AMP bound at the cystathionine β-synthetase 3 (CBS3) nucleotide binding site within the regulatory AMPKγ subunit interacts with the flexible α-linker from the catalytic α subunit to transduce the adenine-binding signal to the kinase domain. The question arises as to how binding of AMP can inhibit dephosphorylation of Thr-172, while at the same time improving access to upstream kinases that phosphorylate the same site. Structural insight explaining the observed inhibition of AMPK by ATP is lacking at present and is identified as a key for understanding the regulation of AMPK activation loop phosphorylation.

Apart from the allosteric mode of regulation, AMPK is part of a kinase cascade. Upstream regulation of AMPK involves one of several kinases capable of phosphorylating AMPK at Thr-172 in the α-subunit. Both liver kinase B1 (LKB1) and Ca^{2+}/Calmodulin-dependent protein kinase kinase 2 (CaMKK2) are firmly established as physiological upstream kinases of AMPK. In addition, transforming growth factor β (TGF-β)-activated kinase 1 (TAK1) has also been reported as AMPK upstream kinase, but did not receive full attention as discussed in detail [2]. The historical origin of the conflict between researchers accepting TAK1 as a possible direct upstream kinase of AMPK and those rejecting this option is explained. Arguments from both sides lead to the conclusion that TAK1 should be accepted as a genuine contextual AMPK upstream kinase. Notably, the same contextual restriction applies to LKB1 and CaMKK2, which, depending on cell type, energy status, and environmental signal, act as alternative AMPK kinases.

As reviewed by Janzen et al., in skeletal muscle, AMPK activity is regulated by glycogen content [3]. Glycogen physically binds AMPK, modifying its conformation to inhibit its activity.

Vice versa, AMPK activity impacts glycogen storage dynamics to modulate exercise metabolism. In a monograph, Thomson summarizes AMPK signal integration in regulating skeletal muscle growth and atrophy [4]. Thomson suggests that activation of AMPKα1 mainly limits muscle growth, for example, by inhibiting protein synthesis, whereas AMPKα2 activation may play a more important role in muscle degradation, for example, through accelerating autophagy. Because a lack of AMPKα1 also inhibits muscle regeneration after injury, AMPKα1 may further have a mandatory function in regulating satellite cell dynamics. In general agreement, Vilchinskaya et al. describe AMPK as a key trigger in disuse-induced skeletal muscle remodelling [5]. In a mouse model overexpressing dominant negative AMPKα1 in skeletal muscle, Egawa et al. confirm the role of AMPK in muscle mass regulation upon unloading and reloading, but do not find evidence for AMPK involvement in fibre type switching [6]. The application of AMPK activating drugs to increase insulin sensitivity for improved glucose uptake in skeletal muscle is also a promising key therapeutic strategy to treat diabetes. Earlier results suggested the possible requirement of a serum factor in the insulin-sensitizing effect of the widely used AMPK activator 5-amino-imidazole-4-carboxamide ribonucleotide (AICAR). Jørgensen et al. clarify this issue by showing in mouse skeletal muscle that the beneficial effect of AICAR stimulation on downstream insulin signalling was not dependent on the presence of a serum factor [7].

Previous studies also reported that AMPK activation improved insulin sensitivity in endothelial cells for modulation of vascular homeostasis. Strembitska et al. investigate insulin-stimulated Akt phosphorylation in response to the AMPK activators AICAR, 991, and A-769662, the latter two chemicals targeting a different part of the AMPK molecule, the ADaM (allosteric drug and metabolism) site [8]. However, Strembitska et al., besides AMPK activation, observed AMPK-independent effects of A-769662 in human umbilical vein endothelial cells (HUVECs) and human aortic endothelial cells (HAECs) [8]. Namely, inhibition of insulin-stimulated Akt phosphorylation and nitric oxide (NO) synthesis by A-769662 was seen in the presence of AMPK inhibitor SBI-0206965. A-769662 also inhibited insulin-stimulated Erk1/2 phosphorylation in mouse embryo fibroblasts (MEFs) and in HAECs, which was independent of AMPK in MEFs, indicating that data obtained using this compound should be interpreted with caution [8]. Contradicting results have also been reported for AMPK-dependent regulation of endothelial NO synthase (eNOS). In endothelial cells, Zippel et al. observe AMPK-dependent inhibition of endothelial NO formation. The data provided suggest that AMPK targets Thr495 of eNOS, the inhibitory site, rather than Ser1177 (which would accelerate NO production) [9]. Notably, Zippel et al. applied genetic models of AMPK deficiency (CRISPR/Cas and mouse knockouts) and mutated eNOS at respective phosphorylation sites before incubating with AMPK in vitro, thus providing strong support for their physiological and mechanistic claims.

Hypertension and kidney disease can be a consequence of suboptimal early-life conditions, that is, by renal programming. Tain and Hsu bring forward the argument that AMPK activators could be applied for renal reprogramming as a protection against disease development [10]. Glosse and Föller review the involvement of AMPK in the regulation of renal transporters [11]. Without a particular focus on the kidney, but with relevance also for renal function, Rowart et al. describe the role of AMPK in the formation of epithelial tight junctions [12]. In particular, the authors discuss the contribution of AMPK in Ca^{2+}-induced assembly of tight junctions.

Over the past decade, AMPK has emerged as a key player in the regulation of whole-body energy homeostasis. AMPK regulates food intake and integrates energy metabolism with several hormones, such as leptin, adiponectin, ghrelin, and insulin [13]. In this review, Wang et al. summarize the role of hypothalamic AMPK in hormonal regulation of energy balance. Nutrient intake, on the other hand, may regulate AMPK activation status. Lyons and Roche discuss the impact of dietary components on AMPK activity [14]. The authors review the evidence of whether specific nutrients and non-nutrient food components modulate AMPK-dependent processes relating to metabolism and inflammation, thus affecting the development of type 2 diabetes and obesity. Pointing out that the reported effects of diet on AMPK are mostly based on animal studies, the authors plead for further investigation in human studies. Resveratrol is one such nutritional substance that has been described as an AMPK activator. Trepiana et al. review the involvement of AMPK in the effects of resveratrol and its derivatives in the context of liver steatosis [15]. Although AMPK activation may only partly explain the preventive and therapeutic effects, the authors conclude that resveratrol represents a potential interesting approach to treat lipid accumulation in liver. Foretz et al. add further support for the potential of AMPK-dependent remodelling of lipid metabolism by providing in vivo evidence for increased fatty acid oxidation and reduced lipid content in mouse liver expressing constitutive active AMPK [16].

Apart from acute effects on the activity of enzymes or localization of proteins, AMPK has also been shown to change gene expression patterns for long-term adaptation involving regulation of transcription factors and chromatin remodelling. Gongol et al. describe AMPK as a key player in epigenetic regulation and discuss the consequent physiological and pathophysiological implications [17]. AMPK is involved in regulation of protein acetylation and itself receives regulation by acetylation, as reviewed by Vancura et al. [18]. Apart from epigenetic and transcriptional regulation, the acetylating and deacetylating events are linked to cellular metabolism, all of which in part is controlled by AMPK. A weighted gene co-expression network analysis was carried out to investigate the interaction of AMPK and autophagy gene products during adipocyte differentiation [19]. In fact, differentiation of cells by definition involves cellular remodelling and thus may generally require autophagy, which could be linked to AMPK. Indeed, AMPK has been recognized as a major driver of autophagy, as reviewed by Tamargo-Gómez and Mariño [20]. Jacquel et al. summarize the evidence that AMPK regulates myeloid differentiation [21]. Because autophagy appears to support myeloid differentiation, the authors suggest investigating the potential of AMPK activators as an anti-leukemic strategy.

Long-term memory depends on the induction of immediate early genes (IEGs). Didier et al. report that AMPK controls the expression of IEGs upon synaptic activation via the cAMP-dependent protein kinase (PKA)/cAMP response element binding (CREB) signalling pathway [22]. Although genetic evidence suggests the requirement of AMPK, the mechanism through which AMPK may regulate PKA activation remains elusive. The authors speculate that AMPK may be required to maintain ATP levels, as a requirement for formation of cyclic AMP. Thus, AMPK may play an indirect role in PKA activation upon synaptic activation.

While many studies focused their attention on the tumour-suppressor effect of AMPK activation, there is now growing evidence that AMPK plays a dual role in cancer, that is, inhibiting growth but enhancing survival. Adding to this discussion, Zhang et al. show that loss of AMPKα2 impairs sonic hedgehog medulloblastoma tumorigenesis [23]. Silwal et al. review the function of AMPK in host defence against infections [24]. As pointed out by the authors, AMPK also plays a dual role, suppressive or supportive for viral infections, depending on the type of virus. The role of AMPK in adaptive and innate immune response to infection of microbial and parasitic infections is also discussed.

Human reproduction represents a less mature field of AMPK research. Martin-Hidalgo et al. review the known cellular roles of AMPK in spermatozoa [25]. The argument is made that AMPK acts as key molecule linking the sperm's energy metabolism and ability to fertilize. In the context of pregnancy complications in humans, Kumagai et al. discuss the possibility of further investigating AMPK activators as a treatment in a subset of conditions [26]. In their perspective, the authors discuss the possibility of AMPK regulation by catechol-*O*-methyltransferase (COMT).

In summary, the current Special Issue provides a representative cross-section of AMPK research and topical reviews. Thanks to the authors submitting their precious work and insights that are presented in this Special Issue, our understanding of AMPK structure, function, and regulation has further progressed. Additionally, it turns out that AMPK biology is more complex than most of us originally anticipated, leading many of the contributing authors to highlight the fact that we still lack information and need to address new questions in subsequent studies. For example, the molecular structure of AMPK, although studied in great detail, does not provide information on the dynamic movements that are inherent to an allosteric enzyme. Moreover, "AMPK" is a heterogeneous mixture of twelve different heterotrimeric complexes (αβγ combinations of α1, α2, β1, β2, γ1, γ2, and γ3) without considering splice variants. The concept emerges that isoforms of AMPK localized at different subcellular compartments may respond to specific cues and regulate only a subset of cellular processes that are now collectively attributed to AMPK. Indeed, AMPK isoform selectivity to specific substrates may arise from a compartmentalized AMPK signalling, rather than from distinct intrinsic kinase substrate specificity. Hence, the spatiotemporal regulation of individual AMPK complexes in various tissues and metabolic conditions awaits further clarification. Furthermore, the development of AMPK activating drugs is constantly progressing behind the scenes, holding more promise than ever for the possible treatment of human disease. AMPK research does not stand still. The knowledge about AMPK accordingly will steadily increase. Besides, the variety of research topics relating to AMPK may continue to evolve. As we are already working on the next edition, we encourage the reader to consider submission of their upcoming AMPK-focused work to the successor Special Issue entitled "AMP-Activated Protein Kinase Signalling 2.0".

Conflicts of Interest: The authors report no conflict of interest.

References

1. Yan, Y.; Zhou, X.E.; Xu, H.E.; Melcher, K. Structure and physiological regulation of AMPK. *Int. J. Mol. Sci.* **2018**, *19*, 3534. [CrossRef]
2. Neumann, D. Is TAK1 a direct upstream kinase of AMPK? *Int. J. Mol. Sci* **2018**, *19*, 2412. [CrossRef]
3. Janzen, N.R.; Whitfield, J.; Hoffman, N.J. Interactive roles for AMPK and glycogen from cellular energy sensing to exercise metabolism. *Int. J. Mol. Sci.* **2018**, *19*, 3344. [CrossRef] [PubMed]
4. Thomson, D.M. The role of AMPK in the regulation of skeletal muscle size, hypertrophy, and regeneration. *Int. J. Mol. Sci.* **2018**, *19*, 3125. [CrossRef] [PubMed]
5. Vilchinskaya, N.A.; Krivoi, I.; Shenkman, B.S. AMP-activated protein kinase as a key trigger for the disuse-induced skeletal muscle remodeling. *Int. J. Mol. Sci.* **2018**, *19*, 3558. [CrossRef] [PubMed]
6. Egawa, T.; Ohno, Y.; Goto, A.; Yokoyama, S.; Hayashi, T.; Goto, K. AMPK mediates muscle mass change but not the transition of myosin heavy chain isoforms during unloading and reloading of skeletal muscles in mice. *Int. J. Mol. Sci.* **2018**, *19*, 2954. [CrossRef] [PubMed]
7. Jorgensen, N.O.; Wojtaszewski, J.F.P.; Kjobsted, R. Serum is not necessary for prior pharmacological activation of AMPK to increase insulin sensitivity of mouse skeletal muscle. *Int. J. Mol. Sci.* **2018**, *19*, 1201. [CrossRef]
8. Strembitska, A.; Mancini, S.J.; Gamwell, J.M.; Palmer, T.M.; Baillie, G.S.; Salt, I.P. A769662 inhibits insulin-stimulated Akt activation in human macrovascular endothelial cells independent of AMP-activated protein kinase. *Int. J. Mol. Sci.* **2018**, *19*, 3886. [CrossRef]
9. Zippel, N.; Loot, A.E.; Stingl, H.; Randriamboavonjy, V.; Fleming, I.; Fisslthaler, B. Endothelial AMP-activated kinase alpha1 phosphorylates eNOS on Thr495 and decreases endothelial NO formation. *Int. J. Mol. Sci.* **2018**, *19*, 2753. [CrossRef]
10. Tain, Y.L.; Hsu, C.N. AMP-activated protein kinase as a reprogramming strategy for hypertension and kidney disease of developmental origin. *Int. J. Mol. Sci.* **2018**, *19*, 1744. [CrossRef]
11. Glosse, P.; Foller, M. AMP-activated protein kinase (AMPK)-dependent regulation of renal transport. *Int. J. Mol. Sci.* **2018**, *19*, 3481. [CrossRef] [PubMed]
12. Rowart, P.; Wu, J.; Caplan, M.J.; Jouret, F. Implications of AMPK in the formation of epithelial tight junctions. *Int. J. Mol. Sci.* **2018**, *19*, 2040. [CrossRef] [PubMed]
13. Wang, B.; Cheng, K.K. Hypothalamic AMPK as a mediator of hormonal regulation of energy balance. *Int. J. Mol. Sci.* **2018**, *19*, 3552. [CrossRef] [PubMed]
14. Lyons, C.L.; Roche, H.M. Nutritional modulation of AMPK-impact upon metabolic-inflammation. *Int. J. Mol. Sci.* **2018**, *19*, 3092. [CrossRef] [PubMed]
15. Trepiana, J.; Milton-Laskibar, I.; Gomez-Zorita, S.; Eseberri, I.; Gonzalez, M.; Fernandez-Quintela, A.; Portillo, M.P. Involvement of 5′-activated protein kinase (AMPK) in the effects of resveratrol on liver steatosis. *Int. J. Mol. Sci.* **2018**, *19*, 3473. [CrossRef] [PubMed]
16. Foretz, M.; Even, P.C.; Viollet, B. AMPK activation reduces hepatic lipid content by increasing fat oxidation in vivo. *Int. J. Mol. Sci.* **2018**, *19*, 2826. [CrossRef] [PubMed]

17. Gongol, B.; Sari, I.; Bryant, T.; Rosete, G.; Marin, T. Ampk: An epigenetic landscape modulator. *Int. J. Mol. Sci.* **2018**, *19*, 3238. [CrossRef]
18. Vancura, A.; Nagar, S.; Kaur, P.; Bu, P.; Bhagwat, M.; Vancurova, I. Reciprocal regulation of AMPK/SNF1 and protein acetylation. *Int. J. Mol. Sci.* **2018**, *19*, 3314. [CrossRef]
19. Ahmed, M.; Hwang, J.S.; Lai, T.H.; Zada, S.; Nguyen, H.Q.; Pham, T.M.; Yun, M.; Kim, D.R. Co-expression network analysis of AMPK and autophagy gene products during adipocyte differentiation. *Int. J. Mol. Sci.* **2018**, *19*, 1808. [CrossRef]
20. Tamargo-Gomez, I.; Marino, G. AMPK: Regulation of metabolic dynamics in the context of autophagy. *Int. J. Mol. Sci.* **2018**, *19*, 3812. [CrossRef]
21. Jacquel, A.; Luciano, F.; Robert, G.; Auberger, P. Implication and regulation of AMPK during physiological and pathological myeloid differentiation. *Int. J. Mol. Sci.* **2018**, *19*, 2991. [CrossRef] [PubMed]
22. Didier, S.; Sauve, F.; Domise, M.; Buee, L.; Marinangeli, C.; Vingtdeux, V. AMP-activated protein kinase controls immediate early genes expression following synaptic activation through the pka/creb pathway. *Int. J. Mol. Sci.* **2018**, *19*, 3716. [CrossRef] [PubMed]
23. Zhang, H.; Kuick, R.; Park, S.S.; Peabody, C.; Yoon, J.; Fernandez, E.C.; Wang, J.; Thomas, D.; Viollet, B.; Inoki, K.; et al. Loss of AMPKalpha2 impairs hedgehog-driven medulloblastoma tumorigenesis. *Int. J. Mol. Sci.* **2018**, *19*, 3287. [CrossRef] [PubMed]
24. Silwal, P.; Kim, J.K.; Yuk, J.M.; Jo, E.K. AMP-activated protein kinase and host defense against infection. *Int. J. Mol. Sci.* **2018**, *19*, 3495. [CrossRef] [PubMed]
25. Martin-Hidalgo, D.; Hurtado de Llera, A.; Calle-Guisado, V.; Gonzalez-Fernandez, L.; Garcia-Marin, L.; Bragado, M.J. AMPK function in mammalian spermatozoa. *Int. J. Mol. Sci.* **2018**, *19*, 3293. [CrossRef] [PubMed]
26. Kumagai, A.; Itakura, A.; Koya, D.; Kanasaki, K. AMP-activated protein (AMPK) in pathophysiology of pregnancy complications. *Int. J. Mol. Sci.* **2018**, *19*, 3076. [CrossRef] [PubMed]

International Journal of
Molecular Sciences

MDPI

Review

Structure and Physiological Regulation of AMPK

Yan Yan [1,2], X. Edward Zhou [1], H. Eric Xu [1,2] and Karsten Melcher [1,*]

[1] Center for Cancer and Cell Biology, Van Andel Research Institute, 333 Bostwick Ave. N.E., Grand Rapids,
    MI 49503, USA; yan.yan@vai.org (Y.Y.); edward.zhou@vai.org (X.E.Z.); eric.xu@vai.org (H.E.X.)
[2] VARI/SIMM Center, Center for Structure and Function of Drug Targets, CAS-Key Laboratory of Receptor
    Research, Shanghai Institute of Materia Medica, Chinese Academy of Sciences, Shanghai 201203, China
*   Correspondence: Karsten.melcher@vai.org; Tel.: +1-616-234-5699

Received: 17 October 2018; Accepted: 6 November 2018; Published: 9 November 2018

Abstract: Adenosine monophosphate (AMP)-activated protein kinase (AMPK) is a heterotrimeric
αβγ complex that functions as a central regulator of energy homeostasis. Energy stress manifests
as a drop in the ratio of adenosine triphosphate (ATP) to AMP/ADP, which activates AMPK's
kinase activity, allowing it to upregulate ATP-generating catabolic pathways and to reduce
energy-consuming catabolic pathways and cellular programs. AMPK senses the cellular energy
state by competitive binding of the three adenine nucleotides AMP, ADP, and ATP to three sites in
its γ subunit, each, which in turn modulates the activity of AMPK's kinase domain in its α subunit.
Our current understanding of adenine nucleotide binding and the mechanisms by which differential
adenine nucleotide occupancies activate or inhibit AMPK activity has been largely informed by crystal
structures of AMPK in different activity states. Here we provide an overview of AMPK structures,
and how these structures, in combination with biochemical, biophysical, and mutational analyses
provide insights into the mechanisms of adenine nucleotide binding and AMPK activity modulation.

Keywords: energy metabolism; AMPK; activation loop; AID; α-linker; β-linker; CBS; LKB1;
CaMKK2; αRIM

1. AMPK Is a Master Regulator of Energy Homeostasis That Is Dysregulated in Disease

AMPK is the primary energy sensor and regulator of energy homeostasis in eukaryotes. It is
activated by energy stress in response to increased ATP consumption (e.g., exercise, cell proliferation,
anabolism) or decreased ATP production (e.g., low glucose levels, oxidative stress, hypoxia), which are
sensed as low ratios of ATP to AMP and ADP. Upon activation, AMPK phosphorylates downstream
targets to directly or indirectly modulate the activities of rate-limiting metabolic enzymes, transcription
and translation factors, proliferation and growth pathways, and epigenetic regulators. Collectively,
this increases oxidative phosphorylation, autophagy, and uptake and metabolism of glucose and fatty
acids, and decreases the synthesis of fatty acids, cholesterol, proteins, and ribosomal RNAs (rRNAs),
as well as decreasing cell growth and proliferation [1–6]. Due to its central roles in metabolism,
AMPK is dysregulated in diabetes, obesity, cardiometabolic disease, and cancer, and it is a promising
pharmacological target [1,2,5,7–10], especially for the treatment of type 2 diabetes [11–13].

2. AMPK Consists of a Stable Core Attached to Moveable Domains

AMPK is a heterotrimeric αβγ protein kinase. In mammals, it is encoded by two alternative α
subunits (α1 and α2), two alternative β subunits (β1 and β2), and three alternative γ subunits (γ1, γ2,
and γ3) that can form up to 12 different αβγ isoforms [14]. The α subunits contain a canonical Ser/Thr
kinase domain (KD), an autoinhibitory domain (AID), an adenine nucleotide sensor segment termed
an α-linker, and a β subunit-interacting C-terminal domain (α-CTD), the latter of which contains
the ST loop, which harbors proposed phosphorylation sites for AKT [15], PKA [16], and GSK [17].

The β subunits are composed of a myristoylated, unstructured N-terminus, a glycogen-binding carbohydrate-binding module (CBM), a scaffolding C-terminal domain (β-CTD) that interacts with both the γ subunit, and the α-CTD, and the extended β-linker loop that connects the CBM with the β-CTD (Figure 1A,B). The three alternative γ subunits consist of N-termini of different lengths and unknown function, followed by a conserved adenine nucleotide-binding domain that contains four cystathione β-synthetase (CBS) AMP/ADP/ATP binding sites (Figure 1). CBS1, 3, and 4 are functional, whereas in CBS2, the ribose-binding Asp residue is replaced by an Arg, and no nucleotide binding has been observed for CBS2 in heterotrimer structures.

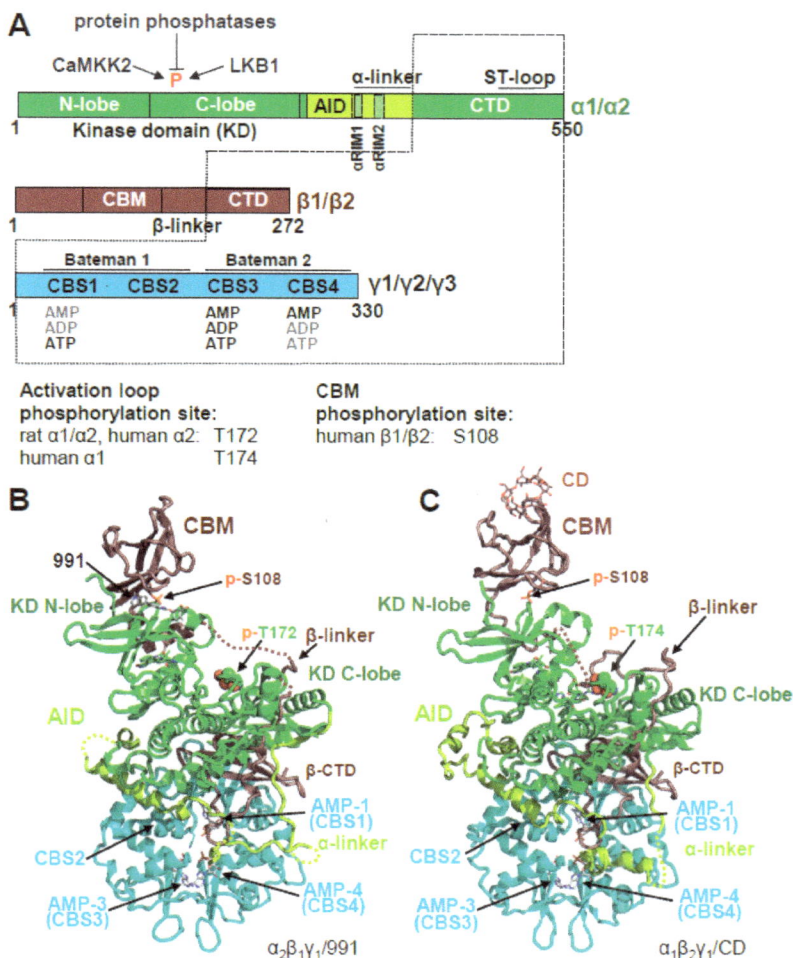

Figure 1. Overall structure of human adenosine monophosphate (AMP)-activated protein kinase (AMPK). (**A**). Domain structure and AMPK isoforms. Activation loop and carbohydrate-binding module (CBM) phosphorylation sites of different isoforms are indicated below the domain map (**B,C**). Crystal structures of phosphorylated, AMP-bound AMPK $\alpha_2\beta_1\gamma_1$/991 ((**B**); PDB: 4CFE) and $\alpha_1\beta_2\gamma_1$/cyclodextrin (CD) ((**C**); PDB: 4RER).

AMPK is a highly dynamic complex with a stable core formed by the γ subunit and the α- and β-CTDs, in which the β-CTD is sandwiched between the α and γ subunits (Figure 1A, core highlighted by dotted lines). Attached to the core are moveable domains whose position is determined by ligand binding and posttranslational modifications. As such, the holo-complex cannot be crystallized in the absence of multiple stabilizing ligands and/or protein engineering. Consequently, the first structures of AMPK consisted of isolated domains, e.g., the KD [18–21], the CBM bound to the glycogen mimic cyclodextrin [22], the yeast and mammalian nucleotide-bound scaffolding cores [23–26], the AID [27], and the yeast KD–AID complex [21] (Figure 2).

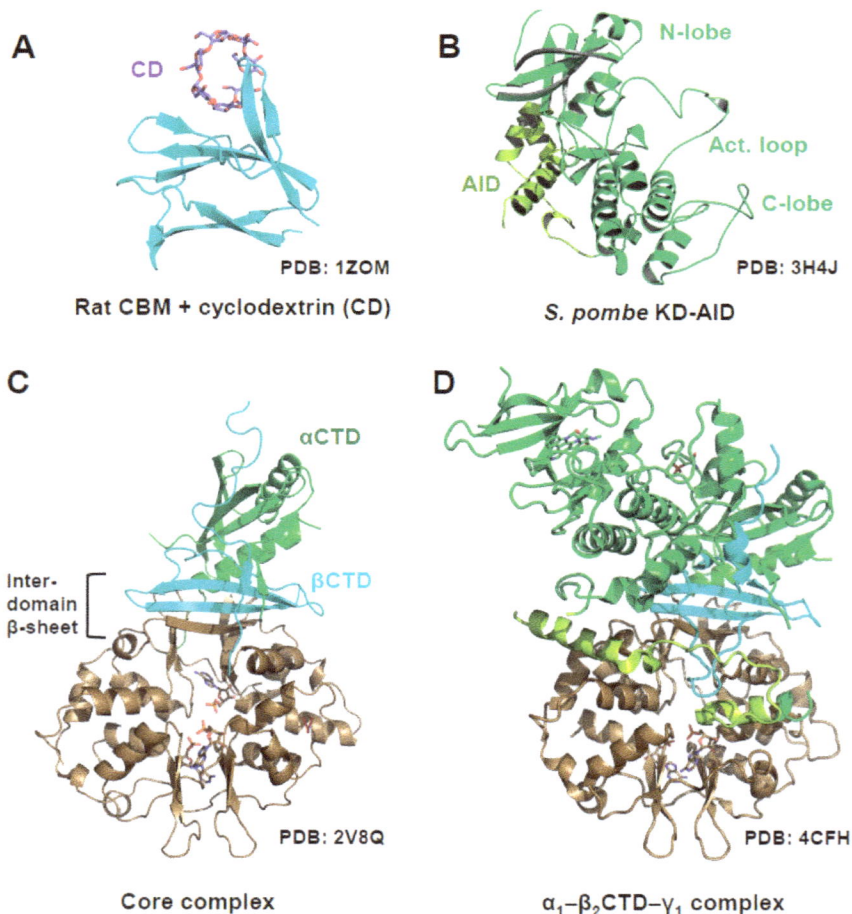

Figure 2. Structure of AMPK domains and subcomplexes. (**A**) Rat CBM bound to cyclodextrin; (**B**) Fission yeast kinase domain–autoinhibitory domain (KD-AID) complex; (**C**) AMP-bound, phosphorylated mammalian AMPK core complex (rat α_1-human β_2-rat γ_1); (**D**) AMP-bound, phosphorylated rat α_1—human β_2CTD—rat γ_1 complex.

Activation Loop Phosphorylation Orchestrates the Catalytic Center for Phosphoryl Transfer

Kinase domains have a highly conserved structure consisting of a smaller N-terminal lobe (N-lobe), composed of a β-sheet and the αB and αC helices, and a larger α-helical C-terminal lobe (C-lobe; see Figures 1B and 2B). The cleft between the lobes is the binding site for substrate peptides and Mg^{2+}–ATP. The two lobes are separated by a flexible hinge at the back that allows them to move towards each other to cycle through substrate-accessible open and catalytically-competent closed conformations as part of the kinase catalytic cycle. Key regulatory elements of the KD are: (i) the activation loop at the entrance of the catalytic cleft; (ii) the αC helix in the N-lobe, which positions the ATP-binding lysine (K47 in human α1) and the Mg^{2+}-binding DFG (Asp-Phe-Gly) loop; and (iii) the peptide substrate-binding catalytic loop in the C-lobe (Figure 3) [28–30].

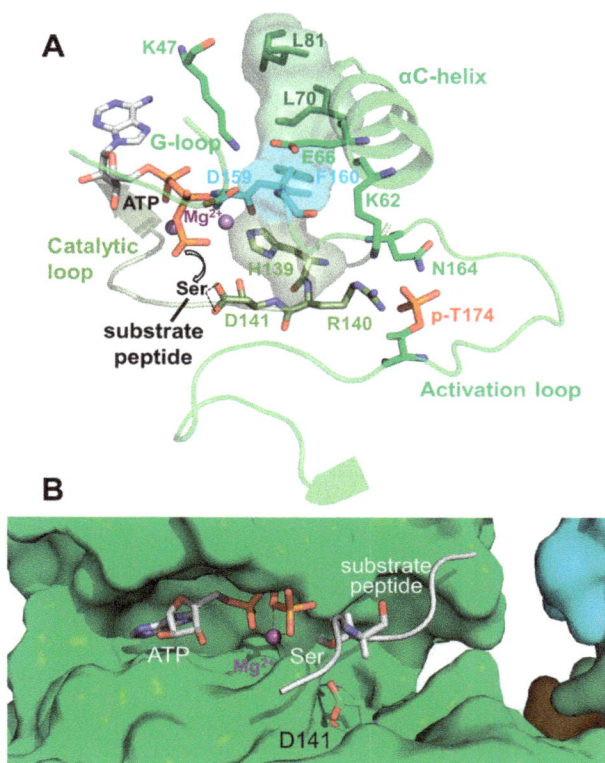

Figure 3. Active protein kinase catalytic cleft. (**A**) Key residues and structural elements of phosphorylated AMP-bound $\alpha_1\beta_2\gamma_1$ AMPK (4RER). Active kinase structures are characterized by a precisely positioned set of motifs for substrate- and adenosine triphosphate (ATP)-binding, in which four residues (L70, L81, H139, F160; shown in stick plus translucent surface presentation) are stacked against each other to form a regulatory spine. In this conformation, the activation loop p-T174 (p-T172 in human α2) positions R140 and D141 from the catalytic loop for peptide substrate binding, and K62 from the αC-helix for aligning the ATP-binding K47 and the Mg^{2+}-binding DFG loop. The AMPK active protein kinase cleft resembles the canonical protein kinase A (PKA) site. To better visualize the active structure, we modeled the serine residue of a substrate peptide and the co-substrate ATP from the structure of PKA (PDB: 1ATP) in the catalytic cleft. Spheres: Mg^{2+} ions. (**B**) Surface presentation of the AMPK catalytic cleft (4RER) overlaid with a stick model of the aligned substrate peptide and ATP from the structure of substrate-bound CDK2 (PDB: 1QMZ). The Ser hydroxyl-positioning AMPK D141 is shown in green stick representation.

AMPK belongs to the RD (Arg-Asp) kinases, in many of which phosphorylation stabilizes the activation loop through a charge interaction between the negatively charged activation loop phosphate, and the positively charged residues from the αC helix (K62 in AMPK α1), the activation loop (N164), and the catalytic loop (R140). This conformation in turn stabilizes the αC helix and positions the arginine (R) and adjacent aspartate (D) of the catalytic loop for substrate binding (Figure 3). The hallmark of active protein kinases is therefore a precisely positioned set of motifs for substrate- and ATP-binding, in which four residues from the catalytic loop (H139), the Mg^{2+}-binding DFG loop (F160), the αC helix (L70) and the αC-αD loop (L81) are stacked against each other [28–30], as found in structures of active AMPK (Figure 3).

3. AMPK Is Activated Both by Direct Allosteric Activation and by Increasing Net Activation Loop Phosphorylation

AMPK activity is regulated at three different levels: at the level of (i) activation loop phosphorylation by upstream kinases, (ii) protection against activation loop dephosphorylation by protein phosphatases, and (iii) at the level of phosphorylation-independent, allosteric kinase activation (Figure 1A). Activation loop phosphorylation increases the AMPK activity by about 100-fold, while allosteric regulation changes AMPK activity up to ten-fold in mammalian cells and about two-fold in recombinant, bacterially produced AMPK [24,31–33]. AMP activates, and ATP inhibits, AMPK through all three mechanisms. ADP more weakly protects against activation loop dephosphorylation, does not allosterically activate AMPK [33–36], and it may not stimulate activation loop phosphorylation [33,37], although the latter is controversial [36].

The two main mammalian AMPK activation loop-phosphorylating kinases are the tumor suppressor LKB1 in complex with STRAD and MO25, and Ca^{2+}/calmodulin-dependent protein kinase kinase β (CaMKK2) [38–42]. While CaMKK2 mediates Ca^{2+}-dependent AMPK phosphorylation, AMP binding to the γ subunit increases activation loop phosphorylation through LKB1 by inducing a conformation that stabilizes formation of a complex between myristoylated AMPK, Axin, and LKB1/STRAD/MO25 [37,43]. However, the structural details of this interaction remain unknown. In addition, activation loop phosphorylation is also modulated by phosphorylation of the ST loop [15–17] and by ubiquitination of AMPK [44] and LKB1 [45].

In addition to adenine nucleotides, glucose, glycogen, and nicotinamide adenine dinucleotides are also important energy metabolites. Glucose has recently been identified as an important AMPK activity regulator, but it does so without direct AMPK binding [43,46]. In contrast, both glycogen and NADPH and NADH can directly bind AMPK: glycogen at the CBM [22,47], and in a reconstituted system, NADPH and NADH at the adenine nucleotide sensor site CBS3 [34,48]. However, the physiological relevance of the glycogen [47,49,50] and NADPH/NADH [34,48] interactions for AMPK activity regulation remains unclear.

Finally, a number of pharmacological activators bind AMPK at a unique site at the interface between CBM and KD (so called allosteric drug and metabolite [ADaM] site), as first shown for Merck compound 991 [51], and derivatives of the Abbot compound A769662 [52]. Binding greatly stabilizes the association of the highly dynamic CBM with the KD [53], an interaction that is also modulated by CBM phosphorylation and carbohydrate binding [49,53]. ADaM site agonists activate AMPK both directly and through increased protection against activation loop dephosphorylation, whose structural details will be covered in detail in a separate article in this issue.

Besides activity regulation, the level of AMPK is regulated by ubiquitination and proteasomal degradation in brown adipose tissue [54], testis [55], certain cancers [55,56], and in the presence of high levels of glucose [57].

3.1. The γ Subunit Contains Three Functional Adenine Nucleotide Binding Sites

The structure of the yeast and mammalian AMPK core scaffolds revealed a disk-shaped γ subunit composed of four CBS sites. Each CBS consists of a strand-helix-strand-strand-helix fold (β1-α1-β2-β3-α2) with long intervening loops (Figure 4A). β1 is often incomplete, but where present, it forms a three-stranded sheet with the two central β-strands (β2 and β3). The β-sheet of one CBS packs parallel with the sheet of a neighboring CBS. The interface between the two sheets forms two clefts, one on the top flat side and one on the bottom flat side of the disk, which are the binding sites for adenine nucleotides (Figure 4C,D). Therefore, each binding site requires a tandem CBS pair to form a functional unit termed the Bateman domain (CBS1 + CBS2 = Bateman domain 1, CBS3 + CBS4 = Bateman domain 2). The structures of the core complexes in the presence of AMP [24,58], ADP [34], or ATP [24,58] revealed adenine nucleotide binding at three sites in mammalian AMPK: CBS1, CBS3, and CBS4.

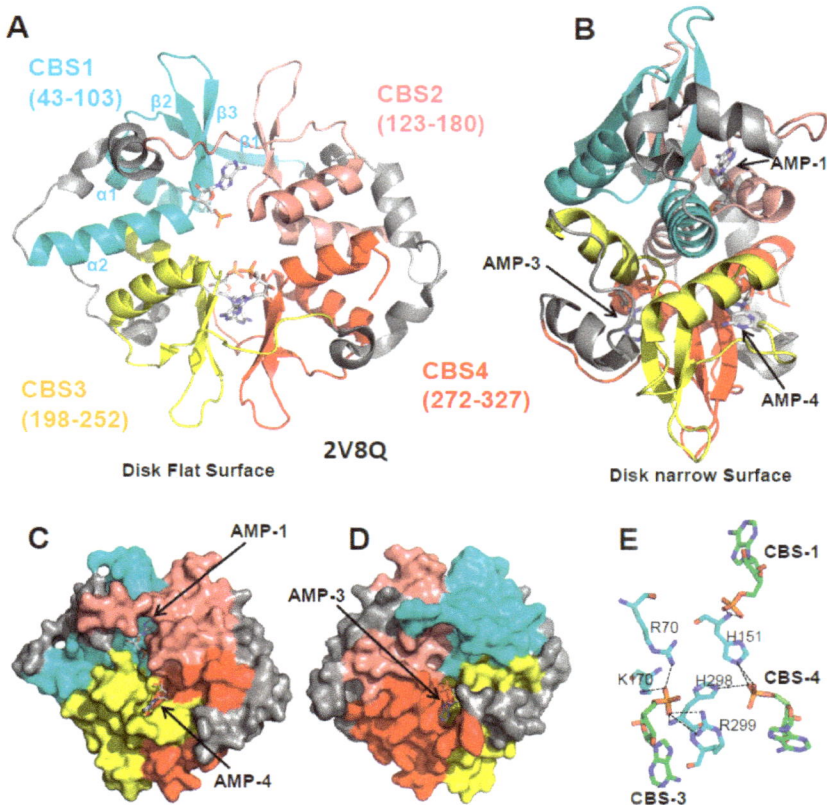

Figure 4. AMP binds three of the four CBS sites of the γ subunit. (**A,B**) Cartoon representation of the γ subunit in two different orientations. AMP molecules are shown in stick representation. The four CBS sites are shown in different colors with the secondary structure elements of CBS1 labeled. (**C,D**) Surface representation of the front and back sides of the disk flat surfaces illustrating the AMP-occupied binding pockets 1, 3, and 4, and the empty CBS2 pocket. (**E**) The phosphate groups (orange) of the three AMP molecules (cyan C atoms) coordinately interact with a set of polar γ subunit residues (green C atoms); O: red, N: blue.

3.2. CBS3 Is the Adenine Nucleotide Sensor Site

While the structure of the core complexes revealed how adenine nucleotides bind the γ subunit, they did not provide information on how the binding signal is transduced to the KD in the α subunit. In 2011, the Gamblin and Carling groups crystallized an AMPK complex containing rat α1, human β2 CTD, and rat γ1 [34]. While this complex is not regulated by protection against activation loop dephosphorylation [51], it retained direct AMPK activation by AMP and ADP. The structure revealed that the α-linker that connects AID and α-CTD directly bound the γ subunit [34], which has been validated in all subsequent AMP-bound AMPK complex structures with a resolved α-linker. A segment of the linker, termed regulatory subunit-interacting motif 2 (αRIM2) [27,59], interacts with AMP at CBS3, suggesting that αRIM2 functions as an adenine nucleotide sensor, and that it mediates the transduction of the adenine-binding signal to the KD [27,34,59]. This function has been validated by several experimental approaches. First, the mutation of either of the two key αRIM2 residues (E362 and R363 in rat α1 and human α2; E364 and R365 in human α1) abolished or largely reduced both AMP-dependent direct AMPK activation [27,49,51] and AMP-dependent protection against activation loop dephosphorylation [49]. Second, AMP increases, and ATP decreases the interaction between isolated α-linker and core AMPK in a reconstituted system, and the AMP increase requires intact E364 and R365 [49]. Third, the AMP-mimetic synthetic AMPK activator C2 activates AMPK α-isotype-selectively (it fully activates α1-containing complexes, but only partially α2 complexes), and this selectivity can be fully reversed by a swap of the αRIM2 regions [60,61].

4. If CBS3 Is the Sensor Site, What Are the Roles of CBS1 and CBS4?

Of the three functional CBS sites, only CBS3 interacts with the α subunit, an interaction that is directly modulated by AMP and ATP. In contrast, CBS1 and CBS4 do not interact with any part of the α- or β-subunit. Moreover, CBS4 binds AMP very tightly [24,58] and it is unlikely to exchange AMP under physiological conditions, yet mutations in CBS4 abolish regulation by AMP [36,58], while mutations in CBS1 have either no [58] or only a small [36] effect on AMPK regulation. Important insight came from a mutational study. When CBS1, CBS4, and the ATP-binding site in the KD are mutated, so that CBS3 remains the only functional adenine nucleotide binding site, it binds AMP only very weakly and with 10–100 times lower affinity than ATP [48]. Since the cellular ATP concentrations are much higher than AMP and ADP concentrations, CBS3 by itself would remain almost completely ATP-bound under both normal and energy stress conditions. However, the phosphates of adenine nucleotides bound to CBS1, 3, and 4 coordinately bind a set of charged and polar amino acids (Figure 4E), so that binding to one site affects binding to the other two sites. Through these coordinated interactions, AMP bound at CBS4, together with additional interactions from αRIM2, stabilizes AMP at CBS3. This increases CBS3's affinity for AMP by two orders of magnitude, and its AMP/ATP binding preference by two to three orders of magnitude [48], allowing CBS3 to sensitively detect physiological energy stress versus non-stress adenine nucleotide levels. Conversely, both CBS3 and CBS1 strongly stabilize AMP-binding at CBS4, so that under physiological conditions CBS4 remains essentially non-exchangeably AMP-bound and CBS1 largely ATP-bound [48].

5. AMP-Binding at CBS3 Destabilizes an Inhibitory AID–KD Interaction

The KD is followed by the AID, a small 48 amino acid domain that inhibits kinase activity about tenfold in the context of a KD–AID fragment [62,63]. Crystal structures of the fission yeast [21] and human [49] AMPK KD–AID fragments revealed a three-helical AID, whose C-terminal helix (α3) directly binds the hinge between the KD N- and C-lobes at the backside of the KD (Figures 2B and 5A). In contrast, in structures of active, AMP-bound AMPK [34,49], the AID is rotated away from the KD and bound to the γ subunit (see structure overlay in Figure 5B). The AID–KD interaction arrests the KD in a unique inactive conformation, in which the ATP binding K47, the Mg^{2+}-binding DFG loop, and the substrate-binding catalytic loop are misaligned, and H139 of the regulatory spine is out of register [21,49] (so called "HRD-out" conformation [30]; Figure 5C). The inhibitory function of the KD–AID interaction was further validated by the mutation of interface residues in either the KD or the AID, all of which made AMPK constitutively active [21,49]. Conversely, binding of the AID to the γ subunit, as seen in structures of AMP-bound AMPK [34,49], allows the KD to adopt the active conformation [27,34,51,59] (Figure 5D). Consistently, mutations in AID-interacting γ subunit residues make AMPK constitutively inactive [59].

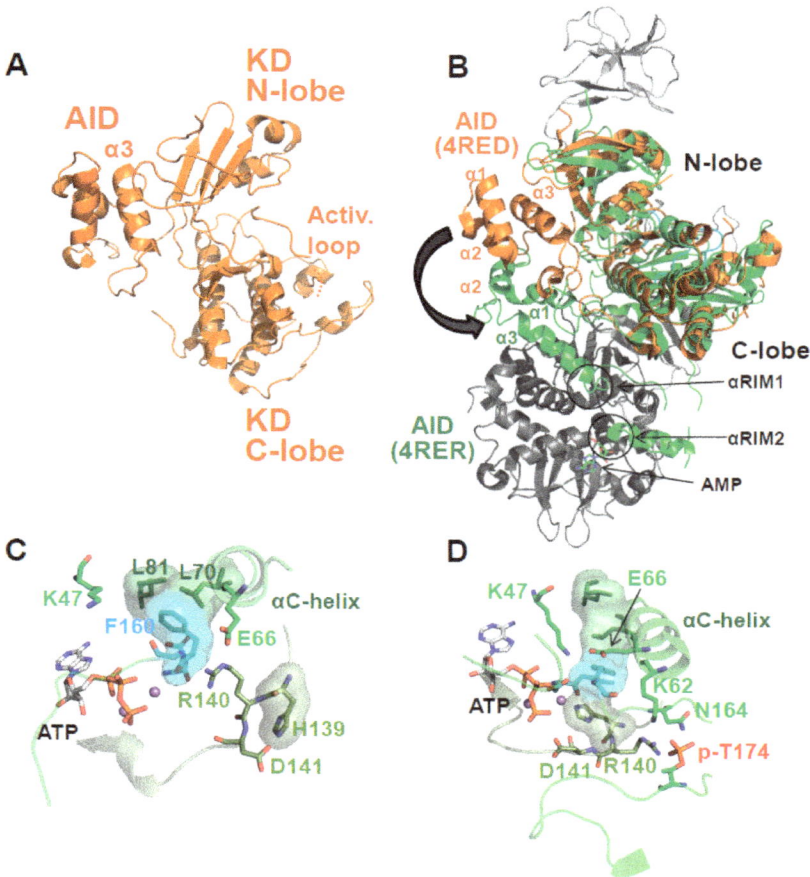

Figure 5. The AID is in equilibrium between KD- and γ-bound conformations. (**A**) Cartoon structure of the human α_1 KD-AID complex. (**B**) Overlay of the inactive KD-AID structure with the structure of active holo-AMPK (α subunit: green; β- and γ-subunits: grey). The arrow indicates the repositioning of the AID in the active structure. (**C,D**) Catalytic center of the inactive (**C**) and active (**D**) AMPK conformation. Stick plus translucent surface presentations indicate the regulatory spine residues L70, L81, H139, and F160. Mg^{2+}-ATP was modeled into both structures for orientation, even though it cannot bind to the inactive structure shown in panel C. Spheres: Mg^{2+} ions.

A Highly Conserved Interaction Network Links αRIM2/CBS3 and AID-αRIM1/CBS2 Binding

The structure of AMP-bound AMPK α_1-β_2CTD-γ_1 [34] first revealed the AID conformation in active AMPK, in which the border of the AID and the N-terminus of the α-linker, termed αRIM1, binds the γ-subunit at the unoccupied CBS2 [27,34,51,59]. Mutational analysis by Ja-Wei Wu's group provided a molecular pathway to link αRIM2 binding of AMP-occupied CBS3 to direct AMPK kinase activation. They first showed that αRIM1/CBS2 interface amino acids corresponding to human α1 I335/M3364 and F342/Y343, and human γ1 R171 and F179 are required for AMP-mediated relief of AMPK autoinhibition [59]. In active, AMP-bound holo-AMPK, the direct interaction of γ1 K170 with both AMP/CBS3 and αRIM2 α1 E364 positions three key residues at the αRIM1 interface. First, the residue following K170, R171, forms Van der Waals interactions and a backbone hydrogen bond with αRIM1 α1 F342. Second, the K170-interacting residues K174 and F175 form Van der Waals bonds with F342 and both Van der Waals and π-stacking interactions with γ1 F179. The latter is the linchpin of the interface and directly interacts with all four αRIM1 residues that are required for the relief of AMPK autoinhibition (I335, M336, Y343, and F342; Figure 6). Similarly, E364, R171, and F179 are also all required for the relief of AMPK autoinhibition [59]. The mutational analysis thus provides strong support that this AMP-stabilized interaction network that is seen in all active structures of holo-AMPK is responsible for shifting the AID equilibrium from the inactive, KD-bound conformation to the active, γ/CBS2-bound conformation.

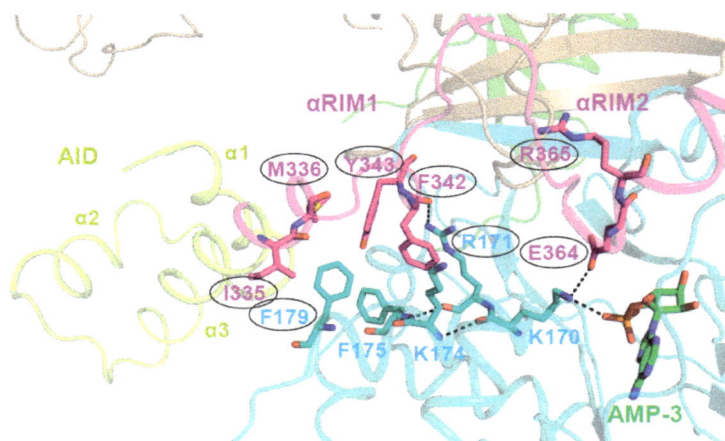

Figure 6. αRIM2/CBS3 and AID-αRIM1/CBS2 interactions are linked. Structure of human AMP-bound AMPK $\alpha_1\beta_2\gamma_1$ (4RER) with key residues shown in a stick presentation; the α-linker is shown in magenta, the γ subunit in cyan, and the AID in light green. AMP bound at CBS3 and αRIM2 E364 directly interact with γ1 K170, which positions the αRIM1-binding residues R171, and indirectly through K174 and F175, F179, thus stabilizing the AID-γ subunit interaction. Consistently, mutations of the αRIM1/γ subunit (and αRIM2/CBS3) interface residues highlighted by oval outlines (human α_1: F342D/Y343D, I335D/M336D, E364, R365; γ_1: R171A, F179D) are constitutively AMP-non-responsive. Dashed lines indicate hydrogen bonds.

ATP binding is thought to disrupt this network. In the structure of the core AMPK complex co-crystallized with ATP [58], ATP was bound to CBS4 and CBS1, which sterically interfered with nucleotide binding at CBS3, and caused rearrangement and disruption of the interaction network [58,59]. However, the physiological relevance of this structure remains unclear, since under physiological conditions, CBS4 does not seem to exchange AMP (see above; [24,48,58]). Therefore, a final understanding of how ATP disrupts the CBS3–α-linker–AID network will require the structure of the holo-AMPK complex, including the α-linker, in ATP-bound conformation.

ADaM site ligands, while not focus of this review, directly activate AMPK by a completely different mechanism. Through binding of both the CBM and the KD [51,52] and stabilization of the CBM–KD interaction [53], the N-terminus of the β-linker at the CBM border adopts a helix that packs parallel to the αC-helix, and it has therefore been named C-interacting helix [51]. This suggested that ADaM site ligands may activate AMPK by stabilizing αC through induced formation of the C-interacting helix, reminiscent of the regulatable αC stabilization of several other protein kinases [64]. Support for this model came from the mutation of H233 in the C-interacting helix, which reduced activation by the ADaM site ligand 991 [51], and by direct demonstration through hydrogen/deuterium exchange mass spectrometry (HDX-MS) that 991 binding strongly and selectively stabilizes αC [48].

6. Regulation of Activation Loop Accessibility

A major regulatory mechanism for AMPK activation by AMP and ADP is the protection of activation loop p-T172 (human α1 T174) against dephosphorylation. p-T172 protection can be demonstrated in a cell-free, reconstituted system independent of the phosphatase used (e.g., PP2C, PP2A, λ-phosphatase), and AMP does not, or only slightly inhibit the dephosphorylation of a different substrate, casein, by PP2Cα [32]. Therefore, reduced dephosphorylation is not due to phosphatase inhibition, but to an AMP/ADP-induced change in the activation loop accessibility. The crystal structure of AMP-bound, phosphorylated AMPK α1–β2CTD–γ1 (PDB: 4CFH) first demonstrated that the activation loop directly interacts with the core of AMPK [34]. Specifically, the stable β-CTD directly bound and stabilized the activation loop (Figure 7). The authors therefore proposed that the core shields the activation loop from phosphatase access. In agreement, mutation of the activation loop-interacting β2 H235 increased p-T172 dephosphorylation in the context of holo-AMPK [34]. However, the construct used in the structure was not regulated by protection against activation loop dephosphorylation [51], indicating that additional parts of AMPK, likely either the β-linker and/or the CBM, were also required for AMP-mediated, and probably ADaM site ligand-mediated protection against activation loop dephosphorylation. Consistently, in structures in which the β-linker is largely resolved (e.g., β2-linker in 4RER [49], β1-linker in 5ISO [65]), p-T172 is clearly protected by the β-linker, especially in the case of the β2-linker. Finally, how can the activation loop in AMP-bound conformation be largely inaccessible to protein phosphatases without affecting accessibility to the T172-phosphorylating upstream protein kinases? Answers to these fundamental questions will likely require the structure of holo-AMPK in the alternative, ATP-bound state and analysis of AMPK's conformational landscape and dynamics in solution.

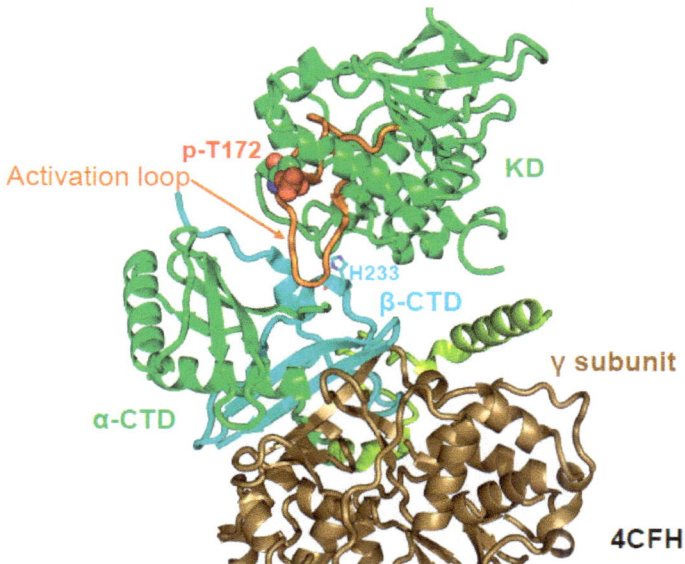

Figure 7. The β-CTD binds and stabilizes the activation loop. Structure of AMP-bound, phosphorylated AMPK α1–β2CTD–γ1 (PDB: 4CFH). The activation loop is highlighted in orange, and p-T172 is shown in sphere presentation.

7. Conclusions and Future Directions

AMPK is a molecular machine consisting of the adenine nucleotide-binding core (γ subunit plus α- and β-CTDs), the catalytic KD, and at least four dynamic domains (AID, CBM, and the α- and β-linkers). We propose that adenine nucleotides, ADaM site ligands, and CBM phosphorylation affect the conformation of the KD through induced movements of the dynamic domains, while phosphorylation of activation loop and S/T loop modulate the KD conformation directly. Through concerted efforts, the mechanism of direct, allosteric AMPK activation through AID movement and αC stabilization is relatively well understood. However, the structural basis of direct inhibition by ATP, of activation loop accessibility regulation through ligands and possibly phosphorylation, and of the AMP-induced interaction with Axin and the LKB1 complex all remain poorly understood. The most important future challenges in AMPK structural biology will therefore be the determination of the structures of holo-AMPK in its inhibited, ATP-bound conformation, and in complex with Axin and LKB/STRAD/MO25.

Author Contributions: The manuscript was written by K.M. with input from all authors.

funding: This research was funded by the Van Andel Research Institute (H.E.X. and K.M.) and the National Institutes of Health (R01 GM129436 to K.M.).

Conflicts of Interest: The authors declare no conflict of interest. The funders had no role in the design of the study; in the collection, analyses, or interpretation of data; in the writing of the manuscript, or in the decision to publish the results.

Abbreviations

ADaM site	Allosteric drug and metabolite-binding site
AID	Autoinhibitory domain
AMPK	AMP-activated protein kinase
αRIM	α-regulatory subunit interaction motif
CaMKK2	Ca^{2+}/calmodulin-dependent protein kinase kinase β
CBM	Carbohydrate-binding module
CBS	Cystathionine β-synthetase
CTD	C-terminal domain
HDX-MS	Hydrogen deuterium exchange mass spectrometry
KD	Kinase domain
LKB1	Liver kinase B1
MO25	Mouse protein-25
PP2A	Protein phosphatase 2A
PP2C	Protein phosphatase 2C
STRAD	STE20-related kinase adaptor

References

1. Yuan, H.X.; Xiong, Y.; Guan, K.L. Nutrient sensing, metabolism, and cell growth control. *Mol. Cell* **2013**, *49*, 379–387. [CrossRef] [PubMed]
2. Garcia, D.; Shaw, R.J. AMPK: Mechanisms of Cellular Energy Sensing and Restoration of Metabolic Balance. *Mol. Cell* **2017**, *66*, 789–800. [CrossRef] [PubMed]
3. Hardie, D.G. AMP-activated protein kinase: An energy sensor that regulates all aspects of cell function. *Genes Dev.* **2011**, *25*, 1895–1908. [CrossRef] [PubMed]
4. Hardie, D.G. Keeping the home fires burning: AMP-activated protein kinase. *J. R. Soc. Interface* **2018**, *15*, 20170774. [CrossRef] [PubMed]
5. Steinberg, G.R.; Kemp, B.E. AMPK in Health and Disease. *Physiol. Rev.* **2009**, *89*, 1025–1078. [CrossRef] [PubMed]
6. Hardie, D.G.; Schaffer, B.E.; Brunet, A. AMPK: An Energy-Sensing Pathway with Multiple Inputs and Outputs. *Trends Cell Biol.* **2016**, *26*, 190–201. [CrossRef] [PubMed]

7. Hardie, D.G. AMPK: A target for drugs and natural products with effects on both diabetes and cancer. *Diabetes* **2013**, *62*, 2164–2172. [CrossRef] [PubMed]

8. Hardie, D.G. Targeting an energy sensor to treat diabetes. *Science* **2017**, *357*, 455–456. [CrossRef] [PubMed]

9. Hardie, D.G.; Ross, F.A.; Hawley, S.A. AMP-activated protein kinase: A target for drugs both ancient and modern. *Chem. Biol.* **2012**, *19*, 1222–1236. [CrossRef] [PubMed]

10. Guigas, B.; Viollet, B. Targeting AMPK: From Ancient Drugs to New Small-Molecule Activators. *EXS* **2016**, *107*, 327–350. [PubMed]

11. Cokorinos, E.C.; Delmore, J.; Reyes, A.R.; Albuquerque, B.; Kjobsted, R.; Jorgensen, N.O.; Tran, J.L.; Jatkar, A.; Cialdea, K.; Esquejo, R.M.; et al. Activation of Skeletal Muscle AMPK Promotes Glucose Disposal and Glucose Lowering in Non-human Primates and Mice. *Cell Metab.* **2017**, *25*, 1147–1159.e10. [CrossRef] [PubMed]

12. Myers, R.W.; Guan, H.P.; Ehrhart, J.; Petrov, A.; Prahalada, S.; Tozzo, E.; Yang, X.; Kurtz, M.M.; Trujillo, M.; Gonzalez Trotter, D.; et al. Systemic pan-AMPK activator MK-8722 improves glucose homeostasis but induces cardiac hypertrophy. *Science* **2017**, *357*, 507–511. [CrossRef] [PubMed]

13. Steneberg, P.; Lindahl, E.; Dahl, U.; Lidh, E.; Straseviciene, J.; Backlund, F.; Kjellkvist, E.; Berggren, E.; Lundberg, I.; Bergqvist, I.; et al. PAN-AMPK activator O304 improves glucose homeostasis and microvascular perfusion in mice and type 2 diabetes patients. *JCI Insight* **2018**, *3*. [CrossRef] [PubMed]

14. Ross, F.A.; MacKintosh, C.; Hardie, D.G. AMP-activated protein kinase: A cellular energy sensor that comes in 12 flavours. *FEBS J.* **2016**, *283*, 2987–3001. [CrossRef] [PubMed]

15. Hawley, S.A.; Ross, F.A.; Gowans, G.J.; Tibarewal, P.; Leslie, N.R.; Hardie, D.G. Phosphorylation by Akt within the ST loop of AMPK-alpha1 down-regulates its activation in tumour cells. *Biochem. J.* **2014**, *459*, 275–287. [CrossRef] [PubMed]

16. Hurley, R.L.; Barre, L.K.; Wood, S.D.; Anderson, K.A.; Kemp, B.E.; Means, A.R.; Witters, L.A. Regulation of AMP-activated protein kinase by multisite phosphorylation in response to agents that elevate cellular cAMP. *J. Biol. Chem.* **2006**, *281*, 36662–36672. [CrossRef] [PubMed]

17. Suzuki, T.; Bridges, D.; Nakada, D.; Skiniotis, G.; Morrison, S.J.; Lin, J.D.; Saltiel, A.R.; Inoki, K. Inhibition of AMPK catabolic action by GSK3. *Mol. Cell* **2013**, *50*, 407–419. [CrossRef] [PubMed]

18. Littler, D.R.; Walker, J.R.; Davis, T.; Wybenga-Groot, L.E.; Finerty, P.J., Jr.; Newman, E.; Mackenzie, F.; Dhe-Paganon, S. A conserved mechanism of autoinhibition for the AMPK kinase domain: ATP-binding site and catalytic loop refolding as a means of regulation. *Acta Crystallogr. Sect. F Struct. Biol. Cryst. Commun.* **2010**, *66*, 143–151. [CrossRef] [PubMed]

19. Nayak, V.; Zhao, K.; Wyce, A.; Schwartz, M.F.; Lo, W.S.; Berger, S.L.; Marmorstein, R. Structure and dimerization of the kinase domain from yeast Snf1, a member of the Snf1/AMPK protein family. *Structure* **2006**, *14*, 477–485. [CrossRef] [PubMed]

20. Handa, N.; Takagi, T.; Saijo, S.; Kishishita, S.; Takaya, D.; Toyama, M.; Terada, T.; Shirouzu, M.; Suzuki, A.; Lee, S.; et al. Structural basis for compound C inhibition of the human AMP-activated protein kinase alpha2 subunit kinase domain. *Acta Crystallogr. D Biol. Crystallogr.* **2011**, *67*, 480–487. [CrossRef] [PubMed]

21. Chen, L.; Jiao, Z.H.; Zheng, L.S.; Zhang, Y.Y.; Xie, S.T.; Wang, Z.X.; Wu, J.W. Structural insight into the autoinhibition mechanism of AMP-activated protein kinase. *Nature* **2009**, *459*, 1146–1149. [CrossRef] [PubMed]

22. Polekhina, G.; Gupta, A.; van Denderen, B.J.; Feil, S.C.; Kemp, B.E.; Stapleton, D.; Parker, M.W. Structural basis for glycogen recognition by AMP-activated protein kinase. *Structure* **2005**, *13*, 1453–1462. [CrossRef] [PubMed]

23. Amodeo, G.A.; Rudolph, M.J.; Tong, L. Crystal structure of the heterotrimer core of Saccharomyces cerevisiae AMPK homologue SNF1. *Nature* **2007**, *449*, 492–495. [CrossRef] [PubMed]

24. Xiao, B.; Heath, R.; Saiu, P.; Leiper, F.C.; Leone, P.; Jing, C.; Walker, P.A.; Haire, L.; Eccleston, J.F.; Davis, C.T.; et al. Structural basis for AMP binding to mammalian AMP-activated protein kinase. *Nature* **2007**, *449*, 496–500. [CrossRef] [PubMed]

25. Townley, R.; Shapiro, L. Crystal structures of the adenylate sensor from fission yeast AMP-activated protein kinase. *Science* **2007**, *315*, 1726–1729. [CrossRef] [PubMed]

26. Jin, X.; Townley, R.; Shapiro, L. Structural insight into AMPK regulation: ADP comes into play. *Structure* **2007**, *15*, 1285–1295. [CrossRef] [PubMed]

27. Chen, L.; Xin, F.J.; Wang, J.; Hu, J.; Zhang, Y.Y.; Wan, S.; Cao, L.S.; Lu, C.; Li, P.; Yan, S.F.; et al. Conserved regulatory elements in AMPK. *Nature* **2013**, *498*, E8–E10. [CrossRef] [PubMed]
28. Kornev, A.P.; Haste, N.M.; Taylor, S.S.; Eyck, L.F. Surface comparison of active and inactive protein kinases identifies a conserved activation mechanism. *Proc. Natl. Acad. Sci. USA* **2006**, *103*, 17783–17788. [CrossRef] [PubMed]
29. Kornev, A.P.; Taylor, S.S. Dynamics-Driven Allostery in Protein Kinases. *Trends Biochem. Sci.* **2015**, *40*, 628–647. [CrossRef] [PubMed]
30. Meharena, H.S.; Chang, P.; Keshwani, M.M.; Oruganty, K.; Nene, A.K.; Kannan, N.; Taylor, S.S.; Kornev, A.P. Deciphering the structural basis of eukaryotic protein kinase regulation. *PLoS Biol.* **2013**, *11*, e1001680. [CrossRef] [PubMed]
31. Sanders, M.J.; Ali, Z.S.; Hegarty, B.D.; Heath, R.; Snowden, M.A.; Carling, D. Defining the mechanism of activation of AMP-activated protein kinase by the small molecule A-769662, a member of the thienopyridone family. *J. Biol. Chem.* **2007**, *282*, 32539–32548. [CrossRef] [PubMed]
32. Davies, S.P.; Helps, N.R.; Cohen, P.T.; Hardie, D.G. 5'-AMP inhibits dephosphorylation, as well as promoting phosphorylation, of the AMP-activated protein kinase. Studies using bacterially expressed human protein phosphatase-2C alpha and native bovine protein phosphatase-2AC. *FEBS Lett.* **1995**, *377*, 421–425. [PubMed]
33. Gowans, G.J.; Hawley, S.A.; Ross, F.A.; Hardie, D.G. AMP is a true physiological regulator of AMP-activated protein kinase by both allosteric activation and enhancing net phosphorylation. *Cell Metab.* **2013**, *18*, 556–566. [CrossRef] [PubMed]
34. Xiao, B.; Sanders, M.J.; Underwood, E.; Heath, R.; Mayer, F.V.; Carmena, D.; Jing, C.; Walker, P.A.; Eccleston, J.F.; Haire, L.F.; et al. Structure of mammalian AMPK and its regulation by ADP. *Nature* **2011**, *472*, 230–233. [CrossRef] [PubMed]
35. Carling, D.; Clarke, P.R.; Zammit, V.A.; Hardie, D.G. Purification and characterization of the AMP-activated protein kinase. Copurification of acetyl-CoA carboxylase kinase and 3-hydroxy-3-methylglutaryl-CoA reductase kinase activities. *Eur. J. Biochem.* **1989**, *186*, 129–136. [PubMed]
36. Oakhill, J.S.; Steel, R.; Chen, Z.P.; Scott, J.W.; Ling, N.; Tam, S.; Kemp, B.E. AMPK is a direct adenylate charge-regulated protein kinase. *Science* **2011**, *332*, 1433–1435. [CrossRef] [PubMed]
37. Zhang, Y.L.; Guo, H.; Zhang, C.S.; Lin, S.Y.; Yin, Z.; Peng, Y.; Luo, H.; Shi, Y.; Lian, G.; Zhang, C.; et al. AMP as a low-energy charge signal autonomously initiates assembly of AXIN-AMPK-LKB1 complex for AMPK activation. *Cell Metab.* **2013**, *18*, 546–555. [CrossRef] [PubMed]
38. Hawley, S.A.; Boudeau, J.; Reid, J.L.; Mustard, K.J.; Udd, L.; Makela, T.P.; Alessi, D.R.; Hardie, D.G. Complexes between the LKB1 tumor suppressor, STRAD alpha/beta and MO25 alpha/beta are upstream kinases in the AMP-activated protein kinase cascade. *J. Biol.* **2003**, *2*, 28. [CrossRef] [PubMed]
39. Hurley, R.L.; Anderson, K.A.; Franzone, J.M.; Kemp, B.E.; Means, A.R.; Witters, L.A. The Ca2+/calmodulin-dependent protein kinase kinases are AMP-activated protein kinase kinases. *J. Biol. Chem.* **2005**, *280*, 29060–29066. [CrossRef] [PubMed]
40. Shaw, R.J.; Kosmatka, M.; Bardeesy, N.; Hurley, R.L.; Witters, L.A.; DePinho, R.A.; Cantley, L.C. The tumor suppressor LKB1 kinase directly activates AMP-activated kinase and regulates apoptosis in response to energy stress. *Proc. Natl. Acad. Sci. USA* **2004**, *101*, 3329–3335. [CrossRef] [PubMed]
41. Woods, A.; Dickerson, K.; Heath, R.; Hong, S.P.; Momcilovic, M.; Johnstone, S.R.; Carlson, M.; Carling, D. Ca2+/calmodulin-dependent protein kinase kinase-beta acts upstream of AMP-activated protein kinase in mammalian cells. *Cell Metab.* **2005**, *2*, 21–33. [CrossRef] [PubMed]
42. Woods, A.; Johnstone, S.R.; Dickerson, K.; Leiper, F.C.; Fryer, L.G.; Neumann, D.; Schlattner, U.; Wallimann, T.; Carlson, M.; Carling, D. LKB1 is the upstream kinase in the AMP-activated protein kinase cascade. *Curr. Biol.* **2003**, *13*, 2004–2008. [CrossRef] [PubMed]
43. Zhang, C.S.; Jiang, B.; Li, M.; Zhu, M.; Peng, Y.; Zhang, Y.L.; Wu, Y.Q.; Li, T.Y.; Liang, Y.; Lu, Z.; et al. The lysosomal v-ATPase-Ragulator complex is a common activator for AMPK and mTORC1, acting as a switch between catabolism and anabolism. *Cell Metab.* **2014**, *20*, 526–540. [CrossRef] [PubMed]
44. Deng, M.; Yang, X.; Qin, B.; Liu, T.; Zhang, H.; Guo, W.; Lee, S.B.; Kim, J.J.; Yuan, J.; Pei, H.; et al. Deubiquitination and Activation of AMPK by USP10. *Mol. Cell* **2016**, *61*, 614–624. [PubMed]
45. Lee, S.W.; Li, C.F.; Jin, G.; Cai, Z.; Han, F.; Chan, C.H.; Yang, W.L.; Li, B.K.; Rezaeian, A.H.; Li, H.Y.; et al. Skp2-dependent ubiquitination and activation of LKB1 is essential for cancer cell survival under energy stress. *Mol. Cell* **2015**, *57*, 1022–1033. [CrossRef] [PubMed]

46. Zhang, C.S.; Hawley, S.A.; Zong, Y.; Li, M.; Wang, Z.; Gray, A.; Ma, T.; Cui, J.; Feng, J.W.; Zhu, M.; et al. Fructose-1,6-bisphosphate and aldolase mediate glucose sensing by AMPK. *Nature* **2017**, *548*, 112–116. [CrossRef] [PubMed]

47. Polekhina, G.; Gupta, A.; Michell, B.J.; van Denderen, B.; Murthy, S.; Feil, S.C.; Jennings, I.G.; Campbell, D.J.; Witters, L.A.; Parker, M.W.; et al. AMPK beta subunit targets metabolic stress sensing to glycogen. *Curr. Biol.* **2003**, *13*, 867–871. [CrossRef]

48. Gu, X.; Yan, Y.; Novick, S.J.; Kovich, A.; Goswami, D.; Ke, J.; Tan, M.H.E.; Wang, L.; Li, X.; de Waal, P.; et al. Deconvoluting AMP-dependent kinase (AMPK) adenine nucleotide binding and sensing. *J. Biol. Chem.* **2017**, *292*, 12653–12666. [CrossRef] [PubMed]

49. Li, X.; Wang, L.; Zhou, X.E.; Ke, J.; de Waal, P.W.; Gu, X.; Tan, M.H.; Wang, D.; Wu, D.; Xu, H.E.; et al. Structural basis of AMPK regulation by adenine nucleotides and glycogen. *Cell Res.* **2015**, *25*, 50–66. [CrossRef] [PubMed]

50. McBride, A.; Ghilagaber, S.; Nikolaev, A.; Hardie, D.G. The glycogen-binding domain on the AMPK beta subunit allows the kinase to act as a glycogen sensor. *Cell Metab.* **2009**, *9*, 23–34. [CrossRef] [PubMed]

51. Xiao, B.; Sanders, M.J.; Carmena, D.; Bright, N.J.; Haire, L.F.; Underwood, E.; Patel, B.R.; Heath, R.B.; Walker, P.A.; Hallen, S.; et al. Structural basis of AMPK regulation by small molecule activators. *Nat. Commun.* **2013**, *4*, 3017. [CrossRef] [PubMed]

52. Calabrese, M.F.; Rajamohan, F.; Harris, M.S.; Caspers, N.L.; Magyar, R.; Withka, J.M.; Wang, H.; Borzilleri, K.A.; Sahasrabudhe, P.V.; Hoth, L.R.; et al. Structural Basis for AMPK Activation: Natural and Synthetic Ligands Regulate Kinase Activity from Opposite Poles by Different Molecular Mechanisms. *Structure* **2014**, *22*, 1161–1172. [CrossRef] [PubMed]

53. Gu, X.; Bridges, M.D.; Yan, Y.; de Waal, P.; Zhou, X.E.; Suino-Powell, K.M.; Xu, H.E.; Hubbell, W.L.; Melcher, K. Conformational heterogeneity of the allosteric drug and metabolite (ADaM) site in AMP-activated protein kinase (AMPK). *J. Biol. Chem.* **2018**, *239*, 16994–17007. [CrossRef] [PubMed]

54. Qi, J.; Gong, J.; Zhao, T.; Zhao, J.; Lam, P.; Ye, J.; Li, J.Z.; Wu, J.; Zhou, H.M.; Li, P. Downregulation of AMP-activated protein kinase by Cidea-mediated ubiquitination and degradation in brown adipose tissue. *EMBO J.* **2008**, *27*, 1537–1548. [CrossRef] [PubMed]

55. Pineda, C.T.; Ramanathan, S.; Fon Tacer, K.; Weon, J.L.; Potts, M.B.; Ou, Y.H.; White, M.A.; Potts, P.R. Degradation of AMPK by a cancer-specific ubiquitin ligase. *Cell* **2015**, *160*, 715–728. [CrossRef] [PubMed]

56. Vila, I.K.; Yao, Y.; Kim, G.; Xia, W.; Kim, H.; Kim, S.J.; Park, M.K.; Hwang, J.P.; Gonzalez-Billalabeitia, E.; Hung, M.C.; et al. A UBE2O-AMPKalpha2 Axis that Promotes Tumor Initiation and Progression Offers Opportunities for Therapy. *Cancer Cell* **2017**, *31*, 208–224. [CrossRef] [PubMed]

57. Lee, J.O.; Lee, S.K.; Kim, N.; Kim, J.H.; You, G.Y.; Moon, J.W.; Jie, S.; Kim, S.J.; Lee, Y.W.; Kang, H.J.; et al. E3 ubiquitin ligase, WWP1, interacts with AMPKalpha2 and down-regulates its expression in skeletal muscle C2C12 cells. *J. Biol. Chem.* **2013**, *288*, 4673–4680. [CrossRef] [PubMed]

58. Chen, L.; Wang, J.; Zhang, Y.Y.; Yan, S.F.; Neumann, D.; Schlattner, U.; Wang, Z.X.; Wu, J.W. AMP-activated protein kinase undergoes nucleotide-dependent conformational changes. *Nat. Struct. Mol. Biol.* **2012**, *19*, 716–718. [CrossRef] [PubMed]

59. Xin, F.J.; Wang, J.; Zhao, R.Q.; Wang, Z.X.; Wu, J.W. Coordinated regulation of AMPK activity by multiple elements in the alpha-subunit. *Cell Res.* **2013**, *23*, 1237–1240. [CrossRef] [PubMed]

60. Hunter, R.W.; Foretz, M.; Bultot, L.; Fullerton, M.D.; Deak, M.; Ross, F.A.; Hawley, S.A.; Shpiro, N.; Viollet, B.; Barron, D.; et al. Mechanism of action of compound-13: An alpha1-selective small molecule activator of AMPK. *Chem. Biol.* **2014**, *21*, 866–879. [CrossRef] [PubMed]

61. Langendorf, C.G.; Ngoei, K.R.; Scott, J.W.; Ling, N.X.; Issa, S.M.; Gorman, M.A.; Parker, M.W.; Sakamoto, K.; Oakhill, J.S.; Kemp, B.E. Structural basis of allosteric and synergistic activation of AMPK by furan-2-phosphonic derivative C2 binding. *Nat. Commun.* **2016**, *7*, 10912. [CrossRef] [PubMed]

62. Crute, B.E.; Seefeld, K.; Gamble, J.; Kemp, B.E.; Witters, L.A. Functional domains of the alpha1 catalytic subunit of the AMP-activated protein kinase. *J. Biol. Chem.* **1998**, *273*, 35347–35354. [CrossRef] [PubMed]

63. Pang, T.; Xiong, B.; Li, J.Y.; Qiu, B.Y.; Jin, G.Z.; Shen, J.K.; Li, J. Conserved alpha-helix acts as autoinhibitory sequence in AMP-activated protein kinase alpha subunits. *J. Biol. Chem.* **2007**, *282*, 495–506. [CrossRef] [PubMed]

64. Palmieri, L.; Rastelli, G. alphaC helix displacement as a general approach for allosteric modulation of protein kinases. *Drug Discov. Today* **2013**, *18*, 407–414. [CrossRef] [PubMed]
65. Willows, R.; Sanders, M.J.; Xiao, B.; Patel, B.R.; Martin, S.R.; Read, J.; Wilson, J.R.; Hubbard, J.; Gamblin, S.J.; Carling, D. Phosphorylation of AMPK by upstream kinases is required for activity in mammalian cells. *Biochem. J.* **2017**, *474*, 3059–3073. [CrossRef] [PubMed]

International Journal of
Molecular Sciences

MDPI

Review

Is TAK1 a Direct Upstream Kinase of AMPK?

Dietbert Neumann

Department of Pathology, CARIM School for Cardiovascular Diseases, Faculty of Health, Medicine and Life Sciences, Maastricht University, 6200 MD Maastricht, The Netherlands; d.neumann@maastrichtuniversity.nl; Tel.: +31-43-387-7167

Received: 29 June 2018; Accepted: 14 August 2018; Published: 15 August 2018

Abstract: Alongside Liver kinase B1 (LKB1) and Ca^{2+}/Calmodulin-dependent protein kinase kinase 2 (CaMKK2), Transforming growth factor-β (TGF-β)-activated kinase 1 (TAK1) has been suggested as a direct upstream kinase of AMP-activated protein kinase (AMPK). Several subsequent studies have reported on the TAK1-AMPK relationship, but the interpretation of the respective data has led to conflicting views. Therefore, to date the acceptance of TAK1 as a genuine AMPK kinase is lagging behind. This review provides with argumentation, whether or not TAK1 functions as a direct upstream kinase of AMPK. Several specific open questions that may have precluded the consensus are discussed based on available data. In brief, TAK1 can function as direct AMPK upstream kinase in specific contexts and in response to a subset of TAK1 activating stimuli. Further research is needed to define the intricate signals that are conditional for TAK1 to phosphorylate and activate AMPKα at T172.

Keywords: TAK1; AMPK; phosphorylation; AMPK kinase

---

1. About AMPK and TAK1

This review addresses questions that are relevant for experts in the field already familiar with AMPK and TAK1. To begin with, I will not discuss whether AMPK and TAK1 are a disparate couple or ideal affiliates, but rather provide entry points for further reading, in case readers are in need of information about AMPK or TAK1. I will not go into any detail with AMPK, because this review is part of the Special Issue on AMPK. Moreover, multiple authors (repeatedly) reviewed AMPK. In addition, there is a growing base of reviews focusing on different aspects of AMPK, such as the functions of AMPK in various tissues, or (patho-)physiological contexts (e.g., [1–4]). For AMPK novices, Hardie provides an excellent overview (e.g., [5–8]). In a nutshell, AMPK is an energy-sensing kinase that functions to maintain cellular and whole body energy balance [9]. AMPK is part of a protein kinase cascade [10]. T172 phosphorylation of the AMPKα subunit activates the kinase, which is dependent on upstream kinases, called AMPK kinases, identified as LKB1 and CaMKK2 [11–14].

TAK1 has been proposed as an alternative third AMPK kinase, which has received varying appreciation. This is the topic of this review. TAK1 is a serine/threonine protein kinase of the mitogen-activated protein kinase kinase kinase (MAP3K) family, playing a crucial role in regulating cell survival, differentiation, apoptosis, and inflammatory responses [15,16]. It forms complexes by binding to its accessory subunits, the TAK1-binding proteins (TAB1, TAB2, TAB3). TAK1 is activated by interleukin-1 (IL-1) and TGF-β receptors, tumour necrosis factor (TNF)-α, Toll-like receptors (TLR), CD40, and the B cell receptor. TAK1 is also involved in activating several intracellular kinases, p38 mitogen-activated protein kinase (p38MAPK), c-Jun N-terminal kinase (JNK), and IκB kinase complex (IKK). Therefore, TAK1 has been described as a regulator of nuclear factor κ-light-chain-enhancer of activated B cells (NF-κB) and MAPKs in proinflammatory signalling. More recently, this picture has been significantly amended, with the roles of TAK1 in tissue homeostasis (reviewed in [17]), as also further discussed below.

2. The Origin of the Debate

In 2006, after the discovery of LKB1 and CaMKK2 as upstream kinases of AMPK, TAK1 was identified as the third kinase capable of activating AMPK [18]. However, different from LKB1 and CaMKK2, TAK1 to date remains a disputed AMPK activating kinase. The reactions of the scientific community range from complete ignorance, to questioning TAK1 as an AMPK kinase, to acceptance without question. In this review, I will provide an overview on TAK1, with respect to its (putative) role as (direct) upstream activating kinase of AMPK.

In yeast, three alternative upstream kinases (Sak1, Tos3, and Elm1) have been described to activate the AMPK ortholog Snf1; knockout of all three kinases replicates the Snf1 knockout phenotype [19,20]. In search for alternative AMPK activating kinases in mammalian cells and by applying a screening approach in yeast, TAK1 was identified as in vitro upstream kinase of AMPK [18]. The in vivo relevance remained unknown, since the authors based their conclusion solely on cell-free and cell-based approaches. Notably, the study included evidence for TAK1 action on AMPK in LKB1-deficient HeLa cells. In the same year, cardiac-specific dominant-negative TAK1 mice were reported to show Wolff-Parkinson-White (WPW)-like phenotype [21], i.e., consistent with the idea that AMPK loss-of-function mutations in AMPKγ2 underlie WPW [22]. In the same study, TAK1 knockout embryos were shown to exhibit defective AMPK signalling. Due to observed midgestation embryonic lethality, the authors subsequently went on to acutely knock out floxed TAK1 alleles in cells using virally delivered Cre. The obtained results again generally supported a role of TAK1 upstream of AMPK, but the authors concluded that LKB1 could have been the intermediate of TAK1 action. The reason for this reservation was that acute loss of TAK1 interfered with the kinase activity of adenovirally delivered LKB1 complex, i.e., consisting of LKB1, mouse protein 25 (MO25) and STE20-related kinase adapter protein (STRAD). It should be noted that LKB1 is considered to be constitutively active upon complex formation with MO25 and STRAD [23,24]. Therefore, the mechanism of LKB1 inhibition, as observed by Xie et al., remains elusive. Accordingly, from these two early publications some discrepancy on the role of TAK1 upstream of AMPK primarily evolved around LKB1, and whether or not it mediates TAK1 effects on AMPK [18,21]. On the other hand, both reports agree on TAK1 as an important regulator of AMPK. In subsequent work on upstream kinases of AMPK, almost all studies have dealt with LKB1 and CaMKK2, which are firmly confirmed without any question. In contrast, the role of TAK1 as direct or indirect AMPK kinase remained obscure.

Until today only few studies further addressed TAK1-AMPK signalling, of which the majority applied chemical tools (such as kinase inhibitors) that are prone to misinterpretation because of possible off-target effects. Moreover, a few reports also indicate signalling of AMPK to TAK1, i.e., turning AMPK into a possible activating kinase of TAK1 [25,26]. In this review, I am focusing on studies using genetic tools and offering clues on the exact role of TAK1 upstream of AMPK, but will also try to integrate controversial findings. As a guide to the reader, I am asking specific remaining open questions that are subsequently either partly or wholly answered, based on scientific evidence.

3. Is TAK1 Capable of Directly Phosphorylating AMPKα at T172 in Cell Free Assays?

In the original paper, Momcilovic et al. used a purified GST-fusion of the isolated Snf1 kinase domain that was directly incubated with an artificial construct of TAK1-TAB1 fusion protein purified from insect cells. Snf1 kinase domain was phosphorylated by a TAK1-TAB1 fusion protein at T210; the site equivalent to T172 in AMPK [18]. In my lab a bacterial co-expression strategy for TAK1 with TAB1 or with TAB2 was developed [27]. TAK1-TAB1 (but not TAK1-TAB2) was active upon co-expression in bacteria, strongly suggesting that the formation of TAK1-TAB1 complexes is sufficient for kinase activation. In contrast, recombinant AMPK heterotrimers after purification from bacteria are not phosphorylated in the α-subunit at T172 [28], but received this modification in presence of either LKB1-MO25-STRAD or TAK1-TAB1 complexes [23,27]. Therefore, mammalian AMPK heterotrimeric complexes can be directly activated by human TAK1-TAB1 complexes, in a process not requiring but resembling LKB1 complex, as shown in cell-free systems.

4. Is TAK1 Activating Cellular AMPK in Absence of LKB1?

This question has been already been addressed in the original work of Momcilovic et al., by using LKB1-deficient HeLa cells transfected with TAK1 and TAB1 plasmids, i.e., revealing that AMPK is activated by the wild-type but not the kinase defective TAK1-TAB1 complex [18]. In my lab, we obtained similar results in HeLa cells using wild type and mutant AMPK [27]. This approach rules out that TAB1 scaffolding is sufficient for AMPK activation. This is an important detail, because TAB1 has been shown to activate p38MAPK employing an unusual autophosphorylation mechanism [29]. In fact, even before TAK1 was suggested as a new upstream kinase, TAB1 was shown to co-immunoprecipitate with AMPK in cardiomyocytes [30]. This latter study also suggested that the association of AMPK with TAB1 did not require prior AMPK activation. Therefore, it seems unlikely that LKB1 action is needed for AMPK-TAB1 interaction, and TAB1 should be able to recruit TAK1 independent of LKB1. Further direct evidence for TAK1-AMPK signalling came from an unexpected pathway: in TNF-related apoptosis-inducing ligand (TRAIL)-treated epithelial cells AMPK was activated by TAK1 [31]. In this study, LKB1 and CaMKK2 were knocked down without affecting the ability of TRAIL to activate AMPK. In contrast, siRNA against TAK1, along with over-expression of kinase-defective TAK1 efficiently interfered with AMPK activation. These data establish TAK1-dependency, as well as LKB1-independency, at least in this particular setting.

5. Is Stimulation of TAK1 Sufficient for Activation of AMPK?

Many researchers doubt whether TAK1 can be considered a genuine upstream kinase, if TAK1 activation is seen in situations where AMPK is not activated. Indeed, Herrero-Martin et al. also observed that TNF-treatment activates TAK1 (as seen by IκB phosphorylation), but did not activate AMPK [31]. In addition, TAK1 activation originally was seen as an intracellular mediator of pro-inflammatory signals (such as TNF-α), giving rise to the development of TAK1 inhibitors for possible treatment of inflammatory disorders [32]. This prevalent view compounded scepticism about TAK1 as a genuine AMPK kinase, since reported AMPK effects are summarized to be the inverse, i.e., anti-inflammatory [3]. Therefore, the role of TAK1 as an upstream kinase of AMPK may be relevant only in certain physiological situations, or in response to specific signals. It should also be noted that LKB1 and CaMKK2 do not share the same input, and may well be active in situations where AMPK is not. In particular, LKB1, a constitutively active kinase upon complex formation, more efficiently phosphorylates AMPK in response to a drop in cellular energy level (through allosteric regulation of AMPK by AMP and ADP). CaMKK2 may be dependent on extracellular input (operating downstream of G-protein coupled receptors), but it is not clear whether transient Ca^{2+} waves, such as those occurring in contracting myocytes, are sufficient to activate AMPK in this cell type. However, both signals, LKB1 and CaMKK2 may also act synergistically [33]. Thus, AMPK phosphorylation at T172 increases through different pathways, downstream of various signals that can be intra- or extracellular. If TAK1 activation per se is insufficient for AMPK activation, TAK1 may still activate AMPK conditionally in response to specific upstream signals.

6. What Is the Cellular Condition Where TAK1 Acts as an Upstream Kinase of AMPK?

In several recent studies, a possible role of TAK1 as upstream mediator of AMPK activation was verified by applying genetic knockdown strategies [34–39]. Although not verifying the role of TAK1 as a direct AMPK kinase, this approach puts TAK1 as an upstream AMPK activating signal into various cellular contexts. Moreover, TRAIL is an example of a distinct extracellular signal that activates TAK1-AMPK signalling [31]. Thus, the question may be asked, whether we can recognise a pattern of cellular challenges or signals where TAK1 is acting as activating AMPK kinase.

In recent literature, TAK1 is interpreted as a regulator of cell death and survival [17], which is well in accordance with the known functions of death ligands, such as TNFα and TRAIL [40]. Notably, TRAIL-induced TAK1-AMPK signalling was shown to induce cytoprotective autophagy in untransformed cells [31], whereas TRAIL induces apoptosis in several cancer cell types. Autophagy is a survival mechanism, which can be elicited by various sublethal stresses as a response to fluctuating external conditions, ranging from extracellular signals, to a change in pH, temperature or oxygen tension [41]. Some of these stresses are not predicted to directly affect cellular energy levels. The known role of AMPK in the control of autophagy in response to nutrient starvation is commonly linked to LKB1 signalling, whereas TRAIL elicits autophagy via TAK1-AMPK [31]. Of note, independent of the upstream signalling pathway, autophagy is an important survival mechanism, providing the cell with building blocks and metabolites. This integrates well with one of AMPKs more general roles; limiting cell proliferation and growth, as well as energy expense, in times of nutrient scarcity, while also enhancing the cell's ability to survive stresses, such as hypoxia and glucose deprivation [42].

As already indicated, TAK1 was shown to activate AMPK in response to various stimuli and different cell types. Receptor activator of NF-κB ligand (RANKL) activated AMPK in osteoclast precursors, and siRNA-mediated TAK1 knockdown blocked RANKL-induced activation of AMPK [34]. RANKL is a member of the TNF superfamily, supporting the idea that a subset of TAK1 activating signals could physiologically activate AMPK. In endothelial cells, Vascular endothelial growth factor (VEGF) stimulated TAK1 and AMPK, whereas TAK1 downregulation by shRNA also inhibited VEGF–stimulated phosphorylation of several kinases, including AMPK [36].

Belinostat promoted reactive oxygen species (ROS) production in PANC-1 cells and increased the ROS induced TAK1/AMPK association resulting in AMPK activation. Anti-oxidants, as well as TAK1 shRNA knockdown, suppressed Belinostat-induced AMPK activation and PANC-1 cell apoptosis [35].

Fasted mice deficient of TAK1 in hepatocytes exhibited severe hepatosteatosis with increased mTORC1 activity, and suppression of autophagy compared with their WT counterparts [43], suggesting reduced AMPK function in these livers. TAK1-deficient hepatocytes exhibited autophagy and suppressed AMPK activity in response to starvation or metformin treatment; however, ectopic activation of AMPK restored autophagy in these cells. These data indicate that TAK1 regulates hepatic lipid metabolism and tumorigenesis via the AMPK/mTORC1 axis [43]. Therefore, it was proposed that TAK1-mediated autophagy in the liver plays a role in preventing excessive lipid accumulation induced by starvation and fat overload [44]. Knockdown of TAK1 decreased the AMPK phosphorylation induced by overexpression of a dominant-negative form of p38α [38], which the authors interpreted as a negative feedback loop. Recent data suggested that TAK1 could be the upstream kinase for AMPK activation by *Helicobacter pylori*, since partial depletion of TAK1 by shRNAs not only inhibited AMPK activation, but also suppressed survival of *H. pylori*-infected gastric epithelial cells [37]. Activation of TAK1 was also found to restrict *Salmonella typhimurium* growth by inducing AMPK activation and autophagy [45]. In this study, TAK1 siRNA led to the inhibition of *S. typhimurium*-induced phosphorylation of AMPK T172, ULK1 S317, and ACC S79. The authors concluded that TAK1 activation leads to AMPK activation, which activates ULK1 by phosphorylating ULK1 S317 and suppressing mTOR activity and ULK1 S757 phosphorylation.

In conclusion, published data indicate TAK1-dependent AMPK activation could be required for induction of autophagy, as a possible survival mechanism in response to acute and specific life-threatening challenges. TAK1-induced autophagy may thus occur in the absence of an energy challenge, such as those elicited through extracellular factors (e.g., TRAIL), or bacterial infections (e.g., *H. pylori*, *S. typhimurium*), and oxidative stress (e.g., Belinostat).

Further conditions promoting TAK1-dependent AMPK activation are likely to be identified.

7. Does AMPK Have a Role in Activating TAK1?

AMPK has been reported to activate TAK1 and mediate pro-inflammatory effects in THP-1 cells [25]. In this study, it was shown that pro-inflammatory signals activated TAK1 signalling, which was then inhibited by AMPKα knockdown. Taking into account the ability of AMPK to bind TAB1 [21], and considering the role of TAB1 in activating TAK1, the interpretation of AMPK as upstream kinase of TAK1 could consequently be challenged. For, example, could the lack of AMPK reduce the availability of TAB1 for subsequent activation of TAK1? Notably, binding of TAB1 to TAK1 in a sequence of molecular events, activates TAK1 by autophosphorylation of T184/T187 [27], and does not require any upstream kinase. Interestingly, the authors of the same study confirmed AMPK-TAK1 interaction in their model, which required both the AMPKα autoinhibitory-domain, and the TAB1-binding domain of TAK1 [25]. The possible AMPK-TAB1 complex formation, and putative requirement of TAB1 as a mediator of AMPK-TAK1 binding in THP-1 cells was not investigated.

In another recent study, AMPKα1 was suggested to participate in renal TAK1 activation and TAK1-dependent signalling induced by angiotensin-II [26]. Angiotensin-II increased the phosphorylation of TAK1 (S412) in renal tissue of AMPKα1+/+ mice but not AMPKα1−/− mice. Notably, S412 is targeted by PKA [46]. Furthermore, the authors also observe that angiotensin-II upregulates the AMPKα1 isoform in renal tissue, and increased TAK1-target gene mRNA and renal protein expression in AMPKα1+/+ mice, but less-so in AMPKα1−/− mice [47]. Using the same argumentation as above, if AMPKα indeed acts as a scaffold for TAB1, one could predict that TAK1 activity is downregulated in AMPKα knockouts.

Therefore, AMPK may be involved in TAK1 activation, but not necessarily as an upstream kinase. Importantly, to date, there is only circumstantial evidence for AMPK to activate TAK1, whereas biochemical proof is available and functional data is accumulating to support TAK1 as a genuine direct AMPK activating kinase.

8. Conclusions

About 12 years after the original publication reporting TAK1 as a 'candidate' AMPK kinase [1], as argued above, the collective data rather confirms the suggested authentic role. Thus, I propose to accept TAK1, in addition to LKB1 and CaMKK2, as the third genuine upstream kinase of AMPK (Figure 1).

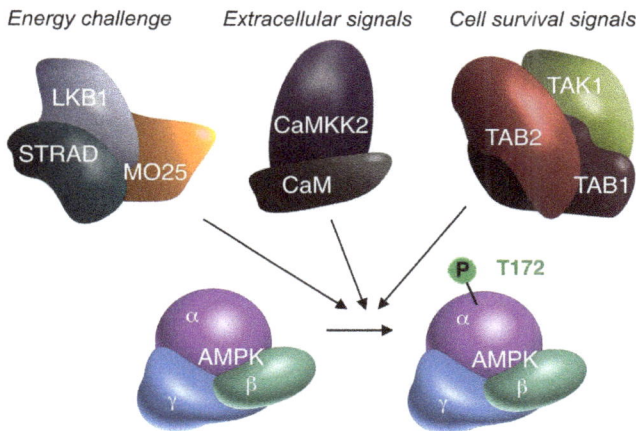

Figure 1. The three alternative AMPK kinases. Biochemical (cell-free), cell biological (in vitro) and animal (in vivo) experimentation suggest that TAK1 can activate AMPKα by phosphorylation of the critical T172 residue. Summative evidence therefore supports TAK1 as an additional AMPK upstream kinase, besides LKB1 and CaMKK2. AMPK may receive (simultaneous) activation from all three upstream kinases. The original signal leading to AMPK activation may differ per upstream kinase, as suggested above. All four kinases are depicted with their accessory subunits, as functional protein complexes. The requirement of TAB1/TAB2/TAB3 for AMPK activation has not been fully elucidated. However, to date TAB1 and/or TAB2 are the most likely candidates. TAB1 may also bind to AMPK independent of TAK1 [30]. MO25: mouse protein 25; STRAD: STE20-related kinase adapter protein; CaM: Calmodulin.

funding: The author has received funding from the Netherlands Organization of Scientific Research VIDI-Innovational Research Grant (NWO-ALW grant no. 864.10.007).

Acknowledgments: I thank Erik Biessen and his team members in the Department of Pathology at Maastricht University for support and the kind atmosphere during the preparation of this manuscript. I also thank the anonymous reviewers for careful reading and Olivia Waring for language editing.

Conflicts of Interest: The author declares no conflict of interest. The funders had no role in the writing of the manuscript and in the decision to publish.

Abbreviations

AMPK	AMP-activated protein kinase
CaMKK2	Ca^{2+}/Calmodulin-dependent protein kinase kinase 2
LKB1	Liver kinase B1
T172	Threonine 172 residue (of AMPKα)
TAB1	TAK1 binding protein 1
TAB2	TAK1 binding protein 2
TAB3	TAK1 binding protein 3
TAK1	Transforming growth factor β-activated protein kinase
TNF-α	Tumour necrosis factor α
TRAIL	Tumor necrosis factor related apoptosis inducing ligand

References

1. Hardie, D.G. Molecular pathways: Is AMPK a friend or a foe in cancer? *Clin. Cancer Res.* **2015**, *21*, 3836–3840. [CrossRef] [PubMed]
2. Lopez, M.; Nogueiras, R.; Tena-Sempere, M.; Dieguez, C. Hypothalamic AMPK: A canonical regulator of whole-body energy balance. *Nat. Rev. Endocrinol.* **2016**, *12*, 421–432. [CrossRef] [PubMed]
3. Day, E.A.; Ford, R.J.; Steinberg, G.R. AMPK as a therapeutic target for treating metabolic diseases. *Trends Endocrinol. Metab.* **2017**, *28*, 545–560. [CrossRef] [PubMed]
4. Salt, I.P.; Hardie, D.G. AMP-activated protein kinase: An ubiquitous signaling pathway with key roles in the cardiovascular system. *Circ. Res.* **2017**, *120*, 1825–1841. [CrossRef] [PubMed]
5. Hardie, D.G.; Ross, F.A.; Hawley, S.A. AMPK: A nutrient and energy sensor that maintains energy homeostasis. *Nat. Rev. Mol. Cell Biol.* **2012**, *13*, 251–262. [CrossRef] [PubMed]
6. Lin, S.C.; Hardie, D.G. AMPK: Sensing glucose as well as cellular energy status. *Cell Metab.* **2018**, *27*, 299–313. [CrossRef] [PubMed]
7. Hardie, D.G.; Schaffer, B.E.; Brunet, A. AMPK: An energy-sensing pathway with multiple inputs and outputs. *Trends Cell Biol.* **2016**, *26*, 190–201. [CrossRef] [PubMed]
8. Hardie, D.G. Keeping the home fires burning: AMP-activated protein kinase. *J. R. Soc. Interface* **2018**, *15*. [CrossRef] [PubMed]
9. Hardie, D.G.; Ashford, M.L. AMPK: Regulating energy balance at the cellular and whole body levels. *Physiology* **2014**, *29*, 99–107. [CrossRef] [PubMed]
10. Carling, D.; Mayer, F.V.; Sanders, M.J.; Gamblin, S.J. AMP-activated protein kinase: Nature's energy sensor. *Nat. Chem. Biol.* **2011**, *7*, 512–518. [CrossRef] [PubMed]
11. Hawley, S.A.; Boudeau, J.; Reid, J.L.; Mustard, K.J.; Udd, L.; Makela, T.P.; Alessi, D.R.; Hardie, D.G. Complexes between the LKB1 tumor suppressor, STRADα/β and MO25α/β are upstream kinases in the AMP-activated protein kinase cascade. *J. Biol.* **2003**, *2*, 28. [CrossRef] [PubMed]
12. Woods, A.; Johnstone, S.R.; Dickerson, K.; Leiper, F.C.; Fryer, L.G.; Neumann, D.; Schlattner, U.; Wallimann, T.; Carlson, M.; Carling, D. LKB1 is the upstream kinase in the AMP-activated protein kinase cascade. *Curr. Biol.* **2003**, *13*, 2004–2008. [CrossRef] [PubMed]
13. Hawley, S.A.; Pan, D.A.; Mustard, K.J.; Ross, L.; Bain, J.; Edelman, A.M.; Frenguelli, B.G.; Hardie, D.G. Calmodulin-dependent protein kinase kinase-β is an alternative upstream kinase for AMP-activated protein kinase. *Cell Metab.* **2005**, *2*, 9–19. [CrossRef] [PubMed]

14. Woods, A.; Dickerson, K.; Heath, R.; Hong, S.P.; Momcilovic, M.; Johnstone, S.R.; Carlson, M.; Carling, D. Ca^{2+}/calmodulin-dependent protein kinase kinase-β acts upstream of AMP-activated protein kinase in mammalian cells. *Cell Metab.* **2005**, *2*, 21–33. [CrossRef] [PubMed]

15. Ajibade, A.A.; Wang, H.Y.; Wang, R.F. Cell type-specific function of TAK1 in innate immune signaling. *Trends Immunol.* **2013**, *34*, 307–316. [CrossRef] [PubMed]

16. Dai, L.; Aye Thu, C.; Liu, X.Y.; Xi, J.; Cheung, P.C. TAK1, more than just innate immunity. *IUBMB Life* **2012**, *64*, 825–834. [CrossRef] [PubMed]

17. Mihaly, S.R.; Ninomiya-Tsuji, J.; Morioka, S. TAK1 control of cell death. *Cell Death Differ.* **2014**, *21*, 1667–1676. [CrossRef] [PubMed]

18. Momcilovic, M.; Hong, S.P.; Carlson, M. Mammalian TAK1 activates SNF1 protein kinase in yeast and phosphorylates AMP-activated protein kinase in vitro. *J. Biol. Chem.* **2006**, *281*, 25336–25343. [CrossRef] [PubMed]

19. Hong, S.P.; Leiper, F.C.; Woods, A.; Carling, D.; Carlson, M. Activation of yeast SNF1 and mammalian AMP-activated protein kinase by upstream kinases. *Proc. Natl. Acad. Sci. USA* **2003**, *100*, 8839–8843. [CrossRef] [PubMed]

20. Sutherland, C.M.; Hawley, S.A.; McCartney, R.R.; Leech, A.; Stark, M.J.; Schmidt, M.C.; Hardie, D.G. Elm1p is one of three upstream kinases for the *Saccharomyces cerevisiae* SNF1 complex. *Curr. Biol.* **2003**, *13*, 1299–1305. [CrossRef]

21. Xie, M.; Zhang, D.; Dyck, J.R.; Li, Y.; Zhang, H.; Morishima, M.; Mann, D.L.; Taffet, G.E.; Baldini, A.; Khoury, D.S.; et al. A pivotal role for endogenous TGF-β-activated kinase-1 in the LKB1/AMP-activated protein kinase energy-sensor pathway. *Proc. Natl. Acad. Sci. USA* **2006**, *103*, 17378–17383. [CrossRef] [PubMed]

22. Grahame Hardie, D. AMP-activated protein kinase: A key regulator of energy balance with many roles in human disease. *J. Intern. Med.* **2014**, *276*, 543–559. [CrossRef] [PubMed]

23. Neumann, D.; Suter, M.; Tuerk, R.; Riek, U.; Wallimann, T. Co-expression of LKB1, MO25α and STRADα in bacteria yield the functional and active heterotrimeric complex. *Mol. Biotechnol.* **2007**, *36*, 220–231. [CrossRef] [PubMed]

24. Alessi, D.R.; Sakamoto, K.; Bayascas, J.R. LKB1-dependent signaling pathways. *Annu. Rev. Biochem.* **2006**, *75*, 137–163. [CrossRef] [PubMed]

25. Kim, S.Y.; Jeong, S.; Jung, E.; Baik, K.H.; Chang, M.H.; Kim, S.A.; Shim, J.H.; Chun, E.; Lee, K.Y. AMP-activated protein kinase-α1 as an activating kinase of TGF-β-activated kinase 1 has a key role in inflammatory signals. *Cell Death Dis.* **2012**, *3*, e357. [CrossRef] [PubMed]

26. Mia, S.; Castor, T.; Musculus, K.; Voelkl, J.; Alesutan, I.; Lang, F. Role of AMP-activated protein kinase α1 in angiotensin-II-induced renal TGFβ-activated kinase 1 activation. *Biochem. Biophys. Res. Commun.* **2016**, *476*, 267–272. [CrossRef] [PubMed]

27. Scholz, R.; Sidler, C.L.; Thali, R.F.; Winssinger, N.; Cheung, P.C.; Neumann, D. Autoactivation of transforming growth factor β-activated kinase 1 is a sequential bimolecular process. *J. Biol. Chem.* **2010**, *285*, 25753–25766. [CrossRef] [PubMed]

28. Neumann, D.; Woods, A.; Carling, D.; Wallimann, T.; Schlattner, U. Mammalian AMP-activated protein kinase: Functional, heterotrimeric complexes by co-expression of subunits in *Escherichia coli*. *Protein Expr. Purif.* **2003**, *30*, 230–237. [CrossRef]

29. Tanno, M.; Bassi, R.; Gorog, D.A.; Saurin, A.T.; Jiang, J.; Heads, R.J.; Martin, J.L.; Davis, R.J.; Flavell, R.A.; Marber, M.S. Diverse mechanisms of myocardial p38 mitogen-activated protein kinase activation: Evidence for MKK-independent activation by a TAB1-associated mechanism contributing to injury during myocardial ischemia. *Circ. Res.* **2003**, *93*, 254–261. [CrossRef] [PubMed]

30. Li, J.; Miller, E.J.; Ninomiya-Tsuji, J.; Russell, R.R., 3rd; Young, L.H. AMP-activated protein kinase activates p38 mitogen-activated protein kinase by increasing recruitment of p38 MAPK to TAB1 in the ischemic heart. *Circ. Res.* **2005**, *97*, 872–879. [CrossRef] [PubMed]

31. Herrero-Martin, G.; Hoyer-Hansen, M.; Garcia-Garcia, C.; Fumarola, C.; Farkas, T.; Lopez-Rivas, A.; Jaattela, M. TAK1 activates AMPK-dependent cytoprotective autophagy in trail-treated epithelial cells. *EMBO J.* **2009**, *28*, 677–685. [CrossRef] [PubMed]

32. Sakurai, H. Targeting of TAK1 in inflammatory disorders and cancer. *Trends Pharmacol. Sci.* **2012**, *33*, 522–530. [CrossRef] [PubMed]

33. Hardie, D.G. AMPK–sensing energy while talking to other signaling pathways. *Cell Metab.* **2014**, *20*, 939–952. [CrossRef] [PubMed]

34. Lee, Y.S.; Kim, Y.S.; Lee, S.Y.; Kim, G.H.; Kim, B.J.; Lee, S.H.; Lee, K.U.; Kim, G.S.; Kim, S.W.; Koh, J.M. AMP kinase acts as a negative regulator of RANKL in the differentiation of osteoclasts. *Bone* **2010**, *47*, 926–937. [CrossRef] [PubMed]

35. Wang, B.; Wang, X.B.; Chen, L.Y.; Huang, L.; Dong, R.Z. Belinostat-induced apoptosis and growth inhibition in pancreatic cancer cells involve activation of TAK1-AMPK signaling axis. *Biochem. Biophys. Res. Commun.* **2013**, *437*, 1–6. [CrossRef] [PubMed]

36. Zippel, N.; Malik, R.A.; Fromel, T.; Popp, R.; Bess, E.; Strilic, B.; Wettschureck, N.; Fleming, I.; Fisslthaler, B. Transforming Growth Factor-β–Activated Kinase 1 Regulates Angiogenesis via AMP-Activated Protein Kinase-α1 and Redox Balance in Endothelial Cells. *Arterioscler. Thromb. Vasc. Biol.* **2013**, *33*, 2792–2799. [CrossRef] [PubMed]

37. Lv, G.; Zhu, H.; Zhou, F.; Lin, Z.; Lin, G.; Li, C. Amp-activated protein kinase activation protects gastric epithelial cells from helicobacter pylori-induced apoptosis. *Biochem. Biophys. Res. Commun.* **2014**, *453*, 13–18. [CrossRef] [PubMed]

38. Jing, Y.; Liu, W.; Cao, H.; Zhang, D.; Yao, X.; Zhang, S.; Xia, H.; Li, D.; Wang, Y.C.; Yan, J.; et al. Hepatic p38α regulates gluconeogenesis by suppressing AMPK. *J. Hepatol.* **2015**, *62*, 1319–1327. [CrossRef] [PubMed]

39. Xu, X.; Sun, J.; Song, R.; Doscas, M.E.; Williamson, A.J.; Zhou, J.; Sun, J.; Jiao, X.; Liu, X.; Li, Y. Inhibition of p70 S6 kinase (S6k1) activity by a77 1726, the active metabolite of leflunomide, induces autophagy through tak1-mediated AMPK and JNK activation. *Oncotarget* **2017**, *8*, 30438–30454. [CrossRef] [PubMed]

40. Flusberg, D.A.; Sorger, P.K. Surviving apoptosis: Life-death signaling in single cells. *Trends Cell Biol.* **2015**, *25*, 446–458. [CrossRef] [PubMed]

41. Kroemer, G.; Marino, G.; Levine, B. Autophagy and the integrated stress response. *Mol. Cell* **2010**, *40*, 280–293. [CrossRef] [PubMed]

42. Zadra, G.; Batista, J.L.; Loda, M. Dissecting the dual role of AMPK in cancer: From experimental to human studies. *Mol. Cancer Res.* **2015**, *13*, 1059–1072. [CrossRef] [PubMed]

43. Inokuchi-Shimizu, S.; Park, E.J.; Roh, Y.S.; Yang, L.; Zhang, B.; Song, J.; Liang, S.; Pimienta, M.; Taniguchi, K.; Wu, X.; et al. TAK1-mediated autophagy and fatty acid oxidation prevent hepatosteatosis and tumorigenesis. *J. Clin. Investig.* **2014**, *124*, 3566–3578. [CrossRef] [PubMed]

44. Seki, E. TAK1-dependent autophagy: A suppressor of fatty liver disease and hepatic oncogenesis. *Mol. Cell. Oncol.* **2014**, *1*, e968507. [CrossRef] [PubMed]

45. Liu, W.; Jiang, Y.; Sun, J.; Geng, S.; Pan, Z.; Prinz, R.A.; Wang, C.; Sun, J.; Jiao, X.; Xu, X. Activation of TGF-β-activated kinase 1 (TAK1) restricts *Salmonella* Typhimurium growth by inducing AMPK activation and autophagy. *Cell Death Dis.* **2018**, *9*, 570. [CrossRef] [PubMed]

46. Kobayashi, Y.; Mizoguchi, T.; Take, I.; Kurihara, S.; Udagawa, N.; Takahashi, N. Prostaglandin E2 enhances osteoclastic differentiation of precursor cells through protein kinase A-dependent phosphorylation of TAK1. *J. Biol. Chem.* **2005**, *280*, 11395–11403. [CrossRef] [PubMed]

47. Mia, S.; Federico, G.; Feger, M.; Pakladok, T.; Meissner, A.; Voelkl, J.; Groene, H.J.; Alesutan, I.; Lang, F. Impact of AMP-activated protein kinase α1 deficiency on tissue injury following unilateral ureteral obstruction. *PLoS ONE* **2015**, *10*, e0135235. [CrossRef] [PubMed]

International Journal of
Molecular Sciences

MDPI

Review

Interactive Roles for AMPK and Glycogen from Cellular Energy Sensing to Exercise Metabolism

Natalie R. Janzen, Jamie Whitfield and Nolan J. Hoffman *

Exercise and Nutrition Research Program, Mary MacKillop Institute for Health Research, Australian Catholic University, Level 5, 215 Spring Street, Melbourne, Victoria 3000, Australia; natalie.janzen@acu.edu.au (N.R.J.); jamie.whitfield@acu.edu.au (J.W.)
* Correspondence: Nolan.Hoffman@acu.edu.au; Tel.: +61-3-9230-8277

Received: 30 August 2018; Accepted: 23 October 2018; Published: 26 October 2018

Abstract: The AMP-activated protein kinase (AMPK) is a heterotrimeric complex with central roles in cellular energy sensing and the regulation of metabolism and exercise adaptations. AMPK regulatory β subunits contain a conserved carbohydrate-binding module (CBM) that binds glycogen, the major tissue storage form of glucose. Research over the past two decades has revealed that the regulation of AMPK is impacted by glycogen availability, and glycogen storage dynamics are concurrently regulated by AMPK activity. This growing body of research has uncovered new evidence of physical and functional interactive roles for AMPK and glycogen ranging from cellular energy sensing to the regulation of whole-body metabolism and exercise-induced adaptations. In this review, we discuss recent advancements in the understanding of molecular, cellular, and physiological processes impacted by AMPK-glycogen interactions. In addition, we appraise how novel research technologies and experimental models will continue to expand the repertoire of biological processes known to be regulated by AMPK and glycogen. These multidisciplinary research advances will aid the discovery of novel pathways and regulatory mechanisms that are central to the AMPK signaling network, beneficial effects of exercise and maintenance of metabolic homeostasis in health and disease.

Keywords: AMP-activated protein kinase; glycogen; exercise; metabolism; cellular energy sensing; energy utilization; liver; skeletal muscle; metabolic disease; glycogen storage disease

1. Introduction

The AMP-activated protein kinase (AMPK) is a heterotrimer composed of a catalytic α subunit and regulatory β and γ subunits, which becomes activated in response to a decrease in cellular energy status. Activation of AMPK results in metabolic adaptations such as increases in glucose uptake and glycolytic flux and fatty acid (FA) oxidation. AMPK activation simultaneously inhibits anabolic processes including protein and FA synthesis. AMPK can also translocate to the nucleus where it regulates transcription factors to increase energy production, meet cellular energy demands and inhibit cell growth and proliferation. Conversely, when energy levels are replete, AMPK activity returns to basal levels, allowing anabolic processes to resume. Given its central roles in cellular metabolic and growth signaling pathways, AMPK remains an appealing target for treating a range of pathologies associated with obesity and aging, including metabolic diseases such as obesity and type 2 diabetes (T2D).

In response to changes in energy supply and demand, glycogen, predominately stored in the liver and skeletal muscle, serves as an important source of energy to maintain metabolic homeostasis. Glycogen is synthesized by the linking of glucose monomers during periods of nutrient excess. In response to energy stress and decreased arterial glucose concentration, rising glucagon levels induce increased hepatic glucose output by promoting the breakdown of glycogen and the conversion of non-glucose substrates into glucose. The newly formed glucose is released into the bloodstream to help restore blood glucose levels. Skeletal muscle glycogen serves as an accessible source of glucose to form adenosine triphosphate (ATP) and to reduce equivalents via glycolytic and oxidative phosphorylation pathways during muscle contraction.

A significant body of evidence demonstrates that AMPK binds glycogen. This physical interaction is mediated by the carbohydrate-binding module (CBM) located within the AMPK β subunit and is thought to allow AMPK to function as a sensor of stored cellular energy. While glycogen is stored in multiple tissues throughout the body, this review will primarily focus on the physical interactions underlying the AMPK β subunit binding to glycogen and its potential functional links to glycogen storage dynamics in the liver and skeletal muscles, as these tissues are central to metabolic and exercise-regulated biological processes. In addition, as the majority of research on this topic has been undertaken in human and mouse model systems, studies in these species will be highlighted. Following a brief background on the regulation of AMPK and glycogen, this review will critically assess the recent advances and focus primarily on studies within the past two decades that have added to our understanding of the physical basis of AMPK-glycogen binding and its potential functional interactions in exercise and metabolism. Key remaining biological questions related to the interactive roles of AMPK and glycogen will be posed along with a discussion of research advancements that are feasible in the next decade with new technologies and experimental models to determine how AMPK-glycogen binding may be therapeutically targeted in health and disease.

2. Roles for AMPK and Glycogen in Metabolism

2.1. AMPK Activation and Signaling

Structural biology-based studies over the past decade have provided new insights into the molecular mechanisms by which AMPK activation is regulated by nucleotides, including changes in AMP:ATP and ADP:ATP ratios (i.e., adenylate energy charge) that occur in response to cellular energy stress [1]. The binding of AMP and ADP to the cystathionine-β-synthase (CBS) domains of the γ subunit promotes AMPK activation through several complementary mechanisms. The binding of AMP promotes AMPK association with liver kinase B1 (LKB1) and the scaffolding protein axin which enhances the effect of T172 phosphorylation [2,3], the primary phosphorylation and activation site in the AMPK α subunit, while simultaneously preventing its dephosphorylation by protein phosphatases [4–7]. This activation is mediated by myristoylation of the G2 site on the N-terminus of the β subunit (Figure 1), which promotes AMPK association with cellular membranes and LKB1 [2]. Furthermore, the binding of AMP, but not ADP, can cause the allosteric activation of AMPK without T172 phosphorylation [5,7–9]. The combination of allosteric activation by nucleotides and increased T172 phosphorylation by LKB1 can increase AMPK activity 1000-fold [10]. Additionally, the phosphorylation of T172 can be regulated by changes in intracellular Ca^{2+} concentrations via the upstream kinase calcium/calmodulin-dependent protein kinase kinase β (CAMKK2) in the absence of changes in adenylate energy charge [11–13].

Once activated, AMPK serves as a metabolic 'switch' to promote catabolic pathways and inhibit anabolic processes. For example, AMPK increases glucose uptake into skeletal muscle by phosphorylating and inhibiting Tre-2, BUB2, CDC16, 1 domain family, members 1 (TBC1D1) and 4 (TBC1D4), promoting glucose transporter 4 (GLUT4) vesicle translocation to the sarcolemmal membrane [14–18]. AMPK also functions in the regulation of lipids, acutely promoting lipid oxidation and inhibiting FA synthesis, primarily through phosphorylation and the inhibition of acetyl-CoA carboxylase (ACC) [19,20]. At the transcriptional level, AMPK phosphorylates and inhibits sterol regulatory element-binding protein 1, a transcription factor that regulates lipid synthesis [21]. Mitochondrial biogenesis is stimulated by AMPK activity through an increase in the peroxisome proliferator-activated receptor gamma coactivator 1-alpha (PGC-1α) transcription, thereby promoting oxidative metabolism [22]. AMPK can also inhibit anabolic pathways by phosphorylating the regulatory associated protein of the mechanistic target of rapamycin (mTOR) (Raptor) and the tuberous sclerosis complex 2 (TSC2), which in turn inactivates mTOR and prevents the phosphorylation of its substrates [23–25].

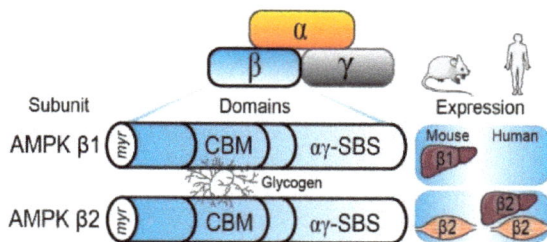

Figure 1. The AMPK is a heterotrimeric protein, consisting of a catalytic α subunit and regulatory β and γ subunits. The β subunit (β1 and β2 isoforms) possesses a glycogen-binding domain (CBM) that mediates AMPK's interaction with glycogen, an N-terminal myristoylation site (myr) and an αγ subunit binding sequence (αγ-SBS) involved in the heterotrimeric complex formation. Tissue expression of the β1 and β2 isoforms varies between humans and mice, as the β2 isoform is predominately expressed in both human liver and skeletal muscles, while mice predominately express the β1 isoform in the liver and the β2 isoform in the skeletal muscles.

2.2. The AMPK β Subunit and Carbohydrate-Binding Module

The AMPK β subunit exists in two isoforms (β1 and β2) and serves as a scaffolding subunit that binds to the AMPK catalytic α and regulatory γ subunits, playing an important role in the physical stability of the heterotrimer (Figure 1) [9,26,27]. In human and mouse skeletal muscles, the β2 isoform is predominantly expressed (Table 1) [28]. In contrast, the liver β subunit isoform expression differs across the mammalian species: the β1 isoform is predominantly expressed in mice, while β2 is predominantly expressed in humans [29,30]. However, despite isoform differences between species, both the β subunit isoforms contain the CBM which mediates physical AMPK-glycogen interaction and binding. Furthermore, the CBM is highly conserved between species, suggesting that the region possesses evolutionary significance and plays similar roles across species [31,32]. The CBM spans residues 68–163 of the β1 subunit and residues 67–163 of the β2 subunit [32,33] and is nearly identical in structure and sequence in both isoforms, with the major difference being the insertion of a threonine at residue 101 in the β2 CBM [34,35]. This insertion is believed to have occurred early in evolutionary history and provides the β2 CBM with a higher affinity for glycogen [33,34]. However, the reason for this divergence in β subunit isoforms is unknown.

Table 1. The AMPK β subunit isoform distribution in human and mouse tissues.

Tissue	B1	B2
Human vastus lateralis	ND	~100%
Human liver	ND	~100%
Mouse extensor digitorum longus	5%	95%
Mouse soleus	18%	78%
Mouse liver	100%	ND

Adapted from References [10,28–30,36]. ND, nondetectable.

2.3. Glycogen Dynamics

A number of proteins are associated with glycogen particles and function as regulators of glycogen synthesis, breakdown, particle size, and degree of branching. Glycogenin initiates glycogen formation and functions as the central protein of the glycogen particle [37]. Glycogen synthase (GS) is the rate-limiting enzyme in glycogen synthesis responsible for attaching UDP-glucose donors together in α-1,4 linkages, the linear links of the glycogen particle. As glucose-6-phosphate (G6P) is a precursor to UDP-glucose, its accumulation is a potent activator of GS, capable of overriding the inhibitory effects of phosphorylation mediated by proteins such as AMPK, glycogen synthase kinase 3, and protein kinase A [38,39]. As its name implies, glycogen branching enzyme (GBE) is responsible for introducing α-1,6 branch points to the growing glycogen particle. The rate-limiting enzyme of glycogen breakdown is glycogen phosphorylase (GP), which is known to be activated by elevated intracellular Ca^{2+}, epinephrine and cAMP concentrations [40,41]. When activated, GP degrades the α-1,4 links of glycogen particles and removes glycosyl units from the non-reducing ends of the glycogen particle [42]. The glycogen debranching enzyme (GDE) assists with the degradation of glycogen and is responsible for breaking the α-1,6 links to allow continued GP activity. Without GDE, GP can only degrade the outer tiers of glycogen particles and stops four glucose residues short of the α-1,6 branch point [39]. Further details regarding glycogen synthesis and breakdown are beyond the scope of this review, and readers are referred to other reviews covering this topic [39,43].

2.4. Glycogen Localization

Recently, there has been an increasing interest regarding the significance of glycogen's subcellular localization in skeletal muscles [44–48]. Glycogen can be concentrated beneath the sarcolemma (subsarcolemmal; SS), between the myofibers along the I band near the mitochondria and sarcoplasmic reticulum (intermyofibrillar; interMF), or within myofibers near the triad junction (intramyofibrillar; intraMF) [44,46]. Depletion of these different glycogen pools impacts muscle function and fatigue, such as impairing Ca^{2+} release and reuptake. Therefore, it has been hypothesized that these different pools of glycogen play a significant role in muscle contraction and fatigue beyond their role as an energy substrate [45,47,48].

Human skeletal muscles contain large stores of glycogen, which can exceed 100 mmol glucosyl units/kg wet weight (~500 mmol/kg dry weight) in the vastus lateralis muscle [49,50] and are primarily concentrated in the interMF space [49]. Conversely, rodents tend to have higher stores of glycogen in the liver compared to the skeletal muscle. For example, mice store about 120 μmol/g wet weight in liver and 15–20 μmol glucose/g wet weight in the type IIA flexor digitorum brevis muscle, with the highest concentrations in the intraMF pool [51]. Interestingly, while the relative contributions of intraMF glycogen to total glycogen are different between humans and mice, the intraMF content as a percentage of the total fiber volume is very similar between species [49,51]. Additionally, substrate utilization is different between species during exercise, as humans rely predominately on intramuscular stores and rodents rely on blood-borne substrates [52–55]. These differences in glycogen storage and utilization between humans and rodents are important considerations in the study design across species when assessing glycogen depletion and/or repletion.

3. Molecular Evidence of AMPK-Glycogen Binding

In 2003, it was first demonstrated that recombinant AMPK β1 CBM bound glycogen using a cell-free assay system [31]. Structural prediction and mutagenesis experiments targeting conserved residues within the CBM thought to mediate glycogen binding demonstrated that W100G and K126Q mutations abolished glycogen binding to the isolated β1 CBM, while W133L, S108E, and G147R mutations partially disrupted glycogen binding. Additionally, the AMPK heterotrimeric complex was found to bind glycogen more tightly than the β1 subunit in isolation; however, the reasons for this differential binding affinity remain unclear [31]. In support of these findings, cell-free assays have also revealed that glycogen has an inhibitory effect on AMPK activity [56]. Furthermore, mutation of critical residues in the β1 CBM (W100G, W133A, K126A, L148A, and T148A) ablated glycogen's inhibition. In these cell-free assays, glycogen with higher branch points had a greater inhibitory effect on AMPK, indicating that glycogen particle size has the capacity to influence AMPK-glycogen interactions [56]. It was also observed that glycogen particles co-localized with the β subunit of AMPK in the cytoplasm of CCL13 cells [57]. A follow-up structural-based study determined the CBM crystal structure in the presence of β-cyclodextrin and confirmed that AMPK indeed interacts with glycogen [32]. Additional experimental approaches such as immunogold cytochemistry have also shown that the AMPK α and β subunits of rat liver tissue are associated with the surface of glycogen particles in situ, providing further molecular evidence supporting this concept of physical AMPK-glycogen interaction [58].

In addition to its role in binding glycogen, the CBM of the β subunit also physically interacts with the kinase domain of the catalytic α subunit, forming a pocket, referred to as the allosteric drug and metabolite (ADaM) site. Small molecule AMPK activators such as A-769662, a β1 subunit specific activator, bind to this site and directly activate AMPK [27,59]. However, to date, any connection of A-769662's subunit specificity in relation to glycogen has been highly speculative and further research is required to establish potential direct links. The ADaM site is stabilized by autophosphorylation of S108 on the CBM and is dissociated when T172 on the α subunit is dephosphorylated [9]. Mutation of S108 to a phosphomimetic glutamic acid (S108E) resulted in reduced glycogen binding [31] and increased AMPK activity in response to AMP and A-769662, even in the presence of a non-phosphorylatable T172A mutation [8]. Conversely, mutation of S108 to a neutral alanine (S108A) had no effect on glycogen binding [31], but reduced AMPK activity in response to AMP and A-769662 [8]. Collectively, these findings infer that glycogen binding may inhibit AMPK activity by disrupting the interaction between the CBM and the kinase domain of the α subunit [1,9,56]. The inhibitory role of the β subunit T148 autophosphorylation on AMPK-glycogen binding has also been a focus of recent research. The mutation of T148 to a phosphomimetic aspartate (T148D) on the β1 subunit inhibits AMPK-glycogen binding in cellular systems [60]. The results from subsequent experiments in isolated rat skeletal muscle suggest that T148 is constitutively phosphorylated both at rest and following electrical stimulation, therefore, preventing glycogen from associating with the AMPK β2 subunit [61]. Further research is necessary to further elucidate the role of T148 in the context of AMPK-glycogen interactions.

Recent research has provided further structural insights into the affinity of the AMPK β subunits for carbohydrates. Isolated β2 CBM has a stronger affinity for carbohydrates than the β1 CBM, binding strongly to both branched and unbranched carbohydrates, with a preference for single α-1,6 branched carbohydrates [34]. One possible explanation for this difference is that a pocket is formed in the CBM by the T101 residue, which is unique to the β2 subunit, therefore, allowing binding to branched carbohydrates [33,34]. In addition, the β1 CBM possesses a threonine at residue 134 which may form a hydrogen bond with the neighboring W133, restricting the ability of the β1 CBM to accommodate carbohydrates, while the β2 CBM possesses a valine which does not bond with W133 [34]. This difference may explain the increased affinity of the β2 subunit for branched carbohydrates even though the 134 residue does not directly contact carbohydrates [34]. These findings indicate that the glycogen structure and branching affect AMPK binding, specifically to β1 subunits, which may dictate the inhibitory effect of glycogen observed in previous studies [34,56]. While the role of AMPK β isoform glycogen binding in the contexts of glycogen structure and branching has been investigated in vitro, it remains to be determined how these characteristics alter the dynamics of AMPK-glycogen binding in vivo.

4. Regulation of Cellular Energy Sensing by AMPK-Glycogen Binding

Several independent lines of evidence suggest that these physical AMPK and glycogen interactions also serve mechanistic functional roles in cellular energy sensing. A number of AMPK substrates are known to be directly involved in glycogen storage and breakdown, highlighting AMPK's role as an important regulator of glycogen metabolism. In vitro, AMPK regulates glycogen synthesis directly via the phosphorylation and inactivation of GS at site 2 [62]. In support of this finding, AMPK α2, but not α1, knockout (KO) mice display blunted phosphorylation of GS at site 2 and higher GS activity in response to stimulation by the AMPK activator 5-aminoimidazole-4-carboxamide ribonucleotide (AICAR) in skeletal muscle [63]. Paradoxically, chronic activation of AMPK also results in an accumulation of glycogen in skeletal and cardiac muscles [38]. While these divergent outcomes appear contradictory, it has been proposed that prolonged AMPK activation leads to glycogen accumulation by increasing glucose uptake and, subsequently, by increasing intracellular G6P, a known allosteric activator of GS. This hypothesis is further supported by recent independent findings using highly specific and potent pharmacological activators demonstrating that skeletal muscle AMPK activation results in increased skeletal muscle glucose uptake and glycogen synthesis in mice and non-human primates [64,65]. This accumulation of G6P overcomes the inhibition of GS by AMPK, thereby increasing GS activity [38]. Furthermore, AMPK activation also shifts fuel utilization towards FA oxidation post-exercise, allowing glucose to be utilized for glycogen resynthesis [66]. In addition to regulating GS activity, phosphorylation of GS at site 2 by AMPK causes GS to localize to the SS and interMF glycogen pools in humans [67]. These findings have been replicated in mouse models, as an R70Q mutation of the AMPK γ1 subunit results in the chronic activation of AMPK and glycogen accumulation in the skeletal muscle interMF region [68]. This has led to the suggestion that AMPK specifically senses and responds to interMF levels of glycogen [69]; however, further research is warranted to verify this hypothesis.

An increase in GP activity has also been observed to be associated with AMPK activation induced by AICAR treatment of isolated rat soleus muscles [70,71]. However, the ability to demonstrate a direct relationship between AMPK and GP activity has been limited by the identification of several AMPK-independent targets of AICAR, including phosphofructokinase, protein kinase C, and heat shock protein 90 [72]. Further research is therefore required to determine if this speculated relationship exists and elucidate the mechanism by which AMPK may regulate GP. In contrast, there is in vitro evidence that GDE binds to residues 68–123 of the AMPK β1 subunit [73]. Mutations in this region that disrupt glycogen binding (W100G and K128Q) do not affect the binding to GDE, indicating that GDE-AMPK binding is not likely mediated by glycogen [73]. AMPK's direct positive effect on glycogen accumulation, its known interaction with glycogen-associated proteins, and its ability to promote energy production through glucose uptake and fat oxidation when glycogen levels are low all support AMPK's role as a cellular energy sensor. Given the limited in vivo data currently available directly linking AMPK to glycogen-associated proteins, additional studies are necessary to further understand the potential direct binding partners and effects of AMPK on the glycogen-associated proteome.

It is important to consider additional factors that may impact physical and functional AMPK-glycogen interactions. In a proteomic screen utilizing purified glycogen from rat liver, AMPK was not included in the proteins detected to be associated with glycogen [74], and this has been replicated in a complementary study in adipocytes [75]. The authors suggested that this may be due to either AMPK protein below the level of detection being able to regulate glycogen or the predominance of the AMPK β1 subunit expression in the tissues studied, as this isoform has a lower affinity for glycogen compared to the β2 subunit [74,75]. In future studies interrogating AMPK and glycogen binding and functional interactions, considerations of the β subunit isoform expression and glycogen localization, as well as sample preparation and experimental variables that may limit the preservation and detection of AMPK-glycogen binding, are warranted in future studies to build upon this strong foundation of molecular and cellular evidence.

5. Linking AMPK and Glycogen to Exercise Metabolism in Physiological Settings

5.1. Regulation of Glycogen Storage by AMPK

In the fifteen years following the discovery of glycogen binding to the CBM on the β subunit, several studies utilizing AMPK isoform knockout (KO) mouse models have provided whole-body physiological evidence of AMPK's interactive functional roles with glycogen. Collectively, studies using AMPK α and β subunit KO mouse models have found that the ablation of AMPK alters liver and skeletal muscle glycogen content, supporting the role of AMPK in the regulation of tissue glycogen dynamics in vivo. Specifically, whole-body β2 KO mice have reduced basal glycogen levels in both liver and skeletal muscles associated with reduced muscle AMPK activity and attenuated maximal and submaximal running capacity compared to wild-type (WT) mice [76]. β2 KO mice also display reduced expression and activity of α1 and α2 subunits as well as compensatory upregulation of the β1 subunit in skeletal muscle [76]. Additional experiments utilizing this β2 KO model have demonstrated its negative impact on the whole-body and tissue metabolism and exercise capacity associated with attenuated AICAR-induced AMPK phosphorylation and glucose uptake in skeletal muscle [77]. As a result of these changes in the AMPK subunit expression and activity, it is difficult to elucidate the precise role of AMPK β2 in glycogen dynamics in this model. Similarly, muscle-specific AMPK β1/β2 KO mice display essentially no T172 phosphorylation in extensor digitorum longus (EDL) and the soleus muscle in response to electrical-stimulated contraction and have vastly reduced exercise capacity, carbohydrate utilization, and glucose uptake during treadmill running [55]. These defects were associated with reduced mitochondrial mRNA expression and reduced mitochondrial protein content [55]. Taken together, these findings suggest an important role of the β subunit in regulating AMPK activity and signaling, cellular glucose uptake and glycogen storage, mitochondrial function, and whole-body exercise capacity and metabolism.

In addition to mouse models targeting the AMPK β subunit(s), recent studies utilizing tissue-specific α1/α2 KO mice have provided support for the functional interactive roles of AMPK and glycogen. Liver-specific AMPK α1/α2 KO mice have an impaired ability to maintain euglycemia during exercise as a result of decreased hepatic glucose output due to decreased glycogenolysis [54]. Specifically, hepatic glycogen content was reduced in KO mice following both fasting and exercise. Phosphorylation of GS was unaffected in KO mice, but a decrease in UDP-glucose pyrophosphorylase 2 content was observed, suggesting reduced glycogen synthesis due to decreased glycogen precursors rather than an altered ability to synthesize glycogen [54]. In addition, when challenged with a long-term fast, these mice had reduced hepatic glycogenolysis and were unable to maintain liver ATP concentration without AMPK activity, providing further support of AMPK's role as an energy sensor [78]. Inducible muscle-specific α1/α2 KO mice have ablated skeletal muscle glycogen resynthesis and FA oxidation following exercise, even though glucose uptake was not affected, suggesting that AMPK functions as a switch to promote fat oxidation in order to preserve glucose for glycogen synthesis [79]. These findings indicate that AMPK can influence glycogen dynamics in physiological settings and that the ablation of AMPK activity reduces hepatic glucose output and is critical for skeletal muscle glycogen supercompensation following exercise. Collectively, studies using genetic models and pharmacological activators to date indicate that AMPK activation regulates glycogen synthesis in striated muscle (i.e., skeletal and cardiac muscles) secondary to increased glucose uptake and G6P accumulation, but not in the liver. Despite these important findings from AMPK transgenic mouse models, the precise role(s) of glycogen binding to the β subunits in the functional regulation of these physiological processes, as opposed to the ablation of the entire α subunits or β subunit(s) containing the CBM, remains to be elucidated.

5.2. Roles for Glycogen Availability in the Regulation of AMPK Activity

A series of physiological studies have demonstrated that low glycogen availability can amplify the AMPK signaling responses and adaptations to exercise. This was originally described in rat skeletal muscles in which AICAR treatment resulted in increased AMPK α2 activity and a markedly reduced glycogen synthase activity in a glycogen-depleted state compared to a glycogen-loaded state [80]. This observation was independent of adenine nucleotide concentrations and has subsequently been replicated in human skeletal muscle following exercise [81]. Additional studies of skeletal muscles have shown reductions in AMPK α1 and α2 association with glycogen, along with increased AMPK α2 activity and translocation to the nucleus following exercise in a glycogen-depleted state [82,83]. Furthermore, the consumption of a high-fat, low-carbohydrate diet followed by one day of a high-carbohydrate diet increases the resting skeletal muscle AMPK α activity in human skeletal muscle compared to a high-carbohydrate diet alone [84], supporting glycogen's inhibitory role on AMPK described in cell-free assays [56]. In a follow-up study, AMPK T172 phosphorylation was increased by exercise to a greater extent in the glycogen-depleted muscle than the normal glycogen repleted state [85]. Similarly, exercise in an overnight carbohydrate-fasted state resulted in increased AMPK T172 phosphorylation and the upregulation of signaling pathways involved in FA oxidation [86], while low glycogen stimulated peroxisome proliferator-activated receptor δ, a transcription factor that regulates fat utilization, in rat skeletal muscle following treadmill running [87]. Reduced glycogen availability is also associated with increases in the regulators of mitochondrial biogenesis, such as p53 and PGC-1α [88,89]. While none of these in vivo studies have directly assessed the functional role of AMPK-glycogen physical interaction, together they provide important physiological insights into how AMPK activity, subcellular localization, and signaling may be regulated by glycogen binding (Figure 2).

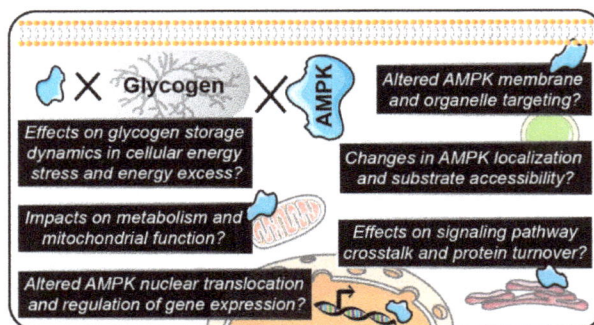

Figure 2. There are several potential alterations in cellular metabolism and signaling as a consequence of dysregulated AMPK-glycogen physical and functional interactions that represent key knowledge gaps in our current understanding and warrant further investigation in future studies. These potential alterations include changes in AMPK localization, translocation, substrates, and signaling pathway crosstalk, and subsequently, alterations in gene expression, cellular metabolism and glycogen storage.

5.3. Metabolic and Glycogen Storage Diseases as Models to Investigate AMPK-Glycogen Binding

Metabolic diseases such as insulin resistance and T2D are associated with impairments in AMPK activity, signaling, and glycogen storage dynamics. Obese patients with T2D have reduced skeletal muscle AMPK, ACC, and TBC1D4 phosphorylation following an acute bout of exercise [90]. In support of these findings, insulin resistance has been associated with suppressed AMPK activity in humans and mice [90,91], although results have been equivocal [92]. The liver-specific AMPK α1/α2 KO mice display an inability to maintain hepatic glucose output during exercise, highlighting the role of AMPK in maintaining euglycemia [54]. Skeletal muscle GS activity has also been demonstrated to be affected by insulin resistance and T2D, as there is increased phosphorylation of GS at site 2, the site phosphorylated by AMPK, which is not seen in healthy controls, resulting in nearly complete GS inactivation and dysregulation of glycogen synthesis [93]. Continued research in metabolic disease populations and rodent models can provide more insight into the significance of dysregulated AMPK and glycogen dynamics.

In addition, glycogen storage diseases provide pathophysiological models that can help provide additional insights into the influence of glycogen dynamics on AMPK. McArdle's disease is characterized by the accumulation of skeletal muscle glycogen due to a deficiency of GP. Individuals with McArdle's disease display higher muscle glycogen both at rest and following exercise compared to healthy controls, and an increased AMPK α2 activity and reduced GS activity in response to exercise [94]. Patients with McArdle's disease also demonstrate increased glucose clearance and ACC phosphorylation, indicating that AMPK activity is increased in order to maintain ATP concentration by promoting glucose uptake and FA oxidation [94]. The inability to break down glycogen, when coupled with retained, albeit reduced, glycogen synthesis, likely results in glycogen accumulation and the failure to utilize this energy source during exercise in this setting of the disease. A mouse model of McArdle's disease containing a p.R50X mutation, a nonsense mutation of nucleotide 148 in exon 1 of the GP gene, showed increased basal AMPK phosphorylation in the tibialis anterior and quadriceps muscles, associated with an increased GLUT4 content and increased AMPK-mediated glucose uptake compared to WT [95]. Following exhaustive exercise, McArdle mice display increased AMPK phosphorylation in the tibialis anterior and EDL muscles, while WT mice display no significant increase in AMPK activity [96]. While increased AMPK activity in McArdle patients and rodent models seems contrary to previous findings, the authors hypothesized that since McArdle disease results in an inability to break down glycogen, there is a subsequent increase in the AMPK activity in order to maintain an energy balance via increased glucose uptake [95,96]. Other rodent models have directly targeted muscle GS, which is affected in patients with Glycogen Storage Disease 0 [97]. Muscle-specific glycogen synthase knock-out models display increased AMPK phosphorylation [98] and markedly reduced glycogen content in skeletal muscle in the basal state, likely due to the retained capacity to break down but an inability to resynthesize glycogen [99,100].

6. Multidisciplinary Techniques and Models to Interrogate Roles for AMPK-Glycogen Interactions

While much remains to be discovered with regard to the molecular and cellular roles and physiological relevance of AMPK-glycogen binding, recent multidisciplinary technical research advances can be used to help address remaining knowledge gaps. For example, global mass spectrometry-based phosphoproteomics have recently revealed a repertoire of new AMPK substrates, providing additional evidence regarding the complexity and interconnection of the AMPK signaling network. A recent phosphoproteomic analysis mapping the human skeletal muscle exercise signaling network before and immediately following a single bout of intense aerobic exercise, in combination with phosphoproteomic analysis of AICAR-stimulated signaling in rat L6 myotubes, identified several novel AMPK substrates [101]. Other recent efforts have predicted and identified novel AMPK substrate phosphorylation sites via chemical genetic screening combined with peptide capture in whole cells [102], as well as affinity proteomics approach to analyzing hepatocyte proteins containing the substrate recognition motif targeted by AMPK phosphorylation [103]. Together, these complementary large-scale approaches have expanded the range of biological functions known to be regulated by AMPK. While additional substrates residing in different subcellular locations and organelles are continuing to be uncovered, the mechanisms underlying AMPK subcellular localization and targeting to substrates residing in these different organelles remains unknown. Future global, unbiased studies such as phosphoproteomics can help identify novel glycogen-associated AMPK substrates, post-translational regulation of glycogen regulatory machinery, AMPK subunit-specific regulation, and subcellular substrate targeting. Furthermore, omics-based approaches will reveal how AMPK-glycogen binding may impact other levels of biological regulation, such as the transcriptome, proteome, metabolome, and lipidome, in the contexts of exercise, metabolism, and beyond [104].

Novel AMPK fluorescence resonance energy transfer (FRET)-based sensors have recently revealed heterogeneous activity and tissue-specific roles for AMPK. These AMPK FRET sensors have permitted the spatiotemporal and dynamic assessment of AMPK activity in single cells [105], 3D cell cultures [106], and transgenic mice [107]. These biosensors build upon traditional methods to interrogate AMPK activity such as kinase assays and immunoblotting, which are limited to targeted measures of mean cellular protein phosphorylation and do not allow the spatiotemporal and dynamic assessment of AMPK activity. Electron microscopy-based approaches have also been used to visualize AMPK-glycogen association in fixed rat liver samples [58]. While improved microscopy technologies and sensors have been used to assess AMPK or glycogen localization, few studies have directly assessed AMPK-glycogen interactions. Utilizing these recent technical advancements will allow for the interrogation of AMPK-glycogen interactions and dynamics across species and physiologically relevant settings (Figure 2).

Despite the large body of research using in vitro models and physiological evidence indicating the potential functional roles for AMPK-glycogen binding, to date, there are no models that have been developed to disrupt and/or examine this physical binding directly in vivo. AMPK subunit KO models, while providing important insights into the functions of AMPK, are limited by the potential compensatory upregulation of other subunit isoforms or the disrupted stability of the AMPK heterotrimer complex (e.g., Reference [76]). In addition, directly assessing the function of β subunit glycogen binding is challenging when additional functions are altered in the presence of subunit deletion, as AMPK activity is impaired when the scaffolding β subunit is removed. The design of novel in vivo models in the future will be informed by previous molecular and cellular findings to allow direct interrogation of the functional relevance of the β subunits and CBM. Generation of novel animal models to specifically target physical AMPK-glycogen binding will provide important advances regarding its physiological significance and capability to be therapeutically targeted in vivo to modulate metabolism and the health benefits of exercise.

Finally, previous studies have primarily utilized centrifugation-based assays to detect and quantify physical AMPK-glycogen association. Novel biotechnological platforms and proximity assays will aid this investigation of AMPK-glycogen binding and AMPK's proximity to glycogen with improved sensitivity and specificity across molecular, cellular, and physiological models. Furthermore, newly developed kinase activity reporters [108] and other non-radioactive activity assays [109] will help provide new measures of intracellular AMPK activity dynamics and complement traditional surrogate measures such as immunoblot analyses of AMPK and ACC phosphorylation. Together these technological advances expand the repertoire of available tools to monitor the range of biological processes regulated by AMPK and further our understanding of the mechanisms and physiological significance underlying AMPK-glycogen interactions.

7. Potential Therapeutic Relevance of Targeting AMPK-Glycogen Binding

Consistent with the therapeutic relevance of the CBM, several lines of evidence demonstrate that the CBM may play a direct functional role in AMPK conformation and activation. The CBM contains the critical S108 autophosphorylation site required for drug-induced AMPK activation in the absence of AMP [8]. Although located on opposite sides of the AMPK heterotrimer, the CBM is conformationally connected to the regulatory AMPK γ subunit and its stabilization is affected by adenine nucleotide binding (e.g., AMP) to the CBS motifs [26]. Despite physical AMPK-glycogen interaction being mediated by the β subunit, mutations in the γ subunit also result in alterations in AMPK activation and glycogen metabolism. The γ2 subunit is known to contain mutations that cause constitutive AMPK activation, resulting in glycogen storage diseases in humans. These mutations result in glycogen accumulation with coexisting deleterious effects on cardiac electrical properties that are characteristic of familial hypertrophic cardiomyopathy and Wolff-Parkinson-White syndrome [110]. In addition, gain of function mutations in the AMPK γ3 subunit predominantly expressed in skeletal muscle result in excess glycogen storage [111] as well as improvements in metabolism via increased mitochondrial biogenesis [112]. Constitutive AMPK activation associated with these γ subunit mutations promotes glycogen synthesis by increasing glucose uptake. As mentioned above, the CBM interacts with the α subunit, forming the ADaM site and stabilizing the kinase domain of the α subunit in its active formation [9,27]. However, when glycogen binds to the CBM, this interaction is destabilized, altering the ADaM site and inhibiting the AMPK activity [9]. For example, isoform-specific allosteric inhibition of AMPK has been shown to be dependent on the β2 subunit CBM in glycogen-containing pancreatic beta cells [113]. The CBM, therefore, functions as both a critical element of AMPK activation as well as a site for the allosteric inhibition by glycogen, highlighting the therapeutic potential of new drugs targeting the ADaM site.

8. Conclusions

AMPK is a central regulator of cellular metabolism and, therefore, possesses significant therapeutic potential for the prevention and treatment of a range of metabolic diseases. A growing body of evidence demonstrates that AMPK physically binds glycogen and this interaction can alter the conformation of AMPK, and subsequently, its activity and downstream signaling. AMPK activity subsequently regulates glycogen metabolism. Recent research has described experimental and physiological settings that impact functional AMPK and glycogen interactions, including AMPK β isoform affinity, glycogen availability, and particle size. Despite our understanding of AMPK's relationship with glycogen, much remains to be elucidated. Further research using new technologies and experimental models can reveal additional mechanisms underlying AMPK and glycogen's interactive roles in cellular energy sensing, exercise, and metabolism. Together, these findings will help provide insights into the physiological and therapeutic relevance of targeting AMPK and glycogen binding in health and disease.

funding: This work is supported by Australian Catholic University (ACU) Faculty of Health Sciences Open Access Publishing Support awarded to N.R.J. and an ACU Research Funding (ACURF) Early Career Researcher Grant awarded to N.J.H.

Acknowledgments: The authors acknowledge that all publications related to AMPK and glycogen could not be discussed in this review due to word limitations and the primary focus on studies published within the past two decades.

Conflicts of Interest: The authors declare no conflict of interest.

Abbreviations

ACC	Acetyl-CoA carboxylase
ADP	Adenosine diphosphate
AICAR	5-aminoimidazole-4-carboxamide ribonucleotide
AMP	Adenosine monophosphate
AMPK	AMP-activated protein kinase
ATP	Adenosine triphosphate
CaMKKβ	Calcium/calmodulin-dependent protein kinase β
cAMP	Cyclic AMP
CBM	Carbohydrate-binding module
CBS	Cystathionine-β-synthase domains
EDL	Extensor digitorum longus
FA	Fatty acid
FRET	Fluorescence resonance energy transfer
G6P	Glucose-6-phosphate
GBE	Glycogen branching enzyme
GDE	Glycogen debranching enzyme
GLUT4	Glucose transporter 4
GP	Glycogen phosphorylase
GS	Glycogen synthase
interMF	Intermyofibrillar
intraMF	Intramyofibrillar
KO	Knock-out
LKB1	Liver kinase B1
mTOR	Mechanistic target of rapamycin
PGC-1α	Peroxisome proliferator-activated receptor gamma coactivator 1-α
Raptor	Regulatory associated protein of mechanistic target of rapamycin

SS	Subsarcolemmal
TBCID1	Tre-2, BUB2, CDC16, 1 domain family, member 1
TBC1D4	Tre-2, BUB2, CDC16, 1 domain family, member 4
TSC2	Tuberous sclerosis complex 2
T2D	Type 2 diabetes
UGP2	UDP-glucose pyrophosphorylase 2
WT	Wild type

References

1. Garcia, D.; Shaw, R.J. AMPK: Mechanisms of cellular energy sensing and restoration of metabolic balance. *Mol. Cell* **2017**, *66*, 789–800. [CrossRef] [PubMed]

2. Oakhill, J.S.; Chen, Z.P.; Scott, J.W.; Steel, R.; Castelli, L.A.; Ling, N.; Macaulay, S.L.; Kemp, B.E. β-Subunit myristoylation is the gatekeeper for initiating metabolic stress sensing by AMP-activated protein kinase (AMPK). *Proc. Natl. Acad. Sci. USA* **2010**, *107*, 19237–19241. [CrossRef] [PubMed]

3. Zhang, Y.L.; Guo, H.; Zhang, C.S.; Lin, S.Y.; Yin, Z.; Peng, Y.; Luo, H.; Shi, Y.; Lian, G.; Zhang, C.; et al. AMP as a low-energy charge signal autonomously initiates assembly of AXIN-AMPK-LKB1 complex for AMPK activation. *Cell Metab.* **2013**, *18*, 546–555. [CrossRef] [PubMed]

4. Garcia-Haro, L.; Garcia-Gimeno, M.A.; Neumann, D.; Beullens, M.; Bollen, M.; Sanz, P. The PP1-R6 protein phosphatase holoenzyme is involved in the glucose-induced dephosphorylation and inactivation of AMP-activated protein kinase, a key regulator of insulin secretion, in MIN6 β cells. *FASEB J.* **2010**, *24*, 5080–5091. [CrossRef] [PubMed]

5. Gowans, G.J.; Hawley, S.A.; Ross, F.A.; Hardie, D.G. AMP is a true physiological regulator of AMP-activated protein kinase by both allosteric activation and enhancing net phosphorylation. *Cell Metab.* **2013**, *18*, 556–566. [CrossRef] [PubMed]

6. Joseph, B.K.; Liu, H.Y.; Francisco, J.; Pandya, D.; Donigan, M.; Gallo-Ebert, C.; Giordano, C.; Bata, A.; Nickels, J.T., Jr. Inhibition of AMP kinase by the protein phosphatase 2A heterotrimer, PP2APpp2r2d. *J. Biol. Chem.* **2015**, *290*, 10588–10598. [CrossRef] [PubMed]

7. Xiao, B.; Sanders, M.J.; Underwood, E.; Heath, R.; Mayer, F.V.; Carmena, D.; Jing, C.; Walker, P.A.; Eccleston, J.F.; Haire, L.F.; et al. Structure of mammalian AMPK and its regulation by ADP. *Nature* **2011**, *472*, 230–233. [CrossRef] [PubMed]

8. Scott, J.W.; Ling, N.; Issa, S.M.; Dite, T.A.; O'Brien, M.T.; Chen, Z.P.; Galic, S.; Langendorf, C.G.; Steinberg, G.R.; Kemp, B.E.; et al. Small molecule drug A-769662 and AMP synergistically activate naive AMPK independent of upstream kinase signaling. *Chem. Biol.* **2014**, *21*, 619–627. [CrossRef] [PubMed]

9. Li, X.; Wang, L.; Zhou, X.E.; Ke, J.; de Waal, P.W.; Gu, X.; Tan, M.H.; Wang, D.; Wu, D.; Xu, H.E.; et al. Structural basis of AMPK regulation by adenine nucleotides and glycogen. *Cell Res.* **2015**, *25*, 50–66. [CrossRef] [PubMed]

10. Kjobsted, R.; Hingst, J.R.; Fentz, J.; Foretz, M.; Sanz, M.N.; Pehmoller, C.; Shum, M.; Marette, A.; Mounier, R.; Treebak, J.T.; et al. AMPK in skeletal muscle function and metabolism. *FASEB J.* **2018**. [CrossRef] [PubMed]

11. Hawley, S.A.; Pan, D.A.; Mustard, K.J.; Ross, L.; Bain, J.; Edelman, A.M.; Frenguelli, B.G.; Hardie, D.G. Calmodulin-dependent protein kinase kinase-β is an alternative upstream kinase for AMP-activated protein kinase. *Cell Metab.* **2005**, *2*, 9–19. [CrossRef] [PubMed]

12. Jensen, T.E.; Rose, A.J.; Jorgensen, S.B.; Brandt, N.; Schjerling, P.; Wojtaszewski, J.F.; Richter, E.A. Possible CaMKK-dependent regulation of AMPK phosphorylation and glucose uptake at the onset of mild tetanic skeletal muscle contraction. *Am. J. Physiol. Endocrinol. Metab.* **2007**, *292*, E1308–E1317. [CrossRef] [PubMed]

13. Woods, A.; Dickerson, K.; Heath, R.; Hong, S.P.; Momcilovic, M.; Johnstone, S.R.; Carlson, M.; Carling, D. Ca²⁺/calmodulin-dependent protein kinase kinase-β acts upstream of AMP-activated protein kinase in mammalian cells. *Cell Metab.* **2005**, *2*, 21–33. [CrossRef] [PubMed]

14. Kjobsted, R.; Munk-Hansen, N.; Birk, J.B.; Foretz, M.; Viollet, B.; Bjornholm, M.; Zierath, J.R.; Treebak, J.T.; Wojtaszewski, J.F. Enhanced muscle insulin sensitivity after contraction/exercise is mediated by AMPK. *Diabetes* **2017**, *66*, 598–612. [CrossRef] [PubMed]

15. Pehmoller, C.; Treebak, J.T.; Birk, J.B.; Chen, S.; Mackintosh, C.; Hardie, D.G.; Richter, E.A.; Wojtaszewski, J.F. Genetic disruption of AMPK signaling abolishes both contraction-and insulin-stimulated TBC1D1 phosphorylation and 14-3-3 binding in mouse skeletal muscle. *Am. J. Physiol. Endocrinol. Metab.* **2009**, *297*, E665–E675. [CrossRef] [PubMed]

16. Vichaiwong, K.; Purohit, S.; An, D.; Toyoda, T.; Jessen, N.; Hirshman, M.F.; Goodyear, L.J. Contraction regulates site-specific phosphorylation of TBC1D1 in skeletal muscle. *Biochem. J.* **2010**, *431*, 311–320. [CrossRef] [PubMed]

17. Whitfield, J.; Paglialunga, S.; Smith, B.K.; Miotto, P.M.; Simnett, G.; Robson, H.L.; Jain, S.S.; Herbst, E.A.F.; Desjardins, E.M.; Dyck, D.J.; et al. Ablating the protein TBC1D1 impairs contraction-induced sarcolemmal glucose transporter 4 redistribution but not insulin-mediated responses in rats. *J. Biol. Chem.* **2017**, *292*, 16653–16664. [CrossRef] [PubMed]

18. Stockli, J.; Meoli, C.C.; Hoffman, N.J.; Fazakerley, D.J.; Pant, H.; Cleasby, M.E.; Ma, X.; Kleinert, M.; Brandon, A.E.; Lopez, J.A.; et al. The RabGAP TBC1D1 plays a central role in exercise-regulated glucose metabolism in skeletal muscle. *Diabetes* **2015**, *64*, 1914–1922. [CrossRef] [PubMed]

19. Hardie, D.G.; Ross, F.A.; Hawley, S.A. AMPK: A nutrient and energy sensor that maintains energy homeostasis. *Nat. Rev. Mol. Cell Biol.* **2012**, *13*, 251–262. [CrossRef] [PubMed]

20. Fullerton, M.D.; Galic, S.; Marcinko, K.; Sikkema, S.; Pulinilkunnil, T.; Chen, Z.P.; O'Neill, H.M.; Ford, R.J.; Palanivel, R.; O'Brien, M.; et al. Single phosphorylation sites in Acc1 and Acc2 regulate lipid homeostasis and the insulin-sensitizing effects of metformin. *Nat. Med.* **2013**, *19*, 1649–1654. [CrossRef] [PubMed]

21. Li, Y.; Xu, S.; Mihaylova, M.M.; Zheng, B.; Hou, X.; Jiang, B.; Park, O.; Luo, Z.; Lefai, E.; Shyy, J.Y.; et al. AMPK phosphorylates and inhibits SREBP activity to attenuate hepatic steatosis and atherosclerosis in diet-induced insulin-resistant mice. *Cell Metab.* **2011**, *13*, 376–388. [CrossRef] [PubMed]

22. Hawley, J.A.; Hargreaves, M.; Joyner, M.J.; Zierath, J.R. Integrative biology of exercise. *Cell* **2014**, *159*, 738–749. [CrossRef] [PubMed]

23. Gwinn, D.M.; Shackelford, D.B.; Egan, D.F.; Mihaylova, M.M.; Mery, A.; Vasquez, D.S.; Turk, B.E.; Shaw, R.J. AMPK phosphorylation of raptor mediates a metabolic checkpoint. *Mol. Cell* **2008**, *30*, 214–226. [CrossRef] [PubMed]

24. Inoki, K.; Zhu, T.; Guan, K.L. Tsc2 mediates cellular energy response to control cell growth and survival. *Cell* **2003**, *115*, 577–590. [CrossRef]

25. Tee, A.R.; Fingar, D.C.; Manning, B.D.; Kwiatkowski, D.J.; Cantley, L.C.; Blenis, J. Tuberous sclerosis complex-1 and -2 gene products function together to inhibit mammalian target of rapamycin (mTOR)-mediated downstream signaling. *Proc. Natl. Acad. Sci. USA* **2002**, *99*, 13571–13576. [CrossRef] [PubMed]

26. Gu, X.; Yan, Y.; Novick, S.J.; Kovach, A.; Goswami, D.; Ke, J.; Tan, M.H.E.; Wang, L.; Li, X.; de Waal, P.W.; et al. Deconvoluting AMP-activated protein kinase (AMPK) adenine nucleotide binding and sensing. *J. Biol. Chem.* **2017**, *292*, 12653–12666. [CrossRef] [PubMed]

27. Xiao, B.; Sanders, M.J.; Carmena, D.; Bright, N.J.; Haire, L.F.; Underwood, E.; Patel, B.R.; Heath, R.B.; Walker, P.A.; Hallen, S.; et al. Structural basis of AMPK regulation by small molecule activators. *Nat. Commun.* **2013**, *4*, 3017. [CrossRef] [PubMed]

28. Olivier, S.; Foretz, M.; Viollet, B. Promise and challenges for direct small molecule AMPK activators. *Biochem. Pharmacol.* **2018**. [CrossRef] [PubMed]

29. Stephenne, X.; Foretz, M.; Taleux, N.; van der Zon, G.C.; Sokal, E.; Hue, L.; Viollet, B.; Guigas, B. Metformin activates AMP-activated protein kinase in primary human hepatocytes by decreasing cellular energy status. *Diabetologia* **2011**, *54*, 3101–3110. [CrossRef] [PubMed]

30. Wu, J.; Puppala, D.; Feng, X.; Monetti, M.; Lapworth, A.L.; Geoghegan, K.F. Chemoproteomic analysis of intertissue and interspecies isoform diversity of AMP-activated protein kinase (AMPK). *J. Biol. Chem.* **2013**, *288*, 35904–35912. [CrossRef] [PubMed]

31. Polekhina, G.; Gupta, A.; Michell, B.J.; van Denderen, B.; Murthy, S.; Feil, S.C.; Jennings, I.G.; Campbell, D.J.; Witters, L.A.; Parker, M.W.; et al. AMPK β subunit targets metabolic stress sensing to glycogen. *Curr. Biol.* **2003**, *13*, 867–871. [CrossRef]

32. Polekhina, G.; Gupta, A.; van Denderen, B.J.; Feil, S.C.; Kemp, B.E.; Stapleton, D.; Parker, M.W. Structural basis for glycogen recognition by AMP-activated protein kinase. *Structure* **2005**, *13*, 1453–1462. [CrossRef] [PubMed]

33. Koay, A.; Woodcroft, B.; Petrie, E.J.; Yue, H.; Emanuelle, S.; Bieri, M.; Bailey, M.F.; Hargreaves, M.; Park, J.T.; Park, K.H.; et al. AMPK β subunits display isoform specific affinities for carbohydrates. *FEBS Lett.* **2010**, *584*, 3499–3503. [PubMed]

34. Mobbs, J.I.; Di Paolo, A.; Metcalfe, R.D.; Selig, E.; Stapleton, D.I.; Griffin, M.D.W.; Gooley, P.R. Unravelling the carbohydrate-binding preferences of the carbohydrate-binding modules of AMP-activated protein kinase. *Chembiochem* **2017**. [CrossRef] [PubMed]

35. Mobbs, J.I.; Koay, A.; Di Paolo, A.; Bieri, M.; Petrie, E.J.; Gorman, M.A.; Doughty, L.; Parker, M.W.; Stapleton, D.I.; Griffin, M.D.; et al. Determinants of oligosaccharide specificity of the carbohydrate-binding modules of AMP-activated protein kinase. *Biochem. J.* **2015**, *468*, 245–257. [CrossRef] [PubMed]

36. O'Neill, H.M. AMPK and exercise: Glucose uptake and insulin sensitivity. *Diabetes Metab. J.* **2013**, *37*, 1–21. [CrossRef] [PubMed]

37. Alonso, M.D.; Lomako, J.; Lomako, W.M.; Whelan, W.J. A new look at the biogenesis of glycogen. *FASEB J.* **1995**, *9*, 1126–1137. [CrossRef] [PubMed]

38. Hunter, R.W.; Treebak, J.T.; Wojtaszewski, J.F.; Sakamoto, K. Molecular mechanism by which AMP-activated protein kinase activation promotes glycogen accumulation in muscle. *Diabetes* **2011**, *60*, 766–774. [CrossRef] [PubMed]

39. Roach, P.J.; Depaoli-Roach, A.A.; Hurley, T.D.; Tagliabracci, V.S. Glycogen and its metabolism: Some new developments and old themes. *Biochem. J.* **2012**, *441*, 763–787. [PubMed]

40. Chasiotis, D.; Sahlin, K.; Hultman, E. Regulation of glycogenolysis in human muscle at rest and during exercise. *J. Appl. Physiol. Respir. Environ. Exerc. Physiol.* **1982**, *53*, 708–715. [CrossRef] [PubMed]

41. Richter, E.A.; Ruderman, N.B.; Gavras, H.; Belur, E.R.; Galbo, H. Muscle glycogenolysis during exercise: Dual control by epinephrine and contractions. *Am. J. Physiol.* **1982**, *242*, E25–E32. [CrossRef] [PubMed]

42. Shearer, J.; Graham, T.E. Novel aspects of skeletal muscle glycogen and its regulation during rest and exercise. *Exerc. Sport Sci. Rev.* **2004**, *32*, 120–126. [CrossRef] [PubMed]

43. Prats, C.; Graham, T.E.; Shearer, J. The dynamic life of the glycogen granule. *J. Biol. Chem.* **2018**, *293*, 7089–7098. [CrossRef] [PubMed]

44. Graham, T.E.; Yuan, Z.; Hill, A.K.; Wilson, R.J. The regulation of muscle glycogen: The granule and its proteins. *Acta Physiol.* **2010**, *199*, 489–498. [CrossRef] [PubMed]

45. Nielsen, J.; Ortenblad, N. Physiological aspects of the subcellular localization of glycogen in skeletal muscle. *Appl. Physiol. Nutr. Metab.* **2013**, *38*, 91–99. [CrossRef] [PubMed]

46. Ortenblad, N.; Nielsen, J.; Saltin, B.; Holmberg, H.C. Role of glycogen availability in sarcoplasmic reticulum Ca^{2+} kinetics in human skeletal muscle. *J. Physiol.* **2011**, *589*, 711–725. [CrossRef] [PubMed]

47. Ortenblad, N.; Westerblad, H.; Nielsen, J. Muscle glycogen stores and fatigue. *J. Physiol.* **2013**, *591*, 4405–4413. [CrossRef] [PubMed]

48. Philp, A.; Hargreaves, M.; Baar, K. More than a store: Regulatory roles for glycogen in skeletal muscle adaptation to exercise. *Am. J. Physiol. Endocrinol. Metab.* **2012**, *302*, E1343–E1351. [CrossRef] [PubMed]

49. Nielsen, J.; Holmberg, H.C.; Schroder, H.D.; Saltin, B.; Ortenblad, N. Human skeletal muscle glycogen utilization in exhaustive exercise: Role of subcellular localization and fibre type. *J. Physiol.* **2011**, *589*, 2871–2885. [CrossRef] [PubMed]

50. Yeo, W.K.; Paton, C.D.; Garnham, A.P.; Burke, L.M.; Carey, A.L.; Hawley, J.A. Skeletal muscle adaptation and performance responses to once a day versus twice every second day endurance training regimens. *J. Appl. Physiol.* **2008**, *105*, 1462–1470. [CrossRef] [PubMed]

51. Nielsen, J.; Cheng, A.J.; Ortenblad, N.; Westerblad, H. Subcellular distribution of glycogen and decreased tetanic Ca^{2+} in fatigued single intact mouse muscle fibres. *J. Physiol.* **2014**, *592*, 2003–2012. [CrossRef] [PubMed]

52. Romijn, J.A.; Coyle, E.F.; Sidossis, L.S.; Gastaldelli, A.; Horowitz, J.F.; Endert, E.; Wolfe, R.R. Regulation of endogenous fat and carbohydrate metabolism in relation to exercise intensity and duration. *Am. J. Physiol.* **1993**, *265*, E380–E391. [CrossRef] [PubMed]

53. van Loon, L.J.; Greenhaff, P.L.; Constantin-Teodosiu, D.; Saris, W.H.; Wagenmakers, A.J. The effects of increasing exercise intensity on muscle fuel utilisation in humans. *J. Physiol.* **2001**, *536*, 295–304. [CrossRef] [PubMed]

54. Hughey, C.C.; James, F.D.; Bracy, D.P.; Donahue, E.P.; Young, J.D.; Viollet, B.; Foretz, M.; Wasserman, D.H. Loss of hepatic AMP-activated protein kinase impedes the rate of glycogenolysis but not gluconeogenic fluxes in exercising mice. *J. Biol. Chem.* **2017**. [CrossRef] [PubMed]

55. O'Neill, H.M.; Maarbjerg, S.J.; Crane, J.D.; Jeppesen, J.; Jorgensen, S.B.; Schertzer, J.D.; Shyroka, O.; Kiens, B.; van Denderen, B.J.; Tarnopolsky, M.A.; et al. AMP-activated protein kinase (AMPK) β1β2 muscle null mice reveal an essential role for AMPK in maintaining mitochondrial content and glucose uptake during exercise. *Proc. Natl. Acad. Sci. USA* **2011**, *108*, 16092–16097. [CrossRef] [PubMed]

56. McBride, A.; Ghilagaber, S.; Nikolaev, A.; Hardie, D.G. The glycogen-binding domain on the AMPK β subunit allows the kinase to act as a glycogen sensor. *Cell Metab.* **2009**, *9*, 23–34. [CrossRef] [PubMed]

57. Hudson, E.R.; Pan, D.A.; James, J.; Lucocq, J.M.; Hawley, S.A.; Green, K.A.; Baba, O.; Terashima, T.; Hardie, D.G. A novel domain in AMP-activated protein kinase causes glycogen storage bodies similar to those seen in hereditary cardiac arrhythmias. *Curr. Biol.* **2003**, *13*, 861–866. [CrossRef]

58. Bendayan, M.; Londono, I.; Kemp, B.E.; Hardie, G.D.; Ruderman, N.; Prentki, M. Association of AMP-activated protein kinase subunits with glycogen particles as revealed in situ by immunoelectron microscopy. *J. Histochem. Cytochem.* **2009**, *57*, 963–971. [CrossRef] [PubMed]

59. Scott, J.W.; van Denderen, B.J.; Jorgensen, S.B.; Honeyman, J.E.; Steinberg, G.R.; Oakhill, J.S.; Iseli, T.J.; Koay, A.; Gooley, P.R.; Stapleton, D.; et al. Thienopyridone drugs are selective activators of AMP-activated protein kinase β1-containing complexes. *Chem. Biol.* **2008**, *15*, 1220–1230. [CrossRef] [PubMed]

60. Oligschlaeger, Y.; Miglianico, M.; Chanda, D.; Scholz, R.; Thali, R.F.; Tuerk, R.; Stapleton, D.I.; Gooley, P.R.; Neumann, D. The recruitment of AMP-activated protein kinase to glycogen is regulated by autophosphorylation. *J. Biol. Chem.* **2015**, *290*, 11715–11728. [CrossRef] [PubMed]

61. Xu, H.; Frankenberg, N.T.; Lamb, G.D.; Gooley, P.R.; Stapleton, D.I.; Murphy, R.M. When phosphorylated at Thr[148], the β2-subunit of AMP-activated kinase does not associate with glycogen in skeletal muscle. *Am. J. Physiol. Cell Physiol.* **2016**, *311*, C35–C42. [CrossRef] [PubMed]

62. Carling, D.; Hardie, D.G. The substrate and sequence specificity of the AMP-activated protein kinase. Phosphorylation of glycogen synthase and phosphorylase kinase. *Biochim. Biophys. Acta* **1989**, *1012*, 81–86. [CrossRef]

63. Jorgensen, S.B.; Nielsen, J.N.; Birk, J.B.; Olsen, G.S.; Viollet, B.; Andreelli, F.; Schjerling, P.; Vaulont, S.; Hardie, D.G.; Hansen, B.F.; et al. The α2-5′AMP-activated protein kinase is a site 2 glycogen synthase kinase in skeletal muscle and is responsive to glucose loading. *Diabetes* **2004**, *53*, 3074–3081. [CrossRef] [PubMed]

64. Cokorinos, E.C.; Delmore, J.; Reyes, A.R.; Albuquerque, B.; Kjobsted, R.; Jorgensen, N.O.; Tran, J.L.; Jatkar, A.; Cialdea, K.; Esquejo, R.M.; et al. Activation of skeletal muscle AMPK promotes glucose disposal and glucose lowering in non-human primates and mice. *Cell Metab.* **2017**. [CrossRef] [PubMed]

65. Myers, R.W.; Guan, H.P.; Ehrhart, J.; Petrov, A.; Prahalada, S.; Tozzo, E.; Yang, X.; Kurtz, M.M.; Trujillo, M.; Gonzalez Trotter, D.; et al. Systemic pan-AMPK activator MK-8722 improves glucose homeostasis but induces cardiac hypertrophy. *Science* **2017**, *357*, 507–511. [CrossRef] [PubMed]

66. Fritzen, A.M.; Lundsgaard, A.M.; Jeppesen, J.; Christiansen, M.L.; Bienso, R.; Dyck, J.R.; Pilegaard, H.; Kiens, B. 5′-AMP activated protein kinase α2 controls substrate metabolism during post-exercise recovery via regulation of pyruvate dehydrogenase kinase 4. *J. Physiol.* **2015**, *593*, 4765–4780. [CrossRef] [PubMed]

67. Prats, C.; Helge, J.W.; Nordby, P.; Qvortrup, K.; Ploug, T.; Dela, F.; Wojtaszewski, J.F. Dual regulation of muscle glycogen synthase during exercise by activation and compartmentalization. *J. Biol. Chem.* **2009**, *284*, 15692–15700. [CrossRef] [PubMed]

68. Barre, L.; Richardson, C.; Hirshman, M.F.; Brozinick, J.; Fiering, S.; Kemp, B.E.; Goodyear, L.J.; Witters, L.A. Genetic model for the chronic activation of skeletal muscle AMP-activated protein kinase leads to glycogen accumulation. *Am. J. Physiol. Endocrinol. Metab.* **2007**, *292*, E802–E811. [CrossRef] [PubMed]

69. Prats, C.; Gomez-Cabello, A.; Hansen, A.V. Intracellular compartmentalization of skeletal muscle glycogen metabolism and insulin signalling. *Exp. Physiol.* **2011**, *96*, 385–390. [CrossRef] [PubMed]

70. Young, M.E.; Leighton, B.; Radda, G.K. Glycogen phosphorylase may be activated by AMP-kinase in skeletal muscle. *Biochem. Soc. Trans.* **1996**, *24*, 268S. [CrossRef] [PubMed]

71. Young, M.E.; Radda, G.K.; Leighton, B. Activation of glycogen phosphorylase and glycogenolysis in rat skeletal muscle by AICAR—An activator of AMP-activated protein kinase. *FEBS Lett.* **1996**, *382*, 43–47. [CrossRef]

72. Daignan-Fornier, B.; Pinson, B. 5-Aminoimidazole-4-carboxamide-1-beta-d-ribofuranosyl 5′-monophosphate (AICAR), a highly conserved purine intermediate with multiple effects. *Metabolites* **2012**, *2*, 292–302. [CrossRef] [PubMed]

73. Sakoda, H.; Fujishiro, M.; Fujio, J.; Shojima, N.; Ogihara, T.; Kushiyama, A.; Fukushima, Y.; Anai, M.; Ono, H.; Kikuchi, M.; et al. Glycogen debranching enzyme association with β-Subunit regulates AMP-activated protein kinase activity. *Am. J. Physiol. Endocrinol. Metab.* **2005**, *289*, E474–E481. [CrossRef] [PubMed]

74. Stapleton, D.; Nelson, C.; Parsawar, K.; McClain, D.; Gilbert-Wilson, R.; Barker, E.; Rudd, B.; Brown, K.; Hendrix, W.; O'Donnell, P.; et al. Analysis of hepatic glycogen-associated proteins. *Proteomics* **2010**, *10*, 2320–2329. [CrossRef] [PubMed]

75. Stapleton, D.; Nelson, C.; Parsawar, K.; Flores-Opazo, M.; McClain, D.; Parker, G. The 3T3-L1 adipocyte glycogen proteome. *Proteome Sci.* **2013**. [CrossRef] [PubMed]

76. Steinberg, G.R.; O'Neill, H.M.; Dzamko, N.L.; Galic, S.; Naim, T.; Koopman, R.; Jorgensen, S.B.; Honeyman, J.; Hewitt, K.; Chen, Z.P.; et al. Whole body deletion of AMP-activated protein kinase β2 reduces muscle AMPK activity and exercise capacity. *J. Biol. Chem.* **2010**, *285*, 37198–37209. [CrossRef] [PubMed]

77. Dasgupta, B.; Ju, J.S.; Sasaki, Y.; Liu, X.; Jung, S.R.; Higashida, K.; Lindquist, D.; Milbrandt, J. The AMPK β2 subunit is required for energy homeostasis during metabolic stress. *Mol. Cell. Biol.* **2012**, *32*, 2837–2848. [CrossRef] [PubMed]

78. Hasenour, C.M.; Ridley, D.E.; James, F.D.; Hughey, C.C.; Donahue, E.P.; Viollet, B.; Foretz, M.; Young, J.D.; Wasserman, D.H. Liver AMP-activated protein kinase is unnecessary for gluconeogenesis but protects energy state during nutrient deprivation. *PLoS ONE* **2017**, *12*, e0170382.

79. Hingst, J.R.; Bruhn, L.; Hansen, M.B.; Rosschou, M.F.; Birk, J.B.; Fentz, J.; Foretz, M.; Viollet, B.; Sakamoto, K.; Faergeman, N.J.; et al. Exercise-induced molecular mechanisms promoting glycogen supercompensation in human skeletal muscle. *Mol. Metab.* **2018**. [CrossRef] [PubMed]

80. Wojtaszewski, J.F.; Jorgensen, S.B.; Hellsten, Y.; Hardie, D.G.; Richter, E.A. Glycogen-dependent effects of 5-aminoimidazole-4-carboxamide (AICA)-riboside on AMP-activated protein kinase and glycogen synthase activities in rat skeletal muscle. *Diabetes* **2002**, *51*, 284–292. [CrossRef] [PubMed]

81. Wojtaszewski, J.F.; MacDonald, C.; Nielsen, J.N.; Hellsten, Y.; Hardie, D.G.; Kemp, B.E.; Kiens, B.; Richter, E.A. Regulation of 5′AMP-activated protein kinase activity and substrate utilization in exercising human skeletal muscle. *Am. J. Physiol. Endocrinol. Metab.* **2003**, *284*, E813–E822. [CrossRef] [PubMed]

82. Steinberg, G.R.; Watt, M.J.; McGee, S.L.; Chan, S.; Hargreaves, M.; Febbraio, M.A.; Stapleton, D.; Kemp, B.E. Reduced glycogen availability is associated with increased AMPKα2 activity, nuclear AMPKα2 protein abundance, and GLUT4 mRNA expression in contracting human skeletal muscle. *Appl. Physiol. Nutr. Metab.* **2006**, *31*, 302–312. [CrossRef] [PubMed]

83. Watt, M.J.; Steinberg, G.R.; Chan, S.; Garnham, A.; Kemp, B.E.; Febbraio, M.A. β-Adrenergic stimulation of skeletal muscle HSL can be overridden by AMPK signaling. *FASEB J.* **2004**, *18*, 1445–1446. [CrossRef] [PubMed]

84. Yeo, W.K.; Lessard, S.J.; Chen, Z.P.; Garnham, A.P.; Burke, L.M.; Rivas, D.A.; Kemp, B.E.; Hawley, J.A. Fat adaptation followed by carbohydrate restoration increases AMPK activity in skeletal muscle from trained humans. *J. Appl. Physiol.* **2008**, *105*, 1519–1526. [CrossRef] [PubMed]

85. Yeo, W.K.; McGee, S.L.; Carey, A.L.; Paton, C.D.; Garnham, A.P.; Hargreaves, M.; Hawley, J.A. Acute signalling responses to intense endurance training commenced with low or normal muscle glycogen. *Exp. Physiol.* **2010**, *95*, 351–358. [CrossRef] [PubMed]

86. Lane, S.C.; Camera, D.M.; Lassiter, D.G.; Areta, J.L.; Bird, S.R.; Yeo, W.K.; Jeacocke, N.A.; Krook, A.; Zierath, J.R.; Burke, L.M.; et al. Effects of sleeping with reduced carbohydrate availability on acute training responses. *J. Appl. Physiol.* **2015**, *119*, 643–655. [CrossRef] [PubMed]

87. Philp, A.; MacKenzie, M.G.; Belew, M.Y.; Towler, M.C.; Corstorphine, A.; Papalamprou, A.; Hardie, D.G.; Baar, K. Glycogen content regulates peroxisome proliferator activated receptor-∂ (PPAR-∂) activity in rat skeletal muscle. *PLoS ONE* **2013**, *8*, e77200. [CrossRef] [PubMed]

88. Bartlett, J.D.; Louhelainen, J.; Iqbal, Z.; Cochran, A.J.; Gibala, M.J.; Gregson, W.; Close, G.L.; Drust, B.; Morton, J.P. Reduced carbohydrate availability enhances exercise-induced p53 signaling in human skeletal muscle: Implications for mitochondrial biogenesis. *Am. J. Physiol. Regul. Integr. Comp. Physiol.* **2013**, *304*, R450–R458. [CrossRef] [PubMed]

89. Psilander, N.; Frank, P.; Flockhart, M.; Sahlin, K. Exercise with low glycogen increases PGC-1α gene expression in human skeletal muscle. *Eur. J. Appl. Physiol.* **2013**, *113*, 951–963. [CrossRef] [PubMed]

90. Sriwijitkamol, A.; Coletta, D.K.; Wajcberg, E.; Balbontin, G.B.; Reyna, S.M.; Barrientes, J.; Eagan, P.A.; Jenkinson, C.P.; Cersosimo, E.; DeFronzo, R.A.; et al. Effect of acute exercise on AMPK signaling in skeletal muscle of subjects with type 2 diabetes: A time-course and dose-response study. *Diabetes* **2007**, *56*, 836–848. [CrossRef] [PubMed]

91. Witters, L.A.; Kemp, B.E. Insulin activation of acetyl-CoA carboxylase accompanied by inhibition of the 5′-AMP-activated protein kinase. *J. Biol. Chem.* **1992**, *267*, 2864–2867. [PubMed]

92. Musi, N.; Fujii, N.; Hirshman, M.F.; Ekberg, I.; Froberg, S.; Ljungqvist, O.; Thorell, A.; Goodyear, L.J. AMP-activated protein kinase (AMPK) is activated in muscle of subjects with type 2 diabetes during exercise. *Diabetes* **2001**, *50*, 921–927. [CrossRef] [PubMed]

93. Hojlund, K.; Staehr, P.; Hansen, B.F.; Green, K.A.; Hardie, D.G.; Richter, E.A.; Beck-Nielsen, H.; Wojtaszewski, J.F. Increased phosphorylation of skeletal muscle glycogen synthase at NH2-terminal sites during physiological hyperinsulinemia in type 2 diabetes. *Diabetes* **2003**, *52*, 1393–1402. [CrossRef] [PubMed]

94. Nielsen, J.N.; Wojtaszewski, J.F.; Haller, R.G.; Hardie, D.G.; Kemp, B.E.; Richter, E.A.; Vissing, J. Role of 5′AMP-activated protein kinase in glycogen synthase activity and glucose utilization: Insights from patients with mcardle's disease. *J. Physiol.* **2002**, *541*, 979–989. [CrossRef] [PubMed]

95. Krag, T.O.; Pinos, T.; Nielsen, T.L.; Duran, J.; Garcia-Rocha, M.; Andreu, A.L.; Vissing, J. Differential glucose metabolism in mice and humans affected by mcardle disease. *Am. J. Physiol. Regul. Integr. Comp. Physiol.* **2016**, *311*, R307–R314. [CrossRef] [PubMed]

96. Nielsen, T.L.; Pinos, T.; Brull, A.; Vissing, J.; Krag, T.O. Exercising with blocked muscle glycogenolysis: Adaptation in the mcardle mouse. *Mol. Genet. Metab.* **2018**, *123*, 21–27. [CrossRef] [PubMed]

97. Kollberg, G.; Tulinius, M.; Gilljam, T.; Ostman-Smith, I.; Forsander, G.; Jotorp, P.; Oldfors, A.; Holme, E. Cardiomyopathy and exercise intolerance in muscle glycogen storage disease 0. *N. Engl. J. Med.* **2007**, *357*, 1507–1514. [CrossRef] [PubMed]

98. Pederson, B.A.; Schroeder, J.M.; Parker, G.E.; Smith, M.W.; DePaoli-Roach, A.A.; Roach, P.J. Glucose metabolism in mice lacking muscle glycogen synthase. *Diabetes* **2005**, *54*, 3466–3473. [CrossRef] [PubMed]

99. Pederson, B.A.; Chen, H.; Schroeder, J.M.; Shou, W.; DePaoli-Roach, A.A.; Roach, P.J. Abnormal cardiac development in the absence of heart glycogen. *Mol. Cell. Biol.* **2004**, *24*, 7179–7187. [CrossRef] [PubMed]

100. Xirouchaki, C.E.; Mangiafico, S.P.; Bate, K.; Ruan, Z.; Huang, A.M.; Tedjosiswoyo, B.W.; Lamont, B.; Pong, W.; Favaloro, J.; Blair, A.R.; et al. Impaired glucose metabolism and exercise capacity with muscle-specific glycogen synthase 1 (gys1) deletion in adult mice. *Mol. Metab.* **2016**, *5*, 221–232. [CrossRef] [PubMed]

101. Hoffman, N.J.; Parker, B.L.; Chaudhuri, R.; Fisher-Wellman, K.H.; Kleinert, M.; Humphrey, S.J.; Yang, P.; Holliday, M.; Trefely, S.; Fazakerley, D.J.; et al. Global phosphoproteomic analysis of human skeletal muscle reveals a network of exercise-regulated kinases and AMPK substrates. *Cell Metab.* **2015**, *22*, 922–935. [CrossRef] [PubMed]

102. Schaffer, B.E.; Levin, R.S.; Hertz, N.T.; Maures, T.J.; Schoof, M.L.; Hollstein, P.E.; Benayoun, B.A.; Banko, M.R.; Shaw, R.J.; Shokat, K.M.; et al. Identification of AMPK phosphorylation sites reveals a network of proteins involved in cell invasion and facilitates large-scale substrate prediction. *Cell Metab.* **2015**, *22*, 907–921. [CrossRef] [PubMed]

103. Ducommun, S.; Deak, M.; Sumpton, D.; Ford, R.J.; Nunez Galindo, A.; Kussmann, M.; Viollet, B.; Steinberg, G.R.; Foretz, M.; Dayon, L.; et al. Motif affinity and mass spectrometry proteomic approach for the discovery of cellular AMPK targets: Identification of mitochondrial fission factor as a new AMPK substrate. *Cell. Signal.* **2015**, *27*, 978–988. [CrossRef] [PubMed]

104. Hoffman, N.J. Omics and exercise: Global approaches for mapping exercise biological networks. *Cold Spring Harb. Perspect. Med.* **2017**. [CrossRef] [PubMed]

105. Tsou, P.; Zheng, B.; Hsu, C.H.; Sasaki, A.T.; Cantley, L.C. A fluorescent reporter of AMPK activity and cellular energy stress. *Cell Metab.* **2011**, *13*, 476–486. [CrossRef] [PubMed]

106. Chennell, G.; Willows, R.J.; Warren, S.C.; Carling, D.; French, P.M.; Dunsby, C.; Sardini, A. Imaging of metabolic status in 3D cultures with an improved AMPK fret biosensor for flim. *Sensors* **2016**, *16*, 1312. [CrossRef] [PubMed]

107. Konagaya, Y.; Terai, K.; Hirao, Y.; Takakura, K.; Imajo, M.; Kamioka, Y.; Sasaoka, N.; Kakizuka, A.; Sumiyama, K.; Asano, T.; et al. A highly sensitive fret biosensor for AMPK exhibits heterogeneous AMPK responses among cells and organs. *Cell Rep.* **2017**, *21*, 2628–2638. [CrossRef] [PubMed]
108. Depry, C.; Mehta, S.; Li, R.; Zhang, J. Visualization of compartmentalized kinase activity dynamics using adaptable BimKARs. *Chem. Biol.* **2015**, *22*, 1470–1479. [CrossRef] [PubMed]
109. Yan, Y.; Gu, X.; Xu, H.E.; Melcher, K. A highly sensitive non-radioactive activity assay for AMP-activated protein kinase (AMPK). *Methods Protoc.* **2018**, *1*, 3. [CrossRef] [PubMed]
110. Arad, M.; Benson, D.W.; Perez-Atayde, A.R.; McKenna, W.J.; Sparks, E.A.; Kanter, R.J.; McGarry, K.; Seidman, J.G.; Seidman, C.E. Constitutively active AMP kinase mutations cause glycogen storage disease mimicking hypertrophic cardiomyopathy. *J. Clin. Investig.* **2002**, *109*, 357–362. [CrossRef] [PubMed]
111. Milan, D.; Jeon, J.T.; Looft, C.; Amarger, V.; Robic, A.; Thelander, M.; Rogel-Gaillard, C.; Paul, S.; Iannuccelli, N.; Rask, L.; et al. A mutation in PRKAG3 associated with excess glycogen content in pig skeletal muscle. *Science* **2000**, *288*, 1248–1251. [CrossRef] [PubMed]
112. Garcia-Roves, P.M.; Osler, M.E.; Holmstrom, M.H.; Zierath, J.R. Gain-of-function R225Q mutation in AMP-activated protein kinase γ3 subunit increases mitochondrial biogenesis in glycolytic skeletal muscle. *J. Biol. Chem.* **2008**, *283*, 35724–35734. [CrossRef] [PubMed]
113. Scott, J.W.; Galic, S.; Graham, K.L.; Foitzik, R.; Ling, N.X.; Dite, T.A.; Issa, S.M.; Langendorf, C.G.; Weng, Q.P.; Thomas, H.E.; et al. Inhibition of AMP-activated protein kinase at the allosteric drug-binding site promotes islet insulin release. *Chem. Biol.* **2015**, *22*, 705–711. [CrossRef] [PubMed]

International Journal of
Molecular Sciences

MDPI

Review

The Role of AMPK in the Regulation of Skeletal Muscle Size, Hypertrophy, and Regeneration

David M. Thomson

Department of Physiology & Developmental Biology, Brigham Young University, Provo, UT 84602, USA;
david_thomson@byu.edu; Tel.: +1-801-422-8709

Received: 28 September 2018; Accepted: 9 October 2018; Published: 11 October 2018

Abstract: AMPK (5′-adenosine monophosphate-activated protein kinase) is heavily involved in skeletal muscle metabolic control through its regulation of many downstream targets. Because of their effects on anabolic and catabolic cellular processes, AMPK plays an important role in the control of skeletal muscle development and growth. In this review, the effects of AMPK signaling, and those of its upstream activator, liver kinase B1 (LKB1), on skeletal muscle growth and atrophy are reviewed. The effect of AMPK activity on satellite cell-mediated muscle growth and regeneration after injury is also reviewed. Together, the current data indicate that AMPK does play an important role in regulating muscle mass and regeneration, with AMPKα1 playing a prominent role in stimulating anabolism and in regulating satellite cell dynamics during regeneration, and AMPKα2 playing a potentially more important role in regulating muscle degradation during atrophy.

Keywords: AMPK; LKB1; autophagy; proteasome; hypertrophy; atrophy; skeletal muscle; AICAR; mTOR; protein synthesis

1. Introduction

5′-adenosine monophosphate-activated protein kinase (AMPK) is an intracellular sensor of ATP consumption that emerged in the late 1990s as a key regulator of skeletal muscle metabolism [1–3]. Its role in the promotion of ATP-producing catabolic processes involved in glucose and fat oxidation is well characterized. Its general identity as a catabolic agent is further illustrated by its stimulation of protein degradation and autophagy [4–6]. Additionally, AMPK inhibits anabolic processes that consume ATP, such as protein synthesis [7]. Given these general actions, AMPK's potential negative effect on skeletal muscle growth has been well-studied over the past 20 years.

In this review, a very brief overview of AMPK structure and function will be presented. Then, AMPK's effect on cell processes that are relevant to the control of cell size, such as protein synthesis, protein degradation and autophagy, will be reviewed. Finally, the known experimental effects of AMPK modulation on skeletal muscle growth and regeneration will be presented.

2. AMPK and Its Activation

2.1. AMPK Structure and Activation

Many excellent sources are available in the literature that provide a thorough review of the molecular and mechanistic details of AMPK structure and activity (e.g., [3,8,9]). Only a brief summary is provided here. Active AMPK is a heterotrimer comprised of three subunits: α, β, and γ. The actual kinase domain is contained within the α subunit, along with the predominant regulatory phosphorylation site, Thr172, which must be phosphorylated to produce any significant activity. The α and γ subunits serve scaffolding and regulatory roles. The γ subunit confers AMP sensitivity to the enzyme through four cystathionine β-synthase (CBS) domains, which can bind AMP, ADP, or ATP.

This interaction with three of these nucleotides confers on AMPK its ability for effectively detecting cellular energy status. During energy stress, when ATP breakdown to ADP accelerates, AMP is generated through the action of adenylate kinase, which transfers a phosphate from one ADP molecule to another, resulting in the production of ATP and AMP. As AMP levels rise, it (and to some degree, ADP) activates AMPK by: (1) increasing AMPK phosphorylation by upstream kinases; (2) decreasing AMPK dephosphorylation by phosphatases; and (3) allosterically activating phosphorylated AMPK [8]. AMPK's response to the decrease in the ATP:AMP ratio are crucial for the cell's ability to maintain appropriate ATP levels because it promotes ATP-generating catabolic processes, while inhibiting ATP-consuming anabolic processes [3,8,9].

Different isoforms exist for each of the AMPK subunits. Two α (α1 and α2), two β (β1 and β2), and three γ (γ1, γ2, and γ3) isoforms result in the possibility of up to 12 distinct AMPK configurations. In human skeletal muscle, however, these configurations are likely limited to α2/β2/γ1 (most abundant), α2/β2/γ3, and α1/β2/γ1 [10]. Of these three, α2/β2/γ3 accounts for the majority of AMPK activation due to high-intensity exercise [11]. In contrast to human muscle, mouse muscle contains β1 trimers (α1β1γ1 and α2β1γ1), although these still only contribute slightly to the overall AMPK activity [12]. While some functional implications of these different configurations have been determined, a full understanding of the full impact of differing trimer contents in tissues and within muscles is still being worked out.

2.2. Upstream AMPK Kinases

In mature skeletal muscle, liver kinase B1 (LKB1) is generally considered the primary AMPK kinase since total AMPK activity is essentially eliminated by muscle-specific LKB1 knockout [13–16]. LKB1 seems, however, to play a more important role in AMPKα2 activation, since AMPKα1 activity is not heavily impacted in skeletal muscle by LKB1 knockout [13,14,17–19].

Calcium/calmodulin-dependent protein kinase (CamKK) [20,21], and transforming growth factor β-activated kinase-1 (TAK1) [19,22] likely also play important roles in the activation of AMPK in skeletal muscle under certain circumstances.

2.3. AMPK Activators in Skeletal Muscle

2.3.1. Exercise

As would be expected given its role as a cellular energy sensor, AMPK is strongly activated in skeletal muscle by repeated muscle contraction [23] and exercise [2,11,24] in both rodents and humans. Activation of AMPKα2-containing trimers by endurance exercise occurs within 5 min of the onset of exercise [25], and likely requires a relatively high intensity effort, usually somewhere above 50% of VO2max [24,26]. AMPK activity returns to baseline levels within 3 h after exercise [26].

While AMPKα2 activity is readily increased by exercise and muscle contraction in rodents [27], increases in AMPKα1 activity after exercise/contraction are less consistent. For example, AMPKα1 activity in mouse quadriceps muscle was approximately four times higher immediately after 90 min of treadmill running at 13–17 m/min [28], but was not activated at all after running at 10–15 m/min for 60 min [29]. Similarly, 30 min of treadmill running at 30% of maximum running capacity activated AMPKα2 in mouse skeletal muscle, but not AMPKα1, while running at 70% of maximum activated both isoforms [30]. In vitro contraction of the extensor digitorum longus (EDL) muscle for 25 min activated AMPKα1, while 20 min of in situ contraction of the tibialis anterior (TA) failed to do so [29]. The data from rodents is confirmed in human studies where cycling for 1 h at 50% and 70% VO2max failed to activate AMPKα1 [24], while a single 30 s sprint [31] or high intensity interval cycling (4 × 30 s bouts of cycle sprints) [32] activated both AMPKα1 and α2 isoforms. Thus, activation of AMPKα1 isoforms by exercise requires greater intensity work and/or duration than for the activation of AMPKα2. As discussed below, this has important implications in relation to AMPK's impact on muscle growth and repair, as AMPKα1 appears to be critical in the regulation of anabolism.

2.3.2. AICAR

The 5-amino-4-imidazolecarboxamide ribonucleoside (AICAR) has been used for nearly 25 years to activate AMPK in various tissues in the body [33], including skeletal muscle [1,34,35]. Upon administration, it is converted into ZMP (AICAR monophosphate), an AMP mimetic that activates AMPK without altering intracellular adenine nucleotide levels. Similar to relatively low-intensity exercise, intraperitoneal injection of AICAR activates AMPKα2 but not AMPKα1 in rat gastrocnemius [7]. Furthermore, AICAR-stimulated glucose uptake is eliminated in AMPKα2 knockout muscle, but not in AMPKα1 knockouts [36], suggesting that at least some of AICAR's metabolic effects are specifically AMPKα2 dependent. Nonetheless, AICAR can activate AMPKα1, since incubation of isolated rat epitrochlearis muscle with 2 mM AICAR activated AMPKα1, albeit to a lesser degree than AMPKα2 [27].

2.3.3. Metformin

Metformin has long been used as a front-line drug in the treatment of insulin resistance and diabetes because of its ability to improve hyperglycemia in an insulin-independent manner. Shortly after AMPK's metabolic actions began to be described, which are similar to those of metformin, it was discovered that at least some of metformin's effects are, indeed, AMPK-dependent, although some are not [37,38]. The activation of AMPK by metformin is mainly indirect, where metformin inhibits mitochondrial oxidative phosphorylation, thereby decreasing ATP production and generating an energetic stress on the cell [38]. Although the liver is considered the principal site of metformin's glucose-regulating effects, chronic, therapeutic dosing of metformin over a 10 week period does increase AMPKα2 (but not AMPKα1) activity in diabetic skeletal muscle [39]. However, it isn't known whether this is a direct effect of the metformin on skeletal muscle since the effect persisted after metformin withdrawal. In mice, a metformin injection modestly increased AMPKα1 and α2 activity, while treatment of isolated epitrochlearis and soleus (SOL) muscle ex vivo with 10 mM (but not 2 mM) metformin markedly activated AMPKα1 and α2 isoforms [40]. However, the relevance of these high concentrations to in vivo metformin action is questionable.

2.3.4. Small Molecule AMPK Activators

A-769662 was the first small molecule AMPK activator described in the literature [41]. It specifically targets β1-containing AMPK trimers, and in skeletal muscle only activates the scarcely-expressed α1β1 complexes [42]. Several additional activators have subsequently been identified, with varying specificities for the different AMPK subunit isoforms. Of them, Ex229 (small molecule 991), PF-739, and MK-8722 have been demonstrated to activate AMPK in skeletal muscle [43], though effects on muscle growth, atrophy, and regeneration are unknown.

3. Regulation of Growth-Related Cell Processes by AMPK

Skeletal muscle growth, in essence, occurs when the rate of protein anabolism exceeds the rate of protein catabolism. Atrophy results when protein catabolism exceeds anabolism [44]. AMPK is known to regulate both processes.

3.1. Effect of AMPK on Protein Synthesis

The first indications that AMPK played a role in the regulation of protein metabolism came in 2002 when it was shown that the fractional rate of protein synthesis in skeletal muscle deceased approximately 45% 1 h after an injection of the AMPK-activating drug, AICAR [7]. This inhibitory effect of AMPK activation on protein synthesis was subsequently observed in cultured muscle cells [45] as well as hepatocytes/liver [46–48], cardiac myocytes [49,50], and cancer cells [51,52], among other cell types and tissues.

AMPK's inhibition of protein synthesis is mediated by regulation of protein translation through the mechanistic target of rapamycin, complex 1 (mTORC1) pathway. Regulation of mTORC1 activity is complex, as it serves as a signaling checkpoint for many environmental inputs including nutrients, energy status and mechanical strain. When activated, mTORC1 drives cell growth in part by stimulating protein synthesis through its phosphorylation of several downstream targets, the best characterized of which are the 70-kDa ribosomal protein S6 kinase (p70S6K1) and eukaryotic initiation factor 4E-binding protein 1 (4E-BP1).

AMPK has been shown to inhibit mTORC1 activity through multiple mechanisms. First, AMPK phosphorylates mTOR, a key component of the mTORC1 complex, at Thr2446 [53], which is thought to impair mTORC1 activity by preventing phosphorylation at Ser2448. This site (Ser2448) was initially thought to promote mTORC1 activity when phosphorylated. Since then, its relevance to mTORC1 activity has been reassessed, and it seems probable that phosphorylation of both sites (i.e., Thr2446 and Ser2448) is inhibitory on mTORC1 activity [54]. Nonetheless, AMPK also inhibits mTORC1 by phosphorylating tuberous sclerosis complex 2 (TSC2). Activation of mTORC1 occurs at the lysosomal membrane through interaction with GTP-bound Rheb [55]. TSC2 acts, in complex with binding partners tuberous sclerosis complex 1 and TBC1 domain family member 7 (TBC1D7) [56], as a GTPase activating protein that converts GTP to GDP, thereby greatly diminishing the ability of Rheb to promote mTOR activity. Finally, AMPK phosphorylates raptor, an mTOR binding partner that is essential for mTORC1 activity. This phosphorylation leads to sequestration of raptor by 14-3-3 proteins, and impaired mTORC1 activity [57].

In addition to its inhibitory action on mTORC1, AMPK also regulates protein synthesis through inhibition of eukaryotic elongation factor 2 (eEF2) activity. Phosphorylation of eEF2 at Thr56 inhibits binding of the elongation factor to the ribosome, thereby slowing elongation rate. Phosphorylation of eEF2 at this site is mediated by eEF2 kinase (eEF2K). AMPK impacts eEF2K activity in two ways. First, p70S6k phosphorylates and inhibits eEF2K (leading to eEF2 activation), and AMPK can prevent this by inhibiting the mTOR pathway, as described above. Secondly, AMPK directly phosphorylates and activates eEF2K, leading to eEF2 inactivation [47,58]. While translation initiation is often considered the rate-limiting step in protein synthesis, control of elongation can, under certain circumstances, be critical in protein synthetic rate [58,59]. For instance, inhibition of eEF2K partially blocks the acute inhibitory effect of contractions on protein synthesis, although this effect does not appear to be regulated by AMPK [60]. Thus, the capacity for eEF2 regulation by AMPK in skeletal muscle remains unclear.

3.2. Effect of AMPK on Catabolic Processes

3.2.1. AMPK and Autophagy

Defective cellular content (organelles, pathogens, etc.) is degraded and recycled through the process of autophagy under low-energy conditions such as nutrient deprivation and exercise. Autophagy involves several subprocesses including engulfment of the target components in an autophagosome, fusion of the autophagasome with a lysosome (forming an autophagolysosome), followed by degradation of the cargo. This is a complex process and a complete description will not be presented here (see reference [61] for an excellent review). However, several key points of autophagy regulation are important in the context of the current topic. Under low-energy conditions, uncoordinated 51-like kinase 1 (ULK1) phosphorylates and activates multiple downstream targets that promote the progression of autophagy, including several autophagy related (ATG) proteins and beclin-1. Under conditions of energy abundance mTORC1 inhibits ULK1 through phosphorylation at Ser757. This, along with its targeting of other autophagy components, leads to mTOR's inhibition of autophagy [61].

AMPK has long been known to regulate autophagy. Initial observations in rat hepatocytes suggested that AICAR-induced AMPK activation inhibited autophagy [62], but subsequent work demonstrated that the AMPK inhibitor, Compound C, and dominant negative AMPK expression also inhibited autophagy, suggesting that AICAR's effects might be AMPK-independent [63]. Since then, AMPK's role in the process remains complicated because its effect seems to be dependent on cell type and metabolic context. Nonetheless, it appears that AMPK generally supports and promotes autophagy [64,65], and this is true in skeletal muscle [66]. It does this through multiple mechanisms. As noted above, AMPK inhibits mTORC1 activity. This relieves mTORC1 inhibition of ULK1, and thereby promotes autophagic flux. Additionally, AMPK directly phosphorylates components of the autophagy regulatory machinery. AMPK phosphorylates ULK1 at several sites [67], and also targets autophagy related protein 9 (ATG9) [4] and beclin-1 [68] downstream of ULK1, promoting autophagy.

3.2.2. AMPK and Ubiquitin-Proteasome Mediated Catabolism

The 26S proteasome degrades proteins that have been tagged for destruction through the attachment of ubiquitin chains. The covalent attachment of ubiquitin to targeted proteins is catalyzed through the action of three enzymes (E1, E2, and E3). E3 actually refers to one of multiple ubiquitin ligases, each of which is specific for the degradation of particular proteins. In skeletal muscle, two E3 enzymes, Atrogin-1 and muscle ring finger-1 (MuRF-1), are known to play a prominent role in proteasomal protein breakdown during muscle atrophy [69].

Expression of the atrophy-related genes Atrogin-1 and MuRF-1 is regulated through members of the forkhead box (FoxO) transcription factors. Anabolic signaling that activates Akt (e.g., via nutritional and hormonal cues mediated by insulin and other growth factors) results in their cytoplasmic localization and subsequent degradation so that they do not induce atrogene transcription. Catabolic stimuli, such as oxidative stress and inflammation, increase MuRF-1 and Atrogin-1 expression in muscle through the mitogen-activated protein kinase (MAPK) p38 as well as nuclear factor-κB (NF-κB) [69].

AMPK stimulates FoxO activity. AICAR injection into mice increases FoxO1 and FoxO3 expression [28,70], although AICAR's upregulation of FoxO1 is not impacted by knockout of AMPKα2 [28]. Treatment of C2C12 myotubes with AICAR results in protein breakdown accompanied by increased expression of FOXO, Atrogin-1, MuRF-1, and two other FoxO target genes, microtubule-associated protein 1A/1B-light chain 3 (LC3), and Bnip3 [71], and these effects are Akt/mTOR independent [6]. AMPK also phosphorylates FoxO3a at a site known to activate the transcription factor and thereby induce generalized protein degradation but this may not necessarily affect its localization in the nucleus [72–74]. AICAR also increases Foxo3 binding to the MuRF-1 and Atrogin-1 promoters [6]. Furthermore, AMPK activation increases nicotinamide adenine dinucleotide (NAD$^+$) concentration which activates the sirtuin 1 (SIRT1) deacetylase. SIRT1-mediated deacetylation of FoxO proteins increases their transcriptional activity [75,76].

4. Influence of AMPK on Skeletal Muscle Size

4.1. AMPK Regulation of Basal Muscle Size

Initial observations in dominant-negative AMPK (AMPK-DN) transgenic mice in which a dominant negative AMPKα2 subunit was overexpressed under the muscle creatine kinase promoter (expressed in heart and skeletal muscle) showed that EDL muscles tended to be larger than in wild-type (WT) mice, suggesting that AMPK might negatively regulate basal muscle mass [77], as would be expected given AMPK's stimulation of catabolism and activation of anabolism. Those initial findings are consistent with later work in which skeletal muscle specific AMPKα1 and α2 double knockout (AMPKα1/α2 dKO) soleus muscles were larger by mass and fiber diameter compared to WT muscles [78]. Myotubes derived via primary muscle cell cultures from these AMPKα1/α2 dKO muscles were likewise larger than from those from WT muscles [78]. On the other hand, muscle-specific AMPKβ1/β2 double knockout (AMPKβ1/β2 dKO) SOL and EDL muscles were reportedly not different in size compared to WT muscles [79]. Why the double knockout of β isoforms did not lead to increased muscle size is not clear, but might be related to the promoter used to drive the Cre-mediated deletion of floxed AMPK. In the case of the AMPKβ1/β2 dKO mice, the muscle creatine kinase (MCK; cardiac and skeletal muscle specific) promoter was used, while in the AMPKα1/α2 dKO mice it was the human skeletal muscle actin (HSA; skeletal muscle specific) promoter. Alteration of AMPK activity in the heart in the AMPKβ1/β2 dKO mice could have influenced the response of skeletal muscle to the knockout since cardiac dysfunction is well-known to induce skeletal muscle atrophy. Thus, not all AMPK deficiency models support the notion that AMPK inhibits basal muscle mass, but these findings are usually derived from AMPK knockout models that are not specific to skeletal muscle. Plantaris muscles from germline AMPKα1 knockout mice are smaller than those from WT mice [80]. The lack of AMPKα1 in all tissues in this model, however, doesn't allow for conclusions regarding the role of AMPK in the muscle specifically, since the lack of AMPKα1 in other tissues may have impacted muscle size (e.g., by decreasing systemic growth factors or other humoral inputs). Furthermore, the lack of AMPKα1 was associated with compensatorily elevated AMPKα2 activity in the muscles, which could have resulted in decreased mass. In agreement with this interpretation, primary cultured myotubes derived from these cells (removed from the systemic environment of the mouse) were larger than WT muscles [80]. Another study found that muscle fibers, especially type IIb fibers, are smaller in whole-body AMPKβ2 knockout mice [81]. The TA muscles from the aforementioned cardiac/skeletal muscle AMPK β1/β2 dKO mice are also smaller vs. WT muscles [82]. Again, the smaller size of these muscle fibers in these studies could be secondary to systemic effects of altered function of other tissues (e.g., heart), though this has not been directly tested.

4.2. Role of AMPK in Skeletal Muscle Hypertrophy

Given AMPK's pro-catabolic and anti-anabolic actions, it was hypothesized that AMPK activity would block overload-induced muscle growth, and the available data generally support this. When comparing the hypertrophic response of rat muscle to synergist ablation-induced overload, AMPK phosphorylation in the hypertrophying muscle was associated with decreased muscle hypertrophy [83], and diminished mTOR pathway signaling [84]. Several subsequent studies have reported negative associations between AMPK phosphorylation or activity and skeletal muscle growth. Indeed, impaired overload hypertrophy in obese rats [85,86], attenuated mTOR phosphorylation in metabolic syndrome patients [87], myotube hypertrophy during differentiation [88], myostatin inhibition of eEF2 and protein synthesis in myotubes [89], and differences in hypertrophy with varied ladder-climbing protocols in rats [90] are all associated inversely with AMPK activity.

Direct pharmacological evidence showing that AMPK inhibits muscle growth has also been demonstrated. An AICAR injection 1 h prior to a bout of resistance exercise-mimicking contractions greatly attenuated the mTOR signaling response to the contraction bout [35], suggesting that AMPK activation would impair the normal increase in protein translation that occurs post resistance exercise. Likewise, continuous perfusion of overloaded plantaris muscles with AICAR after synergist ablation greatly attenuated muscle hypertrophy [91].

Genetic evidence for the inhibitory effect of AMPK on in vivo skeletal muscle hypertrophy was provided by Mounier, et al. [80], who performed synergist ablations on AMPKα1 knockout mice. After 7–21 days of overload, AMPKα1 expression and activity was significantly increased in WT mice (but not AMPKα1-KO mice, as expected). Despite lower basal muscle mass, whole muscle hypertrophy and muscle fiber hypertrophy at 7 and 21 days was greater for the AMPKα1 knockouts. In line with the hypertrophy measurements, mTOR pathway signaling, as assessed by p70S6k and 4E-BP1 phosphorylation was greater, while eEF2 phosphorylation was lower (corresponding to increased eEF2 activity) after overload in the AMPKα1-KO muscles. Importantly, this occurred despite a compensatory increase in AMPKα2 activity basally and at 7 and 21 days after overload in the KO muscles, demonstrating that AMPKα1 is likely the major isoform involved in regulation of overload-induced muscle growth.

On a related note, old age leads to the loss of muscle mass (sarcopenia) and a blunted anabolic response to hypertrophic stimuli. AMPKα2 activation by exercise and AICAR is typically blunted in old age [34,92]. However, AMPK phosphorylation in old overloaded muscle is elevated vs. young overloaded muscles and is negatively correlated to mTOR signaling and hypertrophy [83,84]. Similarly, AMPK phosphorylation 1–3 h after resistance exercise is elevated in old vs. young human muscle, and is associated with delayed mTOR pathway activation [93]. Interestingly, 10 min of continuous electrically stimulated muscle contractions resulted in increased AMPKα2 activity in muscles from young adult (8 months-old) and old (30 months-old) rats, but this response was attenuated by old age [34]. However, under this stimulation protocol, AMPKα1 activity increased after stimulation only in old muscle, and not young, suggesting that old muscle may be hypersensitive to exercise-induced AMPKα1 activation which perhaps contributes to sarcopenia.

4.3. Role of AMPK in the Regulation of Skeletal Muscle Atrophy

The response of the AMPK system to muscle atrophy is unclear. Disuse atrophy of rodent skeletal muscle after 1–4 weeks of hindlimb unloading (HU) has been reported to increase [94] or decrease AMPK phosphorylation [95–97], while others observed no effect of HU on AMPKα1 and AMPKα2 activity or on acetyl-CoA carboxylase (ACC; AMPK target and marker of AMPK activity) phosphorylation [98]. Similarly, AMPK activity is reported by some to increase after 4 and 7 days of denervation in mice and/or rats [99–102], while spinal cord transection does not alter AMPKα2 activity in muscle [103]. In the denervation model, these conflicting results may be due to timing since AMPK phosphorylation in denervated soleus muscles is decreased during at least the first 24 h post-denervation, is not different at 3 days, and is elevated by 7 days post-denervation [104] compared to control muscles. Thus, it appears that AMPK activity in general decreases initially, then increases later on during the adaptation to disuse, at least in the denervation model.

Consistent with a catabolic role for AMPK, HU-induced atrophy of the soleus muscle was partially attenuated in AMPK-DN soleus muscles, potentially through decreased ubiquitin-proteosome activity [98]. It should be noted, however, that in this model, AMPKα1 activity was only mildly decreased by dominant-negative (DN) expression in the transgenic muscles, so the anti-atrophy effect was mainly due to the loss of AMPKα2 activity. Similarly, atrophy in denervated TA muscles from AMPKα2-KO mice was partially blocked compared to WT muscles [101], and this was also associated with decreased autophagic markers, Atrogin-1/MuRF-1 expression and ubiquitination. Akt and 4E-BP1 phosphorylation were unaffected by AMPKα2-KO, suggesting that the attenuation of atrophy was due to decreased protein degradation rather than increased mTOR activity and synthesis. Together, these findings suggest that in contrast to AMPKα1's role in inhibiting skeletal muscle mTOR and hypertrophy, the presence of AMPKα2 plays a more pronounced role in supporting an atrophy response to disuse, and in promoting protein degradation through the ubiquitin-proteasome system.

While the lack of AMPKα2 attenuates atrophy, increased activation of AMPK above normal does not appear to accelerate the loss of muscle mass since daily AICAR injections during 3 days of tibial nerve denervation in rats did not significantly affect skeletal muscle atrophy in soleus and gastrocnemius muscles [105]. Furthermore, 4 weeks of AICAR treatment of mdx mice (a model for Duchenne muscular dystrophy) did not exacerbate atrophy associated with dystrophy, and actually improved muscle function, probably through enhanced autophagic clearing of damaged cell components [106], and/or promotion of a more oxidative muscle phenotype [107].

4.4. Effect of Disruption of LKB1 on Skeletal Muscle Size and Hypertrophy

LKB1 knockout in skeletal muscle results in a nearly complete elimination of basal, exercise, and AICAR-induced AMPKα2 activity [14,18,19] and overall AMPK phosphorylation [13,16,108–110], while it has little [13,14] to no [17,18] effect on AMPKα1 activity, which is an important consideration since AMPKα1 seems to be the major isoform regulating muscle growth [78,80]. LKB1 also phosphorylates several other AMPK family members, at least one of which, sucrose non-fermenting 1 AMPK related kinase (SNARK) is important in the maintenance of muscle mass [111].

The weight of muscles from relatively young mLKB1-KO mice is not statistically different from that of WT muscles [16,19]. However, after approximately 30 weeks of age, muscle mass begins to decline in mLKB1-KO muscles [15]. This atrophy is associated with the development of heart failure, however [15,112], and thus may be primarily due to cardiac cachexia. Consistent with that speculation, muscle weights were, for the most part, similar in a skeletal muscle specific dominant negative LKB1 model, though quadricep muscles were smaller, and diaphragms were larger [113].

Nonetheless, since LKB1 is a primary upstream activator of AMPK in skeletal muscle, McGee et al. [19] hypothesized that the lack of LKB1 in muscle would result in a greater hypertrophic response to overload. However, when plantaris muscles from conditional muscle-specific LKB1 knockout mice (mLKB1-KO; cardiac and skeletal muscle knockout) were overloaded via synergist ablation (of the gastrocnemius muscle), there was no significant difference in the degree of mTOR pathway activation or hypertrophy compared to WT muscles. Importantly, overload increased the activity of AMPKα1 (but not α2) in both WT and KO muscles, showing that this response is regulated at least in part in an LKB1-independent fashion, perhaps via CamKK or TAK1 signaling [19]. Therefore, based on the findings of Monier, et al., that it is the α1 subunit that regulates skeletal muscle mTOR signaling and size [80], the lack of difference in hypertrophic response in the mLKB1-KO muscles is not surprising.

However, when skmLKB1-KO muscles were subjected to an acute bout of intermittent contractions designed to mimic hypertrophy-inducing resistance exercise, mTOR signaling (p70S6k and ribosomal protein S6 phosphorylation) was elevated to a greater extent both basally and immediately post-contraction in knockout vs. WT muscles, as was protein synthesis at 8 h post contraction [110]. AMPK phosphorylation was increased with contractions in WT but not skmLKB1-KO muscles using this contraction protocol, suggesting that the increased mTOR signaling in the knockout muscle could be due to a lack of AMPK activation, but AMPKα1-specific activity was not measured. This suggests that LKB1 can exert catabolic effects under some circumstances, though evidence that this impacts gross muscle hypertrophy is lacking.

Potential effects of LKB1 on muscle atrophy during unloading or denervation are currently unknown.

4.5. Exercise-Induced AMPK Activation and Muscle Hypertrophy

That endurance training interferes with hypertrophy/strength gains has been well-established [114,115]. The accumulation of evidence demonstrating AMPK's anti-anabolic and pro-catabolic effects naturally leads to the question of whether its activation during exercise functionally impairs the ability of muscle to hypertrophy, which, if true, would mechanistically explain the conflict between endurance/hypertrophy responses. In support of this hypothesis, Atherton et al. [116] showed that tissue-autonomous differences in signaling pathway activation may contribute to the inherent differences in gross adaptation that is observed with endurance vs. resistance exercise training. Using in vitro electric stimulation protocols that mimic endurance (low frequency, continuous) and resistance (high-frequency, intermittent) exercise bouts in rat skeletal muscle, they showed that endurance-type stimulation (but not resistance-type stimulation) resulted in AMPK activation and accrual of peroxisome-proliferator-activated receptor γ coactivator-1 α, while resistance-type stimulation (but not endurance-type stimulation) increased phosphorylation of Akt, TSC2, mTOR, downstream mTOR targets, and increased protein synthesis.

In humans, however, the molecular responses to different exercise modalities is less clear and has generally been interpreted as not supporting the hypothesis that physiological AMPK activation (e.g., through endurance exercise training) significantly impacts mTOR signaling and/or protein synthesis [115]. Apró et al. showed in trained male subjects that activation AMPKα2 via 1 h of intense cycling did not significantly impair subsequent activation of mTOR pathway components or mixed muscle fractional protein synthesis after a resistance training bout [117]. However, AMPKα1 was not activated by either exercise bout in this case. Since AMPKα1 is the major AMPK isoform regulating skeletal muscle growth, at least in rodents [80], the lack of an effect of this endurance exercise bout on mTOR or protein synthesis would be expected and does not preclude an AMPK effect if the $\alpha 1$ subunit were actually activated (e.g., by more intense or prolonged exercise than that employed in this study).

Furthermore, while acute AMPK activation immediately after exercise is suppressed in endurance-trained muscle [118,119], chronic endurance exercise training increases basal AMPK$\alpha 1$ protein content and activity. Twelve weeks of treadmill training (90 min/day, 5 days/week) elevated both AMPK$\alpha 1$ and $\alpha 2$ protein content in rat muscle [120]. Similarly, in humans, AMPK$\alpha 1$ (but not AMPK$\alpha 2$) protein concentration [118,121,122] and basal AMPK$\alpha 1$ activity [122] is greater in endurance trained vs. untrained individuals. Thus, the question of whether or not resistance exercise-induced anabolic signaling and hypertrophy are impacted by AMPK$\alpha 1$ activation by endurance exercise training remains unresolved.

4.6. Does Pharmacological AMPK Activation Limit Skeletal Muscle Hypertrophy?

Data showing the effect of pharmacological AMPK activation on load-induced muscle hypertrophy is quite limited. As noted previously, AICAR activation of AMPK attenuates contraction-induced increases in mTOR signaling and overload-induced hypertrophy in rodent muscles [35,91].

Interestingly, metformin treatment of patients with severe burn injury at dosages (850–2550 mg/day for 8 days) previously shown to activate AMPK$\alpha 2$ but not AMPK$\alpha 1$ in skeletal muscle [39] led to a significant increase in protein synthesis [123]. Similarly, in tumor-bearing cachexic rat muscle, metformin treatment rescued protein synthesis and decreased protein degradation while activating AMPK, though isoform-specific activity measures were not taken [124]. This improvement in muscle anabolism may be attributable to the long-appreciated impact of metformin on insulin sensitivity. Improved insulin action at the skeletal muscle would not only improve glucose handling, but protein synthesis as well.

5. Influence of AMPK on Skeletal Muscle Regeneration after Injury

5.1. The Regenerative Process in Skeletal Muscle

Skeletal muscle regeneration after injury is dependent upon the action of muscle stem cells (MuSCs), primarily satellite cells (SCs) which, in uninjured muscle, reside underneath the basal lamina next to mature muscle fibers in a quiescent, mitotically inactive state. Upon muscle damage, these cells activate and proliferate, with their subsequent progeny either engaging in a process of self-renewal to maintain the MuSC pool, or differentiating into myoblasts that then fuse together with other myoblasts or existing myofibers, leading to repair or replacement of the damaged tissue. Many excellent reviews are available for more detail on these events (e.g., [125–127]).

Muscle regeneration is a precisely ordered process that is dependent on the actions and influence of many cellular players at or near the myogenic niche, including SCs, mature muscle fibers, immune cells, fibroblasts, fibroadipogenic progenitors (FAPs), and others [125,126]. Although AMPK likely plays an important role in the regulation of many of these cell types (in macrophages, for instance [128]), the discussion here will be limited to its role in SCs.

5.2. Effect of AMPK on Myogenesis in Culture

The C2C12 adult skeletal muscle myoblast cell line is frequently used as an in vitro culture model for studying the process of myogenesis. C2C12 myoblasts, prior to differentiation in low-serum media, express the $\alpha 2$, $\gamma 2$, and $\gamma 3$ AMPK isoforms, but minimal expression of $\alpha 1$, $\beta 1$, $\beta 2$, and $\gamma 1$ isoforms. Differentiation of the myoblasts into myotubes by exposure to low-serum media, however, induces the expression of all isoforms except for $\gamma 1$ [129]. Consistent with the lack of β isoforms in myoblasts, which should preclude AMPK activation, stimulation of the cells with oligomycin and serum withdrawal activated AMPK much more strongly in myotubes vs. myoblasts [129]. However, other findings show that AICAR is able to activate AMPK in undifferentiated myoblasts [130], suggesting that the lack of β isoforms and AMPK activity in myoblasts is not a generalizable finding.

Activation of AMPK impairs myoblast proliferation. When C2C12 myoblasts are cultured in low glucose conditions (≤5 mM), AMPK is activated leading to impaired differentiation into myotubes. The same phenomenon is true for primary myoblasts, but only at even lower glucose concentrations [131]. Pharmacological AMPK activation with AICAR, metformin and other drugs accomplishes the same impairment in differentiation [130,131].

AMPK also impairs myoblast differentiation in culture. Activation of AMPK with AICAR in differentiating C2C12 myoblasts decreased p21 expression (which normally increases dramatically during differentiation) and cell cycle transition, and decreased myotube formation and myosin heavy chain expression [130]. A similar inhibitory effect of AICAR on primary bovine myoblasts was also observed [132]. Furthermore, transfection of C2C12 myoblasts with CamKKβ, an established AMPK activator, resulted in AMPK activation in myoblasts, cell cycle arrest and impaired proliferation as well as impaired subsequent differentiation, and this effect on proliferation and differentiation was AMPK-dependent since it was blocked by dominant-negative AMPK expression [133]. Together, these in vitro findings suggest that hyperactivation of AMPK in myoblasts blocks muscle proliferation and differentiation.

5.3. Effect of AMPK on Muscle Regeneration In Vivo

Although hyperactivation of AMPK in culture impairs both proliferation and differentiation of myoblasts, the lack of AMPK in SCs in vivo blocks normal muscle regeneration after injury. AMPKα1 is the predominant catalytic isoform in quiescent, activated and differentiating satellite cells [134–136]. Regeneration of damaged muscle is impaired (vs. WT) in both constitutive AMPKα1-KO mice, as well as in mice with AMPKα1-KO induced just before injury [137], and this is associated with decreased satellite cell number and Pax7, myogenic factor 5 (Myf5), and myogenin expression in basal muscles. Furthermore, AMPKα1-KO satellite cells have diminished myogenic capacity when transplanted into WT muscles, showing that the defect in regeneration is mediated by the lack of AMPKα1 in the satellite cells themselves, rather than in other cells in the KO animals (such as fibroblasts, macrophages, etc.) [137].

A similar impairment of regeneration is demonstrated by satellite cell-specific AMPKα1-KO. Fu et al. reported that satellite cells lacking AMPKα1 activate and proliferate more slowly both in culture and in single fiber preparations, and result in a subsequent impairment of muscle regeneration after cardiotoxin injury [136]. SCs, with their scant mitochondria, depend heavily on glycolytic metabolism and, according to the findings of these authors, the lack of AMPK impairs SC activation and proliferation by decreasing Warburg-like glycolysis [136].

Theret et al. also reported that satellite cell-specific AMPKα1-KO impairs muscle regeneration [135]. They showed that when SCs from AMPKα1 knockout mice (but not AMPKα2 knockouts) were collected and differentiated, the lack of AMPKα1 resulted in increased self-renewal instead of differentiation [135]. Similarly, deletion of AMPKα1 in MuSCs in vivo resulted in decreased size of the regenerating fibers along with decreased differentiation and fusion, but increased proliferation of MuSCs. However, in contrast to the report of Fu et al., the impaired regeneration was attributed by these authors to increased lactate dehydrogenase activity and enhanced Warburg-like glycolysis in the AMPKα1-KO SCs. The reason for this discrepancy is not clear, but could be due to different transgenic constructs. Regardless, both studies demonstrate the importance of SC AMPKα1 in allowing for proper regeneration through metabolic regulation. Together with the culture data, the available evidence indicates that AMPK activity must be kept within relatively tight bounds (not too high or too low) for optimal muscle regeneration.

5.4. LKB1's Role in Skeletal Muscle Regeneration

The content of the upstream AMPK kinase, LKB1, increases during myoblast differentiation [138]. Overexpression of LKB1 in C2C12 myoblasts enhances differentiation, while RNAi-mediated knockdown of LKB1 impairs differentiation [138]. While some of this effect is likely due to the action of LKB1 on other targets within the AMPK family, AMPK phosphorylation is also increased substantially during muscle differentiation [138].

The lack of LKB1 in SCs promotes proliferation and self-renewal of the satellite cell pool, but impairs myoblast differentiation [139,140]. The effect on self-renewal is due in part to the activation of the Notch signaling pathway in LKB1-deficient cells, leading to overexpression of Pax7 that appears to be dependent on the decreased AMPK activation in these cells [139]. Other findings indicate that AMPKα1 also regulates self-renewal in a LKB1-independent manner [135]. Furthermore, LKB1's role in SC differentiation is at least partly independent of AMPK through regulation of glycogen synthase kinase (GSK3)/Wnt signaling [140].

6. Conclusions and Future Perspectives

AMPK's role as a signaling nexus for cellular processes that control energy balance has been well established over recent decades. While it certainly is not the only player in the regulation of skeletal muscle development, size, and/or growth, it, and especially the AMPKα1 subunit, has emerged as a key factor that limits muscle size and capacity for hypertrophy. AMPKα2, on the other hand, may play a more substantial role in promoting muscle atrophy than AMPKα1 through its actions on autophagy and protein degradation (summarized in Figure 1). AMPK also limits myogenesis and regeneration after injury, although the loss of AMPKα1 also blocks these processes, showing that some (but not too much) AMPK activity is required for proper regenerative functioning. While many questions regarding AMPK's role in muscle growth and regeneration have been answered, others still remain unanswered. Does AMPKα1-specific activity after endurance exercise interfere with concomitant resistance-training adaptations? What cellular mediators control AMPK's effects on muscle growth and development? How does AMPK activity in neighboring accessory cells support or impair satellite cell function in muscle regeneration? Can pharmacological AMPK activation or inhibition be harnessed to improve hypertrophic and regenerative responses, especially in populations where these are impaired (aging, obesity, diabetes, myopathies, etc.)? What role do LKB1 and other AMPK family members play in these processes? Continuing work in this area will surely shed additional light on these and other important questions.

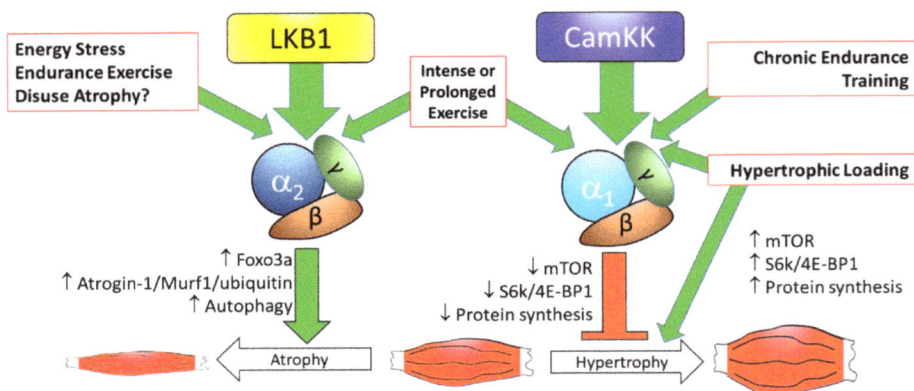

Figure 1. Proposed regulation of skeletal muscle size by 5′-adenosine monophosphate-activated protein kinase (AMPK). Energy stress (decreased ATP/AMP ratio; as in moderately intensive exercise) predominantly activates AMPKα2 via liver kinase B1 (LKB1), while AMPKα1 is only activated by highly-intense or prolonged exercise. Basal AMPKα1 content and activity is also increased by long-term endurance training, perhaps via Calcium/calmodulin-dependent protein kinase (CamKK) action, or other AMPK kinases. AMPKα2 stimulates catabolic processes by increasing Foxo3a, Atrogin-1 and MuRF-1 expression/activity and increasing autophagy, leading, under certain circumstance, to muscle atrophy, but has little effect on protein anabolism. AMPKα1 impairs mTOR signaling, slows protein synthesis, and blocks hypertrophy. Hypertrophic loading (i.e., resistance exercise) stimulates mechanistic target of rapamycin (mTOR) signaling, protein synthesis, and hypertrophy, but also activates AMPKα1 independent of LKB1 (perhaps via CamKK or other means), limiting the hypertrophic growth. ↑: increase expression or activity; ↓: decreased expression or activity.

Conflicts of Interest: The author declares no conflict of interest.

Abbreviations

4E-BP1	eukaryotic initiation factor 4E binding protein 1
ACC	acetyl-CoA carboxylase
AICAR	5-amino-4-imidazolecarboxamide ribonucleoside
AMPK	AMP-activated protein kinase
CamKK	calcium/calmodulin-dependent protein kinase
CBS	cystathionine β-synthase
dKO	double knockout
DN	dominant negative
EDL	extensor digitorum longus
eEF2	eukaryotic elongation factor
eEF2K	eEF2 kinase
FAP	fibroadipogenic progenitor
FoxO	forkhead box
GSK3	glycogen synthase kinase 3
HU	hindlimb unloading
HSA	human skeletal muscle actin
KO	knockout
LC3	microtubule-associated protein 1A/1B-light chain 3
LKB1	liver kinase B1
MAPK	mitogen activated protein kinase
MCK	muscle creatine kinase

mLKB1-KO	muscle-specific LKB1 knockout
mTORC1	mechanistic target of rapamycin, complex 1
MuRF-1	muscle ring finger-1
Myf5	myogenic factor 5
NF-κB	nuclear factor kappa B
SC	satellite cell
SIRT-1	sirtuin 1
skmLKB1-KO	skeletal muscle-specific LKB1 knockout
SNARK	sucrose non-fermenting 1 AMPK related kinase
SOL	soleus
TA	tibialis anterior
TAK1	transforming growth factor β-activated protein kinase
TBC1D7	TBC1 domain family member 7
TGFβ	transforming growth factor β
TSC2	tuberous sclerosis complex 2
ULK1	uncoordinated 51-like kinase 1
Wnt	wingless/integrated
WT	wild-type

References

1. Merrill, G.F.; Kurth, E.J.; Hardie, D.G.; Winder, W.W. Aica riboside increases amp-activated protein kinase, fatty acid oxidation, and glucose uptake in rat muscle. *Am. J. Physiol.* **1997**, *273*, E1107–E1112. [CrossRef] [PubMed]
2. Winder, W.W.; Hardie, D.G. Inactivation of acetyl-coa carboxylase and activation of amp-activated protein kinase in muscle during exercise. *Am. J. Physiol.* **1996**, *270*, E299–E304. [CrossRef] [PubMed]
3. Winder, W.W.; Thomson, D.M. Cellular energy sensing and signaling by amp-activated protein kinase. *Cell. Biochem. Biophys.* **2007**, *47*, 332–347. [CrossRef] [PubMed]
4. Weerasekara, V.K.; Panek, D.J.; Broadbent, D.G.; Mortenson, J.B.; Mathis, A.D.; Logan, G.N.; Prince, J.T.; Thomson, D.M.; Thompson, J.W.; Andersen, J.L. Metabolic-stress-induced rearrangement of the 14-3-3zeta interactome promotes autophagy via a ulk1- and ampk-regulated 14-3-3zeta interaction with phosphorylated atg9. *Mol. Cell. Biol.* **2014**, *34*, 4379–4388. [CrossRef] [PubMed]
5. Sanchez, A.M.; Csibi, A.; Raibon, A.; Cornille, K.; Gay, S.; Bernardi, H.; Candau, R. Ampk promotes skeletal muscle autophagy through activation of forkhead foxo3a and interaction with ulk1. *J. Cell. Biochem.* **2012**, *113*, 695–710. [CrossRef] [PubMed]
6. Romanello, V.; Guadagnin, E.; Gomes, L.; Roder, I.; Sandri, C.; Petersen, Y.; Milan, G.; Masiero, E.; Del Piccolo, P.; Foretz, M.; et al. Mitochondrial fission and remodelling contributes to muscle atrophy. *EMBO J.* **2010**, *29*, 1774–1785. [CrossRef] [PubMed]
7. Bolster, D.R.; Crozier, S.J.; Kimball, S.R.; Jefferson, L.S. Amp-activated protein kinase suppresses protein synthesis in rat skeletal muscle through down-regulated mammalian target of rapamycin (mtor) signaling. *J. Biol. Chem.* **2002**, *277*, 23977–23980. [CrossRef] [PubMed]
8. Hardie, D.G. Keeping the home fires burning: Amp-activated protein kinase. *J. R. Soc. Interface* **2018**, *15*. [CrossRef] [PubMed]
9. Carling, D. Ampk signalling in health and disease. *Curr. Opin. Cell Biol.* **2017**, *45*, 31–37. [CrossRef] [PubMed]
10. Wojtaszewski, J.F.; Birk, J.B.; Frosig, C.; Holten, M.; Pilegaard, H.; Dela, F. 5'amp activated protein kinase expression in human skeletal muscle: Effects of strength training and type 2 diabetes. *J. Physiol.* **2005**, *564*, 563–573. [CrossRef] [PubMed]
11. Birk, J.B.; Wojtaszewski, J.F. Predominant α2/β2/γ3 ampk activation during exercise in human skeletal muscle. *J. Physiol.* **2006**, *577*, 1021–1032. [CrossRef] [PubMed]
12. Treebak, J.T.; Birk, J.B.; Hansen, B.F.; Olsen, G.S.; Wojtaszewski, J.F. A-769662 activates ampk β1-containing complexes but induces glucose uptake through a pi3-kinase-dependent pathway in mouse skeletal muscle. *Am. J. Physiol. Cell Physiol.* **2009**, *297*, C1041–C1052. [CrossRef] [PubMed]

13. Koh, H.J.; Arnolds, D.E.; Fujii, N.; Tran, T.T.; Rogers, M.J.; Jessen, N.; Li, Y.; Liew, C.W.; Ho, R.C.; Hirshman, M.F.; et al. Skeletal muscle-selective knockout of lkb1 increases insulin sensitivity, improves glucose homeostasis, and decreases trb3. *Mol. Cell. Biol.* **2006**, *26*, 8217–8227. [CrossRef] [PubMed]

14. Sakamoto, K.; McCarthy, A.; Smith, D.; Green, K.A.; Grahame Hardie, D.; Ashworth, A.; Alessi, D.R. Deficiency of lkb1 in skeletal muscle prevents ampk activation and glucose uptake during contraction. *EMBO J.* **2005**, *24*, 1810–1820. [CrossRef] [PubMed]

15. Thomson, D.M.; Hancock, C.R.; Evanson, B.G.; Kenney, S.G.; Malan, B.B.; Mongillo, A.D.; Brown, J.D.; Hepworth, S.; Fillmore, N.; Parcell, A.C.; et al. Skeletal muscle dysfunction in muscle-specific lkb1 knockout mice. *J. Appl. Physiol. (1985)* **2010**, *108*, 1775–1785. [CrossRef] [PubMed]

16. Thomson, D.M.; Porter, B.B.; Tall, J.H.; Kim, H.J.; Barrow, J.R.; Winder, W.W. Skeletal muscle and heart lkb1 deficiency causes decreased voluntary running and reduced muscle mitochondrial marker enzyme expression in mice. *Am. J. Physiol. Endocrinol. Metab.* **2007**, *292*, E196–E202. [CrossRef] [PubMed]

17. Tanner, C.B.; Madsen, S.R.; Hallowell, D.M.; Goring, D.M.; Moore, T.M.; Hardman, S.E.; Heninger, M.R.; Atwood, D.R.; Thomson, D.M. Mitochondrial and performance adaptations to exercise training in mice lacking skeletal muscle lkb1. *Am. J. Physiol. Endocrinol. Metab.* **2013**, *305*, E1018–E1029. [CrossRef] [PubMed]

18. Jeppesen, J.; Maarbjerg, S.J.; Jordy, A.B.; Fritzen, A.M.; Pehmoller, C.; Sylow, L.; Serup, A.K.; Jessen, N.; Thorsen, K.; Prats, C.; et al. Lkb1 regulates lipid oxidation during exercise independently of ampk. *Diabetes* **2013**, *62*, 1490–1499. [CrossRef] [PubMed]

19. McGee, S.L.; Mustard, K.J.; Hardie, D.G.; Baar, K. Normal hypertrophy accompanied by phosphoryation and activation of amp-activated protein kinase α1 following overload in lkb1 knockout mice. *J. Physiol.* **2008**, *586*, 1731–1741. [CrossRef] [PubMed]

20. Jensen, T.E.; Rose, A.J.; Jorgensen, S.B.; Brandt, N.; Schjerling, P.; Wojtaszewski, J.F.; Richter, E.A. Possible camkk-dependent regulation of ampk phosphorylation and glucose uptake at the onset of mild tetanic skeletal muscle contraction. *Am. J. Physiol. Endocrinol. Metab.* **2007**, *292*, E1308–E1317. [CrossRef] [PubMed]

21. Ferey, J.L.; Brault, J.J.; Smith, C.A.; Witczak, C.A. Constitutive activation of camkkα signaling is sufficient but not necessary for mtorc1 activation and growth in mouse skeletal muscle. *Am. J. Physiol. Endocrinol. Metab.* **2014**, *307*, E686–E694. [CrossRef] [PubMed]

22. Hindi, S.M.; Sato, S.; Xiong, G.; Bohnert, K.R.; Gibb, A.A.; Gallot, Y.S.; McMillan, J.D.; Hill, B.G.; Uchida, S.; Kumar, A. Tak1 regulates skeletal muscle mass and mitochondrial function. *JCI Insight* **2018**, *3*, e98441. [CrossRef] [PubMed]

23. Vavvas, D.; Apazidis, A.; Saha, A.K.; Gamble, J.; Patel, A.; Kemp, B.E.; Witters, L.A.; Ruderman, N.B. Contraction-induced changes in acetyl-coa carboxylase and 5′-amp-activated kinase in skeletal muscle. *J. Biol. Chem.* **1997**, *272*, 13255–13261. [CrossRef] [PubMed]

24. Fujii, N.; Hayashi, T.; Hirshman, M.F.; Smith, J.T.; Habinowski, S.A.; Kaijser, L.; Mu, J.; Ljungqvist, O.; Birnbaum, M.J.; Witters, L.A.; et al. Exercise induces isoform-specific increase in 5′amp-activated protein kinase activity in human skeletal muscle. *Biochem. Biophys. Res. Commun.* **2000**, *273*, 1150–1155. [CrossRef] [PubMed]

25. Stephens, T.J.; Chen, Z.P.; Canny, B.J.; Michell, B.J.; Kemp, B.E.; McConell, G.K. Progressive increase in human skeletal muscle ampkα2 activity and acc phosphorylation during exercise. *Am. J. Physiol. Endocrinol. Metab.* **2002**, *282*, E688–E694. [CrossRef] [PubMed]

26. Wojtaszewski, J.F.; Nielsen, P.; Hansen, B.F.; Richter, E.A.; Kiens, B. Isoform-specific and exercise intensity-dependent activation of 5′-amp-activated protein kinase in human skeletal muscle. *J. Physiol.* **2000**, *528*, 221–226. [CrossRef] [PubMed]

27. Musi, N.; Hayashi, T.; Fujii, N.; Hirshman, M.F.; Witters, L.A.; Goodyear, L.J. Amp-activated protein kinase activity and glucose uptake in rat skeletal muscle. *Am. J. Physiol. Endocrinol. Metab.* **2001**, *280*, E677–E684. [CrossRef] [PubMed]

28. Jorgensen, S.B.; Wojtaszewski, J.F.; Viollet, B.; Andreelli, F.; Birk, J.B.; Hellsten, Y.; Schjerling, P.; Vaulont, S.; Neufer, P.D.; Richter, E.A.; et al. Effects of α-ampk knockout on exercise-induced gene activation in mouse skeletal muscle. *FASEB J.* **2005**, *19*, 1146–1148. [CrossRef] [PubMed]

29. Dzamko, N.; Schertzer, J.D.; Ryall, J.G.; Steel, R.; Macaulay, S.L.; Wee, S.; Chen, Z.P.; Michell, B.J.; Oakhill, J.S.; Watt, M.J.; et al. Ampk-independent pathways regulate skeletal muscle fatty acid oxidation. *J. Physiol.* **2008**, *586*, 5819–5831. [CrossRef] [PubMed]

30. Maarbjerg, S.J.; Jorgensen, S.B.; Rose, A.J.; Jeppesen, J.; Jensen, T.E.; Treebak, J.T.; Birk, J.B.; Schjerling, P.; Wojtaszewski, J.F.; Richter, E.A. Genetic impairment of ampkα2 signaling does not reduce muscle glucose uptake during treadmill exercise in mice. *Am. J. Physiol. Endocrinol. Metab.* **2009**, *297*, E924–E934. [CrossRef] [PubMed]

31. Chen, Z.P.; McConell, G.K.; Michell, B.J.; Snow, R.J.; Canny, B.J.; Kemp, B.E. Ampk signaling in contracting human skeletal muscle: Acetyl-coa carboxylase and no synthase phosphorylation. *Am. J. Physiol. Endocrinol. Metab.* **2000**, *279*, E1202–E1206. [CrossRef] [PubMed]

32. Gibala, M.J.; McGee, S.L.; Garnham, A.P.; Howlett, K.F.; Snow, R.J.; Hargreaves, M. Brief intense interval exercise activates ampk and p38 mapk signaling and increases the expression of pgc-1α in human skeletal muscle. *J. Appl. Physiol. (1985)* **2009**, *106*, 929–934. [CrossRef] [PubMed]

33. Corton, J.M.; Gillespie, J.G.; Hawley, S.A.; Hardie, D.G. 5-aminoimidazole-4-carboxamide ribonucleoside. A specific method for activating amp-activated protein kinase in intact cells? *Eur. J. Biochem.* **1995**, *229*, 558–565. [CrossRef] [PubMed]

34. Hardman, S.E.; Hall, D.E.; Cabrera, A.J.; Hancock, C.R.; Thomson, D.M. The effects of age and muscle contraction on ampk activity and heterotrimer composition. *Exp. Gerontol.* **2014**, *55*, 120–128. [CrossRef] [PubMed]

35. Thomson, D.M.; Fick, C.A.; Gordon, S.E. Ampk activation attenuates s6k1, 4e-bp1, and eef2 signaling responses to high-frequency electrically stimulated skeletal muscle contractions. *J. Appl. Physiol. (1985)* **2008**, *104*, 625–632. [CrossRef] [PubMed]

36. Jorgensen, S.B.; Viollet, B.; Andreelli, F.; Frosig, C.; Birk, J.B.; Schjerling, P.; Vaulont, S.; Richter, E.A.; Wojtaszewski, J.F. Knockout of the α2 but not α1 5'-amp-activated protein kinase isoform abolishes 5-aminoimidazole-4-carboxamide-1-β-4-ribofuranosidebut not contraction-induced glucose uptake in skeletal muscle. *J. Biol. Chem.* **2004**, *279*, 1070–1079. [CrossRef] [PubMed]

37. Zhou, G.; Myers, R.; Li, Y.; Chen, Y.; Shen, X.; Fenyk-Melody, J.; Wu, M.; Ventre, J.; Doebber, T.; Fujii, N.; et al. Role of amp-activated protein kinase in mechanism of metformin action. *J. Clin. Investig.* **2001**, *108*, 1167–1174. [CrossRef] [PubMed]

38. Rena, G.; Hardie, D.G.; Pearson, E.R. The mechanisms of action of metformin. *Diabetologia* **2017**, *60*, 1577–1585. [CrossRef] [PubMed]

39. Musi, N.; Hirshman, M.F.; Nygren, J.; Svanfeldt, M.; Bavenholm, P.; Rooyackers, O.; Zhou, G.; Williamson, J.M.; Ljunqvist, O.; Efendic, S.; et al. Metformin increases amp-activated protein kinase activity in skeletal muscle of subjects with type 2 diabetes. *Diabetes* **2002**, *51*, 2074–2081. [CrossRef] [PubMed]

40. Oshima, R.; Yamada, M.; Kurogi, E.; Ogino, Y.; Serizawa, Y.; Tsuda, S.; Ma, X.; Egawa, T.; Hayashi, T. Evidence for organic cation transporter-mediated metformin transport and 5'-adenosine monophosphate-activated protein kinase activation in rat skeletal muscles. *Metabolism* **2015**, *64*, 296–304. [CrossRef] [PubMed]

41. Cool, B.; Zinker, B.; Chiou, W.; Kifle, L.; Cao, N.; Perham, M.; Dickinson, R.; Adler, A.; Gagne, G.; Iyengar, R.; et al. Identification and characterization of a small molecule ampk activator that treats key components of type 2 diabetes and the metabolic syndrome. *Cell Metab.* **2006**, *3*, 403–416. [CrossRef] [PubMed]

42. Scott, J.W.; van Denderen, B.J.; Jorgensen, S.B.; Honeyman, J.E.; Steinberg, G.R.; Oakhill, J.S.; Iseli, T.J.; Koay, A.; Gooley, P.R.; Stapleton, D.; et al. Thienopyridone drugs are selective activators of amp-activated protein kinase β1-containing complexes. *Chem. Biol.* **2008**, *15*, 1220–1230. [CrossRef] [PubMed]

43. Olivier, S.; Foretz, M.; Viollet, B. Promise and challenges for direct small molecule ampk activators. *Biochem. Pharmacol.* **2018**, *153*, 147–158. [CrossRef] [PubMed]

44. McGlory, C.; van Vliet, S.; Stokes, T.; Mittendorfer, B.; Phillips, S.M. The impact of exercise and nutrition on the regulation of skeletal muscle mass. *J. Physiol.* **2018**. [CrossRef] [PubMed]

45. Williamson, D.L.; Bolster, D.R.; Kimball, S.R.; Jefferson, L.S. Time course changes in signaling pathways and protein synthesis in c2c12 myotubes following ampk activation by aicar. *Am. J. Physiol. Endocrinol. Metab.* **2006**, *291*, E80–E89. [CrossRef] [PubMed]

46. Dubbelhuis, P.F.; Meijer, A.J. Hepatic amino acid-dependent signaling is under the control of amp-dependent protein kinase. *FEBS Lett.* **2002**, *521*, 39–42. [CrossRef]

47. Horman, S.; Browne, G.; Krause, U.; Patel, J.; Vertommen, D.; Bertrand, L.; Lavoinne, A.; Hue, L.; Proud, C.; Rider, M. Activation of amp-activated protein kinase leads to the phosphorylation of elongation factor 2 and an inhibition of protein synthesis. *Curr. Biol.* **2002**, *12*, 1419–1423. [CrossRef]

48. Reiter, A.K.; Bolster, D.R.; Crozier, S.J.; Kimball, S.R.; Jefferson, L.S. Repression of protein synthesis and mtor signaling in rat liver mediated by the ampk activator aminoimidazole carboxamide ribonucleoside. *Am. J. Physiol. Endocrinol. Metab.* **2005**, *288*, E980–E988. [CrossRef] [PubMed]

49. McLeod, L.E.; Proud, C.G. Atp depletion increases phosphorylation of elongation factor eef2 in adult cardiomyocytes independently of inhibition of mtor signalling. *FEBS Lett.* **2002**, *531*, 448–452. [CrossRef]

50. Chan, A.Y.; Soltys, C.L.; Young, M.E.; Proud, C.G.; Dyck, J.R. Activation of amp-activated protein kinase inhibits protein synthesis associated with hypertrophy in the cardiac myocyte. *J. Biol. Chem.* **2004**, *279*, 32771–32779. [CrossRef] [PubMed]

51. Xiang, X.; Saha, A.K.; Wen, R.; Ruderman, N.B.; Luo, Z. Amp-activated protein kinase activators can inhibit the growth of prostate cancer cells by multiple mechanisms. *Biochem. Biophys. Res. Commun.* **2004**, *321*, 161–167. [CrossRef] [PubMed]

52. Fay, J.R.; Steele, V.; Crowell, J.A. Energy homeostasis and cancer prevention: The amp-activated protein kinase. *Cancer Prev. Res. (Phila)* **2009**, *2*, 301–309. [CrossRef] [PubMed]

53. Cheng, S.W.; Fryer, L.G.; Carling, D.; Shepherd, P.R. Thr2446 is a novel mammalian target of rapamycin (mtor) phosphorylation site regulated by nutrient status. *J. Biol. Chem.* **2004**, *279*, 15719–15722. [CrossRef] [PubMed]

54. Figueiredo, V.C.; Markworth, J.F.; Cameron-Smith, D. Considerations on mtor regulation at serine 2448: Implications for muscle metabolism studies. *Cell. Mol. Life Sci.* **2017**, *74*, 2537–2545. [CrossRef] [PubMed]

55. Long, X.; Lin, Y.; Ortiz-Vega, S.; Yonezawa, K.; Avruch, J. Rheb binds and regulates the mtor kinase. *Curr. Biol.* **2005**, *15*, 702–713. [CrossRef] [PubMed]

56. Dibble, C.C.; Elis, W.; Menon, S.; Qin, W.; Klekota, J.; Asara, J.M.; Finan, P.M.; Kwiatkowski, D.J.; Murphy, L.O.; Manning, B.D. Tbc1d7 is a third subunit of the tsc1-tsc2 complex upstream of mtorc1. *Mol. Cell* **2012**, *47*, 535–546. [CrossRef] [PubMed]

57. Gwinn, D.M.; Shackelford, D.B.; Egan, D.F.; Mihaylova, M.M.; Mery, A.; Vasquez, D.S.; Turk, B.E.; Shaw, R.J. Ampk phosphorylation of raptor mediates a metabolic checkpoint. *Mol. Cell* **2008**, *30*, 214–226. [CrossRef] [PubMed]

58. Rose, A.J.; Richter, E.A. Regulatory mechanisms of skeletal muscle protein turnover during exercise. *J. Appl. Physiol. (1985)* **2009**, *106*, 1702–1711. [CrossRef] [PubMed]

59. Nielsen, P.J.; McConkey, E.H. Evidence for control of protein synthesis in hela cells via the elongation rate. *J. Cell. Physiol.* **1980**, *104*, 269–281. [CrossRef] [PubMed]

60. Rose, A.J.; Alsted, T.J.; Jensen, T.E.; Kobbero, J.B.; Maarbjerg, S.J.; Jensen, J.; Richter, E.A. A ca(2+)-calmodulin-eef2k-eef2 signalling cascade, but not ampk, contributes to the suppression of skeletal muscle protein synthesis during contractions. *J. Physiol.* **2009**, *587*, 1547–1563. [CrossRef] [PubMed]

61. Bento, C.F.; Renna, M.; Ghislat, G.; Puri, C.; Ashkenazi, A.; Vicinanza, M.; Menzies, F.M.; Rubinsztein, D.C. Mammalian autophagy: How does it work? *Annu. Rev. Biochem.* **2016**, *85*, 685–713. [CrossRef] [PubMed]

62. Samari, H.R.; Seglen, P.O. Inhibition of hepatocytic autophagy by adenosine, aminoimidazole-4-carboxamide riboside, and n6-mercaptopurine riboside. Evidence for involvement of amp-activated protein kinase. *J. Biol. Chem.* **1998**, *273*, 23758–23763. [CrossRef] [PubMed]

63. Meley, D.; Bauvy, C.; Houben-Weerts, J.H.; Dubbelhuis, P.F.; Helmond, M.T.; Codogno, P.; Meijer, A.J. Amp-activated protein kinase and the regulation of autophagic proteolysis. *J. Biol. Chem.* **2006**, *281*, 34870–34879. [CrossRef] [PubMed]

64. Herzig, S.; Shaw, R.J. Ampk: Guardian of metabolism and mitochondrial homeostasis. *Nat. Rev. Mol. Cell Biol.* **2018**, *19*, 121–135. [CrossRef] [PubMed]

65. Lee, J.W.; Park, S.; Takahashi, Y.; Wang, H.G. The association of ampk with ulk1 regulates autophagy. *PLoS ONE* **2010**, *5*, e15394. [CrossRef] [PubMed]

66. Martin-Rincon, M.; Morales-Alamo, D.; Calbet, J.A.L. Exercise-mediated modulation of autophagy in skeletal muscle. *Scand. J. Med. Sci. Sports* **2018**, *28*, 772–781. [CrossRef] [PubMed]

67. Kim, J.; Kundu, M.; Viollet, B.; Guan, K.L. Ampk and mtor regulate autophagy through direct phosphorylation of ulk1. *Nat. Cell Biol.* **2011**, *13*, 132–141. [CrossRef] [PubMed]

68. Zhang, D.; Wang, W.; Sun, X.; Xu, D.; Wang, C.; Zhang, Q.; Wang, H.; Luo, W.; Chen, Y.; Chen, H.; et al. Ampk regulates autophagy by phosphorylating becn1 at threonine 388. *Autophagy* **2016**, *12*, 1447–1459. [CrossRef] [PubMed]

69. Rom, O.; Reznick, A.Z. The role of e3 ubiquitin-ligases murf-1 and mafbx in loss of skeletal muscle mass. *Free Radic. Biol. Med.* **2016**, *98*, 218–230. [CrossRef] [PubMed]

70. Nystrom, G.J.; Lang, C.H. Sepsis and ampk activation by aicar differentially regulate foxo-1, -3 and -4 mrna in striated muscle. *Int J. Clin. Exp. Med.* **2008**, *1*, 50–63. [PubMed]

71. Nakashima, K.; Yakabe, Y. Ampk activation stimulates myofibrillar protein degradation and expression of atrophy-related ubiquitin ligases by increasing foxo transcription factors in c2c12 myotubes. *Biosci. Biotechnol. Biochem.* **2007**, *71*, 1650–1656. [CrossRef] [PubMed]

72. Greer, E.L.; Dowlatshahi, D.; Banko, M.R.; Villen, J.; Hoang, K.; Blanchard, D.; Gygi, S.P.; Brunet, A. An ampk-foxo pathway mediates longevity induced by a novel method of dietary restriction in *C. elegans. Curr. Biol.* **2007**, *17*, 1646–1656. [CrossRef] [PubMed]

73. Sanchez, A.M.; Candau, R.B.; Csibi, A.; Pagano, A.F.; Raibon, A.; Bernardi, H. The role of amp-activated protein kinase in the coordination of skeletal muscle turnover and energy homeostasis. *Am. J. Physiol. Cell Physiol.* **2012**, *303*, C475–C485. [CrossRef] [PubMed]

74. Greer, E.L.; Oskoui, P.R.; Banko, M.R.; Maniar, J.M.; Gygi, M.P.; Gygi, S.P.; Brunet, A. The energy sensor amp-activated protein kinase directly regulates the mammalian foxo3 transcription factor. *J. Biol. Chem.* **2007**, *282*, 30107–30119. [CrossRef] [PubMed]

75. Canto, C.; Gerhart-Hines, Z.; Feige, J.N.; Lagouge, M.; Noriega, L.; Milne, J.C.; Elliott, P.J.; Puigserver, P.; Auwerx, J. Ampk regulates energy expenditure by modulating nad+ metabolism and sirt1 activity. *Nature* **2009**, *458*, 1056–1060. [CrossRef] [PubMed]

76. Canto, C.; Jiang, L.Q.; Deshmukh, A.S.; Mataki, C.; Coste, A.; Lagouge, M.; Zierath, J.R.; Auwerx, J. Interdependence of ampk and sirt1 for metabolic adaptation to fasting and exercise in skeletal muscle. *Cell. Metab.* **2010**, *11*, 213–219. [CrossRef] [PubMed]

77. Mu, J.; Barton, E.R.; Birnbaum, M.J. Selective suppression of amp-activated protein kinase in skeletal muscle: Update on 'lazy mice'. *Biochem. Soc. Trans.* **2003**, *31*, 236–241. [CrossRef] [PubMed]

78. Lantier, L.; Mounier, R.; Leclerc, J.; Pende, M.; Foretz, M.; Viollet, B. Coordinated maintenance of muscle cell size control by amp-activated protein kinase. *FASEB J.* **2010**, *24*, 3555–3561. [CrossRef] [PubMed]

79. O'Neill, H.M.; Maarbjerg, S.J.; Crane, J.D.; Jeppesen, J.; Jorgensen, S.B.; Schertzer, J.D.; Shyroka, O.; Kiens, B.; van Denderen, B.J.; Tarnopolsky, M.A.; et al. Amp-activated protein kinase (ampk) β1β2 muscle null mice reveal an essential role for ampk in maintaining mitochondrial content and glucose uptake during exercise. *Proc. Natl. Acad. Sci. USA* **2011**, *108*, 16092–16097. [CrossRef] [PubMed]

80. Mounier, R.; Lantier, L.; Leclerc, J.; Sotiropoulos, A.; Pende, M.; Daegelen, D.; Sakamoto, K.; Foretz, M.; Viollet, B. Important role for ampkα1 in limiting skeletal muscle cell hypertrophy. *FASEB J.* **2009**, *23*, 2264–2273. [CrossRef] [PubMed]

81. Steinberg, G.R.; O'Neill, H.M.; Dzamko, N.L.; Galic, S.; Naim, T.; Koopman, R.; Jorgensen, S.B.; Honeyman, J.; Hewitt, K.; Chen, Z.P.; et al. Whole body deletion of amp-activated protein kinase β2 reduces muscle ampk activity and exercise capacity. *J. Biol. Chem.* **2010**, *285*, 37198–37209. [CrossRef] [PubMed]

82. Thomas, M.M.; Wang, D.C.; D'Souza, D.M.; Krause, M.P.; Layne, A.S.; Criswell, D.S.; O'Neill, H.M.; Connor, M.K.; Anderson, J.E.; Kemp, B.E.; et al. Muscle-specific ampk β1β2-null mice display a myopathy due to loss of capillary density in nonpostural muscles. *FASEB J.* **2014**, *28*, 2098–2107. [CrossRef] [PubMed]

83. Thomson, D.M.; Gordon, S.E. Diminished overload-induced hypertrophy in aged fast-twitch skeletal muscle is associated with ampk hyperphosphorylation. *J. Appl. Physiol. (1985)* **2005**, *98*, 557–564. [CrossRef] [PubMed]

84. Thomson, D.M.; Gordon, S.E. Impaired overload-induced muscle growth is associated with diminished translational signalling in aged rat fast-twitch skeletal muscle. *J. Physiol.* **2006**, *574*, 291–305. [CrossRef] [PubMed]

85. Paturi, S.; Gutta, A.K.; Kakarla, S.K.; Katta, A.; Arnold, E.C.; Wu, M.; Rice, K.M.; Blough, E.R. Impaired overload-induced hypertrophy in obese zucker rat slow-twitch skeletal muscle. *J. Appl. Physiol. (1985)* **2010**, *108*, 7–13. [CrossRef] [PubMed]

86. Katta, A.; Kakarla, S.K.; Manne, N.D.; Wu, M.; Kundla, S.; Kolli, M.B.; Nalabotu, S.K.; Blough, E.R. Diminished muscle growth in the obese zucker rat following overload is associated with hyperphosphorylation of ampk and dsrna-dependent protein kinase. *J. Appl. Physiol. (1985)* **2012**, *113*, 377–384. [CrossRef] [PubMed]

87. Layne, A.S.; Nasrallah, S.; South, M.A.; Howell, M.E.; McCurry, M.P.; Ramsey, M.W.; Stone, M.H.; Stuart, C.A. Impaired muscle ampk activation in the metabolic syndrome may attenuate improved insulin action after exercise training. *J. Clin. Endocrinol. Metab.* **2011**, *96*, 1815–1826. [CrossRef] [PubMed]

88. Egawa, T.; Ohno, Y.; Goto, A.; Ikuta, A.; Suzuki, M.; Ohira, T.; Yokoyama, S.; Sugiura, T.; Ohira, Y.; Yoshioka, T.; et al. Aicar-induced activation of ampk negatively regulates myotube hypertrophy through the hsp72-mediated pathway in c2c12 skeletal muscle cells. *Am. J. Physiol. Endocrinol. Metab.* **2014**, *306*, E344–E354. [CrossRef] [PubMed]

89. Deng, Z.; Luo, P.; Lai, W.; Song, T.; Peng, J.; Wei, H.K. Myostatin inhibits eef2k-eef2 by regulating ampk to suppress protein synthesis. *Biochem. Biophys. Res. Commun.* **2017**, *494*, 278–284. [CrossRef] [PubMed]

90. Luciano, T.F.; Marques, S.O.; Pieri, B.L.; de Souza, D.R.; Araujo, L.V.; Nesi, R.T.; Scheffer, D.L.; Comin, V.H.; Pinho, R.A.; Muller, A.P.; et al. Responses of skeletal muscle hypertrophy in wistar rats to different resistance exercise models. *Physiol. Res.* **2017**, *66*, 317–323. [PubMed]

91. Gordon, S.E.; Lake, J.A.; Westerkamp, C.M.; Thomson, D.M. Does amp-activated protein kinase negatively mediate aged fast-twitch skeletal muscle mass? *Exerc. Sport Sci. Rev.* **2008**, *36*, 179–186. [CrossRef] [PubMed]

92. Reznick, R.M.; Zong, H.; Li, J.; Morino, K.; Moore, I.K.; Yu, H.J.; Liu, Z.X.; Dong, J.; Mustard, K.J.; Hawley, S.A.; et al. Aging-associated reductions in amp-activated protein kinase activity and mitochondrial biogenesis. *Cell Metab.* **2007**, *5*, 151–156. [CrossRef] [PubMed]

93. Drummond, M.J.; Dreyer, H.C.; Pennings, B.; Fry, C.S.; Dhanani, S.; Dillon, E.L.; Sheffield-Moore, M.; Volpi, E.; Rasmussen, B.B. Skeletal muscle protein anabolic response to resistance exercise and essential amino acids is delayed with aging. *J. Appl. Physiol. (1985)* **2008**, *104*, 1452–1461. [CrossRef] [PubMed]

94. Hilder, T.L.; Baer, L.A.; Fuller, P.M.; Fuller, C.A.; Grindeland, R.E.; Wade, C.E.; Graves, L.M. Insulin-independent pathways mediating glucose uptake in hindlimb-suspended skeletal muscle. *J. Appl. Physiol. (1985)* **2005**, *99*, 2181–2188. [CrossRef] [PubMed]

95. Han, B.; Zhu, M.J.; Ma, C.; Du, M. Rat hindlimb unloading down-regulates insulin like growth factor-1 signaling and amp-activated protein kinase, and leads to severe atrophy of the soleus muscle. *Appl. Physiol. Nutr. Metab.* **2007**, *32*, 1115–1123. [CrossRef] [PubMed]

96. Liu, J.; Peng, Y.; Cui, Z.; Wu, Z.; Qian, A.; Shang, P.; Qu, L.; Li, Y.; Liu, J.; Long, J. Depressed mitochondrial biogenesis and dynamic remodeling in mouse tibialis anterior and gastrocnemius induced by 4-week hindlimb unloading. *IUBMB Life* **2012**, *64*, 901–910. [CrossRef] [PubMed]

97. Cannavino, J.; Brocca, L.; Sandri, M.; Grassi, B.; Bottinelli, R.; Pellegrino, M.A. The role of alterations in mitochondrial dynamics and pgc-1α over-expression in fast muscle atrophy following hindlimb unloading. *J. Physiol.* **2015**, *593*, 1981–1995. [CrossRef] [PubMed]

98. Egawa, T.; Goto, A.; Ohno, Y.; Yokoyama, S.; Ikuta, A.; Suzuki, M.; Sugiura, T.; Ohira, Y.; Yoshioka, T.; Hayashi, T.; et al. Involvement of ampk in regulating slow-twitch muscle atrophy during hindlimb unloading in mice. *Am. J. Physiol. Endocrinol. Metab.* **2015**, *309*, E651–E662. [CrossRef] [PubMed]

99. Gao, H.; Li, Y.F. Distinct signal transductions in fast- and slow-twitch muscles upon denervation. *Physiol. Rep.* **2018**, *6*. [CrossRef] [PubMed]

100. Paul, P.K.; Gupta, S.K.; Bhatnagar, S.; Panguluri, S.K.; Darnay, B.G.; Choi, Y.; Kumar, A. Targeted ablation of traf6 inhibits skeletal muscle wasting in mice. *J. Cell Biol.* **2010**, *191*, 1395–1411. [CrossRef] [PubMed]

101. Guo, Y.; Meng, J.; Tang, Y.; Wang, T.; Wei, B.; Feng, R.; Gong, B.; Wang, H.; Ji, G.; Lu, Z. Amp-activated kinase α2 deficiency protects mice from denervation-induced skeletal muscle atrophy. *Arch. Biochem. Biophys.* **2016**, *600*, 56–60. [CrossRef] [PubMed]

102. Ribeiro, C.B.; Christofoletti, D.C.; Pezolato, V.A.; de Cassia Marqueti Durigan, R.; Prestes, J.; Tibana, R.A.; Pereira, E.C.; de Sousa Neto, I.V.; Durigan, J.L.; da Silva, C.A. Leucine minimizes denervation-induced skeletal muscle atrophy of rats through akt/mtor signaling pathways. *Front. Physiol.* **2015**, *6*, 73. [CrossRef] [PubMed]

103. Dreyer, H.C.; Glynn, E.L.; Lujan, H.L.; Fry, C.S.; DiCarlo, S.E.; Rasmussen, B.B. Chronic paraplegia-induced muscle atrophy downregulates the mtor/s6k1 signaling pathway. *J. Appl. Physiol. (1985)* **2008**, *104*, 27–33. [CrossRef] [PubMed]

104. Stouth, D.W.; Manta, A.; Ljubicic, V. Protein arginine methyltransferase expression, localization, and activity during disuse-induced skeletal muscle plasticity. *Am. J. Physiol. Cell Physiol.* **2018**, *314*, C177–C190. [CrossRef] [PubMed]

105. Paulsen, S.R.; Rubink, D.S.; Winder, W.W. Amp-activated protein kinase activation prevents denervation-induced decline in gastrocnemius glut-4. *J. Appl. Physiol. (1985)* **2001**, *91*, 2102–2108. [CrossRef] [PubMed]

106. Pauly, M.; Daussin, F.; Burelle, Y.; Li, T.; Godin, R.; Fauconnier, J.; Koechlin-Ramonatxo, C.; Hugon, G.; Lacampagne, A.; Coisy-Quivy, M.; et al. Ampk activation stimulates autophagy and ameliorates muscular dystrophy in the mdx mouse diaphragm. *Am. J. Pathol.* **2012**, *181*, 583–592. [CrossRef] [PubMed]

107. Ljubicic, V.; Jasmin, B.J. Amp-activated protein kinase at the nexus of therapeutic skeletal muscle plasticity in duchenne muscular dystrophy. *Trends Mol. Med.* **2013**, *19*, 614–624. [CrossRef] [PubMed]

108. Thomson, D.M.; Brown, J.D.; Fillmore, N.; Condon, B.M.; Kim, H.J.; Barrow, J.R.; Winder, W.W. Lkb1 and the regulation of malonyl-coa and fatty acid oxidation in muscle. *Am. J. Physiol. Endocrinol. Metab.* **2007**, *293*, E1572–E1579. [CrossRef] [PubMed]

109. Brown, J.D.; Hancock, C.R.; Mongillo, A.D.; Benjamin Barton, J.; DiGiovanni, R.A.; Parcell, A.C.; Winder, W.W.; Thomson, D.M. Effect of lkb1 deficiency on mitochondrial content, fibre type and muscle performance in the mouse diaphragm. *Acta Physiol. (Oxf.)* **2011**, *201*, 457–466. [CrossRef] [PubMed]

110. Moore, T.M.; Mortensen, X.M.; Ashby, C.K.; Harris, A.M.; Kump, K.J.; Laird, D.W.; Adams, A.J.; Bray, J.K.; Chen, T.; Thomson, D.M. The effect of caffeine on skeletal muscle anabolic signaling and hypertrophy. *Appl. Physiol. Nutr. Metab.* **2017**, *42*, 621–629. [CrossRef] [PubMed]

111. Lessard, S.J.; Rivas, D.A.; So, K.; Koh, H.J.; Queiroz, A.L.; Hirshman, M.F.; Fielding, R.A.; Goodyear, L.J. The ampk-related kinase snark regulates muscle mass and myocyte survival. *J. Clin. Investig.* **2016**, *126*, 560–570. [CrossRef] [PubMed]

112. Jessen, N.; Koh, H.J.; Folmes, C.D.; Wagg, C.; Fujii, N.; Lofgren, B.; Wolf, C.M.; Berul, C.I.; Hirshman, M.F.; Lopaschuk, G.D.; et al. Ablation of lkb1 in the heart leads to energy deprivation and impaired cardiac function. *Biochim. Biophys. Acta* **2010**, *1802*, 593–600. [CrossRef] [PubMed]

113. Miura, S.; Kai, Y.; Tadaishi, M.; Tokutake, Y.; Sakamoto, K.; Bruce, C.R.; Febbraio, M.A.; Kita, K.; Chohnan, S.; Ezaki, O. Marked phenotypic differences of endurance performance and exercise-induced oxygen consumption between ampk and lkb1 deficiency in mouse skeletal muscle: Changes occurring in the diaphragm. *Am. J. Physiol. Endocrinol. Metab.* **2013**, *305*, E213–E229. [CrossRef] [PubMed]

114. Hickson, R.C. Interference of strength development by simultaneously training for strength and endurance. *Eur. J. Appl. Physiol. Occup. Physiol.* **1980**, *45*, 255–263. [CrossRef] [PubMed]

115. Coffey, V.G.; Hawley, J.A. Concurrent exercise training: Do opposites distract? *J. Physiol.* **2017**, *595*, 2883–2896. [CrossRef] [PubMed]

116. Atherton, P.J.; Babraj, J.; Smith, K.; Singh, J.; Rennie, M.J.; Wackerhage, H. Selective activation of ampk-pgc-1α or pkb-tsc2-mtor signaling can explain specific adaptive responses to endurance or resistance training-like electrical muscle stimulation. *FASEB J.* **2005**, *19*, 786–788. [CrossRef] [PubMed]

117. Apro, W.; Moberg, M.; Hamilton, D.L.; Ekblom, B.; van Hall, G.; Holmberg, H.C.; Blomstrand, E. Resistance exercise-induced s6k1 kinase activity is not inhibited in human skeletal muscle despite prior activation of ampk by high-intensity interval cycling. *Am. J. Physiol. Endocrinol. Metab.* **2015**, *308*, E470–E481. [CrossRef] [PubMed]

118. Nielsen, J.N.; Mustard, K.J.; Graham, D.A.; Yu, H.; MacDonald, C.S.; Pilegaard, H.; Goodyear, L.J.; Hardie, D.G.; Richter, E.A.; Wojtaszewski, J.F. 5'-amp-activated protein kinase activity and subunit expression in exercise-trained human skeletal muscle. *J. Appl. Physiol. (1985)* **2003**, *94*, 631–641. [CrossRef] [PubMed]

119. McConell, G.K.; Lee-Young, R.S.; Chen, Z.P.; Stepto, N.K.; Huynh, N.N.; Stephens, T.J.; Canny, B.J.; Kemp, B.E. Short-term exercise training in humans reduces ampk signalling during prolonged exercise independent of muscle glycogen. *J. Physiol.* **2005**, *568*, 665–676. [CrossRef] [PubMed]

120. Tarini, V.A.; Carnevali, L.C., Jr.; Arida, R.M.; Cunha, C.A.; Alves, E.S.; Seelaender, M.C.; Schmidt, B.; Faloppa, F. Effect of exhaustive ultra-endurance exercise in muscular glycogen and both α1 and α2 ampk protein expression in trained rats. *J. Physiol. Biochem.* **2013**, *69*, 429–440. [CrossRef] [PubMed]

121. Langfort, J.; Viese, M.; Ploug, T.; Dela, F. Time course of glut4 and ampk protein expression in human skeletal muscle during one month of physical training. *Scand. J. Med. Sci. Sports* **2003**, *13*, 169–174. [CrossRef] [PubMed]

122. Frosig, C.; Jorgensen, S.B.; Hardie, D.G.; Richter, E.A.; Wojtaszewski, J.F. 5'-amp-activated protein kinase activity and protein expression are regulated by endurance training in human skeletal muscle. *Am. J. Physiol. Endocrinol. Metab.* **2004**, *286*, E411–E417. [CrossRef] [PubMed]

123. Gore, D.C.; Wolf, S.E.; Sanford, A.; Herndon, D.N.; Wolfe, R.R. Influence of metformin on glucose intolerance and muscle catabolism following severe burn injury. *Ann. Surg.* **2005**, *241*, 334–342. [CrossRef] [PubMed]

124. Oliveira, A.G.; Gomes-Marcondes, M.C. Metformin treatment modulates the tumour-induced wasting effects in muscle protein metabolism minimising the cachexia in tumour-bearing rats. *BMC Cancer* **2016**, *16*, 418. [CrossRef] [PubMed]

125. Wosczyna, M.N.; Rando, T.A. A muscle stem cell support group: Coordinated cellular responses in muscle regeneration. *Dev. Cell* **2018**, *46*, 135–143. [CrossRef] [PubMed]

126. Deyhle, M.R.; Hyldahl, R.D. The role of t lymphocytes in skeletal muscle repair from traumatic and contraction-induced injury. *Front. Physiol.* **2018**, *9*, 768. [CrossRef] [PubMed]

127. Cornelison, D. "Known unknowns": Current questions in muscle satellite cell biology. *Curr. Top. Dev. Biol.* **2018**, *126*, 205–233. [PubMed]

128. Mounier, R.; Theret, M.; Arnold, L.; Cuvellier, S.; Bultot, L.; Goransson, O.; Sanz, N.; Ferry, A.; Sakamoto, K.; Foretz, M.; et al. Ampkα1 regulates macrophage skewing at the time of resolution of inflammation during skeletal muscle regeneration. *Cell Metab.* **2013**, *18*, 251–264. [CrossRef] [PubMed]

129. Niesler, C.U.; Myburgh, K.H.; Moore, F. The changing ampk expression profile in differentiating mouse skeletal muscle myoblast cells helps confer increasing resistance to apoptosis. *Exp. Physiol.* **2007**, *92*, 207–217. [CrossRef] [PubMed]

130. Williamson, D.L.; Butler, D.C.; Alway, S.E. Ampk inhibits myoblast differentiation through a pgc-1α-dependent mechanism. *Am. J. Physiol. Endocrinol. Metab.* **2009**, *297*, E304–E314. [CrossRef] [PubMed]

131. Fulco, M.; Cen, Y.; Zhao, P.; Hoffman, E.P.; McBurney, M.W.; Sauve, A.A.; Sartorelli, V. Glucose restriction inhibits skeletal myoblast differentiation by activating sirt1 through ampk-mediated regulation of nampt. *Dev. Cell* **2008**, *14*, 661–673. [CrossRef] [PubMed]

132. Miyake, M.; Takahashi, H.; Kitagawa, E.; Watanabe, H.; Sakurada, T.; Aso, H.; Yamaguchi, T. Ampk activation by aicar inhibits myogenic differentiation and myostatin expression in cattle. *Cell Tissue Res.* **2012**, *349*, 615–623. [CrossRef] [PubMed]

133. Ye, C.; Zhang, D.; Zhao, L.; Li, Y.; Yao, X.; Wang, H.; Zhang, S.; Liu, W.; Cao, H.; Yu, S.; et al. Camkk2 suppresses muscle regeneration through the inhibition of myoblast proliferation and differentiation. *Int. J. Mol. Sci.* **2016**, *17*. [CrossRef] [PubMed]

134. Fu, X.; Zhao, J.X.; Zhu, M.J.; Foretz, M.; Viollet, B.; Dodson, M.V.; Du, M. Amp-activated protein kinase α1 but not α2 catalytic subunit potentiates myogenin expression and myogenesis. *Mol. Cell. Biol.* **2013**, *33*, 4517–4525. [CrossRef] [PubMed]

135. Theret, M.; Gsaier, L.; Schaffer, B.; Juban, G.; Ben Larbi, S.; Weiss-Gayet, M.; Bultot, L.; Collodet, C.; Foretz, M.; Desplanches, D.; et al. Ampkα1-ldh pathway regulates muscle stem cell self-renewal by controlling metabolic homeostasis. *EMBO J.* **2017**, *36*, 1946–1962. [CrossRef] [PubMed]

136. Fu, X.; Zhu, M.J.; Dodson, M.V.; Du, M. Amp-activated protein kinase stimulates warburg-like glycolysis and activation of satellite cells during muscle regeneration. *J. Biol. Chem.* **2015**, *290*, 26445–26456. [CrossRef] [PubMed]

137. Fu, X.; Zhu, M.; Zhang, S.; Foretz, M.; Viollet, B.; Du, M. Obesity impairs skeletal muscle regeneration through inhibition of ampk. *Diabetes* **2016**, *65*, 188–200. [PubMed]

138. Mian, I.; Pierre-Louis, W.S.; Dole, N.; Gilberti, R.M.; Dodge-Kafka, K.; Tirnauer, J.S. Lkb1 destabilizes microtubules in myoblasts and contributes to myoblast differentiation. *PLoS ONE* **2012**, *7*, e31583. [CrossRef] [PubMed]

139. Shan, T.; Zhang, P.; Xiong, Y.; Wang, Y.; Kuang, S. Lkb1 deletion upregulates pax7 expression through activating notch signaling pathway in myoblasts. *Int. J. Biochem. Cell Biol.* **2016**, *76*, 31–38. [CrossRef] [PubMed]

140. Shan, T.; Zhang, P.; Liang, X.; Bi, P.; Yue, F.; Kuang, S. Lkb1 is indispensable for skeletal muscle development, regeneration, and satellite cell homeostasis. *Stem Cells* **2014**, *32*, 2893–2907. [CrossRef] [PubMed]

International Journal of
Molecular Sciences

MDPI

Review

AMP-Activated Protein Kinase as a Key Trigger for the Disuse-Induced Skeletal Muscle Remodeling

Natalia A. Vilchinskaya [1], Igor I. Krivoi [2] and Boris S. Shenkman [1,*]

[1]   Myology Laboratory, Institute of Biomedical Problems RAS, Moscow 123007, Russia;
     vilchinskayanatalia@gmail.com
[2]   Department of General Physiology, St. Petersburg State University, St. Petersburg 199034, Russia;
     iikrivoi@gmail.com
*   Correspondence: bshenkman@mail.ru

Received: 1 October 2018; Accepted: 9 November 2018; Published: 12 November 2018

Abstract: Molecular mechanisms that trigger disuse-induced postural muscle atrophy as well as myosin phenotype transformations are poorly studied. This review will summarize the impact of 5′ adenosine monophosphate -activated protein kinase (AMPK) activity on mammalian target of rapamycin complex 1 (mTORC1)-signaling, nuclear-cytoplasmic traffic of class IIa histone deacetylases (HDAC), and myosin heavy chain gene expression in mammalian postural muscles (mainly, soleus muscle) under disuse conditions, i.e., withdrawal of weight-bearing from ankle extensors. Based on the current literature and the authors' own experimental data, the present review points out that AMPK plays a key role in the regulation of signaling pathways that determine metabolic, structural, and functional alternations in skeletal muscle fibers under disuse.

Keywords: AMPK; HDAC4/5; p70S6K; MyHC I(β), motor endplate remodeling; soleus muscle; mechanical unloading; hindlimb suspension

1. Introduction

Skeletal muscle is a highly plastic organ, which is able to change its structure and metabolism depending on the mode of contractile activity. Such conditions as hypokinesia, immobilization, paralysis, and weightlessness can lead to a complex of atrophic changes (most pronounced in postural muscles), resulting from a significant reduction in muscle mass and contractile function [1,2]. Skeletal muscle disuse also leads to a reduction in muscle stiffness and slow-to-fast myosin phenotype transformations [2–7]. Muscle atrophy observed during muscle inactivation under conditions of real and simulated microgravity, joint immobilization, or spinal isolation is associated with an increase in proteolytic processes and a decrease in protein synthesis [4,8–11]. Myosin phenotype shift occurs as a result of a decrease in the gene expression of the slow isoform of myosin heavy chain (MyHC) and an increase in the expression of the fast MyHC isoforms [12–14].

To study the mechanisms of muscle disuse atrophy, a variety of experimental models with the different rate of reduction in muscle electrical and contractile activity are used. In this sense, one of the most suitable models is a rodent hindlimb suspension (HS) technique, which prevents the hindlimbs from touching any supporting surface, resulting in a cessation of rat soleus neuromuscular activity [15–17]. Similar effects are observed during dry immersion in human skeletal muscle [1,18,19]. These models not only provide an almost complete cessation of the soleus muscle contractile activity, simulating the effects of weightlessness, but also allow the experimentalist to avoid invasive procedures (denervation, spinal isolation, administration of toxins, etc.). Hence, this review will mainly summarize the data obtained in HS and dry immersion models.

Despite a large number of studies aimed at the analysis of disuse muscle atrophy, the triggering mechanisms of its development within a few hours/days after withdrawal of weight-bearing from postural muscles are still poorly studied [1,15,16]. The earliest effects of unloading (the first 24 h) on postural muscle include: (1) Depolarization of the sarcolemma due to an inactivation of the α2 subunit of the Na,K-ATPase [20,21], (2) disintegration of cholesterol rafts [22], and (3) translocation of the neuronal NO-synthase from the subsarcolemmal compartment to the cytoplasm [23]. However, the mechanisms of development of these changes and, most importantly, the dependence of these mechanisms on molecular triggers determined by the level of muscle contractile activity/inactivity remain unknown.

It seems natural that the reduction/cessation of electrical and, accordingly, contractile activity of the muscle can lead to changes in the basic physiological mechanisms that directly depend on the activity of muscle fibers.

1. Changes in electrogenic signaling mechanisms due to termination of electrical activity (hypothetical decrease in Na^+ concentration inside the muscle fibers, the expected temporary cessation of Ca^{2+} flow through voltage-dependent L-type calcium channels).

2. Mechanosensory molecular changes due to termination of the mechanical action of extracellular matrix structures on mechanosensory molecules (changes in the state of integrins, etc. [24], termination of the active state of actin stress-fibers, inactivation of mechanosensitive channels, inactivation of mechanosensory myofibrillar proteins).

3. Changes in energy metabolism as a result of termination of ATP expenditure (changes in the ratio of ATP/ADP/AMP and PCr/Cr, accumulation of glycogen and reactive oxygen species).

It is not yet possible to trace the entire complex of processes that link the cessation of electrical activity and the elimination of mechanical loading of muscle fibers with the development of early molecular events during mechanical unloading. As for the consequences of the cessation of metabolic energy expenditure during unloading, the situation is somewhat different. In the 1990s, it was hypothesized that mechanical unloading leads to a change in the balance of high-energy phosphate compounds towards the accumulation of fully phosphorylated compounds [25]. Wakatsuki and co-authors have shown the accumulation of phosphocreatine in the soleus muscle of rats after 10 days of HS [26]. In 1987, Henriksen and Tischler reported a 25% increase in glycogen content in rat soleus during the first three days of HS [27].

If the described changes in energy metabolism are the potential triggers for signaling processes leading to the development of postural muscle atrophy, reduced intrinsic muscle stiffness, and myosin phenotype shift, there should be a specific sensor of the state of energy metabolism. Such a sensor has long been known. It is 5′ adenosine monophosphate -activated protein kinase (AMPK), the cell's main energy sensor reacting to the changes in the ratio of high-energy phosphates (Figure 1). Therefore, the termination of electrical activity in soleus muscle at the initial stage of gravitational unloading should affect AMPK activity.

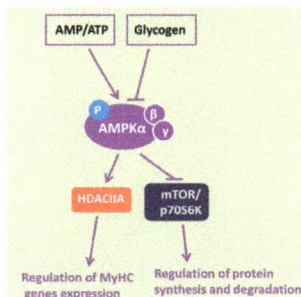

Figure 1. Key physiological regulators of 5′ adenosine monophosphate -activated protein kinase (AMPK) activity in skeletal muscle and physiologically-relevant AMPK targets. (original scheme)

2. AMPK Is a Key Energy Sensor and Metabolic Regulator of Signaling Pathways in Skeletal Muscle Fibers

AMPK is involved in transmission of extracellular and intracellular signals by phosphorylation of various substrates in many metabolic reactions in skeletal muscle. AMRK is a heterotrimeric complex consisting of three proteins: α-subunit, which has its own kinase activity, and two regulatory subunits, β and γ [28].

AMPK activity is regulated both allosterically and by post-translational modifications (phosphorylation). Allosteric activation of AMPK is carried out with the help of AMP and its analogues. Phosphorylation AMPK at Thr172 of the α-subunit leads to its activation. This phosphorylation is regulated by calcium-/calmodulin-dependent kinase kinase 2 (CaMKK2) and liver kinase B1 (LKB1). Further activation of AMPK occurs due to conformational changes occurring upon binding of AMP or ADP to the γ-subunit of AMPK, promoting Thr172 phosphorylation at the α-subunit. The combined effect of Thr172 phosphorylation of the α-subunit of AMPK and allosteric regulation leads to more than a 1000-fold increase in AMPK activity, which makes AMPK highly sensitive to alternations in the energy status of the cell [29].

Activation of AMPK may also occur under the action of extracellular regulatory factors, such as interleukin-6 (IL-6) and brain-derived neurotrophic factor (BDNF) [30]. Interestingly, AMPK is involved in the regulation of IL-6 expression in skeletal muscles [31]. One of the factors of AMPK activation is nitric oxide (NO), the production of which is determined by the activity of neuronal and endothelial NO synthases [32,33]. There is evidence that AMPK activation can result from mechanical stretch via components of the dystrophin-glycoprotein complex (at least in cardiomyocytes) [34].

AMPK can also be phosphorylated on Ser485/491 sites by protein kinase D and some isoforms of protein kinase C [35], which leads to inhibition of AMPK activity. A decrease in AMPK activity is associated with increased glycogen content, as well as accumulation of ATP and creatine phosphate [36,37]. AMPK has a number of molecular targets in skeletal muscle. It is known that AMPK can activate Na,K-ATPase [38,39] and phosphorylate neuronal NO-synthase [40]. AMPK is also involved in the regulation of protein synthesis and degradation [36]. AMPK is a negative regulator of protein synthesis in skeletal muscle. This kinase can inhibit the key regulator of protein synthesis, the mammalian target of rapamycin complex 1 (mTORC1), through phosphorylation of TSC2 [41] and raptor [42]. AMPK can also be involved in the degradation of myofibrillar proteins [43]. Nakashima and co-authors have shown that AMPK participates in the degradation of myofibrillar proteins through the activation of forkhead box proteins (FOXO) transcription factors and subsequent up-regulation of muscle-specific E3 ubiquitin-ligases atrogin-1/MAFbx and MuRF-1 [44].

In addition, AMPK, as a key energy sensor of the cell regulating energy metabolism, participates in the initiation of autophagy [45]. AMPK can directly phosphorylate Unc-51-like kinase (ULK-1) across several sites as well as activate autophagy by inhibiting mTOR activity [46,47].

In recent years, it has been shown that AMPK can influence the expression of a number of genes by phosphorylation of class IIA histone deacetylases (HDAC4, HDAC5, HDAC7), leading to their exclusion from the nucleus and activation of gene expression [48–50].

Thus, according to modern concepts, AMPK activity is mainly determined by the state of energy metabolism: AMPK activity increases with increased ATP consumption, AMP accumulation, and glycogen depletion, and decreases with the accumulation of ATP and glycogen in muscle fibers. Activated AMPK phosphorylates and retains class IIA HDACs outside the myonuclei (thereby contributing to the expression of a number of genes) and inhibits the activity of mTORC1 and its primary targets (Figure 1). Dephosphorylated AMPK, on the contrary, promotes HDACs nuclear import and transcriptional suppression of gene expression, while reducing the degree of mTORC1 suppression.

3. AMPK Activity under Conditions of Mechanical Unloading

Until recently, AMPK activity/phosphorylation in skeletal muscle under unloading conditions has been poorly studied. Moreover, the literature reveals contradictory results concerning AMPK Thr172 phosphorylation at various time-points in different models of disuse. It has been shown that four- and seven-day denervation resulted in a significant increase in AMPK phosphorylation in rodent skeletal muscles [51,52], while deletion of AMPKα2 significantly attenuated denervation-induced skeletal muscle wasting and protein degradation [51]. In human skeletal muscle with recent complete cervical spinal cord injury, AMPKα2 protein abundance decreased by 25% during the first year after injury, without significant change in AMPKα1 content. Furthermore, AMPK phosphorylation on Thr172 was significantly decreased during the first year post-spinal cord injury in human vastus lateralis muscle [53]. Thirty-day space flight and subsequent recovery did not affect AMPK Thr172 phosphorylation in murine longissimus dorsi muscle [54]. In terms of fiber-type composition, murine longissimus dorsi is similar to soleus muscle. However, it should be noted that it is not quite correct to compare the data obtained from rat and mouse soleus muscle. It is known that rat soleus muscle, as well as human soleus, comprises about 85% of slow-twitch fibers expressing the slow isoform of MyHC, while mouse soleus consists of approximately 40% slow-twitch fibers [55]. Obviously, this fact determines the essential features of metabolism in the mouse soleus muscle and its response to gravitational unloading. Therefore, unloading-induced changes in mouse soleus can significantly differ from that of rats and humans.

Vilchinskaya et al. (2015) have shown for the first time that short-term (three days) gravitational unloading via dry immersion leads to a significant decrease in the level of AMPK Thr172 phosphorylation in human soleus muscle [56]. The literature data on the AMPK activity in rat soleus following HS are inconsistent. AMPK activity, which is usually assessed by the level of Thr172 phosphorylation [57–59], was reported to be reduced in rat soleus after two weeks of HS [60], whereas Hilder and co-authors showed that 14-day HS results in a significant increase in AMPK Thr172 phosphorylation in rat soleus [61]. At the same time, Egawa and others did not find any changes in AMPKα1 и AMPKα2 activity in mouse soleus after 14-day HS, however, the level of ACC phosphorylation was upregulated [62,63].

A significant reduction in AMPK Thr172 phosphorylation was previously observed in rat soleus at the early stage (24 h) of hindlimb unloading [11]. A recent study has also demonstrated a significant decrease in AMPK Thr172 phosphorylation in rat soleus muscle already after 6- and 12-h mechanical unloading [64]. In an inactive skeletal muscle, a rapid accumulation of completely phosphorylated high-energy phosphates can occur, resulting in reduced AMPK activity. Thus, AMPK dephosphorylation at the early stage of gravitational unloading can be caused by a decrease in postural muscle energy consumption and a corresponding change in the ratio of phosphorylated and dephosphorylated adenine nucleotides (ATP, ADP, AMP).

It is known that binding of AMPK to glycogen results in reduced AMPK activity [37]. Therefore, it is possible that a decrease in the activity of AMPK at the initial stages of mechanical unloading may be associated with glycogen accumulation. Indeed, glycogen concentration in rat soleus muscle during the first three days of HS is significantly increased [27].

After seven-day HS, AMPK phosphorylation does not differ from that of control [11], which correlates well with the restoration of electromyographic activity of rat soleus following six to seven days of HS [15]. Moreover, 14-day HS resulted in a significant increase in AMPK Thr172 phosphorylation in rat soleus [61,65]. It is notable that after 14-day HS, the increase of AMPK activity (judging by ACC phosphorylation) [62,63] is less pronounced in murine vs rat soleus.

Now, it is difficult to establish the precise mechanisms that cause an increase in AMPK activity by 14 days of HS. However, some assumptions can be made concerning potential signaling mechanisms leading to this phenomenon. There is evidence that the concentration of interleukin-6 (IL-6) is significantly increased in rodent skeletal muscle after 5- and 14-day HS [66,67] as well as in human skeletal muscle following 60-day bed rest [68]. It is known that IL-6 can increase AMPK activity in rodent skeletal muscle [69], and it is likely that increased concentration of IL-6 during long-term gravitational unloading promotes AMPK hyperphosphorylation. BDNF was also shown to increase AMPK phosphorylation in isolated rat extensor digitorum muscle [30]. Therefore, an increase in BDNF mRNA expression in the spinal cord and soleus muscle of rats after 14 days of HS [70] could be the cause of the increased AMPK Thr172 phosphorylation during this period.

Thus, HS experiments show complex time-course changes in AMPK activity in the rodent soleus muscle under mechanical unloading. It is important to note that the level of AMPK phosphorylation is significantly reduced during the first day of HS, a period that precedes atrophy development. Additionally, the most likely cause of such a decrease is a shift in the ratio of phosphorylated and dephosphorylated adenine nucleotides (AMP/ATP and ADP/ATP ratios).

4. The Role of AMPK in the Regulation of mTOR/p70S6K and Akt/FOXO3/MuRF-1/MAFbx/ Atrogin-1 Signaling Pathways under Gravitational Unloading

Anabolic processes in skeletal muscle fibers are regulated by a number of signaling pathways, the most important of which is the mammalian target of rapamycin complex 1 (mTORC1) signaling pathway. mTORC1, through its downstream targets (p70S6K, 4E-BP1), stimulates mRNA translation initiation on a ribosome. Activation of protein synthesis following resistance exercise is associated with the increased level of p70S6K (Thr389) phosphorylation, leading to subsequent phosphorylation of ribosomal protein S6 and initiation of protein synthesis. Some authors reported a decrease in p70S6K (Thr389) phosphorylation in rat soleus after four to five days of HS [71,72]. Other studies showed that even 7–10 days of HS did not affect p70S6K phosphorylation, the level of which decreased only following 14 days of unloading [73–76].

It is well known that activated AMPK has an inhibitory effect on the anabolic processes via suppression of mTORC1 and its key substrate, p70S6K [41–43,77–80]. Summing up the available data, we can assume that changes in p70S6K and AMPK phosphorylation in rat soleus muscle during the first two weeks of HS show reciprocal relations and complex dynamics. It is important to note that during the early stage of HS (one to three days), an increase in p70S6K Thr389 phosphorylation is accompanied by a decrease in the level of AMPK Thr172 phosphorylation. However, after seven days of HS, there is no difference between these parameters and control values [11]. Two-week HS results in an opposite effect: A decrease in p70S6K phosphorylation is accompanied by an increase in AMPK phosphorylation. These data are consistent with the report by Sugiura and co-authors (2005) that showed no changes in p70S6K phosphorylation after 10 days of HS [74], whereas 14-day HS led to a significant decrease in p70S6K phosphorylation as compared to control levels [71,72,75,76,81]. Interestingly, according to Hilder et al. (2005) and Zhang et al. (2018), the level of AMPK phosphorylation following 14-day unloading is significantly increased [61,65]. The high level of AMPK phosphorylation is accompanied by a decrease in phosphorylation of not only p70S6K, but also Akt and FOXO3, which can lead to an upregulation of muscle-specific E3-ubiquitin ligases, MuRF-1 and MAFbx/atrogin-1 [65]. These results are consistent with earlier studies showing the ability of AMPK to stimulate FOXO3 dephosphorylation and the expression of E3-ubiquitin ligases [44,82]. Time-course changes in the level of p70S6K and AMPK phosphorylation during mechanical unloading suggest that an increase in p70S6K phosphorylation at the first day of HS may be due to the low AMPK activity. This hypothesis has been recently tested by Vilchinskaya and co-authors [83]. Pretreatment of rats with 5-aminoimidazole-4-carboxamide ribonucleotide (AICAR) several days before and during 24-h HS prevented both a decrease in AMPK Th172 phosphorylation as well as an increase in p70S6K Thr389 phosphorylation [83]. This result fully confirmed the hypothesis and once again demonstrated the role of AMPK as a negative regulator of mTORC1-signaling [41,42]. The results obtained by Vilchinskaya et al. (2017) suggest that AMPK dephosphorylation in postural soleus muscle at the initial stage of mechanical unloading is one of the reasons for a paradoxical increase in the level of p70S6K phosphorylation. Increased p70S6K phosphorylation is usually considered to be a consequence of (1) inactivation of endogenous mTORC1 inhibitor tuberous sclerosis complex (TSC1/2) due to

AMPK dephosphorylation [41] and (2) accumulation of sphingolipid ceramide [84]. Interestingly, it was previously shown that activation of AMPK with AICAR prevents ceramide accumulation in skeletal muscle fibers [85]. Therefore, it can be assumed that AMPK dephosphorylation can contribute to the accumulation of ceramide in rat soleus under gravitational unloading [86–88]. Hsieh and co-authors (2014) observed an interesting effect of p70S6K hyperphosphorylation. The authors showed that activated p70S6K promotes phosphorylation of insulin receptor substrate (IRS1) on Ser636-639, which leads to a reduction in IRS-1 activity and, accordingly, dephosphorylation of downstream protein kinase Akt (Figure 2).

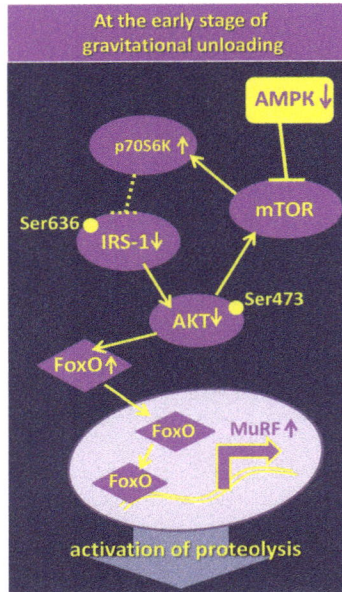

Figure 2. Hypothetical role of AMPK in the activation of signaling pathways regulating the expression of E3-ubiquitin ligases during gravitational unloading.

Dephosphorylation of Akt on Ser473, as a rule, causes an increased expression of E3-ubiquitin ligases (MuRF-1 and MAFbx/atrogin-1) [88]. It is well known that even short-term (one to three days) gravitational unloading leads to Akt dephosphorylation and upregulation of E3-ubiquitin ligases [10,66]. According to a number of authors, gravitational unloading results in a significant decrease in IRS-1 content [54,89]. Based on these literature data, it can be assumed that AMPK dephosphorylation during the first day of unloading leads to an increase in p70S6K phosphorylation resulting in the increased E3-ubiquitin ligases' expression and enhanced proteolysis. Recent experiments with inhibition of p70S6K phosphorylation during the first day of gravitational unloading do not contradict this hypothesis: The use of rapamycin (mTORC1 inhibitor) resulted in a significant decrease in MuRF-1 and MAFbx/atrogin-1 expression [88].

Thus, AMPK dephosphorylation at the initial period of HS leads to an increase in the level of p70S6K phosphorylation, which can contribute to the subsequent upregulation of proteolytic enzymes.

AMPK is known to activate autophagy by ULK phosphorylation (see above). Therefore, one would expect to see a reduction in autophagy markers at the initial period of unloading when AMPK activity is downregulated. However, it has been recently shown that most of autophagy markers, except for ULK, were upregulated in rat soleus following 6- and 12-h HS [64]. The authors of the cited report emphasize that this state of autophagy markers is not consistent with a reduced level of AMPK phosphorylation. They explain this phenomenon by the possible unequal phosphorylation of different AMPK isoforms. It is clear that this issue needs further investigation.

At a later stage of unloading, an increase in AMPK phosphorylation is accompanied by a decrease in p70S6K phosphorylation and activation of the proteolytic signaling system [65].

At the same time, no significant changes in phosphorylation of Akt and p70S6K and the rate of protein synthesis were found in the soleus of transgenic mice overexpressing the dominant-negative mutant of AMPK following 14-day HS [62]. Thus, the lack of AMPK activity in the transgenic mice did not affect anabolic signaling following a relatively long period of unloading (14 days). However, 14-day unloading of these transgenic mice induced a significant increase in the expression of both markers of the ubiquitin-proteasome system and ULK, a marker of autophagy. The differences between dominant-negative AMPK mutants and wild-type mice were significant [62]. There were also significant differences in the severity of atrophic changes. The weight of soleus muscle in dominant-negative AMPK mutants after 14-day HS was significantly reduced, but to a lesser extent than that in wild-type mice (difference of 10–15%) [62,63]. Thus, AMPK can contribute to the development of muscle atrophy during unloading. This contribution to muscle atrophy seems to be carried out via the activation of catabolic processes rather than through the suppression of anabolic regulation.

This fact allows us to suggest that, at later stages of gravitational unloading, AMPK activity can act as a key signaling node of protein homeostasis in skeletal muscle.

5. AMPK Is Involved in the Regulation of Myosin Phenotype under Mechanical Unloading

Unloading-induced changes in skeletal muscle myosin phenotype have been observed by a number of researchers. It is well established that there is an increase in the expression of fast-type myosin isoforms and a decrease in the expression of slow-type myosin isoform in skeletal muscles of space-flown rodents as well as astronauts [3,4,90,91]. HS results in a significant increase in the content of fast-twitch (type II) fibers and a significant decrease in the proportion of slow-twitch (type I) fibers in rat soleus muscle [7,92–95]. In samples of soleus muscle, taken from astronauts after a six-month spaceflight, there was a decrease in the proportion of fibers expressing slow MyHC isoform and an increase in the proportion of fibers expressing fast MyHC isoforms [96]. Mechanisms of molecular regulation of myosin phenotype transformation remain largely unexplored. It is known that AMPK can affect the expression of a number of genes by phosphorylation of class IIA histone deacetylases (HDAC4, HDAC5, HDAC7), which leads to HDACs dissociation from gene promoters and removal from the nucleus, thereby allowing for increased gene expression [48–50]. It was earlier shown that at the first day of mechanical unloading, mRNA expression of the slow MyHC isoform in rat soleus is reduced [13]. One of the ways of MyHC expression regulation in muscle fibers is associated with phosphorylation of histone deacetylase 4 (HDAC4) [7,97]. However, it remains unclear what mechanisms are implicated in such a rapid decrease in gene expression. Until recently, there have been no studies on the role of AMPK in the regulation of myosin phenotype in muscle fibers under conditions of gravitational unloading. Can AMPK be involved in the regulation of MyHC gene expression in skeletal muscle fibers? To answer this question, Vilchinskaya et al. (2017) carried out an experiment with AICAR pretreatment of Wistar male rats [98]. There was a significant decrease in MyHC I (β) pre-mRNA expression and a pronounced tendency to a decrease in the mature MyHC I(β) mRNA expression in rat soleus muscle after 24 h of HS. These results are in good agreement with the data previously obtained in skeletal muscles of Sprague-Dawley rats under similar conditions [13]. However, when exposed to AICAR, there were not any significant decreases in the MyHC I(β) pre-mRNA and mature mRNA expression. Since one of the possible mechanisms of AMPK gene expression regulation is linked to histone deacetylase 4 and 5 (HDAC4/HDAC5) phosphorylation status [49,50], it was hypothesized that a significant decrease in AMPK Thr172 phosphorylation after one-day HS would result in HDAC4 and HDAC5 nuclear accumulation [83]. This hypothesis is supported by a previous report that showed a significant deacetylation of histone H3 at the MyHC I (β) gene in rat soleus following seven-day HS [99]. In addition, it has been previously shown that HDAC4 is predominantly localized to the nuclei in fast-twitch fibers in contrast to the sarcoplasm in slow-twitch fibers [100]. Indeed, in our experiment with 24-h HS, HDAC4 accumulation in the nuclear fraction was found; however, in the AICAR-pretreated group, the accumulation of HDAC4 in the nuclei did not occur, which correlates well with data on AMPK phosphorylation and confirms the hypothesis of AMPK-dependent control of nucleocytoplasmic trafficking of HDAC4 [98,101]. Recently, Yoshihara and co-authors have found nuclear accumulation of HDAC4 in rat gastrocnemius muscle following 10 days of ankle joint immobilization. This accumulation (as in the experiment of Vilchinskaya and co-authors) was accompanied by a decrease in the level of AMPK phosphorylation [102]. As for HDAC5, even a slight increase in AMPK activity in the control AICAR-pretreated rat was accompanied by a decrease in HDAC5 content in the nuclear fraction. This phenomenon could be associated with HDAC5 phosphorylation via AMPK. However, 24-h HS resulted in a significant decrease in the nuclear HDAC5 content. Such a reduction in HDAC5 content in the nuclear fraction could be associated with HDAC5 degradation or HDAC5 phosphorylation and subsequent nuclear export. HDAC5 can also be a target for protein kinase D (PKD) [103,104]. Since the AMPK activity is significantly reduced within the first day of unloading (see above), it appears that HDAC5 nuclear export during unloading is not directly linked to AMPK. It has been shown that a decrease in AMPK activity can result in an upregulation of protein kinase D (PKD) phosphorylation [103]. Indeed, one-day HS resulted in a significant increase in PKD Ser916 phosphorylation, however, in the HS+AICAR group, PKD phosphorylation did not differ from the control levels. Interestingly, such a decrease in PKD phosphorylation in the HS+AICAR group vs. the HS group can significantly attenuate the loss of nuclear HDAC5. It is noteworthy that an increase in PKD phosphorylation in unloaded animals did not affect the content of nuclear HDAC4. Obviously, under unloading conditions, HDAC4 does not appear to be a target of PKD. This study was the first to observe the reciprocal relationship between AMPK and PKD in an inactivated skeletal muscle. In addition, gravitational unloading led to an increase in the level of negative AMPK phosphorylation on Ser485/491, which is known to be associated with PKD [35]. However, allosteric activation of AMPK by AICAR reduced the intensity of negative phosphorylation, possibly due to a reduction in PKD phosphorylation.

AMPK is known to stimulate the expression of peroxisome proliferator-activated receptor gamma coactivator 1-alpha (PGC1α), the most important regulator of the expression and activity of signaling proteins. In particular, PGC1α is involved in the control of the expression of the slow isoform of MyHC [105]. A significant decrease in PGC1α expression was found in mouse soleus after four days of hindlimb unloading [106]. However, Vilchinskaya et al. (2017) found no changes in PGC1α mRNA expression after 24-h HS [98]. AMPK activation by AICAR pretreatment also did not induce any changes in PGC1α expression. Obviously, the impact of AMPK on the expression of the slow isoform of MyHC during the first day of unloading is carried out primarily through the trafficking of HDAC4, without the involvement of PGC1α.

Thus, the results obtained by Vilchinskaya and the co-authors (2017) clearly show that AMPK Thr172 dephosphorylation within the first day of gravitational unloading has a significant effect on the regulation of myosin phenotype in rat postural muscle. In particular, AMPK Thr172 dephosphorylation led to a decrease in MyHC I(β) pre-mRNA and mRNA expression. The study of Vilchinskaya et al. (2017) indicates that HDAC4 is not a target of PKD at the early stages of unloading, and, probably, HDAC4 nuclear import results from a decrease in the AMPK activity (Figure 3). It is possible that an increase in PKD activity can lead to HDAC5 nuclear export.

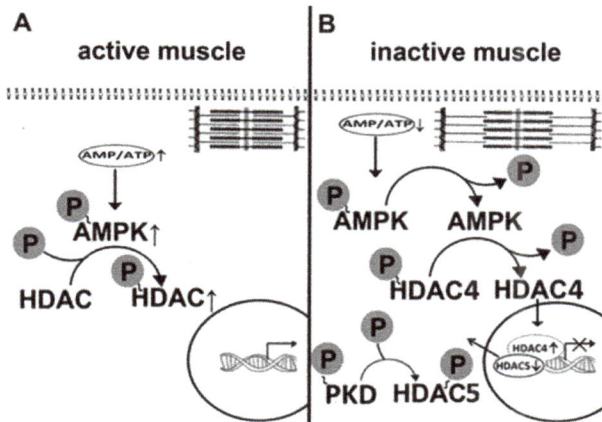

Figure 3. The role of AMPK in class IIa histone deacetylases traffic in rat soleus muscle at the initial stages of gravitational unloading (modified from [98]). (**A**): active muscle; (**B**): inactive muscle. Time-course studies on the MyHCI (β) expression demonstrate that a reduction in MyHCI (β) mRNA expression begins on the first day of unloading and then steadily decreases during, at least, two weeks of hindlimb unloading [13,14]. As shown in dominant-negative AMPK mutants, AMPK has no effect on the expression of slow MyHC after 14-day unloading [63]. It is clear that an experiment with transgenic animals does not allow for tracing the effect of AMPK on the expression of MyHC at the different time-points of unloading. However, the results of this experiment indicate that the decrease in the expression of slow isoform of MyHC after the completion of the initial stage of hindlimb unloading is determined not by AMPK, but by some other mechanisms, for example, via inhibition of the calcineurin/ Nuclear factor of activated T-cells (NFAT) signaling pathway [14,107].

6. AMPK and Disuse-Induced Motor Endplate Remodeling

Reliable neuromuscular transmission is essential for normal bodily functions. Motor endplate is a highly specialized sarcolemma region in which the transmission of activity from motor neurons to striated muscle fiber is realized. Endplate structure, including features of nicotinic acetylcholine receptors' (nAChRs) localization and distribution, are among the factors crucial for maintenance of highly efficient neuromuscular transmission [108,109].

Ultrastructure of the neuromuscular junction strongly depends on the motor activity, and exhibits high morphological and functional plasticity. Different modes of increased activity are manifested in the morphological remodeling, such as the expansion of the neuromuscular junction size. Reduced patterns of neuromuscular activity also trigger endplate remodeling. Alterations in neuromuscular junction stability and integrity progressively increase with age [110–114] and myodystrophy [115–117], in animal models of muscle injures [118,119], after denervation [120] and prolonged (four weeks) HS [121]. Although the molecular mechanisms underlying the structural and functional endplate plasticity are intensively studied, they are not completely clear [113,115,116,122–125].

Currently, many facts point to the involvement of AMPK in neuromuscular junction remodeling. AMPK is linked to a variety of cellular processes and is also considered a crucial activator of autophagy and its downstream target, ULK1 [126]. Autophagy is involved in neuromuscular junction preservation during aging, and impairments in autophagy exacerbate synaptic structure degeneration [111]. AMPK and AMPK-activated autophagy are among the most important factors that maintain neuromuscular junction stability [111,127,128]. Pharmacological activation of AMPK by AICAR administration improves integrity of neuromuscular junctions and prevents skeletal muscle pathology in a mouse model of severe spinal muscular atrophy [128]. Additionally, AICAR treatment has been shown to stimulate autophagy and ameliorate muscular dystrophy in the mdx mice, a model of Duchenne muscular dystrophy, suggesting AMPK as a powerful therapeutic target [129,130]. Also, AMPK may play a role in activating PGC-1α, a key regulatory protein in skeletal muscle adaptation to physical activity. PGC-1α has been reported to play a major role in maintaining neuromuscular junction integrity [122,123].

AMPK also plays a beneficial role in the regulation of skeletal muscle cholesterol synthesis as well as sarcolemma cholesterol levels [131,132]. Direct molecular interaction between membrane cholesterol and the nAChRs has been shown [133]. Moreover, cholesterol and lipid rafts contribute to the orchestration of nAChRs clustering at the endplate region [134,135]. In addition, AMPK can affect the Na,K-ATPase activity [38,39]. The targeting and activity of the Na,K-ATPase are also influenced by the cholesterol environment [136–139], and reciprocal interactions between cholesterol and the α2 Na,K-ATPase isozyme has been suggested [22,140]. Notably, both the nAChRs and the α2 Na,K-ATPase isozyme are enriched at the endplate region, co-localized, and co-immunoprecipitated, suggesting that these proteins exist as a functional multimolecular complex to regulate electrogenesis and to maintain the effectiveness of neuromuscular transmission [141–143].

The loss of the α2 Na,K-ATPase isozyme electrogenic activity accompanied by disturbances in lipid rafts and endplate structure stability were observed in rat soleus muscle even after 6–12 h of HS [21,22,64]. Such acute disuse decreased the endplate area and increased the density of the nAChRs distribution. These changes were accompanied by decreased phosphorylation of AMPK and its substrate, ACC, and increased autophagy [64]. Autophagy is known to be involved in the nAChRs turnover regulation [127] and has a major impact on neuromuscular synaptic function [111]. So, an increase in autophagy after acute HS can reflect an adaptive response to compensate for the endplate area loss through increasing the density of the nAChRs distribution [64].

Notably, pretreatment of the rats with AMPK activator, AICAR, followed by HS, stabilized the resting membrane potential, endplates area, and the nAChRs distribution density, indicating that AMPK activation can prevent disuse-induced endplate structural and functional reorganization (unpublished observation).

In summary, these novel findings indicate that endplate functional and structural characteristics rapidly (within hours) respond to skeletal muscle disuse. Decreased phosphorylation of AMPK accompanied by increased autophagy is the earliest disuse-induced remodeling event preceding the overt skeletal muscle atrophy.

7. Conclusions

AMPK demonstrates multidirectional changes in a mammalian soleus muscle during gravitational unloading: A deep decrease in the level of phosphorylation and kinase activity at the early stages of the process and a significant increase in activity at the later stages. Time-course changes in AMPK activity under unloading are nonlinear and require a more detailed analysis at each stage of the process. The experiments discussed in this review show that changes in AMPK activity under unloading conditions can have a significant impact on the key signaling pathways and molecular structures in skeletal muscle fiber. As a result of reduction in AMPK activity at the initial stage of unloading, paradoxical hyperphosphorylation of p70S6K occurs, which, according to in vitro experiments, can lead to the activation of proteolytic processes in muscle fiber [83,87]. A decrease in the level of phosphorylation and kinase activity of AMPK at the early stages of unloading also affects the change in nucleocytoplasmic traffic of class IIA HDACs, resulting in a decrease in the expression of the myh7 gene (slow isoform of MyHC) and, possibly, a number of other genes controlling energy metabolism. At the later time-points of unloading, an increase in the level of AMPK phosphorylation/activity is accompanied by a decrease in the activity of p70S6K [76], FOXO3 dephosphorylation, and an increase in the expression of the key enzymes of the ubiquitin-proteasome pathway [65], which, obviously, should lead to decreased muscle protein synthesis and enhanced proteolysis. Unfortunately, to date, the precise physiological mechanisms, both intramuscular and systemic, underlying such a complex and nonlinear nature of AMPK activity during gravitational unloading are not known. It is possible that glycogen accumulation [27] or a change in the ratio of dephosphorylated and phosphorylated high-energy phosphates could lead to a deep AMPK dephosphorylation. The cause of a gradual increase in AMPK phosphorylation in rat soleus during the first week of HS remains unclear. It is possible that this increase is due to a gradual increase in the electromyographic activity of the postural muscle [15]. It is difficult to explain a significant increase in AMPK phosphorylation in rat soleus by the end of the second week of HS. Possibly, such systemic factors as BDNF and/or interleukin-6 could play a role in these processes. All these questions have yet to be answered.

Thus, recent studies have revealed the key role of AMPK in the processes of deep remodeling of signaling pathways, leading to changes in metabolism, structure, and function of postural muscle fibers under unloading conditions. Further studies are needed to elucidate new signaling mechanisms that trigger, determine, and limit atrophy development and intrinsic muscle stiffness, as well as myosin phenotype changes, in a mammalian postural muscle under mechanical unloading.

Funding: The work was supported by the Russian Science Foundation (RSF) grant no. 18-15-00107.

Acknowledgments: The authors express their deep gratitude to Timur M. Mirzoev for the participation in the manuscript preparation and creative criticism of the paper.

Conflicts of Interest: The authors declare no conflict of interest.

Abbreviations

AMPK	5′ adenosine monophosphate -activated protein kinase
AICAR	5-aminoimidazole-4-carboxamide ribonucleotide
BGPA	β-Guanidinopropionic acid
HS	hindlimb suspension
mTOR	mammalian/mechanistic target of rapamycin
p70S6K	ribosomal protein S6 kinase beta-1
CaMKKβ	Ca(2+)/calmodulin-dependent protein kinase kinase β
LKB1	liver kinase B1
BDNF	brain-derived neurotrophic factor
MuRF-1	Muscle RING-finger protein-1
FOXO	forkhead box proteins
IRS-1	insulin receptor substrate 1
MyHC	myosin heavy chain isoforms
nAChRs	nicotinic acetylcholine receptors
ULK-1	Unc-51-like kinase
PGC1α	peroxisome proliferator-activated receptor gamma coactivator 1-alpha

References

1. Grigor'ev, A.I.; Kozlovskaia, I.B.; Shenkman, B.S. The role of support afferents in organisation of the tonic muscle system. *Rossiiskii fiziologicheskii zhurnal imeni IM Sechenova* **2004**, *90*, 508–521.
2. Kozlovskaia, I.B.; Grigor'eva, L.S.; Gevlich, G.I. Comparative analysis of the effect of weightlessness and its model on the velocity-strength properties and tonus of human skeletal muscles. *Kosm. Biol. Aviakosm. Med.* **1984**, *18*, 22–26. [PubMed]
3. Oganov, V.S.; Skuratova, S.A.; Murashko, L.M.; Guba, F.; Takach, O. Effect of short-term space flights on physiological properties and composition of myofibrillar proteins of the skeletal muscles of rats. *Kosm. Biol. Aviakosm. Med.* **1988**, *22*, 50–54. [PubMed]
4. Baldwin, K.M.; Haddad, F.; Pandorf, C.E.; Roy, R.R.; Edgerton, V.R. Alterations in muscle mass and contractile phenotype in response to unloading models: Role of transcriptional/pretranslational mechanisms. *Front. Physiol.* **2013**, *4*, 284. [CrossRef] [PubMed]
5. Kandarian, S.C.; Stevenson, E.J. Molecular events in skeletal muscle during disuse atrophy. *Exerc. Sport Sci. Rev.* **2002**, *30*, 111–116. [CrossRef] [PubMed]
6. Adams, G.R.; Baldwin, K.M. Age dependence of myosin heavy chain transitions induced by creatine depletion in rat skeletal muscle. *J. Appl. Physiol.* **1995**, *78*, 368–371. [CrossRef] [PubMed]

7. Shenkman, B.S. From Slow to Fast: Hypogravity-Induced Remodeling of Muscle Fiber Myosin Phenotype. *Acta Nat.* **2016**, *8*, 47–59.

8. Chopard, A.; Hillock, S.; Jasmin, B.J. Molecular events and signalling pathways involved in skeletal muscle disuse-induced atrophy and the impact of countermeasures. *J. Cell Mol. Med.* **2009**, *13*, 3032–3050. [CrossRef] [PubMed]

9. Bodine, S.C. Disuse-induced muscle wasting. *Int. J. Biochem. Cell Biol.* **2013**, *45*, 2200–2208. [CrossRef] [PubMed]

10. Kachaeva, E.V.; Shenkman, B.S. Various jobs of proteolytic enzymes in skeletal muscle during unloading: Facts and speculations. *J. Biomed. Biotechnol.* **2011**, *2012*, 493618. [CrossRef] [PubMed]

11. Mirzoev, T.; Tyganov, S.; Vilchinskaya, N.; Lomonosova, Y.; Shenkman, B. Key Markers of mTORC1-Dependent and mTORC1-Independent Signaling Pathways Regulating Protein Synthesis in Rat Soleus Muscle during Early Stages of Hindlimb Unloading. *Cell. Physiol. Biochem.* **2016**, *39*, 1011–1020. [CrossRef] [PubMed]

12. Stevens, L.; Sultan, K.R.; Peuker, H.; Gohlsch, B.; Mounier, Y.; Pette, D. Time dependent changes in myosin heavy chain mRNA and protein isoforms 94 in unloaded soleus muscle of rat. *Am. J. Physiol. Cell. Physiol.* **1999**, *277*, 1044–1049. [CrossRef] [PubMed]

13. Giger, J.M.; Bodell, P.W.; Zeng, M.; Baldwin, K.M.; Haddad, F. Rapid muscle atrophy response to unloading: Pretranslational processes involving MHC and actin. *J. Appl. Physiol.* **2009**, *107*, 1204–1212. [CrossRef] [PubMed]

14. Lomonosova, Y.N.; Turtikova, O.V.; Shenkman, B.S. Reduced expression of MyHC slow isoform in rat soleus during unloading is accompanied by alterations of endogenous inhibitors of calcineurin/NFAT signaling pathway. *J. Muscle Res. Cell Motil.* **2016**, *37*, 7–16. [CrossRef] [PubMed]

15. Alford, E.K.; Roy, R.R.; Hodgson, J.A.; Edgerton, V.R. Electromyography of rat soleus, medial gastrocnemius, and tibialis anterior during hind limb suspension. *Exp. Neurol.* **1987**, *96*, 635–649. [CrossRef]

16. Kawano, F.; Matsuoka, Y.; Oke, Y.; Higo, Y.; Terada, M.; Wang, X.D.; Nakai, N.; Fukuda, H.; Imajoh-Ohmi, S.; Ohira, Y. Role(s) of nucleoli and phosphorylation of ribosomal protein S6 and/or HSP27 in the regulation of muscle mass. *Am. J. Physiol. Cell Physiol.* **2007**, *293*, 35–44. [CrossRef] [PubMed]

17. De-Doncker, L.; Kasri, M.; Picquet, F.; Falempin, M. Physiologically adaptive changes of the L5afferent neurogram and of the rat soleus EMG activity during 14 days of hindlimb unloading and recovery. *J. Exp. Biol.* **2005**, *208*, 4585–4592. [CrossRef] [PubMed]

18. Kozlovskaya, I.; Dmitrieva, I.; Grigorieva, L.; Kirenskaya, A.; Kreidich, Y. Gravitational mechanisms in the motor system studies in real and simulated weightlessness. In *Stance and Motion Facts and Concepts*; Gurfinkel, V.S., Loffe, M.E., Massion, J., Eds.; Springer: New York, NY, USA, 1988; ISBN 978-1-4899-0823-0.

19. Navasiolava, N.M.; Custaud, M.A.; Tomilovskaya, E.S.; Larina, I.M.; Mano, T.; Gauquelin-Koch, G.; Gharib, C.; Kozlovskaya, I.B. Long-term dry immersion: Review and prospects. *Eur. J. Appl. Physiol.* **2011**, *111*, 1235–1260. [CrossRef] [PubMed]

20. Kravtsova, V.V.; Matchkov, V.V.; Bouzinova, E.V.; Vasiliev, A.N.; Razgovorova, I.A.; Heiny, J.A.; Krivoi, I.I. Isoform-specific Na,K-ATPase alterations precede disuse-induced atrophy of rat soleus muscle. *Biomed. Res. Int.* **2015**, 720172. [CrossRef] [PubMed]

21. Kravtsova, V.V.; Petrov, A.M.; Matchkov, V.V.; Bouzinova, E.V.; Vasiliev, A.N.; Benziane, B.; Zefirov, A.L.; Chibalin, A.V.; Heiny, J.A.; Krivoi, I.I. Distinct α2 Na,K-ATPase membrane pools are differently involved in early skeletal muscle remodeling during disuse. *J. Gen. Physiol.* **2016**, *147*, 175–188. [CrossRef] [PubMed]

22. Petrov, A.M.; Kravtsova, V.V.; Matchkov, V.V.; Vasiliev, A.N.; Zefirov, A.L.; Chibalin, A.V.; Heiny, J.A.; Krivoi, I.I. Membrane lipid rafts are disturbed in the response of rat skeletal muscle to short-term disuse. *Am. J. Physiol. Cell Physiol.* **2017**, *312*, 627–C637. [CrossRef] [PubMed]

23. Lechado, I.; Terradas, A.; Vitadello, M.; Traini, L.; Namuduri, A.V.; Gastaldello, S.; Gorza, L. Sarcolemmal loss of active nNOS (Nos1) is an oxidative stress-dependent, early event driving disuse atrophy. *J. Pathol.* **2018**. [CrossRef]

24. Gordon, S.E.; Flück, M.; Booth, F.W. Selected Contribution: Skeletal muscle focal adhesion kinase, paxillin, and serum response factor are loading dependent. *J. Appl. Physiol.* **2001**, *90*, 1174–1183. [CrossRef] [PubMed]

25. Ohira, Y.; Yasui, W.; Kariya, F.; Wakatsuki, T.; Nakamura, K.; Asakura, T.; Edgerton, V.R. Metabolic adaptation of skeletal muscles to gravitational unloading. *Acta Astronaut.* **1994**, *33*, 113–117. [CrossRef]

26. Wakatsuki, T.; Ohira, Y.; Yasui, W.; Nakamura, K.; Asakura, T.; Ohno, H.; Yamamoto, M. Responses of contractile properties in rat soleus to high-energy phosphates and/or unloading. *Jpn. J. Physiol.* **1994**, *44*, 193–204. [CrossRef] [PubMed]

27. Henriksen, E.J.; Tischler, M.E. Time Course of the Response of Carbohydrate Metabolism to Unloading of the Soleus. *Metabolism* **1988**, *37*, 201–208. [CrossRef]

28. Witczak, C.A.; Sharoff, C.G.; Goodyear, L.J. AMP-activated protein kinase in skeletal muscle: From structure and localization to its role as a master regulator of cellular metabolism. *Cell Mol. Life Sci.* **2008**, *65*, 3737–3755. [CrossRef] [PubMed]

29. Mounier, R.; Théret, M.; Lantier, L.; Foretz, M.; Viollet, B. Expanding roles for AMPK in skeletal muscle plasticity. *Trends Endocrinol. Metab.* **2015**, *26*, 275–286. [CrossRef] [PubMed]

30. Matthews, V.B.; Aström, M.B.; Chan, M.H.; Bruce, C.R.; Krabbe, K.S.; Prelovsek, O.; Akerström, T.; Yfanti, C.; Broholm, C.; Mortensen, O.H.; et al. Brain-derived neurotrophic factor is produced by skeletal muscle cells in response to contraction and enhances fat oxidation via activation of AMP-activated protein kinase. *Diabetologia* **2009**, *52*, 1409–1418. [CrossRef] [PubMed]

31. Glund, S.; Treebak, J.T.; Long, Y.C.; Barres, R.; Viollet, B.; Wojtaszewski, J.F.; Zierath, J.R. Role of adenosine 5′-monophosphate-activated protein kinase in interleukin-6 release from isolated mouse skeletal muscle. *Endocrinology* **2009**, *150*, 600–606. [CrossRef] [PubMed]

32. Lira, V.A.; Soltow, Q.A.; Long, J.H.; Betters, J.L.; Sellman, J.E.; Criswell, D.S. Nitric oxide increases GLUT4 expression and regulates AMPK signaling in skeletal muscle. *Am. J. Physiol. Endocrinol. Metab.* **2007**, *293*, 1062–1068. [CrossRef] [PubMed]

33. Lira, V.A.; Brown, D.L.; Lira, A.K.; Kavazis, A.N.; Soltow, Q.A.; Zeanah, E.H.; Criswell, D.S. Nitric oxide and AMPK cooperatively regulate PGC-1α in skeletal muscle. *J. Physiol.* **2010**, *588*, 3551–3566. [CrossRef] [PubMed]

34. Garbincius, J.F.; Michele, D.E. Dystrophin-glycoprotein complex regulates muscle nitric oxide production through mechanoregulation of AMPK signaling. *Proc. Natl. Acad. Sci. USA* **2015**, *112*, 13663–13668. [CrossRef] [PubMed]

35. Coughlan, K.A.; Valentine, R.J.; Sudit, B.S.; Allen, K.; Dagon, Y.; Kahn, B.B.; Ruderman, N.B.; Saha, A.K. PKD1 Inhibits AMPKα2 through Phosphorylation of Serine 491 and Impairs Insulin Signaling in Skeletal Muscle Cells. *Biochem. J.* **2016**, *473*, 4681–4697. [CrossRef] [PubMed]

36. Hardie, D.G.; Ross, F.A.; Hawley, S.A. AMP-activated protein kinase: A target for drugs both ancient and modern. *Chem. Biol.* **2012**, *19*, 1222–1236. [CrossRef] [PubMed]

37. Koay, A.; Woodcroft, B.; Petrie, E.J.; Yue, H.; Emanuelle, S.; Bieri, M.; Bailey, M.F.; Hargreaves, M.; Park, J.T.; Park, K.H.; et al. AMPK beta subunits display isoform specific affinities for carbohydrates. *FEBS Lett.* **2010**, *584*, 3499–3503. [CrossRef] [PubMed]

38. Ingwersen, M.S.; Kristensen, M.; Pilegaard, H.; Wojtaszewski, J.F.; Richter, E.A.; Juel, C. Na, K-ATPase activity in mouse muscle is regulated by AMPK and PGC-1α. *J. Membr. Biol.* **2011**, *242*, 1–10. [CrossRef] [PubMed]

39. Benziane, B.; Bjornholm, M.; Pirkmajer, S.; Austin, R.L.; Kotova, O.; Viollet, B.; Zierath, J.R.; Chibalin, A.V. Activation of AMP-activated protein kinase stimulates Na$^+$,K$^+$-ATPase activity in skeletal muscle cells. *J. Biol. Chem.* **2012**, *287*, 23451–23463. [CrossRef] [PubMed]

40. Chen, Z.P.; McConell, G.K.; Belinda, J.; Snow, R.J.; Canny, B.J.; Kemp, B.E. AMPK signaling in contracting human skeletal muscle: Acetyl-CoA carboxylase and NO synthase phosphorylation. *Am. J. Physiol. Endocrinol. Metab.* **2000**, *279*, E1202–E1206. [CrossRef] [PubMed]

41. Inoki, K.; Zhu, T.; Guan, K.L. TSC2 mediates cellular energy response to control cell growth and survival. *Cell* **2003**, *115*, 577–590. [CrossRef]

42. Gwinn, D.M.; Shackelford, D.B.; Egan, D.F.; Mihaylova, M.M.; Mery, A.; Vasquez, D.S.; Turk, B.E.; Shaw, R.J. AMPK phosphorylation of raptor mediates a metabolic checkpoint. *Mol. Cell* **2008**, *30*, 214–226. [CrossRef] [PubMed]

43. Nystrom, G.J.; Lang, C.H. Sepsis and AMPK Activation by AICAR Differentially Regulate FoxO-1, -3 and -4 mRNA in Striated Muscle. *Int. J. Clin. Exp. Med.* **2008**, *1*, 50–63. [PubMed]

44. Nakashima, K.; Yakabe, Y.; Ishida, A.; Katsumata, M. Effects of orally administered glycine on myofibrillar proteolysis and expression of proteolytic-related genes of skeletal muscle in chicks. *Amino Acids* **2008**, *35*, 451–456. [CrossRef] [PubMed]

45. Kim, J.; Kundu, M.; Viollet, B.; Guan, K. AMPK and mTOR regulate autophagy via direct phosphorylation of ULK1. *Nat. Cell Biol.* **2011**, *13*, 132–141. [CrossRef] [PubMed]
46. Egan, D.F.; Shackelford, D.B.; Mihaylova, M.M.; Gelino, S.; Kohnz, R.A.; Mair, W.; Vasquez, D.S.; Joshi, A.; Gwinn, D.M.; Taylor, R.; et al. Phosphorylation of ULK1 (hATG1) by AMP-activated protein kinase connects energy sensing to mitophagy. *Science* **2011**, *331*, 456–461. [CrossRef] [PubMed]
47. Dunlop, E.A.; Tee, A.R. The kinase triad, AMPK, mTORC1 and ULK1, maintains energy and nutrient homoeostasis. *Biochem. Soc. Trans.* **2013**, *41*, 939–943. [CrossRef] [PubMed]
48. Mihaylova, M.M.; Shaw, R.J. The AMPK signalling pathway coordinates cell growth, autophagy and metabolism. *Nat. Cell Biol.* **2011**, *13*, 1016–1023. [CrossRef] [PubMed]
49. Röckl, K.S.; Hirshman, M.F.; Brandauer, J.; Fujii, N.; Witters, L.A.; Goodyear, L.J. Skeletal muscle adaptation to exercise training: AMP-activated protein kinase mediates muscle fiber type shift. *Diabetes* **2007**, *56*, 2062–2069. [CrossRef] [PubMed]
50. McGee, S.L.; Hargreaves, M. AMPK-mediated regulation of transcription in skeletal muscle. *Clin. Sci. (London)* **2010**, *118*, 507–518. [CrossRef] [PubMed]
51. Guo, Y.; Meng, J.; Tang, Y.; Wang, T.; Wei, B.; Feng, R.; Gong, B.; Wang, H.; Ji, G.; Lu, Z. AMP-activated kinase α2 deficiency protects mice from denervation-induced skeletal muscle atrophy. *Arch. Biochem. Biophys.* **2016**, *600*, 56–60. [CrossRef] [PubMed]
52. Gao, H.; Li, Y.F. Distinct signal transductions in fast- and slow- twitch muscles upon denervation. *Physiol. Rep.* **2018**, *6*. [CrossRef] [PubMed]
53. Kostovski, E.; Boon, H.; Hjeltnes, N.; Lundell, L.S.; Ahlsén, M.; Chibalin, A.V.; Krook, A.; Iversen, P.O.; Widegren, U. Altered content of AMP-activated protein kinase isoforms in skeletal muscle from spinal cord injured subjects. *Am. J. Physiol. Endocrinol. Metab.* **2013**, *305*, E1071–E1080. [CrossRef] [PubMed]
54. Mirzoev, T.M.; Vil'chinskaia, N.A.; Lomonosova, I.N.; Nemirovskaia, T.L.; Shenkman, B.S. Effect of 30-day space flight and subsequent readaptation on the signaling processes in m. longissimus dorsi of mice. *Aviakosm. Ekolog. Med.* **2014**, *48*, 12–15. [PubMed]
55. Augusto, V.; Padovani, R.C.; Campos, G.E.R. Skeletal muscle fiber types in C57BL6J mice. *Braz. J. Morphol. Sci.* **2004**, *21*, 89–94.
56. Vilchinskaya, N.A.; Mirzoev, T.M.; Lomonosova, Y.N.; Kozlovskaya, I.B.; Shenkman, B.S. Human muscle signaling responses to 3-day head-out dry immersion. *J. Musculoskelet. Neuronal Interact.* **2015**, *15*, 286–293. [PubMed]
57. Hawley, S.A.; Boudeau, J.; Reid, J.L.; Mustard, K.J.; Udd, L.; Mäkelä, T.P.; Alessi, D.R.; Hardie, D.G. Complexes between the LKB1 tumor suppressor, STRAD alpha/beta and MO25 alpha/beta are upstream kinases in the AMP-activated protein kinase cascade. *J. Biol.* **2003**, *2*, 28. [CrossRef] [PubMed]
58. Hawley, S.A.; Pan, D.A.; Mustard, K.J.; Ross, L.; Bain, J.; Edelman, A.M.; Frenguelli, B.G.; Hardie, D.G. Calmodulin-dependent protein kinase kinase-beta is an alternative upstream kinase for AMP-activated protein kinase. *Cell Metab.* **2005**, *2*, 9–19. [CrossRef] [PubMed]
59. Woods, A.; Vertommen, D.; Neumann, D.; Turk, R.; Bayliss, J.; Schlattner, U.; Wallimann, T.; Carling, D.; Rider, M.H. Identification of phosphorylation sites in AMP-activated protein kinase (AMPK) for upstream AMPK kinases and study of their roles by site-directed mutagenesis. *J. Biol. Chem.* **2003**, *278*, 28434–28442. [CrossRef] [PubMed]
60. Han, B.; Zhu, M.J.; Ma, C.; Du, M. Rat hindlimb unloading down-regulates insulin like growth factor-1 signaling and AMP-activated protein kinase, and leads to severe atrophy of the soleus muscle. *Appl. Physiol. Nutr. Metab.* **2007**, *32*, 1115–1123. [CrossRef] [PubMed]
61. Hilder, T.L.; Baer, L.A.; Fuller, P.M.; Fuller, C.A.; Grindeland, R.E.; Wade, C.E.; Graves, L.M. Insulinindependent pathways mediating glucose uptake in hindlimbsuspended skeletal muscle. *J. Appl. Physiol.* **2005**, *99*, 2181–2188. [CrossRef] [PubMed]
62. Egawa, T.; Goto, A.; Ohno, Y.; Yokoyama, S.; Ikuta, A.; Suzuki, M.; Sugiura, T.; Ohira, Y.; Yoshioka, T.; Hayashi, T.; et al. Involvement of AMPK in regulating slow-twitch muscle atrophy during hindlimb unloading in mice. *Am. J. Physiol. Endocrinol. Metab.* **2015**, *309*, 651–662. [CrossRef] [PubMed]
63. Egawa, T.; Ohno, Y.; Goto, A.; Yokoyama, S.; Hayashi, T.; Goto, K. AMPK Mediates Muscle Mass Change But Not the Transition of Myosin Heavy Chain Isoforms during Unloading and Reloading of Skeletal Muscles in Mice. *Int. J. Mol. Sci.* **2018**, *19*, 2954. [CrossRef] [PubMed]

64. Chibalin, A.V.; Benziane, B.; Zakyrjanova, G.F.; Kravtsova, V.V.; Krivoi, I.I. Early endplate remodeling and skeletal muscle signaling events following rat hindlimb suspension. *J. Cell. Physiol.* **2018**, *233*, 6329–6336. [CrossRef] [PubMed]

65. Zhang, S.F.; Zhang, Y.; Li, B.; Chen, N. Physical inactivity induces the atrophy of skeletal muscle of rats through activating AMPK/FoxO3 signal pathway. *Eur. Rev. Med. Pharmacol. Sci.* **2018**, *22*, 199–209. [PubMed]

66. Grano, M.; Mori, G.; Minielli, V.; Barou, O.; Colucci, S.; Giannelli, G.; Alexandre, C.; Zallone, A.Z.; Vico, L. Rat hindlimb unloading by tail suspension reduces osteoblast differentiation, induces IL-6 secretion, and increases bone resorption in ex vivo cultures. *Calcif. Tissue Int.* **2002**, *70*, 176–185. [CrossRef] [PubMed]

67. Yakabe, M.; Ogawa, S.; Ota, H.; Iijima, K.; Eto, M.; Ouchi, Y.; Akishita, M. Inhibition of interleukin-6 decreases atrogene expression and ameliorates tail suspension-induced skeletal muscle atrophy. *PLoS ONE* **2018**, *13*, e0191318. [CrossRef] [PubMed]

68. Mutin-Carnino, M.; Carnino, A.; Roffino, S.; Chopard, A. Effect of muscle unloading, reloading and exercise on inflammation during a head-down bed rest. *Int. J. Sports Med.* **2014**, *35*, 28–34. [CrossRef] [PubMed]

69. Ruderman, N.B.; Keller, C.; Richard, A.M.; Saha, A.K.; Luo, Z.; Xiang, X.; Giralt, M.; Ritov, V.B.; Menshikova, E.V.; Kelley, D.E.; et al. Interleukin-6 regulation of AMP-activated protein kinase. Potential role in the systemic response to exercise and prevention of the metabolic syndrome. *Diabetes* **2006**, *55*, S48–54. [CrossRef] [PubMed]

70. Yang, W.; Zhang, H. Effects of hindlimb unloading on neurotrophins in the rat spinal cord and soleus muscle. *Brain Res.* **2016**, *1630*, 1–9. [CrossRef] [PubMed]

71. Dupont, E.; Cieniewski-Bernard, C.; Bastide, B.; Stevens, L. Electrostimulation during hindlimb unloading modulates PI3K-AKT downstream targets without preventing soleus atrophy and restores slow phenotype through ERK. *Am. J. Physiol. Regul. Integr. Comp. Physiol.* **2011**, *300*, 408–417. [CrossRef] [PubMed]

72. Bajotto, G.; Sato, Y.; Kitaura, Y.; Shimomura, Y. Effect of branched-chain amino acid supplementation during unloading on regulatory components of protein synthesis in atrophied soleus muscles. *Eur. J. Appl. Physiol.* **2011**, *111*, 1815–1828. [CrossRef] [PubMed]

73. Childs, T.E.; Spangenburg, E.E.; Vyas, D.R.; Booth, F.W. Temporal alterations in protein signaling cascades during recovery from muscle atrophy. *Am. J. Physiol. Cell. Physiol.* **2003**, *285*, 391–398. [CrossRef] [PubMed]

74. Sugiura, T.; Abe, N.; Nagano, M.; Goto, K.; Sakuma, K.; Naito, H.; Yoshioka, T.; Powers, S.K. Changes in PKB/Akt and calcineurin signaling during recovery in atrophied soleus muscle induced by unloading. *Am. J. Physiol. Regul. Integr. Comp. Physiol.* **2005**, *288*, 1273–1278. [CrossRef] [PubMed]

75. Gwag, T.; Lee, K.; Ju, H.; Shin, H.; Lee, J.W.; Choi, I. Stress and signaling responses of rat skeletal muscle to brief endurance exercise during hindlimb unloading: A catch-up process for atrophied muscle. *Cell Physiol. Biochem.* **2009**, *24*, 537–546. [CrossRef] [PubMed]

76. Lysenko, E.A.; Turtikova, O.V.; Kachaeva, E.V.; Ushakov, I.B.; Shenkman, B.S. Time course of ribosomal kinase activity during hindlimb unloading. *Dokl. Biochem. Biophys.* **2010**, *434*, 223–226. [CrossRef] [PubMed]

77. Hamilton, D.L.; Philp, A.; MacKenzie, M.G.; Patton, A.; Towler, M.C.; Gallagher, I.J.; Bodine, S.C.; Baar, K. Molecular brakes regulating mTORC1 activation in skeletal muscle following synergist ablation. *Am. J. Physiol. Endocrinol. Metab.* **2014**, *307*, E365–E373. [CrossRef] [PubMed]

78. Xu, J.; Ji, J.; Yan, X.H. Cross-talk between AMPK and mTOR in regulating energy balance. *Crit. Rev. Food. Sci. Nutr.* **2012**, *52*, 373–381. [CrossRef] [PubMed]

79. McGee, S.L.; Mustard, K.J.; Hardie, D.G.; Baar, K. Normal hypertrophy accompanied by phosphoryation and activation of AMP-activated protein kinase alpha1 following overload in LKB1 knockout mice. *J. Physiol.* **2008**, *586*, 1731–1741. [CrossRef] [PubMed]

80. Mounier, R.; Lantier, L.; Leclerc, J.; Sotiropoulos, A.; Pende, M.; Daegelen, D.; Sakamoto, K.; Foretz, M.; Viollet, B. Important role for AMPKalpha1 in limiting skeletal muscle cell hypertrophy. *FASEB J.* **2009**, *23*, 2264–2273. [CrossRef] [PubMed]

81. Hornberger, T.A.; Hunter, R.B.; Kandarian, S.C.; Esser, K.A. Regulation of translation factors during hindlimb unloading and denervation of skeletal muscle in rats. *Am. J. Physiol. Cell Physiol.* **2001**, *281*, 179–187. [CrossRef] [PubMed]

82. Krawiec, B.J.; Nystrom, G.J.; Frost, R.A.; Jefferson, L.S.; Lang, C.H. AMP-activated protein kinase agonists increase mRNA content of the muscle-specific ubiquitin ligases MAFbx and MuRF1 in C2C12 cells. *Am. J. Physiol. Endocrinol. Metab.* **2007**, *292*, E1555–E1567. [CrossRef] [PubMed]

83. Vilchinskaya, N.A.; Mochalova, E.P.; Belova, S.P.; Shenkman, B.S. Dephosphorylation of AMP-activated protein kinase in a postural muscle: A key signaling event on the first day of functional unloading. *Biophysics* **2016**, *61*, 1019–1025. [CrossRef]

84. Hsieh, C.T.; Chuang, J.H.; Yang, W.C.; Yin, Y.; Lin, Y. Ceramide inhibits insulin-stimulated Akt phosphorylation through activation of Rheb/mTORC1/S6K signaling in skeletal muscle. *Cell. Signal.* **2014**, *26*, 1400–1408. [CrossRef] [PubMed]

85. Erickson, K.A.; Smith, M.E.; Anthonymuthu, T.S.; Evanson, M.J.; Brassfield, E.S.; Hodson, A.E.; Bressler, M.A.; Tucker, B.J.; Thatcher, M.O.; Prince, J.T.; et al. AICAR inhibits ceramide biosynthesis in skeletal muscle. *Diabetol. Metab. Syndr.* **2012**, *4*, 45. [CrossRef] [PubMed]

86. Bryndina, I.G.; Shalagina, M.N.; Ovechkin, S.V.; Ovchinina, N.G. Sphingolipids in skeletal muscles of C57Bl/6 mice after short-term simulated microgravity. *Rossiiskii fiziologicheskii zhurnal imeni IM Sechenova* **2014**, *100*, 1280–1286.

87. Salaun, E.; Lefeuvre-Orfila, L.; Cavey, T.; Martin, B.; Turlin, B.; Ropert, M.; Loreal, O.; Derbré, F. Myriocin prevents muscle ceramide accumulation but not muscle fiber atrophy during short-term mechanical unloading. *J. Appl. Physiol.* **2016**, *120*, 178–187. [CrossRef] [PubMed]

88. Shenkman, B.; Vilchinskaya, N.; Mochalova, E.; Belova, S.; Nemirovskaya, T. Signaling events at the early stage of muscle disuse. *FEBS J.* **2017**, *284*, 367. [CrossRef]

89. Nakao, R.; Hirasaka, K.; Goto, J.; Ishidoh, K.; Yamada, C.; Ohno, A.; Okumura, Y.; Nonaka, I.; Yasutomo, K.; Baldwin, K.M.; et al. Ubiquitin ligase Cbl-b is a negative regulator for insulin-like growth factor 1 signaling during muscle atrophy caused by unloading. *Mol. Cell. Biol.* **2009**, *29*, 4798–4811. [CrossRef] [PubMed]

90. Shenkman, B.S.; Nemirovskaya, T.L. Calcium-dependent signaling mechanisms and soleus fiber remodeling under gravitational unloading. *J. Muscle Res. Cell Motil.* **2008**, *29*, 221–230. [CrossRef] [PubMed]

91. Fitts, R.H.; Riley, D.R.; Widrick, J.J. Physiology of a microgravity environment invited review: Microgravity and skeletal muscle. *J. Appl. Physiol.* **2000**, *89*, 823–839. [CrossRef] [PubMed]

92. Templeton, G.H.; Sweeney, H.L.; Timson, B.F.; Padalino, M.; Dudenhoeffer, G.A. Changes in fiber composition of soleus muscle during rat hindlimb suspension. *J. Appl. Physiol.* **1988**, *65*, 1191–1195. [CrossRef] [PubMed]

93. Desplanches, D.; Mayet, M.H.; Sempore, B.; Flandrois, R. Structural and functional responses to prolonged hindlimb suspension in rat muscle. *J. Appl. Physiol.* **1987**, *63*, 558–563. [CrossRef] [PubMed]

94. Desplanches, D.; Kayar, S.R.; Sempore, B.; Flandrois, R.; Hoppeler, H. Rat soleus muscle ultrastructure after hindlimb suspension. *J. Appl. Physiol.* **1990**, *69*, 504–508. [CrossRef] [PubMed]

95. Riley, D.A.; Slocum, G.R.; Bain, J.L.; Sedlak, F.R.; Sedlak, F.R.; Sowa, T.E.; Mellender, J.W. Rat hindlimb unloading: Soleus histochemistry, ultrastructure, and electromyography. *J. Appl. Physiol.* **1990**, *69*, 58–66. [CrossRef] [PubMed]

96. Trappe, S.; Costill, D.; Gallagher, P.; Creer, A.; Peters, J.R.; Evans, H.; Riley, D.A.; Fitts, R.H. Exercise in space: Human skeletal muscle after 6 months aboard the International Space Station. *J. Appl. Physiol.* **2009**, *106*, 1159–1168. [CrossRef] [PubMed]

97. Liu, Y.; Randall, W.R.; Martin, F.; Schneider, M.F. Activity-dependent and -independent nuclear fluxes of HDAC4 mediated by different kinases in adult skeletal muscle. *J. Cell Biol.* **2005**, *168*, 887–897. [CrossRef] [PubMed]

98. Vilchinskaya, N.A.; Mochalova, E.P.; Nemirovskaya, T.L.; Mirzoev, T.M.; Turtikova, O.V.; Shenkman, B.S. Rapid decline in MyHC I(β) mRNA expression in rat soleus during hindlimb unloading is associated with AMPK dephosphorylation. *J. Physiol.* **2017**, *595*, 7123–7134. [CrossRef] [PubMed]

99. Pandorf, C.E.; Haddad, F.; Wright, C.; Bodell, P.W.; Baldwin, K.M. Differential epigenetic modifications of histones at the myosin heavy chain genes in fast and slow skeletal muscle fibers and in response to muscle unloading. *Am. J. Physiol. Cell Physiol.* **2009**, *297*, 6–16. [CrossRef] [PubMed]

100. Cohen, T.J.; Choi, M.C.; Kapur, M.; Lira, V.A.; Lira, V.A.; Yan, Z.; Yao, T.P. HDAC4 regulates muscle fiber type-specific gene expression programs. *Mol. Cells.* **2015**, *38*, 343–348. [CrossRef] [PubMed]

101. Vilchinskaya, N.A.; Turtikova, O.V.; Shenkman, B.S. Regulation of the Nuclear–Cytoplasmic Traffic of Class IIa Histone Deacetylases in Rat Soleus Muscle at the Early Stage of Gravitational Unloading. *Biol. Membr.* **2017**, *34*, 109–115. [CrossRef]

102. Yoshihara, T.; Machida, S.; Kurosaka, Y.; Kakigi, R.; Sugiura, T.; Naito, H. Immobilization induces nuclear accumulation of HDAC4 in rat skeletal muscle. *J. Physiol. Sci.* **2016**, *66*, 337–343. [CrossRef] [PubMed]

103. McGee, S.L.; Swinton, C.; Morrison, S.; Gaur, V.; Gaur, V.; Campbell, D.E.; Jorgensen, S.B.; Kemp, B.E.; Baar, K.; Steinberg, G.R.; Hargreaves, M. Compensatory regulation of HDAC5 in muscle maintains metabolic adaptive responses and metabolism in response to energetic stress. *Faseb, J.* **2014**, *28*, 3384–3395. [CrossRef] [PubMed]

104. Ya, F.; Rubin, C.S. Protein kinase D: Coupling extracellular stimuli to the regulation of cell physiology. *EMBO Rep.* **2011**, *12*, 785–796.

105. Gan, Z.; Rumsey, J.; Hazen, B.C.; Lai, L.; Leone, T.C.; Vega, R.B.; Xie, H.; Conley, KE.; Auwerx, J.; Smith, S.R.; et al. Nuclear receptor/microRNA circuitry links muscle fiber type to energy metabolism. *J. Clin. Investig.* **2013**, *123*, 2564–2575. [CrossRef] [PubMed]

106. Cannavino, J.; Brocca, L.; Sandri, M.; Bottinelli, R.; Pellegrino, M.A. PGC1-α over-expression prevents metabolic alterations and soleus muscle atrophy in hindlimb unloaded mice. *J. Physiol.* **2014**, *592*, 4575–4589. [CrossRef] [PubMed]

107. Sharlo, C. A.; Lomonosova, Y. N.; Turtikova, O.V.; Mitrofanova, O.V.; Kalamkarov, G.R.; Bugrova, A.E.; Shevchenko, T.F. The Role of GSK-3β Phosphorylation in the Regulation of Slow Myosin Expression in Soleus Muscle during Functional Unloading. *Biochem. (Mosc.) Suppl. Ser. A Membr. Cell Biol.* **2018**, *1*, 85–91. [CrossRef]

108. Wood, S.J.; Slater, C.R. Safety factor at the neuromuscular junction. *Prog. Neurobiol.* **2001**, *64*, 393–429. [CrossRef]

109. Ruff, R.L. Endplate contributions to the safety factor for neuromuscular transmission. *Muscle Nerve* **2011**, *44*, 854–861. [CrossRef] [PubMed]

110. Cheng, A.; Morsch, M.; Murata, Y.; Ghazanfari, N.; Reddel, S.W.; Phillips, W.D. Sequence of age-associated changes to the mouse neuromuscular junction and the protective effects of voluntary exercise. *PLoS ONE* **2013**, *8*, e67970. [CrossRef] [PubMed]

111. Carnio, S.; LoVerso, F.; Baraibar, M.A.; Longa, E.; Khan, M.M.; Maffei, M.; Reischl, M.; Canepari, M.; Loefler, S.; Kern, H.; et al. Autophagy impairment in muscle induces neuromuscular junction degeneration and precocious aging. *Cell Rep.* **2014**, *8*, 1509–1521. [CrossRef] [PubMed]

112. Willadt, S.; Nash, M.; Slater, C.R. Age-related fragmentation of the motor endplate is not associated with impaired neuromuscular transmission in the mouse diaphragm. *Sci. Rep.* **2016**, *6*, 24849. [CrossRef] [PubMed]

113. Lee, K.M.; Chand, K.K.; Hammond, L.A.; Lavidis, N.A.; Noakes, P.G. Functional decline at the aging neuromuscular junction is associated with altered laminin-α4 expression. *Aging* **2017**, *9*, 880–899. [CrossRef] [PubMed]

114. Hughes, D.C.; Marcotte, G.R.; Marshall, A.G.; West, D.W.D.; Baehr, L.M.; Wallace, M.A.; Saleh, P.M.; Bodine, S.C.; Baar, K. Age-related differences in dystrophin: Impact on force transfer proteins, membrane integrity, and neuromuscular junction stability. *J. Gerontol. A Biol. Sci. Med. Sci.* **2017**, *72*, 640–648. [CrossRef] [PubMed]

115. Rudolf, R.; Khan, M.M.; Labeit, S.; Deschenes, M.R. Degeneration of neuromuscular junction in age and dystrophy. *Front. Aging Neurosci.* **2014**, *6*, 99. [CrossRef] [PubMed]

116. Pratt, S.J.; Valencia, A.P.; Le, G.K.; Shah, S.B.; Lovering, R.M. Pre- and postsynaptic changes in the neuromuscular junction in dystrophic mice. *Front. Physiol.* **2015**, *6*, 252. [CrossRef] [PubMed]

117. van der Pijl, E.M.; van Putten, M.; Niks, E.H.; Verschuuren, J.J.; Aartsma-Rus, A.; Plomp, J.J. Characterization of neuromuscular synapse function abnormalities in multiple Duchenne muscular dystrophy mouse models. *Eur. J. Neurosci.* **2016**, *43*, 1623–1635. [CrossRef] [PubMed]

118. Falk, D.J.; Todd, A.G.; Lee, S.; Soustek, M.S.; ElMallah, M.K.; Fuller, D.D.; Notterpek, L.; Byrne, B.J. Peripheral nerve and neuromuscular junction pathology in Pompe disease. *Hum. Mol. Genet.* **2015**, *24*, 625–636. [CrossRef] [PubMed]

119. Tu, H.; Zhang, D.; Corrick, R.M.; Muelleman, R.L.; Wadman, M.C.; Li, Y.L. Morphological regeneration and functional recovery of neuromuscular junctions after tourniquet-induced injuries in mouse hindlimb. *Front. Physiol.* **2017**, *8*, 207. [CrossRef] [PubMed]

120. Yampolsky, P.; Pacifici, P.G.; Witzemann, V. Differential muscle-driven synaptic remodeling in the neuromuscular junction after denervation. *Eur. J. Neurosci.* **2010**, *31*, 646–658. [CrossRef] [PubMed]

121. Deschenes, M.R.; Wilson, M.H. Age-related differences in synaptic plasticity following muscle unloading. *J. Neurobiol.* **2003**, *57*, 246–256. [CrossRef] [PubMed]

122. Nishimune, H.; Stanford, J.A.; Mori, Y. Role of exercise in maintaining the integrity of the neuromuscular junction. *Muscle Nerve* **2014**, *49*, 315–324. [CrossRef] [PubMed]

123. Gonzalez-Freire, M.; de Cabo, R.; Studenski, S.A.; Ferrucci, L. The neuromuscular junction: Aging at the crossroad between nerves and muscle. *Front. Aging Neurosci.* **2014**, *6*, 208. [CrossRef] [PubMed]

124. Tintignac, L.A.; Brenner, H.R.; Rüegg, M.A. Mechanisms regulating neuromuscular junction development and function and causes of muscle wasting. *Physiol. Rev.* **2015**, *95*, 809–852. [CrossRef] [PubMed]

125. Wang, J.; Wang, F.; Zhang, P.; Liu, H.; He, J.; Zhang, C.; Fan, M.; Chen, X. PGC-1α over-expression suppresses the skeletal muscle atrophy and myofiber-type composition during hindlimb unloading. *Biosci. Biotechnol. Biochem.* **2017**, *81*, 500–513. [CrossRef] [PubMed]

126. Martin-Rincon, M.; Morales-Alamo, D.; Calbet, J.A.L. Exercise-mediated modulation of autophagy in skeletal muscle. *Scand. J. Med. Sci. Sports* **2018**, *28*, 772–781. [CrossRef] [PubMed]

127. Khan, M.M.; Strack, S.; Wild, F.; Hanashima, A.; Gasch, A.; Brohm, K.; Reischl, M.; Carnio, S.; Labeit, D.; Sandri, M.; et al. Role of autophagy, SQSTM1, SH3GLB1, and TRIM63 in the turnover of nicotinic acetylcholine receptors. *Autophagy* **2014**, *10*, 123–136. [CrossRef] [PubMed]

128. Cerveró, C.; Montull, N.; Tarabal, O.; Piedrafita, L.; Esquerda, J.E.; Calderó, J. Chronic treatment with the AMPK agonist AICAR prevents skeletal muscle pathology but fails to improve clinical outcome in a mouse model of severe spinal muscular atrophy. *Neurotherapeutics* **2016**, *13*, 198–216. [CrossRef] [PubMed]

129. Pauly, M.; Daussin, F.; Burelle, Y.; Li, T.; Godin, R.; Fauconnier, J.; Koechlin-Ramonatxo, C.; Hugon, G.; Lacampagne, A.; Coisy-Quivy, M.; et al. AMPK activation stimulates autophagy and ameliorates muscular dystrophy in the mdx mouse diaphragm. *Am. J. Pathol.* **2012**, *181*, 583–592. [CrossRef] [PubMed]

130. Ljubicic, V.; Jasmin, B.J. AMP-activated protein kinase at the nexus of therapeutic skeletal muscle plasticity in Duchenne muscular dystrophy. *Trends Mol. Med.* **2013**, *19*, 614–624. [CrossRef] [PubMed]

131. Habegger, K.M.; Hoffman, N.J.; Ridenour, C.M.; Brozinick, J.T.; Elmendorf, J.S. AMPK enhances insulin-stimulated GLUT4 regulation via lowering membrane cholesterol. *Endocrinology* **2012**, *153*, 2130–2141. [CrossRef] [PubMed]

132. Ambery, A.G.; Tackett, L.; Penque, B.A.; Brozinick, J.T.; Elmendorf, J.S. Exercise training prevents skeletal muscle plasma membrane cholesterol accumulation, cortical actin filament loss, and insulin resistance in C57BL/6J mice fed a western-style high-fat diet. *Physiol. Rep.* **2017**, *5*, e13363. [CrossRef] [PubMed]

133. Brannigan, G.; Hénin, J.; Law, R.; Eckenhoff, R.; Klein, M.L. Embedded cholesterol in the nicotinic acetylcholine receptor. *Proc. Natl. Acad. Sci. USA* **2008**, *105*, 14418–14423. [CrossRef] [PubMed]

134. Zhu, D.; Xiong, W.C.; Mei, L. Lipid rafts serve as a signaling platform for nicotinic acetylcholine receptor clustering. *J. Neurosci.* **2006**, *26*, 4841–4851. [CrossRef] [PubMed]

135. Willmann, R.; Pun, S.; Stallmach, L.; Sadasivam, G.; Santos, A.F.; Caroni, P.; Fuhrer, C. Cholesterol and lipid microdomains stabilize the postsynapse at the neuromuscular junction. *EMBO J.* **2006**, *25*, 4050–4060. [CrossRef] [PubMed]

136. Chen, Y.; Li, X.; Ye, Q.; Tian, J.; Jing, R.; Xie, Z. Regulation of α1 Na/K-ATPase expression by cholesterol. *J. Biol. Chem.* **2011**, *286*, 15517–15524. [CrossRef] [PubMed]

137. Kapri-Pardes, E.; Katz, A.; Haviv, H.; Mahmmoud, Y.; Ilan, M.; Khalfin-Penigel, I.; Carmeli, S.; Yarden, O.; Karlish, S.J.D. Stabilization of the α2 isoform of Na,K-ATPase by mutations in a phospholipid binding pocket. *J. Biol. Chem.* **2011**, *286*, 42888–42899. [CrossRef] [PubMed]

138. Haviv, H.; Habeck, M.; Kanai, R.; Toyoshima, C.; Karlish, S.J. Neutral phospholipids stimulate Na,K-ATPase activity: A specific lipid-protein interaction. *J. Biol. Chem.* **2013**, *288*, 10073–10081. [CrossRef] [PubMed]

139. Cornelius, F.; Habeck, M.; Kanai, R.; Toyoshima, C.; Karlish, S.J. General and specific lipid-protein interactions in Na,K-ATPase. *Biochim. Biophys. Acta* **2015**, *1848*, 1729–1743. [CrossRef] [PubMed]

140. Kravtsova, V.V.; Petrov, A.M.; Vasiliev, A.N.; Zefirov, A.L.; Krivoi, I.I. Role of cholesterol in the maintenance of endplate electrogenesis in rat diaphragm. *Bull. Exp. Biol. Med.* **2015**, *158*, 298–300. [CrossRef] [PubMed]

141. Heiny, J.A.; Kravtsova, V.V.; Mandel, F.; Radzyukevich, T.L.; Benziane, B.; Prokofiev, A.V.; Pedersen, S.E.; Chibalin, A.V.; Krivoi, I.I. The nicotinic acetylcholine receptor and the Na,K-ATPase α2 isoform interact to regulate membrane electrogenesis in skeletal muscle. *J. Biol. Chem.* **2010**, *285*, 28614–28626. [CrossRef] [PubMed]

142. Chibalin, A.V.; Heiny, J.A.; Benziane, B.; Prokofiev, A.V.; Vasiliev, A.N.; Kravtsova, V.V.; Krivoi, I.I. Chronic nicotine exposure modifies skeletal muscle Na,K-ATPase activity through its interaction with the nicotinic acetylcholine receptor and phospholemman. *PLoS ONE* **2012**, *7*, e33719. [CrossRef] [PubMed]

143. Matchkov, V.V.; Krivoi, I.I. Specialized functional diversity and interactions of the Na,K-ATPase. *Front. Physiol.* **2016**, *7*, 179. [CrossRef] [PubMed]

International Journal of
Molecular Sciences

MDPI

Article

AMPK Mediates Muscle Mass Change But Not the Transition of Myosin Heavy Chain Isoforms during Unloading and Reloading of Skeletal Muscles in Mice

Tatsuro Egawa [1,2,3,*], Yoshitaka Ohno [4], Ayumi Goto [1,2], Shingo Yokoyama [4], Tatsuya Hayashi [2] and Katsumasa Goto [1,4]

[1] Department of Physiology, Graduate School of Health Sciences, Toyohashi SOZO University, Toyohashi, Aichi 440-8511, Japan; ayumi.goto8@gmail.com (A.G.); gotok@sepia.ocn.ne.jp (K.G.)
[2] Laboratory of Sports and Exercise Medicine, Graduate School of Human and Environmental Studies, Kyoto University, Kyoto 606-8501, Japan; tatsuya@kuhp.kyoto-u.ac.jp
[3] Laboratory of Health and Exercise Sciences, Graduate School of Human and Environmental Studies, Kyoto University, Kyoto 606-8501, Japan
[4] Laboratory of Physiology, School of Health Sciences, Toyohashi SOZO University, Toyohashi, Aichi 440-8511, Japan; yohno@sozo.ac.jp (Y.O.); s-yokoyama@sozo.ac.jp (S.Y.)
* Correspondence: egawa.tatsuro.4u@kyoto-u.ac.jp; Tel.: +81-75-753-6613, Fax: +81-75-753-6885

Received: 27 August 2018; Accepted: 26 September 2018; Published: 27 September 2018

Abstract: $5'$AMP-activated protein kinase (AMPK) plays an important role in the regulation of skeletal muscle mass and fiber-type distribution. However, it is unclear whether AMPK is involved in muscle mass change or transition of myosin heavy chain (MyHC) isoforms in response to unloading or increased loading. Here, we checked whether AMPK controls muscle mass change and transition of MyHC isoforms during unloading and reloading using mice expressing a skeletal-muscle-specific dominant-negative AMPKα1 (AMPK-DN). Fourteen days of hindlimb unloading reduced the soleus muscle weight in wild-type and AMPK-DN mice, but reduction in the muscle mass was partly attenuated in AMPK-DN mice. There was no difference in the regrown muscle weight between the mice after 7 days of reloading, and there was concomitantly reduced AMPKα2 activity, however it was higher in AMPK-DN mice after 14 days reloading. No difference was observed between the mice in relation to the levels of slow-type MyHC I, fast-type MyHC IIa/x, and MyHC IIb isoforms following unloading and reloading. The levels of 72-kDa heat-shock protein, which preserves muscle mass, increased in AMPK-DN-mice. Our results indicate that AMPK mediates the progress of atrophy during unloading and regrowth of atrophied muscles following reloading, but it does not influence the transition of MyHC isoforms.

Keywords: atrophy; regrowth; sirtuin 1 (SIRT1); peroxisome proliferator-activated receptor gamma coactivator 1-α (PGC1α); heat shock protein; fiber-type

1. Introduction

The skeletal muscle is the largest organ in the body and plays a crucial role in metabolism. Loss of skeletal muscle function and mass leads to disorders such as sarcopenia and insulin resistance [1]. Muscle loading is a vital process in the regulation of skeletal muscle properties. Increased loading induced by mechanical stretch or strength exercises leads to muscle hypertrophy and regrowth of atrophied skeletal muscles [2–4]. By contrast, unloading, as well as inactivity, causes skeletal muscle atrophy, especially that of the antigravitational slow-twitch muscles [5,6].

$5'$AMP-activated protein kinase (AMPK) is a central regulator of cellular metabolism and energy homeostasis in mammalian tissues. AMPK plays an important role in the regulation of skeletal muscle mass; AMPK inhibits hypertrophy of skeletal muscle cells [7–9] and rodent skeletal muscle [7,10], and AMPK activity negatively correlates the degree of hypertrophy in rat skeletal muscle [11,12]. We have previously shown that AMPK regulates unloading-induced skeletal muscle atrophy [13]. However, it remains unclear whether AMPK plays a role in the regrowth of atrophied skeletal muscles in response to increased loading.

The skeletal muscle fibers in mammals can be roughly divided into slow- and fast-twitch types, which are further classified into type I, type IIa, type IIx, and type IIb. The four muscle fibers, respectively, contain protein isoforms of myosin heavy chain (MyHC) I, IIa, IIx, and IIb [14]. The slow-to-fast transition of MyHC isoforms is observed in unloading-associated atrophied slow soleus muscles [4,15]. In this regard, AMPK is a potential regulator of skeletal muscle fiber-type distribution. Training-induced increases in MyHC IIa/x isoforms is attenuated in AMPKα2-deficient mice [16]. Chronic administration of an AMPK activator promotes a switch to type I fibers in the skeletal muscles of rodents [16–18]. Furthermore, AMPK triggers oxidative adaptation and is involved in training-induced fiber-type shift [17]. However, to date, it remains unclear whether AMPK is involved in the transition of MyHC isoforms during unloading and reloading.

AMPK controls skeletal muscle plasticity through a variety of molecular responses. Interaction of AMPK with sirtuin 1 (SIRT1) and peroxisome proliferator-activated receptor gamma coactivator 1-α (PGC1α) comprises a pivotal regulatory network in metabolic homeostasis [19]. In terms of muscle mass regulation, the AMPK–SIRT1 axis acts as a sensor of nutrient availability and regulates muscle development [20], and pharmacological AMPK activation inhibits muscle cell growth through a PGC1α-dependent mechanism [21]. In addition, we recently showed that the interaction of AMPK with 72-kDa heat-shock protein (HSP72) is associated with hypertrophy as well as atrophy of skeletal muscles [9,13]. Although AMPK's regulation of muscle mass is thus clear, no previous study, to our knowledge, has examined the association between AMPK and SIRT1, PGC1α, and HSP72 during unloading-induced atrophy or regrowth following atrophy.

Here, we investigated the role of AMPK in the changes in muscle mass and transition of MyHC isoforms, and its associated molecular responses during unloading and reloading. We performed 14-day hindlimb suspension and 14-day ambulation recovery procedures with wild-type littermate (WT) mice and transgenic mice that overexpressed a muscle-specific dominant-negative mutant of AMPKα1 (AMPK-DN) (Figure 1) [22]. AMPK is a heterotrimeric kinase, consisting of a catalytic α-subunit and two regulatory subunits, β and γ. Two distinct α-isoforms (α1 and α2) exist in skeletal muscle. These mice exhibit almost complete depletion in AMPKα2 activity and moderate depletion in AMPKα1 activity [22–25].

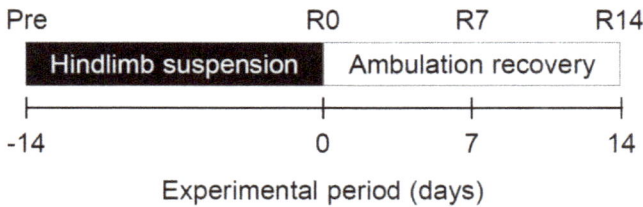

Figure 1. Summary of the experimental protocol. The hindlimbs of both dominant-negative mutant of AMPK (AMPK-DN) and wild-type littermate (WT) mice were continuously suspended for 14 days. After 14 days, the mice were allowed ambulation recovery. Pre: before hindlimb suspension; R0, R7, and R14: 0, 7, and 14 days after ambulation recovery; *n* = 8 per group.

2. Results

2.1. 5′AMP-Activated Protein Kinase (AMPK) Activity

Changes in isoform-specific AMPK activities during unloading and reloading are shown in Figure 2. AMPKα1 activity was lower in AMPK-DN mice than that of WT mice during the overall experimental period (genotype effect, $p < 0.05$), but there was no time effect in AMPKα1 activity (Figure 2A). AMPKα2 activity was almost completely suppressed in AMPK-DN mice compared with that in WT mice during the overall experimental period (genotype effect, $p < 0.05$) (Figure 2B). AMPKα2 activity decreased in response to 7 days of reloading, and remained suppressed after 14 days of reloading (Figure 2B).

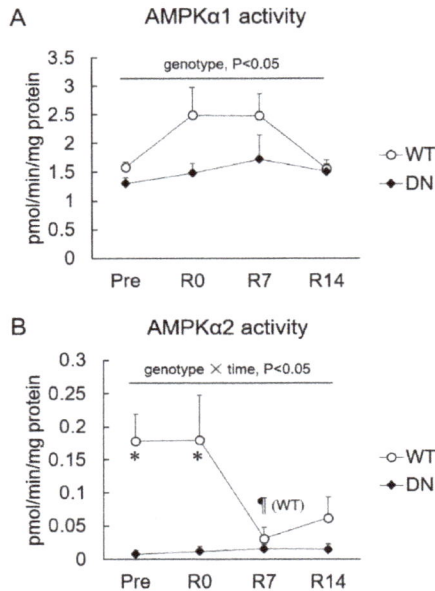

Figure 2. Changes in the isoform-specific 5′AMP-activated protein kinase (AMPK) activity after 14 days of hindlimb unloading, and at 0, 7, and 14 days of reloading. (**A**) AMPKα1 activity. (**B**) AMPKα2 activity. Values are means ± SE; $n = 7$–8 per group. Statistical results of two-way ANOVA (genotype, time, and genotype × time) are described in the Figure. *, significant difference between genotypes at same time point. ¶, significant difference from R0 independent of genotype, unless special mention in Figure.

2.2. Body Weight and Soleus Muscle Weight

Changes in body weight and soleus muscle weight relative to the body weight during unloading and reloading are shown in Figure 3. The body weight was lower in AMPK-DN mice than that in WT mice during the overall experimental period (genotype effect, $p < 0.05$) (Figure 3A). The body weight of both mice decreased in response to 14 days of unloading and was restored after 7 days of reloading (Figure 3A). The soleus muscle weight relative to the body weight also decreased following the 14-day unloading, and recovered after 7 and 14 days of reloading for both groups, but the reduction was attenuated in AMPK-DN mice compared with that in WT mice (Figure 3B). There was no difference in the muscle weight of both mice groups after 7 days of reloading, but it was higher in AMPK-DN mice after 14 days reloading (Figure 3B).

Int. J. Mol. Sci. **2018**, *19*, 2954

A
body weight

B
muscle weight/body weight

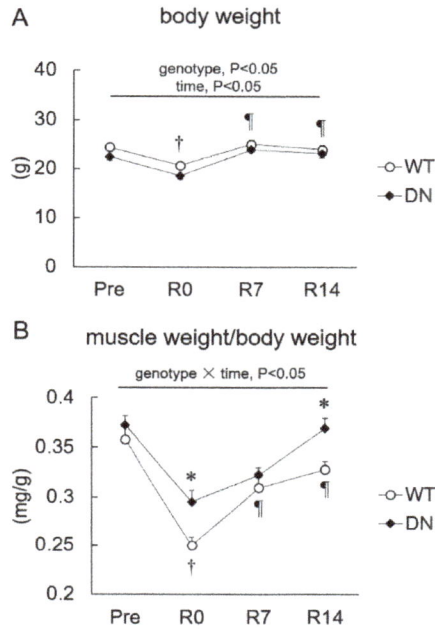

Figure 3. Changes in the body weight and soleus weight after 14 days of hindlimb unloading, and at 0, 7, and 14 days of reloading. (**A**) Body weight. (**B**) Relative soleus weight to body weight. Values are means ± SE; $n = 8$ per group. Statistical results of two-way ANOVA (genotype, time, and genotype × time) are described in the Figure. *, significant difference between genotypes at same time point. †, significant difference from Pre independent of genotype. ¶, significant difference from R0 independent of genotype.

2.3. Levels of Slow- and Fast-Type Myosin Heavy Chain (MyHC) Isoforms

We examined the levels of slow-type I and fast-type IIa, IIx, and IIb MyHC isoforms in the soleus muscle (Figure 4). Owing to a technical limitation, IIa and IIx MyHC phenotypes were expressed as IIa/x. MyHC I levels decreased following 14 days of unloading, but there was no difference between their levels in WT and AMPK-DN mice (Figure 4A). The MyHC IIa/x level was lower in AMPK-DN mice than that in WT mice during the overall experimental period (genotype effect, $p < 0.05$), but no change was observed following unloading and reloading (Figure 3B). The MyHC IIb level was higher in AMPK-DN mice than that in WT mice during the overall experimental period (genotype effect, $p < 0.05$) (Figure 4C). The levels did not change by unloading, but increased after seven days of reloading in both mice groups (Figure 4C).

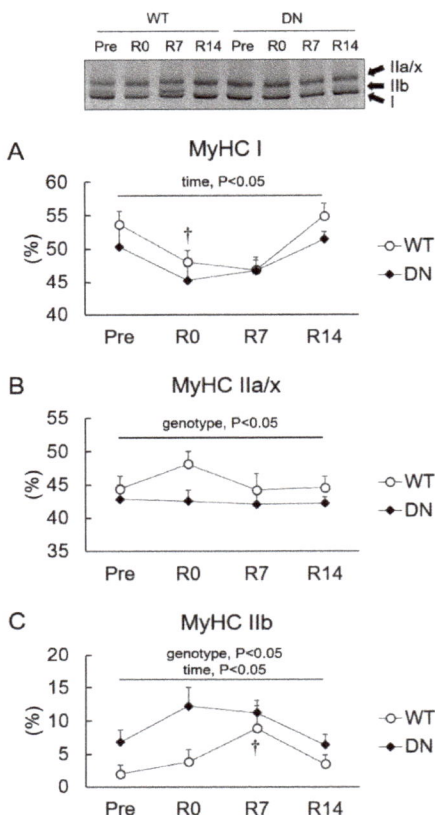

Figure 4. Changes in relative levels of myosin heavy chain (MyHC) isoforms in the soleus muscles after 14 days of hindlimb unloading, and at 0, 7, and 14 days of reloading. (**A**) MyHC I. (**B**) MyHC IIa/x. (**C**) MyHC IIb. Representative image is shown. Values are means ± SE; $n = 8$ per group. Statistical results of two-way ANOVA (genotype, time, and genotype × time) are described in the Figure. †, significant difference from Pre independent of genotype.

2.4. Sirtuin 1 (SIRT1) Activity and Peroxisome Proliferator-Activated Receptor Gamma Coactivator 1-α (PGC1α) mRNA Levels

To clarify the role of SIRT1 and PGC1α in the changes in the AMPK-mediated phenotype during muscle mass change, SIRT1 activity and PGC1α mRNA levels were examined (Figure 5). SIRT1 activity increased following 7 days of reloading, but it was similar in the WT and AMPK-DN mice (Figure 5A). No difference was observed in the PGC1α mRNA level between WT and AMPK-DN mice during the overall experimental period (Figure 5B). The PGC1α mRNA level was not changed after 14 days of unloading, but it decreased in response to 7 days of reloading in both mice groups (Figure 5B).

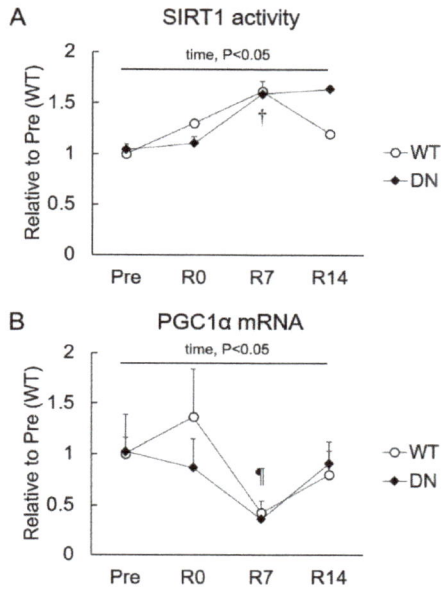

Figure 5. Changes in sirtuin 1 (SIRT1) activity and peroxisome proliferator-activated receptor gamma coactivator 1-alpha (PGC1α) mRNA expression after 14 days of hindlimb unloading, and at 0, 7, and 14 days of reloading. (**A**) SIRT1 activity; $n = 3$ per group. (**B**) PGC1α mRNA; $n = 8$ per group. Values are means ± SE. Statistical results of two-way ANOVA (genotype, time, and genotype × time) are described in the Figure. †, significant difference from Pre independent of genotype. ¶, significant difference from R0 independent of genotype.

2.5. 72-kDa Heat-Shock Protein (HSP72) Levels

HSP72 levels were determined to examine the relationship between HSP72 and AMPK-mediated skeletal muscle atrophy and regrowth (Figure 6). The HSP72 level was higher in AMPK-DN mice than that in WT mice during the overall experimental period (genotype effect, $p < 0.05$) (Figure 6B). HSP72 levels increased following 7 days of reloading, which remained unchanged throughout the 14 days of reloading in both mice groups (Figure 6).

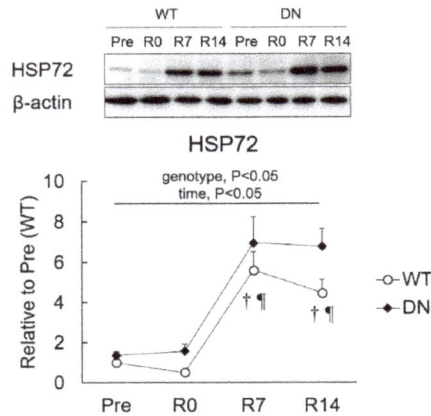

Figure 6. Changes in the 72-kDa heat shock protein (HSP72) expression after 14 days of hindlimb unloading, and at 0, 7, and 14 days of reloading. Representative immunoblots are shown. Values are means ± SE. Statistical results of two-way ANOVA (genotype, time, and genotype × time) are described in the Figure. †, significant difference from Pre independent of genotype. ¶, significant difference from R0 independent of genotype.

3. Discussion

The present study reports the following novel findings with respect to the role of AMPK in muscle mass change and fiber-type shift during unloading and reloading. First, AMPKα2 activity is suppressed in response to the reloading procedure (Figure 2). Second, the regrowth of soleus muscle weight in response to reloading was accelerated in AMPK-DN mice after seven days reloading (Figure 3). Third, AMPK-DN mice showed a higher proportion of MyHC IIb than WT mice, and the slow-to-fast transition of MyHC isoforms was identical in WT and AMPK-DN mice (Figure 4). Fourth, no difference was observed between the mice in response to unloading- and reloading-induced changes of SIRT1 activity and PGC1α mRNA levels (Figure 5). Fifth, AMPK-DN mice showed higher levels of HSP72 (Figure 6) in the soleus muscles than WT mice.

It is accepted that diminished loading leads to skeletal muscle atrophy, and increased loading following unloading induces regrowth [2–6]. Fourteen days of hindlimb unloading showed a 30% decrease in the soleus muscle mass of WT mice, whereas it was attenuated in AMPK-DN mice (Figure 3A). Such atrophic responses of the soleus muscles were in accordance with those seen in our previous study, which used the same procedures [13]. Our findings did not show the upregulation of AMPKα1 and α2 activity in response to 14 days of unloading (Figure 2). A previous study reported that AMPK signaling was activated at the early phase (three days) of unloading and returned to the basal level at 7 days [26]. This may be why we could not detect the upregulation of AMPK activity after 14 days of unloading.

The difference in muscle weight disappeared following seven days reloading. Interestingly, the difference was expanded again after 14 days reloading (Figure 3B). This suggests that the lack of AMPK activity promotes regrowth of atrophied skeletal muscles, especially in the latter phase of recovery. AMPK is known to be a negative regulator of muscle hypertrophy, and this is supported by our findings that AMPKα2 activity is greatly suppressed during the reloading (hypertrophy) phase (Figure 2B). A previous report has shown that AMPK phosphorylation was suppressed by seven days of reloading following unloading in mouse heart muscle [27]. Therefore, reduction of AMPK activity (mainly α2) might contribute to progress in skeletal muscle regrowth. To our knowledge, this is the first study to report on the association of AMPK with regrowth from unloading-induced atrophy of skeletal muscles.

The soleus muscles of adult mice have high levels of slow-type I and fast-type IIa MyHC, and low levels of fast-type IIx or IIb MyHC [4]. Unloading results in a slow-to-fast transition of MyHC isoforms [4,15], whereas reloading reverses it [28]. We saw that 14 days of unloading decreased MyHC I levels and tended to increase MyHC IIa and IIb levels (Figure 4), suggesting that a slow-to-fast transition of MyHC isoforms occurred. In addition, the 14 days of reloading induced a fast-to-slow transition of MyHC. However, the transition occurred identically in WT and AMPK-DN mice. This suggests that AMPK does not play a role in unloading- and reloading-induced transition of MyHC isoforms. On the other hand, we found that AMPK-DN mice exhibited an increased proportion of MyHC IIb (Figure 4C) and decreased proportion of MyHC IIa/x (Figure 4B). These results indicate that lack of AMPK activity (mainly α2) leads to slow-to-fast transition of muscle fiber type, suggesting that AMPK is associated with a regulation of skeletal muscle fiber-type distribution, as has been suggested [16–18].

SIRT1 is a key modulator of cell proliferation, hormone response, stress response, apoptosis, and cell metabolism [29]. AMPK and SIRT1 both regulate each other and share many common target molecules, including PGC1α [30]. Previous studies have shown that AMPK regulates muscle formation through SIRT1- or PGC1α-dependent mechanisms [20,21]. Moreover, SIRT1 and PGC1α both protect skeletal muscles from denervation-induced atrophy [31,32], suggesting that they both are important regulators of skeletal muscle mass. Here, no difference was observed between the WT and AMPK-DN mice in response to the unloading- and reloading-induced change of PGC1α mRNA levels (Figure 5B), and no significant difference in SIRT1 activity between the mice (Figure 5A). However, AMPK-DN mice exhibited higher SIRT1 activity after 14 days of reloading compared with that in WT mice, although this was not statistically significant. Moreover, considering that PGC1α expression and activity are controlled by posttranslational modifications [33], more detailed experiments are needed to make clear the involvement of SIRT1 and PGC1α in AMPK-mediated regrowth of atrophied muscle.

HSP72 is one of the most prominent members of the HSP family. Previous studies have shown that HSP72 levels increase under hypertrophic conditions [34,35], whereas they decrease under atrophic conditions [13,29]. Moreover, we have previously shown that an HSP-dependent mechanism underlies the AMPK-mediated inhibition of skeletal muscle hypertrophy [9]. Thus, HSP72 possibly plays an important role in muscle mass regulation. In the present study, we found that the HSP72 levels in the soleus muscles of AMPK-DN mice were higher than those in WT mice (Figure 6). In this regard, it has been shown that overexpressing HSP72 in transgenic mice prevents immobilization-induced skeletal muscle atrophy [36], and improves skeletal muscle recovery from unloading-induced atrophy [37]. We therefore hypothesized that the protection from atrophy and acceleration of regrowth of skeletal muscles in AMPK-DN mice was partly attributable to the high levels of HSP72.

Previously, we have shown that AMPK controlled hypertrophy and atrophy of skeletal muscle through protein degradation systems [9]. In addition, it has been shown that AMPK maintained muscle cell size through protein synthesis pathways [10]. Skeletal muscle mass is ultimately determined by balancing protein synthesis and degradation, and thus our findings that the suppression of AMPK activity attenuated unloading-induced atrophy and accelerated regrowth of skeletal muscle are probably attributed to changes in protein turnover systems, for example, mammalian target of rapamycin signaling, autophagy, and the ubiquitin-proteasome system. Therefore, further investigations measuring protein turnover systems are required to validate AMPK-mediated muscle mass regulation during unloading and reloading.

In conclusion, our current results indicate that AMPK mediates the progress of atrophy of skeletal muscles during unloading, and regrowth of atrophied muscles following reloading. To the best of our knowledge, this is the first report to show the effect of AMPK on skeletal muscle adaptations following recovery from unloading-induced atrophy. Our findings contribute to understanding the complex molecular responses during loading-associated skeletal muscle adaptations.

4. Materials and Methods

4.1. Animals

Transgenic mice expressing a dominant-negative mutant of AMPKα1 in the skeletal muscles [22] were purchased from the Laboratory Animal Resource Bank at the National Institute of Biomedical Innovation (Resource No. nbio085, Osaka, Japan). Male 12–16 week old AMPK-DN mice ($n = 32$) and WT littermate mice ($n = 32$) with C57BL/6NCr background were used. All mice were housed in an animal room maintained at 22–24 °C with a 12-h light–dark cycle, and were fed a standard laboratory diet with water given ad libitum. All animal-related protocols were performed in accordance with the Guide for the Care and Use of Laboratory Animals as adopted and promulgated by the National Institutes of Health (Bethesda, MD, USA), and were approved by the Animal Use Committee at Toyohashi SOZO University (A2012002, approved 7 August 2012; A2013003, approved 6 August 2013; and A2014003, approved 27 August 2014).

4.2. Procedure of Hindlimb Suspension and Ambulation Recovery

The hindlimbs of both AMPK-DN and WT mice were continuously suspended for 14 days as described previously [38]. After 14 days, the mice were allowed ambulation recovery. Eight mice of each strain were killed at baseline (untreated pre-experimental control: Pre), and at 0 (R0), 7 (R7) and 14 (R14) days of ambulation recovery (Figure 1). Their soleus muscles were dissected under anesthesia with intraperitoneal injection of sodium pentobarbital (50 mg/kg). The muscles were trimmed of excess fat and connective tissues, weighed, frozen in liquid nitrogen, and stored at −80 °C. The left soleus muscle was cross-sectionally sliced into halves at the mid-belly region, and the proximal half of the left soleus muscle was immediately frozen in 2-methylbutane cooled with liquid nitrogen, and stored at −80 °C for immunohistochemical analyses. The distal half of the left soleus muscle was used for real-time RT-PCR analysis, and the right soleus muscle was used for western blotting.

4.3. Sample Preparation and Western Blotting

Sample preparation and western blotting were performed as described previously [34,39]. Briefly, the muscles were homogenized in ice-cold lysis buffer (CelLytic MT, Sigma–Aldrich, St. Louis, MO, USA) containing a protease/phosphatase inhibitor (5872, Cell Signaling Technology, Danvers, MA, USA). The homogenates were then centrifuged at 16,000× g for 15 min at 4 °C. The supernatant was collected and solubilized in Laemmli's sample buffer containing mercaptoethanol and was then boiled. Protein samples (10 µg) were separated by SDS-PAGE using a 10% polyacrylamide gel, after which the proteins were transferred onto polyvinylidene difluoride membranes. Next, the membranes were blocked for 1 h at room temperature in Blocking One-P (Nacalai Tesque, Kyoto, Japan), and then incubated overnight at 4 °C with primary antibodies: HSP72 (ADI-SPA-812, Enzo Life Sciences, Farmingdale, NY, USA), and β-actin (4967, Cell Signaling Technology); and diluted in Tris-buffered saline with 0.1% Tween 20 (TBS-T). The membranes were then washed with TBS-T and treated with anti-rabbit IgG (7074, Cell Signaling Technology) for 1 h at room temperature. After the final wash with TBS-T, protein bands were visualized using chemiluminescence (Wako Pure Chemical Industries, Osaka, Japan). The intensity of the signals was quantified using ImageJ (National Institutes of Health, Bethesda, MD, USA). The level of β-actin was evaluated as an internal control.

4.4. Real-Time RT-PCR Analyses

Real-time RT-PCR analyses were performed as was described previously [35]. Briefly, total RNA was extracted from muscles using the miRNeasy Mini kit (Qiagen, Hilden, Germany) according to the manufacturer's protocol. For the detection of mRNA, the RNA was reverse-transcribed to cDNA using PrimeScript RT Master Mix (Takara Bio, Otsu, Japan), and then synthesized cDNA was applied to real-time RT-PCR (Thermal Cycler Dice Real Time System IIMRQ, Takara Bio) using Takara SYBR Premix Ex Taq II (Takara Bio). Relative fold change of expression was calculated by the comparative CT method. To normalize the amount of total RNA present in each reaction, S18 ribosomal RNA (18S rRNA) was used as an internal standard. The following primers were used: PGC1α (Ppargc1a), 5′-GCTGCATGGTTCTGAGTGCTAAG-3′ (forward) and 5′-AGCCGTGACCACTGACAACGAG-3′ (reverse); 18S rRNA, 3′-ACTCAACACGGGAAACCTCA-5′ (forward) and 3′-AACCAGACAAATCGCTCCAC-5′ (reverse).

4.5. Myosin Heavy Chain (MyHC) Isoform Detection

Analysis of MyHC isoform (I, IIa, IIx, and IIb) levels was performed using a previously described method, albeit with a modification [40,41]. Briefly, the homogenate sample proteins (5 μg) were separated by SDS-PAGE using a 7% polyacrylamide gel at 120 V for 19 h in a temperature-controlled chamber at 4 °C. After electrophoresis, the gels were stained with Oriole™ Fluorescent Gel Stain (Bio-Rad Laboratories, Hercules, CA, USA). The gels were visualized using Light-Capture (AE-6971, ATTO Corporation, Tokyo, Japan) and analyzed using ImageJ.

4.6. Sirtuin 1 (SIRT1) Activity Assay

SIRT1 activity was determined using the SIRT1 Fluorometric Assay Kit (AS-72155, AnaSpec, Fremont, CA, USA) according to the manufacturer's instructions. Fluorescence was measured using a fluorometric reader (Fluoroskan FL, ThermoFisher Scientific, Waltham, MA, USA) with excitation at 490 nm and emission at 520 nm.

4.7. 5′AMP-Activated Protein Kinase (AMPK) Activity Assay

The kinase activities of α1-containing AMPK complex (AMPKα1) and α2-containing AMPK complex (AMPKα2) were measured as described previously [39]. The supernatants from the muscle homogenates (50 μg protein) were incubated with either the anti-α1 or -α2 antibody [42] and Protein A Sepharose beads (Amersham Biosciences, Uppsala, Sweden) at 4 °C overnight. The beads were subjected to the kinase reaction using the SAMS peptide as a substrate.

4.8. Statistical Analyses

All values were expressed as means ± SE. Statistical significance was analyzed using two-way analysis of variance (ANOVA), with genotype (WT and AMPK-DN) and time (Pre, R0, R7, and R14) as the main factors. If there was significant time effect, post hoc multiple-comparison tests were performed among groups (Pre, R0, R7, and R14). If there were any significant interactions (genotype × times), post hoc simple-effects tests were performed. Post hoc analyses were conducted with Tukey–Kramer's test. The differences between groups were considered statistically significant at $p < 0.05$.

Author Contributions: T.E. performed experiments, analyzed the data, and contributed with drafting the manuscript. Y.O. conceived and designed the research. A.G. performed experiments and analyzed the data. S.Y. performed experiments and analyzed the data. T.H. conceived and designed the research, and drafted the manuscript. K.G. conceived and designed the research, analyzed the data, and drafted the manuscript. All authors interpreted the results, contributed to the discussion, edited and revised the manuscript, and read and approved the final version of the manuscript.

Acknowledgments: This study was supported in part by JSPS KAKENHI (Tatsuro Egawa, 26560371 and 18H03148; Yoshitaka Ohno, 16K12942 and 18K10796; Ayumi Goto, 16H07182 and 18K17934; Shingo Yokoyama, 16K16450; Katsumasa Goto, 16K13022, 17K01762, and 18H03160); JSPS Fellows (Ayumi Goto, 14J00286); the Council for Science, Technology and Innovation; and SIP (Funding agency: Bio-oriented Technology Research Advancement Institution, NARO) (Tatsuya Hayashi, 14533567). Additional research grants were provided by the Nakatomi Foundation (TE); the Foundation for Dietary Scientific Research (TE); Takeda Research Support (TH, TKDS20170531015); the Science Research Promotion Fund from the Promotion and Mutual Aid Corporation for Private Schools of Japan; and Graduate School of Health Sciences, Toyohashi SOZO University (KG).

Conflicts of Interest: The authors declare no conflict of interest.

Abbreviations

AMPK	5′AMP-activated protein kinase
AMPK-DN	dominant-negative mutant of AMPKα1
AMPKα1	α1-containing AMPK complex
AMPKα2	α2-containing AMPK complex
ANOVA	analysis of variance
HSP72	72-kDa heat shock protein
MyHC	myosin heavy chain
PGC1α	peroxisome proliferator-activated receptor gamma coactivator 1-alpha
TBS-T	Tris buffered saline with 0.1% Tween 20

References

1. Kalyani, R.R.; Corriere, M.; Ferrucci, L. Age-related and disease-related muscle loss: The effect of diabetes, obesity, and other diseases. *Lancet Diabetes Endocrinol.* **2014**, *2*, 819–829. [CrossRef]
2. Goldspink, G. Changes in muscle mass and phenotype and the expression of autocrine and systemic growth factors by muscle in response to stretch and overload. *J. Anat.* **1999**, *194*, 323–334. [CrossRef] [PubMed]
3. Selsby, J.T.; Rother, S.; Tsuda, S.; Pracash, O.; Quindry, J.; Dodd, S.L. Intermittent hyperthermia enhances skeletal muscle regrowth and attenuates oxidative damage following reloading. *J. Appl. Physiol.* **2007**, *102*, 1702–1707. [CrossRef] [PubMed]
4. Yokoyama, S.; Ohno, Y.; Egawa, T.; Yasuhara, K.; Nakai, A.; Sugiura, T.; Ohira, Y.; Yoshioka, T.; Okita, M.; Origuchi, T.; et al. Heat shock transcription factor 1-associated expression of slow myosin heavy chain in mouse soleus muscle in response to unloading with or without reloading. *Acta Physiol.* **2016**, *217*, 325–337. [CrossRef] [PubMed]
5. Thomason, D.B.; Booth, F.W. Atrophy of the soleus muscle by hindlimb unweighting. *J. Appl. Physiol.* **1990**, *68*, 1–12. [CrossRef] [PubMed]
6. Ohira, Y.; Yoshinaga, T.; Ohara, M.; Kawano, F.; Wang, X.D.; Higo, Y.; Terada, M.; Matsuoka, Y.; Roy, R.R.; Edgerton, V.R. The role of neural and mechanical influences in maintaining normal fast and slow muscle properties. *Cells Tissues Organs* **2006**, *182*, 129–142. [CrossRef] [PubMed]
7. Lantier, L.; Mounier, R.; Leclerc, J.; Pende, M.; Foretz, M.; Viollet, B. Coordinated maintenance of muscle cell size control by AMP-activated protein kinase. *FASEB J.* **2010**, *24*, 3555–3561. [CrossRef] [PubMed]
8. Nakashima, K.; Yakabe, Y. AMPK activation stimulates myofibrillar protein degradation and expression of atrophy-related ubiquitin ligases by increasing FOXO transcription factors in C2C12 myotubes. *Biosci. Biotechnol. Biochem.* **2007**, *71*, 1650–1656. [CrossRef] [PubMed]
9. Egawa, T.; Ohno, Y.; Goto, A.; Ikuta, A.; Suzuki, M.; Ohira, T.; Yokoyama, S.; Sugiura, T.; Ohira, Y.; Yoshioka, T.; et al. AICAR-induced activation of AMPK negatively regulates myotube hypertrophy through the HSP72-mediated pathway in C2C12 skeletal muscle cells. *Am. J. Physiol. Endocrinol. Metab.* **2014**, *306*, E344–E354. [CrossRef] [PubMed]
10. Mounier, R.; Lantier, L.; Leclerc, J.; Sotiropoulos, A.; Pende, M.; Daegelen, D.; Sakamoto, K.; Foretz, M.; Viollet, B. Important role for AMPKalpha1 in limiting skeletal muscle cell hypertrophy. *FASEB J.* **2009**, *23*, 2264–2273. [CrossRef] [PubMed]
11. Paturi, S.; Gutta, A.K.; Kakarla, S.K.; Katta, A.; Arnold, E.C.; Wu, M.; Rice, K.M.; Blough, E.R. Impaired overload-induced hypertrophy in obese Zucker rat slow-twitch skeletal muscle. *J. Appl. Physiol.* **2010**, *108*, 7–13. [CrossRef] [PubMed]
12. Thomson, D.M.; Gordon, S.E. Diminished overload-induced hypertrophy in aged fast-twitch skeletal muscle is associated with AMPK hyperphosphorylation. *J. Appl. Physiol.* **2005**, *98*, 557–564. [CrossRef] [PubMed]
13. Egawa, T.; Goto, A.; Ohno, Y.; Yokoyama, S.; Ikuta, A.; Suzuki, M.; Sugiura, T.; Ohira, Y.; Yoshioka, T.; Hayashi, T.; et al. Involvement of AMPK in regulating slow-twitch muscle atrophy during hindlimb unloading in mice. *Am. J. Physiol. Endocrinol. Metab.* **2015**, *309*, E651–E662. [CrossRef] [PubMed]
14. Pette, D.; Staron, R.S. Myosin isoforms, muscle fiber types, and transitions. *Microsc. Res. Tech.* **2000**, *50*, 500–509. [CrossRef]

15. Haddad, F.; Qin, A.X.; Zeng, M.; McCue, S.A.; Baldwin, K.M. Interaction of hyperthyroidism and hindlimb suspension on skeletal myosin heavy chain expression. *J. Appl. Physiol.* **1998**, *85*, 2227–2236. [CrossRef] [PubMed]

16. Rockl, K.S.; Hirshman, M.F.; Brandauer, J.; Fujii, N.; Witters, L.A.; Goodyear, L.J. Skeletal muscle adaptation to exercise training: AMP-activated protein kinase mediates muscle fiber type shift. *Diabetes* **2007**, *56*, 2062–2069. [CrossRef] [PubMed]

17. Narkar, V.A.; Downes, M.; Yu, R.T.; Embler, E.; Wang, Y.X.; Banayo, E.; Mihaylova, M.M.; Nelson, M.C.; Zou, Y.; Juguilon, H.; et al. AMPK and PPARdelta agonists are exercise mimetics. *Cell* **2008**, *134*, 405–415. [CrossRef] [PubMed]

18. Suwa, M.; Nakano, H.; Kumagai, S. Effects of chronic AICAR treatment on fiber composition, enzyme activity, UCP3, and PGC-1 in rat muscles. *J. Appl. Physiol.* **2003**, *95*, 960–968. [CrossRef] [PubMed]

19. Canto, C.; Auwerx, J. PGC-1alpha, SIRT1 and AMPK, an energy sensing network that controls energy expenditure. *Curr. Opin. Lipidol.* **2009**, *20*, 98–105. [CrossRef] [PubMed]

20. Fulco, M.; Cen, Y.; Zhao, P.; Hoffman, E.P.; McBurney, M.W.; Sauve, A.A.; Sartorelli, V. Glucose restriction inhibits skeletal myoblast differentiation by activating SIRT1 through AMPK-mediated regulation of Nampt. *Dev. Cell* **2008**, *14*, 661–673. [CrossRef] [PubMed]

21. Williamson, D.L.; Butler, D.C.; Alway, S.E. AMPK inhibits myoblast differentiation through a PGC-1alpha-dependent mechanism. *Am. J. Physiol. Endocrinol. Metab.* **2009**, *297*, E304–E314. [CrossRef] [PubMed]

22. Miura, S.; Kai, Y.; Kamei, Y.; Bruce, C.R.; Kubota, N.; Febbraio, M.A.; Kadowaki, T.; Ezaki, O. Alpha2-AMPK activity is not essential for an increase in fatty acid oxidation during low-intensity exercise. *Am. J. Physiol. Endocrinol. Metab.* **2009**, *296*, E47–E55. [CrossRef] [PubMed]

23. Kano, Y.; Poole, D.C.; Sudo, M.; Hirachi, T.; Miura, S.; Ezaki, O. Control of microvascular PO(2) kinetics following onset of muscle contractions: Role for AMPK. *Am. J. Physiol. Regul. Integr. Comp. Physiol.* **2011**, *301*, R1350–R1357. [CrossRef] [PubMed]

24. Tadaishi, M.; Miura, S.; Kai, Y.; Kawasaki, E.; Koshinaka, K.; Kawanaka, K.; Nagata, J.; Oishi, Y.; Ezaki, O. Effect of exercise intensity and AICAR on isoform-specific expressions of murine skeletal muscle PGC-1alpha mRNA: A role of beta(2)-adrenergic receptor activation. *Am. J. Physiol. Endocrinol. Metab.* **2011**, *300*, E341–E349. [CrossRef] [PubMed]

25. Fujii, N.; Hirshman, M.F.; Kane, E.M.; Ho, R.C.; Peter, L.E.; Seifert, M.M.; Goodyear, L.J. AMP-activated protein kinase alpha2 activity is not essential for contraction- and hyperosmolarity-induced glucose transport in skeletal muscle. *J. Biol. Chem.* **2005**, *280*, 39033–39041. [CrossRef] [PubMed]

26. Cannavino, J.; Brocca, L.; Sandri, M.; Grassi, B.; Bottinelli, R.; Pellegrino, M.A. The role of alterations in mitochondrial dynamics and PGC-1alpha over-expression in fast muscle atrophy following hindlimb unloading. *J. Physiol.* **2015**, *593*, 1981–1995. [CrossRef] [PubMed]

27. Zhong, G.; Li, Y.; Li, H.; Sun, W.; Cao, D.; Li, J.; Zhao, D.; Song, J.; Jin, X.; Song, H.; et al. Simulated Microgravity and Recovery-Induced Remodeling of the Left and Right Ventricle. *Front. Physiol.* **2016**, *7*, 274. [CrossRef] [PubMed]

28. Miyazaki, M.; Hitomi, Y.; Kizaki, T.; Ohno, H.; Katsumura, T.; Haga, S.; Takemasa, T. Calcineurin-mediated slow-type fiber expression and growth in reloading condition. *Med. Sci. Sports Exerc.* **2006**, *38*, 1065–1072. [CrossRef] [PubMed]

29. Chang, H.C.; Guarente, L. SIRT1 and other sirtuins in metabolism. *Trends Endocrinol. Metab.* **2014**, *25*, 138–145. [CrossRef] [PubMed]

30. Ruderman, N.B.; Xu, X.J.; Nelson, L.; Cacicedo, J.M.; Saha, A.K.; Lan, F.; Ido, Y. AMPK and SIRT1: A long-standing partnership? *Am. J. Physiol. Endocrinol. Metab.* **2010**, *298*, E751–E760. [CrossRef] [PubMed]

31. Lee, D.; Goldberg, A.L. SIRT1 protein, by blocking the activities of transcription factors FoxO1 and FoxO3, inhibits muscle atrophy and promotes muscle growth. *J. Biol. Chem.* **2013**, *288*, 30515–30526. [CrossRef] [PubMed]

32. Sandri, M.; Lin, J.; Handschin, C.; Yang, W.; Arany, Z.P.; Lecker, S.H.; Goldberg, A.L.; Spiegelman, B.M. PGC-1alpha protects skeletal muscle from atrophy by suppressing FoxO$_3$ action and atrophy-specific gene transcription. *Proc. Natl. Acad. Sci. USA* **2006**, *103*, 16260–16265. [CrossRef] [PubMed]

33. Fernandez-Marcos, P.J.; Auwerx, J. Regulation of PGC-1alpha, a nodal regulator of mitochondrial biogenesis. *Am. J. Clin. Nutr.* **2011**, *93*, 884S–890S. [CrossRef] [PubMed]

34. Ohno, Y.; Yamada, S.; Sugiura, T.; Ohira, Y.; Yoshioka, T.; Goto, K. A possible role of NF-kappaB and HSP72 in skeletal muscle hypertrophy induced by heat stress in rats. *Gener. Physiol. Biophys.* **2010**, *29*, 234–242. [CrossRef]

35. Yasuhara, K.; Ohno, Y.; Kojima, A.; Uehara, K.; Beppu, M.; Sugiura, T.; Fujimoto, M.; Nakai, A.; Ohira, Y.; Yoshioka, T.; et al. Absence of heat shock transcription factor 1 retards the regrowth of atrophied soleus muscle in mice. *J. Appl. Physiol.* **2011**, *111*, 1142–1149. [CrossRef] [PubMed]

36. Senf, S.M.; Dodd, S.L.; McClung, J.M.; Judge, A.R. Hsp70 overexpression inhibits NF-kappaB and Foxo3a transcriptional activities and prevents skeletal muscle atrophy. *FASEB J.* **2008**, *22*, 3836–3845. [CrossRef] [PubMed]

37. Miyabara, E.H.; Nascimento, T.L.; Rodrigues, D.C.; Moriscot, A.S.; Davila, W.F.; AitMou, Y.; deTombe, P.P.; Mestril, R. Overexpression of inducible 70-kDa heat shock protein in mouse improves structural and functional recovery of skeletal muscles from atrophy. *Pflugers Arch.* **2012**, *463*, 733–741. [CrossRef] [PubMed]

38. Goto, A.; Ohno, Y.; Ikuta, A.; Suzuki, M.; Ohira, T.; Egawa, T.; Sugiura, T.; Yoshioka, T.; Ohira, Y.; Goto, K. Up-regulation of adiponectin expression in antigravitational soleus muscle in response to unloading followed by reloading, and functional overloading in mice. *PLoS ONE* **2013**, *8*, e81929. [CrossRef] [PubMed]

39. Egawa, T.; Hamada, T.; Kameda, N.; Karaike, K.; Ma, X.; Masuda, S.; Iwanaka, N.; Hayashi, T. Caffeine acutely activates 5′ adenosine monophosphate-activated protein kinase and increases insulin-independent glucose transport in rat skeletal muscles. *Metabolism* **2009**, *58*, 1609–1617. [CrossRef] [PubMed]

40. Talmadge, R.J.; Roy, R.R. Electrophoretic separation of rat skeletal muscle myosin heavy-chain isoforms. *J. Appl. Physiol.* **1993**, *75*, 2337–2340. [CrossRef] [PubMed]

41. Masuda, S.; Hayashi, T.; Hashimoto, T.; Taguchi, S. Correlation of dystrophin-glycoprotein complex and focal adhesion complex with myosin heavy chain isoforms in rat skeletal muscle. *Acta Physiol.* **2009**, *195*, 483–494. [CrossRef] [PubMed]

42. Toyoda, T.; Hayashi, T.; Miyamoto, L.; Yonemitsu, S.; Nakano, M.; Tanaka, S.; Ebihara, K.; Masuzaki, H.; Hosoda, K.; Inoue, G.; et al. Possible involvement of the alpha1 isoform of 5′ AMP-activated protein kinase in oxidative stress-stimulated glucose transport in skeletal muscle. *Am. J. Physiol. Endocrinol. Metab.* **2004**, *287*, E166–E173. [CrossRef] [PubMed]

International Journal of
Molecular Sciences

MDPI

Article

Serum Is Not Necessary for Prior Pharmacological Activation of AMPK to Increase Insulin Sensitivity of Mouse Skeletal Muscle

Nicolas O. Jørgensen, Jørgen F. P. Wojtaszewski * and Rasmus Kjøbsted

Section of Molecular Physiology, Department of Nutrition, Exercise and Sports, Faculty of Science,
University of Copenhagen, DK-2100 Copenhagen, Denmark; nioj@nexs.ku.dk (N.O.J.);
rasmus.kjobsted@nexs.ku.dk (R.K.)
* Correspondence: jwojtaszewski@nexs.ku.dk; Tel.: +45-3532-1625

Received: 23 March 2018; Accepted: 10 April 2018; Published: 15 April 2018

Abstract: Exercise, contraction, and pharmacological activation of AMP-activated protein kinase (AMPK) by 5-aminoimidazole-4-carboxamide ribonucleotide (AICAR) have all been shown to increase muscle insulin sensitivity for glucose uptake. Intriguingly, improvements in insulin sensitivity following contraction of isolated rat and mouse skeletal muscle and prior AICAR stimulation of isolated rat skeletal muscle seem to depend on an unknown factor present in serum. One study recently questioned this requirement of a serum factor by showing serum-independency with muscle from old rats. Whether a serum factor is necessary for prior AICAR stimulation to increase insulin sensitivity of mouse skeletal muscle is not known. Therefore, we investigated the necessity of serum for this effect of AICAR in mouse skeletal muscle. We found that the ability of prior AICAR stimulation to improve insulin sensitivity of mouse skeletal muscle did not depend on the presence of serum during AICAR stimulation. Although prior AICAR stimulation did not enhance proximal insulin signaling, insulin-stimulated phosphorylation of Tre-2/BUB2/CDC16-domain family member 4 (TBC1D4) Ser711 was greater in prior AICAR-stimulated muscle compared to all other groups. These results imply that the presence of a serum factor is not necessary for prior AMPK activation by AICAR to enhance insulin sensitivity of mouse skeletal muscle.

Keywords: exercise; glucose uptake; AMP-activated protein kinase; TBC1D4; AS160

---

1. Introduction

Skeletal muscle accounts for the vast majority of whole body glucose disposal in response to insulin [1]. Because muscle insulin resistance is a major cause of metabolic diseases such as type 2 diabetes [2], identifying molecular mechanisms involved in the regulation of muscle insulin sensitivity is central for the development of pharmacological therapies. Interestingly, exercise in the form of running and swimming, as well as contraction of isolated muscle has been shown to increase insulin sensitivity for glucose uptake in healthy and insulin resistant skeletal muscle [3–11]. Recently, we have provided genetic evidence to support that the insulin-sensitizing effect of exercise, contraction, and 5-aminoimidazole-4-carboxaminde ribonucleotide (AICAR) stimulation is dependent of AMP-activated protein kinase (AMPK) in skeletal muscle [11,12].

Improved muscle insulin sensitivity after contraction, AICAR stimulation, and presumably exercise further seems to depend on an unknown humoral factor present in serum [13,14]. Initial findings point towards one (or several) serum protein(s), as isolated muscle stimulated to contract in trypsin-treated serum does not exhibit enhanced insulin sensitivity [15]. Importantly, it has been shown that the origin of the serum factor is of no importance and is not specific for the individual species as serum collected from fasting and resting humans and rats promotes contraction-induced improvements in insulin sensitivity of rat skeletal muscle equally well [15]. Although much effort has been devoted to uncover the identity of the serum factor(s) [15–18], this has yet to be identified.

Evidence implies that a serum factor is necessary for improving insulin sensitivity after contraction of isolated rodent skeletal muscle [13–16,19] as well as after AICAR stimulation of rat skeletal muscle [14]. One study recently questioned this requirement of a serum factor by demonstrating improved insulin sensitivity after prior AICAR stimulation of isolated skeletal muscle from old rats in the absence of serum [20]. Therefore, we investigated whether the effect of AICAR on insulin sensitivity for glucose uptake in isolated mouse extensor digitorum longus (EDL) was dependent on the presence of serum, as this is currently unknown. Additionally, we examined AMPK and insulin signaling in collected muscle samples that may support the molecular and mechanistic signature of AICAR-induced improvements of muscle insulin sensitivity.

2. Results

2.1. Acute Serum Stimulation Does Not Affect Basal or AICAR-Stimulated Glucose Uptake in Mouse Skeletal Muscle

Glucose uptake increased in EDL muscle in response to acute AICAR stimulation (Figure 1A). Neither basal nor AICAR-stimulated glucose uptake was affected by the presence of serum (Figure 1A,B).

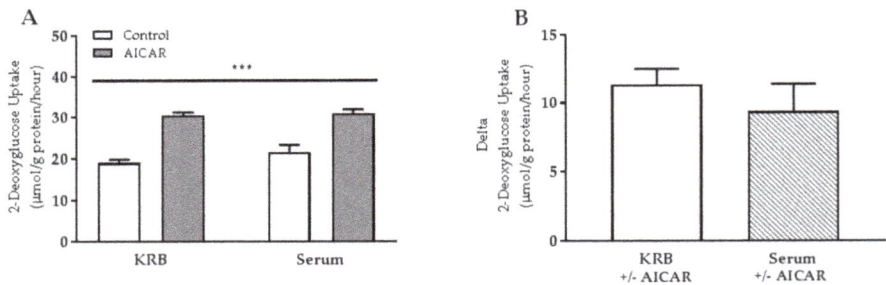

Figure 1. Acute serum- and 5-aminoimidazole-4-carboxamide ribonucleotide (AICAR)-stimulated glucose uptake in isolated mouse skeletal muscle. (**A**) 2-deoxyglucose (2-DG) uptake in isolated extensor digitorum longus (EDL) muscle from C57Bl/6 mice in response to 50 min of serum and/or AICAR stimulation; (**B**) Delta 2-DG uptake (AICAR minus control) in Krebs Ringer buffer (KRB) and serum-stimulated muscles. Data were analyzed by a two-way repeated-measures analysis of variance (ANOVA) and a Student's *t*-test, respectively. *** $p < 0.001$ indicates main effect of AICAR. Values are means $\pm$ SEM. $n = 8$ in all groups.

Alongside the increase in glucose uptake, acute AICAR stimulation increased phosphorylation of AMPK Thr172, acetyl-CoA carboxylase (ACC) Ser212, Tre-2/BUB2/CDC16-domain family member 1 (TBC1D1) Ser231, and Tre-2/BUB2/CDC16-domain family member 4 (TBC1D4) Ser711 compared to control muscles (Figure 2A–D). Of these phosphorylation sites, only phosphorylation of AMPK Thr172 increased in skeletal muscle when incubated in serum (Figure 2A) compared to the standard serum-free incubation buffer. As downstream targets of AMPK was unaffected by the presence of serum, this may indicate that serum does not affect AMPK activity in incubated skeletal muscle. Besides a small decrease of AMPKα2 protein content in serum-incubated muscles, total protein abundance of the measured proteins was not affected by serum and AICAR stimulation.

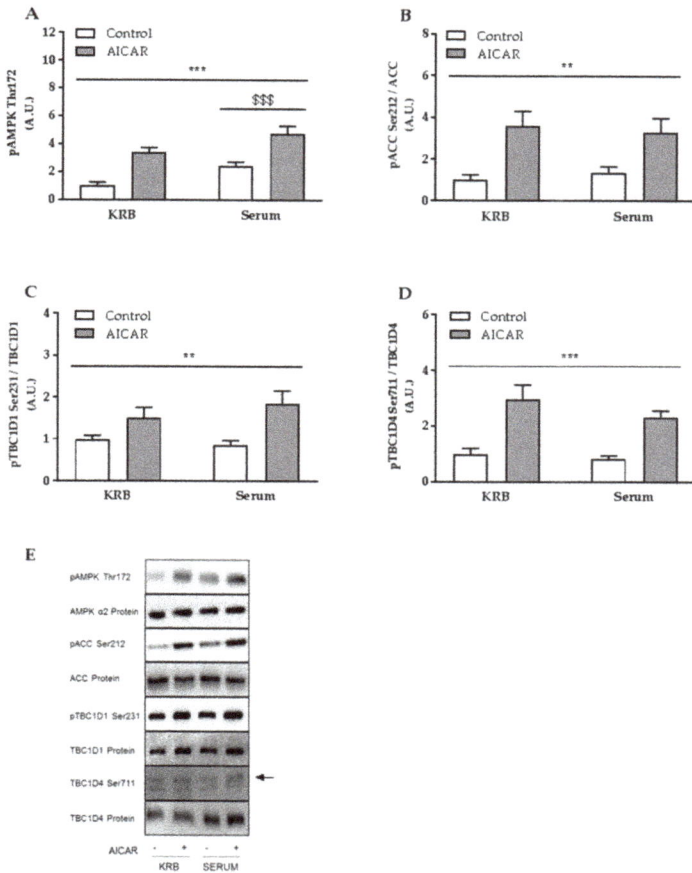

Figure 2. AMP-activated protein kinase (AMPK) signaling in isolated mouse skeletal muscle following acute serum and AICAR stimulation. (**A**) Phosphorylation of AMPK Thr172; (**B**) Acetyl-CoA carboxylase (ACC) Ser212; (**C**) Tre-2/BUB2/CDC16-domain family member 1 (TBC1D1) Ser231; and (**D**) Tre-2/BUB2/CDC16-domain family member 4 (TBC1D4) Ser711, in isolated EDL muscle from C57Bl/6 mice in response to 50 min of serum and/or AICAR stimulation; (**E**) Representative immunoblots. Data were analyzed by a two-way repeated-measures ANOVA. *** $p < 0.001$ and ** $p < 0.01$ indicate main effect of AICAR. $^{\$\$\$}$ $p < 0.001$ indicates main effect of serum. Values are means $\pm$ SEM. $n = 8$ in all groups. A.U., arbitrary units.

2.2. The Absence of Serum Does Not Influence the Ability of Prior AICAR Stimulation to Increase Mouse Muscle Insulin Sensitivity

As acute AICAR stimulation increased glucose uptake and AMPK-related downstream signaling similarly in EDL muscle incubated in the presence or absence of serum, we tested whether serum was in fact necessary for prior AICAR stimulation to increase muscle insulin sensitivity. We found that insulin sensitivity was increased 6 h after prior AICAR stimulation in mouse EDL muscle regardless of whether or not serum was present during AICAR stimulation (Figure 3A). Thus, the incremental increase in insulin-stimulated glucose uptake was significantly higher in prior AICAR-stimulated muscle independent of serum presence (Figure 3B).

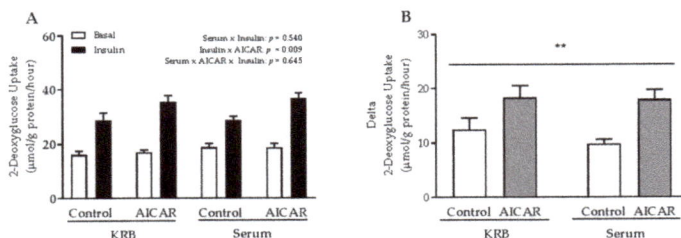

Figure 3. Enhanced insulin sensitivity after prior AICAR stimulation in isolated mouse skeletal muscle incubated in the absence or presence of serum. (**A**) Basal and submaximal insulin-stimulated 2-DG uptake in isolated EDL muscle from C57Bl/6 mice 6 h after prior AICAR stimulation in KRB or serum. (**B**) Delta 2-DG uptake (insulin minus basal) in prior control and AICAR-stimulated muscles. Data were analyzed by a three-way repeated-measures ANOVA and a two-way ANOVA, respectively. Possible interactions between groups are indicated in the figure. ** $p < 0.01$ indicates main effect of AICAR. Values are means ± SEM. $n = 4$–6 and $n = 8$–12 in serum and KRB group, respectively.

2.3. Increased Muscle Insulin Sensitivity Coincides with Elevated AMPK Signaling

Since we have previously reported intracellular signaling in prior serum- and AICAR-stimulated mouse EDL muscle [12], we decided to evaluate intracellular signaling only in EDL muscle incubated without serum. Previously we have shown that AMPK signaling is elevated in isolated muscle 6 h into recovery from acute AICAR and serum stimulation [12]. Concomitantly, we found that phosphorylation of AMPK Thr172, ACC Ser212, and TBC1D1 Ser231 was also increased 6 h after prior AICAR stimulation in muscle incubated without serum (Figure 4A–C). No change in total protein expression of AMPKα2, ACC, TBC1D1, Glucose transporter 4 (GLUT4), and Hexokinase II (HK-II) was found 6 h into recovery from acute AICAR stimulation.

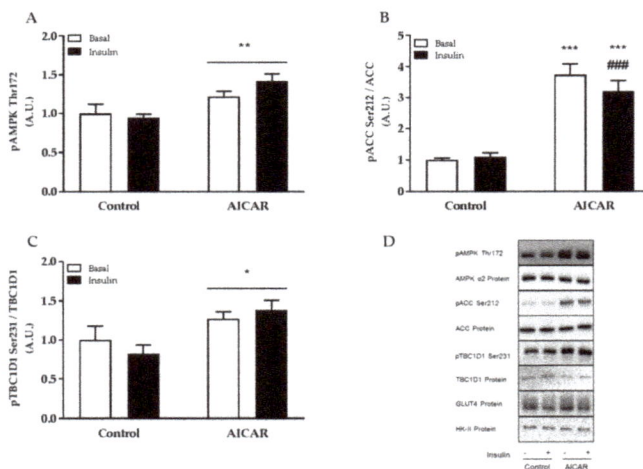

Figure 4. AMPK signaling in isolated mouse skeletal muscle after prior AICAR stimulation in serum-free incubation buffer. (**A**) Phosphorylation of AMPK Thr172; (**B**) ACC Ser212; and (**C**) TBC1D1 Ser231 in isolated EDL muscle from C57Bl/6 mice 6 h after prior AICAR stimulation in KRB; (**D**) Representative immunoblots. Data were analyzed by a two-way repeated-measures ANOVA. ** $p < 0.01$ and * $p < 0.05$ indicate main effect of AICAR. *** $p < 0.001$ indicates effect of AICAR within group. ### $p < 0.001$ indicates effect of insulin within AICAR. Values are means ± SEM. $n = 8$ and $n = 12$ in control and AICAR group, respectively. A.U., arbitrary units.

2.4. Insulin-Stimulated Phosphorylation of Akt Thr308 and Ser473 Is Not Affected by Prior AICAR Stimulation

Several observations indicate that improvements in muscle insulin sensitivity following exercise, contraction, and AICAR stimulation occur in the absence of elevated proximal insulin signaling [7,11,12,14,21–23]. In line, we found that submaximal insulin-stimulated phosphorylation of Akt Thr308 and Ser473 was similar between control and prior AICAR-stimulated muscles incubated without serum (Figure 5A,B). No change in total protein expression of Akt2 was found 6 h into recovery from acute AICAR stimulation.

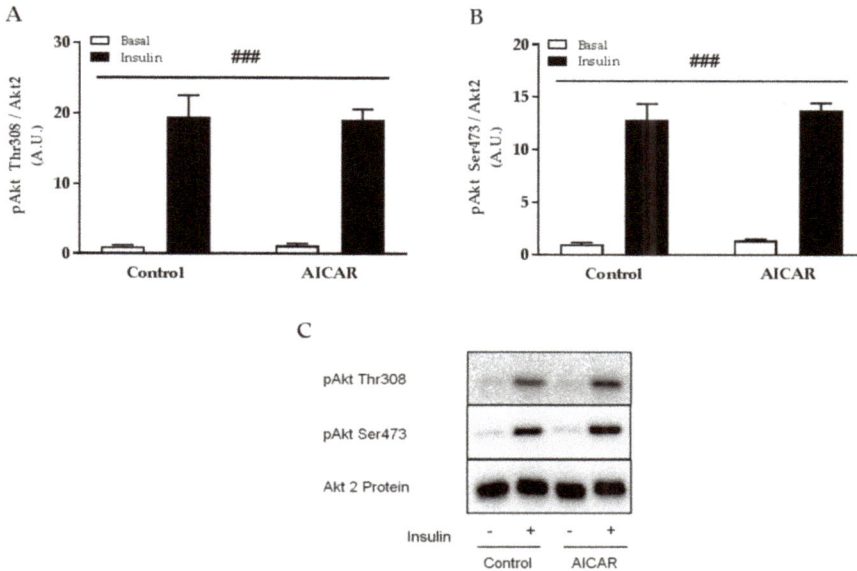

Figure 5. Akt signaling in isolated mouse skeletal muscle after prior AICAR stimulation in serum-free incubation buffer. Insulin-stimulated phosphorylation of (**A**) Akt Thr308 and (**B**) Ser473 in isolated EDL muscle from C57Bl/6 mice 6 h after prior AICAR stimulation in KRB. (**C**) Representative immunoblots. Data were analyzed by a two-way repeated-measures ANOVA. $^{\#\#\#}$ $p < 0.001$ indicates main effect of insulin. Values are means $\pm$ SEM. $n = 8$ and $n = 11$ in control and AICAR group, respectively. A.U., arbitrary units.

2.5. Insulin-Stimulated Phosphorylation of TBC1D4 Ser711 Is Elevated in Prior AICAR-Stimulated Muscle

Phosphorylation of TBC1D4 has been shown to be important for insulin-stimulated glucose uptake in skeletal muscle [24,25] and evidence suggests that TBC1D4 may relay improvements in insulin sensitivity of muscle previously stimulated with AICAR and serum [12]. We found that insulin-stimulated phosphorylation of TBC1D4 Ser595 and Thr649 was similar between control and prior AICAR-stimulated muscles incubated without serum (Figure 6A,B). In contrast, phosphorylation of AMPK downstream target TBC1D4 Ser711 was significantly higher in prior AICAR- and insulin-stimulated muscle compared to all other groups (Figure 6C), signifying a potential role of TBC1D4 Ser711 for regulating muscle insulin sensitivity. No change in total protein expression of TBC1D4 was found 6 h into recovery from acute AICAR stimulation.

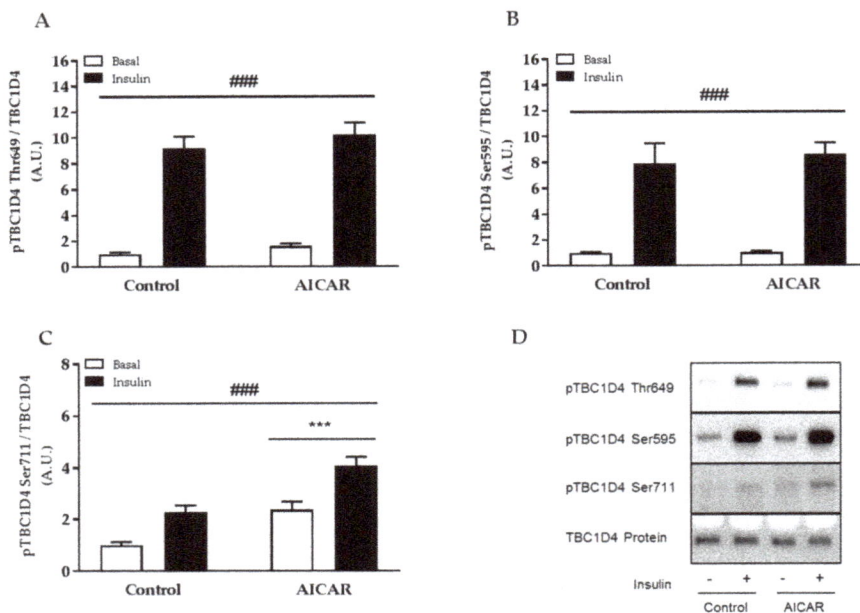

Figure 6. TBC1D4 signaling in isolated mouse skeletal muscle after prior AICAR stimulation in serum-free incubation buffer. Insulin-stimulated phosphorylation of (**A**) TBC1D4 Thr649; (**B**) Ser595; and (**C**) Ser711 in isolated EDL muscle from C57Bl/6 mice 6 h after prior AICAR stimulation in KRB; (**D**) Representative immunoblots. Data were analyzed by a two-way repeated-measures ANOVA. ### $p < 0.001$ indicates main effect of insulin. *** $p < 0.001$ indicates main effect of AICAR. Values are means $\pm$ SEM. $n = 8$ and $n = 12$ in control and AICAR group, respectively. A.U., arbitrary units.

3. Discussion

Here, we demonstrate that improved insulin sensitivity after prior AICAR stimulation of isolated mouse skeletal muscle does not require the presence of serum. Additionally, we show that the insulin-sensitizing effect of AICAR occurs independently of elevated proximal insulin signaling but coincides with elevated insulin-stimulated phosphorylation of TBC1D4 Ser711, a known downstream target of AMPK. Thus, our data suggest that a serum factor is not needed for prior pharmacological activation of AMPK to enhance insulin sensitivity of mouse skeletal muscle in contrast to previous assumptions [12].

It has previously been suggested that the presence of a serum factor is necessary for improved insulin sensitivity after prior AICAR stimulation of rat epitrochlearis muscle [14]. The data presented herein oppose the findings by Fisher et al. [14], though the muscles studied differed with regards to type, species, and gender. However, a recent study reported that prior AICAR stimulation improved insulin sensitivity of rat epitrochlearis muscle in the absence of serum [20], emphasizing that the discrepancies observed between the present study and that of Fisher et al. [14] are not due to differences in muscle type or species. Since Oki et al. [20] and Fisher et al. [14] investigated muscle from old (24 months-old) and young (likely ~6 weeks-old) rats, respectively, the observed difference could be due to an effect of age somehow affecting the necessity of a serum factor to mediate improvements in muscle insulin sensitivity after prior AICAR stimulation. However, the mice used in the present study were young, suggesting that, at least in mice, the ability for pharmacological AMPK activation to increase skeletal muscle insulin sensitivity in the absence of serum is not restricted to aged muscle.

Since we investigated muscle from female mice in the present study, data presented here and in the study by Fisher et al. [14], where a serum factor was found necessary for AICAR-induced improvement of insulin sensitivity in muscle from young male rats, suggests that a gender difference may be responsible for the observed discrepancy. Whether a serum factor is indeed needed for prior AICAR stimulation to improve insulin sensitivity of young male mouse muscle is not known at present. Interestingly though, in a recent study, serum was found necessary for prior contraction to improve insulin sensitivity of isolated skeletal muscle from young male mice [19]. As such, we cannot exclude that gender-related differences in skeletal muscle may exist and therefore influence whether or not a serum factor has to be present for AICAR to improve muscle insulin sensitivity.

We found that acute serum stimulation of isolated mouse muscle did not affect glucose uptake or phosphorylation of TBC1D4. This is in contrast to another study showing that acute serum stimulation increases glucose uptake in isolated rat skeletal muscle as well as phosphorylation of Akt and TBC1D4 [23]. These findings were likely due to the presence of insulin in serum, as the authors reported a similar increase in glucose uptake and phosphorylation of TBC1D4 when incubating rat muscle in serum-free buffer with an insulin concentration equivalent to that found in the used serum [23]. We speculate that the inconsistency between this and the aforementioned study with regards to the acute effects of serum stimulation may relate to the use of serum from different species. Thus, although serum was obtained from healthy male rats [23] and humans (the present study) in the fasted and rested condition, fasting insulin concentrations in rat serum are typically twice as high of that found in human serum [26,27]. This may be the cause of the observed differences in glucose uptake and cellular signaling between the two studies.

Several studies have reported that the increase in muscle insulin sensitivity after exercise, contraction, and AICAR stimulation occurs independently of enhanced proximal insulin signaling (e.g., from insulin binding to Akt activity) [7,11,12,14,21–23]. In line, we found that insulin-stimulated phosphorylation of Akt Thr308 and Ser473 was similar between control and prior AICAR-stimulated muscles although glucose uptake was not.

We have previously reported that AMPK downstream signaling is increased in isolated muscle 6 h after prior AICAR and serum stimulation [12]. Furthermore, this increase seems to depend on a persistent increase in AMPK α2β2γ3 activity, which likely regulates muscle insulin sensitivity [12]. In prior AICAR- but non-serum-stimulated muscle, we also found a persistent increase in phosphorylation of AMPK Thr172 and downstream targets ACC Ser212 and TBC1D1 Ser231 indicating that prior AICAR stimulation improves insulin sensitivity similarly in serum and non-serum treated muscles. Interestingly, elevated phosphorylation of AMPK Thr172, ACC Ser212 and TBC1D1 Ser231 is not found in insulin-sensitized rodent skeletal muscle after prior contraction [11,23] even though activity of the AMPK α2β2γ3 complex is increased [11]. Thus, despite that differences in phosphorylation of AMPK targets are observed between prior contracted and AICAR-stimulated skeletal muscle, increased activity of the AMPK α2β2γ3 complex is found during both conditions supporting the notion that AICAR and contraction improve muscle insulin sensitivity via the AMPK α2β2γ3 complex.

TBC1D4 is a Rab GTPase-activating protein involved in the regulation of GLUT4 trafficking [28]. In skeletal muscle, TBC1D4 is phosphorylated by Akt, which seems important for increasing glucose uptake in response to insulin [24,25,29]. TBC1D4 is also targeted at Ser711 by AMPK during exercise, contraction, and acute AICAR stimulation [11,30]. Importantly, phosphorylation of TBC1D4 Ser711 seems to be regulated directly by the AMPK α2β2γ3 complex [11] but it does not seem to affect muscle glucose uptake per se [30]. Several findings have pointed towards a role of TBC1D4 in regulating muscle insulin sensitivity given its function as a point of convergence for exercise (AMPK) and insulin (Akt) signaling. Indeed, recent evidence from our muscle-specific AMPK transgenic mouse models supports the notion of an AMPK-TBC1D4 signaling axis involved in the regulation of muscle insulin sensitivity as both improvements in insulin-stimulated glucose uptake and phosphorylation of TBC1D4 Ser711 are abrogated in AMPK-deficient muscle after prior in situ contraction as well as prior AICAR stimulation of serum-incubated muscle [11,12]. In accordance, we found that insulin-stimulated phosphorylation of TBC1D4 was increased at Ser711 in prior AICAR-stimulated muscle concomitant with enhanced insulin sensitivity.

Taken together, improved insulin sensitivity of mouse skeletal muscle after prior pharmacological activation of AMPK by AICAR does not require the presence of a serum factor. This is in contrast to findings in prior contracted and AICAR-stimulated skeletal muscle from young male rats in which one (or several) unknown serum factor(s) seems important to enhance insulin sensitivity [13–16,23]. Moreover, considerable evidence points toward an important role of elevated phosphorylation of TBC1D4 Ser711 for enhancing muscle insulin sensitivity after AMPK activating stimuli, signifying the importance of the upstream regulator AMPK.

4. Materials and Methods

4.1. Animals

All animal experiments were approved by the Danish Animal Experiments Inspectorate (#2014-15-2934-01037, approved 4 March 2014) and complied with the EU convention for the protection of vertebra used for scientific purposes (Council of Europe, Treaty 123/170, Strasbourg, France, 1985/1998). Animals used in this study were C57Bl/6J female mice from Taconic (Ejby, Denmark). Young mice (19.9 ± 1.9 g [means ± SD]) were maintained on a 12:12 hour light-dark cycle with free access to standard rodent chow and water. Serum was obtained from a healthy man (Body Mass Index: 24.1 kg/m^2, 34 years of age) in the overnight fasted and rested state (blood glucose concentration = 5.4 mmol/L). The serum was collected by antecubital venous catheter and kept frozen at −26 °C until used and was not refrozen for later use. Collection of human serum was approved by the Ethics Committee of Copenhagen (#H-3-2012-140, approved 29 November 2012) and complied with the ethical guidelines of the Declaration of Helsinki II. Informed consent was obtained from the serum donor before entering the study.

4.2. Muscle Incubations

Fed animals were anesthetized by an intraperitoneal injection of Pentobarbital (10 mg/100 g body weight) before EDL muscles were isolated and suspended in incubation chambers containing Krebs Ringer buffer (KRB) as previously described [12]. In short, EDL muscles were incubated for 50 min in the absence or presence of 1 mmol/L AICAR (Toronto Research Chemicals, Toronto, ON, Canada) in KRB or 100% human serum. Subsequent to AICAR stimulation, muscles were allowed to recover for 6 h in KRB supplemented with 5 mmol/L of D-glucose after which they were incubated in KRB with or without a submaximal insulin concentration (100 µU/mL, 30 min). 2-deoxyglucose (2-DG) uptake was measured during the last 10 min of the 30 min stimulation period by adding 1 mmol/L [^{3}H]2-DG (0.028 MBq/mL) and 7 mmol/L [^{14}C]mannitol (0.0083 MBq/mL) to the incubation medium. For glucose uptake measurements in response to acute AICAR and serum stimulation, EDL muscles were incubated in KRB or 100% human serum for 50 min with or without 1 mmol/L AICAR. Following stimulation all muscles were washed in KRB for 1 min before 2-DG uptake was measured during a 10 min incubation period. For measurements of acute AICAR-stimulated muscle glucose uptake, AICAR was present in the incubation medium throughout the entire incubation period. For all incubations, 2-DG uptake was determined as previously described [11].

4.3. Muscle Processing, Sodium Dodecyl Sulfate Polyacrylamide Gel Electrophoresis (SDS-PAGE), and Western Blot Analyses

Muscles were homogenized as previously described [12] and lysates were collected and frozen in liquid nitrogen for subsequent analyses. The bicinchoninic acid method was used to determine total protein abundance in muscle lysates. Lysates were boiled in Laemmli buffer and subjected to SDS-PAGE and immunoblotting as previously described [12].

4.4. Antibodies

Primary antibodies against Akt2 (#3063), pAkt-Ser473 (#9271), pAkt-Thr308 (#9275), pAMPKα-Thr172 (#2531), pACC-Ser79/212 (#3661), pTBC1D4-Ser588 (mouse: Ser595) (#8730), and pTBC1D4-Thr642 (mouse: Thr649) (#8881) were from Cell Signaling Technology (Danvers, MA, USA) Antibody against pTBC1D1-Ser231 (#NRG-1848963) was from Millipore (Burlington, MA, USA), AMPKα2 antibody (#SC-19131) and Hexokinase II were from Santa Cruz (Dallas, TX, USA)(#SC-6521) while GLUT4 antibody (#PA1-1065) was from Thermo Fisher Scientific (Waltham, MA, USA). ACC protein was detected using horseradish peroxidase-conjugated streptavidin from Dako (Glostrup, Denmark), (#P0397). TBC1D1 and TBC1D4 protein as well as phosphorylation of TBC1D4-Ser711 were detected using antibodies as previously described [30,31].

4.5. Statistics

Data are presented as the means ± SEM unless stated otherwise. Results on cellular signaling are presented in figures as relative to the basal or control group levels within the given experiment. An unpaired Student's *t*-test (Figure 1B) as well as three-way (Figure 3A) and two-way (remaining figures) ANOVA with and without repeated measures were used to assess statistical differences. The Student–Newman–Keuls test was used for post hoc testing. Main effects are indicated with lines comprising the affected groups and symbols in Figure 4B represent post hoc test corrected *p*-values. *, #, and $ indicate effects of AICAR, insulin, and serum, respectively. Statistical significance was defined as $p < 0.05$.

Acknowledgments: The authors would like to thank Laurie J. Goodyear (Joslin Diabetes Center and Harvard Medical School, Boston, MA, USA) for the donation of antibody. This study was funded by the Danish Council for Independent Research Medical Sciences and the Novo Nordisk Foundation.

Author Contributions: Nicolas O. Jørgensen performed experiments, analyzed the data and contributed with drafting the manuscript. Jørgen F. P. Wojtaszewski conceived and designed the research and contributed with drafting the manuscript. Rasmus Kjøbsted conceived and designed the research, performed the experiments, analyzed the data, and drafted the manuscript. All authors interpreted the results, contributed to the discussion, edited and revised the manuscript and read and approved the final version of the manuscript.

Conflicts of Interest: The authors declare no conflict of interest.

Abbreviations

2-DG	2-deoxyglucose
ACC	Acetyl-CoA carboxylase
AICAR	5-aminoimidazole-4-carboxamide ribonucleotide
AMPK	AMP-activated protein kinase
ANOVA	Analysis of variance
A.U.	Arbitrary units
EDL	Extensor digitorum longus
GLUT4	Glucose transporter 4
HK-II	Hexokinase II
KRB	Krebs Ringer buffer
SDS-PAGE	Sodium dodecyl sulfate polyacrylamide gel electrophoresis
TBC1D1	Tre-2/BUB2/CDC16-domain family member 1
TBC1D4	Tre-2/BUB2/CDC16-domain family member 4

References

1. DeFronzo, R.A. The triumvirate: Beta-cell, muscle, liver. A collusion responsible for NIDDM. *Diabetes* **1988**, *37*, 667–687. [CrossRef] [PubMed]
2. DeFronzo, R.; Tripathy, D. Skeletal muscle insulin resistance is the primary defect in type 2 diabetes. *Diabetes Care* **2009**, *32*, S157–S163. [CrossRef] [PubMed]

3. Richter, E.A.; Garetto, L.P.; Goodman, M.N.; Ruderman, N.B. Enhanced muscle glucose metabolism after exercise: Modulation by local factors. *Am. J. Physiol.* **1984**, *246*, E476–E482. [CrossRef] [PubMed]

4. Cartee, G.D.; Young, D.A.; Sleeper, M.D.; Zierath, J.; Wallberg-Henriksson, H.; Holloszy, J.O. Prolonged increase in insulin-stimulated glucose transport in muscle after exercise. *Am. J. Physiol.* **1989**, *256*, E494–E499. [CrossRef] [PubMed]

5. Richter, E.A.; Mikines, K.J.; Galbo, H.; Kiens, B. Effect of exercise on insulin action in human skeletal muscle. *J. Appl. Physiol.* **1989**, *66*, 876–885. [CrossRef] [PubMed]

6. Wojtaszewski, J.F.P.; Hansen, B.F.; Kiens, B.; Richter, E.A. Insulin signaling in human skeletal muscle: Time course and effect of exercise. *Diabetes* **1997**, *46*, 1775–1781. [CrossRef] [PubMed]

7. Wojtaszewski, J.F.; Hansen, B.F.; Gade; Kiens, B.; Markuns, J.F.; Goodyear, L.J.; Richter, E.A. Insulin signaling and insulin sensitivity after exercise in human skeletal muscle. *Diabetes* **2000**, *49*, 325–331. [CrossRef] [PubMed]

8. Pehmøller, C.; Brandt, N.; Birk, J.B.; Høeg, L.D.; Sjøberg, K.A.; Goodyear, L.J.; Kiens, B.; Richter, E.A.; Wojtaszewski, J.F.P. Exercise alleviates lipid-induced insulin resistance in human skeletal muscle-signaling interaction at the level of TBC1 domain family member 4. *Diabetes* **2012**, *61*, 2743–2752. [CrossRef] [PubMed]

9. Castorena, C.M.; Arias, E.B.; Sharma, N.; Cartee, G.D. Post-exercise Improvement in Insulin-Stimulated Glucose Uptake Occurs Concomitant with Greater AS160 Phosphorylation in Muscle from Normal and Insulin Resistant Rats. *Diabetes* **2014**, *63*, 1–37. [CrossRef] [PubMed]

10. Hamada, T.; Arias, E.B.; Cartee, G.D. Increased submaximal insulin-stimulated glucose uptake in mouse skeletal muscle after treadmill exercise. *J. Appl. Physiol.* **2006**, *101*, 1368–1376. [CrossRef] [PubMed]

11. Kjøbsted, R.; Munk-Hansen, N.; Birk, J.B.; Foretz, M.; Viollet, B.; Björnholm, M.; Zierath, J.R.; Treebak, J.T.; Wojtaszewski, J.F.P. Enhanced Muscle Insulin Sensitivity After Contraction/Exercise Is Mediated by AMPK. *Diabetes* **2017**, *66*, 598–612. [CrossRef] [PubMed]

12. Kjøbsted, R.; Treebak, J.T.; Fentz, J.; Lantier, L.; Viollet, B.; Birk, J.B.; Schjerling, P.; Björnholm, M.; Zierath, J.R.; Wojtaszewski, J.F.P. Prior AICAR stimulation increases insulin sensitivity in mouse skeletal muscle in an AMPK-dependent manner. *Diabetes* **2015**, *64*, 2042–2055. [CrossRef] [PubMed]

13. Cartee, G.D.; Holloszy, J.O. Exercise increases susceptibility of muscle glucose transport to activation by various stimuli. *Am. J. Physiol.* **1990**, *258*, E390–E393. [CrossRef] [PubMed]

14. Fisher, J.S.; Gao, J.; Han, D.-H.; Holloszy, J.O.; Nolte, L.A. Activation of AMP kinase enhances sensitivity of muscle glucose transport to insulin. *Am. J. Physiol. Endocrinol. Metab.* **2002**, *282*, E18–E23. [CrossRef] [PubMed]

15. Gao, J.; Gulve, E.A.; Holloszy, J.O. Contraction-induced increase in muscle insulin sensitivity: Requirement for a serum factor. *Am. J. Physiol.* **1994**, *266*, E186–E192. [CrossRef] [PubMed]

16. Dumke, C.L.; Kim, J.; Arias, E.B.; Cartee, G.D. Role of kallikrein-kininogen system in insulin-stimulated glucose transport after muscle contractions. *J. Appl. Physiol.* **2002**, *92*, 657–664. [CrossRef] [PubMed]

17. Schweitzer, G.G.; Castorena, C.M.; Hamada, T.; Funai, K.; Arias, E.B.; Cartee, G.D. The B2 receptor of bradykinin is not essential for the post-exercise increase in glucose uptake by insulin-stimulated mouse skeletal muscle. *Physiol. Res.* **2011**, *60*, 511–519. [PubMed]

18. Schweitzer, G.G.; Cartee, G.D. Postexercise skeletal muscle glucose transport is normal in kininogen-deficient rats. *Med. Sci. Sports Exerc.* **2011**, *43*, 1148–1153. [CrossRef] [PubMed]

19. Zhang, X.; Hiam, D.; Hong, Y.H.; Zulli, A.; Hayes, A.; Rattigan, S.; McConell, G.K. Nitric oxide is required for the insulin sensitizing effects of contraction in mouse skeletal muscle. *J. Physiol.* **2017**, *595*, 7427–7439. [CrossRef] [PubMed]

20. Oki, K.; Arias, E.B.; Kanzaki, M.; Cartee, G.D. Prior treatment with the AMPK activator AICAR induces subsequently enhanced glucose uptake in isolated skeletal muscles from 24 month-old rats. *Appl. Physiol. Nutr. Metab.* **2018**. [CrossRef] [PubMed]

21. Bonen, A.; Tan, M.H.; Watson-Wright, W.M. Effects of exercise on insulin binding and glucose metabolism in muscle. *Can. J. Physiol. Pharmacol.* **1984**, *62*, 1500–1504. [CrossRef] [PubMed]

22. Funai, K.; Schweitzer, G.G.; Sharma, N.; Kanzaki, M.; Cartee, G.D. Increased AS160 phosphorylation, but not TBC1D1 phosphorylation, with increased postexercise insulin sensitivity in rat skeletal muscle. *Am. J. Physiol. Endocrinol. Metab.* **2009**, *297*, E242–E251. [CrossRef] [PubMed]

23. Funai, K.; Schweitzer, G.G.; Castorena, C.M.; Kanzaki, M.; Cartee, G.D. In vivo exercise followed by in vitro contraction additively elevates subsequent insulin-stimulated glucose transport by rat skeletal muscle. *Am. J. Physiol. Endocrinol. Metab.* **2010**, *298*, E999–E1010. [CrossRef] [PubMed]

24. Kramer, H.F.; Witczak, C.A.; Taylor, E.B.; Fujii, N.; Hirshman, M.F.; Goodyear, L.J. AS160 regulates insulin- and contraction-stimulated glucose uptake in mouse skeletal muscle. *J. Biol. Chem.* **2006**, *281*, 31478–31485. [CrossRef] [PubMed]

25. Chen, S.; Wasserman, D.H.; MacKintosh, C.; Sakamoto, K. Mice with AS160/TBC1D4-Thr649Ala knockin mutation are glucose intolerant with reduced insulin sensitivity and altered GLUT4 trafficking. *Cell Metab.* **2011**, *13*, 68–79. [CrossRef] [PubMed]

26. Burrows, S. Insulin response in glucose-tolerance tests. *Am. J. Clin. Pathol.* **1967**, *47*, 709–713. [CrossRef] [PubMed]

27. Tsai, A.C. Serum insulin concentration, insulin release and degradation, glucose tolerance and in vivo insulin sensitivity in cholesterol-fed rats. *J. Nutr.* **1977**, *107*, 546–551. [CrossRef] [PubMed]

28. Cartee, G.D.; Wojtaszewski, J.F.P. Role of Akt substrate of 160 kDa in insulin-stimulated and contraction-stimulated glucose transport. *Appl. Physiol. Nutr. Metab.* **2007**, *32*, 557–566. [CrossRef] [PubMed]

29. Kramer, H.F.; Witczak, C.A.; Fujii, N.; Jessen, N.; Taylor, E.B.; Arnolds, D.E.; Sakamoto, K.; Hirshman, M.F.; Goodyear, L.J. Distinct signals regulate AS160 phosphorylation in response to insulin, AICAR, and contraction in mouse skeletal muscle. *Diabetes* **2006**, *55*, 2067–2076. [CrossRef] [PubMed]

30. Treebak, J.T.; Taylor, E.B.; Witczak, C.A.; An, D.; Toyoda, T.; Koh, H.-J.; Xie, J.; Feener, E.P.; Wojtaszewski, J.F.P.; Hirshman, M.F.; et al. Identification of a novel phosphorylation site on TBC1D4 regulated by AMP-activated protein kinase in skeletal muscle. *Am. J. Physiol. Cell Physiol.* **2010**, *298*, C377–C385. [CrossRef] [PubMed]

31. Treebak, J.T.; Pehmøller, C.; Kristensen, J.M.; Kjøbsted, R.; Birk, J.B.; Schjerling, P.; Richter, E.A.; Goodyear, L.J.; Wojtaszewski, J.F.P. Acute exercise and physiological insulin induce distinct phosphorylation signatures on TBC1D1 and TBC1D4 proteins in human skeletal muscle. *J. Physiol.* **2014**, *592*, 351–375. [CrossRef] [PubMed]

International Journal of
Molecular Sciences

MDPI

Article

A769662 Inhibits Insulin-Stimulated Akt Activation in Human Macrovascular Endothelial Cells Independent of AMP-Activated Protein Kinase

Anastasiya Strembitska [1], Sarah J. Mancini [1], Jonathan M. Gamwell [1,2], Timothy M. Palmer [1,3], George S. Baillie [1] and Ian P. Salt [1,*]

[1] Institute of Cardiovascular and Medical Sciences, College of Medical, Veterinary and Life Sciences, University of Glasgow, Glasgow G12 8QQ, UK; a.strembitska.1@research.gla.ac.uk (A.S.); sarah.mancini@glasgow.ac.uk (S.J.M.); jonathan.gamwell@rdm.ox.ac.uk (J.M.G.); tim.palmer@hyms.ac.uk (T.M.P.); george.baillie@glasgow.ac.uk (G.S.B.)
[2] Radcliffe Department of Medicine, University of Oxford, Oxford OX3 9DU, UK
[3] Centre for Atherothrombosis and Metabolic Disease, Hull York Medical School, University of Hull, Hull HU6 7RX, UK
* Correspondence: ian.salt@glasgow.ac.uk; Tel.: +44-(0)141-330-2049; Fax: +44-(0)141-330-5481

Received: 16 November 2018; Accepted: 3 December 2018; Published: 5 December 2018

Abstract: Protein kinase B (Akt) is a key enzyme in the insulin signalling cascade, required for insulin-stimulated NO production in endothelial cells (ECs). Previous studies have suggested that AMP-activated protein kinase (AMPK) activation stimulates NO synthesis and enhances insulin-stimulated Akt activation, yet these studies have largely used indirect activators of AMPK. The effects of the allosteric AMPK activator A769662 on insulin signalling and endothelial function was therefore examined in cultured human macrovascular ECs. Surprisingly, A769662 inhibited insulin-stimulated NO synthesis and Akt phosphorylation in human ECs from umbilical veins (HUVECs) and aorta (HAECs). In contrast, the AMPK activators compound **991** and AICAR had no substantial inhibitory effect on insulin-stimulated Akt phosphorylation in ECs. Inhibition of AMPK with SBI-0206965 had no effect on the inhibition of insulin-stimulated Akt phosphorylation by A769662, suggesting the inhibitory action of A769662 is AMPK-independent. A769662 decreased IGF1-stimulated Akt phosphorylation yet had no effect on VEGF-stimulated Akt signalling in HUVECs, suggesting that A769662 attenuates early insulin/IGF1 signalling. The effects of A769662 on insulin-stimulated Akt phosphorylation were specific to human ECs, as no effect was observed in the human cancer cell lines HepG2 or HeLa, as well as in mouse embryonic fibroblasts (MEFs). A769662 inhibited insulin-stimulated Erk1/2 phosphorylation in HAECs and MEFs, an effect that was independent of AMPK in MEFs. Therefore, despite being a potent AMPK activator, A769662 has effects unlikely to be mediated by AMPK in human macrovascular ECs that reduce insulin sensitivity and eNOS activation.

Keywords: AMP-activated protein kinase; protein kinase B; Akt; insulin signalling; A769662; endothelial function

---

1. Introduction

Endothelial cells (ECs) are essential for modulation of vascular homeostasis and signal transduction [1], including the production and regulation of vascular tone, modulation of inflammatory responses and maintenance of an anti-atherogenic phenotype of vascular smooth muscle cells (VSMCs) [1]. Being the downstream target of phosphoinositide (PI) 3-kinase (PI3K), protein kinase B (Akt) is one of the key kinases regulating cell survival, cell-cycle progression and metabolism [2].

Under physiological conditions, insulin [3], VEGF [4] or IGF1 [5] stimulation rapidly leads to increased Akt activity by increasing Ser473 and Thr308 phosphorylation.

In healthy subjects, insulin acts as a vasodilator, stimulating calcium-independent NO synthesis in cultured human aortic ECs (HAECs) through Akt-mediated phosphorylation of endothelial nitric oxide synthase (eNOS) Ser1177 and Ser615 [6]. However, in murine models of diabetes and people with diabetes, impaired insulin-stimulated blood flow and NO bioavailability have been demonstrated [7]. Decreased NO bioavailability leads to endothelial dysfunction in animal models and human subjects, increasing risk of cardiovascular events and pro-inflammatory signalling in the vasculature [8].

AMP-activated protein kinase (AMPK) is a heterotrimeric serine/threonine protein kinase sensitive to intracellular AMP levels which acts as a master regulator of energy metabolism [9,10]. AMPK is activated in response to an increase in the intracellular (AMP + ADP):ATP ratio and simultaneous AMPKα Thr172 phosphorylation by the upstream kinases liver kinase B1 or calcium/calmodulin-dependent protein kinase kinase β [9,10]. Activated AMPK inhibits anabolic processes which consume ATP [9,10], reduces inflammation [11,12], inhibits endothelial cell proliferation [13], stimulates mitochondrial biogenesis [14], and increases insulin sensitivity [9,15], making it an attractive pharmacological target for diabetes and cardiovascular pathologies [10]. There are 12 differentially expressed AMPK isoforms, each composed of catalytic α1/α2, and regulatory β1/β2 and γ1/γ2/γ3 subunits [9,10], allowing a specialised cellular and systemic response to different metabolic stimuli [10].

Previous studies have reported that AMPK activation improved insulin sensitivity and energy homeostasis, and attenuated inflammatory signalling in insulin-sensitive tissues, such as muscle [14], liver [16,17] and adipose tissue [11,18] as well as ECs [12,15]. Several AMPK activators have been shown to increase NO synthesis in HAECs in an AMPK-dependent manner [10]. Nevertheless, it is unclear whether all AMPK activators improve insulin sensitivity and vascular function, as previous studies have largely been conducted using a variety of compounds that activate AMPK by altering cellular nucleotide ratios, including rosiglitazone, resveratrol, metformin and canagliflozin or mimic AMP, such as AICAR (5-amino-4-imidazolecarboxamide ribonucleoside) [12,19–22].

In the last 15 years, there has been considerable effort to develop AMPK-selective small-molecule activators, resulting in the development of A769662 [23], the discovery of salicylate as an AMPK activator and, more recently, the development of compound **991**—a compound 5–10-fold more potent than A769662 [24]. A769662 is an allosteric activator that binds complexes containing AMPKβ1, also inhibiting AMPKα Thr172 dephosphorylation [24,25]. Since being first described as an AMPK activator, some AMPK-independent effects of A769662 have been reported [26–29], yet it remains a commonly utilised pharmacological tool for selective AMPK activation. The effects of A769662 on insulin signalling and insulin-stimulated Akt/eNOS axis activation in primary human ECs were therefore determined.

2. Results

2.1. Insulin-Stimulated Signalling and NO Production are Reduced by A769662 in Human Endothelial Cells

Stimulation of HUVECs with concentrations of insulin above 0.1 µM robustly stimulated phosphorylation of Akt at Ser473 and Thr308 (Supplementary Figure S1), similar to concentrations of insulin previously demonstrated to be required for insulin-stimulated NO synthesis in cultured endothelial cells [30]. All subsequent experiments were therefore conducted using 1 µM insulin. To examine the effect of A769662 on HUVEC insulin signalling, cells were preincubated in the presence or absence of A769662 prior to stimulation with insulin and NO synthesis and the extent of Akt phosphorylation assessed. Insulin modestly increased NO synthesis, yet insulin-stimulated NO synthesis was lost upon preincubation with A769662 (Figure 1A).

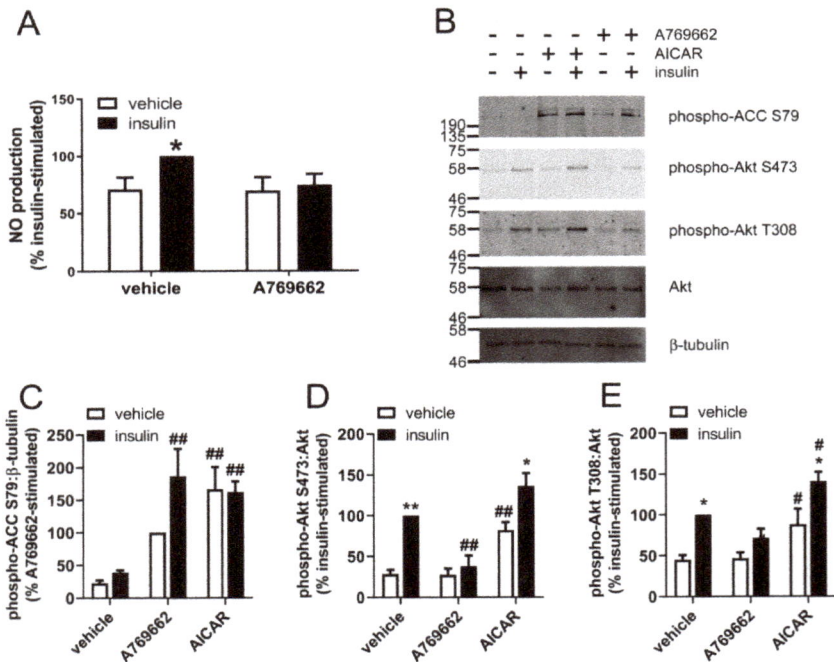

Figure 1. The effect of A769662 on insulin-stimulated NO production and Akt phosphorylation in HUVECs. (**A**) HUVECs were stimulated with A769662 (50 µM, 45 min) prior to insulin (1 µM, 15 min), conditioned media collected and NO production assessed. Data (mean ± SEM) shown from three independent replicates. (**B–E**) HUVECs were stimulated with A769662 (50 µM, 45 min) or AICAR (2 mM, 45 min) prior to insulin (51 µM, 15 min) and cell lysates prepared. Proteins were resolved by SDS-PAGE and immunoblotted with the antibodies indicated. (**B**) Representative immunoblots from three biological replicates with molecular weight markers indicated. Phospho-Akt Thr308 protein levels were assessed by stripping and re-probing the membranes. Densitometric quantification of (**C**) ACC, (**D**) Akt Ser473 and (**E**) Akt Thr308 phosphorylation normalised to β-tubulin or Akt (mean ± SEM). * $p < 0.05$, ** $p < 0.01$, relative to absence of insulin. # $p < 0.05$, ## $p < 0.01$ relative to absence of AMPK activator.

A769662 (50 µM, 45 min) increased AMPK activity, assessed by immunoblotting of AMPK-specific ACC (acetyl CoA carboxylase) Ser79 phosphorylation (Figure 1B,C), yet insulin had no effect on ACC phosphorylation. The inhibition of NO production was associated with markedly reduced insulin-stimulated phosphorylation of Akt at Ser473 and Thr308 (Figure 1D,E). In contrast, the AMPK activator AICAR, which is converted to the AMP mimetic ZMP in cells, increased basal and insulin-stimulated Akt Ser473 and Thr308 phosphorylation (Figure 1D,E), despite activating AMPK to a similar degree as assessed by ACC Ser79 phosphorylation (Figure 1C). Furthermore, the direct AMPK activator compound **991**, which allosterically activates AMPK at the same site as A769662 [31], had no effect on basal or insulin-stimulated Akt Thr308 phosphorylation, and only modestly reduced insulin-stimulated Akt Ser473 phosphorylation by 10% despite activating AMPK to a similar extent (Figure 2).

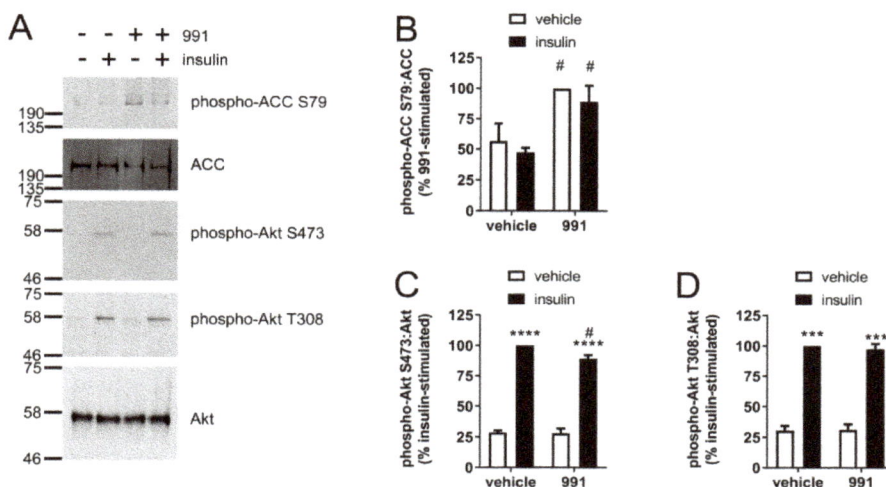

Figure 2. The effect of compound **991** on insulin-stimulated Akt phosphorylation in HUVECs. HUVECs were stimulated with compound **991** (5 µM, 45 min) prior to insulin (1 µM, 15 min) and cell lysates prepared. Proteins were resolved by SDS-PAGE and immunoblotted with the antibodies indicated. (**A**) Representative immunoblots from three biological replicates with molecular weight markers indicated. Phospho-Akt Thr308 protein levels were assessed by stripping and re-probing the membranes. Densitometric quantification of (**B**) ACC Ser79, (**C**) Akt Ser473 and (**D**) Akt Thr308 phosphorylation normalised to ACC or Akt (mean ± SEM). *** $p < 0.001$, **** $p < 0.0001$ relative to absence of insulin. # $p < 0.05$ relative to absence of compound **991**.

To examine the relationship between AMPK activation and inhibition of insulin-stimulated Akt phosphorylation by A769662, the concentration dependence of either effect of A769662 was assessed. Significant A769662-mediated stimulation of ACC phosphorylation was achieved with 50–100 µM A769662 (Figure 3A,B). A769662 decreased both insulin-stimulated Akt Ser473 and Thr308 phosphorylation in HUVECs in a concentration-dependent manner (Figure 3A), whereby 100 µM A769662 significantly inhibited Akt S473 phosphorylation and the statistical significance of insulin-stimulated Akt phosphorylation at either site was lost at concentrations above 10 µM A769662 (Figure 3C,D).

Furthermore, the time dependence of the inhibitory effect of A769662 on insulin-stimulated Akt phosphorylation was assessed. A769662 (50 µM) rapidly stimulated ACC phosphorylation in HUVECs within 5 min, an effect that was sustained for at least 2 h (Figure 3E,F). The inhibitory effect of A769662 on insulin-stimulated Akt Ser473 phosphorylation occurred similarly rapidly within 5 min and was sustained for 1 h (Figure 3E,G).

To determine whether the inhibition of insulin-stimulated Akt phosphorylation in ECs was AMPK-dependent, similar experiments were conducted after prior incubation in the SBI-0206965, which has recently been described as a selective inhibitor of AMPK [32]. Preincubation with 30 µM SBI-0206965 completely inhibited A769662-stimulated ACC Ser79 phosphorylation (Figure 4A,B), yet had no effect on the inhibition of insulin-stimulated Akt phosphorylation at Ser473 or Thr308 (Figure 4A,C,D), further indicating that the inhibitory effect of A769662 was AMPK-independent.

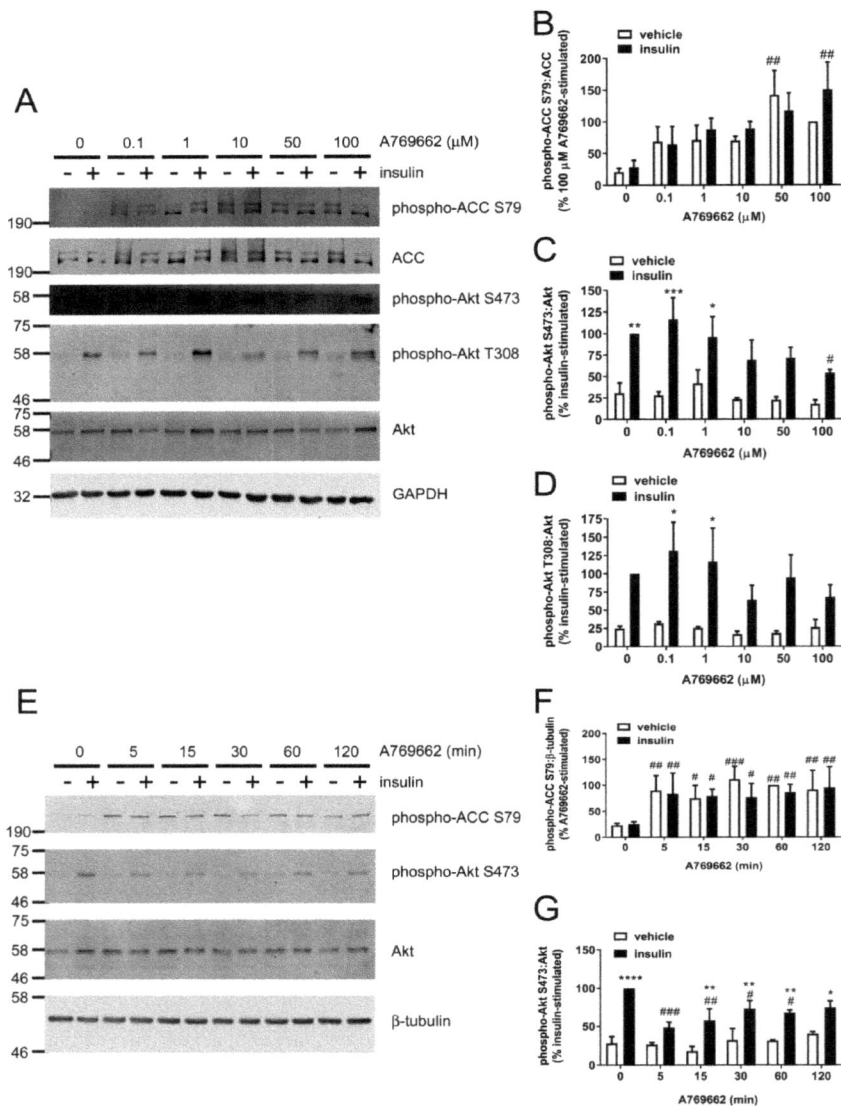

Figure 3. A769662 decreases Akt phosphorylation in a concentration- and time-dependent manner in HUVECs. HUVECs were stimulated with (**A–D**) the indicated concentrations of A769662 for 45 min or (**E–G**) 50 μM A769662 for the indicated durations prior to insulin (1 μM, 15 min). Cell lysates were prepared and immunoblotted with the antibodies indicated. (**A,E**) Representative immunoblots from three independent biological replicates with molecular weight markers indicated. Total ACC protein level was assessed by stripping and re-probing the membranes. Densitometric quantification of (**B,F**) ACC Ser79, (**C,G**) Akt Ser473 and (**D**) Akt Thr308 phosphorylation (mean ± SEM). * $p < 0.05$, ** $p < 0.01$, *** $p < 0.001$, **** $p < 0.0001$ relative to absence of insulin. # $p < 0.05$, ## $p < 0.01$, ### $p < 0.001$ relative to absence of A769662.

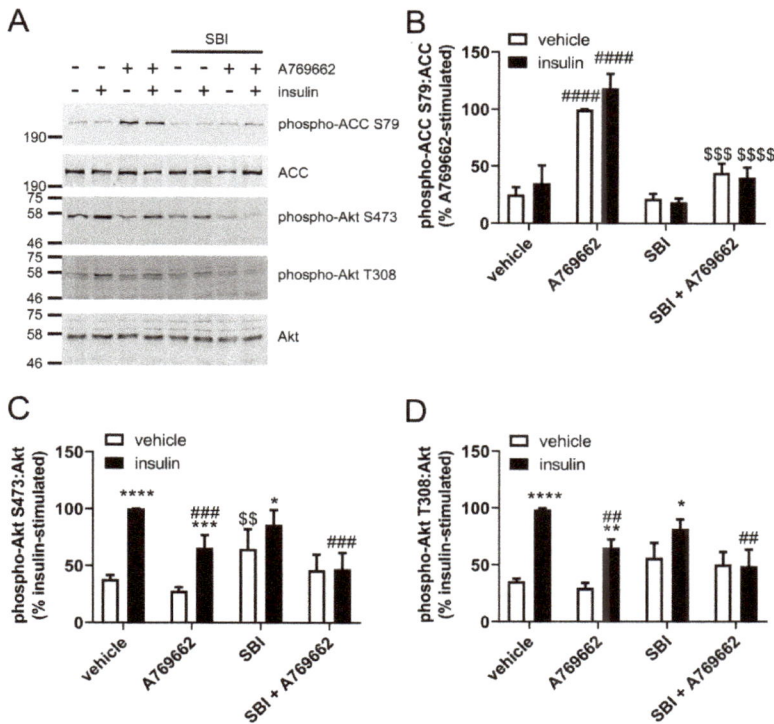

Figure 4. SBI-0206965 inhibits A769662-stimulated AMPK activation without altering inhibition of insulin-stimulated Akt phosphorylation in HUVECs. HUVECs were stimulated with A769662 (50 μM, 45 min) prior to insulin (1 μM, 15 min) after preincubation in the presence or absence of SBI-0206965 (SBI, 30 μM, 30 min) and cell lysates prepared. Proteins were resolved by SDS-PAGE and immunoblotted with the antibodies indicated. (**A**) Representative immunoblots from six biological replicates with molecular weight markers indicated. Densitometric quantification of (**B**) ACC, (**C**) Akt Ser473 and (**D**) Akt Thr308 phosphorylation normalised to total ACC or Akt (mean ± SEM). * $p < 0.05$, ** $p < 0.01$, *** $p < 0.001$, **** $p < 0.0001$ relative to absence of insulin. ## $p < 0.01$, ### $p < 0.001$, #### relative to absence of A769662. \$\$ $p < 0.01$, \$\$\$ $p < 0.001$, \$\$\$\$ $p < 0.0001$ relative to absence of SBI-0206965.

2.2. Insulin Signalling and Insulin-Stimulated NO Production are Significantly Decreased in A769662-Treated HAECs

To determine whether this inhibition of insulin signalling was conserved in ECs from other regions of the vasculature, insulin-stimulated NO synthesis and signalling were assessed in HAECs. A769662 (50 μM) markedly inhibited insulin-stimulated NO synthesis (Figure 5A), an effect associated with reduced insulin-stimulated phosphorylation of Akt Ser473 (Figure 5B,C). To examine whether insulin signalling through an alternative pathway independent of PI3K was influenced by A769662, insulin-stimulated Erk1/2 phosphorylation was assessed. Intriguingly, A769662 significantly inhibited insulin-stimulated Erk1/2 phosphorylation (Figure 5B,D).

To examine whether the inhibitory action of A769662 on insulin-stimulated Akt phosphorylation was observed in non-endothelial cell lines, similar experiments were conducted in the HeLa tumour and HepG2 hepatoma cell lines. Preincubation with A769662 (50 μM, 45 min) or compound **991** (5 μM, 60 min) significantly increased ACC Ser79 phosphorylation in both HeLa (Supplementary Figure S2) and HepG2 cells (Supplementary Figure S3) to a similar extent. Unlike HUVECs, preincubation with A769662 had no statistically significant effect on basal or insulin-stimulated Akt Ser473 phosphorylation in either cell line, although the statistical significance of the effect of insulin was lost upon preincubation with A769662 in HeLa cells (Supplementary Figure S2C,D). Compound **991** did not affect insulin-stimulated Akt Ser473 and Thr308 phosphorylation in HepG2 cells (Supplementary Figure S3C,D) yet intriguingly did increase insulin-stimulated Akt Ser473 and Thr308 phosphorylation in HeLa cells (Supplementary Figure S2C,D). The inhibition of insulin-stimulated Akt Ser473 phosphorylation by preincubation with A769662 therefore seems to be restricted to ECs, and it is not observed in insulin-sensitive human cell lines.

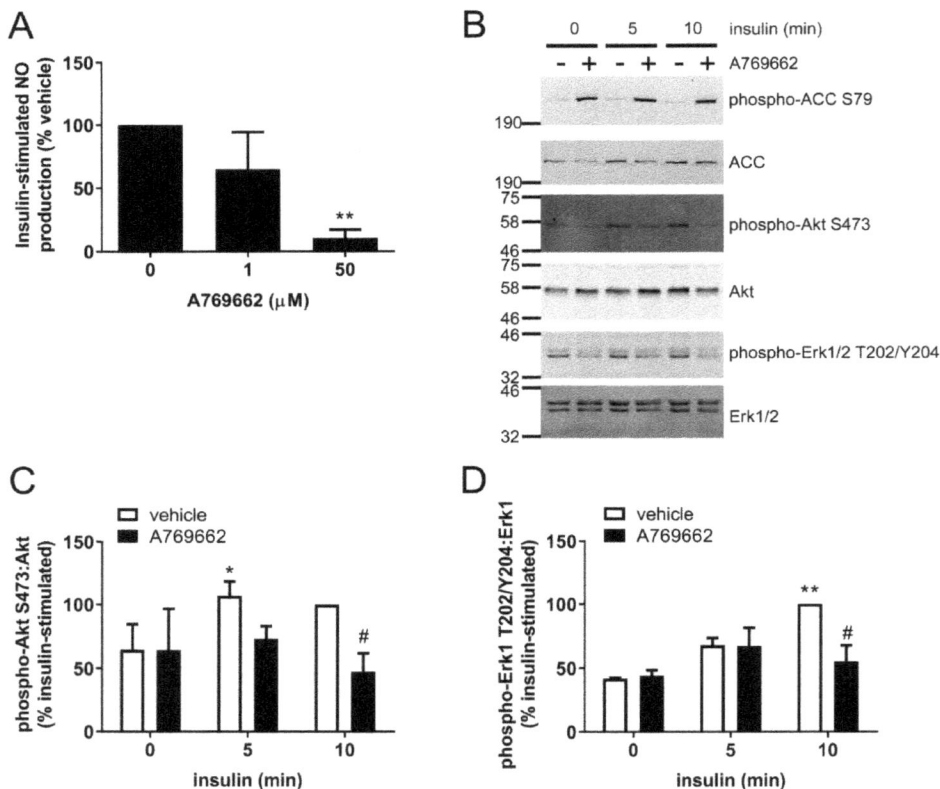

Figure 5. The effect of A769662 on insulin-stimulated NO synthesis and signalling in HAECs. (**A**) HAECs were stimulated with the indicated concentrations of A769662 prior to insulin stimulation (1 μM, 15 min). Conditioned media was collected and NO production assessed. (**B–D**) HAECs were incubated with 50 μM A769662 for 45 min prior to insulin stimulation (1 μM) for the indicated durations. Cell lysates were prepared and immunoblotted with the antibodies indicated. (**B**) Representative immunoblots from three independent biological replicates with molecular weight markers indicated. (**C,D**) Densitometric quantification (mean ± SEM) of (**C**) Akt Ser473 or (**D**) Erk1 Thr202/Tyr204 phosphorylation normalised to total Akt or Erk1 levels respectively (mean ± SEM). * $p < 0.05$, ** $p < 0.01$ relative to absence of insulin. # $p < 0.05$, relative to absence of A769662.

2.3. A769662 Inhibits Insulin-Stimulated Erk1/2 Phosphorylation in an AMPK-Independent Manner

To further assess the AMPK-dependence of the inhibition of insulin-stimulated Akt and Erk1/2 phosphorylation by A769662, SV40-immortalised wild-type (WT) or AMPK knockout (KO) mouse embryonic fibroblasts (MEFs) [33] were stimulated with insulin after prior incubation in the presence or absence of A769662 (Figure 6). A769662 (100 μM, 30 min) robustly stimulated ACC phosphorylation in WT MEFs and ACC phosphorylation was undetectable in KO MEFs (Figure 6A). Insulin stimulated Akt Ser473 and Erk1/2 phosphorylation to a similar extent in cells from either genotype, yet A769662 had no effect on insulin-stimulated Akt Ser473 phosphorylation in either genotype (Figure 6B). In contrast, insulin-stimulated Erk1/2 phosphorylation was significantly inhibited by preincubation with A769662 in cells of either genotype (Figure 6C), indicating an AMPK-independent effect.

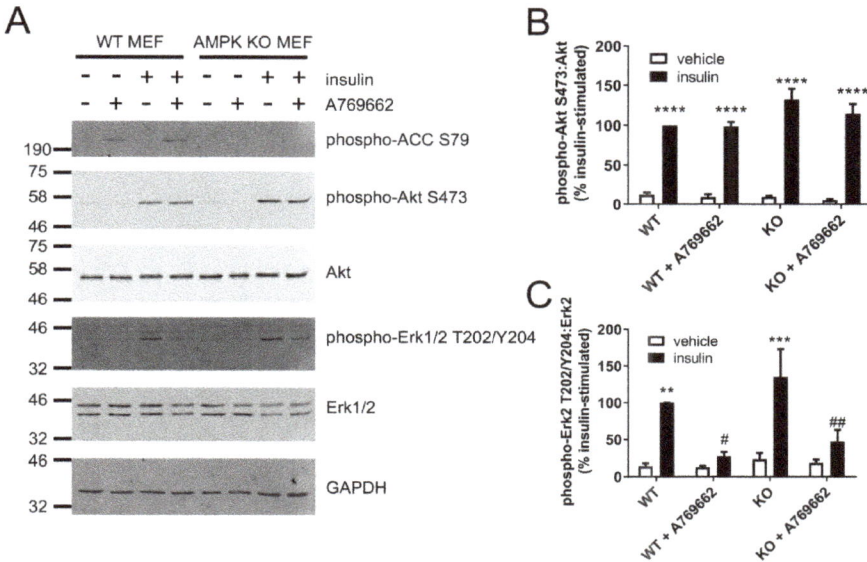

Figure 6. The effect of A769662 on insulin-stimulated Akt and Erk1/2 phosphorylation in wild-type and AMPK knockout MEFs. Wild-type (WT) and AMPK knockout (KO) MEFs were pre-incubated with A769662 (100 μM, 30 min) prior to insulin stimulation (1 μM, 15 min). Cell lysates were prepared, proteins resolved by SDS-PAGE and immunoblotted with the antibodies indicated. (**A**) Representative immunoblots from four independent biological replicates with molecular weight markers shown. (**B,C**) Densitometric quantification of (**B**) Akt Ser473 or (**C**) Erk2 Thr202/Tyr204 phosphorylation normalised to total Akt or Erk2 levels respectively (mean ± SEM). ** $p < 0.01$, *** $p < 0.001$, **** $p < 0.0001$ relative to absence of insulin. # $p < 0.05$, ## $p < 0.01$ relative to absence of A769662.

2.4. A769662 Inhibits IGF1-Stimulated Akt Ser473 Phosphorylation but Has No Effect on VEGF Signalling

To examine whether A769662 inhibits Akt phosphorylation in response to growth factors other than insulin in ECs, the effect of A769662 on IGF-1 and VEGF were assessed in HUVECs. VEGF (10 ng/mL, 10 min) stimulated a significant increase in Erk1 Thr202/Tyr204 phosphorylation, yet there was only a trend towards an increase in Akt Ser473 phosphorylation (Figure 7). In contrast, IGF1 (25 ng/mL, 10 min) significantly stimulated Akt Ser473 phosphorylation and tended to increase Erk1 Thr202/Tyr204 phosphorylation (Figure 7). Preincubation with A769662 significantly increased ACC Ser79 phosphorylation (Figure 7B) and modestly inhibited IGF1-stimulated Akt phosphorylation (Figure 7C). Although previous studies from our laboratory have shown VEGF stimulated AMPK activity after 5 min [34,35], stimulation with VEGF for 10 min only tended to increase ACC Ser79 phosphorylation in HUVECs (Figure 7B). This disparity may reflect the difference in incubation time, as in previous studies, VEGF stimulated transient ACC Ser79 phosphorylation that reached a maximum at 5 min and decreased rapidly after this point [34]. Furthermore, in contrast to the inhibitory action of A769662 on insulin-stimulated Erk1/2 phosphorylation observed in HAECs (Figure 5), IGF-1 significantly stimulated Erk1/2 phosphorylation only in the presence of A769662 (Figure 7D).

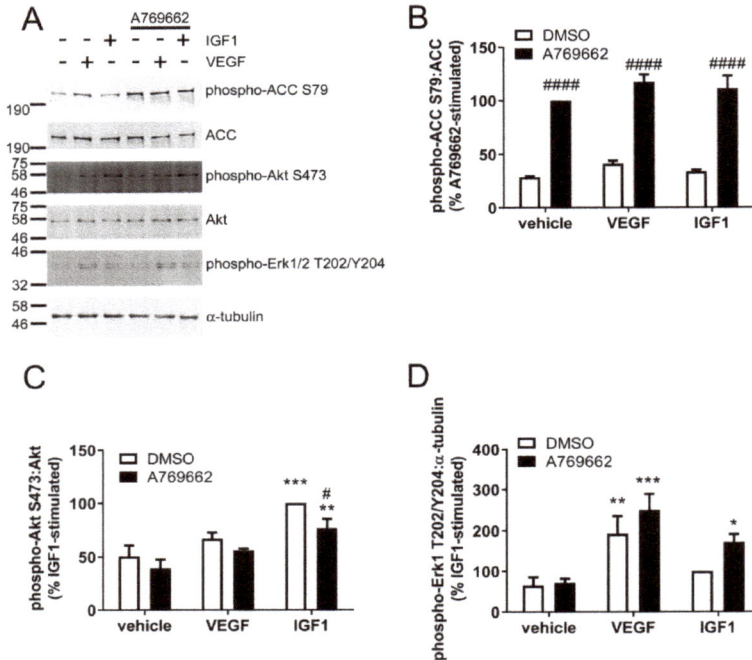

Figure 7. A769662 tends to attenuate IGF1-stimulated Ser473 Akt phosphorylation but has no effect on VEGF signalling in HUVECs. HUVECs were stimulated with A769662 (50 μM 45 min) prior to VEGF (10 ng/mL, 10 min) or IGF1 (25 ng/mL, 10 min). Cell lysates were prepared, proteins resolved by SDS-PAGE and immunoblotted with the antibodies indicated. (**A**) Representative immunoblots from four independent biological replicates with molecular weight markers indicated. Total ACC protein level was assessed by stripping and re-probing the membranes. (**B–D**) Densitometric quantification of (**B**) ACC Ser79, (**C**) Akt Ser473 or (**D**) Erk1 Thr202/Tyr204 phosphorylation normalised to total ACC, Akt or α-tubulin levels respectively (mean ± SEM). * $p < 0.05$, ** $p < 0.01$, *** $p < 0.001$ relative to absence of IGF1/VEGF. # $p < 0.05$, #### $p < 0.0001$ relative to absence of A769662.

2.5. AMPK Complexes Containing α1 and β1 Isoforms Contribute the Majority of Total Cellular AMPK Activity in HAECs

A769662 and compound **991** selectively activate AMPK complexes containing the β1 regulatory subunit [31]. It has previously been demonstrated that HepG2 cells principally express AMPKβ1, whereas HeLa cells express both AMPKβ1 and AMPKβ2 [36], yet the proportion of AMPKβ1/β2 complexes in endothelial cells has not been reported. The activities of AMPK complexes containing specific AMPKβ and α isoforms was therefore assessed in HAECs. Complexes containing AMPKβ1 accounted for approximately 60% of the total cellular AMPK activity and AMPKβ2 the remaining 40%

in HAECs. As previously reported, complexes containing AMPKα1 represent the majority (~95%) of total cellular AMPK activity in HAECs [21] (Figure 8).

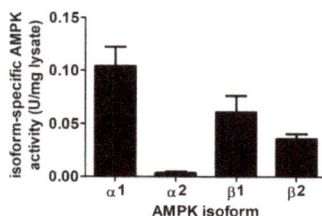

Figure 8. Activities of complexes containing specific AMPK isoforms in HAECs. AMPK was immunoprecipitated from HAEC lysates (100 µg) and AMPK activity assessed in immunoprecipitates by incorporation of ^{32}P from [γ-^{32}P]ATP into SAMS peptide. AMPK activity is presented as U/mg lysate protein (1 U = 1 nmol ^{32}P incorporated into SAMS peptide/min). Results shown are mean ± SEM activity from three independent experiments.

3. Discussion

This study demonstrates that A769662 inhibited the effects of insulin on Akt phosphorylation and NO synthesis in ECs, but had no substantive effect on Akt phosphorylation in MEFs, HeLa or HepG2 cells. This action of A769662 was not attenuated when AMPK activity was inhibited, and was not recapitulated by the alternative AMPK activators compound **991** and AICAR, suggesting this is an AMPK-independent action of A769662. In addition, A769662 inhibited insulin-stimulated Erk1 phosphorylation in ECs and MEFs, an effect that was still apparent in MEFs lacking AMPK activity. Taken together, these data suggest an EC-specific action of A769662 on early insulin signalling that is independent of AMPK.

Previous studies have demonstrated multiple beneficial effects of AMPK activation on insulin signalling, lipid and plasma glucose levels [17,23,37,38]. In ECs, AMPK activation has been reported to stimulate NO synthesis and angiogenesis while inhibiting pro-inflammatory signalling and reactive oxygen species synthesis [10]. Most of these studies have used AMPK activators that either indirectly activate AMPK through the inhibition of mitochondrial ATP synthesis, such as metformin, berberine and resveratrol or mimic AMP, such as AICAR [39]. As a consequence, these AMPK activators also have numerous other effects that are not mediated by AMPK. In contrast, A769662 and compound **991** are direct allosteric activators of AMPK that selectively or show a bias toward activating AMPK complexes containing the β1 regulatory subunit isoform [31]. Few studies have examined the endothelial effects of A769662, although it has been reported to inhibit antioxidant gene expression [40,41], viability and proliferation [13,32] and proinflammatory signalling [12,42] in human ECs. Furthermore, compound **991** has been also reported to inhibit proinflammatory signalling in ECs [12].

It is, therefore, surprising that A769662 inhibited insulin-stimulated NO synthesis in HUVECs and HAECs, as many studies report that AMPK activation stimulates NO synthesis [10]. Indeed, A769662 stimulates activating eNOS Ser1177 phosphorylation in hearts and AICAR and resveratrol have been reported to improve impaired insulin-mediated vascular responses in rodents [43–45]. Endothelial AMPK activation is not always associated with eNOS phosphorylation or NO synthesis, however [46,47], and the lack of insulin-stimulated NO synthesis in A769662-stimulated HAECs and HUVECs is likely to be a consequence of reduced insulin-stimulated Akt phosphorylation, since Akt is required for the activation of eNOS by insulin [48]. A769662 also decreased IGF1-stimulated Akt phosphorylation. Insulin and IGF1 both activate Akt [49], differing only in the initial step, as they bind to the insulin receptor (IR) and IGF1 receptor (IGF1R) respectively. In addition, IGF1R and IR share enough structural similarity to allow interchangeable binding of IGF1 and insulin [49,50]. Although IGF1 has a higher affinity for IGF1R, IGF1 may still bind IR and vice-versa. This may explain why A769662 seems to have a more modest effect on IGF1-stimulated Akt Ser473 phosphorylation, as IGF1 signalling may recruit distinct populations of IR substrate (IRS) proteins compared to insulin [51]. On the other hand, unlike insulin and IGF1, VEGF signalling does not utilise IRSs to recruit PI3K and was unaffected by A769662 preincubation [52]. As both Akt Ser473 and Thr308 phosphorylation were inhibited by A769662 in ECs, this suggests that A769662 does not simply inhibit the kinases that phosphorylate those sites, mammalian target of rapamycin complex 2 (mTORC2) or phosphoinositide-dependent protein kinase-1 (PDK1) respectively [2]. In addition, A769662 also inhibited insulin-stimulated Erk1/2 phosphorylation in ECs and MEFs, which is activated by a pathway separate to that of Akt after insulin receptor activation [53]. Importantly, A769662 also inhibited insulin-stimulated Erk1/2 phosphorylation in MEFs lacking AMPK, demonstrating that this effect is AMPK-independent. A769662 has previously been demonstrated to have no direct effect on the activity of Akt or PDK1, although it did inhibit the Erk1 kinase, mitogen-activated protein kinase kinase-1 (MKK1), in vitro [54]. In contrast, IGF-1-stimulated Erk1 phosphorylation was accentuated by A769662 in HUVECs, such that direct inhibition of MKK1 by A769662 may not simply underlie the inhibition of insulin-stimulated Erk1/2 phosphorylation. Taken together, these data indicate that A769662 inhibits at the level of the insulin/IGF-1 receptor or a receptor-associated protein, thereby inhibiting both pathways.

The inhibition of insulin-stimulated Akt phosphorylation by A769662 in ECs was not recapitulated by either compound 991 or AICAR, both of which activate AMPK to a similar degree. Compound **991** allosterically activates AMPK at a similar site to A769662 [31], whereas AICAR is phosphorylated to the AMP-mimetic, ZMP within cells [39]. Indeed, AICAR increased basal and insulin-stimulated Akt phosphorylation in ECs, in agreement with previous reports [15,21]. As both Akt Ser473 and Thr308 phosphorylation increased, this likely reflects stimulation of PI3K or earlier signalling events by AICAR, independent of AMPK activation. Given that the inhibition of Akt phosphorylation was still observed in ECs in which AMPK activity had been completely inhibited by SBI-0206965 and the lack of effect of compound **991** despite the similar mechanism by which compound **991** activates AMPK, these data indicate an AMPK-independent action of A769662 on insulin-stimulated Akt phosphorylation, although it occurs with a similar concentration to that required for AMPK activation.

Intriguingly, the inhibitory effect of A769662 on insulin-stimulated Akt activation was limited to ECs, as it was not observed in MEFs, HeLa or HepG2 cells, although the statistical significance of stimulation by insulin was lost in HeLa cells preincubated with A769662. A769662 (100 μM) has been demonstrated previously to inhibit basal Akt Ser473 phosphorylation in prostate cancer cell lines, yet stimulated Akt Ser473 phosphorylation at a lower concentration (50 μM) [55]. Indeed, previous studies have shown no effect of A769662 on insulin-stimulated Akt activation in rat adult cardiomyocytes, human myotubes or L6 cells [29,56,57]. In contrast to this neutral effect reported in striated muscle cells, A769662 has also been reported to stimulate glucose uptake in muscle cells by increasing PI3K association with IRS1, suggesting an AMPK-independent effect that increased Akt activity [58]. In addition, high concentrations of A769662 stimulated Akt phosphorylation in a manner sensitive to the PI3K inhibitor, wortmannin in CHO cells expressing the δ-opioid receptor [59]. It is clear, therefore, that A769662 influences Akt differentially in different cell types, providing further evidence that these effects are not mediated by AMPK.

In addition, we demonstrate that preincubation of HeLa cells, but not HepG2 cells or ECs, with compound **991** increased insulin-stimulated Akt phosphorylation. These data argue for a cell-type specific/selective effect of A769662 on insulin signalling, whereby only A769662 markedly reduces insulin signalling in ECs and insulin-stimulated Erk1/2 phosphorylation in ECs and MEFs. This cell-type selectivity could be related to differential expression of IR/IGF1R between cell types. HUVECs have been previously reported to have approximately 400,000 IGF1R/cell and 40,000 IR/cell [30]. Furthermore, according to the updated version of Human Protein Atlas (https://www.proteinatlas. org/) [60], HepG2 and HeLa cells have been reported to express 21.9 and 27.5 IGF1R transcripts per million (TPM), respectively and 15.9 and 1.9 IR TPM respectively [61]. Higher IR levels could preserve insulin-stimulated Akt and Erk1 phosphorylation by increasing the number of activated receptors and signal intensity. The high relative IR expression levels in HepG2 cells may therefore explain why A769662 had no effect on attenuation of insulin signalling in this cell line, whereas in HUVECs and HAECs, which have lower IR expression levels, there was such a marked effect of A769662.

It is unlikely that the differential effects of A769662 in the different cell types can be explained by differences in AMPK isoform expression, due to the selectivity of A769662 and compound **991** [31]. Both compounds would be expected to activate AMPK complexes containing β1, and complexes containing AMPKβ1 were found to contribute ~60% of the total cellular AMPK activity in HAECs. It has previously been reported that HepG2 cells principally express AMPKβ1, whereas HeLa cells express both AMPKβ1 and AMPKβ2 [36]. This further indicates that the inhibition of insulin signalling in ECs by A769662 is an AMPK-independent effect and unrelated to differential actions on specific pools of AMPK within ECs, as all the cell types investigated express abundant AMPKβ1 levels. As AMPK-independent effects of A769662, including inhibition of 26S proteasome activity in MEFs, voltage-gated Na^+ channels in rat neurons and the Na^+/K^+ ATPase in L6 cells [26,28,29] have been reported previously, inhibition of insulin signalling in ECs may similarly be considered AMPK-independent. In addition, A769662 was recently reported to promote vasorelaxation in rabbit and rat arteries by reducing cytosolic Ca^{2+} levels, by an endothelium-dependent yet AMPK-independent manner [62].

In conclusion, A769662 decreases insulin-stimulated NO synthesis and Akt Ser473 phosphorylation in HUVECs and HAECs in a manner likely to be independent of AMPK. Furthermore, A769662 decreases insulin-stimulated Erk1/2 phosphorylation in a manner that is AMPK-independent in MEFs and HAECs. Taken together, these data demonstrate that caution should be exercised when interpreting data obtained using A769662 as a tool in cultured human endothelial cells.

4. Materials and Methods

4.1. Materials

Cryopreserved HUVECs, HAECs and MV2 medium were purchased from Promocell (Heidelberg, Germany). SV40-immortalised wild-type and AMPKα1 and AMPKα2 knock-out MEFs were kindly provided by Dr. B. Viollet (Institut Cochin, Paris, France) and have been described previously [33]. HeLa and HepG2 cells were obtained from ATCC (Manassas, VA, USA). IGF1, VEGF, SBI-0206965 and porcine insulin were purchased from Sigma Aldrich (St. Louis, MO, USA). AICAR was purchased from Toronto Research Chemicals Inc. (Ontario, ON, Canada). Compound **991** was synthetised by MRC Technology. A769662 and mouse anti-β-tubulin (#11307) antibodies were purchased from Abcam (Cambridge, UK). Rabbit anti-phospho-Akt Thr308 (#13038), anti-phospho-Akt Ser473 (#4058), anti-ERK1/2 (#9102), anti-phospho-ACC Ser79 (#3661), anti-ACC (#3676), anti-α-tubulin (#2144) and mouse anti-Akt (#2920) and anti-phospho-p44/42 Erk1/2 (Thr202/Tyr204) (#9106) antibodies were from New England Biolabs UK (Hitchin, UK). Sheep anti-AMPKα1, anti-AMPKα2, anti-AMPKβ1 and anti-AMPKβ2 antibodies used for immunoprecipitation of AMPK complexes containing specific subunit isoforms were a kind gift from Professor D.G. Hardie (University of Dundee) and have been described previously [39,63,64]. IRdye680 or 800-labelled donkey anti-mouse IgG (#926-32212) and anti-rabbit IgG (#926-68023 and #926-32213) antibodies were from LI-COR Biosciences (Lincoln, UK). Medium 199, RPMI 1640 and DMEM (4.5 g/L glucose) were from Life Technologies (Paisley, UK).

4.2. Cell Culture and Experimental Design

HAECs and HUVECs were cultured in MV2 medium supplemented with EC growth factor mix, 5% (*v/v*) serum (Promocell). HUVECs and HAECs were utilised between passages 3 and 6. MEFs and HeLa cells were cultured in DMEM supplemented with 10% (*v/v*) foetal calf serum. HepG2 cells were cultured in RPMI 1640 supplemented with 10% (*v/v*) foetal calf serum, 1 mM sodium pyruvate. Once cells reached 90–100% confluence, HUVECs/HAECs were incubated in serum-free Medium 199 and HeLa cells and HepG2 cells were incubated with serum-free DMEM for 2 h and then with AMPK activators prior to stimulation with insulin (1 µM), VEGF (10 ng/mL) or IGF1 (25 ng/mL). Cells were placed on ice and washed with PBS prior to lysis in Triton X-100-based lysis buffer (50 mM Tris-HCl, pH 7.4 at 4 °C, 50 mM NaF, 1 mM $Na_4P_2O_7$, 1 mM EDTA, 1 mM EGTA, 1% (v/v) Triton-X-100, 250 mM mannitol, 1 mM DTT, 1 mM Na_3VO_4, 0.1 mM benzamidine, 0.1 mM PMSF, 5 µg/mL SBTI). Cell lysates were scraped into microcentrifuge tubes and incubated on ice for 20 min, centrifuged (5 min, 21,910× *g*, 4 °C) and the subsequent supernatants stored at −20 °C. Lysate protein concentrations were determined using Bradford or BCA methods, as previously described [6,21].

4.3. SDS-Polyacrylamide Gel Electrophoresis and Immunoblotting

Cell lysate proteins were resolved by SDS-PAGE and immunoblotted with antibodies diluted in 50% (v/v) LI-COR blocking buffer in TBS containing 0.1% (v/v) Tween-20 as described previously [13]. Proteins were visualised using infrared dye-labelled secondary antibodies on a LI-COR Odyssey infrared imaging system and analysed using the ImageJ software for densitometric quantification of band intensity. In some cases, indicated in the figure legends, immunoblots were stripped in 0.2 M NaOH for 10 min, washed multiple times with Tris-buffered saline (TBS) until the pH returned to 7.2, and then blocked in TBS supplemented with 5% (w/v) milk powder prior to probing with primary antibodies. In all other cases, immunoblots for phospho- and total protein levels of ACC, Akt and Erk1/2 were obtained on immunoblots conducted concurrently in parallel.

4.4. NO Assay

HAECs and HUVECs were cultured in 6-well plates until 100% confluent and serum-starved for 2 h in Medium 199 and then in the presence of Krebs-Ringer-Hepes (KRH) buffer for 20 min. The medium was replaced and cells incubated in the presence or absence of stimuli in KRH for a further 15–20 min. Samples of medium (50 µL) were taken at different intervals and 200 µL methanol added to each. Samples were centrifuged (21,910× *g*, 4 °C, 20 min), and supernatants were stored at −20 °C. NO concentration was determined using a Sievers 280A NO Meter (Sievers, Boulder, CO, USA) as described previously [21].

4.5. AMPK Assay

AMPK specific isoforms were immunoprecipitated from HAECs using sheep anti-AMPKα1, anti-AMPKα2 [63], anti-AMPKβ1 [39] or anti-AMPKβ2 [64] antibodies bound to protein G-Sepharose (1 µg antibody and 5 µL packed volume protein G-Sepharose/immunoprecipitation) in IP buffer (50 mM Tris-HCl (pH 7.4 at 4 °C), 150 mM NaCl, 50 mM NaF, 5 mM $Na_4P_2O_7$, 1 mM EDTA, 1 mM EGTA, 1% (v/v) Triton-X-100, 1% (v/v) glycerol, 1 mM DTT, 0.1 mM benzamidine, 1 mM PMSF, 5 µg/mL SBTI, 1 mM Na_3VO_4). Immunoprecipitates were then washed into HBD buffer (50 mM HEPES-NaOH (pH 7.4), 0.02% (v/v) Brij-35, 1 mM DTT) and assayed for AMPK activity using SAMS substrate peptide as described previously [12,35].

4.6. Statistical Analysis

All data is expressed as a relative change (mean ± SEM) from control baseline or normalised to group of reference for each experiment. Prism software (GraphPad Software, San Diego, CA, USA) was used to perform one or two-way ANOVA (with Tukey's or Dunnett's post hoc multiple comparison tests), where appropriate, using $p < 0.05$ as significant.

Supplementary Materials: The following are available online at http://www.mdpi.com/1422-0067/19/12/3886/s1.

Author Contributions: Conceptualisation, A.S. and I.P.S.; Methodology, A.S. and I.P.S.; Investigation, A.S., S.J.M., J.M.G. and I.P.S.; Validation, A.S., S.J.M. and J.M.G.; Formal Analysis, A.S., S.J.M. and I.P.S.; Writing—Original Draft Preparation, A.S. and I.P.S.; Writing—Review & Editing, I.P.S., T.M.P. and G.S.B.; Supervision, I.P.S.; Funding Acquisition I.P.S. and T.M.P. All listed authors approved the final version of the manuscript.

funding: This work was supported by the British Heart Foundation studentship (FS/14/61/31284 to A.S.), a British Heart Foundation Project Grant (PG/13/82/30483 to I.P.S. and T.M.P.) and a Diabetes UK equipment grant (BDA11/0004309 to I.P.S. and T.M.P.).

Conflicts of Interest: The authors declare no conflict of interest. The funders had no role in the design of the study; in the collection, analyses, or interpretation of data; in the writing of the manuscript, and in the decision to publish the results.

References

1. Zhao, Y.; Vanhoutte, P.M.; Leung, S.W.S. Vascular nitric oxide: Beyond eNOS. *J. Pharmacol. Sci.* **2015**, *12*, 83–94. [CrossRef] [PubMed]
2. Manning, B.D.; Toker, A. AKT/PKB Signaling: Navigating the Network. *Cell* **2017**, *169*, 381–405. [CrossRef] [PubMed]
3. Jiang, Z.Y.; He, Z.; King, B.L.; Kuroki, T.; Opland, D.M.; Suzuma, K.; Suzuma, I.; Ueki, K.; Kulkarni, R.N.; Kahn, C.R. Characterization of multiple signaling pathways of insulin in the regulation of vascular endothelial growth factor expression in vascular cells and angiogenesis. *J. Biol. Chem.* **2003**, *278*, 31964–31971. [CrossRef] [PubMed]
4. Takahashi, S.; Mendelsohn, M.E. Synergistic activation of endothelial nitric-oxide synthase (eNOS) by HSP90 and Akt. Calcium-independent eNOS activation involves formation of an HSP90-Akt-CaM-bound eNOS complex. *J. Biol. Chem.* **2003**, *278*, 30821–30827. [CrossRef] [PubMed]
5. Michell, B.J.; Griffiths, J.E.; Mitchelhill, K.I.; Rodriguez-Crespo, I.; Tiganis, T.; Bozinovski, S.; Montellano, P.R.O.; Kemp, B.E.; Pearson, R.B. The Akt kinase signals directly to endothelial nitric oxide synthase. *Curr. Biol.* **1999**, *9*, 845–848. [CrossRef]
6. Ritchie, S.A.; Kohlhaas, C.F.; Boyd, A.R.; Yalla, K.C.; Walsh, K.; Connell, J.M.; Salt, I.P. Insulin-stimulated phosphorylation of endothelial nitric oxide synthase at serine-615 contributes to nitric oxide synthesis. *Biochem. J.* **2010**, *426*, 85–90. [CrossRef] [PubMed]
7. Reynolds, L.J.; Credeur, D.P.; Manrique, C.; Padilla, J.; Fadel, P.J.; Thyfault, J.P. Obesity, type 2 diabetes, and impaired insulin-stimulated blood flow: Role of skeletal muscle NO synthase and endothelin-1. *J. Appl. Physiol.* **2017**, *122*, 38–47. [CrossRef] [PubMed]
8. Tabit, C.E.; Chung, W.B.; Vita, J.A. Endothelial dysfunction in diabetes mellitus: Molecular mechanisms and clinical implications. *Rev. Endocr. Metab. Disord.* **2010**, *11*, 61–74. [CrossRef]
9. Day, E.A.; Ford, R.J.; Steinberg, G.R. AMPK as a Therapeutic Target for Treating Metabolic Diseases. *Trends Endocrinol. MeTab.* **2017**, *28*, 545–560. [CrossRef]
10. Salt, I.P.; Hardie, D.G. AMP-Activated Protein Kinase: An Ubiquitous Signaling Pathway with Key Roles in the Cardiovascular System. *Circ. Res.* **2017**, *120*, 1825–1841. [CrossRef]
11. Mancini, S.J.; White, A.D.; Bijland, S.; Rutherford, C.; Graham, D.; Richter, E.A.; Viollet, B.; Touyz, R.M.; Palmer, T.M.; Salt, I.P. Activation of AMP-activated protein kinase rapidly suppresses multiple pro-inflammatory pathways in adipocytes including IL-1 receptor-associated kinase-4 phosphorylation. *Mol. Cell. Endocrinol.* **2017**, *440*, 44–56. [CrossRef] [PubMed]
12. Mancini, S.J.; Boyd, D.; Katwan, O.J.; Strembitska, A.; Almabrouk, T.A.; Kennedy, S.; Palmer, T.M.; Salt, I.P. Canagliflozin inhibits interleukin-1β-stimulated cytokine and chemokine secretion in vascular endothelial cells by AMP-activated protein kinase-dependent and independent mechanisms. *Sci. Rep.* **2018**, *8*, 5276. [CrossRef] [PubMed]
13. Peyton, K.J.; Liu, X.M.; Yu, Y.; Yates, B.; Durante, W. Activation of AMP-Activated Protein Kinase Inhibits the Proliferation of Human Endothelial Cells. *J. Pharmacol. Exp. Ther.* **2012**, *342*, 827–834. [CrossRef] [PubMed]
14. Li, C.; Reif, M.M.; Craige, S.M.; Kant, S.; Keaney, J.F. Endothelial AMPK activation induces mitochondrial biogenesis and stress adaptation via eNOS-dependent mTORC1 signaling. *Nitric Oxide-Biol. Chem.* **2016**, *55–56*, 45–53. [CrossRef] [PubMed]
15. Ido, Y.; Carling, D.; Ruderman, N. Hyperglycemia-induced apoptosis in human umbilical vein endothelial cells: Inhibition by the AMP-activated protein kinase activation. *Diabetes* **2002**, *51*, 159–167. [CrossRef] [PubMed]
16. Li, Y.; Xu, S.; Mihaylova, M.M.; Zheng, B.; Hou, X.; Jiang, B.; Park, O.; Luo, Z.; Lefai, E.; Shyy, J.Y.J.; et al. AMPK phosphorylates and inhibits SREBP activity to attenuate hepatic steatosis and atherosclerosis in diet-induced insulin-resistant mice. *Cell MeTab.* **2011**, *13*, 376–388. [CrossRef] [PubMed]

17. Ford, R.J.; Fullerton, M.D.; Pinkosky, S.L.; Day, E.A.; Scott, J.W.; Oakhill, J.S.; Bujak, A.L.; Smith, B.K.; Crane, J.D.; Blümer, R.M.; et al. Metformin and salicylate synergistically activate liver AMPK, inhibit lipogenesis and improve insulin sensitivity. *Biochem. J.* **2015**, *468*, 125–132. [CrossRef] [PubMed]

18. Mottillo, E.P.; Desjardins, E.M.; Crane, J.D.; Smith, B.K.; Green, A.E.; Ducommun, S.; Henriksen, T.I.; Rebalka, I.A.; Razi, A.; Sakamoto, K.; et al. Lack of Adipocyte AMPK Exacerbates Insulin Resistance and Hepatic Steatosis through Brown and Beige Adipose Tissue Function. *Cell Metab.* **2016**, *24*, 118–129. [CrossRef]

19. Boyle, J.G.; Logan, P.J.; Ewart, M.A.; Reihill, J.A.; Ritchie, S.A.; Connell, J.M.; Cleland, S.J.; Salt, I.P. Rosiglitazone Stimulates Nitric Oxide Synthesis in Human Aortic Endothelial Cells via AMP-activated Protein Kinase. *J. Biol. Chem.* **2008**, *283*, 11210–11217. [CrossRef]

20. Csiszar, A.; Labinskyy, N.; Podlutsky, A.; Kaminski, P.M.; Wolin, M.S.; Zhang, C.; Mukhopadhyay, P.; Pacher, P.; Hu, F.; De Cabo, R.; et al. Vasoprotective effects of resveratrol and SIRT1: Attenuation of cigarette smoke-induced oxidative stress and proinflammatory phenotypic alterations. *Am. J. Physiol. Heart Circ. Physiol.* **2008**, *294*, H2721–H2735. [CrossRef]

21. Morrow, V.A.; Foufelle, F.; Connell, J.M.C.; Petrie, J.R.; Gould, G.W.; Salt, I.P. Direct activation of AMP-activated protein kinase stimulates nitric-oxide synthesis in human aortic endothelial cells. *J. Biol. Chem.* **2003**, *278*, 31629–31639. [CrossRef] [PubMed]

22. Davis, B.J.; Xie, Z.; Viollet, B.; Zou, M.H. Activation of the AMP-activated kinase by antidiabetes drug metformin stimulates nitric oxide synthesis in vivo by promoting the association of heat shock protein 90 and endothelial nitric oxide synthase. *Diabetes* **2006**, *55*, 496–505. [CrossRef] [PubMed]

23. Cool, B.; Zinker, B.; Chiou, W.; Kifle, L.; Cao, N.; Perham, M.; Dickinson, R.; Adler, A.; Gagne, G.; Iyengar, R.; et al. Identification and characterization of a small molecule AMPK activator that treats key components of type 2 diabetes and the metabolic syndrome. *Cell MeTab.* **2006**, *3*, 403–416. [CrossRef] [PubMed]

24. Bultot, L.; Jensen, T.E.; Lai, Y.C.; Madsen, A.L.; Collodet, C.; Kviklyte, S.; Deak, M.; Yavari, A.; Foretz, M.; Ghaffari, S.; et al. Benzimidazole derivative small-molecule 991 enhances AMPK activity and glucose uptake induced by AICAR or contraction in skeletal muscle. *Am. J. Physiol. Endocrinol. MeTab.* **2016**, *311*, E706–E719. [CrossRef] [PubMed]

25. Ducommun, S.; Ford, R.J.; Bultot, L.; Deak, M.; Bertrand, L.; Kemp, B.E.; Steinberg, G.R.; Sakamoto, K. Enhanced activation of cellular AMPK by dual-small molecule treatment: AICAR and A769662. *Am. J. Physiol. Endocrinol. MeTab.* **2014**, *306*, E688–E696. [CrossRef] [PubMed]

26. Asiedu, M.N.; Han, C.; Dib-Hajj, S.D.; Waxman, S.G.; Price, T.J.; Dussor, G. The AMPK activator A769662 blocks voltage-gated sodium channels: Discovery of a novel pharmacophore with potential utility for analgesic development. *PLoS ONE* **2017**, *12*, e0169882. [CrossRef] [PubMed]

27. Vlachaki Walker, J.M.; Robb, J.L.; Cruz, A.M.; Malhi, A.; Weightman Potter, P.G.; Ashford, M.L.; McCrimmon, R.J.; Ellacott, K.L.; Beall, C. AMP-activated protein kinase (AMPK) activator A-769662 increases intracellular calcium and ATP release from astrocytes in an AMPK-independent manner. *Diabetes Obes. MeTab.* **2017**, *19*, 997–1005. [CrossRef]

28. Moreno, D.; Knecht, E.; Viollet, B.; Sanz, P. A769662, a novel activator of AMP-activated protein kinase, inhibits non-proteolytic components of the 26S proteasome by an AMPK-independent mechanism. *FEBS Lett.* **2008**, *582*, 2650–2654. [CrossRef]

29. Benziane, B.; Björnholm, M.; Lantier, L.; Viollet, B.; Zierath, J.R.; Chibalin, A.V. AMP-activated protein kinase activator A-769662 is an inhibitor of the Na^+-K^+-ATPase. *Am. J. Physiol. Cell Physiol.* **2009**, *297*, C1554–C1567. [CrossRef]

30. Zeng, G.; Quon, M.J. Insulin-stimulated production of nitric oxide is inhibited by Wortmannin: Direct measurement in vascular endothelial cells. *J. Clin. Investig.* **1996**, *98*, 894–898. [CrossRef]

31. Xiao, B.; Sanders, M.J.; Carmena, D.; Bright, N.J.; Haire, L.F.; Underwood, E.; Patel, B.R.; Heath, R.B.; Walker, P.A.; Hallen, S.; et al. Structural basis of AMPK regulation by small molecule activators. *Nat Commun.* **2013**, *4*, 1–17. [CrossRef] [PubMed]

32. Dite, T.A.; Langendorf, C.G.; Hoque, A.; Galic, S.; Rebello, R.J.; Ovens, A.J.; Lindqvist, L.M.; Ngoei, K.R.; Ling, N.X.; Furic, L.; et al. AMP-activated protein kinase selectively inhibited by the type II inhibitor SBI-0206965. *J. Biol. Chem.* **2018**, *293*, 8874–8885. [CrossRef] [PubMed]

33. Jørgensen, S.B.; Viollet, B.; Andreelli, F.; Frøsig, C.; Birk, J.B.; Schjerling, P.; Vaulont, S.; Richter, E.A.; Wojtaszewski, J.F. Knockout of the $\alpha2$ but Not α_1 5'-AMP-activated Protein Kinase Isoform Abolishes

5-Aminoimidazole-4-carboxamide-1-β-4-ribofuranoside- but Not Contraction-induced Glucose Uptake in Skeletal Muscle. *J. Biol. Chem.* **2004**, *279*, 1070–1079. [CrossRef] [PubMed]

34. Reihill, J.A.; Ewart, M.A.; Salt, I.P. The role of AMP-activated protein kinase in the functional effects of vascular endothelial growth factor-A and -B in human aortic endothelial cells. *Vasc. Cell.* **2011**, *3*, 9. [CrossRef] [PubMed]

35. Heathcote, H.R.; Mancini, S.J.; Strembitska, A.; Jamal, K.; Reihill, J.A.; Palmer, T.M.; Gould, G.W.; Salt, I.P. Protein kinase C phosphorylates AMP-activated protein kinase α1 Ser487. *Biochem. J.* **2016**, *473*, 4681–4697. [CrossRef] [PubMed]

36. Scott, J.W.; Galic, S.; Graham, K.L.; Foitzik, R.; Ling, N.X.; Dite, T.A.; Issa, S.M.; Langendorf, C.G.; Weng, Q.P.; Thomas, H.E.; et al. Inhibition of AMP-Activated Protein Kinase at the Allosteric Drug-Binding Site Promotes Islet Insulin Release. *Chem. Biol.* **2015**, *22*, 705–711. [CrossRef] [PubMed]

37. Lee, Y.S.; Kim, W.S.; Kim, K.H.; Yoon, M.J.; Cho, H.J.; Shen, Y.; Ye, J.M.; Lee, C.H.; Oh, W.K.; Kim, C.T.; et al. Berberine, a natural plant product, activates AMP-activated protein kinase with beneficial metabolic effects in diabetic and insulin-resistant states. *Diabetes* **2006**, *55*, 2256–2264. [CrossRef] [PubMed]

38. Zang, M.; Zuccollo, A.; Hou, X.; Nagata, D.; Walsh, K.; Herscovitz, H.; Brecher, P.; Ruderman, N.B.; Cohen, R.A. AMP-activated protein kinase is required for the lipid-lowering effect of metformin in insulin-resistant human HepG2 cells. *J. Biol. Chem.* **2004**, *279*, 47898–47905. [CrossRef]

39. Hawley, S.A.; Ross, F.A.; Chevtzoff, C.; Green, K.A.; Evans, A.; Fogarty, S.; Towler, M.C.; Brown, L.J.; Ogunbayo, O.A.; Evans, A.M.; et al. Use of cells expressing γ subunit variants to identify diverse mechanisms of AMPK activation. *Cell MeTab.* **2010**, *11*, 554–565. [CrossRef]

40. Liu, X.; Peyton, K.J.; Shebib, A.R.; Wang, H.; Korthuis, R.J.; Durante, W. Activation of AMPK stimulates heme oxygenase-1 gene expression and human endothelial cell survival. *Am. J. Physiol. Heart Circ. Physiol.* **2011**, *300*, H84–H93. [CrossRef]

41. Dang, Y.; Ling, S.; Duan, J.; Ma, J.; Ni, R.; Xu, J.W. Bavachalcone-Induced Manganese Superoxide Dismutase Expression through the AMP-Activated Protein Kinase Pathway in Human Endothelial Cells. *Pharmacology* **2015**, *95*, 105–110. [CrossRef] [PubMed]

42. Rutherford, C.; Speirs, C.; Williams, J.J.L.; Ewart, M.A.; Mancini, S.J.; Hawley, S.A.; Delles, C.; Viollet, B.; Costa-Pereira, A.P.; Baillie, G.S.; et al. Phosphorylation of Janus kinase 1 (JAK1) by AMP-activated protein kinase (AMPK) links energy sensing to anti-inflammatory signaling. *Sci. Signal.* **2016**, *9*, ra109. [CrossRef] [PubMed]

43. Kim, A.S.; Miller, E.J.; Wright, T.M.; Li, J.; Qi, D.; Atsina, K.; Zaha, V.; Sakamoto, K.; Young, L.H. A small molecule AMPK activator protects the heart against ischemia-reperfusion injury. *J. Mol. Cell. Cardiol.* **2011**, *51*, 24–32. [CrossRef] [PubMed]

44. Bradley, E.A.; Zhang, L.; Genders, A.J.; Richards, S.M.; Rattigan, S.; Keske, M.A. Enhancement of insulin-mediated rat muscle glucose uptake and microvascular perfusion by 5-aminoimidazole-4-carboxamide-1-β-d-ribofuranoside. *Cardiovasc. Diabetol.* **2015**, *14*, 91. [CrossRef] [PubMed]

45. Liu, Z.; Jiang, C.; Zhang, J.; Liu, B.; Du, Q. Resveratrol inhibits inflammation and ameliorates insulin resistant endothelial dysfunction via regulation of AMP-activated protein kinase and sirtuin 1 activities. *J. Diabetes* **2016**, *8*, 324–335. [CrossRef]

46. Stahmann, N.; Woods, A.; Spengler, K.; Heslegrave, A.; Bauer, R.; Krause, S.; Viollet, B.; Carling, D.; Heller, R. Activation of AMP-activated protein kinase by vascular endothelial growth factor mediates endothelial angiogenesis independently of nitric-oxide synthase. *J. Biol. Chem.* **2010**, *285*, 10638–10652. [CrossRef]

47. Mount, P.F.; Hill, R.E.; Fraser, S.A.; Levidiotis, V.; Katsis, F.; Kemp, B.E.; Power, D.A. Acute renal ischemia rapidly activates the energy sensor AMPK but does not increase phosphorylation of eNOS-Ser[1177]. *Am. J. Physiol. Physiol.* **2005**, *289*, F1103–F1115. [CrossRef]

48. Montagnani, M.; Chen, H.; Barr, V.A.; Quon, M.J. Insulin-stimulated Activation of eNOS Is Independent of Ca2+ but Requires Phosphorylation by Akt at Ser1179. *J. Biol. Chem.* **2001**, *276*, 30392–30398. [CrossRef]

49. Van Heemst, D. Insulin, IGF-1 and longevity. *Aging Dis.* **2010**, *1*, 147–157. [CrossRef]

50. Varewijck, A.J.; Janssen, J.A. Insulin and its analogues and their affinities for the IGF1 receptor. *Endocr. Relat. Cancer* **2012**, *19*, F63–F75. [CrossRef]

51. Cai, W.; Sakaguchi, M.; Kleinridders, A.; Gonzalez-Del Pino, G.; Dreyfuss, J.M.; O'Neill, B.T.; Ramirez, A.K.; Pan, H.; Winnay, J.N.; Boucher, J.; et al. Domain-dependent effects of insulin and IGF-1 receptors on signalling and gene expression. *Nat. Commun.* **2017**, *8*, 14892. [CrossRef] [PubMed]

52. Olsson, A.K.; Dimberg, A.; Kreuger, J.; Claesson-Welsh, L. VEGF receptor signalling—In control of vascular function. *Nat. Rev. Mol. Cell. Biol.* **2006**, *7*, 359–371. [CrossRef] [PubMed]

53. King, G.L.; Park, K.; Li, Q. Selective insulin resistance and the development of cardiovascular diseases in diabetes: The 2015 Edwin Bierman Award Lecture. *Diabetes* **2016**, *65*, 1462–1471. [CrossRef] [PubMed]

54. Göransson, O.; McBride, A.; Hawley, S.A.; Ross, F.A.; Shpiro, N.; Foretz, M.; Viollet, B.; Hardie, D.G.; Sakamoto, K. Mechanism of action of A-769662, a valuable tool for activation of AMP-activated protein kinase. *J. Biol. Chem.* **2007**, *282*, 32549–32560. [CrossRef] [PubMed]

55. Choudhury, Y.; Yang, Z.; Ahmad, I.; Nixon, C.; Salt, I.P.; Leung, H.Y. AMP-activated protein kinase (AMPK) as a potential therapeutic target independent of PI3K/Akt signaling in prostate cancer. *Oncoscience* **2014**, *1*, 446. [CrossRef] [PubMed]

56. Timmermans, A.D.; Balteau, M.; Gélinas, R.; Renguet, E.; Ginion, A.; de Meester, C.; Sakamoto, K.; Balligand, J.L.; Bontemps, F.; Vanoverschelde, J.L.; et al. A-769662 potentiates the effect of other AMP-activated protein kinase activators on cardiac glucose uptake. *Am. J. Physiol. Heart Circ. Physiol.* **2014**, *306*, H1619–H1630. [CrossRef] [PubMed]

57. Green, C.J.; Pedersen, M.; Pedersen, B.K.; Scheele, C. Elevated NF-κB activation is conserved in human myocytes cultured from obese type 2 diabetic patients and attenuated by AMP-activated protein kinase. *Diabetes* **2011**, *60*, 2810–2819. [CrossRef]

58. Treebak, J.T.; Birk, J.B.; Hansen, B.F.; Olsen, G.S.; Wojtaszewski, J.F.P. A-769662 activates AMPK beta1-containing complexes but induces glucose uptake through a PI3-kinase-dependent pathway in mouse skeletal muscle. *Am. J. Physiol. Cell Physiol.* **2009**, *297*, C1041–C1052. [CrossRef]

59. Olianas, M.C.; Dedoni, S.; Onali, P. Signalling pathways mediating phosphorylation and inactivation of glycogen synthase kinase-3β by the recombinant human δ-opioid receptor stably expressed in Chinese hamster ovary cells. *Neuropharmacology* **2011**, *60*, 1326–1336. [CrossRef]

60. Human Protein Atlas. Available online: https://www.proteinatlas.org/ENSG00000017427-IGF1/cell (accessed on 6 June 2018).

61. Uhlen, M.; Oksvold, P.; Fagerberg, L.; Lundberg, E.; Jonasson, K.; Forsberg, M.; Zwahlen, M.; Kampf, C.; Wester, K.; Hober, S.; et al. Towards a knowledge-based Human Protein Atlas. *Nat. Biotechnol.* **2010**, *28*, 1248–1250. [CrossRef]

62. Huang, Y.; Smith, C.A.; Chen, G.; Sharma, B.; Miner, A.S.; Barbee, R.W.; Ratz, P.H. The AMP-Dependent Protein Kinase (AMPK) activator A-769662 causes arterial relaxation by reducing cytosolic free calcium independently of an increase in AMPK phosphorylation. *Front. Pharmacol.* **2017**, *8*, 1–14. [CrossRef] [PubMed]

63. Woods, A.; Salt, I.; Scott, J.; Hardie, D.G.; Carling, D. The alpha1 and alpha2 isoforms of the AMP-activated protein kinase have similar activities in rat liver but exhibit differences in substrate specificity in vitro. *FEBS Lett.* **1996**, *397*, 347–351. [CrossRef]

64. Durante, P.E.; Mustard, K.J.; Park, S.H.; Winder, W.W.; Hardie, D.G. Effects of endurance training on activity and expression of AMP-activated protein kinase isoforms in rat muscles. *Am. J. Physiol. Endocrinol. Metab.* **2002**, *283*, E178–E186. [CrossRef] [PubMed]

International Journal of
Molecular Sciences

MDPI

Article

Endothelial AMP-Activated Kinase α1 Phosphorylates eNOS on Thr495 and Decreases Endothelial NO Formation

Nina Zippel [1], Annemarieke E. Loot [1], Heike Stingl [1,2], Voahanginirina Randriamboavonjy [1,2], Ingrid Fleming [1,2] and Beate Fisslthaler [1,2,*]

[1] Institute for Vascular Signalling, Centre for Molecular Medicine, Johann Wolfgang Goethe University, 60590 Frankfurt, Germany; NZippel@gmx.de (N.Z.); a.loot@certe.nl (A.E.L.); Stingl@vrc.uni-frankfurt.de (H.S.); Voahangy@vrc.uni-frankfurt.de (V.R.); fleming@em.uni-frankfurt.de (I.F.)
[2] DZHK (German Centre for Cardiovascular Research) partner site RhineMain, Theodor Stern Kai 7, 60590 Frankfurt, Germany
* Correspondence: fisslthaler@vrc.uni-frankfurt.de; Tel.: +49-69-6301-6994

Received: 2 August 2018; Accepted: 11 September 2018; Published: 13 September 2018

Abstract: AMP-activated protein kinase (AMPK) is frequently reported to phosphorylate Ser1177 of the endothelial nitric-oxide synthase (eNOS), and therefore, is linked with a relaxing effect. However, previous studies failed to consistently demonstrate a major role for AMPK on eNOS-dependent relaxation. As AMPK also phosphorylates eNOS on the inhibitory Thr495 site, this study aimed to determine the role of AMPKα1 and α2 subunits in the regulation of NO-mediated vascular relaxation. Vascular reactivity to phenylephrine and acetylcholine was assessed in aortic and carotid artery segments from mice with global (AMPKα$^{-/-}$) or endothelial-specific deletion (AMPKα$^{\Delta EC}$) of the AMPKα subunits. In control and AMPKα1-depleted human umbilical vein endothelial cells, eNOS phosphorylation on Ser1177 and Thr495 was assessed after AMPK activation with thiopental or ionomycin. Global deletion of the AMPKα1 or α2 subunit in mice did not affect vascular reactivity. The endothelial-specific deletion of the AMPKα1 subunit attenuated phenylephrine-mediated contraction in an eNOS- and endothelium-dependent manner. In in vitro studies, activation of AMPK did not alter the phosphorylation of eNOS on Ser1177, but increased its phosphorylation on Thr495. Depletion of AMPKα1 in cultured human endothelial cells decreased Thr495 phosphorylation without affecting Ser1177 phosphorylation. The results of this study indicate that AMPKα1 targets the inhibitory phosphorylation Thr495 site in the calmodulin-binding domain of eNOS to attenuate basal NO production and phenylephrine-induced vasoconstriction.

Keywords: endothelial nitric-oxide synthase; vasodilation; phenylephrine; vasoconstriction; endothelial cells; ionomycin

---

1. Introduction

AMP-activated protein kinase (AMPK) is activated in response to intracellular energy depletion, e.g., during insulin resistance when cellular glucose uptake is limited—especially in contracting skeletal muscle [1] or in cultured cells in the absence of extracellular glucose or hypoxia [2]. Once activated, AMPK acts to conserve energy by stimulating glucose uptake and mitochondrial biosynthesis, as well as by stimulating autophagy to provide substrates for metabolism. At the same time, AMPK inhibits anabolic pathways, such as cholesterol biosynthesis and fatty-acid synthesis, which are not essential for survival (for a recent review, see Reference [3]). In addition to activation by energy-sensitive stimuli, AMPK can also be stimulated following cell exposure to cytokines, growth factors, and mechanical stimuli [4]. In endothelial cells, AMPK was implicated in the inhibition of cell activation [5,6], as well as in angiogenesis in vitro [7] and in vivo [8]. These effects are, at least partially, attributed to the phosphorylation and stimulation of the endothelial nitric-oxide (NO) synthase (eNOS) by AMPK. This claim was backed up by reports of AMPK-dependent phosphorylation of eNOS (on Ser1177) following the exposure of cultured endothelial cells to agonists such as the vascular endothelial growth factor (VEGF) [9] and adiponectin [10], or pharmacological agents including peroxisome proliferator-activated receptor (PPAR) agonists [11] and statins [12]. Similar reports were also published using AMPK activators such as 5-aminoimidazole-4-carboxamide ribonucleotide (AICAR) [13] and metformin [14,15], or natural polyphenols like amurensin G [16] or resveratrol [17]. However, the effects are generally weak and much less impressive than the stimulation seen in response to hypoxia [7], shear stress [18–20], and thrombin [21] which result in robust AMPK activation.

Evidence for a link between AMPK- and NO-dependent alterations in vascular reactivity is also not consistent and depends on the model studied. For example, in resistance arteries in rat hindlimb and cremaster muscles, AICAR induces an NO- and endothelium-independent relaxation [22]. In mice, small-molecule AMPK activators, PT-1 or A769662, elicit the vasodilation of mesenteric arteries by decreasing intracellular Ca^{2+} levels and inducing depolymerization of the actin cytoskeleton [23,24]. In other studies, AICAR was reported to impair the relaxation elicited by sodium nitroprusside (SNP), indicating a general effect on smooth-muscle contractility [25]. In genetic models, the situation is not any clearer as the deletion of the AMPKα1 subunit did not affect acetylcholine (ACh)-induced NO production and relaxation unless mice were treated with angiotensin II over seven days [26]. Also, in isolated phenylephrine-contracted rings of murine aorta, AICAR elicited a profound dose-dependent relaxation that was independent of either the endothelium or the inhibition of eNOS, and mediated by the AMPKα1 subunit in smooth-muscle cells [27]. The most thorough study investigating the role of endothelial AMPKα subunits on vascular function and blood pressure reported hypertension in endothelial-specific AMPKα1 knockout mice; however, in the mesentery artery, the effect was attributed to the opening of charybdotoxin-sensitive potassium channels and smooth-muscle hyperpolarization [28]. The global deletion of the AMPKα2 subunit was also reported to attenuate the ACh-induced relaxation of murine aorta. This effect was attributed to eNOS uncoupling via an AMPKα2-mediated proteasomal degradation of the GTP cyclohydrolase [29], which generates the eNOS cofactor, tetrahydrobiopterin. Also, other researchers failed to detect any evidence for the AMPK-dependent activation of eNOS [30,31]. In this study, we set out to make a more thorough analysis of the effects of AMPKα1 and AMPKα2 deletion on NO-mediated vascular function. We also carefully studied changes in eNOS phosphorylation in cultured and native endothelial cells.

2. Results

2.1. Consequences of Global AMPKα Deletion on Vascular Responsiveness

The maximal KCl- and phenylephrine-induced contractions of isolated aortic rings were indistinguishable between wild-type mice and their corresponding AMPKα1$^{-/-}$ littermates (Figure 1A,B). However, there was a tendency toward an attenuated contraction in the aortic rings from the AMPKα1$^{-/-}$ mice and −log half maximal effective concentration (pEC$_{50}$) values were −6.899 ± 0.082 and −6.711 ± 0.099 (n = 7, not significant (n.s.)) in rings from wild-type and AMPKα1$^{-/-}$ mice, respectively. The endothelium- and NO-dependent relaxation elicited by ACh (Figure 1C), as well as the endothelium-independent relaxation elicited by SNP (Figure 1D), was identical in vessels from both strains. When experiments were repeated using carotid arteries, samples from AMPKα1$^{-/-}$ mice demonstrated a slightly weaker contractile response to KCl than the wild-type mice, as well as a slightly attenuated response to phenylephrine (pEC$_{50}$ values were −6.393 ± 0.065 and −5.895 ± 0.093, respectively, (n = 7, n.s.) in rings from wild-type and AMPKα1$^{-/-}$ mice (Figure S1). Again, there was no apparent difference in relaxant responsiveness to ACh or SNP.

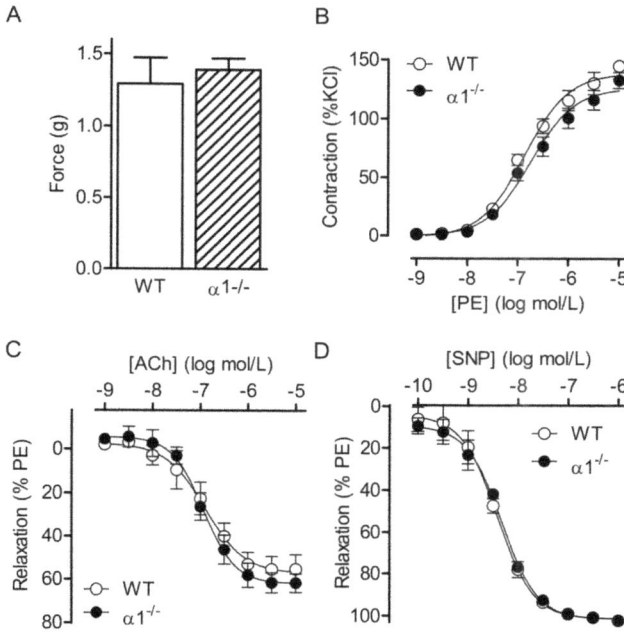

Figure 1. Consequences of global AMP-activated protein kinase (AMPK) α1 deletion on vascular reactivity. Vascular reactivity in aortic rings from wild-type (WT) and AMPKα1$^{-/-}$ (α1$^{-/-}$) mice. (**A**) Responsiveness of endothelium-intact aortic rings to KCl (80 mmol/L). (**B–D**) Concentration–response curves to (**B**) phenylephrine (PE), (**C**) acetylcholine (ACh), and (**D**) sodium nitroprusside (SNP). The graphs summarize data obtained from seven animals in each group.

Similar experiments using arteries from AMPKα2$^{-/-}$ mice gave essentially the same results, i.e., no significant difference in either the agonist-induced contraction or relaxation of the aorta in either the presence or absence of a functional endothelium (Figure 2).

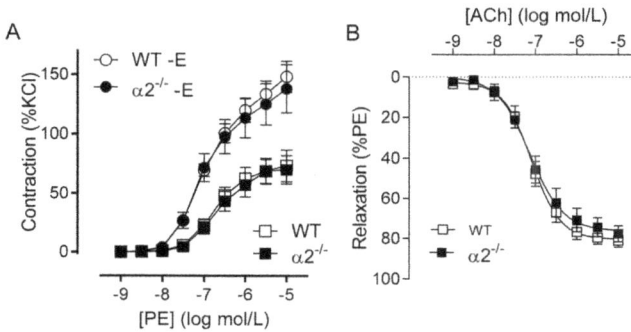

Figure 2. Consequences of global AMPKα2 deletion on vascular reactivity. Vascular reactivity in aortic rings from wild-type (WT) and AMPKα2$^{-/-}$ (α2$^{-/-}$) mice: (**A**) contractile response to phenylephrine (PE) in the presence and absence (−E) of endothelium. (**B**) Concentration-dependent relaxation due to acetylcholine. The graphs summarize data obtained from seven animals in each group.

2.2. Consequence of Endothelial-Specific AMPKα Deletion on Vascular Responsiveness

As the global deletion of AMPK seemed to affect vascular smooth-muscle contraction rather than NO-mediated relaxation, animals lacking the AMPKα1 or AMPKα2 subunits specifically in endothelial cells (i.e., AMPKα1$^{\Delta EC}$ and AMPKα2$^{\Delta EC}$ mice) were generated, and the specificity of the deletion verified in isolated cluster of differentiation 144 (CD144)-positive pulmonary endothelial cells (Figure S2A). The deletion of endothelial AMPKα1 did not influence the expression level of AMPKα1 in whole aortic lysates (Figure S2B).

Endothelial-specific deletion of AMPKα1 did not affect the N$^\omega$-nitro-L-arginine methyl ester (L-NAME)-induced contraction of aortic rings (Figure 3A), which is an index of basal Ca^{2+}-independent NO production under isometric stretch conditions [32], or that induced by KCl (not shown). However, the ACh-induced relaxation was slightly improved by endothelial-specific AMPKα1 deletion (Figure 3B) with pEC$_{50}$ values for ACh of -7.217 ± 0.095 and -7.360 ± 0.076 ($n = 14$; n.s.) in rings from wild-type and AMPKα1$^{-/-}$ mice, respectively. An increased production of NO was evident as a markedly impaired contraction of aortic rings from AMPKα1$^{\Delta EC}$ mice compared to rings from wild-type mice to phenylephrine that was abolished by L-NAME (Figure 3C, Table 1). Similarly, removal of the endothelium with 3-[(3-cholamidopropyl)dimethylammonio]-1-propanesulfonate (CHAPS) also abrogated the improved relaxation that was dependent on AMPKα1 deletion (Figure 3D, Table 1).

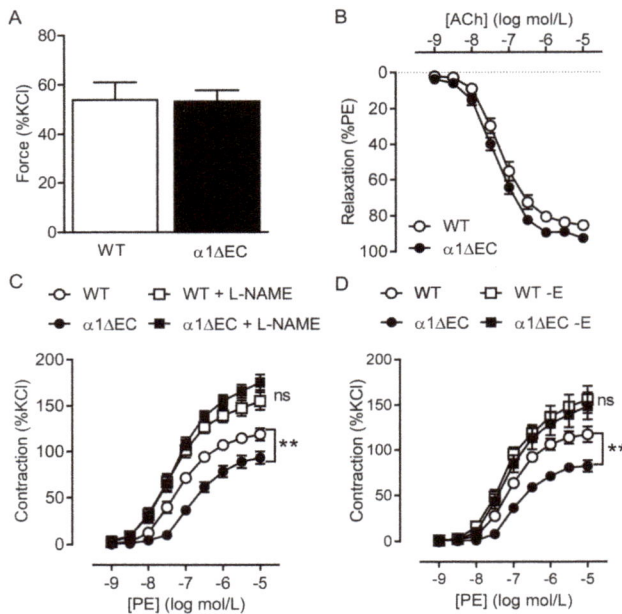

Figure 3. Consequences of endothelial-specific deletion of AMPKα1 on vascular reactivity. (**A**) Effect of N$^\omega$-nitro-L-arginine methyl ester (L-NAME; 300 μmol/L) on the tone of aortic rings from wild-type (WT) and AMPKα1$^{\Delta EC}$ (α1ΔEC) mice pre-contracted to 30% of the maximal KCl-induced contraction by phenylephrine. (**B**) Concentration-dependent relaxation due to acetylcholine in aortic rings pre-constricted with phenylephrine from wild-type (WT) and AMPKα1$^{\Delta EC}$ (α1ΔEC) mice. (**C,D**) Concentration-dependent contraction of aortic rings from wild-type (WT) and AMPKα1$^{\Delta EC}$ (α1ΔEC) mice due to phenylephrine. Experiments were performed in the absence and presence of L-NAME (300 μmol/L, (**C**) and in the presence and absence (−E) of functional endothelium (**D**); $n = 10$ to 16, ** $p < 0.01$ AMPKα1$^{\Delta EC}$ versus wild type.

Table 1. The $-\log$ half maximal effective concentration (pEC_{50}) values relating to the consequences of endothelial-specific deletion of AMP activated protein kinase (AMPK) $\alpha1$ on vascular response to phenylephrine. Experiments were performed in endothelium intact rings the presence of solvent or N^{ω}-nitro-L-arginine methyl ester (L-NAME; 300 µmol/L), as well as in endothelium-denuded ($-E$) samples from the same animals; $n = 10\text{--}16$.

pEC_{50} Values	Solvent		L-NAME		$-E$	
Wild type	-7.04 ± 0.13		-7.43 ± 0.10		-7.21 ± 0.05	
AMPK$^{\Delta EC}$	-6.77 ± 0.05	*	-7.31 ± 0.09	§§	-7.13 ± 0.05	§§

* $p < 0.05$ versus wild type; §§ $p < 0.001$ versus solvent.

Endothelial-specific deletion of the AMPK$\alpha2$ subunit had no consequence on the phenylephrine-induced contraction, ACh-induced relaxation, or the SNP-induced relaxation of isolated aortic rings from wild type versus the respective AMPK$\alpha2^{\Delta EC}$ littermates (Figure S3).

2.3. Vascular Responses to AMPK Activators

One reason for the lack of consequence of AMPK$\alpha1$ deletion on agonist-induced relaxation may be related to the fact that the ACh-induced phosphorylation and activation of eNOS is, like that of other agonists, largely regulated by the activity of Ca^{2+}/calmodulin-dependent kinase II [33]. Therefore, responses to two potential AMPK activators, i.e., resveratrol [34] and amurensin G [16], as well as two reportedly specific small-molecule AMPK activators, 991 and PT-1, were studied.

Resveratrol elicited the almost complete relaxation of aortic rings from wild-type and AMPK$\alpha1^{\Delta EC}$ mice (Figure S4A), but these responses were insensitive to NOS inhibition, and therefore, unrelated to its activation. Amurensin G is reported to activate AMPK in endothelial cells and increase eNOS phosphorylation [16]. While it was able to elicit the NOS inhibitor-sensitive relaxation of aortic rings from wild-type mice, it was equally effective and equally sensitive to NOS inhibition in aortic rings from corresponding AMPK$\alpha1^{\Delta EC}$ mice (Figure S4B). Thus, amurensin G exerted its relaxation in an eNOS-dependent manner and the activity of both AMPK activators was AMPK$\alpha1$-independent. The situation was similar when PT-1 and 991 were studied. The compounds elicited phosphorylation of AMPK in endothelial cells of murine aortic rings from wild-type cells (Figure S4E). Although these compounds elicited vascular relaxation, the responses were slow, and although they were sensitive to NOS inhibition, the effects were comparable in aortic rings from wild-type and AMPK$\alpha1^{\Delta EC}$ mice (Figure S4C,D).

2.4. AMPK$\alpha1$ and eNOS Phosphorylation

The activity of eNOS is reciprocally regulated by its phosphorylation on the activator site, Ser1177 [35,36], and the inhibitory site, Thr495 [37–39]. The next step was, therefore, to analyze the ability of AMPK to phosphorylate eNOS on these two residues in vitro. Wild-type (Myc-tagged) eNOS or eNOS mutants in which Thr495 was substituted with alanine or aspartate (Thr495A, Thr495D), or Ser1177 was substituted with alanine or aspartate (Ser1177A, Ser1177D) were overexpressed in HEK293 cells and used as the substrate for in vitro kinase assays for recombinant AMPK$\alpha1$. While the phosphorylation of wild-type eNOS and Ser1177 mutants was clearly detectable, there was only minimal phosphorylation of the Thr495 mutants (Figure 4A). These findings could be confirmed using phospho-specific antibodies to assess eNOS phosphorylation on Ser1177 and Thr495 on immunoprecipitated eNOS from human endothelial cells (Figure 4B).

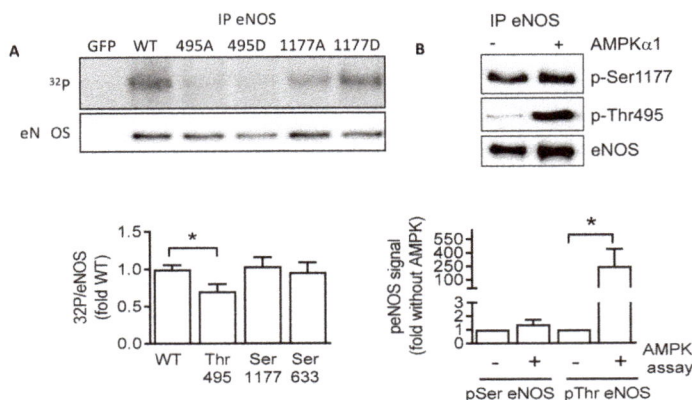

Figure 4. Endothelial nitric-oxide synthase (eNOS) is a substrate of AMPK in vitro. (**A**) Wild-type eNOS, as well as Thr495 and Ser1177 mutants, was overexpressed in HEK293 cells, then immunoprecipitated and used as substrate for AMPKα1 in in vitro kinase assays. The upper panel shows the autoradiograph of eNOS proteins. The lower panel shows the Western blot of the immunoprecipitated (IP) eNOS protein used as input. The graph summarizes the data from four independent experiments. (**B**) Wild-type eNOS (Flag-tagged) overexpressed in human umbilical vein endothelial cells was immunoprecipitated and used as a substrate for AMPKα1. Phosphorylation was assessed using specific antibodies for phosphorylated Ser1177 (p-Ser1177) and Thr945 eNOS (p-Thr495). The graph summarizes the data from five independent kinase reactions. * $p < 0.05$.

To transfer the observations to a more physiological system, changes in eNOS phosphorylation were studied in response to thiopental in cultured endothelial cells. Thiopental elicited the rapid and pronounced phosphorylation of AMPK (Figure 5A) in primary cultures of human endothelial cells. At the same time, thiopental decreased the phosphorylation of eNOS on Ser1177 and increased eNOS phosphorylation on Thr495 (Figure 5B). In contrast, the Ca^{2+}-elevating agonist and NO-dependent vasodilator, bradykinin, elicited a significant increase in the phosphorylation of Ser1177, and had no significant effect on Thr495 phosphorylation.

Next, AMPKα1 was deleted in human endothelial cells using a "clustered regularly interspaced short palindromic repeats" (CRISPR)/CRISPR-associated protein 9 (Cas9)-based approach. After each passaging, protein expression levels of AMPK α1 were analyzed with Western blotting. After passages 4–5, the protein was no longer detectable ($n = 6$, Figure 6), and agonist-induced changes in eNOS phosphorylation were assessed. Given that bradykinin and ACh receptors are rapidly lost during culture, cells were stimulated with the Ca^{2+} ionophore, ionomycin. In AMPKα1-expressing cells, ionomycin elicited the phosphorylation of eNOS on Ser1177 and the dephosphorylation of Thr495, followed by a rapid re-phosphorylation on Thr495, similar to the effects of other Ca^{2+}-elevating agonists [39]. In AMPKα1-depleted cells, basal Thr495 phosphorylation of eNOS was significantly impaired, and the re-phosphorylation after 2 min was also less pronounced than in control cells. The ionomycin-induced phosphorylation of eNOS Ser1177 was not affected by the depletion of AMPKα1 (Figure 6).

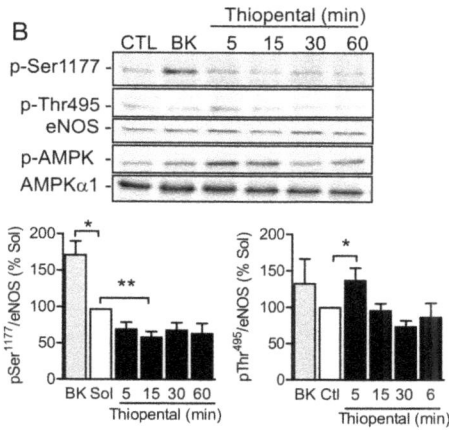

Figure 5. Effect of AMPK activation on eNOS activity and phosphorylation. Cultured human endothelial cells were incubated with solvent (Sol), thiopental (1 mmol/L, 5–60 min), or bradykinin (BK; 1 µmol/L, 2 min). Thereafter, the phosphorylation of (**A**) AMPK and (**B**) phosphorylation of eNOS at Ser1177 and Thr495 were assessed. Bar graphs summarize the data obtained in four to five different cell batches; * $p < 0.05$, ** $p < 0.01$ versus solvent treatment.

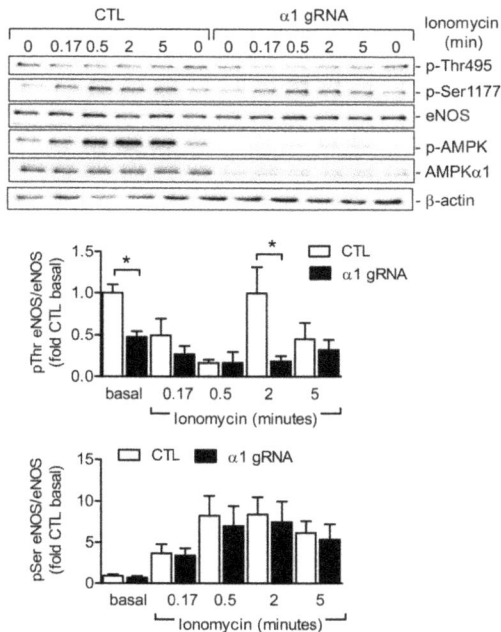

Figure 6. Effect of AMPKα1 deletion on the ionomycin-induced phosphorylation of eNOS on Thr495 and Ser1177. The AMPKα1 subunit was deleted in cultured human cells using the "clustered regularly interspaced short palindromic repeats" (CRISPR)/CRISPR-associated protein 9 (Cas9) system and AMPKα1-specific guide RNAs (gRNAs), and was stimulated with ionomycin (100 nmol/L) for up to 5 min. Representative Western blots are shown of six independent experiments. Bar graphs summarize the evaluation of p-eNOS to total eNOS ($n = 6$); * $p < 0.05$ versus control guide RNA.

3. Discussion

The results of the present study revealed that global deletion of the AMPKα1 or AMPKα2 subunits in healthy animals had no major impact on the relaxant function of isolated endothelium-intact murine aortae or carotid arteries. A small decrease in the contractile response to phenylephrine was apparent in global AMPKα1-deficient arteries, which was much more pronounced following endothelial-specific deletion of the AMPKα1 subunit. Moreover, the attenuated contractile response observed in arteries from AMPKα1$^{\Delta EC}$ mice was sensitive to L-NAME and removal of the endothelium, indicating that an increase in NO production by the endothelium underlies the effects observed. Mechanistically, the activity of AMPKα1 could be linked to the phosphorylation of eNOS on the inhibitory Thr495 site.

There are numerous reports describing the effects on AMPK activators on the phosphorylation of eNOS on Ser1177, which suggests that AMPK acts as an eNOS activator [13,40]. However, most studies linking AMPK with eNOS Ser1177 relied on compound C to inhibit AMPK, or AICAR, phenformin, or resveratrol to activate AMPK. This is a concern as the specificity of these pharmacological tools is questionable, with AMPK-dependent and -independent effects being attributed to both activators and inhibitors [41]. In the present study, the previously reported AMPK activators, amurensin G [16] and resveratrol [34], were studied together with the reportedly more specific small-molecule activators of AMPK, PT-1 [42] and 991 [43], in vascular reactivity studies. No endothelial-specific and AMPKα1-dependent effects were detected using any of the substances tested. Most studies using genetically modified models that reported effects on vascular reactivity focused on global AMPKα-deficient mice, and the defects were usually attributed to vascular smooth-muscle cells [44,45]. Effects on vascular function in global AMPKα1$^{-/-}$ mice were only observed in exercising or angiotensin-II-treated mice. The protective effects of voluntary exercise on vascular function were attributed to AMPKα1 via an effect on eNOS [26,46]. However, the only study to investigate changes in vascular reactivity in vessels from mice lacking AMPKα subunits specifically in endothelial cells linked changes in blood pressure with a carybdotoxin-sensitive potassium channel and endothelial-cell hyperpolarization [28].

The majority of reports describing AMPK-mediated effects on vascular function in disease models, as well as in healthy mice, focused on the AMPKα2 subunit, which suppresses reduced nicotinamide adenine dinucleotide phosphate (NADPH) oxidase activity and the production of reactive oxygen species to inhibit 26S proteasomal activity [47]. One consequence of this was the stabilization of GTP cyclohydrolase, the key sepiapterin biosynthetic enzyme that generates the essential eNOS cofactor, tetrahydrobiopterin [29]. It was, therefore, somewhat surprising that no major alterations in NO-mediated relaxation due to ACh could be detected in arteries from animals constitutively lacking the AMPKα2 subunit. Moreover, endothelial-specific deletion of AMPKα2 also failed to affect vascular NO production.

The enzyme eNOS can be phosphorylated on serine, threonine, and tyrosine residues, findings which highlight the potential role of phosphorylation in regulating eNOS activity. There are numerous putative phosphorylation sites, but most is known about the functional consequences of phosphorylation of a serine residue (human eNOS sequence: Ser1177) in the reductase domain and a threonine residue (human eNOS sequence Thr495) within the calmodulin (CaM)-binding domain. Maximal eNOS activity is linked with the simultaneous dephosphorylation of Thr495 and phosphorylation of Ser1177 [39,48].

In unstimulated cultured endothelial cells, Ser1177 is not phosphorylated, but it is rapidly phosphorylated after the application of fluid shear stress [35], VEGF [49] or bradykinin [39]. The kinases involved in this process vary with the stimuli applied. For example, while shear stress elicits the phosphorylation of Ser1177 by protein kinase A (PKA), insulin, estrogen, and VEGF mainly phosphorylate eNOS in endothelial cells via protein kinase B (Akt) [50]. The bradykinin-, Ca^{2+} ionophore-, and thapsigargin-induced phosphorylation of Ser1177 is mediated by Ca^{2+}/calmodulin-dependent kinase II (CaMKII) [39]. Thr495, on the other hand, is constitutively phosphorylated in all endothelial cells investigated to date, and it is a negative regulatory site, i.e., phosphorylation is associated with a decrease in enzyme activity [38,39]. The constitutively active kinase that phosphorylates eNOS Thr495 is most probably protein kinase C (PKC) [38], even though there is some confusion regarding the specific isoform(s) involved. AMPK can, however, also phosphorylate Thr495 [37]. The results of this study clearly indicate a role for endothelial cell AMPKα1 in the negative regulation of NO production and vascular tone, and as such, are in line with a previous study that reported an increased NO component to total relaxation in the mesenteric arteries of AMPKα1$^{\Delta EC}$ mice compared to wild type [28], this correlated to an enhanced eNOS Thr495 phosphorylation in mesenteric arteries compared to the aorta in wild type mice [51]. Our study also goes further to demonstrate that, in in vitro kinase assays, AMPKα1 clearly phosphorylated eNOS on Thr495, an effect that was prevented by the mutation of Thr495 to Ala or Asp. Also, in AMPKα1-depleted human endothelial cells, basal eNOS phosphorylation on Thr495 was decreased and its re-phosphorylation in response to agonist stimulation was significantly delayed, an effect that can account for the increase in NO generation by AMPKα1-deficient endothelial cells. At this stage, it is not possible to rule out a role for AMPK in the regulation of Ser1177 phosphorylation, as the higher basal phosphorylation of this residue in the transduced cells studied may have masked AMPK-dependent effects. However, the functional studies using vessels from AMPKα1 knockout mice clearly hint at an inhibitory rather than a stimulatory effect of AMPK on eNOS activity. The link between eNOS Thr495 phosphorylation and NO production can be explained by interference with the binding of CaM to the CaM-binding domain. Indeed, in endothelial cells stimulated with agonists such as bradykinin, histamine, or a Ca^{2+} ionophore, substantially more CaM binds to eNOS when Thr495 is dephosphorylated [39]. Analysis of the crystal structure of the eNOS CaM-binding domain with CaM indicates that the phosphorylation of eNOS Thr495 not only causes electrostatic repulsion of nearby glutamate residues within CaM, but may also affect eNOS Glu498, and thus, induce a conformational change within eNOS itself [52]. AMPK activation was also linked with the phosphorylation of eNOS on Ser1177 in isolated endothelial cells [13,37,53], but contrasted somewhat with the lack of effect on endothelium-dependent vascular reactivity [27]. In the present study, only a small increase in Ser1177 phosphorylation was detected in vitro using different cellular sources of eNOS (i.e., HEK cells or human endothelial cells).

In cultured endothelial cells, we found thiopental to be an effective AMPK activator and could demonstrate that AMPKα1 phosphorylates eNOS on Thr495, an observation that fits well with an earlier report [37]. This phosphorylation step is generally associated with eNOS inhibition due to the decreased binding of Ca^{2+}/calmodulin to the enzyme [39], and implies that the activation of AMPK in isolated vessels would act to decrease relaxation and increase vascular tone, which is exactly the response that was observed in the vascular reactivity studies.

In addition to direct phosphorylation, there are various signaling pathways described for AMPK to influence eNOS activity. AMPK was previously reported to prevent the estradiol-induced phosphorylation of eNOS by preventing the association of eNOS with heat-shock protein 90 (Hsp90), which is generally required for kinase binding to the eNOS signalosome [54]. Any link between AMPK and Hsp90 was not addressed in the current study given the clear effect of AMPKα1 on eNOS phosphorylation in vitro. Direct effects on eNOS activity may not be the only way via which AMPK activation can affect NO signaling. Indeed, AMPKα1 activation could affect the bioavailability of NO by improving mitochondrial function and stimulating the transcriptional regulation of anti-inflammatory enzymes, such as superoxide dismutase 2, to alter the production of reactive oxygen species [55].

In summary, endothelial AMPKα subunits have no direct activating effect on eNOS in vivo. Rather, since AMPKα1 phosphorylates eNOS on the inhibitory Thr495 site, AMPK activation attenuates NO production. No link between AMPKα2 and phenylephrine- or ACh-induced changes in vascular tone were detected. Moreover, while some of AMPK activators tested did affect vascular tone, the effects were independent of the endothelial-specific deletion of AMPKα1.

4. Materials and Methods

4.1. Materials

The antibodies used were directed against p-Ser1177 (Cell signaling, Cat. No. 9571) and p-Thr495 eNOS (Cell signaling, Cat. No. 9574), eNOS (BD Transduction, 610296), p-Thr172 AMPK (Cell signaling, Cat. No. 2535), AMPKα2 (Cell signaling, Cat. No. 2757), β-actin (Sigma, Cat. No. A5441), Flag (Sigma, Cat. No. F3165), and c-Myc (Santa Cruz, Cat. No. SC-40). The AMPKα1 antibody was generated by Eurogentec by injecting rabbits with the AMPKα1-specific peptide H_2N–CRA RHT LDE LNPQKS KHQ–$CONH_2$. All other substances were obtained from Sigma-Aldrich (Munich, Germany). $^{32}P\gamma$-ATP was obtained from Hartmann Analytics (Braunschweig, Germany).

4.2. Animals

AMPKα1$^{-/-}$ or AMPKα2$^{-/-}$ mice (kindly provided by Benoit Viollet, Paris via the European Mouse Mutant Archive, Munich, Germany) were bred heterozygous and housed at the Goethe University Hospital and knockouts or their respective wild-type littermates were used. AMPKα1$^{flox/flox}$ and α2$^{flox/flox}$ mice with *loxP* sites flanking a coding exon (provided by Benoit Viollet) were crossed with transgenic mouse lines overexpressing Cre recombinase under control of the vascular endothelial (VE)-cadherin promoter to generate the appropriate endothelial-specific AMPKα deletion; Cre$^{+/-}$ mice are referred throughout as AMPKα1^{ΔEC} and AMPKα2^{ΔEC} mice and Cre$^{-/-}$ mice are referred as their respective WT littermates. The investigation conforms to the Guide for the Care and Use of Laboratory Animals published by the European Commission Directive 86/609/EEC. For the isolation of tissues, mice were euthanized with 4% isoflurane in air and subsequent exsanguination.

4.3. Vascular Reactivity Measurements

Aortae and carotid arteries were prepared free of adhering tissue and cut into 2.0-mm segments. Aortic rings were mounted in standard 10-mL organ bath chambers, stretched to 1 g tension and responses were measured in g. Carotid artery rings were mounted in 5-mL wire myograph chambers (DMT, Aarhus, Denmark), stretched to 90% of their diameter at 100 mmHg, and responses were measured in mN/mm segment length. Contractile responses to a high K$^+$ buffer (80 mmol/L KCl) or cumulatively increasing concentrations of phenylephrine were assessed. Relaxation to cumulatively increasing concentrations of ACh, resveratrol (Sigma, Munich, Germany), 2-chloro-5-[[5-[[5-(4,5-dimethyl-2-nitrophenyl)-2-furanyl]methylene]-4,5-dihydro-4-oxo-2-thiazolyl] amino]benzoic acid (PT-1; Tocris, Biotechne, Wiesbaden, Germany), amurensin G (kindly provided by K.W. Kang, Seoul, Korea), 5-((6-chloro-5-(1-methyl-1H-indol-5-yl)-1H-benzo [d]imidazol-2-yl)oxy)-2-methylbenzoic acid (991; SpiroChem AG, Switzerland), or SNP was assessed in segments pre-contracted with phenylephrine to 80% of their maximal contraction due to KCl in the presence and absence of L-NAME. Relaxation was expressed as the percentage of phenylephrine precontraction. Removal of the endothelium was performed by intraluminal application of CHAPS (0.5%, 30 s) into the aortae.

4.4. Cell Culture

Human endothelial cells: Human umbilical vein endothelial cells were isolated and cultured as previously described [56] and used up to passage 2. The use of human material in this study conforms to the principles outlined in the Declaration of Helsinki, and the isolation of endothelial cells was approved in written form by the ethics committee of Goethe University. For lentiviral and adenoviral transduction, human umbilical vein endothelial cells (Promocell, Heidelberg, Germany) were used and cultured up to passage 8 in endothelial growth medium 2 (Promocell, Heidelberg, Germany).

Murine pulmonary endothelial cells: Mouse lungs were freshly processed as previously described [18].

4.5. Adenoviral Transduction of Human Umbilical Vein Endothelial Cells

Adenoviral particles expressing the C-terminal Flag-tagged human full-length eNOS were used to transduce cultured umbilical vein endothelial cells as described previously [57].

4.6. In Vitro Kinase Assay

The eNOS wild-type or mutant proteins with C-terminal myc or Flag tags were overexpressed by transfection in HEK cells or adenoviral transduction in human umbilical vein endothelial cells, and after two days, cells were lysed and eNOS was immunoprecipitated by c-myc or Flag immunoprecipitation (IP). The immunoprecipitated proteins were used as a substrate for kinase assays with purified AMPKα1/β1/γ1 subunits (Merck Millipore, Darmstadt, Germany, Cat No. 1480) [20]. The lysates were separated by SDS-PAGE and blotted with antibodies specific for the phosphorylation sites of eNOS. Alternatively, ^{32}PγATP was used to radioactively label the protein. Proteins were separated by SDS-PAGE, and the gel was exposed to X-ray film after drying.

4.7. CRISPR/Cas9-Mediated Knock-Down of AMPKα1

Human umbilical vein endothelial cells (Promocell, Heidelberg, Germany) were transduced with lentiviral particles mediating the expression of Cas9 (Lenti-Cas9-2A-Blast was provided by Jason Moffat (Addgene plasmid # 73310)) and selected by blasticidin (10 μg/mL). Thereafter, a second lentiviral transduction with guide RNAs directed against AMPKα1 (Addgene numbers 76253 and 75254 provided by David Root, Cambridge, MA, USA) was performed, and puromycin (1 μg/mL) was used to select for double-transduced cells. The efficiency of the knockdown was analyzed by Western blotting.

4.8. Immunoblotting

Cells were lysed in Triton X-100 lysis buffer (Tris/HCl pH 7.5; 50 mmol/L; NaCl, 150 mmol/L; ethyleneglycoltetraacetic acid (EGTA), 2 mmol/L; ethylenediaminetetraacetic acid (EDTA) 2 mmol/L; Triton X-100, 1% (v/v); NaF, 25 mmol/L; $Na_4P_2O_7$, 10 mmol/L; 2 μg/mL each of leupeptin, pepstatin A, antipain, aprotinin, chymostatin, and trypsin inhibitor, and phenylmethylsulfonyl fluoride (PMSF), 40 μg/mL). Detergent-soluble proteins were heated with SDS-PAGE sample buffer and separated by SDS-PAGE, and specific proteins were detected by immunoblotting. To assess the phosphorylation of proteins, either equal amounts of protein from each sample were loaded twice and one membrane incubated with the phospho-specific antibody and the other with an antibody recognizing total protein, or blots were reprobed with the appropriate antibody.

4.9. Statistical Analyses

Data are expressed as mean ± standard error of the mean (SEM). Statistical evaluation was done using Student's *t*-test for unpaired data or ANOVA for repeated measures where appropriate. Values of $p < 0.05$ were considered statistically significant.

Supplementary Materials: Supplementary materials can be found at http://www.mdpi.com/1422-0067/19/9/2753/s1. Figure S1. Vascular function in carotid arteries from wild-type (WT) and AMPKα1$^{-/-}$ mice. (A) Contraction induced by KCl (80 mmol/L), (B) concentration response curves to phenylephrine (PE), and relaxation curves to (C) acetylcholine (ACh) or (D) sodium nitroprusside (SNP) in PE-contracted vessels. The graphs summarize data obtained from 7 animals in each group. Figure S2. Endothelial cell specific deletion of AMPKα1. (A) AMPKα1 expression in freshly isolated pulmonary endothelial cells from AMPKα1^{ΔEC} or Cre$^{-/-}$ (wild-type; WT) mice. (B) Expression of eNOS, AMPKα1 and AMPKα2 in aortic ring lysates from WT or AMPKα1^{ΔEC} (ΔEC) mice. (A) The blots presented are representative of 12 additional experiments using 2 mice per group. Figure S3. Effect of endothelial specific deletion of AMPKα2 on vascular reactivity of aortic rings (A) Dose dependent contraction to PE of wild-type (open symbols) or AMPKα2^{ΔEC} mice (closed symbols). (B) Relaxation curves of aortic rings to acetylcholine (ACh) after PE constriction of wild-type (open symbols) or AMPKα2^{ΔEC} mice (closed symbols). (C) Dose-dependent relaxation to SNP. The graphs summarize data obtained from 6 animals in each group. Figure S4. Effect of AMPK activators on the relaxation of aortic rings. (A,B) Concentration dependent effects of resveratrol (A) and amurensin G (B) on vascular tone in phenylephrine preconstricted aortic rings from wild-type (WT) and AMPKα1$^{\Delta EC}$ (α1$^{\Delta EC}$) mice; $n = 6$ animals in each group. (C,D) Time-dependent effects of PT-1 (30 μmol/L) and 991 (30 μmol/L) on vascular tone in phenylephrine preconstricted aortic rings from wild-type (WT) and AMPKα1$^{\Delta EC}$ (α1$^{\Delta EC}$) mice; $n = 4$ animals in each group. (E) Effects of the AMPK activators on the phosphorylation of AMPK (on Thr172) and ACC (Ser79) in endothelial cells isolated from aortic rings from wild-type mice. Experiments were performed in the absence (Basal) and presence of 991 (30 μmol/L), AICAR (0.5 mmol/L) or PT-1 (30 μmol/L) for 60 min. Comparable results were obtained in 3 additional independent experiments.

Author Contributions: N.Z., A.E.L., H.S., V.R., and B.F. performed the experiments and interpreted the data. I.F. and B.F. planned the study and wrote the manuscript.

funding: This work was supported by the Deutsche Forschungsgemeinschaft (SFB834/2 A9, SFB834/3 A5 and Exzellenzcluster 147 "Cardio-Pulmonary Systems").

Acknowledgments: The authors are indebted to Isabel Winter, Katharina Bruch, Mechtild Pipenbrock-Gyamfi, and Katharina Herbig for expert technical assistance.

Conflicts of Interest: The authors have no relationships to disclose.

Abbreviations

AMPK	AMP-activated protein kinase
eNOS	endothelial nitric-oxide synthase
AICAR	5-aminoimidazole-4-carboxamide ribonucleotide
VEGF	vascular endothelial growth factor
PPAR	peroxisome proliferator-activated receptor
SNP	sodium nitroprusside
ACh	acetylcholine
PE	phenylephrine
L-NAME	N$^{\omega}$-nitro-L-arginine methyl ester
CRISPR	clustered regularly interspaced short palindromic repeats
Cas9	CRISPR-associated protein 9

References

1. Musi, N.; Fujii, N.; Hirshman, M.F.; Ekberg, I.; Froberg, S.; Ljungqvist, O.; Thorell, A.; Goodyear, L.J. AMP-activated protein kinase (AMPK) is activated in muscle of subjects with type 2 diabetes during exercise. *Diabetes* **2001**, *50*, 921–927. [CrossRef] [PubMed]

2. Laderoute, K.R.; Amin, K.; Calaoagan, J.M.; Knapp, M.; Le, T.; Orduna, J.; Foretz, M.; Viollet, B. 5′-AMP-activated protein kinase (AMPK) is induced by low-oxygen and glucose deprivation conditions found in solid-tumor microenvironments. *Mol. Cell. Biol.* **2006**, *26*, 5336–5347. [CrossRef] [PubMed]

3. Herzig, S.; Shaw, R.J. AMPK: Guardian of metabolism and mitochondrial homeostasis. *Nat. Rev. Mol. Cell Biol.* **2018**, *19*, 121–135. [CrossRef] [PubMed]

4. Fisslthaler, B.; Fleming, I. Activation and signaling by the AMP-activated protein kinase in endothelial cells. *Circ. Res.* **2009**, *105*, 114–127. [CrossRef] [PubMed]

5. Bess, E.; Fisslthaler, B.; Fromel, T.; Fleming, I. Nitric oxide-induced activation of the AMP-activated protein kinase α2 subunit attenuates IκB kinase activity and inflammatory responses in endothelial cells. *PLoS ONE* **2011**, *6*, e20848. [CrossRef] [PubMed]

6. Cacicedo, J.M.; Yagihashi, N.; Keaney, J.F.; Ruderman, N.B.; Ido, Y. AMPK inhibits fatty acid-induced increases in NF-κB transactivation in cultured human umbilical vein endothelial cells. *Biochem. Biophys. Res. Commun.* **2004**, *324*, 1204–1209. [CrossRef] [PubMed]

7. Nagata, D.; Mogi, M.; Walsh, K. AMP-activated protein kinase (AMPK) signaling in endothelial cells is essential for angiogenesis in response to hypoxic stress. *J. Biol. Chem.* **2003**, *278*, 31000–31006. [CrossRef] [PubMed]

8. Yu, J.W.; Deng, Y.P.; Han, X.; Ren, G.F.; Cai, J.; Jiang, G.J. Metformin improves the angiogenic functions of endothelial progenitor cells via activating AMPK/eNOS pathway in diabetic mice. *Cardiovasc. Diabetol.* **2016**, *15*, 88. [CrossRef] [PubMed]

9. Reihill, J.A.; Ewart, M.A.; Hardie, D.G.; Salt, I.P. AMP-activated protein kinase mediates VEGF-stimulated endothelial NO production. *Biochem. Biophys. Res. Commun.* **2007**, *354*, 1084–1088. [CrossRef] [PubMed]

10. Cheng, K.K.; Lam, K.S.; Wang, Y.; Huang, Y.; Carling, D.; Wu, D.; Wong, C.; Xu, A. Adiponectin-induced endothelial nitric oxide synthase activation and nitric oxide production are mediated by APPL1 in endothelial cells. *Diabetes* **2007**, *56*, 1387–1394. [CrossRef] [PubMed]

11. Boyle, J.G.; Logan, P.J.; Ewart, M.A.; Reihill, J.A.; Ritchie, S.A.; Connell, J.M.; Cleland, S.J.; Salt, I.P. Rosiglitazone stimulates nitric oxide synthesis in human aortic endothelial cells via AMP-activated protein kinase. *J. Biol. Chem.* **2008**, *283*, 11210–11217. [CrossRef] [PubMed]

12. Rossoni, L.; Wareing, M.; Wenceslau, C.; Al-Abri, M.; Cobb, C.; Austin, C. Acute simvastatin increases endothelial nitric oxide synthase phosphorylation via AMP-activated protein kinase and reduces contractility of isolated rat mesenteric resistance arteries. *Clin. Sci. (Lond.)* **2011**, *121*, 449–458. [CrossRef] [PubMed]

13. Morrow, V.A.; Foufelle, F.; Connell, J.M.; Petrie, J.R.; Gould, G.W.; Salt, I.P. Direct activation of AMP-activated protein kinase stimulates nitric-oxide synthesis in human aortic endothelial cells. *J. Biol. Chem.* **2003**, *278*, 31629–31639. [CrossRef] [PubMed]

14. Davis, B.J.; Xie, Z.; Viollet, B.; Zou, M.H. Activation of the AMP-activated kinase by antidiabetes drug metformin stimulates nitric oxide synthesis in vivo by promoting the association of heat shock protein 90 and endothelial nitric oxide synthase. *Diabetes* **2006**, *55*, 496–505. [CrossRef] [PubMed]

15. Zou, M.H.; Kirkpatrick, S.S.; Davis, B.J.; Nelson, J.S.; Wiles, W.G.; Schlattner, U.; Neumann, D.; Brownlee, M.; Freeman, M.B.; Goldman, M.H. Activation of the AMP-activated protein kinase by the anti-diabetic drug metformin in vivo. Role of mitochondrial reactive nitrogen species. *J. Biol. Chem.* **2004**, *279*, 43940–43951. [CrossRef] [PubMed]

16. Hien, T.T.; Oh, W.K.; Quyen, B.T.; Dao, T.T.; Yoon, J.H.; Yun, S.Y.; Kang, K.W. Potent vasodilation effect of amurensin G is mediated through the phosphorylation of endothelial nitric oxide synthase. *Biochem. Pharmacol.* **2012**, *84*, 1437–1450. [CrossRef] [PubMed]

17. Li, X.; Dai, Y.; Yan, S.; Shi, Y.; Li, J.; Liu, J.; Cha, L.; Mu, J. Resveratrol lowers blood pressure in spontaneously hypertensive rats via calcium-dependent endothelial NO production. *Clin. Exp. Hypertens.* **2016**, *38*, 287–293. [CrossRef] [PubMed]

18. Fleming, I.; Fisslthaler, B.; Dixit, M.; Busse, R. Role of PECAM-1 in the shear-stress-induced activation of Akt and the endothelial nitric oxide synthase (eNOS) in endothelial cells. *J. Cell Sci.* **2005**, *118*, 4103–4111. [CrossRef] [PubMed]

19. Dixit, M.; Bess, E.; Fisslthaler, B.; Hartel, F.V.; Noll, T.; Busse, R.; Fleming, I. Shear stress-induced activation of the AMP-activated protein kinase regulates FoxO1a and angiopoietin-2 in endothelial cells. *Cardiovasc. Res.* **2008**, *77*, 160–168. [CrossRef] [PubMed]

20. Fisslthaler, B.; Fleming, I.; Keseru, B.; Walsh, K.; Busse, R. Fluid shear stress and NO decrease the activity of the hydroxy-methylglutaryl coenzyme A reductase in endothelial cells via the AMP-activated protein kinase and FoxO1. *Circ. Res.* **2007**, *100*, e12–e21. [CrossRef] [PubMed]

21. Thors, B.; Halldorsson, H.; Jonsdottir, G.; Thorgeirsson, G. Mechanism of thrombin mediated eNOS phosphorylation in endothelial cells is dependent on ATP levels after stimulation. *Biochim. Biophys. Acta* **2008**, *1783*, 1893–1902. [CrossRef] [PubMed]

22. Bradley, E.A.; Eringa, E.C.; Stehouwer, C.D.; Korstjens, I.; van Nieuw Amerongen, G.P.; Musters, R.; Sipkema, P.; Clark, M.G.; Rattigan, S. Activation of AMP-activated protein kinase by 5-aminoimidazole-4-carboxamide-1-beta-D-ribofuranoside in the muscle microcirculation increases nitric oxide synthesis and microvascular perfusion. *Arterioscler. Thromb. Vasc. Biol.* **2010**, *30*, 1137–1142. [CrossRef] [PubMed]

23. Schneider, H.; Schubert, K.M.; Blodow, S.; Kreutz, C.P.; Erdogmus, S.; Wiedenmann, M.; Qiu, J.; Fey, T.; Ruth, P.; Lubomirov, L.T.; et al. AMPK Dilates Resistance Arteries via Activation of SERCA and BKCa Channels in Smooth Muscle. *Hypertension* **2015**, *66*, 108–116. [CrossRef] [PubMed]

24. Schubert, K.M.; Qiu, J.; Blodow, S.; Wiedenmann, M.; Lubomirov, L.T.; Pfitzer, G.; Pohl, U.; Schneider, H. The AMP-Related Kinase (AMPK) Induces Ca^{2+}-Independent Dilation of Resistance Arteries by Interfering With Actin Filament Formation. *Circ. Res.* **2017**, *121*, 149–161. [CrossRef] [PubMed]

25. Davis, B.; Rahman, A.; Arner, A. AMP-activated kinase relaxes agonist induced contractions in the mouse aorta via effects on PKC signaling and inhibits NO-induced relaxation. *Eur. J. Pharmacol.* **2012**, *695*, 88–95. [CrossRef] [PubMed]

26. Schuhmacher, S.; Foretz, M.; Knorr, M.; Jansen, T.; Hortmann, M.; Wenzel, P.; Oelze, M.; Kleschyov, A.L.; Daiber, A.; Keaney, J.F.; et al. α1 AMP-activated protein kinase preserves endothelial function during chronic angiotensin II treatment by limiting Nox2 upregulation. *Arterioscler. Thromb. Vasc. Biol.* **2011**, *31*, 560–566. [CrossRef] [PubMed]

27. Goirand, F.; Solar, M.; Athea, Y.; Viollet, B.; Mateo, P.; Fortin, D.; Leclerc, J.; Hoerter, J.; Ventura-Clapier, R.; Garnier, A. Activation of AMP kinase alpha1 subunit induces aortic vasorelaxation in mice. *J. Physiol.* **2007**, *581*, 1163–1171. [CrossRef] [PubMed]

28. Enkhjargal, B.; Godo, S.; Sawada, A.; Suvd, N.; Saito, H.; Noda, K.; Satoh, K.; Shimokawa, H. Endothelial AMP-activated protein kinase regulates blood pressure and coronary flow responses through hyperpolarization mechanism in mice. *Arterioscler. Thromb. Vasc. Biol.* **2014**, *34*, 1505–1513. [CrossRef] [PubMed]

29. Wang, S.; Xu, J.; Song, P.; Viollet, B.; Zou, M.H. In vivo activation of AMP-activated protein kinase attenuates diabetes-enhanced degradation of GTP cyclohydrolase I. *Diabetes* **2009**, *58*, 1893–1901. [CrossRef] [PubMed]

30. Stahmann, N.; Woods, A.; Carling, D.; Heller, R. Thrombin activates AMP-activated protein kinase in endothelial cells via a pathway involving Ca^{2+}/calmodulin-dependent protein kinase kinase beta. *Mol. Cell. Biol.* **2006**, *26*, 5933–5945. [CrossRef] [PubMed]

31. Stahmann, N.; Woods, A.; Spengler, K.; Heslegrave, A.; Bauer, R.; Krause, S.; Viollet, B.; Carling, D.; Heller, R. Activation of AMP-activated protein kinase by vascular endothelial growth factor mediates endothelial angiogenesis independently of nitric-oxide synthase. *J. Biol. Chem.* **2010**, *285*, 10638–10652. [CrossRef] [PubMed]

32. Fleming, I.; Bauersachs, J.; Fisslthaler, B.; Busse, R. Calcium-independent activation of endothelial nitric oxide synthase in response to tyrosine phosphatase inhibitors and fluid shear stress. *Circ. Res.* **1998**, *82*, 686–695. [CrossRef] [PubMed]

33. Schneider, J.C.; El, K.D.; Chereau, C.; Lanone, S.; Huang, X.L.; De Buys Roessingh, A.S.; Mercier, J.C.; Dall'Ava-Santucci, J.; Dinh-Xuan, A.T. Involvement of Ca^{2+}/calmodulin-dependent protein kinase II in endothelial NO production and endothelium-dependent relaxation. *Am. J. Physiol. Heart Circ. Physiol.* **2003**, *284*, H2311–H2319. [CrossRef] [PubMed]

34. Zang, M.; Xu, S.; Maitland-Toolan, K.A.; Zuccollo, A.; Hou, X.; Jiang, B.; Wierzbicki, M.; Verbeuren, T.J.; Cohen, R.A. Polyphenols stimulate AMP-activated protein kinase, lower lipids, and inhibit accelerated atherosclerosis in diabetic LDL receptor-deficient mice. *Diabetes* **2006**, *55*, 2180–2191. [CrossRef] [PubMed]

35. Dimmeler, S.; Fleming, I.; Fisslthaler, B.; Hermann, C.; Busse, R.; Zeiher, A.M. Activation of nitric oxide synthase in endothelial cells by Akt-dependent phosphorylation. *Nature* **1999**, *399*, 601–605. [CrossRef] [PubMed]

36. Fulton, D.; Mcgiff, J.C.; Quilley, J. Pharmacological evaluation of an epoxide as the putative hyperpolarizing factor mediating the nitric oxide-independent vasodilator effect of bradykinin in the rat heart. *J. Pharm. Exp. Ther.* **1999**, *287*, 497–503.

37. Chen, Z.P.; Mitchelhill, K.I.; Michell, B.J.; Stapleton, D.; Rodriguez-Crespo, I.; Witters, L.A. AMP-activated protein kinase phosphorylation of endothelial no synthase. *FEBS Lett.* **1999**, 285–289. [CrossRef]

38. Michell, B.J.; Chen, Z.p.; Tiganis, T.; Stapleton, D.; Katsis, F.; Power, D.A.; Sim, A.T.; Kemp, B.E. Coordinated control of endothelial nitric-oxide synthase phosphorylation by protein kinase C and the cAMP-dependent protein kinase. *J. Biol. Chem.* **2001**, *276*, 17625–17628. [CrossRef] [PubMed]

39. Fleming, I.; Fisslthaler, B.; Dimmeler, S.; Kemp, B.E.; Busse, R. Phosphorylation of Thr(495) regulates Ca(2+)/calmodulin-dependent endothelial nitric oxide synthase activity. *Circ. Res.* **2001**, *88*, E68–E75. [CrossRef] [PubMed]

40. Gaskin, F.S.; Kamada, K.; Yusof, M.; Korthuis, R.J. 5'-AMP-activated protein kinase activation prevents postischemic leukocyte-endothelial cell adhesive interactions. *Am. J. Physiol. Heart Circ. Physiol.* **2007**, *292*, H326–H332. [CrossRef] [PubMed]

41. Huang, Y.; Smith, C.A.; Chen, G.; Sharma, B.; Miner, A.S.; Barbee, R.W.; Ratz, P.H. The AMP-dependent protein kinase (AMPK) activator A-769662 causes arterial relaxation by reducing cytosolic free calcium independently of an increase in AMPK phosphorylation. *Front. Pharmacol.* **2017**, *8*. [CrossRef] [PubMed]

42. Jensen, T.E.; Ross, F.A.; Kleinert, M.; Sylow, L.; Knudsen, J.R.; Gowans, G.J.; Hardie, D.G.; Richter, E.A. PT-1 selectively activates AMPKγ1 complexes in mouse skeletal muscle, but activates all three + subunit

complexes in cultured human cells by inhibiting the respiratory chain. *Biochem. J.* **2015**, *467*, 461–472. [CrossRef] [PubMed]

43. Bultot, L.; Jensen, T.E.; Lai, Y.C.; Madsen, A.L.B.; Collodet, C.; Kviklyte, S.; Deak, M.; Yavari, A.; Foretz, M.; Ghaffari, S.; et al. Benzimidazole derivative small-molecule 991 enhances AMPK activity and glucose uptake induced by AICAR or contraction in skeletal muscle. *Am. J. Physiol. Endocrinol. Metab.* **2016**, *311*, E706–E719. [CrossRef] [PubMed]

44. Wang, S.; Liang, B.; Viollet, B.; Zou, M.H. Inhibition of the AMP-activated protein kinase-alpha2 accentuates agonist-induced vascular smooth muscle contraction and high blood pressure in mice. *Hypertension* **2011**, *57*, 1010–1017. [CrossRef] [PubMed]

45. Sun, G.Q.; Li, Y.B.; Du, B.; Meng, Y. Resveratrol via activation of AMPK lowers blood pressure in DOCA-salt hypertensive mice. *Clin. Exp. Hypertens.* **2015**, *37*, 616–621. [CrossRef] [PubMed]

46. Kröller-Schön, S.; Jansen, T.; Hauptmann, F.; Schüler, A.; Heeren, T.; Hausding, M.; Oelze, M.; Viollet, B.; Keaney, J.F.; Wenzel, P.; et al. α1AMP-activated protein kinase mediates vascular protective effects of exercise. *Arterioscler. Thromb. Vasc. Biol.* **2012**, *32*, 1634–1641. [CrossRef] [PubMed]

47. Wang, S.; Zhang, M.; Liang, B.; Xu, J.; Xie, Z.; Liu, C.; Viollet, B.; Yan, D.; Zou, M.H. AMPKalpha2 deletion causes aberrant expression and activation of NAD(P)H oxidase and consequent endothelial dysfunction in vivo: Role of 26S proteasomes. *Circ. Res.* **2010**, *106*, 1117–1128. [CrossRef] [PubMed]

48. Mount, P.F.; Kemp, B.E.; Power, D.A. Regulation of endothelial and myocardial NO synthesis by multi-site eNOS phosphorylation. *J. Mol. Cell. Cardiol.* **2007**, *42*, 271–279. [CrossRef] [PubMed]

49. Michell, B.J.; Griffiths, J.E.; Mitchelhill, K.I.; Rodriguez-Crespo, I.; Tiganis, T.; Bozinovski, S.; de Montellano, P.R.O.; Kemp, B.E.; Pearson, R.B. The Akt kinase signals directly to endothelial nitric oxide synthase. *Curr. Biol.* **1999**, *9*, 845–848. [CrossRef]

50. Gallis, B.; Corthals, G.L.; Goodlett, D.R.; Ueba, H.; Kim, F.; Presnell, S.R.; Figeys, D.; Harrison, D.G.; Berk, B.C.; Aebersold, R.; et al. Identification of flow-dependent endothelial nitric-oxide synthase phosphorylation sites by mass spectrometry and regulation of phosphorylation and nitric oxide production by the phosphatidylinositol 3-Kinase inhibitor LY294002. *J. Biol. Chem.* **1999**, *274*, 30101–30108. [CrossRef] [PubMed]

51. Ohashi, J.; Sawada, A.; Nakajima, S.; Noda, K.; Takaki, A.; Shimokawa, H. Mechanisms for enhanced endothelium-derived hyperpolarizing factor-mediated responses in microvessels in mice. *Circ. J.* **2012**, *76*, 1768–1779. [CrossRef] [PubMed]

52. Aoyagi, M.; Arvai, A.S.; Tainer, J.A.; Getzoff, E.D. Structural basis for endothelial nitric oxide synthase binding to calmodulin. *EMBO J.* **2003**, *22*, 766–775. [CrossRef] [PubMed]

53. Chen, Z.; Peng, I.C.; Sun, W.; Su, M.I.; Hsu, P.H.; Fu, Y.; Zhu, Y.; DeFea, K.; Pan, S.; Tsai, M.D.; et al. AMP-activated protein kinase functionally phosphorylates endothelial nitric oxide synthase Ser633. *Circ. Res.* **2009**, *104*, 496–505. [CrossRef] [PubMed]

54. Schulz, E.; Anter, E.; Zou, M.H.; Keaney, J.F., Jr. Estradiol-mediated endothelial nitric oxide synthase association with heat shock protein 90 requires adenosine monophosphate-dependent protein kinase. *Circulation* **2005**, *111*, 3473–3480. [CrossRef] [PubMed]

55. Zippel, N.; Malik, R.A.; Fromel, T.; Popp, R.; Bess, E.; Strilic, B.; Wettschureck, N.; Fleming, I.; Fisslthaler, B. Transforming growth factor-beta-activated kinase 1 regulates angiogenesis via AMP-activated protein kinase-α1 and redox balance in endothelial cells. *Arterioscler. Thromb. Vasc. Biol.* **2013**, *33*, 2792–2799. [CrossRef] [PubMed]

56. Busse, R.; Lamontagne, D. Endothelium-derived bradykinin is responsible for the increase in calcium produced by angiotensin-converting enzyme inhibitors in human endothelial cells. *Naunyn Schmiedebergs Arch. Pharmacol.* **1991**, *344*, 126–129. [CrossRef] [PubMed]

57. Michaelis, U.R.; Falck, J.R.; Schmidt, R.; Busse, R.; Fleming, I. Cytochrome P4502C9-derived epoxyeicosatrienoic acids induce the expression of cyclooxygenase-2 in endothelial cells. *Arterioscler. Thromb. Vasc. Biol.* **2005**, *25*, 321–326. [CrossRef] [PubMed]

International Journal of
Molecular Sciences

MDPI

Review

AMP-Activated Protein Kinase as a Reprogramming Strategy for Hypertension and Kidney Disease of Developmental Origin

You-Lin Tain [1,2] and Chien-Ning Hsu [3,*]

[1] Departments of Pediatrics, Kaohsiung Chang Gung Memorial Hospital and Chang Gung University College of Medicine, Kaohsiung 833, Taiwan; tainyl@hotmail.com
[2] Institute for Translational Research in Biomedicine, Kaohsiung Chang Gung Memorial Hospital and Chang Gung University College of Medicine, Kaohsiung 833, Taiwan
[3] Department of Pharmacy, Kaohsiung Chang Gung Memorial Hospital, Kaohsiung 833, Taiwan
* Correspondence: chien_ning_hsu@hotmail.com; Tel.: +886-975-368-975; Fax: +886-7733-8009

Received: 2 June 2018; Accepted: 12 June 2018; Published: 12 June 2018

Abstract: Suboptimal early-life conditions affect the developing kidney, resulting in long-term programming effects, namely renal programming. Adverse renal programming increases the risk for developing hypertension and kidney disease in adulthood. Conversely, reprogramming is a strategy aimed at reversing the programming processes in early life. AMP-activated protein kinase (AMPK) plays a key role in normal renal physiology and the pathogenesis of hypertension and kidney disease. This review discusses the regulation of AMPK in the kidney and provides hypothetical mechanisms linking AMPK to renal programming. This will be followed by studies targeting AMPK activators like metformin, resveratrol, thiazolidinediones, and polyphenols as reprogramming strategies to prevent hypertension and kidney disease. Further studies that broaden our understanding of AMPK isoform- and tissue-specific effects on renal programming are needed to ultimately develop reprogramming strategies. Despite the fact that animal models have provided interesting results with regard to reprogramming strategies targeting AMPK signaling to protect against hypertension and kidney disease with developmental origins, these results await further clinical translation.

Keywords: AMP-activated protein kinase; developmental origins of health and disease (DOHaD); hypertension; kidney disease; nutrient-sensing signals; oxidative stress; renin-angiotensin system

1. Introduction

Hypertension and kidney disease have a significant impact on morbidity and mortality worldwide. Hypertension and kidney disease can be the cause and consequence of one another. Importantly, both disorders can originate in early life. Kidneys play a key role in blood pressure (BP) regulation. The developing kidney is highly vulnerable to environmental effects in fetal and infantile life, leading to long-term programming effects on the morphology and functioning of the kidney [1,2]. Adverse renal programming increases the risk for developing hypertension and kidney disease in adulthood [3]. This notion has become a globally recognized concept as developmental origins of health and disease (DOHaD) [4]. Conversely, the DOHaD concept also allows reprogramming [5], a strategy aimed at reversing the initial programming processes prior to the onset of hypertension and kidney disease, in order to shift therapeutic interventions from adulthood to early life. A growing body of evidence suggests that AMP-activated protein kinase (AMPK) plays a decisive role in the normal renal physiology and pathogenesis of hypertension and kidney disease [6,7]. Based on the two aspects of the DOHaD concept, this review will first present the evidence for the link between AMPK signaling and programming mechanisms that may lead to hypertension and kidney disease of developmental origin, with a focus on the kidney. This will be followed by potential pharmacological interventions targeting AMPK signaling that may serve as reprogramming strategies to halt the growing epidemic of hypertension and kidney disease.

2. AMP-Activated Protein Kinase in the Renal System

2.1. The Structure and Function of AMP-Activated Protein Kinase

AMPK is a phylogenetically conserved, ubiquitously expressed serine/threonine protein kinase. AMPK is a heterotrimer, composed of an α (α1, α2) catalytic subunit, a regulatory and structurally crucial β (β1, β2) subunit, and a regulatory γ (γ1, γ2, γ3) subunit [8–10]. These isoforms are encoded by distinct genes and differentially expressed, and have unique tissue-specific expression profiles, creating the potential to generate a diverse collection of 12 αβγ heterotrimer combinations. AMPK has a diverse range of biological functions, including cellular energy homeostasis, glucose metabolism, lipid metabolism, protein synthesis, redox regulation, mitochondria biogenesis, autophagy, ion transport, tumor suppression, anti-inflammation, and nitric oxide (NO) synthesis [8–11]. The structure and function of these different isoforms have been reviewed in detail previously [8–10], and it is not within the scope of the current review to thoroughly outline these further. However, for the purposes of the discussion below, it is important to note that activation of AMPK leads to the biological functions that are linked to renal pathophysiology.

2.2. Regulation of AMP-Activated Protein Kinase AMP-Activated Protein Kinase in the Kidney

AMPK is strongly expressed in the kidney, where it is involved in diverse physiological and pathologic processes, including sensing cellular energy status, sodium and ion transport, podocyte function, BP control, the epithelial-to-mesenchymal transition, and NO production [7–11]. In the rat kidney, α1 and β1 subunits are predominant [7,11]. Except for the muscle-specific γ3 isoform, both γ1 and γ2 subunits are similarly expressed in the kidney. However, little is known regarding the differences in AMPK subunit expression between different cell types within the kidney.

The activity of AMPK is mainly regulated by the AMP and adenosine triphosphate (ATP) ratio. AMPK is activated both allosterically and by post-translational modifications. The most well-defined mechanisms of AMPK activation are phosphorylation at αThr^{172} by upstream AMPK kinases and by AMP or adenosine diphosphate (ADP) binding to the γ subunit. So far, at least three kinases and three phosphatases have been identified as upstream AMPK-activating kinases, including liver kinase B1 (LKB1), TGFβ-activated kinase 1 (TAK1), Ca^{2+}-/calmodulin-dependent protein kinase β (CaMKKβ), protein phosphatase 2A (PP2A), protein phosphatase 2C (PP2C), and Mg^{2+}-/Mn^{2+}-dependent protein phosphatase 1E (PPM1E) [12]. Additionally, AMPK can also be regulated by intracellular calcium and oxidant signaling, as well as extracellular signaling like hormones and cytokines [13]. Furthermore, AMPK is the target of a growing number of pharmacological activators [14].

AMPK has transcriptional effects on numerous enzymes that mediate cellular energy metabolism. AMPK can induce mitochondrial biogenesis by activating the peroxisome proliferator-activated receptor-γ (PPARγ) coactivator-1α (PGC-1α), either directly or through the silent information regulator transcript 1 (SIRT1) [14,15]. Additionally, AMPK and SIRT1 can mediate phosphorylation and deacetylation of PGC-1α, respectively [15], to regulate the expression of PPAR target genes. As reviewed elsewhere, several PPAR target genes contribute to renal programming and hypertension of developmental origins, such as *Sod2*, *Nrf2*, *Sirt7*, *Ren*, *Nos2*, *Nos3*, and *Sgk1* [16]. Another important downstream effect of AMPK is the inhibition of the mammalian target of rapamycin (mTOR). Both AMPK and mTOR can oppositely regulate unc-51-like kinase 1/2 (ULK1/2) activity by phosphorylation to mediate autophagy, a cellular catabolic process in which key organelles are transported to lysosomes for degradation. In addition to activating ULK1/2, AMPK can promote autophagy through SIRT1 by de-acetylating several autophagy-related proteins [17]. Given that autophagy is involved in the pathogenesis of many kidney diseases, and that AMPK regulates autophagic protection against kidney injury, AMPK is becoming a potential target for kidney disease therapy [18]. Additionally, AMPK has been shown to exert anti-inflammatory, antioxidant, and anti-apoptosis effects. Moreover, AMPK regulates many sodium and ion transport proteins in the renal tubular cells, including the epithelial Na^+ channel (ENaC) [19], the Na^+–K^+–$2Cl^-$ cotransporter (NKCC2) [20], Na^+/K^+-ATPase (NaKATPase) [21], the vacuolar H^+-ATPase (V-ATPase) [22], and others [7,11,23]. The activation of AMPK and its biochemical pathways are illustrated in Figure 1.

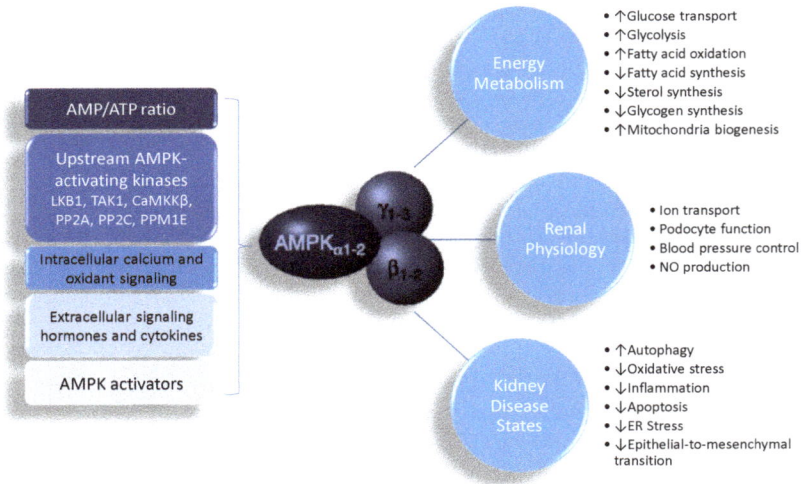

Figure 1. Schematic representation of AMP-activated protein kinase (AMPK) actions and its biochemical functions in the kidney. ↑ = increased. ↓ = decreased.

2.3. AMP-Activated Protein Kinase in Hypertension and Kidney Disease

Emerging evidence suggests that dysfunction in the AMPK signaling pathway is involved with the development of various cardiovascular diseases, including hypertension [6]. Despite several AMPK activators having been assessed in a number of human studies, interventions necessary to provide a reprogramming strategy and prove causation remain undeveloped. It is for this reason that much of our knowledge of potential mechanisms of renal programming, the impacts of AMPK in renal programming, and reprogramming strategies targeting AMPK signaling come from studies using animal models.

In a genetic hypertension model of a spontaneously hypertensive rat (SHR), AMPK activation was reduced in the aorta of the SHR, while 5-aminoimidazole-4carboxamide riboside (AICAR), a direct AMPK activator, lowered BP [24]. Our previous report showed that metformin, a known AMPK activator, blocks the development of hypertension in SHRs, which is associated with increased renal NO production [25]. Like metformin, activation of AMPK by perinatal resveratrol supplementation has been shown to mitigate the development of hypertension in adult SHRs [26]. Additionally, we recently found that maternal plus post-weaning high-fat diets induced hypertension and reduced renal cortical protein levels of phosphorylated AMPK2α in offspring kidneys, which was prevented by resveratrol therapy [27]. Similarly, AMPK activator metformin was reported to protect adult offspring against the developmental programming of hypertension induced by a maternal plus post-weaning high-fat diet [28]. However, the reprogramming effects of the AMPK activator have not been fully assessed in other developmental programming models for hypertension. Therefore, further investigation is

Int. J. Mol. Sci. **2018**, *19*, 1744

needed to reveal the precise role of AMPK in hypertension of developmental origins, especially the reprogramming effects of AMPK activation.

Apart from its role in the development of hypertension, as mentioned above, there have been many studies on the effects of AMPK in kidney diseases, notably diabetic nephropathy, autosomal dominant polycystic kidney disease, subtotal nephrectomy, lupus nephritis, and renal fibrosis [7]. Although AMPK signals have been studied in established kidney diseases, so far there remains a lack of data on the role of AMPK in renal programming and kidney disease of developmental origin.

3. Common Mechanisms Link AMP-Activated Protein Kinase to Renal Programming

Despite the fact that several organ systems can be programmed in response to adverse environmental exposures in early life, renal programming is considered to be decisive in the development of hypertension, as well as kidney disease [3,5,29,30]. Thus far, a number of proposed mechanisms, including dysregulation of the renin–angiotensin system (RAS), impaired sodium transporters, a nutrient-sensing signal, and oxidative stress have been linked to renal programming [3,5,29,30]. Each mechanism related to AMPK signaling will be discussed in turn.

3.1. Renin–Angiotensin System

The RAS is a central regulator of BP and renal function. The RAS consists of two opposing axes: the angiotensin converting enzyme (ACE)-angiotensin (Ang) II type 1 receptor (AT1R) classical axis, mediated primarily by Ang II; and the ACE2-angiotensin-(1–7)-Mas receptor axis, mediated mainly by Angiotensin-(1–7) [31]. In contrast to ACE, ACE2 appears to adjust the angiotensin II type 2 receptor (AT2R) and the angiotensin (1–7) receptor Mas in a way that opposes the development of hypertension [31]. Over-activation of the classical RAS leads to hypertension and kidney disease [31]. Notably, this hormone-signaling pathway controls kidney development [32]. Both RAS axes and the above-mentioned RAS components have been linked to fetal programming [33–36]. There is a biphasic response with reduced classical RAS expression at birth that increases with age. Early-life renal programming might activate the classical RAS, leading to hypertension and kidney disease development in later life. Conversely, early blockade of the classical RAS has been shown to prevent the development of hypertension and kidney disease [37–39]. Decreased renal AMPK expression has been found in uninephretomized rats with the activation of the RAS [40], which was prevented by the blockage of the RAS. AMPKα2 knockout mice expressed high ACE levels, resulting in vasoconstriction [41]. Conversely, prenatal metformin therapy has been shown to restore the maternal high-fructose plus post-weaning high-fat diet-induced increases of RAS components *Ren*, *Atp6ap2*, *Agt*, *Ace*, and *Agtr1a* in the kidney cortex, resulting in protection from hypertension [28]. Given that resveratrol, an indirect AMPK activator, was reported to exert its protective effects in association with increased expression of the AT2R and Mas receptors [42], further studies are required to determine whether AMPK has a role in activation of the ACE2-angiotensin-(1–7)-Mas receptor axis and AT2R to prevent hypertension, and whether this contributes to the reprogramming effects of AMPK activators. Nevertheless, the detailed mechanisms underlying the modulation of RAS by AMPK and its contributions to protection from programmed hypertension kidney disease still await for further study in different models of developmental programming.

3.2. Sodium Transporters

Hypertension and kidney disease of developmental origin have been associated with enhanced sodium reabsorption, attributed to the increased expression of sodium transporters [1,5,29,30]. Several adverse environmental impacts during early life leading to predisposition toward impaired sodium transporters have been reported, including maternal low-protein diet, maternal high-fat diet, maternal exposure to continuous light, and prenatal glucocorticoid exposure [1,5,29]. Several sodium transporters have been identified in the programming processes, including Na^+/Cl^- cotransporter (NCC), type 3 sodium hydrogen exchanger (NHE3), NKCC2, and NaKATPase. We previously showed that a maternal high-fructose diet plus a postnatal high-salt diet increased renal cortical protein levels of NKCC2, NHE3, and NCC in a two-hit model of programmed hypertension [43]. Notably, AMPK regulates several sodium transporters, such as NKCC2 and NaKATPase, which may account for its beneficial effects on hypertension and kidney disease. However, there are as yet no studies examining the role of AMPK in sodium transporters in the kidneys from animal models of programmed hypertension and kidney disease.

3.3. Nutrient-Sensing Signals

Nutrient-sensing signals regulate fetal metabolism and development in response to maternal nutritional input. AMPK is a well-known nutrient-sensing signal [44]. In addition to AMPK, known nutrient-sensing signals exist in the kidney, including SIRT, PPARs, PGC-1α, and mTOR [44]. The interplay between AMPK and other nutrient-sensing signals, driven by early-life input, can regulate PPARs and their target genes, thereby promoting programmed hypertension and kidney disease [16,45]. Our previous work demonstrated that resveratrol, an AMPK activator, prevents the development of hypertension programmed by maternal plus post-weaning high-fructose diets, via regulation of nutrient-sensing signals [46], supporting the notion that nutrient-sensing signals might be a common mechanism underlying the pathogenesis of hypertension and kidney disease of developmental origin. Since AMPK is a crucial hub for the nutrient-sensing signals network, further studies are required to determine whether AMPK has a role in the regulation of renal programming, and whether AMPK activators can serve as reprogramming strategies to prevent the developmental programming of hypertension and kidney disease.

3.4. Oxidative Stress

Oxidative stress is an oxidative shift characterized by an imbalance between oxidants (e.g., reactive oxygen species (ROS)) and antioxidants, in favor of oxidants. The developing fetus is highly vulnerable to oxidative stress damage, due to its low antioxidant capacity [47]. As reviewed elsewhere [29,45], numerous pre- and peri-natal inputs have been linked to renal programming attributed to oxidative stress, including imbalanced maternal nutrition, maternal diabetes, preeclampsia, prenatal hypoxia, maternal nicotine exposure, maternal inflammation, prenatal glucocorticoid exposure, and a high-fat maternal diet. Conversely, some reprogramming interventions have targeted antioxidants in order to reduce oxidative stress, and accordingly, prevent hypertension and kidney disease of developmental origin [30]. As nutrient-sensing is interconnected with redox regulation, AMPK has a key role in regulating antioxidant defense during oxidative stress. AMPK has been reported to upregulate several antioxidant genes, such as superoxide dismutase (SOD), uncoupling protein 2 (UCP2), and nuclear factor erythroid-2-related factor (NRF2) [48]. Additionally, AMPK activation was shown to suppress nicotinamide adenine dinucleotide phosphate (NADPH) oxidase, a primary source of ROS [49]. Furthermore, AMPK promotes autophagy. Since mitochondria are another major source of ROS within cells, activation of mitochondrial autophagy driven by AMPK is also beneficial for reducing oxidative stress. Thus, these findings suggest that the interplay between AMPK and oxidative stress contributes to programmed hypertension and kidney disease.

All of these observations demonstrate a close link between AMPK and other hypothetical mechanisms involved in renal programming. Nevertheless, there remains no definite conclusion that AMPK plays a central role on mediating other mechanisms leading to hypertension and developmental kidney disease.

4. Reprogramming Strategy Targeting AMP-Activated Protein Kinase Signaling

Reprogramming strategies to counterbalance the programming processes that have been employed range from nutritional intervention and lifestyle modification to pharmacological therapy [1,5,50]. Currently, a variety of therapeutic regimens have been reported to either activate or inhibit AMPK activity and its downstream signaling pathway. Since AMPK inhibition, either by AMPK silencing or AMPK inhibitors (e.g., compound C), contributes to hypertension [51,52], treatment modalities for AMPK activation have become more attractive reprogramming strategies. Both indirect and direct AMPK activators have been studied in established hypertension and kidney disease [6,13,18,53]. Modulators that cause AMP or calcium accumulation without a direct interaction with AMPK are classified as "indirect AMPK activators". Several kinds of indirect AMPK activators have been studied in relation to cardiovascular and kidney disease [13], including metformin, resveratrol, thiazolidinediones (TZDs), polyphenols, berberine, ginsenoside, α-lipoic acid, quercetin, and so on. On the other hand, direct AMPK activators induce conformational changes in the AMPK complex through direct interaction with a specific subunit of AMPK. While some are potent pan-activators (e.g., AICAR) for all 12 heterotrimetric AMPK complexes, others show isoform-specific activations for the α1 (e.g., compound-13), β1 (e.g., PF-06409577 and PF-249), β1/β2 (e.g., GSK621), or γ1 isoforms (e.g., PT-1) [13,48]. However, present knowledge of AMPK activators in the kidney is significantly less advanced than that for other organs, such as the liver, muscles, and heart. So far, only one report has shown that selective AMPKβ1 activators PF-06409577 and PF-249 protect against kidney damage in a rat model of diabetic nephropathy [54]. In the current review, we will primarily be limited to pharmacological therapies aimed at AMPK signaling as a reprogramming strategy to prevent hypertension and kidney disease of developmental origin. Of note, pharmacotherapies will be narrowly restricted to those beginning prior to the onset of hypertension and kidney disease. SHRs, for example, reveal a rise in BP starting from six weeks of age, and a steep increase between 6 and 24 weeks. We therefore restrict our discussion mainly to therapies starting before six weeks of age, using the SHR model. These AMPK activation modalities are listed in Table 1 [25–28,46,55–62]. Because the field of DOHaD research is beginning to emerge, this list is by no means complete and is expected to grow rapidly.

Table 1. Reprogramming strategy targeted on AMPK signaling in animal models of programmed hypertension and kidney disease.

Animal Models	Gender/Species	Age at Evaluation	Dose and Period of Treatment	Reprogramming Effects	Ref.
Metformin					
SHR [1]	Male SHR	12 weeks	Metformin (500 mg/kg/day) between 4 to 12 weeks of age	Prevented hypertension	[25]
Maternal high-fructose plus post-weaning high-fat diet	Male SD [2] rats	12 weeks	Metformin (500 mg/kg/day) for 3 weeks during pregnancy	Attenuated hypertension;	[28]
Resveratrol and other polyphenols					
SHR	Male SHR	11 weeks	Resveratrol (50 mg/L) in drinking water between 3–11 weeks of age	Attenuated hypertension	[55]
SHR	Male and female SHR	12 weeks	Resveratrol (4g/kg of diet) between gestational day 0.5 and postnatal day 21	Attenuated hypertension	[26]
SHR	Male SHR	13 weeks	Resveratrol (50 mg/L) in drinking water between 3–13 weeks of age	Attenuated hypertension	[56]
Prenatal hypoxia and postnatal high-fat diet	Male SD rats	12 weeks	Resveratrol (4g/kg of diet) between 3–12 weeks of age	Prevented hypertension	[57]
Maternal plus post-weaning high-fructose diets	Male SD rats	12 weeks	Resveratrol (50 mg/L) in drinking water from weaning to three months of age	Prevented hypertension	[46]
Maternal plus post-weaning high-fat diets	Male SD rats	16 weeks	0.5% resveratrol in drinking water between 2 and 4 months of age	Prevented hypertension	[27]
SHR	Male SHR	7 weeks	Magnolol (100 mg/kg/day) between 4 to 7 weeks of age	Attenuated hypertension	[58]
SHR	Male SHR	20 weeks	Berberine (100 mg/kg/day) between 3 to 20 weeks of age	Attenuated hypertension and kidney damage	[59]

Table 1. *Cont.*

Animal Models	Gender/Species	Age at Evaluation	Dose and Period of Treatment	Reprogramming Effects	Ref.
High-salt stroke-prone SHR	Male stroke-prone SHR	16 weeks	Genistein (0.06% wt/wt diet) between 7 to 16 weeks of age	Attenuated hypertension and kidney damage	[60]
Thiazolidinediones					
SHR	Male SHR	7 weeks	Pioglitazone (10 mg/kg/day) between 5 to 7 weeks of age	Attenuated hypertension	[61]
SHR	Male SHR	13 weeks	Rosiglitazone (150 mg/kg/day) between 5 to 13 weeks of age	Attenuated hypertension	[62]

[1] SHR: Spontaneously hypertensive rat; [2] SD rats: Sprague–Dawley rats.

4.1. Metformin

Metformin, the most commonly prescribed first-line antidiabetic drug in the world, exerts its beneficial effects primarily by AMPK activation. Despite a growing body of evidence indicating the protective effects of metformin in established cardiovascular and kidney diseases [63,64], only a few studies have been conducted to explore its reprogramming effects on programmed hypertension and kidney disease. Early metformin treatment in the pre-hypertensive stage blocks the development of hypertension in SHRs [25]. Additionally, maternal metformin therapy protects adult offspring against the developmental programming of hypertension induced by a maternal plus post-weaning high-fat diet [28]. However, a concern raised by these studies is that AMPK-independent effects on metformin were also reported. A better understanding of the AMPK-dependent and -independent mechanisms responsible for the protective effects of metformin on programmed hypertension and kidney disease is therefore warranted.

4.2. Resveratrol and Other Polyphenols

Polyphenols are a large group of phytochemicals found in plant-based food. Resveratrol is a naturally occurring polyphenol phytoalexin. It has been considered to have cardiovascular protective effects, including against hypertension [65]. Mechanisms of activation of AMPK by resveratrol appear to elevate AMP levels and inhibit mitochondrial ATP production. Early resveratrol therapy mitigates the development of hypertension in adult SHRs of both sexes [26,55,56]. Using the prenatal hypoxia and postnatal high-fat diet rat model, post-weaning resveratrol treatment protects adult offspring against programmed hypertension [57]. Our previous report showed that a maternal plus post-weaning high-fat diet induced hypertension and reduced protein levels of phosphorylated AMPK2α in the offspring kidney cortex, which resveratrol therapy prevented [27]. Additionally, early post-weaning resveratrol therapy was reported to prevent the development of hypertension of adult offspring exposed to maternal and post-weaning high-fructose consumption [46].

Interestingly, increased AMP levels and ATP depletion leading to uric acid production have been demonstrated as key mediators in the pathogenesis of fructose-induced metabolic syndrome and hypertension [66]. Our previous study showed that the mechanisms underlying the development of hypertension in offspring exposed to maternal high-fructose consumption are different from those in adult rats fed with a high-fructose diet [67]. This may explain why resveratrol increases AMP levels to activate AMPK, resulting in a beneficial effect on the offspring's BP. However, the reprogramming effect of maternal resveratrol on kidney disease has not been fully assessed in developmental programming models. Of note is that resveratrol has multifaceted biological functions; however, to what extent does its reprogramming effect on hypertension and kidney disease can be attributed to AMPK activation deserves further elucidation.

In addition to resveratrol, several polyphenols are capable of activating AMPK, including quercetin, genistein, epigallocatechin gallate, anthocyanin, magnolol, berberine, and so on. Early treatment with magnolol, berberine, or genistein offered protective effects against programmed hypertension in adult SHRs [58–60], whereas quercetin did not [68]. While many polyphenols are antioxidants and exert beneficial effects on oxidative stress-related disorders [69], evidence for their reprogramming effects on hypertension and kidney disease as AMPK activators is equivocal. A better understanding of the differential mechanisms of various polyphenols in the prevention and treatment of programmed hypertension and kidney disease is therefore warranted.

4.3. Thiazolidinediones

Thiazolidinediones (TZDs) are a class of insulin-sensitizing drugs, including pioglitazone, rosiglitazone, and troglitazone. TZDs exert their effects mainly by activating PPARγ. They are also known to act in part through AMPK activation. TZDs that activate AMPK are associated with the accumulation of AMP. Treatment with pioglitazone and rosiglitazone prior to the onset of hypertension can be protective in SHRs, by attenuating the development of hypertension [61,62]. However, another study showed that perinatal pioglitazone treatment fails to confer antihypertensive or renoprotective effects in adult fawn-hooded hypertensive rats [70]. Thus, further examination is required to understand the protective effects of TZDs in programmed hypertension and kidney disease, which are exerted mainly via AMPK or PPARγ signaling pathway.

4.4. Others

Despite progress made in recent years in discovering direct AMPK activators [13,53], little is known regarding their reprogramming effects on hypertension and kidney disease of developmental origins. The first direct AMPK activator, AICAR, has been reported to lower BP in adult SHRs [24]. However, the reprogramming effect of AICAR on programmed hypertension and kidney disease has not been examined yet in developmental programming models. Additionally, oxidative modification of the AMPKα subunit appears to be a major mechanism by which AMPK is activated under conditions of oxidative stress. Therefore, any modulators that induce intracellular ROS generation might serve as AMPK activators. Furthermore, it is of great importance to understand the interplay between the AMPK signaling and other mechanisms underlying renal programming; the application of reprogramming strategies targeting the above-mentioned mechanisms is also feasible for early intervention.

5. Conclusions and Future Perspectives

Hypertension and kidney disease in adult life can be programmed by early-life input. This concept opens a new window for preventing or delaying the onset of hypertension and kidney disease via reprogramming. Studies in short-lived animals, with controlled interventions across their lifespan, have provided interesting results from reprogramming interventions to prevent programmed hypertension and kidney disease via targeting AMPK signaling.

Regardless of recent advances in pharmacotherapies for hypertension and kidney disease, only a few studies have targeted their potential for reprogramming. In the current review, the beneficial effects of these treatments are all coming from indirect AMPK activators that are known to act in both an AMPK-dependent and -independent manner. There remains a lack of data regarding AMPK isoforms, specific knockdown models, or direct AMPK activators for renal programming and developmental programming of hypertension. Using modern genomic techniques [71,72], the identification of nephron segment-specific pathways regulated by AMPK isoforms will be an emerging area of interest. Additionally, another question raised from the current review is that the follow-up periods after the cessation of treatment in most cited reprogramming studies was relatively short. Of note is that the reprogramming effects of some perinatal interventions seem to persist beyond six months of age in female, but not male, SHRs [73,74]. Since sex differences appear in AMPK signaling [75], we must determine the long-term effects of AMPK activators in different programming models, and whether there is a sex-dependent response. Furthermore, a major concern in the translation from animal model to human use of AMPK activators is the activators' still-unknown adverse effects. Because pharmacological activation of AMPK is required to reach a specific target tissue, off-target effects may counter the therapeutic effects we are aiming for. For example, one possible effect of AMPK being off-target is the stimulation of the hypothalamus to increase food intake, despite the fact that our target organ is the kidney.

The evidence of the reprogramming effects of AMPK is just the beginning of the field. It is worth noting that instead of fully elucidating the potential mechanisms, these studies pointed out several key mechanisms linking AMPK to renal programming. It is clear that better understanding of the isoform- and tissue-specific effects of AMPK for programmed hypertension and kidney disease are required before a reprogramming strategy targeting AMPK could be translated from animal studies to human trials.

Author Contributions: C.-N.H.: contributed to concept generation, data interpretation, drafting of the manuscript, critical revision of the manuscript, and approval of the article; Y.-L.T.: contributed to concept generation, data interpretation, critical revision of the manuscript, and approval of the article.

Acknowledgments: This work was supported by the Grants CMRPG8F0023, CMRPG8G0672, and CMRPG8H0081 from Chang Gung Memorial Hospital, Kaohsiung, Taiwan.

Conflicts of Interest: The authors declare no conflict of interest.

Abbreviations

ACE	Angiotensin converting enzyme
AICAR	5-aminoimidazole-4carboxamide riboside
AMPK	AMP-activated protein kinase
AT1R	Angiotensin II type 1 receptor
CaMKKβ	Ca^{2+}-/calmodulin-dependent protein kinase β
DOHaD	Developmental origins of health and disease
LKB1	Liver kinase B1
mTOR	Mammalian target of rapamycin
NCC	Na^+/Cl^- cotransporter
NHE3	Type 3 sodium hydrogen exchanger
NKCC2	Na-K-2Cl cotransporter
PGC-1α	Peroxisome proliferator-activated receptor γ coactivator-1α
PPAR	Peroxisome proliferator-activated receptor
PPM1E	Mg^{2+}-/Mn^{2+}-dependent protein phosphatase 1E
PP2A	Protein phosphatase 2A
PP2C	protein phosphatase 2C
RAS	Renin-angiotensin system
SD	Sprague–Dawley
SHR	Spontaneously hypertensive rat
SIRT	Silent information regulator transcript
TAK1	TGFβ-activated kinase 1
Ulk1	Unc-51-like kinase 1

References

1. Chong, E.; Yosypiv, I.V. Developmental programming of hypertension and kidney disease. *Int. J. Nephrol.* **2012**, *2012*, 15. [CrossRef] [PubMed]
2. Luyckx, V.A.; Bertram, J.F.; Brenner, B.M.; Fall, C.; Hoy, W.E.; Ozanne, S.E.; Vikse, B.E. Effect of fetal and child health on kidney development and long-term risk of hypertension and kidney disease. *Lancet* **2013**, *382*, 273–283. [CrossRef]

3. Kett, M.M.; Denton, K.M. Renal programming: Cause for concern? *Am. J. Physiol. Regul. Integr. Comp. Physiol.* **2011**, *300*, R791–R803. [CrossRef] [PubMed]
4. Haugen, A.C.; Schug, T.T.; Collman, G.; Heindel, J.J. Evolution of DOHaD: The impact of environmental health sciences. *J. Dev. Orig. Health Dis.* **2015**, *6*, 55–64. [CrossRef] [PubMed]
5. Tain, Y.L.; Joles, J.A. Reprogramming: A preventive strategy in hypertension focusing on the kidney. *Int. J. Mol. Sci.* **2015**, *17*, 23. [CrossRef] [PubMed]
6. Xu, Q.; Si, L.Y. Protective effects of AMP-activated protein kinase in the cardiovascular system. *J. Cell Mol. Med.* **2010**, *14*, 2604–2613. [CrossRef] [PubMed]
7. Rajani, R.; Pastor-Soler, N.M.; Hallows, K.R. Role of AMP-activated protein kinase in kidney tubular transport, metabolism, and disease. *Curr. Opin. Nephrol. Hypertens.* **2017**, *26*, 375–383. [CrossRef] [PubMed]
8. Hardie, D.G.; Ross, F.A.; Hawley, S.A. AMPK: A nutrient and energy sensor that maintains energy homeostasis. *Nat. Rev. Mol. Cell Biol.* **2012**, *13*, 251–262. [CrossRef] [PubMed]
9. Grahame Hardie, D. AMP-activated protein kinase: A key regulator of energy balance with many roles in human disease. *J. Intern. Med.* **2014**, *276*, 543–559. [CrossRef] [PubMed]
10. Moreira, D.; Silvestre, R.; Cordeiro-da-Silva, A.; Estaquier, J.; Foretz, M.; Viollet, B. AMP-activated Protein Kinase as a Target For Pathogens: Friends Or Foes? *Curr. Drug Targets* **2016**, *17*, 942–953. [CrossRef] [PubMed]
11. Fraser, S.; Mount, P.; Hill, R.; Levidiotis, V.; Katsis, F.; Stapleton, D.; Kemp, B.E.; Power, D.A. Regulation of the energy sensor AMP-activated protein kinase in the kidney by dietary salt intake and osmolality. *Am. J. Physiol. Renal Physiol.* **2005**, *288*, F578–F586. [CrossRef] [PubMed]
12. Jeon, S.M. Regulation and function of AMPK in physiology and diseases. *Exp. Mol. Med.* **2016**, *48*, e245. [CrossRef] [PubMed]
13. Kim, J.; Yang, G.; Kim, Y.; Kim, J.; Ha, J. AMPK activators: Mechanisms of action and physiological activities. *Exp. Mol. Med.* **2016**, *48*, e224. [CrossRef] [PubMed]
14. Sugden, M.C.; Caton, P.W.; Holness, M.J. PPAR control: It's SIRTainly as easy as PGC. *J. Endocrinol.* **2010**, *204*, 93–104. [CrossRef] [PubMed]
15. Finck, B.N.; Kelly, D.P. Peroxisome proliferator-activated receptor gamma coactivator-1 (PGC-1) regulatory cascade in cardiac physiology and disease. *Circulation* **2007**, *115*, 2540–2548. [CrossRef] [PubMed]
16. Tain, Y.L.; Hsu, C.N.; Chan, J.Y. PPARs Link Early Life Nutritional insults to later programmed hypertension and metabolic syndrome. *Int. J. Mol. Sci.* **2015**, *17*, 20. [CrossRef] [PubMed]
17. Dutta, D.; Calvani, R.; Bernabei, R.; Leeuwenburgh, C.; Marzetti, E. Contribution of impaired mitochondrial autophagy to cardiac aging: Mechanisms and therapeutic opportunities. *Circ. Res.* **2012**, *110*, 1125–1138. [CrossRef] [PubMed]
18. Allouch, S.; Munusamy, S. AMP-activated Protein Kinase as a Drug Target in Chronic Kidney Disease. *Curr. Drug Targets* **2018**, *19*, 709–720. [CrossRef] [PubMed]
19. Carattino, M.D.; Edinger, R.S.; Grieser, H.J.; Wise, R.; Neumann, D.; Schlattner, U.; Johnson, J.P.; Kleyman, T.R.; Hallows, K.R. Epithelial sodium channel inhibition by AMP-activated protein kinase in oocytes and polarized renal epithelial cells. *J. Biol. Chem.* **2005**, *280*, 17608–17616. [CrossRef] [PubMed]
20. Fraser, S.A.; Gimenez, I.; Cook, N.; Jennings, I.; Katerelos, M.; Katsis, F.; Levidiotis, V.; Kemp, B.E.; Power, D.A. Regulation of the renal-specific Na$^+$–K$^+$–2Cl$^-$ co-transporter NKCC2 by AMP-activated protein kinase (AMPK). *Biochem. J.* **2007**, *405*, 85–93. [CrossRef] [PubMed]
21. Vadasz, I.; Dada, L.A.; Briva, A.; Trejo, H.E.; Welch, L.C.; Chen, J.; Toth, P.T.; Lecuona, E.; Witters, L.A.; Schumacker, P.T.; et al. AMP-activated protein kinase regulates CO$_2$-induced alveolar epithelial dysfunction in rats and human cells by promoting Na,K-ATPase endocytosis. *J. Clin. Invest.* **2008**, *118*, 752–762. [CrossRef] [PubMed]
22. Hallows, K.R.; Alzamora, R.; Li, H.; Gong, F.; Smolak, C.; Neumann, D.; Pastor-Soler, N.M. AMP-activated protein kinase inhibits alkaline pH and PKA-induced apical vacuolar H$^+$-ATPase accumulation in epididymal clear cells. *Am. J. Physiol. Cell Physiol.* **2009**, *296*, C672–C681. [CrossRef] [PubMed]
23. Pastor-Soler, N.M.; Hallows, K.R. AMP-activated protein kinase regulation of kidney tubular transport. *Curr. Opin. Nephrol. Hypertens.* **2012**, *21*, 523–533. [CrossRef] [PubMed]
24. Ford, R.J.; Teschke, S.R.; Reid, E.B.; Durham, K.K.; Kroetsch, J.T.; Rush, J.W. AMP-activated protein kinase activator AICAR acutely lowers blood pressure and relaxes isolated resistance arteries of hypertensive rats. *J. Hypertens.* **2012**, *30*, 725–733. [CrossRef] [PubMed]

25. Tsai, C.M.; Kuo, H.C.; Hsu, C.N.; Huang, L.T.; Tain, Y.L. Metformin reduces asymmetric dimethylarginine and prevents hypertension in spontaneously hypertensive rats. *Transl. Res.* **2014**, *164*, 452–459. [CrossRef] [PubMed]

26. Care, A.S.; Sung, M.M.; Panahi, S.; Gragasin, F.S.; Dyck, J.R.; Davidge, S.T.; Bourque, S.L. Perinatal Resveratrol Supplementation to Spontaneously Hypertensive Rat Dams Mitigates the Development of Hypertension in Adult Offspring. *Hypertension* **2016**, *67*, 1038–1044. [CrossRef] [PubMed]

27. Tain, Y.L.; Lin, Y.J.; Sheen, J.M.; Lin, I.C.; Yu, H.R.; Huang, L.T.; Hsu, C.N. Resveratrol prevents the combined maternal plus postweaning high-fat-diets-induced hypertension in male offspring. *J. Nutr. Biochem.* **2017**, *48*, 120–127. [CrossRef] [PubMed]

28. Tain, Y.L.; Wu, K.L.H.; Lee, W.C.; Leu, S.; Chan, J.Y.H. Prenatal Metformin Therapy Attenuates Hypertension of Developmental Origin in Male Adult Offspring Exposed to Maternal High-Fructose and Post-Weaning High-Fat Diets. *Int. J. Mol. Sci.* **2018**, *19*, 1066. [CrossRef] [PubMed]

29. Tain, Y.L.; Hsu, C.N. Developmental Origins of Chronic Kidney Disease: Should We Focus on Early Life? *Int. J. Mol. Sci.* **2017**, *18*, 381. [CrossRef] [PubMed]

30. Tain, Y.L.; Chan, S.H.H.; Chan, J.Y.H. Biochemical basis for pharmacological intervention as a reprogramming strategy against hypertension and kidney disease of developmental origin. *Biochem. Pharmacol.* **2018**, *153*, 82–90. [CrossRef] [PubMed]

31. Te Riet, L.; van Esch, J.H.; Roks, A.J.; van den Meiracker, A.H.; Danser, A.H. Hypertension: Renin-angiotensin-aldosterone system alterations. *Circ. Res.* **2015**, *116*, 960–975. [CrossRef] [PubMed]

32. Yosypiv, I.V. Renin-angiotensin system in ureteric bud branching morphogenesis: Insights into the mechanisms. *Pediatr. Nephrol.* **2011**, *26*, 1499–1512. [CrossRef] [PubMed]

33. Bogdarina, I.; Welham, S.; King, P.J.; Burns, S.P.; Clark, A.J. Epigenetic modification of the renin-angiotensin system in the fetal programming of hypertension. *Circ. Res.* **2007**, *100*, 520–526. [CrossRef] [PubMed]

34. Chappell, M.C.; Marshall, A.C.; Alzayadneh, E.M.; Shaltout, H.A.; Diz, D.I. Update on the Angiotensin converting enzyme 2-Angiotensin (1-7)-MAS receptor axis: Fetal programing, sex differences, and intracellular pathways. *Front. Endocrinol.* **2014**, *4*, 201. [CrossRef] [PubMed]

35. Siragy, H.M. The angiotensin II type 2 receptor and the kidney. *J. Renin Angiotensin Aldosterone Syst.* **2010**, *11*, 33–36. [CrossRef] [PubMed]

36. Ali, Q.; Dhande, I.; Samuel, P.; Hussain, T. Angiotensin type 2 receptor null mice express reduced levels of renal angiotensin II type 2 receptor/angiotensin (1-7)/Mas receptor and exhibit greater high-fat diet-induced kidney injury. *J. Renin Angiotensin Aldosterone Syst.* **2016**, *17*. [CrossRef] [PubMed]

37. Sherman, R.C.; Langley-Evans, S.C. Early administration of angiotensin-converting enzyme inhibitor captopril, prevents the development of hypertension programmed by intrauterine exposure to a maternal low-protein diet in the rat. *Clin. Sci.* **1998**, *94*, 373–381. [CrossRef] [PubMed]

38. Hsu, C.N.; Lee, C.T.; Huang, L.T.; Tain, Y.L. Aliskiren in early postnatal life prevents hypertension and reduces asymmetric dimethylarginine in offspring exposed to maternal caloric restriction. *J. Renin Angiotensin Aldosterone Syst.* **2015**, *16*, 506–513. [CrossRef] [PubMed]

39. Hsu, C.N.; Wu, K.L.; Lee, W.C.; Leu, S.; Chan, J.Y.; Tain, Y.L. Aliskiren administration during early postnatal life sex-specifically alleviates hypertension programmed by maternal high fructose consumption. *Front. Physiol.* **2016**, *7*, 299. [CrossRef] [PubMed]

40. Yang, K.K.; Sui, Y.; Zhou, H.R.; Shen, J.; Tan, N.; Huang, Y.M.; Li, S.S.; Pan, Y.H.; Zhang, X.X.; Zhao, H.L. Cross-talk between AMP-activated protein kinase and renin-angiotensin system in uninephrectomised rats. *J. Renin Angiotensin Aldosterone Syst.* **2016**, *17*. [CrossRef] [PubMed]

41. Kohlstedt, K.; Trouvain, C.; Boettger, T.; Shi, L.; Fisslthaler, B.; Fleming, I. AMP-activated protein kinase regulates endothelial cell angiotensin-converting enzyme expression via p53 and the post-transcriptional regulation of microRNA-143/145. *Circ. Res.* **2013**, *112*, 1150–1158. [CrossRef] [PubMed]

42. Kim, E.N.; Kim, M.Y.; Lim, J.H.; Kim, Y.; Shin, S.J.; Park, C.W.; Kim, Y.S.; Chang, Y.S.; Yoon, H.E.; Choi, B.S. The protective effect of resveratrol on vascular aging by modulation of the renin-angiotensin system. *Atherosclerosis* **2018**, *270*, 123–131. [CrossRef] [PubMed]

43. Tain, Y.L.; Lee, W.C.; Leu, S.; Wu, K.; Chan, J. High salt exacerbates programmed hypertension in maternal fructose-fed male offspring. *Nutr. Metab. Cardiovasc. Dis.* **2015**, *25*, 1146–1151. [CrossRef] [PubMed]

44. Efeyan, A.; Comb, W.C.; Sabatini, D.M. Nutrient-sensing mechanisms and pathways. *Nature* **2015**, *517*, 302–310. [CrossRef] [PubMed]

45. Tain, Y.L.; Hsu, C.N. Interplay between oxidative stress and nutrient sensing signaling in the developmental origins of cardiovascular disease. *Int. J. Mol. Sci.* **2017**, *18*, 841. [CrossRef] [PubMed]

46. Tain, Y.L.; Lee, W.C.; Wu, K.; Leu, S.; Chan, J.Y.H. Resveratrol prevents the development of hypertension programmed by maternal plus post-weaning high-fructose consumption through modulation of oxidative stress, nutrient-sensing signals, and gut microbiota. *Mol. Nutr. Food Res.* **2018**, in press. [CrossRef] [PubMed]

47. Dennery, P.A. Oxidative stress in development: Nature or nurture? *Free Radic. Biol. Med.* **2010**, *49*, 1147–1151. [CrossRef] [PubMed]

48. Trewin, A.J.; Berry, B.J.; Wojtovich, A.P. Exercise and Mitochondrial Dynamics: Keeping in Shape with ROS and AMPK. *Antioxidants* **2018**, *7*, 7. [CrossRef] [PubMed]

49. Song, P.; Zou, M.H. Regulation of NAD(P)H oxidases by AMPK in cardiovascular systems. *Free Radic. Biol. Med.* **2012**, *52*, 1607–1619. [CrossRef] [PubMed]

50. Nüsken, E.; Dötsch, J.; Weber, L.T.; Nüsken, K.D. Developmental Programming of Renal Function and Re-Programming Approaches. *Front. Pediatr.* **2018**, *6*, 36. [CrossRef] [PubMed]

51. Cao, X.; Luo, T.; Luo, X.; Tang, Z. Resveratrol prevents Ang II-induced hypertension via AMPK activation and RhoA/ROCK suppression in mice. *Hypertens. Res.* **2014**, *37*, 803–810. [CrossRef] [PubMed]

52. Ford, R.J.; Rush, J.W. Endothelium-dependent vasorelaxation to the AMPK activator AICAR is enhanced in aorta from hypertensive rats and is NO and EDCF dependent. *Am. J. Physiol. Heart Circ. Physiol.* **2011**, *300*, H64–H75. [CrossRef] [PubMed]

53. Olivier, S.; Foretz, M.; Viollet, B. Promise and challenges for direct small molecule AMPK activators. *Biochem. Pharmacol.* **2018**, *153*, 147–158. [CrossRef] [PubMed]

54. Salatto, C.T.; Miller, R.A.; Cameron, K.O.; Cokorinos, E.; Reyes, A.; Ward, J.; Calabrese, M.F.; Kurumbail, R.G.; Rajamohan, F.; Kalgutkar, A.S.; et al. Selective Activation of AMPK β1-Containing Isoforms Improves Kidney Function in a Rat Model of Diabetic Nephropathy. *J. Pharmacol. Exp. Ther.* **2017**, *361*, 303–311. [CrossRef] [PubMed]

55. Javkhedkar, A.A.; Banday, A.A. Antioxidant resveratrol restores renal sodium transport regulation in SHR. *Physiol. Rep.* **2015**, *3*, e12618. [CrossRef] [PubMed]

56. Bhatt, S.R.; Lokhandwala, M.F.; Banday, A.A. Resveratrol prevents endothelial nitric oxide synthase uncoupling and attenuates development of hypertension in spontaneously hypertensive rats. *Eur. J. Pharmacol.* **2011**, *667*, 258–264. [CrossRef] [PubMed]

57. Rueda-Clausen, C.F.; Morton, J.S.; Dolinsky, V.W.; Dyck, J.R.; Davidge, S.T. Synergistic effects of prenatal hypoxia and postnatal high-fat diet in the development of cardiovascular pathology in young rats. *Am. J. Physiol. Regul. Integr. Comp. Physiol.* **2012**, *303*, R418–R426. [CrossRef] [PubMed]

58. Liang, X.; Xing, W.; He, J.; Fu, F.; Zhang, W.; Su, F.; Liu, F.; Ji, L.; Gao, F.; Su, H.; et al. Magnolol administration in normotensive young spontaneously hypertensive rats postpones the development of hypertension: Role of increased PPARγ, reduced TRB3 and resultant alleviative vascular insulin resistance. *PLoS ONE* **2015**, *10*, e0120366. [CrossRef] [PubMed]

59. Guo, Z.; Sun, H.; Zhang, H.; Zhang, Y. Anti-hypertensive and renoprotective effects of berberine in spontaneously hypertensive rats. *Clin. Exp. Hypertens.* **2015**, *37*, 332–339. [CrossRef] [PubMed]

60. Cho, T.M.; Peng, N.; Clark, J.T.; Novak, L.; Roysommuti, S.; Prasain, J.; Wyss, J.M. Genistein attenuates the hypertensive effects of dietary NaCl in hypertensive male rats. *Endocrinology* **2007**, *148*, 5396–5402. [CrossRef] [PubMed]

61. Dovinová, I.; Barancik, M.; Majzunova, M.; Zorad, S.; Gajdosechová, L.; Gresová, L.; Cacanyiova, S.; Kristek, F.; Balis, P.; Chan, J.Y. Effects of PPARγ agonist pioglitazone on redox-sensitive cellular signaling in young spontaneously hypertensive rats. *PPAR Res.* **2013**, *2013*, 11. [CrossRef] [PubMed]

62. Wu, L.; Wang, R.; de Champlain, J.; Wilson, T.W. Beneficial and deleterious effects of rosiglitazone on hypertension development in spontaneously hypertensive rats. *Am. J. Hypertens.* **2004**, *17*, 749–756. [CrossRef] [PubMed]

63. Wang, Y.W.; He, S.J.; Feng, X.; Cheng, J.; Luo, Y.T.; Tian, L.; Huang, Q. Metformin: A review of its potential indications. *Drug Des. Dev. Ther.* **2017**, *11*, 2421–2429. [CrossRef] [PubMed]

64. De Broe, M.E.; Kajbaf, F.; Lalau, J.D. Renoprotective Effects of Metformin. *Nephron* **2018**, *138*, 261–274. [CrossRef] [PubMed]

65. Li, H.; Xia, N.; Förstermann, U. Cardiovascular effects and molecular targets of resveratrol. *Nitric Oxide* **2012**, *26*, 102–110. [CrossRef] [PubMed]

66. Johnson, R.J.; Sanchez-Lozada, L.G.; Nakagawa, T. The effect of fructose on renal biology and disease. *J. Am. Soc. Nephrol.* **2010**, *21*, 2036–2039. [CrossRef] [PubMed]

67. Tain, Y.L.; Leu, S.; Wu, K.L.; Lee, W.C.; Chan, J.Y. Melatonin prevents maternal fructose intake-induced programmed hypertension in the offspring: Roles of nitric oxide and arachidonic acid metabolites. *J. Pineal Res.* **2014**, *57*, 80–89. [CrossRef] [PubMed]

68. Carlstrom, J.; Symons, J.D.; Wu, T.C.; Bruno, R.S.; Litwin, S.E.; Jalili, T. A quercetin supplemented diet does not prevent cardiovascular complications in spontaneously hypertensive rats. *J. Nutr.* **2007**, *137*, 628–633. [CrossRef] [PubMed]

69. Goszcz, K.; Duthie, G.G.; Stewart, D.; Leslie, S.J.; Megson, I.L. Bioactive polyphenols and cardiovascular disease: Chemical antagonists, pharmacological agents or xenobiotics that drive an adaptive response? *Br. J. Pharmacol.* **2017**, *174*, 1209–1225. [CrossRef] [PubMed]

70. Koeners, M.P.; Wesseling, S.; Sánchez, M.; Braam, B.; Joles, J.A. Perinatal Inhibition of NF-κB has long-term antihypertensive and renoprotective effects in fawn-hooded hypertensive rats. *Am. J. Hypertens.* **2016**, *29*, 123–131. [CrossRef] [PubMed]

71. Lee, J.W.; Chou, C.L.; Knepper, M.A. Deep Sequencing in Microdissected Renal Tubules Identifies Nephron Segment-Specific Transcriptomes. *J. Am. Soc. Nephrol.* **2015**, *26*, 2669–2677. [CrossRef] [PubMed]

72. Gonzalez-Vicente, A.; Hopfer, U.; Garvin, J.L. Developing Tools for Analysis of Renal Genomic Data: An Invitation to Participate. *J. Am. Soc. Nephrol.* **2017**, *28*, 3438–3440. [CrossRef] [PubMed]

73. Koeners, M.P.; van Faassen, E.E.; Wesseling, S.; de Sain-van der Velden, M.; Koomans, H.A.; Braam, B.; Joles, J.A. Maternal supplementation with citrulline increases renal nitric oxide in young spontaneously hypertensive rats and has long-term antihypertensive effects. *Hypertension* **2007**, *50*, 1077–1084. [CrossRef] [PubMed]

74. Koeners, M.P.; Braam, B.; Joles, J.A. Perinatal inhibition of NF-kappaB has long-term antihypertensive effects in spontaneously hypertensive rats. *J. Hypertens.* **2011**, *29*, 1160–1166. [CrossRef] [PubMed]

75. Mukai, Y.; Ozaki, H.; Serita, Y.; Sato, S. Maternal fructose intake during pregnancy modulates hepatic and hypothalamic AMP-activated protein kinase signalling in a sex-specific manner in offspring. *Clin. Exp. Pharmacol. Physiol.* **2014**, *41*, 331–337. [CrossRef] [PubMed]

International Journal of
Molecular Sciences

MDPI

Review

AMP-Activated Protein Kinase (AMPK)-Dependent Regulation of Renal Transport

Philipp Glosse [1] and Michael Föller [2,*

[1]   Institute of Agricultural and Nutritional Sciences, Martin Luther University Halle-Wittenberg,
      D-06120 Halle (Saale), Germany; philipp.glosse@landw.uni-halle.de
[2]   Institute of Physiology, University of Hohenheim, D-70599 Stuttgart, Germany
*   Correspondence: michael.foeller@uni-hohenheim.de; Tel.: +49-711-459-24566

Received: 26 September 2018; Accepted: 30 October 2018; Published: 6 November 2018

Abstract: AMP-activated kinase (AMPK) is a serine/threonine kinase that is expressed in most cells and activated by a high cellular AMP/ATP ratio (indicating energy deficiency) or by Ca^{2+}. In general, AMPK turns on energy-generating pathways (e.g., glucose uptake, glycolysis, fatty acid oxidation) and stops energy-consuming processes (e.g., lipogenesis, glycogenesis), thereby helping cells survive low energy states. The functional element of the kidney, the nephron, consists of the glomerulus, where the primary urine is filtered, and the proximal tubule, Henle's loop, the distal tubule, and the collecting duct. In the tubular system of the kidney, the composition of primary urine is modified by the reabsorption and secretion of ions and molecules to yield final excreted urine. The underlying membrane transport processes are mainly energy-consuming (active transport) and in some cases passive. Since active transport accounts for a large part of the cell's ATP demands, it is an important target for AMPK. Here, we review the AMPK-dependent regulation of membrane transport along nephron segments and discuss physiological and pathophysiological implications.

Keywords: transporter; carrier; pump; membrane; energy deficiency

1. Introduction

The 5'-adenosine monophosphate (AMP)–activated protein kinase (AMPK) is a serine/threonine protein kinase that is evolutionarily conserved and functions as an intracellular energy sensor in mammalian cells [1–5]. It is a central regulator of energy homeostasis and affects many important cellular functions including growth, differentiation, autophagy, and metabolism [1,2,6]. During energy depletion when cellular AMP levels are high relative to the adenosine triphosphate (ATP) concentration, AMPK activates energy-providing pathways including glucose uptake, glycolysis, or fatty acid oxidation [7–10]. Simultaneously, processes consuming ATP (e.g., gluconeogenesis, lipogenesis, or protein synthesis) are inhibited [7–10].

Being expressed in most mammalian cells, AMPK is a heterotrimeric protein consisting of a catalytic α (α1 or α2), scaffolding β (β1 or β2), and a regulatory nucleotide-binding γ (γ1, γ2, or γ3) subunit with the expression pattern differing from cell type to cell type [1,2,11–14]. Induction of AMPK activity involves phosphorylation of the conserved threonine residue Thr172 within the activation loop of the α subunit's kinase domain by various protein kinases including the tumor suppressor liver kinase B1 (LKB1), Ca^{2+}/calmodulin–dependent protein kinase kinase β (CaMKKβ), and transforming growth factor beta-activated kinase 1 [1,15–28]. AMPK activation in cellular energy depletion is primarily mediated by an increase in the AMP/ATP or ADP/ATP ratio [8,29,30]. Thus, AMP or ADP binding to the subunit at cystathionine-beta-synthase repeats results in conformational changes that allows for the phosphorylation at Thr172 by LKB1. This results in an enhancement of AMPK activity by >100-fold [1,8,12,15,31–36]. Moreover, AMP or ADP binding prevents dephosphorylation at Thr172 by protein phosphatases [8,12,37,38]. Additionally, binding of AMP, but not ADP, activates AMPK allosterically [8,11,12,37]. Conversely, ATP binding to the cystathionine-beta-synthase domain results in AMPK dephosphorylation by protein phosphatases [1,8,39].

Besides LKB1-associated regulation of AMPK phosphorylation, an alternative Ca^{2+}-involving activation mechanisms independent of AMP exists [6,12,40,41]. Protein kinase CaMKKβ phosphorylates AMPK at Thr172 in response to elevated intracellular Ca^{2+} levels which may be caused by mediators such as thrombin or ghrelin [6,12,23,40,42,43]. Intracellular Ca^{2+} store depletion detected by the Ca^{2+}-sensing protein stromal interacting molecule-1 leads to store-operated Ca^{2+} entry (SOCE) involving the Ca^{2+} release-activated Ca^{2+} channel Orai1 [44–49]. Orai1-mediated SOCE impacts on many cellular functions including cell proliferation, differentiation, migration, and cytokine production [44,50–55]. SOCE is involved in a sort of feedback mechanism involving AMPK: SOCE activates AMPK through CaMKKβ. AMPK in turn inhibits SOCE [45]. Moreover, AMPK inhibits SOCE by regulating Orai1 membrane abundance (at least in UMR106 cells) [44,56].

AMPK is a major regulator of whole body energy homeostasis [10,12], impacting on a variety of organs including liver [57–61], skeletal [62–66] and cardiac muscle [67–73], kidney [74–77], and bone [78–80]. In the kidney, AMPK regulates epithelial transport, podocyte function, blood pressure, epithelial-to-mesenchymal transition, autophagy as well as nitric oxide synthesis [75,76,81–83]. Not surprisingly, AMPK is highly relevant for renal pathophysiology, including ischemia, diabetic renal hypertrophy, polycystic kidney disease, chronic kidney disease, and hypertension [40,67,74–76]. This review summarizes the contribution of AMPK to the regulation of renal transport and hence to the final composition of excreted urine. Moreover, pathophysiological implications are discussed.

2. AMPK and Renal Tubular Transport

The kidney is particularly relevant for fluid, electrolyte, and acid–base homeostasis. In addition, it is an endocrine organ producing different hormones such as erythropoietin, Klotho, and calcitriol, the active form of vitamin D [84–86]. The kidneys are made up of about 1 million nephrons, their functional elements. A nephron comprises the glomerulus surrounded by the Bowman´s capsule, the proximal tubule, Henle's loop, distal tubule, and the collecting duct. The primary urine is filtered in the glomerulus. Its composition is similar to plasma. In general, large molecules and particularly proteins >6000 Dalton are normally filtered to a low extent, if at all. The renal tubular system modifies the primary urine by reabsorbing or secreting ions and molecules, ultimately yielding the final urine [85–87]. Epithelial transport is mainly dependent on ATP-dependent pumps (primary-active), secondary-or tertiary-active transporters, as well as carriers and channels (passive, facilitated diffusion). Since active transport consumes energy by definition, it is not surprising that it is subject to regulation by AMPK. Moreover, even passive transport involving glucose transporter (GLUT) carriers is controlled by AMPK [74,75].

2.1. Na⁺/K⁺-ATPase

The ubiquitously expressed Na^+/K^+-ATPase is a primary active ATP-driven pump that mediates the basolateral extrusion of $3Na^+$ in exchange of $2K^+$, thereby establishing a transmembrane Na^+ gradient, which is the prerequisite for secondary active Na^+-dependent transport (e.g., through Na^+-dependent glucose cotransporter 1 and 2 (SGLT1/2), Na^+/H^+ exchanger isoform 1 (NHE1), Na^+-coupled phosphate transporter (NaPi-IIa), or Na^+-K^+-$2Cl^-$ cotransporter (NKCC2), as discussed below) [75,88–94]. Almost one-third of the body's energy is consumed by this pump [95]. Therefore, it does make sense that it is regulated by AMPK [74–76,94]: AMPK inhibits Na^+/K^+-ATPase in airway epithelial cells by promoting its endocytosis [96–100]. However, AMPK stimulates Na^+/K^+-ATPase membrane expression in skeletal muscle cells [101] and in renal epithelia [102], thereby counteracting renal ischemia-induced Na^+/K^+-ATPase endocytosis [103]. Interestingly, AMPKβ1 deficiency was found not to alter outcome in an ischemic kidney injury model in mice [104]. Hence, the effect of AMPK on Na^+/K^+-ATPase appears to be highly tissue-specific [74,75].

2.2. Proximal Tubule

A wide variety of luminal Na^+-dependent cotransporters, which are secondary active, are involved in epithelial transport in the proximal tubule. Secondary active transporters utilize the energy of the transmembrane Na^+ gradient generated by the primary active ATP-consuming Na^+/K^+-ATPase to facilitate transport of a substrate against its concentration gradient [105,106]. These transporters and the basolateral Na^+/K^+-ATPase consume substantial amounts of total cellular energy [74,75,107]. Hence, AMPK has been demonstrated to be an important regulator of proximal tubule transport [74,75].

2.2.1. Glucose Transport

Since glucose is freely filtered by the glomerulus, glucose concentration in primary urine is similar to the plasma glucose concentration, whereas excreted urine is usually free of glucose [108–110]. The sugar is reabsorbed in the proximal tubule by the Na^+-dependent glucose cotransporter 1 and 2 (SGLT1 and 2), the different expression patterns and properties of which ensure total glucose reabsorption as long as the plasma glucose concentration is not abnormally high [89,108]. SGLT2 has a high transport capacity but low affinity for glucose and is predominantly expressed in the kidney, while SGLT1 is also expressed in other tissues including the small intestine. SGLT2 contributes to the reabsorption of up to 90% of filtered glucose [108,109,111,112]. On the other hand, AMPK-regulated SGLT1 [7,92,113] has a low transport capacity but high affinity for glucose and reabsorbs the remaining glucose [108–110,114,115]. Glucose leaves the basolateral membrane through passive glucose carriers GLUT1 and GLUT2 [108,116–118]. AMPK activates SGLT1-dependent glucose transport, presumably by stimulating membrane insertion of the cotransporter as observed in colorectal Caco-2 cells [92,119]. In line with this, AMPK activation is associated with increased *SGLT1* expression and glucose uptake in cardiomyocytes [113,120]. Although the AMPK-dependent regulation of SGLT1 in the proximal tubule has not explicitly been addressed, it is tempting to speculate that it is similar to other cell types [92,113,119,120]. The regulation of SGLT by AMPK is a doubled-edged sword: on the one hand, SGLT1-dependent reabsorption of glucose in proximal tubular cells requires energy which is generated by β-oxidation of fatty acids to a large extent [121,122]. On the other hand, it prevents the loss of energy-rich glucose [122,123], thereby maintaining the Na^+/K^+-ATPase-facilitated Na^+ gradient for Na^+-dependent transport and many other cellular processes [75,76]. SGLT1-mediated glucose uptake is linked to the GLUT1-dependent efflux at the basolateral side [108,116]. GLUT1 activity is stimulated by AMPK in various cell types [124–131]. Therefore, it is conceivable that renal GLUT1 might also be regulated by AMPK in order to save energy-providing glucose. In line with this, Baldwin et al. (1997) showed enhanced glucose uptake via GLUT1 in baby hamster kidney cells treated with AMPK activator 5-aminoimidazole-4-carboxamide ribonucleotide (AICAR) [132]. Moreover, Sokolovska et al. (2010) reported that metformin, another pharmacological AMPK activator, increased *GLUT1* gene expression in rat kidneys [133]. Also, AMPK activation was associated with enhanced activity of GLUT2. These studies, however, found reduced SGLT1 membrane abundance upon AMPK activation, at least in the case of murine intestinal tissue [134,135].

2.2.2. Na^+/H^+ Exchanger Isoform 1

The ubiquitous Na^+/H^+ exchanger isoform 1 (NHE1) participates in cell volume and pH regulation by extruding one cytosolic H^+ in exchange for one extracellular Na^+ [136,137]. NHE1 is expressed in all parts of the nephron, including the proximal tubule. However, it cannot be detected in the macula densa and intercalated cells of the distal nephron [136,138,139]. In the proximal tubule, NHE1 is particularly important for HCO_3^- reabsorption [140]. In hypoxia, anaerobic glycolysis is predominant, which results in intracellular accumulation of lactate and H^+ [90]. Acidosis, however, inhibits glycolysis [90,141,142] and would jeopardize cellular energy generation. AMPK-dependent stimulation of NHE1 activity in human embryonic kidney (HEK) cells therefore helps cells keep up anaerobic glycolysis in oxygen deficiency, as demonstrated by Rotte et al. (2010) [90]. Given that NHE1 is needed for proximal tubular HCO_3^- reabsorption [140], AMPK may help retain HCO_3^-, thereby alleviating acidosis in energy deficiency and hypoxia.

2.2.3. Creatine Transporter

In some organs with high metabolic activity, including skeletal muscle, heart, and brain, creatine is used to refuel cellular ATP levels [143–145]. In the proximal tubule, creatine, a small molecule that is freely filtered, is also reabsorbed through secondary active Na^+-dependent creatine transporter (CRT) (SLC6A8) [7,75,143,146]. AMPK has been demonstrated to downregulate CRT activity and apical membrane expression in a polarized mouse S3 proximal tubule cell line, presumably through mammalian target of rapamycin signaling [147]. The AMPK-dependent inhibition of CRT may help reduce unnecessary energy expenditure [75]. Conversely, AMPK stimulates CRT-mediated creatine transport in cardiomyocytes [148,149]. This again demonstrates that AMPK effects are tissue-specific [148].

2.2.4. Na^+-Coupled Phosphate Transporter IIa

Inorganic phosphate is mainly reabsorbed by the secondary active Na^+-coupled phosphate transporter (NaPi-IIa) (SLC34A1) in the proximal tubule [93,150–152]. Employing electrophysiological recordings in *Xenopus* oocytes, it was shown that AMPK inhibits NaPi-IIa [93]. Kinetics analysis revealed that AMPK decreases NaPi-IIa membrane expression rather than changing its properties.

The regulation of phosphate metabolism by AMPK is not restricted to NaPi-IIa: Recently, AMPK was demonstrated to control the formation of bone-derived hormone fibroblast growth factor 23 (FGF23) [56], which induces renal phosphate excretion by extracellular-signal regulated kinases 1/2 (ERK1/2)-mediated degradation of membrane NaPi-IIa [150]. AMPK inhibits FGF23 production in cell culture and in mice [56]. Despite markedly elevated FGF23 serum levels in AMPKα1-deficient mice, renal phosphate excretion was not different from wild-type animals [56]. The same holds true for cellular localization of NaPi-IIa and renal ERK1/2 [56]. Thus, it is possible that AMPK deficiency is paralleled with some FGF23 resistance.

2.3. Loop of Henle

2.3.1. Na^+-K^+-$2Cl^-$ Cotransporter

The Na^+-K^+-$2Cl^-$ cotransporter (NKCC2), expressed in the thick ascending limb (TAL) of the loop of Henle and macula densa, is required for the generation of a hypertonic medullary interstitium, a mechanism needed for concentrating urine [75,76,88,91]. NKCC2 is a direct substrate of AMPK which phosphorylates it at its stimulatory serine residue Ser-126 [153]. Moreover, exposure of murine macula densa-like cells to low salt leads to AMPK activation and increased NKCC2 phosphorylation [154]. In addition, increased subapical expression (and apparent reduced apical expression) of NKCC2 in the medullary TAL of the loop of Henle along with elevated urinary Na^+ excretion in AMPKβ1-deficient mice on a normal salt diet were observed [155]. This is in line with AMPK being an important regulator of NKCC2-mediated salt retention in the medullary TAL of Henle [155]. Efe et al. (2016) recently observed markedly increased outer medullary expression of NKCC2 in rats treated with the AMPK activator metformin [156]. However, according to a recent in vivo study by Udwan et al. (2017), a low salt diet induced upregulation of NKCC2 surface expression in mouse kidneys but left AMPK activity unchanged [157]. Therefore, the exact role of AMPK in stimulating NKCC2 remains to be established.

2.3.2. Renal Outer Medullary K^+ Channel

The apical renal outer medullary K^+ channel (ROMK) is required for NKCC2 to work properly, as it allows the recirculation of K^+ ions taken up by NKCC2 into the lumen [75,88]. AMPK is an inhibitor of ROMK by downregulating both channel activity and membrane abundance of the channel protein in a heterologous expression system using *Xenopus* oocytes [158]. In vivo studies revealed that the AMPK effect on ROMK is relevant for the renal excretion of K^+ after an acute K^+ challenge, as upregulation of renal ROMK1 protein expression and the ability of K^+ elimination were more pronounced in AMPKα1-deficient than in wild-type mice [158].

2.4. Distal Tubule

2.4.1. Cystic Fibrosis Transmembrane Conductance Regulator

The ATP-gated and cyclic AMP (cAMP)-dependent Cl^- channel cystic fibrosis transmembrane conductance regulator (CFTR) participates in Cl^- secretion and is broadly known for its role in cystic fibrosis, the pathophysiology of which is due to channel malfunction [74–76,159]. In the kidney, CFTR contributes to Cl^- secretion in the distal tubule and the principal cells of the cortical and medullary collecting ducts [74,75,160]. AMPK has been demonstrated to inhibit CFTR-dependent Cl^- conductance in *Xenopus* oocytes [159] and to decrease CFTR channel activity in the lung [161,162] and colon [163]. cAMP-stimulated cell proliferation and CFTR-dependent Cl^- secretion play a decisive role for epithelial cyst enlargement in autosomal dominant polycystic kidney disease (ADPKD) [164]. In line with this, AMPK activation inhibits CFTR in Madin-Darby canine kidney (MDCK) cells [165] as well as decreases cystogenesis in murine models of ADPKD [165,166], suggesting a potential role for pharmacological AMPK activation in the treatment of ADPKD [165,166].

2.4.2. Ca^{2+} Transport

Most Ca^{2+} is reabsorbed by passive paracellular diffusion along with other ions and water through tight junctions in the proximal tubule and the more distal parts of the nephron [88,167]. Conversely, only 5–10% of filtered Ca^{2+} is reabsorbed by transcellular transport involving the apical transient receptor potential vanilloid 5 channel TRPV5 in the distal convoluted tubule [88]: Ca^{2+} enters the cell through TRPV5, whereas basolateral Ca^{2+} efflux is accomplished by the Na^+/Ca^{2+} exchanger (NCX) and the Ca^{2+}-ATPase [88,167,168]. AMPK has been shown to inhibit NCX and decrease Orai1-mediated SOCE in murine dendritic cells [169]. Therefore, it is tempting to speculate that Ca^{2+} reabsorption may be downregulated in the distal tubule in ATP deficiency [169,170]. Indeed, AMPK downregulates Orai1-dependent SOCE in T-lymphocytes [171], endothelial cells [45], and in osteoblast-like cells [56]. Since renal Orai1 activity contributes to kidney fibrosis [172], AMPK-mediated Orai1 downregulation may also be therapeutically desirable.

2.5. Collecting Duct

2.5.1. Epithelial Na^+ Channel

In the collecting duct, fine tuning of Na^+ and K^+ homeostasis is accomplished by epithelial Na^+ channel (ENaC) and ROMK K^+ channel. Both channels are controlled by the renin-angiotensin-aldosterone system [173–175] regulating extracellular volume and hence arterial blood pressure [173–177]. Na^+ reabsorption by ENaC in the late distal convoluted tubule and cortical collecting duct principal cells is a highly energy-demanding process, as it utilizes the electrochemical driving force generated by the basolateral Na^+/K^+-ATPase [74–76,176,178]. AMPK inhibits epithelial Na^+ transport in various tissues, including lung [96,179], colonic [180], and renal cortical collecting duct cells [180–183]. In line with this, AMPKα1-deficient mice exhibit increased renal ENaC expression [180]. In detail, AMPK downregulates ENaC surface expression by inducing the binding of the ubiquitin ligase neural precursor cell expressed developmentally downregulated protein 4-2 (Nedd4-2) to ENaC subunits, resulting in ENaC ubiquitination with subsequent endocytosis and degradation [177,180,184]. In line with this, activation of AMPK enhances the tubuloglomerular feedback and induces urinary diuresis and Na^+ excretion in rats [185]. However, AMPKα1$^{-/-}$ mice with genetic kidney-specific AMPKα2 deletion exhibit a moderate increase in diuresis and natriuresis, possibly because NKCC2 activity is insufficient despite upregulated ENaC activity [186]. Taken together, AMPK activity limits ENaC-dependent energy-consuming Na$^+$ reabsorption [177,180,181,185].

2.5.2. Voltage-Gated K$^+$ Channel

The voltage-gated K$^+$ channel (KCNQ1) is important for the cardiovascular system as well as for electrolyte and fluid homeostasis and is expressed in the distal nephron including the collecting duct [170,187–189]. Its exact role is ill-defined, although a contribution to cell volume regulation is postulated [75,187]. Similar to ENaC, AMPK inhibits KCNQ1 via Nedd4-2, as demonstrated in collecting duct principal cells of rat ex vivo kidney slices [187], MDCK cells [190], and *Xenopus* oocytes [191].

2.5.3. Vacuolar H$^+$-ATPase

The primary active vacuolar H$^+$-ATPase (V-ATPase) is located at the apical membrane of proximal tubule cells and collecting duct type A intercalated cells. It contributes to the regulation of acid–base homeostasis by secreting H$^+$ ions into the tubular lumen [76,192,193]. AMPK inhibits the protein kinase A (PKA)-dependent membrane expression of V-ATPase in collecting duct intercalated cells of rat ex vivo kidney slices [193]. Moreover, epididymal proton-secreting clear cells, developmentally related to intercalated cells, exhibit reduced apical membrane abundance of V-ATPase after in vivo perfusion with the AMPK activator 5-aminoimidazole-4-carboxamide-1-beta-D-ribofuranoside (AICAR) into rats [194]. It appears to be likely that energy deficiency limits highly energy-consuming primary active H$^+$ excretion in the proximal tubule, whereas secondary active NHE1-dependent H$^+$ secretion is maintained, thereby keeping up at least anaerobic glycolysis [192]. The opposing effects of AMPK and PKA on V-ATPase expression and activity in kidney intercalated cells can be explained by different phosphorylation sites, as AMPK and PKA phosphorylate the A subunit at Ser-384 and Ser-175, respectively [195,196]. McGuire and Forgac (2018) further demonstrated that AMPK increases lysosomal V-ATPase assembly and activity in HEK293T cells under conditions of energy depletion [197]. In cells depleted of energy, acidification of autophagic intracellular compartments by V-ATPases enables the lysosomal degradation of proteins and lipids to generate energy substrates for ATP production [197,198]. Thus, it appears to be likely that AMPK-regulated V-ATPase activity depends on its concrete cellular localization and function [197].

2.5.4. Water and Urea Handling

AMPK also regulates renal urea and water handling [76,199]. In the inner medullary collecting duct, osmotic gradients are generated by NKCC2 and urea transporter UT-A1 and water is reabsorbed through aquaporin 2 (AQP2) [76,156,199,200]. The concentration of urine requires the antidiuretic hormone vasopressin, which binds to vasopressin type 2 receptors of collecting duct principal cells, resulting in cAMP-mediated activation of PKA and subsequent phosphorylation and apical membrane insertion of AQP2 and UT-A1 [76,156,199]. Congenital nephrogenic diabetes insipidus (NDI) is a disease primarily caused by mutations of vasopressin type 2 receptors that is characterized by renal resistance to vasopressin and limited urine concentrating capacity [156,201]. According to two in vivo studies using rodent models of congenital NDI, the metformin-stimulated AMPK activation ameliorates the ability of the kidney to concentrate urine by increasing the phosphorylation and apical membrane expression of inner medullary AQP2 and UT-A1 [156,202]. In contrast, an ex vivo treatment of rat kidney slices with AICAR led to reduced apical membrane insertion of AQP2 [203]. Moreover, AMPK antagonizes the desmopressin-induced AQP2 phosphorylation in vitro, thus also suggesting an inhibitory function of AMPK on AQP2 regulation [203]. It appears likely that AMPK-independent effects of the pharmacological AMPK agonists contribute to this discrepancy [156,202,203]. Thus, further studies are clearly required.

3. Conclusions and Perspectives

A growing list of studies indicates the pivotal role of AMPK as a metabolic-sensing regulator of a multitude of transport processes in the kidney [7,74–76,170]. Particularly, AMPK activation under conditions of energy deficiency is expected to differentially modulate renal epithelial ion transport in order to preserve cellular energy homeostasis (Figure 1) [7,74–76,94,170]. Alongside the above discussed function of AMPK in kidney tubular transport, a variety of other transport proteins, which are expressed in the kidney as well, are regulated by AMPK in extrarenal tissues [7,94,170,204] that are reviewed elsewhere [170] and [7] and summarized in Table 1. Future studies are required to focus on the therapeutic value of pharmacological AMPK manipulation to combat kidney disease [74–76,205,206].

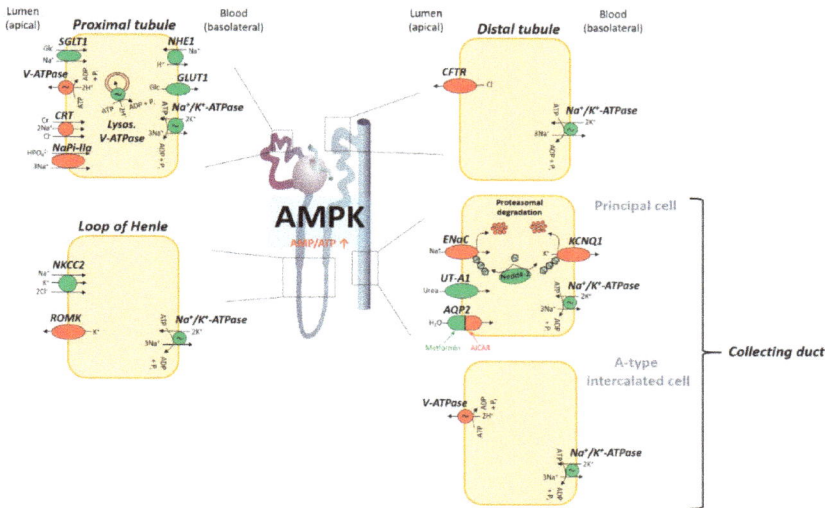

Figure 1. Tentative model illustrating AMPK-dependent effects on renal transport along the nephron. Cellular energy depletion (e.g., during hypoxia) leads to an elevated AMP/ATP ratio and subsequent AMPK activation. AMPK in turn regulates a multitude of active and passive epithelial transport processes along the renal tubular system in order to maintain cellular energy homeostasis. Ion channels, transport proteins, and ATPases that are activated upon AMPK stimulation are depicted as green icons, whereas red coloring indicates AMPK-dependent inhibition (see text for details). AMP, 5′-adenosine monophosphate; AMPK, AMP-activated protein kinase; SGLT1, Na$^+$-dependent glucose cotransporter 1; V-ATPase, vacuolar H$^+$-ATPase; CRT, creatine transporter; NaPi-IIa, Na$^+$-coupled phosphate transporter IIa; NHE1, Na$^+$/H$^+$ exchanger isoform 1; GLUT1, glucose transporter 1; NKCC2, Na$^+$-K$^+$-2Cl$^-$ cotransporter; ROMK, renal outer medullary K$^+$ channel; CFTR, cystic fibrosis transmembrane conductance regulator; ENaC, epithelial Na$^+$ channel; KCNQ1, voltage-gated K$^+$ channel; Nedd4-2, neural precursor cell expressed developmentally downregulated protein 4-2; UT-A1, urea transporter A1; AQP2, aquaporin 2.

Table 1. Overview of transport proteins regulated by AMPK in extrarenal tissues and evidence for renal expression.

Ion Channel/Transporter and Method of Modifying AMPK Activity	AMPK Effect	Cell Type of Studied AMPK Effect/Ref.	Evidence for Renal Expression/Ref.
Heterologous expression systems			
Kir2.1	Reduction of channel activity and membrane abundance via Nedd4-2 mediated endocytosis	*Xenopus* oocytes [207]	Human proximal tubular cells [208]
Kv1.5	Reduction of channel activity and membrane abundance via Nedd4-2 mediated endocytosis	*Xenopus* oocytes [209]	Human kidney biopsies [210]
Kv11.1 (hERG)	Reduction of channel activity and membrane abundance via Nedd4-2 mediated endocytosis	*Xenopus* oocytes [211]	Human proximal and distal convoluted tubule [212]
SMIT	Reduction of channel activity	*Xenopus* oocytes [213]	Rat kidney medulla [214]
BGT1	Reduction of channel activity	*Xenopus* oocytes [213]	Human kidney inner medulla [215] and mouse kidney medulla (basolateral membranes of collecting ducts and TAL of Henle) [216]
EAAT3	Reduction of channel activity and membrane abundance	*Xenopus* oocytes [217]	Mouse renal proximal tubule [218]
NCX	Reduction of channel activity and membrane abundance	*Xenopus* oocytes [169]	Rat distal convoluted tubule [219]
$K_{2P}10.1$ (TREK-2)	Inhibition of channel activity via phosphorylation at Ser-326 and Ser-359	HEK293 cells [220]	Human proximal tubule [221]
$K_{Ca}1.1$	Increase in channel activity and membrane abundance	*Xenopus* oocytes [222]	Human clear cell renal cell carcinoma (ccRCC) and healthy kidney cortex [223]
Pharmacological Manipulation			
$K_{Ca}1.1$	Inhibition of channel activity	Rat carotid body type I cells [224]	
Kir6.2	Upregulation of channel activity Up- or down-regulation of channel activity	Rat cardiomyocytes [225] Rat pancreatic beta-cells [226,227]	Rat renal tubular epithelial cells [228]
KCa3.1	Reduction of channel activity	Human airway epithelial cells [229]	Human proximal tubular cells [230]
MCT1 and MCT4	Upregulation of mRNA expression	Rat skeletal muscle [231]	MCT1: basolateral membrane of mouse proximal tubular epithelial cells [232] MCT4: human ccRCC [233]
PepT1	Downregulation of channel activity and brush-border membrane abundance	Caco-2 cells [234]	Rat renal proximal tubule [235]
Orai1	Downregulation of cell membrane abundance and SOCE	Rat UMR106 osteoblast-like cells [56]	Rat glomerular mesangial cells [236]
Genetically Modified Mouse Models			
Orai1		Mouse T-lymphocytes [171] Mouse dendritic cells [169]	

Int. J. Mol. Sci. **2018**, *19*, 3481

Author Contributions: P.G. and M.F. wrote this review.

funding: The authors were supported by the Deutsche Forschungsgemeinschaft (DFG; Fo695/2-1).

Conflicts of Interest: The authors declare no conflict of interest.

Abbreviations

ADP	Adenosine diphosphate
ADPKD	Autosomal dominant polycystic kidney disease
AMPK	5′-adenosine monophosphate (AMP)–activated protein kinase
AQP2	Aquaporin 2
ATP	Adenosine triphosphate
BGT1	Betaine/γ-aminobutyric acid (GABA) transporter 1
CaMKKβ	Ca^{2+}/calmodulin–dependent protein kinase kinase β
cAMP	Cyclic adenosine monophosphate
ccRCC	Clear cell renal cell carcinoma
CFTR	Cystic fibrosis transmembrane conductance regulator
CRT	Creatine transporter
EAAT3	Excitatory amino acid transporter 3
ENaC	Epithelial Na^+ channel
ERK1/2	Extracellular-signal regulated kinases 1/2
FGF23	Fibroblast growth factor 23
GLUT	Glucose transporter
HEK	Human embryonic kidney cells
hERG	Human ether-a-go-go-related gene
Kca	Ca^{2+} activated K^+ channels
KCNQ1	Voltage-gated K^+ channel
Kir	Inwardly rectifying K^+ channels
Kv	Voltage gated K^+ channels
LKB1	Liver kinase B1
MCT	Monocarboxylate transporters
MDCK	Madin-Darby canine kidney cells
NaPi-IIa	Na^+-coupled phosphate transporter
NCX	Na^+/Ca^{2+} exchanger
NDI	Nephrogenic diabetes insipidus
Nedd4-2	Neural precursor cell expressed developmentally down-regulated protein 4-2
NHE1	Na^+/H^+ exchanger isoform 1
NKCC2	Na^+-K^+-$2Cl^-$ cotransporter
PepT1	H^+-coupled di- and tripeptide transporter 1
PKA	Protein kinase A
ROMK	Renal outer medullary K^+ channel
SGLT	Na^+-dependent glucose cotransporter
SMIT	Na^+ coupled myoinositol transporter
SOCE	Store-operated Ca^{2+} entry
TAL	Thick ascending limb
TREK-2	Tandem pore domain K^+ channel 2
TRPV5	Transient receptor potential vanilloid 5 channel
UT	Urea transporter
V-ATPase	Vacuolar H^+-ATPase

References

1. Ramesh, M.; Vepuri, S.B.; Oosthuizen, F.; Soliman, M.E. Adenosine Monophosphate-Activated Protein Kinase (AMPK) as a Diverse Therapeutic Target: A Computational Perspective. *Appl. Biochem. Biotechnol.* **2016**, *178*, 810–830. [CrossRef] [PubMed]

2. Mihaylova, M.M.; Shaw, R.J. The AMPK signalling pathway coordinates cell growth, autophagy and metabolism. *Nat. Cell Biol.* **2011**, *13*, 1016–1023. [CrossRef] [PubMed]

3. Hardie, D.G. The AMP-activated protein kinase pathway—New players upstream and downstream. *J. Cell Sci.* **2004**, *117*, 5479–5487. [CrossRef] [PubMed]

4. Viollet, B. AMPK: Lessons from transgenic and knockout animals. *Front. Biosci.* **2009**, *14*, 19–44. [CrossRef]

5. Viollet, B.; Andreelli, F.; Jørgensen, S.B.; Perrin, C.; Flamez, D.; Mu, J.; Wojtaszewski, J.F.P.; Schuit, F.C.; Birnbaum, M.; Richter, E.; et al. Physiological role of AMP-activated protein kinase (AMPK): Insights from knockout mouse models. *Biochem. Soc. Trans.* **2003**, *31*, 216–219. [CrossRef] [PubMed]

6. Hardie, D.G.; Schaffer, B.E.; Brunet, A. AMPK: An Energy-Sensing Pathway with Multiple Inputs and Outputs. *Trends Cell Biol.* **2016**, *26*, 190–201. [CrossRef] [PubMed]

7. Dërmaku-Sopjani, M.; Abazi, S.; Faggio, C.; Kolgeci, J.; Sopjani, M. AMPK-sensitive cellular transport. *J. Biochem.* **2014**, *155*, 147–158. [CrossRef] [PubMed]

8. Hardie, D.G. AMPK—Sensing energy while talking to other signalling pathways. *Cell Metab.* **2014**, *20*, 939–952. [CrossRef] [PubMed]

9. Hardie, D.G.; Carling, D.; Gamblin, S.J. AMP-activated protein kinase: Also regulated by ADP? *Trends Biochem. Sci.* **2011**, *36*, 470–477. [CrossRef] [PubMed]

10. Hardie, D.G.; Ross, F.A.; Hawley, S.A. AMPK: A nutrient and energy sensor that maintains energy homeostasis. *Nat. Rev. Mol. Cell Biol.* **2012**, *13*, 251–262. [CrossRef] [PubMed]

11. Ross, F.A.; Jensen, T.E.; Hardie, D.G. Differential regulation by AMP and ADP of AMPK complexes containing different γ subunit isoforms. *Biochem. J.* **2016**, *473*, 189–199. [CrossRef] [PubMed]

12. Hardie, D.G.; Lin, S.-C. AMP-activated protein kinase—Not just an energy sensor. *F1000Research* **2017**, *6*, 1724. [CrossRef] [PubMed]

13. Thornton, C.; Snowden, M.A.; Carling, D. Identification of a novel AMP-activated protein kinase β subunit isoform that is highly expressed in skeletal muscle. *J. Biol. Chem.* **1998**, *273*, 12443–12450. [CrossRef] [PubMed]

14. Viollet, B.; Andreelli, F.; Jørgensen, S.B.; Perrin, C.; Geloen, A.; Flamez, D.; Mu, J.; Lenzner, C.; Baud, O.; Bennoun, M.; et al. The AMP-activated protein kinase α2 catalytic subunit controls whole-body insulin sensitivity. *J. Clin. Investig.* **2003**, *111*, 91–98. [CrossRef] [PubMed]

15. Hawley, S.A.; Davison, M.; Woods, A.; Davies, S.P.; Beri, R.K.; Carling, D.; Hardie, D.G. Characterization of the AMP-activated Protein Kinase Kinase from Rat Liver and Identification of Threonine 172 as the Major Site at Which It Phosphorylates AMP-activated Protein Kinase. *J. Biol. Chem.* **1996**, *271*, 27879–27887. [CrossRef] [PubMed]

16. Hong, S.-P.; Leiper, F.C.; Woods, A.; Carling, D.; Carlson, M. Activation of yeast Snf1 and mammalian AMP-activated protein kinase by upstream kinases. *Proc. Natl. Acad. Sci. USA* **2003**, *100*, 8839–8843. [CrossRef] [PubMed]

17. Hawley, S.A.; Boudeau, J.; Reid, J.L.; Mustard, K.J.; Udd, L.; Mäkelä, T.P.; Alessi, D.R.; Hardie, D.G. Complexes between the LKB1 tumor suppressor, STRAD α/β and MO25 α/β are upstream kinases in the AMP-activated protein kinase cascade. *J. Biol.* **2003**, *2*, 28. [CrossRef] [PubMed]

18. Woods, A.; Johnstone, S.R.; Dickerson, K.; Leiper, F.C.; Fryer, L.G.D.; Neumann, D.; Schlattner, U.; Wallimann, T.; Carlson, M.; Carling, D. LKB1 Is the Upstream Kinase in the AMP-Activated Protein Kinase Cascade. *Curr. Biol.* **2003**, *13*, 2004–2008. [CrossRef] [PubMed]

19. Shaw, R.J.; Kosmatka, M.; Bardeesy, N.; Hurley, R.L.; Witters, L.A.; DePinho, R.A.; Cantley, L.C. The tumor suppressor LKB1 kinase directly activates AMP-activated kinase and regulates apoptosis in response to energy stress. *Proc. Natl. Acad. Sci. USA* **2004**, *101*, 3329–3335. [CrossRef] [PubMed]

20. Herrero-Martín, G.; Høyer-Hansen, M.; García-García, C.; Fumarola, C.; Farkas, T.; López-Rivas, A.; Jäättelä, M. TAK1 activates AMPK-dependent cytoprotective autophagy in TRAIL-treated epithelial cells. *EMBO J.* **2009**, *28*, 677–685. [CrossRef] [PubMed]

21. Momcilovic, M.; Hong, S.-P.; Carlson, M. Mammalian TAK1 activates Snf1 protein kinase in yeast and phosphorylates AMP-activated protein kinase in vitro. *J. Biol. Chem.* **2006**, *281*, 25336–25343. [CrossRef] [PubMed]

22. Fujiwara, Y.; Kawaguchi, Y.; Fujimoto, T.; Kanayama, N.; Magari, M.; Tokumitsu, H. Differential AMP-activated Protein Kinase (AMPK) Recognition Mechanism of Ca^{2+}/Calmodulin-dependent Protein Kinase Kinase Isoforms. *J. Biol. Chem.* **2016**, *291*, 13802–13808. [CrossRef] [PubMed]

23. Hawley, S.A.; Pan, D.A.; Mustard, K.J.; Ross, L.; Bain, J.; Edelman, A.M.; Frenguelli, B.G.; Hardie, D.G. Calmodulin-dependent protein kinase kinase-β is an alternative upstream kinase for AMP-activated protein kinase. *Cell Metab.* **2005**, *2*, 9–19. [CrossRef] [PubMed]

24. Hurley, R.L.; Anderson, K.A.; Franzone, J.M.; Kemp, B.E.; Means, A.R.; Witters, L.A. The Ca^{2+}/calmodulin-dependent protein kinase kinases are AMP-activated protein kinase kinases. *J. Biol. Chem.* **2005**, *280*, 29060–29066. [CrossRef] [PubMed]

25. Burkewitz, K.; Zhang, Y.; Mair, W.B. AMPK at the nexus of energetics and aging. *Cell Metab.* **2014**, *20*, 10–25. [CrossRef] [PubMed]

26. Neumann, D. Is TAK1 a Direct Upstream Kinase of AMPK? *Int. J. Mol. Sci.* **2018**, *19*, 2412. [CrossRef] [PubMed]

27. Zhu, X.; Dahlmans, V.; Thali, R.; Preisinger, C.; Viollet, B.; Voncken, J.W.; Neumann, D. AMP-activated Protein Kinase Up-regulates Mitogen-activated Protein (MAP) Kinase-interacting Serine/Threonine Kinase 1a-dependent Phosphorylation of Eukaryotic Translation Initiation Factor 4E. *J. Biol. Chem.* **2016**, *291*, 17020–17027. [CrossRef] [PubMed]

28. Viollet, B.; Foretz, M. Revisiting the mechanisms of metformin action in the liver. *Ann. Endocrinol.* **2013**, *74*, 123–129. [CrossRef] [PubMed]

29. Sakamoto, K.; Göransson, O.; Hardie, D.G.; Alessi, D.R. Activity of LKB1 and AMPK-related kinases in skeletal muscle: Effects of contraction, phenformin, and AICAR. *Am. J. Physiol. Endocrinol. Metab.* **2004**, *287*, E310–E317. [CrossRef] [PubMed]

30. Sakamoto, K.; McCarthy, A.; Smith, D.; Green, K.A.; Grahame Hardie, D.; Ashworth, A.; Alessi, D.R. Deficiency of LKB1 in skeletal muscle prevents AMPK activation and glucose uptake during contraction. *EMBO J.* **2005**, *24*, 1810–1820. [CrossRef] [PubMed]

31. Cheung, P.C.F.; Salt, I.P.; Davies, S.P.; Hardie, D.G.; Carling, D. Characterization of AMP-activated protein kinase γ-subunit isoforms and their role in AMP binding. *Biochem. J.* **2000**, *346*, 659–669. [CrossRef] [PubMed]

32. Sanders, M.J.; Grondin, P.O.; Hegarty, B.D.; Snowden, M.A.; Carling, D. Investigating the mechanism for AMP activation of the AMP-activated protein kinase cascade. *Biochem. J.* **2007**, *403*, 139–148. [CrossRef] [PubMed]

33. Xiao, B.; Sanders, M.J.; Underwood, E.; Heath, R.; Mayer, F.V.; Carmena, D.; Jing, C.; Walker, P.A.; Eccleston, J.F.; Haire, L.F.; et al. Structure of mammalian AMPK and its regulation by ADP. *Nature* **2011**, *472*, 230–233. [CrossRef] [PubMed]

34. Oakhill, J.S.; Chen, Z.-P.; Scott, J.W.; Steel, R.; Castelli, L.A.; Ling, N.; Macaulay, S.L.; Kemp, B.E. β-Subunit myristoylation is the gatekeeper for initiating metabolic stress sensing by AMP-activated protein kinase (AMPK). *Proc. Natl. Acad. Sci. USA* **2010**, *107*, 19237–19241. [CrossRef] [PubMed]

35. Oakhill, J.S.; Steel, R.; Chen, Z.-P.; Scott, J.W.; Ling, N.; Tam, S.; Kemp, B.E. AMPK is a direct adenylate charge-regulated protein kinase. *Science* **2011**, *332*, 1433–1435. [CrossRef] [PubMed]

36. Viollet, B.; Mounier, R.; Leclerc, J.; Yazigi, A.; Foretz, M.; Andreelli, F. Targeting AMP-activated protein kinase as a novel therapeutic approach for the treatment of metabolic disorders. *Diabetes Metab.* **2007**, *33*, 395–402. [CrossRef] [PubMed]

37. Gowans, G.J.; Hawley, S.A.; Ross, F.A.; Hardie, D.G. AMP Is a True Physiological Regulator of AMP-Activated Protein Kinase by Both Allosteric Activation and Enhancing Net Phosphorylation. *Cell Metab.* **2013**, *18*, 556–566. [CrossRef] [PubMed]

38. Davies, S.P.; Helps, N.R.; Cohen, P.T.; Hardie, D.G. 5'-AMP inhibits dephosphorylation, as well as promoting phosphorylation, of the AMP-activated protein kinase. Studies using bacterially expressed human protein phosphatase-2C α and native bovine protein phosphatase-2A c. *FEBS Lett.* **1995**, *377*, 421–425. [PubMed]

39. Chen, L.; Wang, J.; Zhang, Y.-Y.; Yan, S.F.; Neumann, D.; Schlattner, U.; Wang, Z.-X.; Wu, J.-W. AMP-activated protein kinase undergoes nucleotide-dependent conformational changes. *Nat. Struct. Mol. Biol.* **2012**, *19*, 716–718. [CrossRef] [PubMed]

40. Garcia, D.; Shaw, R.J. AMPK: Mechanisms of Cellular Energy Sensing and Restoration of Metabolic Balance. *Mol. Cell.* **2017**, *66*, 789–800. [CrossRef] [PubMed]

41. Woods, A.; Dickerson, K.; Heath, R.; Hong, S.-P.; Momcilovic, M.; Johnstone, S.R.; Carlson, M.; Carling, D. Ca^{2+}/calmodulin-dependent protein kinase kinase-β acts upstream of AMP-activated protein kinase in mammalian cells. *Cell Metab.* **2005**, *2*, 21–33. [CrossRef] [PubMed]

42. Stahmann, N.; Woods, A.; Carling, D.; Heller, R. Thrombin activates AMP-activated protein kinase in endothelial cells via a pathway involving Ca^{2+}/calmodulin-dependent protein kinase kinase β. *Mol. Cell. Biol.* **2006**, *26*, 5933–5945. [CrossRef] [PubMed]

43. Yang, Y.; Atasoy, D.; Su, H.H.; Sternson, S.M. Hunger states switch a flip-flop memory circuit via a synaptic AMPK-dependent positive feedback loop. *Cell* **2011**, *146*, 992–1003. [CrossRef] [PubMed]

44. Lang, F.; Eylenstein, A.; Shumilina, E. Regulation of Orai1/STIM1 by the kinases SGK1 and AMPK. *Cell Calcium* **2012**, *52*, 347–354. [CrossRef] [PubMed]

45. Sundivakkam, P.C.; Natarajan, V.; Malik, A.B.; Tiruppathi, C. Store-operated Ca^{2+} entry (SOCE) induced by protease-activated receptor-1 mediates STIM1 protein phosphorylation to inhibit SOCE in endothelial cells through AMP-activated protein kinase and p38β mitogen-activated protein kinase. *J. Biol. Chem.* **2013**, *288*, 17030–17041. [CrossRef] [PubMed]

46. Zhang, B.; Yan, J.; Umbach, A.T.; Fakhri, H.; Fajol, A.; Schmidt, S.; Salker, M.S.; Chen, H.; Alexander, D.; Spichtig, D.; et al. NFκB-sensitive Orai1 expression in the regulation of FGF23 release. *J. Mol. Med.* **2016**, *94*, 557–566. [CrossRef] [PubMed]

47. Prakriya, M.; Feske, S.; Gwack, Y.; Srikanth, S.; Rao, A.; Hogan, P.G. Orai1 is an essential pore subunit of the CRAC channel. *Nature* **2006**, *443*, 230–233. [CrossRef] [PubMed]

48. Zhang, S.L.; Kozak, J.A.; Jiang, W.; Yeromin, A.V.; Chen, J.; Yu, Y.; Penna, A.; Shen, W.; Chi, V.; Cahalan, M.D. Store-dependent and -independent modes regulating Ca^{2+} release-activated Ca^{2+} channel activity of human Orai1 and Orai3. *J. Biol. Chem.* **2008**, *283*, 17662–17671. [CrossRef] [PubMed]

49. Tiruppathi, C.; Ahmmed, G.U.; Vogel, S.M.; Malik, A.B. Ca^{2+} signaling, TRP channels, and endothelial permeability. *Microcirculation* **2006**, *13*, 693–708. [CrossRef] [PubMed]

50. Yu, F.; Sun, L.; Machaca, K. Constitutive recycling of the store-operated Ca^{2+} channel Orai1 and its internalization during meiosis. *J. Cell Biol.* **2010**, *191*, 523–535. [CrossRef] [PubMed]

51. Baryshnikov, S.G.; Pulina, M.V.; Zulian, A.; Linde, C.I.; Golovina, V.A. Orai1, a critical component of store-operated Ca^{2+} entry, is functionally associated with Na$^+$/Ca^{2+} exchanger and plasma membrane Ca^{2+} pump in proliferating human arterial myocytes. *Am. J. Physiol. Cell Physiol.* **2009**, *297*, C1103–C1112. [CrossRef] [PubMed]

52. Johnstone, L.S.; Graham, S.J.L.; Dziadek, M.A. STIM proteins: Integrators of signalling pathways in development, differentiation and disease. *J. Cell. Mol. Med.* **2010**, *14*, 1890–1903. [CrossRef] [PubMed]

53. Yang, S.; Zhang, J.J.; Huang, X.-Y. Orai1 and STIM1 are critical for breast tumor cell migration and metastasis. *Cancer Cell* **2009**, *15*, 124–134. [CrossRef] [PubMed]

54. Stathopulos, P.B.; Ikura, M. Store operated calcium entry: From concept to structural mechanisms. *Cell Calcium* **2017**, *63*, 3–7. [CrossRef] [PubMed]

55. Ambudkar, I.S.; de Souza, L.B.; Ong, H.L. TRPC1, Orai1, and STIM1 in SOCE: Friends in tight spaces. *Cell Calcium* **2017**, *63*, 33–39. [CrossRef] [PubMed]

56. Glosse, P.; Feger, M.; Mutig, K.; Chen, H.; Hirche, F.; Hasan, A.A.; Gaballa, M.M.S.; Hocher, B.; Lang, F.; Foller, M. AMP-activated kinase is a regulator of fibroblast growth factor 23 production. *Kidney Int.* **2018**, *94*, 491–501. [CrossRef] [PubMed]

57. Hasenour, C.M.; Berglund, E.D.; Wasserman, D.H. Emerging role of AMP-activated protein kinase in endocrine control of metabolism in the liver. *Mol. Cell. Endocrinol.* **2013**, *366*, 152–162. [CrossRef] [PubMed]

58. Li, Y.; Xu, S.; Mihaylova, M.M.; Zheng, B.; Hou, X.; Jiang, B.; Park, O.; Luo, Z.; Lefai, E.; Shyy, J.Y.-J.; et al. AMPK phosphorylates and inhibits SREBP activity to attenuate hepatic steatosis and atherosclerosis in diet-induced insulin-resistant mice. *Cell Metab.* **2011**, *13*, 376–388. [CrossRef] [PubMed]

59. Foretz, M.; Viollet, B. Activation of AMPK for a Break in Hepatic Lipid Accumulation and Circulating Cholesterol. *EBioMedicine* **2018**, *31*, 15–16. [CrossRef] [PubMed]

60. Merlen, G.; Gentric, G.; Celton-Morizur, S.; Foretz, M.; Guidotti, J.-E.; Fauveau, V.; Leclerc, J.; Viollet, B.; Desdouets, C. AMPKα1 controls hepatocyte proliferation independently of energy balance by regulating Cyclin A2 expression. *J. Hepatol.* **2014**, *60*, 152–159. [CrossRef] [PubMed]

61. Foretz, M.; Viollet, B. Regulation of hepatic metabolism by AMPK. *J. Hepatol.* **2011**, *54*, 827–829. [CrossRef] [PubMed]

62. Kjøbsted, R.; Hingst, J.R.; Fentz, J.; Foretz, M.; Sanz, M.-N.; Pehmøller, C.; Shum, M.; Marette, A.; Mounier, R.; Treebak, J.T.; et al. AMPK in skeletal muscle function and metabolism. *FASEB J.* **2018**, *32*, 1741–1777. [CrossRef] [PubMed]

63. Mounier, R.; Théret, M.; Lantier, L.; Foretz, M.; Viollet, B. Expanding roles for AMPK in skeletal muscle plasticity. *Trends Endocrinol. Metab.* **2015**, *26*, 275–286. [CrossRef] [PubMed]

64. Kjøbsted, R.; Munk-Hansen, N.; Birk, J.B.; Foretz, M.; Viollet, B.; Björnholm, M.; Zierath, J.R.; Treebak, J.T.; Wojtaszewski, J.F.P. Enhanced Muscle Insulin Sensitivity After Contraction/Exercise Is Mediated by AMPK. *Diabetes* **2017**, *66*, 598–612. [CrossRef] [PubMed]

65. Cokorinos, E.C.; Delmore, J.; Reyes, A.R.; Albuquerque, B.; Kjøbsted, R.; Jørgensen, N.O.; Tran, J.-L.; Jatkar, A.; Cialdea, K.; Esquejo, R.M.; et al. Activation of Skeletal Muscle AMPK Promotes Glucose Disposal and Glucose Lowering in Non-human Primates and Mice. *Cell Metab.* **2017**, *25*, 1147–1159. [CrossRef] [PubMed]

66. Fentz, J.; Kjøbsted, R.; Birk, J.B.; Jordy, A.B.; Jeppesen, J.; Thorsen, K.; Schjerling, P.; Kiens, B.; Jessen, N.; Viollet, B.; et al. AMPKα is critical for enhancing skeletal muscle fatty acid utilization during in vivo exercise in mice. *FASEB J.* **2015**, *29*, 1725–1738. [CrossRef] [PubMed]

67. Arad, M.; Seidman, C.E.; Seidman, J.G. AMP-Activated Protein Kinase in the Heart: Role during Health and Disease. *Circ. Res.* **2007**, *100*, 474–488. [CrossRef] [PubMed]

68. Voelkl, J.; Alesutan, I.; Primessnig, U.; Feger, M.; Mia, S.; Jungmann, A.; Castor, T.; Viereck, R.; Stöckigt, F.; Borst, O.; et al. AMP-activated protein kinase α1-sensitive activation of AP-1 in cardiomyocytes. *J. Mol. Cell. Cardiol.* **2016**, *97*, 36–43. [CrossRef] [PubMed]

69. Liao, Y.; Takashima, S.; Maeda, N.; Ouchi, N.; Komamura, K.; Shimomura, I.; Hori, M.; Matsuzawa, Y.; Funahashi, T.; Kitakaze, M. Exacerbation of heart failure in adiponectin-deficient mice due to impaired regulation of AMPK and glucose metabolism. *Cardiovasc. Res.* **2005**, *67*, 705–713. [CrossRef] [PubMed]

70. Russell, R.R.; Li, J.; Coven, D.L.; Pypaert, M.; Zechner, C.; Palmeri, M.; Giordano, F.J.; Mu, J.; Birnbaum, M.J.; Young, L.H. AMP-activated protein kinase mediates ischemic glucose uptake and prevents postischemic cardiac dysfunction, apoptosis, and injury. *J. Clin. Investig.* **2004**, *114*, 495–503. [CrossRef] [PubMed]

71. Gélinas, R.; Mailleux, F.; Dontaine, J.; Bultot, L.; Demeulder, B.; Ginion, A.; Daskalopoulos, E.P.; Esfahani, H.; Dubois-Deruy, E.; Lauzier, B.; et al. AMPK activation counteracts cardiac hypertrophy by reducing O-GlcNAcylation. *Nat. Commun.* **2018**, *9*, 374. [CrossRef] [PubMed]

72. Chen, K.; Kobayashi, S.; Xu, X.; Viollet, B.; Liang, Q. AMP activated protein kinase is indispensable for myocardial adaptation to caloric restriction in mice. *PLoS ONE* **2013**, *8*, e59682. [CrossRef] [PubMed]

73. Zhang, P.; Hu, X.; Xu, X.; Fassett, J.; Zhu, G.; Viollet, B.; Xu, W.; Wiczer, B.; Bernlohr, D.A.; Bache, R.J.; et al. AMP activated protein kinase-α2 deficiency exacerbates pressure-overload-induced left ventricular hypertrophy and dysfunction in mice. *Hypertension* **2008**, *52*, 918–924. [CrossRef] [PubMed]

74. Hallows, K.R.; Mount, P.F.; Pastor-Soler, N.M.; Power, D.A. Role of the energy sensor AMP-activated protein kinase in renal physiology and disease. *Am. J. Physiol. Renal. Physiol.* **2010**, *298*, F1067–F1077. [CrossRef] [PubMed]

75. Pastor-Soler, N.M.; Hallows, K.R. AMP-activated protein kinase regulation of kidney tubular transport. *Curr. Opin. Nephrol. Hypertens.* **2012**, *21*, 523–533. [CrossRef] [PubMed]

76. Rajani, R.; Pastor-Soler, N.M.; Hallows, K.R. Role of AMP-activated protein kinase in kidney tubular transport, metabolism, and disease. *Curr. Opin. Nephrol. Hypertens.* **2017**, *26*, 375–383. [CrossRef] [PubMed]

77. Lee, M.-J.; Feliers, D.; Mariappan, M.M.; Sataranatarajan, K.; Mahimainathan, L.; Musi, N.; Foretz, M.; Viollet, B.; Weinberg, J.M.; Choudhury, G.G.; et al. A role for AMP-activated protein kinase in diabetes-induced renal hypertrophy. *Am. J. Physiol. Renal. Physiol.* **2007**, *292*, F617–F627. [CrossRef] [PubMed]

78. Jeyabalan, J.; Shah, M.; Viollet, B.; Chenu, C. AMP-activated protein kinase pathway and bone metabolism. *J. Endocrinol.* **2012**, *212*, 277–290. [CrossRef] [PubMed]

79. McCarthy, A.D.; Cortizo, A.M.; Sedlinsky, C. Metformin revisited: Does this regulator of AMP-activated protein kinase secondarily affect bone metabolism and prevent diabetic osteopathy. *World J. Diabetes* **2016**, *7*, 122–133. [CrossRef] [PubMed]

80. Kanazawa, I. Interaction between bone and glucose metabolism. *Endocr. J.* **2017**, *64*, 1043–1053. [CrossRef] [PubMed]

81. Tain, Y.-L.; Hsu, C.-N. AMP-Activated Protein Kinase as a Reprogramming Strategy for Hypertension and Kidney Disease of Developmental Origin. *Int. J. Mol. Sci.* **2018**, *19*, 1744. [CrossRef] [PubMed]

82. Tsai, C.-M.; Kuo, H.-C.; Hsu, C.-N.; Huang, L.-T.; Tain, Y.-L. Metformin reduces asymmetric dimethylarginine and prevents hypertension in spontaneously hypertensive rats. *Transl. Res.* **2014**, *164*, 452–459. [CrossRef] [PubMed]

83. Allouch, S.; Munusamy, S. AMP-activated Protein Kinase as a Drug Target in Chronic Kidney Disease. *Curr. Drug Targets* **2018**, *19*, 709–720. [CrossRef] [PubMed]

84. Curthoys, N.P.; Moe, O.W. Proximal tubule function and response to acidosis. *Clin. J. Am. Soc. Nephrol.* **2014**, *9*, 1627–1638. [CrossRef] [PubMed]

85. Wallace, M.A. Anatomy and Physiology of the Kidney. *AORN J.* **1998**, *68*, 799–820. [CrossRef]

86. Mount, D.B. Thick ascending limb of the loop of Henle. *Clin. J. Am. Soc. Nephrol.* **2014**, *9*, 1974–1986. [CrossRef] [PubMed]

87. Zhang, J.L.; Rusinek, H.; Chandarana, H.; Lee, V.S. Functional MRI of the kidneys. *J. Magn. Reson. Imaging* **2013**, *37*, 282–293. [CrossRef] [PubMed]

88. Blaine, J.; Chonchol, M.; Levi, M. Renal control of calcium, phosphate, and magnesium homeostasis. *Clin. J. Am. Soc. Nephrol.* **2015**, *10*, 1257–1272. [CrossRef] [PubMed]

89. Lee, Y.J.; Han, H.J. Regulatory mechanisms of Na^+/glucose cotransporters in renal proximal tubule cells. *Kidney Int. Suppl.* **2007**, S27–S35. [CrossRef] [PubMed]

90. Rotte, A.; Pasham, V.; Eichenmüller, M.; Bhandaru, M.; Föller, M.; Lang, F. Upregulation of Na^+/H^+ exchanger by the AMP-activated protein kinase. *Biochem. Biophys. Res. Commun.* **2010**, *398*, 677–682. [CrossRef] [PubMed]

91. Palmer, L.G.; Schnermann, J. Integrated control of Na transport along the nephron. *Clin. J. Am. Soc. Nephrol.* **2015**, *10*, 676–687. [CrossRef] [PubMed]

92. Sopjani, M.; Bhavsar, S.K.; Fraser, S.; Kemp, B.E.; Föller, M.; Lang, F. Regulation of Na^+-coupled glucose carrier SGLT1 by AMP-activated protein kinase. *Mol. Membr. Biol.* **2010**, *27*, 137–144. [CrossRef] [PubMed]

93. Dërmaku-Sopjani, M.; Almilaji, A.; Pakladok, T.; Munoz, C.; Hosseinzadeh, Z.; Blecua, M.; Sopjani, M.; Lang, F. Down-regulation of the Na^+-coupled phosphate transporter NaPi-IIa by AMP-activated protein kinase. *Kidney Blood Press. Res.* **2013**, *37*, 547–556. [CrossRef] [PubMed]

94. Hallows, K.R. Emerging role of AMP-activated protein kinase in coupling membrane transport to cellular metabolism. *Curr. Opin. Nephrol. Hypertens.* **2005**, *14*, 464–471. [CrossRef] [PubMed]

95. Noske, R.; Cornelius, F.; Clarke, R.J. Investigation of the enzymatic activity of the Na^+, K^+-ATPase via isothermal titration microcalorimetry. *Biochim. Biophys. Acta* **2010**, *1797*, 1540–1545. [CrossRef] [PubMed]

96. Woollhead, A.M.; Scott, J.W.; Hardie, D.G.; Baines, D.L. Phenformin and 5-aminoimidazole-4-carboxamide-1-β-D-ribofuranoside (AICAR) activation of AMP-activated protein kinase inhibits transepithelial Na^+ transport across H441 lung cells. *J. Physiol.* **2005**, *566*, 781–792. [CrossRef] [PubMed]

97. Woollhead, A.M.; Sivagnanasundaram, J.; Kalsi, K.K.; Pucovsky, V.; Pellatt, L.J.; Scott, J.W.; Mustard, K.J.; Hardie, D.G.; Baines, D.L. Pharmacological activators of AMP-activated protein kinase have different effects on Na^+ transport processes across human lung epithelial cells. *Br. J. Pharmacol.* **2007**, *151*, 1204–1215. [CrossRef] [PubMed]

98. Vadász, I.; Dada, L.A.; Briva, A.; Trejo, H.E.; Welch, L.C.; Chen, J.; Tóth, P.T.; Lecuona, E.; Witters, L.A.; Schumacker, P.T.; et al. AMP-activated protein kinase regulates CO_2-induced alveolar epithelial dysfunction in rats and human cells by promoting Na, K-ATPase endocytosis. *J. Clin. Investig.* **2008**, *118*, 752–762. [CrossRef] [PubMed]

99. Gusarova, G.A.; Dada, L.A.; Kelly, A.M.; Brodie, C.; Witters, L.A.; Chandel, N.S.; Sznajder, J.I. α1-AMP-activated protein kinase regulates hypoxia-induced Na, K-ATPase endocytosis via direct phosphorylation of protein kinase C zeta. *Mol. Cell. Biol.* **2009**, *29*, 3455–3464. [CrossRef] [PubMed]

100. Gusarova, G.A.; Trejo, H.E.; Dada, L.A.; Briva, A.; Welch, L.C.; Hamanaka, R.B.; Mutlu, G.M.; Chandel, N.S.; Prakriya, M.; Sznajder, J.I. Hypoxia leads to Na, K-ATPase downregulation via Ca^{2+} release-activated Ca(2+) channels and AMPK activation. *Mol. Cell. Biol.* **2011**, *31*, 3546–3556. [CrossRef] [PubMed]

101. Benziane, B.; Björnholm, M.; Pirkmajer, S.; Austin, R.L.; Kotova, O.; Viollet, B.; Zierath, J.R.; Chibalin, A.V. Activation of AMP-activated protein kinase stimulates Na^+, K^+-ATPase activity in skeletal muscle cells. *J. Biol. Chem.* **2012**, *287*, 23451–23463. [CrossRef] [PubMed]

102. Alves, D.S.; Farr, G.A.; Seo-Mayer, P.; Caplan, M.J. AS160 associates with the Na^+, K^+-ATPase and mediates the adenosine monophosphate-stimulated protein kinase-dependent regulation of sodium pump surface expression. *Mol. Biol. Cell* **2010**, *21*, 4400–4408. [CrossRef] [PubMed]

103. Seo-Mayer, P.W.; Thulin, G.; Zhang, L.; Alves, D.S.; Ardito, T.; Kashgarian, M.; Caplan, M.J. Preactivation of AMPK by metformin may ameliorate the epithelial cell damage caused by renal ischemia. *Am. J. Physiol. Renal. Physiol.* **2011**, *301*, F1346–F1357. [CrossRef] [PubMed]

104. Mount, P.F.; Gleich, K.; Tam, S.; Fraser, S.A.; Choy, S.-W.; Dwyer, K.M.; Lu, B.; van Denderen, B.; Fingerle-Rowson, G.; Bucala, R.; et al. The outcome of renal ischemia-reperfusion injury is unchanged in AMPK-β1 deficient mice. *PLoS ONE* **2012**, *7*, e29887. [CrossRef] [PubMed]

105. Fitzgerald, G.A.; Mulligan, C.; Mindell, J.A. A general method for determining secondary active transporter substrate stoichiometry. *eLife* **2017**, *6*. [CrossRef] [PubMed]

106. Forrest, L.R.; Krämer, R.; Ziegler, C. The structural basis of secondary active transport mechanisms. *Biochim. Biophys. Acta* **2011**, *1807*, 167–188. [CrossRef] [PubMed]

107. Mandel, L.J.; Balaban, R.S. Stoichiometry and coupling of active transport to oxidative metabolism in epithelial tissues. *Am. J. Physiol.* **1981**, *240*, F357–F371. [CrossRef] [PubMed]

108. Mather, A.; Pollock, C. Glucose handling by the kidney. *Kidney Int. Suppl.* **2011**, *79*, S1–S6. [CrossRef] [PubMed]

109. Bakris, G.L.; Fonseca, V.A.; Sharma, K.; Wright, E.M. Renal sodium-glucose transport: Role in diabetes mellitus and potential clinical implications. *Kidney Int.* **2009**, *75*, 1272–1277. [CrossRef] [PubMed]

110. Wright, E.M.; Hirayama, B.A.; Loo, D.F. Active sugar transport in health and disease. *J. Intern. Med.* **2007**, *261*, 32–43. [CrossRef] [PubMed]

111. Hawley, S.A.; Ford, R.J.; Smith, B.K.; Gowans, G.J.; Mancini, S.J.; Pitt, R.D.; Day, E.A.; Salt, I.P.; Steinberg, G.R.; Hardie, D.G. The Na^+/Glucose Cotransporter Inhibitor Canagliflozin Activates AMPK by Inhibiting Mitochondrial Function and Increasing Cellular AMP Levels. *Diabetes* **2016**, *65*, 2784–2794. [CrossRef] [PubMed]

112. You, G.; Lee, W.-S.; Barros, E.J.G.; Kanai, Y.; Huo, T.-L.; Khawaja, S.; Wells, R.G.; Nigam, S.K.; Hediger, M.A. Molecular Characteristics of Na^+-coupled Glucose Transporters in Adult and Embryonic Rat Kidney. *J. Biol. Chem.* **1995**, *270*, 29365–29371. [CrossRef] [PubMed]

113. Banerjee, S.K.; Wang, D.W.; Alzamora, R.; Huang, X.N.; Pastor-Soler, N.M.; Hallows, K.R.; McGaffin, K.R.; Ahmad, F. SGLT1, a novel cardiac glucose transporter, mediates increased glucose uptake in PRKAG2 cardiomyopathy. *J. Mol. Cell. Cardiol.* **2010**, *49*, 683–692. [CrossRef] [PubMed]

114. Wright, E.M. Renal Na^+-glucose cotransporters. *Am. J. Physiol. Renal. Physiol.* **2001**, *280*, F10–F18. [CrossRef] [PubMed]

115. Pajor, A.M.; Wright, E.M. Cloning and functional expression of a mammalian Na^+/nucleoside cotransporter. A member of the SGLT family. *J. Biol. Chem.* **1992**, *267*, 3557–3560. [PubMed]

116. Linden, K.C.; DeHaan, C.L.; Zhang, Y.; Glowacka, S.; Cox, A.J.; Kelly, D.J.; Rogers, S. Renal expression and localization of the facilitative glucose transporters GLUT1 and GLUT12 in animal models of hypertension and diabetic nephropathy. *Am. J. Physiol. Renal. Physiol.* **2006**, *290*, F205–F213. [CrossRef] [PubMed]

117. Dominguez, J.H.; Camp, K.; Maianu, L.; Garvey, W.T. Glucose transporters of rat proximal tubule: Differential expression and subcellular distribution. *Am. J. Physiol.* **1992**, *262*, F807–F812. [CrossRef] [PubMed]

118. Thorens, B.; Lodish, H.F.; Brown, D. Differential localization of two glucose transporter isoforms in rat kidney. *Am. J. Physiol.* **1990**, *259*, C286–C294. [CrossRef] [PubMed]

119. Castilla-Madrigal, R.; Barrenetxe, J.; Moreno-Aliaga, M.J.; Lostao, M.P. EPA blocks TNF-α-induced inhibition of sugar uptake in Caco-2 cells via GPR120 and AMPK. *J. Cell. Physiol.* **2018**, *233*, 2426–2433. [CrossRef] [PubMed]

120. Di Franco, A.; Cantini, G.; Tani, A.; Coppini, R.; Zecchi-Orlandini, S.; Raimondi, L.; Luconi, M.; Mannucci, E. Sodium-dependent glucose transporters (SGLT) in human ischemic heart: A new potential pharmacological target. *Int. J. Cardiol.* **2017**, *243*, 86–90. [CrossRef] [PubMed]

121. Portilla, D. Energy metabolism and cytotoxicity. *Semin. Nephrol.* **2003**, *23*, 432–438. [CrossRef]

122. Le Hir, M.; Dubach, U.C. Peroxisomal and mitochondrial β-oxidation in the rat kidney: Distribution of fatty acyl-coenzyme A oxidase and 3-hydroxyacyl-coenzyme A dehydrogenase activities along the nephron. *J. Histochem. Cytochem.* **1982**, *30*, 441–444. [CrossRef] [PubMed]

123. Uchida, S.; Endou, H. Substrate specificity to maintain cellular ATP along the mouse nephron. *Am. J. Physiol.* **1988**, *255*, F977–F983. [CrossRef] [PubMed]

124. Fryer, L.G.D.; Foufelle, F.; Barnes, K.; Baldwin, S.A.; Woods, A.; Carling, D. Characterization of the role of the AMP-activated protein kinase in the stimulation of glucose transport in skeletal muscle cells. *Biochem. J.* **2002**, *363*, 167–174. [CrossRef] [PubMed]

125. Al-Bayati, A.; Lukka, D.; Brown, A.E.; Walker, M. Effects of thrombin on insulin signalling and glucose uptake in cultured human myotubes. *J. Diabetes Complicat.* **2016**, *30*, 1209–1216. [CrossRef] [PubMed]

126. Andrade, B.M.; Cazarin, J.; Zancan, P.; Carvalho, D.P. AMP-activated protein kinase upregulates glucose uptake in thyroid PCCL3 cells independent of thyrotropin. *Thyroid* **2012**, *22*, 1063–1068. [CrossRef] [PubMed]

127. Takeno, A.; Kanazawa, I.; Notsu, M.; Tanaka, K.-I.; Sugimoto, T. Glucose uptake inhibition decreases expressions of receptor activator of nuclear factor-kappa B ligand (RANKL) and osteocalcin in osteocytic MLO-Y4-A2 cells. *Am. J. Physiol. Endocrinol. Metab.* **2018**, *314*, E115–E123. [CrossRef] [PubMed]

128. Wang, Y.; Zhang, Y.; Wang, Y.; Peng, H.; Rui, J.; Zhang, Z.; Wang, S.; Li, Z. WSF-P-1, a novel AMPK activator, promotes adiponectin multimerization in 3T3-L1 adipocytes. *Biosci. Biotechnol. Biochem.* **2017**, *81*, 1529–1535. [CrossRef] [PubMed]

129. Yamada, S.; Kotake, Y.; Sekino, Y.; Kanda, Y. AMP-activated protein kinase-mediated glucose transport as a novel target of tributyltin in human embryonic carcinoma cells. *Metallomics* **2013**, *5*, 484–491. [CrossRef] [PubMed]

130. Yu, H.; Zhang, H.; Dong, M.; Wu, Z.; Shen, Z.; Xie, Y.; Kong, Z.; Dai, X.; Xu, B. Metabolic reprogramming and AMPKα1 pathway activation by caulerpin in colorectal cancer cells. *Int. J. Oncol.* **2017**, *50*, 161–172. [CrossRef] [PubMed]

131. Abbud, W.; Habinowski, S.; Zhang, J.Z.; Kendrew, J.; Elkairi, F.S.; Kemp, B.E.; Witters, L.A.; Ismail-Beigi, F. Stimulation of AMP-activated protein kinase (AMPK) is associated with enhancement of Glut1-mediated glucose transport. *Arch. Biochem. Biophys.* **2000**, *380*, 347–352. [CrossRef] [PubMed]

132. Baldwin, S.A.; Barros, L.F.; Griffiths, M.; Ingram, J.; Robbins, E.C.; Streets, A.J.; Saklatvala, J. Regulation of GLUTI in response to cellular stress. *Biochem. Soc. Trans.* **1997**, *25*, 954–958. [CrossRef] [PubMed]

133. Sokolovska, J.; Isajevs, S.; Sugoka, O.; Sharipova, J.; Lauberte, L.; Svirina, D.; Rostoka, E.; Sjakste, T.; Kalvinsh, I.; Sjakste, N. Influence of metformin on GLUT1 gene and protein expression in rat streptozotocin diabetes mellitus model. *Arch. Physiol. Biochem.* **2010**, *116*, 137–145. [CrossRef] [PubMed]

134. Walker, J.; Jijon, H.B.; Diaz, H.; Salehi, P.; Churchill, T.; Madsen, K.L. 5-aminoimidazole-4-carboxamide riboside (AICAR) enhances GLUT2-dependent jejunal glucose transport: A possible role for AMPK. *Biochem. J.* **2005**, *385*, 485–491. [CrossRef] [PubMed]

135. Sakar, Y.; Meddah, B.; Faouzi, M.A.; Cherrah, Y.; Bado, A.; Ducroc, R. Metformin-induced regulation of the intestinal D-glucose transporters. *J. Physiol. Pharmacol.* **2010**, *61*, 301–307. [PubMed]

136. Parker, M.D.; Myers, E.J.; Schelling, J.R. Na$^+$–H$^+$ exchanger-1 (NHE1) regulation in kidney proximal tubule. *Cell. Mol. Life Sci.* **2015**, *72*, 2061–2074. [CrossRef] [PubMed]

137. Odunewu, A.; Fliegel, L. Acidosis-mediated regulation of the NHE1 isoform of the Na$^+$/H$^+$ exchanger in renal cells. *Am. J. Physiol. Renal. Physiol.* **2013**, *305*, F370–F381. [CrossRef] [PubMed]

138. Biemesderfer, D.; Reilly, R.F.; Exner, M.; Igarashi, P.; Aronson, P.S. Immunocytochemical characterization of Na$^+$-H$^+$ exchanger isoform NHE-1 in rabbit kidney. *Am. J. Physiol.* **1992**, *263*, F833–F840. [CrossRef] [PubMed]

139. Peti-Peterdi, J.; Chambrey, R.; Bebok, Z.; Biemesderfer, D.; St John, P.L.; Abrahamson, D.R.; Warnock, D.G.; Bell, P.D. Macula densa Na$^+$/H$^+$ exchange activities mediated by apical NHE2 and basolateral NHE4 isoforms. *Am. J. Physiol. Renal. Physiol.* **2000**, *278*, F452–F463. [CrossRef] [PubMed]

140. Baum, M.; Moe, O.W.; Gentry, D.L.; Alpern, R.J. Effect of glucocorticoids on renal cortical NHE-3 and NHE-1 mRNA. *Am. J. Physiol.* **1994**, *267*, F437–F442. [CrossRef] [PubMed]

141. Hue, L.; Beauloye, C.; Marsin, A.-S.; Bertrand, L.; Horman, S.; Rider, M.H. Insulin and Ischemia Stimulate Glycolysis by Acting on the Same Targets Through Different and Opposing Signaling Pathways. *J. Mol. Cell. Cardiol.* **2002**, *34*, 1091–1097. [CrossRef] [PubMed]

142. Marsin, A.-S.; Bouzin, C.; Bertrand, L.; Hue, L. The stimulation of glycolysis by hypoxia in activated monocytes is mediated by AMP-activated protein kinase and inducible 6-phosphofructo-2-kinase. *J. Biol. Chem.* **2002**, *277*, 30778–30783. [CrossRef] [PubMed]

143. Wyss, M.; Kaddurah-Daouk, R. Creatine and creatinine metabolism. *Physiol. Rev.* **2000**, *80*, 1107–1213. [CrossRef] [PubMed]

144. García-Delgado, M.; Peral, M.J.; Cano, M.; Calonge, M.L.; Ilundáin, A.A. Creatine transport in brush-border membrane vesicles isolated from rat kidney cortex. *J. Am. Soc. Nephrol.* **2001**, *12*, 1819–1825. [PubMed]

145. Wallimann, T.; Wyss, M.; Brdiczka, D.; Nicolay, K.; Eppenberger, H.M. Intracellular compartmentation, structure and function of creatine kinase isoenzymes in tissues with high and fluctuating energy demands: The 'phosphocreatine circuit' for cellular energy homeostasis. *Biochem. J.* **1992**, *281*, 21–40. [CrossRef] [PubMed]

146. Neumann, D.; Schlattner, U.; Wallimann, T. A molecular approach to the concerted action of kinases involved in energy homoeostasis. *Biochem. Soc. Trans.* **2003**, *31*, 169–174. [CrossRef] [PubMed]

147. Li, H.; Thali, R.F.; Smolak, C.; Gong, F.; Alzamora, R.; Wallimann, T.; Scholz, R.; Pastor-Soler, N.M.; Neumann, D.; Hallows, K.R. Regulation of the creatine transporter by AMP-activated protein kinase in kidney epithelial cells. *Am. J. Physiol. Renal. Physiol.* **2010**, *299*, F167–F177. [CrossRef] [PubMed]

148. Darrabie, M.D.; Arciniegas, A.J.L.; Mishra, R.; Bowles, D.E.; Jacobs, D.O.; Santacruz, L. AMPK and substrate availability regulate creatine transport in cultured cardiomyocytes. *Am. J. Physiol. Endocrinol. Metab.* **2011**, *300*, E870–E876. [CrossRef] [PubMed]

149. Santacruz, L.; Arciniegas, A.J.L.; Darrabie, M.; Mantilla, J.G.; Baron, R.M.; Bowles, D.E.; Mishra, R.; Jacobs, D.O. Hypoxia decreases creatine uptake in cardiomyocytes, while creatine supplementation enhances HIF activation. *Physiol. Rep.* **2017**, *5*. [CrossRef] [PubMed]

150. Erben, R.G.; Andrukhova, O. FGF23-Klotho signaling axis in the kidney. *Bone* **2017**, *100*, 62–68. [CrossRef] [PubMed]

151. Biber, J.; Hernando, N.; Forster, I.; Murer, H. Regulation of phosphate transport in proximal tubules. *Pflugers Arch.* **2009**, *458*, 39–52. [CrossRef] [PubMed]

152. Murer, H.; Forster, I.; Biber, J. The sodium phosphate cotransporter family SLC34. *Pflugers Arch.* **2004**, *447*, 763–767. [CrossRef] [PubMed]

153. Fraser, S.A.; Gimenez, I.; Cook, N.; Jennings, I.; Katerelos, M.; Katsis, F.; Levidiotis, V.; Kemp, B.E.; Power, D.A. Regulation of the renal-specific Na^+-K^+-$2Cl^-$ co-transporter NKCC2 by AMP-activated protein kinase (AMPK). *Biochem. J.* **2007**, *405*, 85–93. [CrossRef] [PubMed]

154. Cook, N.; Fraser, S.A.; Katerelos, M.; Katsis, F.; Gleich, K.; Mount, P.F.; Steinberg, G.R.; Levidiotis, V.; Kemp, B.E.; Power, D.A. Low salt concentrations activate AMP-activated protein kinase in mouse macula densa cells. *Am. J. Physiol. Renal. Physiol.* **2009**, *296*, F801–F809. [CrossRef] [PubMed]

155. Fraser, S.A.; Choy, S.-W.; Pastor-Soler, N.M.; Li, H.; Davies, M.R.P.; Cook, N.; Katerelos, M.; Mount, P.F.; Gleich, K.; McRae, J.L.; et al. AMPK couples plasma renin to cellular metabolism by phosphorylation of ACC1. *Am. J. Physiol. Renal. Physiol.* **2013**, *305*, F679–F690. [CrossRef] [PubMed]

156. Efe, O.; Klein, J.D.; LaRocque, L.M.; Ren, H.; Sands, J.M. Metformin improves urine concentration in rodents with nephrogenic diabetes insipidus. *JCI Insight* **2016**, *1*. [CrossRef] [PubMed]

157. Udwan, K.; Abed, A.; Roth, I.; Dizin, E.; Maillard, M.; Bettoni, C.; Loffing, J.; Wagner, C.A.; Edwards, A.; Feraille, E. Dietary sodium induces a redistribution of the tubular metabolic workload. *J. Physiol.* **2017**, *595*, 6905–6922. [CrossRef] [PubMed]

158. Siraskar, B.; Huang, D.Y.; Pakladok, T.; Siraskar, G.; Sopjani, M.; Alesutan, I.; Kucherenko, Y.; Almilaji, A.; Devanathan, V.; Shumilina, E.; et al. Downregulation of the renal outer medullary K^+ channel ROMK by the AMP-activated protein kinase. *Pflugers Arch.* **2013**, *465*, 233–245. [CrossRef] [PubMed]

159. Hallows, K.R.; Raghuram, V.; Kemp, B.E.; Witters, L.A.; Foskett, J.K. Inhibition of cystic fibrosis transmembrane conductance regulator by novel interaction with the metabolic sensor AMP-activated protein kinase. *J. Clin. Investig.* **2000**, *105*, 1711–1721. [CrossRef] [PubMed]

160. Morales, M.M.; Falkenstein, D.; Lopes, A.G. The Cystic Fibrosis Transmembrane Regulator (CFTR) in the kidney. *An. Acad. Bras. Ciênc.* **2000**, *72*, 399–406. [CrossRef] [PubMed]

161. Hallows, K.R.; McCane, J.E.; Kemp, B.E.; Witters, L.A.; Foskett, J.K. Regulation of channel gating by AMP-activated protein kinase modulates cystic fibrosis transmembrane conductance regulator activity in lung submucosal cells. *J. Biol. Chem.* **2003**, *278*, 998–1004. [CrossRef] [PubMed]

162. King, J.D.; Fitch, A.C.; Lee, J.K.; McCane, J.E.; Mak, D.-O.D.; Foskett, J.K.; Hallows, K.R. AMP-activated protein kinase phosphorylation of the R domain inhibits PKA stimulation of CFTR. *Am. J. Physiol. Cell Physiol.* **2009**, *297*, C94–C101. [CrossRef] [PubMed]

163. Kongsuphol, P.; Hieke, B.; Ousingsawat, J.; Almaca, J.; Viollet, B.; Schreiber, R.; Kunzelmann, K. Regulation of Cl⁻ secretion by AMPK in vivo. *Pflugers Arch.* **2009**, *457*, 1071–1078. [CrossRef] [PubMed]

164. Li, H.; Findlay, I.A.; Sheppard, D.N. The relationship between cell proliferation, Cl-secretion, and renal cyst growth: A study using CFTR inhibitors. *Kidney Int.* **2004**, *66*, 1926–1938. [CrossRef] [PubMed]

165. Takiar, V.; Nishio, S.; Seo-Mayer, P.; King, J.D.; Li, H.; Zhang, L.; Karihaloo, A.; Hallows, K.R.; Somlo, S.; Caplan, M.J. Activating AMP-activated protein kinase (AMPK) slows renal cystogenesis. *Proc. Natl. Acad. Sci. USA* **2011**, *108*, 2462–2467. [CrossRef] [PubMed]

166. Yuajit, C.; Muanprasat, C.; Gallagher, A.-R.; Fedeles, S.V.; Kittayaruksakul, S.; Homvisasevongsa, S.; Somlo, S.; Chatsudthipong, V. Steviol retards renal cyst growth through reduction of CFTR expression and inhibition of epithelial cell proliferation in a mouse model of polycystic kidney disease. *Biochem. Pharmacol.* **2014**, *88*, 412–421. [CrossRef] [PubMed]

167. Jeon, U.S. Kidney and calcium homeostasis. *Electrolyte Blood Press.* **2008**, *6*, 68–76. [CrossRef] [PubMed]

168. Na, T.; Peng, J.-B. TRPV5: A Ca^{2+} channel for the fine-tuning of Ca^{2+} reabsorption. *Handb. Exp. Pharmacol.* **2014**, *222*, 321–357. [PubMed]

169. Nurbaeva, M.K.; Schmid, E.; Szteyn, K.; Yang, W.; Viollet, B.; Shumilina, E.; Lang, F. Enhanced Ca^{2+} entry and Na^+/Ca^{2+} exchanger activity in dendritic cells from AMP-activated protein kinase-deficient mice. *FASEB J.* **2012**, *26*, 3049–3058. [CrossRef] [PubMed]

170. Lang, F.; Föller, M. Regulation of ion channels and transporters by AMP-activated kinase (AMPK). *Channels (Austin)* **2014**, *8*, 20–28. [CrossRef] [PubMed]

171. Bhavsar, S.K.; Schmidt, S.; Bobbala, D.; Nurbaeva, M.K.; Hosseinzadeh, Z.; Merches, K.; Fajol, A.; Wilmes, J.; Lang, F. AMPKα1-sensitivity of Orai1 and Ca^{2+} entry in T-lymphocytes. *Cell. Physiol. Biochem.* **2013**, *32*, 687–698. [CrossRef] [PubMed]

172. Mai, X.; Shang, J.; Liang, S.; Yu, B.; Yuan, J.; Lin, Y.; Luo, R.; Zhang, F.; Liu, Y.; Lv, X.; et al. Blockade of Orai1 Store-Operated Calcium Entry Protects against Renal Fibrosis. *J. Am. Soc. Nephrol.* **2016**, *27*, 3063–3078. [CrossRef] [PubMed]

173. Shigaev, A.; Asher, C.; Latter, H.; Garty, H.; Reuveny, E. Regulation of sgk by aldosterone and its effects on the epithelial Na⁺ channel. *Am. J. Physiol. Renal. Physiol.* **2000**, *278*, F613–F619. [CrossRef] [PubMed]

174. Zaika, O.; Mamenko, M.; Staruschenko, A.; Pochynyuk, O. Direct activation of ENaC by angiotensin II: Recent advances and new insights. *Curr. Hypertens. Rep.* **2013**, *15*, 17–24. [CrossRef] [PubMed]

175. Staruschenko, A. Regulation of transport in the connecting tubule and cortical collecting duct. *Compr. Physiol.* **2012**, *2*, 1541–1584. [PubMed]

176. Bhalla, V.; Hallows, K.R. Mechanisms of ENaC regulation and clinical implications. *J. Am. Soc. Nephrol.* **2008**, *19*, 1845–1854. [CrossRef] [PubMed]

177. Bhalla, V.; Oyster, N.M.; Fitch, A.C.; Wijngaarden, M.A.; Neumann, D.; Schlattner, U.; Pearce, D.; Hallows, K.R. AMP-activated kinase inhibits the epithelial Na⁺ channel through functional regulation of the ubiquitin ligase Nedd4-2. *J. Biol. Chem.* **2006**, *281*, 26159–26169. [CrossRef] [PubMed]

178. Hager, H.; Kwon, T.H.; Vinnikova, A.K.; Masilamani, S.; Brooks, H.L.; Frøkiaer, J.; Knepper, M.A.; Nielsen, S. Immunocytochemical and immunoelectron microscopic localization of α-, β-, and γ-ENaC in rat kidney. *Am. J. Physiol. Renal. Physiol.* **2001**, *280*, F1093–F1106. [CrossRef] [PubMed]

179. Myerburg, M.M.; King, J.D.; Oyster, N.M.; Fitch, A.C.; Magill, A.; Baty, C.J.; Watkins, S.C.; Kolls, J.K.; Pilewski, J.M.; Hallows, K.R. AMPK agonists ameliorate sodium and fluid transport and inflammation in cystic fibrosis airway epithelial cells. *Am. J. Respir. Cell Mol. Biol.* **2010**, *42*, 676–684. [CrossRef] [PubMed]

180. Almaça, J.; Kongsuphol, P.; Hieke, B.; Ousingsawat, J.; Viollet, B.; Schreiber, R.; Amaral, M.D.; Kunzelmann, K. AMPK controls epithelial Na⁺ channels through Nedd4-2 and causes an epithelial phenotype when mutated. *Pflugers Arch.* **2009**, *458*, 713–721. [CrossRef] [PubMed]

181. Carattino, M.D.; Edinger, R.S.; Grieser, H.J.; Wise, R.; Neumann, D.; Schlattner, U.; Johnson, J.P.; Kleyman, T.R.; Hallows, K.R. Epithelial sodium channel inhibition by AMP-activated protein kinase in oocytes and polarized renal epithelial cells. *J. Biol. Chem.* **2005**, *280*, 17608–17616. [CrossRef] [PubMed]

182. Yu, H.; Yang, T.; Gao, P.; Wei, X.; Zhang, H.; Xiong, S.; Lu, Z.; Li, L.; Wei, X.; Chen, J.; et al. Caffeine intake antagonizes salt sensitive hypertension through improvement of renal sodium handling. *Sci. Rep.* **2016**, *6*, 25746. [CrossRef] [PubMed]

183. Weixel, K.M.; Marciszyn, A.; Alzamora, R.; Li, H.; Fischer, O.; Edinger, R.S.; Hallows, K.R.; Johnson, J.P. Resveratrol inhibits the epithelial sodium channel via phopshoinositides and AMP-activated protein kinase in kidney collecting duct cells. *PLoS ONE* **2013**, *8*, e78019. [CrossRef] [PubMed]

184. Ho, P.-Y.; Li, H.; Pavlov, T.S.; Tuerk, R.D.; Tabares, D.; Brunisholz, R.; Neumann, D.; Staruschenko, A.; Hallows, K.R. β1Pix exchange factor stabilizes the ubiquitin ligase Nedd4-2 and plays a critical role in ENaC regulation by AMPK in kidney epithelial cells. *J. Biol. Chem.* **2018**, *293*, 11612–11624. [CrossRef] [PubMed]

185. Huang, D.Y.; Gao, H.; Boini, K.M.; Osswald, H.; Nürnberg, B.; Lang, F. In vivo stimulation of AMP-activated protein kinase enhanced tubuloglomerular feedback but reduced tubular sodium transport during high dietary NaCl intake. *Pflugers Arch.* **2010**, *460*, 187–196. [CrossRef] [PubMed]

186. Lazo-Fernández, Y.; Baile, G.; Meade, P.; Torcal, P.; Martínez, L.; Ibañez, C.; Bernal, M.L.; Viollet, B.; Giménez, I. Kidney-specific genetic deletion of both AMPK α-subunits causes salt and water wasting. *Am J. Physiol. Renal. Physiol.* **2017**, *312*, F352–F365. [CrossRef] [PubMed]

187. Alzamora, R.; Gong, F.; Rondanino, C.; Lee, J.K.; Smolak, C.; Pastor-Soler, N.M.; Hallows, K.R. AMP-activated protein kinase inhibits KCNQ1 channels through regulation of the ubiquitin ligase Nedd4-2 in renal epithelial cells. *Am. J. Physiol. Renal. Physiol.* **2010**, *299*, F1308–F1319. [CrossRef] [PubMed]

188. Vallon, V.; Grahammer, F.; Richter, K.; Bleich, M.; Lang, F.; Barhanin, J.; Völkl, H.; Warth, R. Role of KCNE1-dependent K$^+$ fluxes in mouse proximal tubule. *J. Am. Soc. Nephrol.* **2001**, *12*, 2003–2011. [PubMed]

189. Vallon, V.; Grahammer, F.; Volkl, H.; Sandu, C.D.; Richter, K.; Rexhepaj, R.; Gerlach, U.; Rong, Q.; Pfeifer, K.; Lang, F. KCNQ1-dependent transport in renal and gastrointestinal epithelia. *Proc. Natl. Acad. Sci. USA* **2005**, *102*, 17864–17869. [CrossRef] [PubMed]

190. Andersen, M.N.; Krzystanek, K.; Jespersen, T.; Olesen, S.-P.; Rasmussen, H.B. AMP-activated protein kinase downregulates Kv7.1 cell surface expression. *Traffic* **2012**, *13*, 143–156. [CrossRef] [PubMed]

191. Alesutan, I.; Föller, M.; Sopjani, M.; Dërmaku-Sopjani, M.; Zelenak, C.; Fröhlich, H.; Velic, A.; Fraser, S.; Kemp, B.E.; Seebohm, G.; et al. Inhibition of the heterotetrameric K$^+$ channel KCNQ1/KCNE1 by the AMP-activated protein kinase. *Mol. Membr. Biol.* **2011**, *28*, 79–89. [CrossRef] [PubMed]

192. Al-Bataineh, M.M.; Gong, F.; Marciszyn, A.L.; Myerburg, M.M.; Pastor-Soler, N.M. Regulation of proximal tubule vacuolar H$^+$-ATPase by PKA and AMP-activated protein kinase. *Am. J. Physiol. Renal. Physiol.* **2014**, *306*, F981–F995. [CrossRef] [PubMed]

193. Gong, F.; Alzamora, R.; Smolak, C.; Li, H.; Naveed, S.; Neumann, D.; Hallows, K.R.; Pastor-Soler, N.M. Vacuolar H$^+$-ATPase apical accumulation in kidney intercalated cells is regulated by PKA and AMP-activated protein kinase. *Am. J. Physiol. Renal. Physiol.* **2010**, *298*, F1162–F1169. [CrossRef] [PubMed]

194. Hallows, K.R.; Alzamora, R.; Li, H.; Gong, F.; Smolak, C.; Neumann, D.; Pastor-Soler, N.M. AMP-activated protein kinase inhibits alkaline pH- and PKA-induced apical vacuolar H$^+$-ATPase accumulation in epididymal clear cells. *Am. J. Physiol. Cell Physiol.* **2009**, *296*, C672–C681. [CrossRef] [PubMed]

195. Alzamora, R.; Al-bataineh, M.M.; Liu, W.; Gong, F.; Li, H.; Thali, R.F.; Joho-Auchli, Y.; Brunisholz, R.A.; Satlin, L.M.; Neumann, D.; et al. AMP-activated protein kinase regulates the vacuolar H$^+$-ATPase via direct phosphorylation of the A subunit (ATP6V1A) in the kidney. *Am. J. Physiol. Renal. Physiol.* **2013**, *305*, F943–F956. [CrossRef] [PubMed]

196. Alzamora, R.; Thali, R.F.; Gong, F.; Smolak, C.; Li, H.; Baty, C.J.; Bertrand, C.A.; Auchli, Y.; Brunisholz, R.A.; Neumann, D.; et al. PKA regulates vacuolar H$^+$-ATPase localization and activity via direct phosphorylation of the a subunit in kidney cells. *J. Biol. Chem.* **2010**, *285*, 24676–24685. [CrossRef] [PubMed]

197. McGuire, C.M.; Forgac, M. Glucose starvation increases V-ATPase assembly and activity in mammalian cells through AMP kinase and phosphatidylinositide 3-kinase/Akt signaling. *J. Biol. Chem.* **2018**, *293*, 9113–9123. [CrossRef] [PubMed]

198. Collins, M.P.; Forgac, M. Regulation of V-ATPase Assembly in Nutrient Sensing and Function of V-ATPases in Breast Cancer Metastasis. *Front. Physiol* **2018**, *9*, 902. [CrossRef] [PubMed]

199. Sands, J.M.; Klein, J.D. Physiological insights into novel therapies for nephrogenic diabetes insipidus. *Am. J. Physiol. Renal. Physiol.* **2016**, *311*, F1149–F1152. [CrossRef] [PubMed]

200. Denton, J.S.; Pao, A.C.; Maduke, M. Novel diuretic targets. *Am. J. Physiol. Renal. Physiol.* **2013**, *305*, F931–F942. [CrossRef] [PubMed]

201. Bech, A.P.; Wetzels, J.F.M.; Nijenhuis, T. Effects of sildenafil, metformin, and simvastatin on ADH-independent urine concentration in healthy volunteers. *Physiol. Rep.* **2018**, *6*, e13665. [CrossRef] [PubMed]

202. Klein, J.D.; Wang, Y.; Blount, M.A.; Molina, P.A.; LaRocque, L.M.; Ruiz, J.A.; Sands, J.M. Metformin, an AMPK activator, stimulates the phosphorylation of aquaporin 2 and urea transporter A1 in inner medullary collecting ducts. *Am. J. Physiol. Renal. Physiol.* **2016**, *310*, F1008–F1012. [CrossRef] [PubMed]

203. Al-bataineh, M.M.; Li, H.; Ohmi, K.; Gong, F.; Marciszyn, A.L.; Naveed, S.; Zhu, X.; Neumann, D.; Wu, Q.; Cheng, L.; et al. Activation of the metabolic sensor AMP-activated protein kinase inhibits aquaporin-2 function in kidney principal cells. *Am. J. Physiol. Renal. Physiol.* **2016**, *311*, F890–F900. [CrossRef] [PubMed]

204. Andersen, M.N.; Rasmussen, H.B. AMPK: A regulator of ion channels. *Commun. Integr. Biol.* **2012**, *5*, 480–484. [CrossRef] [PubMed]

205. Nickolas, T.L.; Jamal, S.A. Bone kidney interactions. *Rev. Endocr. Metab. Disord.* **2015**, *16*, 157–163. [CrossRef] [PubMed]

206. Graciolli, F.G.; Neves, K.R.; Barreto, F.; Barreto, D.V.; Dos Reis, L.M.; Canziani, M.E.; Sabbagh, Y.; Carvalho, A.B.; Jorgetti, V.; Elias, R.M.; et al. The complexity of chronic kidney disease-mineral and bone disorder across stages of chronic kidney disease. *Kidney Int.* **2017**, *91*, 1436–1446. [CrossRef] [PubMed]

207. Alesutan, I.; Munoz, C.; Sopjani, M.; Dërmaku-Sopjani, M.; Michael, D.; Fraser, S.; Kemp, B.E.; Seebohm, G.; Föller, M.; Lang, F. Inhibition of Kir2.1 (KCNJ2) by the AMP-activated protein kinase. *Biochem. Biophys. Res. Commun.* **2011**, *408*, 505–510. [CrossRef] [PubMed]

208. Derst, C.; Karschin, C.; Wischmeyer, E.; Hirsch, J.R.; Preisig-Müller, R.; Rajan, S.; Engel, H.; Grzeschik, K.-H.; Daut, J.; Karschin, A. Genetic and functional linkage of Kir5.1 and Kir2.1 channel subunits. *FEBS Lett.* **2001**, *491*, 305–311. [CrossRef]

209. Mia, S.; Munoz, C.; Pakladok, T.; Siraskar, G.; Voelkl, J.; Alesutan, I.; Lang, F. Downregulation of Kv1.5 K channels by the AMP-activated protein kinase. *Cell. Physiol. Biochem.* **2012**, *30*, 1039–1050. [CrossRef] [PubMed]

210. Bielanska, J.; Hernandez-Losa, J.; Perez-Verdaguer, M.; Moline, T.; Somoza, R.; Cajal, S.; Condom, E.; Ferreres, J.; Felipe, A. Voltage-Dependent Potassium Channels Kv1.3 and Kv1.5 in Human Cancer. *Curr. Cancer Drug Targets* **2009**, *9*, 904–914. [CrossRef] [PubMed]

211. Almilaji, A.; Munoz, C.; Elvira, B.; Fajol, A.; Pakladok, T.; Honisch, S.; Shumilina, E.; Lang, F.; Föller, M. AMP-activated protein kinase regulates hERG potassium channel. *Pflugers Arch.* **2013**, *465*, 1573–1582. [CrossRef] [PubMed]

212. Wadhwa, S.; Wadhwa, P.; Dinda, A.K.; Gupta, N.P. Differential expression of potassium ion channels in human renal cell carcinoma. *Int. Urol. Nephrol.* **2009**, *41*, 251–257. [CrossRef] [PubMed]

213. Munoz, C.; Sopjani, M.; Dërmaku-Sopjani, M.; Almilaji, A.; Föller, M.; Lang, F. Downregulation of the osmolyte transporters SMIT and BGT1 by AMP-activated protein kinase. *Biochem. Biophys. Res. Commun.* **2012**, *422*, 358–362. [CrossRef] [PubMed]

214. Yamauchi, A.; Nakanishi, T.; Takamitsu, Y.; Sugita, M.; Imai, E.; Noguchi, T.; Fujiwara, Y.; Kamada, T.; Ueda, N. In vivo osmoregulation of Na/myo-inositol cotransporter mRNA in rat kidney medulla. *J. Am. Soc. Nephrol.* **1994**, *5*, 62–67. [PubMed]

215. Rasola, A.; Galietta, L.J.; Barone, V.; Romeo, G.; Bagnasco, S. Molecular cloning and functional characterization of a GABA/betaine transporter from human kidney. *FEBS Lett.* **1995**, *373*, 229–233. [CrossRef]

216. Zhou, Y.; Holmseth, S.; Hua, R.; Lehre, A.C.; Olofsson, A.M.; Poblete-Naredo, I.; Kempson, S.A.; Danbolt, N.C. The betaine-GABA transporter (BGT1, slc6a12) is predominantly expressed in the liver and at lower levels in the kidneys and at the brain surface. *Am. J. Physiol. Renal. Physiol.* **2012**, *302*, F316–F328. [CrossRef] [PubMed]

217. Sopjani, M.; Alesutan, I.; Dërmaku-Sopjani, M.; Fraser, S.; Kemp, B.E.; Föller, M.; Lang, F. Down-regulation of Na$^+$-coupled glutamate transporter EAAT3 and EAAT4 by AMP-activated protein kinase. *J. Neurochem.* **2010**, *113*, 1426–1435. [CrossRef] [PubMed]

218. Hu, Q.X.; Ottestad-Hansen, S.; Holmseth, S.; Hassel, B.; Danbolt, N.C.; Zhou, Y. Expression of Glutamate Transporters in Mouse Liver, Kidney, and Intestine. *J. Histochem. Cytochem.* **2018**, *66*, 189–202. [CrossRef] [PubMed]

219. Schmitt, R.; Ellison, D.H.; Farman, N.; Rossier, B.C.; Reilly, R.F.; Reeves, W.B.; Oberbäumer, I.; Tapp, R.; Bachmann, S. Developmental expression of sodium entry pathways in rat nephron. *Am. J. Physiol. Renal. Physiol.* **1999**, *276*, F367–F381. [CrossRef]

220. Kréneisz, O.; Benoit, J.P.; Bayliss, D.A.; Mulkey, D.K. AMP-activated protein kinase inhibits TREK channels. *J. Physiol.* **2009**, *587*, 5819–5830. [CrossRef] [PubMed]

221. Gu, W.; Schlichthörl, G.; Hirsch, J.R.; Engels, H.; Karschin, C.; Karschin, A.; Derst, C.; Steinlein, O.K.; Daut, J. Expression pattern and functional characteristics of two novel splice variants of the two-pore-domain potassium channel TREK-2. *J. Physiol.* **2002**, *539*, 657–668. [CrossRef] [PubMed]

222. Föller, M.; Jaumann, M.; Dettling, J.; Saxena, A.; Pakladok, T.; Munoz, C.; Ruth, P.; Sopjani, M.; Seebohm, G.; Rüttiger, L.; et al. AMP-activated protein kinase in BK-channel regulation and protection against hearing loss following acoustic overstimulation. *FASEB J.* **2012**, *26*, 4243–4253. [CrossRef] [PubMed]

223. Rabjerg, M.; Oliván-Viguera, A.; Hansen, L.K.; Jensen, L.; Sevelsted-Møller, L.; Walter, S.; Jensen, B.L.; Marcussen, N.; Köhler, R. High expression of KCa3.1 in patients with clear cell renal carcinoma predicts high metastatic risk and poor survival. *PLoS ONE* **2015**, *10*, e0122992. [CrossRef] [PubMed]

224. Wyatt, C.N.; Mustard, K.J.; Pearson, S.A.; Dallas, M.L.; Atkinson, L.; Kumar, P.; Peers, C.; Hardie, D.G.; Evans, A.M. AMP-activated protein kinase mediates carotid body excitation by hypoxia. *J. Biol. Chem.* **2007**, *282*, 8092–8098. [CrossRef] [PubMed]

225. Yoshida, H.; Bao, L.; Kefaloyianni, E.; Taskin, E.; Okorie, U.; Hong, M.; Dhar-Chowdhury, P.; Kaneko, M.; Coetzee, W.A. AMP-activated protein kinase connects cellular energy metabolism to KATP channel function. *J. Mol. Cell. Cardiol.* **2012**, *52*, 410–418. [CrossRef] [PubMed]

226. Lim, A.; Park, S.-H.; Sohn, J.-W.; Jeon, J.-H.; Park, J.-H.; Song, D.-K.; Lee, S.-H.; Ho, W.-K. Glucose deprivation regulates KATP channel trafficking via AMP-activated protein kinase in pancreatic β-cells. *Diabetes* **2009**, *58*, 2813–2819. [CrossRef] [PubMed]

227. Chang, T.-J.; Chen, W.-P.; Yang, C.; Lu, P.-H.; Liang, Y.-C.; Su, M.-J.; Lee, S.-C.; Chuang, L.-M. Serine-385 phosphorylation of inwardly rectifying K⁺ channel subunit (Kir6.2) by AMP-dependent protein kinase plays a key role in rosiglitazone-induced closure of the K(ATP) channel and insulin secretion in rats. *Diabetologia* **2009**, *52*, 1112–1121. [CrossRef] [PubMed]

228. Tan, X.-H.; Zheng, X.-M.; Yu, L.-X.; He, J.; Zhu, H.-M.; Ge, X.-P.; Ren, X.-L.; Ye, F.-Q.; Bellusci, S.; Xiao, J.; et al. Fibroblast growth factor 2 protects against renal ischaemia/reperfusion injury by attenuating mitochondrial damage and proinflammatory signalling. *J. Cell. Mol. Med.* **2017**, *21*, 2909–2925. [CrossRef] [PubMed]

229. Klein, H.; Garneau, L.; Trinh, N.T.N.; Privé, A.; Dionne, F.; Goupil, E.; Thuringer, D.; Parent, L.; Brochiero, E.; Sauvé, R. Inhibition of the KCa3.1 channels by AMP-activated protein kinase in human airway epithelial cells. *Am. J Physiol. Cell Physiol.* **2009**, *296*, C285–C295. [CrossRef] [PubMed]

230. Huang, C.; Shen, S.; Ma, Q.; Chen, J.; Gill, A.; Pollock, C.A.; Chen, X.-M. Blockade of KCa3.1 ameliorates renal fibrosis through the TGF-β1/Smad pathway in diabetic mice. *Diabetes* **2013**, *62*, 2923–2934. [CrossRef] [PubMed]

231. Takimoto, M.; Takeyama, M.; Hamada, T. Possible involvement of AMPK in acute exercise-induced expression of monocarboxylate transporters MCT1 and MCT4 mRNA in fast-twitch skeletal muscle. *Metab. Clin. Exp.* **2013**, *62*, 1633–1640. [CrossRef] [PubMed]

232. Becker, H.M.; Mohebbi, N.; Perna, A.; Ganapathy, V.; Capasso, G.; Wagner, C.A. Localization of members of MCT monocarboxylate transporter family Slc16 in the kidney and regulation during metabolic acidosis. *Am. J. Physiol. Renal. Physiol.* **2010**, *299*, F141–F154. [CrossRef] [PubMed]

233. Fisel, P.; Kruck, S.; Winter, S.; Bedke, J.; Hennenlotter, J.; Nies, A.T.; Scharpf, M.; Fend, F.; Stenzl, A.; Schwab, M.; et al. DNA methylation of the SLC16A3 promoter regulates expression of the human lactate transporter MCT4 in renal cancer with consequences for clinical outcome. *Clin. Cancer Res.* **2013**, *19*, 5170–5181. [CrossRef] [PubMed]

234. Pieri, M.; Christian, H.C.; Wilkins, R.J.; Boyd, C.A.R.; Meredith, D. The apical (hPepT1) and basolateral peptide transport systems of Caco-2 cells are regulated by AMP-activated protein kinase. *Am. J. Physiol. Gastrointest. Liver Physiol.* **2010**, *299*, G136–G143. [CrossRef] [PubMed]

235. Shen, H.; Smith, D.E.; Yang, T.; Huang, Y.G.; Schnermann, J.B.; Brosius, F.C. Localization of PEPT1 and PEPT2 proton-coupled oligopeptide transporter mRNA and protein in rat kidney. *Am. J. Physiol. Renal. Physiol.* **1999**, *276*, F658–F665. [CrossRef]
236. Shen, B.; Zhu, J.; Zhang, J.; Jiang, F.; Wang, Z.; Zhang, Y.; Li, J.; Huang, D.; Ke, D.; Ma, R.; et al. Attenuated mesangial cell proliferation related to store-operated Ca^{2+} entry in aged rat: The role of STIM 1 and Orai 1. *Age* **2013**, *35*, 2193–2202. [CrossRef] [PubMed]

International Journal of
Molecular Sciences

MDPI

Review

Implications of AMPK in the Formation of Epithelial Tight Junctions

Pascal Rowart [1], Jingshing Wu [2], Michael J. Caplan [2] and François Jouret [1,3,*]

[1] Groupe Interdisciplinaire de Génoprotéomique Appliquée (GIGA), Cardiovascular Sciences, University of Liège (ULiège), Avenue de L'Hôpital 11, 4000 Liège, Belgium; pascal.rowart@uliege.be

[2] Department of Cellular and Molecular Physiology, Yale School of Medicine, New Haven, CT 06520, USA; jingshing.wu@yale.edu (J.W.); michael.caplan@yale.edu (M.J.C.)

[3] Division of Nephrology, Centre Hospitalier Universitaire de Liège (CHU of Liège), University of Liège (CHU ULiège), 13-B4000 Liège, Belgium

* Correspondence: francois.jouret@chuliege.be; Tel.: +32-(4)-366-25-40; Fax: +32-(4)-366-21-37

Received: 12 June 2018; Accepted: 9 July 2018; Published: 13 July 2018

Abstract: Tight junctions (TJ) play an essential role in the epithelial barrier. By definition, TJ are located at the demarcation between the apical and baso-lateral domains of the plasma membrane in epithelial cells. TJ fulfill two major roles: (i) TJ prevent the mixing of membrane components; and (ii) TJ regulate the selective paracellular permeability. Disruption of TJ is regarded as one of the earliest hallmarks of epithelial injury, leading to the loss of cell polarity and tissue disorganization. Many factors have been identified as modulators of TJ assembly/disassembly. More specifically, in addition to its role as an energy sensor, adenosine monophosphate-activated protein kinase (AMPK) participates in TJ regulation. AMPK is a ubiquitous serine/threonine kinase composed of a catalytic α-subunit complexed with regulatory β-and γ-subunits. AMPK activation promotes the early stages of epithelial TJ assembly. AMPK phosphorylates the adherens junction protein afadin and regulates its interaction with the TJ-associated protein zonula occludens (ZO)-1, thereby facilitating ZO-1 distribution to the plasma membrane. In the present review, we detail the signaling pathways up-and down-stream of AMPK activation at the time of Ca^{2+}-induced TJ assembly.

Keywords: AMPK; tight junctions; epithelial cells; ZO-1; par complex; MDCK; nectin-afadin; adherent junctions

---

The correct establishment and maintenance of cell-cell contacts and cell polarity in multicellular organisms are crucial for normal cell physiology and tissue homeostasis [1,2]. Epithelial cells form barriers that protect and separate multicellular organisms from the external environment. Such compartmentalization provides different internal and external environments with specialized functions [3,4]. In addition to its compartmentalization role, epithelial cell membrane integrity plays a major role in the defense against pathological organisms, as well as against disease development. Hence, the loss of cell polarity and membrane disruption are observed in cancer, acute kidney injury, apoptosis, and infection, as well as at the initial stages of some central nervous system neoplasia [5–9].

To maintain these two distinct regions and protect the organism, epithelial cells are sealed together by a junctional complex formed by four main components located along the apico-basal axis, as follows: (1) TJ; (2) adherens junctions (AJ); (3) gap junctions; and (4) desmosomes. Disturbances in the formation and maintenance of TJ are observed in many pathological conditions, including cancer [10,11]. For this reason, a better understanding of their regulation may lead to novel targeted therapies [12]. Many factors have been identified as modulators of the assembly and disassembly of TJ. AMP-Activated protein kinase (AMPK) has emerged as one of these. Indeed, epithelial cells are involved in electrolyte and fluid transport across the apical and basal membrane, which consume a major part of an epithelial cell's internal energy currency, namely its stores of

Adenosine triphosphate (ATP) [13]. AMPK is the main energy sensor in all eukaryotic cells that regulate their levels of ATP. During cellular stresses such as hypoxia, starvation, glucose deprivation, or muscle contraction, the ratio of ADP/ATP or AMP/ATP will change. To restore the cellular energy balance, AMPK will promote catabolic pathways and inhibit anabolic pathways [14]. Interestingly enough, the depletion of ATP results in the rapid dislocation of cellular tight junctions (TJ), whereas ATP repletion induces a recovery of tight-junction integrity [15]. As an example, in case of kidney ischemia, TJ disassembly between proximal tubule cells allows the paracellular backleak of the ultrafiltrate into the interstitium, which in turn aggravates the renal hypoperfusion [16]. Zhang et al. were the first to demonstrate a role of AMPK in the regulation of epithelial tight junction assembly and disassembly. Note that the role of AMPK in TJ regulation appears to be independent of intracellular [ATP] levels. These observations opened new investigations into the mechanisms through which AMPK serves at the crossroads between the regulation of cellular energy and TJ homeostasis [17].

1. TJ Are Multiprotein Complexes

The concept of TJ (also known as *zonula occludens*) emerged in 1963 with Farquhar and Palade's experiments demonstrating, by electron microscopy, the regular occurrence in various rat and pig epithelia of a characteristic junctional complex whose components bear a relationship to each other and to the lumen of the organ [18]. Briefly, an individual TJ strand is associated laterally with another TJ strand on adjacent cells to form paired strands, where the extracellular space at this region is completely obliterated [19]. TJ are the most apical constituent of the junctional complex with AJ immediately underneath [20]. TJ can be thought of as gaskets that define and seal the most apical border of cell-cell contacts. For several years following their morphological observation, TJ were further investigated and major progress was achieved by Stevenson et al. in 1986. They explored the biochemical structure of TJ and identified ZO1 as one of their protein components [21], which has gone on to serve as a canonical marker for the assessment of TJ formation.

TJ have two main roles. First, they demarcate the apical and basolateral domains of polarized cells by acting like a fence [22]. TJ prevent membrane proteins from diffusing freely between the two membrane compartments. This function is absolutely required in order for apical and basolateral domains to maintain their distinct lipid and protein compositions [23]. Second, they create a physiological and structural paracellular barrier that regulates the selective passage and exchange of molecules [24–27]. In many situations, various materials are selectively transported across cellular sheets, and this occurs either by direct transcellular transport or by paracellular flux through TJ. The selective passage of these components through TJ is mediated by aqueous pores whose structures have yet to be fully defined [28]. In addition to these functions, TJ are connected to signaling pathways that communicate with the cell cytoplasm and subcellular components [29]. To be able to perform all of these diverse functions, TJ must possess a complicated architecture based upon a multiprotein complex that is composed of more than 40 proteins that are classified either as transmembrane proteins or as cytoplasmic proteins bound to the actin cytoskeleton [30] (Figure 1).

1.1. Transmembrane Proteins

The three main transmembrane proteins are claudins [31], occludins [32], and Junctional Adhesion Molecules (JAMs) [33]. Both claudins and occludins contain four transmembrane regions (Tetraspans) with their amino-and carboxyl-terminal ends directed to the cytoplasm [34]. There are two isoforms of occludin generated by alternative splicing [35], whereas claudins comprise more than 24 members [36]. The extended C-terminus of claudins and occludins have been shown to be essential for interactions with the soluble cytoskeletal ZO proteins. These interactions mediate the association of the ZO proteins with the plasma membrane, which is an obligate step in the formation of TJ [37]. In vivo studies revealed that claudins are more necessary for the structural integrity of TJ than occludins. In occludin knock-out (KO) mice, TJ were morphologically intact [38] in comparison to claudin-KO mice, which showed the disappearance of TJ strands [39,40]. These results indicated that claudin is required for the formation of TJ strands, and suggested that occludin is rather required for TJ stability. The last category of transmembrane proteins are JAMs. They have a single transmembrane domain and their extracellular portion is folded into two immunoglobulin-like domains. Three isoforms are currently known: JAM-A [33], JAM-B, and JAM-C [41,42]. The three JAMs are co-distributed in epithelial cells with the ZO-1 protein. Evidence of a role for JAMs in TJ formation is supported by several studies. Overexpression of JAM proteins enhances the recruitment of TJ components and leads to the increased accumulation of ZO-1 and occludin [43].

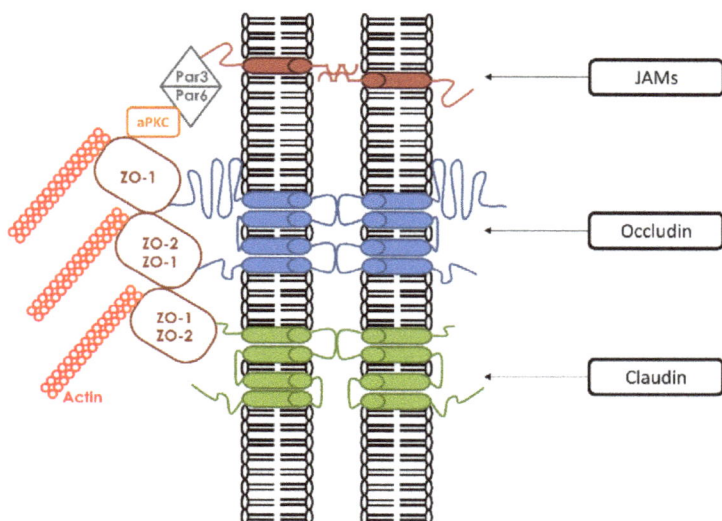

Figure 1. Molecular components of tight junctions. They are composed of three families of transmembrane proteins that include Occludins, Claudins, and Junctional Adhesion Molecule (JAMs). Every transmembrane protein is associated with cytoplasmic adaptor proteins such as the zonula occludens proteins ZO-1 and ZO-2. These interactions are mediated through their cytosolic tails. Other interactions occur between ZO proteins and additional cytoplasmic proteins (Par6, Par3, and aPKC). ZO proteins are also connected with the actin cytoskeleton.

1.2. Cytoplasmic Proteins

The "cytoplasmic plaque of TJ" serves as a link between the transmembrane TJ proteins and the actin cytoskeleton [44]. Most of the cytoplasmic proteins can attach to the TJ plaque via PDZ-domains. A PDZ domain is a common structural domain that interacts with stereotypical sequences embedded within the C-terminal regions of transmembrane proteins. PDZ domain-containing proteins can interact with other PDZ domain-containing proteins and through these multiplexed associations can anchor TJ membrane proteins to the cytoskeleton. PDZ domains are implicated in a variety of signaling mechanisms [45]. The two most important PDZ-containing proteins identified at the TJ plaque are the zonula-occludens-(ZO)1 proteins, which belong to the membrane-associated guanylate kinase family (MAGUK), and the partitioning defective proteins (Par), members of the Par3/aPKC/Par6 polarity complex.

The MAGUK family includes three structurally related proteins: ZO-1, ZO-2, and ZO-3. ZO-1 was the first to be associated with TJ [21,46]. They all share a similar structural organization, with an N-terminal region containing three PDZ domains. In vitro as well as in vivo analyses showed that the first PDZ domain (PDZ1) of the three ZO proteins has binding affinities for the C-terminal domains of claudins [37]. This PDZ-dependent interaction with ZO proteins promotes the proper targeting of claudins to the TJ. Furthermore, ZO-1/ZO-2 knock-down (KD) cells show disruptions in claudin localization associated with barrier dysfunction [47–49]. The second PDZ domain (PDZ2) is responsible for dimerization with other ZO proteins [50]. The third PDZ domain (PDZ3) is associated with the interaction with JAM-A [43]. Surprisingly, these three PDZ domains are not sufficient for the recruitment of ZO proteins to TJ [51]. In addition to these three PDZ domains, ZO proteins also have other regions that are required for their recruitment to TJ. These include SH3 and GUK domains, which can interact with afadin and occludin, respectively [52].

Given the fact that occludin-deficient cells are able to form normal TJ, with the appropriate distribution of ZO-1 [53], alternative interactions must necessarily be involved. One possibility would be the interaction of ZO proteins with α-catenin, a cytoplasmic actin-binding protein that associates with the β-catenin/E-cadherin complex at AJ [54,55]. However, α-catenin-deficient cells are able to recruit ZO-1 to the plasma membrane, indicating that this interaction is not critical. The final hypothesis focuses on the nectin/afadin complex and the specific interaction of the proline-rich regions of afadin with the SH3 domain of ZO proteins. Interaction between afadin and ZO-1 during the formation of cell-cell junctions in MDCK cells has been reported [56]. This interaction is principally observed before TJ are formed. During and after the formation of TJ, ZO proteins appear to be dissociated from afadin, and afadin becomes associated with nectin at AJ [57–60]. The association between TJ components and the AJ complex (α-catenin/β-catenin/E-cadherin and nectin/afadin) thus appears to catalyze the deposition of TJ proteins to the cell surface in the early steps of TJ formation. In summary, ZO proteins are essential for TJ formation, as well as for the linking of TJ membrane proteins to the actin cytoskeleton [51,61].

Besides the MAGUK family, members of the Par family play key roles in TJ assembly. Observations on the asymmetric divisions occurring in the *C. elegans* zygote led to the discovery of six Par proteins by Kemphues et al. in 1988, which are essential for the partitioning of early determinants and the development of embryonic polarity [62]. Only Par3 (Also known as Bazooka) and Par6 were found to be colocalized in *C. elegans* embryos. They both contain a PDZ domain and are able to bind to each other [63]. Par6 contains both N-terminal and C-terminal regions and three conserved domains for their interactions with other members of this complex. Its first domain PB1 (Phox/Bem 1) is located at the N-terminal region and is essential for the interaction of Par6 with atypical protein kinase C (aPKC). The second is Cdc42/Rac interaction binding (CRIB) and can be directly modulated by the cell division control protein 42 (Cdc42). The third one is a PDZ domain located at the C-terminal region. Accumulating evidence showed that Par3 and Par6 function together with aPKC [64]. The PB1 domain of Par6 binds the PB1 domain of aPKC to form a heterodimer. Par3 also contains N-terminal and C-terminal regions separated by three central PDZ domains. This tripartite Par3/aPKC/Par6 is known as the "Par complex" and is conserved from worms to vertebrates [65,66]. This interaction is a membrane targeting signal. In the Par complex, Par3 associates with the Par6/aPKC heterodimer by a PDZ-PDZ domain interaction at the onset of epithelial polarization [67]. Several molecules, such as nectin and JAMs, can bind the PDZ1 domain of Par3.

2. AMPK Is a Key Regulator of Energy Balance

Each accomplish energy-requiring tasks through the hydrolysis of ATP into ADP, which serves as their immediate source of energy [68]. Maintaining an adequate supply of energy is an essential requirement for survival, which means that ATP levels must be kept at a sufficient concentration. The main sensor of cellular energy status is the AMP-activated protein kinase (AMPK). When ATP levels fall, its main function is to switch off anabolic and biosynthetic pathways that consume ATP and to switch on catabolic pathways that produce ATP [69] (Figure 2). When the overall energy levels in cells decrease due to increased demands or decreased availability of substrates, AMPK gets activated through a combination of phosphorylation by upstream kinases and/or direct activation by AMP and ADP [70,71].

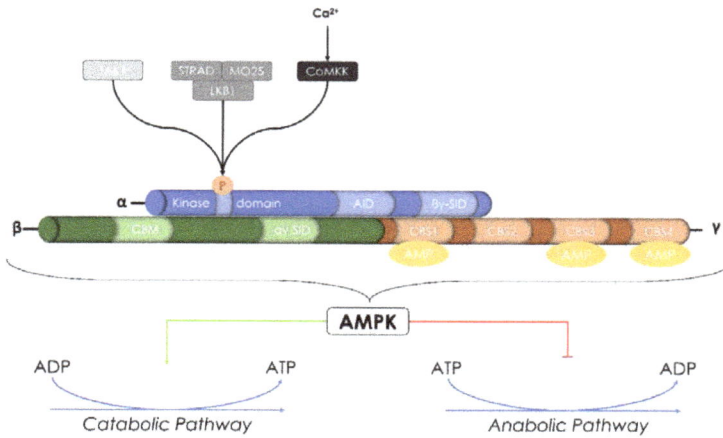

Figure 2. Structure and activation of AMP-activated protein kinase (AMPK). AMPK is a heterotrimeric α-β-γ serine/threonine kinase. It is made up of a catalytic α-subunit complexed with regulatory β-and γ-subunits. It can be activated through the phosphorylation of Thr-172 by two main upstream kinases: Ca^{2+}-Calmodulin Kinase Kinase (CaMKK) and Liver Kinase B1 (LKB1). Transforming growth factor-β-activated kinase (TAK1) was also described as a new AMPK regulatory kinase. In addition to its activation by upstream kinases, AMPK can also be allosterically activated by AMP. Once activated, AMPK responds to changes in the level of Adenosine triphosphate (ATP) by switching off either anabolic and biosynthetic pathways consuming ATP or switching on catabolic pathways that produce ATP.

2.1. AMPK: Structure and Regulation

AMPK is a heterotrimeric serine/threonine kinase. It is made up of a catalytic α-subunit complexed with regulatory β-and γ-subunits [72]. There are 12 unique heterotrimeric combinations of AMPK. In mammals, the α-subunit is encoded by two isoforms, and the β-and γ-subunits are encoded by two and three isoforms, respectively (α1, α2, β1, β2, γ1, γ2, and γ3). All these isoforms have differential tissue-specific expression and activity [73–75].

The α-subunit possesses an *N*-terminal kinase domain that mediates its catalytic activity and a C-terminal subunit-interacting domain that plays a role in the interaction with β-and γ-subunits (βγ-subunit interacting domain-βγ-SID) [76]. The α1-subunit is expressed in many organs (kidney, heart, brain, spleen, liver, lung, and skeletal muscle), unlike α2-subunit, which is essentially expressed primarily in skeletal muscle [77]. In addition to their different tissue/organ expression, α1- and α2-subunits are differentially expressed within the cell. Indeed, α1 is predominantly expressed in the cytosol, whereas α2 is localized to the nucleus in periods of high energy demand [78]. The β-subunit contains a central glycogen-binding domain CBM (carbohydrate-binding module) that permits the interaction of AMPK with glycogen particles and a C-terminal region essential for the assembly of the α β γ complex [79]. The γ-subunit contains four cystathionine-β-synthase (CBS) tandem sequence repeats that fold to form two "Bateman domains" and can bind AMP or ATP to regulate the AMPK activation [80] (Figure 2).

Phosphorylation of Thr-172 in the α-subunit catalytic loop is the main pathway that produces the activation of AMPK. Nevertheless, there are three major mechanisms responsible for the AMPK activation: (i) upstream kinases [71]; (ii) the increase of [AMP] and/or [ADP] [81,82]; and (iii) direct binding to the γ-subunit of AMP for the allosteric activation of AMPK [83]. Beside its activation by allosteric AMP binding and upstream kinases, AMPK has been reported to have an autophosphorylation ability at β-subunit Thr-148 [84,85].

2.2. AMPK: Upstream Kinase and Substrates

Two major upstream AMPK-regulatory kinases have been discovered that are both serine/threonine kinases. The first one is the liver-kinase-B1 (LKB1) and the second one is the Ca^{2+}/calmodulin-dependent kinase kinase (CaMKKβ). LKB1 and CaMKK can activate AMPK in response to energy stress as signaled by elevated AMP levels or to increases of intracellular $[Ca^{2+}]$ levels in an AMP-independent manner, respectively. Other studies have also demonstrated that transforming growth factor-β-activated kinase (TAK1) may represent a third AMPK kinase.

LKB1 phosphorylates and activates AMPK in vitro following increased cellular [AMP] levels [86]. LKB1 activity requires the binding of the scaffolding-related adaptor mouse protein 25 (MO25) and the pseudokinase STe-20 Related ADaptor (STRAD) via the formation of the holoenzyme complex [86]. In cells lacking the expression of LKB1, the activation of AMPK in response to the increase of the AMP/ATP ratio is abolished, suggesting that LKB1 is required for the AMPK phosphorylation when [AMP] increases in the cell [86]. Other studies demonstrated that, in certain circumstances, AMPK can be activated, even in the absence of LKB1 [87,88]. Hence, CaMKK emerged as another main AMPK kinase [89]. In contrast to LKB1, the AMPK phosphorylation by CaMKK does not require a disturbance of the ATP/AMP ratio, but rather an increase in intracellular Ca^{2+} [90]. The addition of the Ca^{2+} ionophore A23187 activates AMPK-via the phosphorylation of Thr-172-approximately 10-fold more in cells expressing a kinase-inactive mutant of LKB1 compared to wild-type cells. Conversely, the AMPK activation by Ca^{2+} ionophore A23187 was abolished by the CaMKK inhibitor STO-609 [91]. Other studies confirmed this result and observed that the overexpression of CaMKK increases AMPK activity, whereas the inhibition of CaMKK reduces AMPK activity [92]. These results suggest a physiological role of LKB1 and CaMKK as AMPK regulatory kinases in mammalian cells [92].

AMPK is a modulator of several pathways [93–95]. For example, AMPK negatively regulates two enzymes involved in lipid synthesis: HMGCR (3-hydroxy-3-methylglutaryl coenzyme A reductase) [96] and ACC (acetyl-CoA carboxylase) [97]. AMPK also exerts a potent effect on glucose metabolism. Glucose uptake is facilitated through the translocation of glucose transporter 4 (GLUT4) to the cell membrane and also through the regulation of *GLUT4* gene expression in response to AMPK activation [98]. AMPK also regulates glycogen metabolism. AMPK activation phosphorylates and decreases the activation of glycogen synthase (GS), thus reducing glycogen synthesis [99]. Thus, AMPK activation reduces the production of stored forms of metabolic energy and diminishes the activity of energy utilizing pathways, while it increases the capacity of cells to import energetic precursors and to produce ATP. Kishton et al. showed that AMPK actively restrained aerobic glycolysis in cells through the inhibition of mTORC1, while promoting oxidative metabolism and mitochondrial Complex I activity producing ATP [100]. The inhibition of AMPK-related kinase 5 (ARK5), an upstream regulator of AMPK, leads to a collapse of cellular ATP levels. Proteomics highlighted the down-regulation of multiple subunits of complexes I, III, and IV of the mitochondrial respiratory chain following the depletion of ARK5 [101]. These studies suggest a role of AMPK in the production of ATP by the mitochondrial respiratory chain. AMPK facilitates the assembly of TJ.

3. AMPK and ZO-1

Zhang et al. [17] demonstrated in 2006 that AMPK could regulate the assembly of epithelial TJ in the MDCK cell line. The authors used a Ca^{2+} switch-based model described in 1978 [102] to decipher the role of AMPK in TJ assembly. Cell-cell adhesion, as well as TJ integrity in polarized epithelial cells, is rapidly lost when the Ca^{2+} is removed from the extracellular medium. On the other hand, the re-addition of Ca^{2+} into the culture medium induces the rapid assembly of cell-cell contacts and subsequent TJ formation. Depletion of Ca^{2+} from the medium causes ZO-1 to translocate from the cell periphery to the cytoplasm. Upon the re-addition of Ca^{2+}, ZO-1 moves back to the TJ. With this model, the authors showed that AMPK is phosphorylated during the Ca^{2+}-induced TJ assembly, while the total amount of AMPK remains unchanged. They also examined the AMPK activity by measuring the phosphorylated form of ACC, one of the principle AMPK substrates. They found an eight-fold increase in pACC following a Ca^{2+}-switch. It is important to note that such AMPK phosphorylation and activation was not attributable to changes in cellular [ATP] levels during a Ca^{2+}-switch. To evaluate the potential effect of AMPK in TJ formation, the authors monitored the time course of ZO-1 relocation to cell-cell junctions with or without the chemical AMPK activator AICAR (which acts an AMP mimic) added at the time of the Ca^{2+} switch. The amount of ZO-1 relocated to TJ in the presence of AICAR during the Ca^{2+} switch was higher in comparison to a classic Ca^{2+} switch. Furthermore, the addition of AICAR in Ca^{2+}-depleted medium was sufficient to activate AMPK and to accelerate ZO-1 relocation to TJ [17]. They also measured the paracellular flux of 70-kDa dextran in MDCK monolayers during the Ca^{2+}-switch in the presence or absence of AICAR. The presence of AICAR led to a slight but statistically significant decrease in the dextran flux rate across the monolayers. Similarly to Zhang et al., Peng et al. also measured the paracellular flux by measuring inulin in intestinal epithelial cells with or without AICAR. Incubation with AICAR led to a significant decrease in the flux rate across the cell monolayers. These effects were also abolished by the AMPK inhibitor, Compound C. These studies indicated that the backleak effect is decreased by the activation of AMPK, independently of [ADP/ATP] changes. In MDCK cells expressing dominant negative AMPK, the early initiation of TJ assembly was compromised. Still, normal-appearing TJ could eventually form in AMPK-deficient cells over time, suggesting that AMPK activation supports the initiation of TJ formation, but other factors, including extracellular Ca^{2+}, are required for the long-term stabilization of TJ. Shortly after the publication of the Zhang et al. study, Zheng et al. confirmed these findings and in addition showed that the activation of AMPK in response to the initiation of junction formation requires LKB1. They also generated MDCK cell lines expressing a kinase-dead mutant form of AMPKα1 and monitored the effect of Ca^{2+}-switch on TER, a measurement for the paracellular barrier function and integrity of TJ [103]. Expression of kinase-dead AMPK significantly decreased the peak level of TER, meaning that the formation of functional TJ is suppressed in the absence of AMPK [104].

4. AMPK and Afadin-Nectin System

In the above-detailed model, the nectin-afadin system is required for the deposition of junction components induced by AMPK activation. The involvement of the nectin-afadin complex in cell adhesion has been described in AJ formation [105]. Still, afadin KD cells are not able to induce the relocalization of ZO-1 and occludin at TJ sites following the addition of AICAR in low Ca^{2+} medium. AMPK therefore appears to be connected to afadin in the Ca^{2+}-independent TJ formation. Furthermore, immunoprecipitation between afadin and AMPK revealed that afadin is a direct substrate of AMPK. The addition of Compound-C inhibited the phosphorylation of afadin, whereas the afadin signal was increased without the AMPK inhibitor [106]. Since afadin directly binds to ZO-1 [56], the authors investigated whether AMPK activation increases the interaction between afadin and ZO-1, thereby facilitating the assembly of TJ. Immunoprecipitation, with or without AICAR during a Ca^{2+} switch, revealed an enhanced interaction between these two proteins after Ca^{2+}-switch and even more in the case of AICAR exposure. These results suggested that AMPK activation might facilitate TJ assembly by phosphorylating afadin and inducing its association with ZO-1.

5. AMPK Effectors

Recent research investigated the potential of AMPK effectors to preserve the epithelial architecture. The first study focused on the multimodular polarity scaffold protein GIV (G-alpha interacting vesicle associated protein) [107]. This protein has been demonstrated to regulate cell polarity and morphogenesis [108], as well as cell-cell junction formation through its ability to bind Par3 [109] and the Cadherin-catenin complex [110]. A role for AMPK-mediated phosphorylation of GIV at serine 245 when [ATP] levels decreased was suggested. The phosphorylation of GIV at ps 245 triggered its localization to TJ by increasing its ability to bind TJ-associated microtubules and AJ-localized protein complexes. The addition of Compound-C inhibited AMPK-mediated phosphorylation of GIV and induced the destabilization of TJ and the reduction of TER. On the other hand, metformin (an AMPK activator) and AICAR triggered GIV phosphorylation and stabilized TJ, with subsequent enhanced TER [107].

In 2005, butyrate emerged as a new candidate to promote enhanced intestinal barrier function as reflected by increases in TER in vitro [111]. A few years later, Peng and co-workers explored whether the effect of butyrate on the intestinal epithelial barrier is related to AMPK. Using a model of Ca^{2+} switch, with or without the addition of butyrate, they demonstrated that the amount of pAMPK, as well as pACC, increased after the treatment with butyrate in a time-dependent manner. The addition of AICAR in the culture medium increased TER, whereas Compound-C abolished this effect. Butyrate also promoted a faster relocalization of ZO-1 and occludin at the cell periphery and tightened the intestinal barrier [112]. These results further support a role for AMPK activation in TJ formation. Additional studies using sodium butyrate further support the AMPK activation by CaMKK due to the increasing of the intracellular concentration of Ca^{2+}. Furthermore, the AMPK activation also increases the phosphorylation of PKCβ, a key player in TJ regulation [65,113]. This study underscores the putative interplay between the AMPK and PKC family in the formation of TJ [114].

Another model involving porcine intestinal epithelial cells investigated the effect of L-glutamine (Gln) in the preservation of TJ. Gln was described as a critically important nutrient for the maintenance of intestinal mucosal barrier integrity in humans and animals [115]. Indeed, the depletion of Gln results in a decreased abundance of TJ-associated proteins and increased intestinal paracellular permeability, whereas the addition of Gln resulted in increased TER, enhanced TJ-protein abundance, and the localization of TJ proteins to the plasma membrane [116]. TJ proteins, such as claudin-1, claudin-4, and ZO-1, are localized more abundantly at TJ sites in the presence of Gln. In addition, the abundance of pAMPK was further enhanced by the addition of Gln. The beneficial effect of Gln, as well as the phosphorylation of AMPK, was abrogated in a low Ca^{2+} medium and with the use of STO-609 (CaMKK inhibitor). Gln increased the intracellular [ATP] levels, but these were not affected by STO-609, meaning that AMPK acts as a TJ regulator via the CaMKK pathway in a model of Ca^{2+}-induced TJ formation [117]. Our recent work further supports a similar role for CaMKK in the activation of AMPK during a Ca^{2+} switch. This study found that the pharmacological inhibition of CaMKK or the direct inhibition of AMPK by Compound-C hampered AMPK phosphorylation and ZO-1 relocation to the TJ during a Ca^{2+} switch in MDCK cells, whereas the inactivation of LKB1 by shRNA did not significantly influence these processes [118].

Park et al. focused their research on the beneficial effect of Theaflavins (TFs), a polyphenol pigment in black tea, known to have anti-hyperglycemic, antioxidant, and anti-inflammatory effects. They previously found that TFs induced AMPK activation. By measuring the fluorescein transport across epithelial cells, they observed a decrease in its transport by pre-treatment with TFs, thereby revealing a reduction of paracellular permeability. Moreover, the TJ-related proteins claudin-1, occludin, and ZO-1 were significantly increased at TJ. Furthermore, Compound-C restored the fluorescein transport and inhibited the action of TFs. The authors highlighted the AMPK-mediated expression of claudin-1, occludin, and ZO-1 at TJ in intestinal cells by TFs [119]. Another natural agent, i.e., Forskolin, has been demonstrated to have beneficial effects on the AMPK-mediated TJ formation. This compound increased the phosphorylation and activation of AMPK with a comparable effect to the addition of 2 Deoxy-D-Glucose (2-DG) in a placenta epithelial cell culture. Forskolin treatment markedly enhanced the assembly of TJ strands, with higher ZO-1 relocation at TJ, while only weak ZO-1 staining was observed in control cells. The authors also used dominant negative AMPK transfected cells or Compound-C to inhibit AMPK activation and examined whether or not Forskolin-induced TJ assembly is mediated by AMPK activation. In both cases, the Forskolin effect was abrogated, resulting in lower ZO-1 relocation at TJ [120]. Thus, the activation of AMPK by Forskolin may enhance TJ formation.

Int. J. Mol. Sci. **2018**, *19*, 2040

AMPKα-null *Drosophila* die before reaching adulthood, while the transgenic expression of wild-type AMPK in AMPKα-null mutants allowed them to successfully develop into adults. A detailed examination of the embryonic epithelial structure of AMPKα-null mutants revealed a major disorganization of apico-basal polarity. They also assessed whether AMPK is necessary for cell polarity in mammalian cells. 2-DG treatment of unpolarized epithelial cells, such as the LS174T line, induced major changes in cell shape with the formation of a polarized actin cytoskeleton and a brush-border-like structure. Interestingly, this actin polarization was suppressed by the AMPK-specific inhibitor Compound C [121]. Furthermore, the authors have shown that AMPKα mutation in *Drosophila* embryos leads to the abnormal distribution of epithelial polarity markers. The consequent loss of polarity along with over-proliferative aberration could promote cancers [122]. In addition, AMPK has been shown to suppress tumorigenesis and the Warburg effect [123]. Therefore, one may speculate that AMPK-mediated TJ strengthening may help inhibit adenocarcinoma and tumorigenesis. Additional studies involving AMPK KO mice showed higher intestinal permeability when compared with WT mice, as indicated by decreased TER and increased paracellular FITC-dextran permeability, indicating a leaking gut. To investigate the integrity of TJ, ZO-1 immunofluorescence staining was analyzed and ZO-1 labeled at the tip of villi was impaired in AMPK KO mice [124].

Another role of AMPK in TJ formation and maintenance could be hypothetically linked to dietary methionine restriction (MR). MR has been found to modify the protein composition of TJ complexes in epithelial cells [125]. In addition, the stimulation of *S*-adenosyl-l-methionine, a key intermediate of methionine metabolism, led to the consumption of both Met and ATP [126] and AMPK activation [127]. These two studies may suggest a new role of AMPK in TJ maintenance in the case of MR. Along with other nutritional regimens modulating AMPK, Zinc has the potential to function as a TJ modifier and selective enhancer of epithelial barrier function [128] by regulating claudin-3 and occludin [129]. Since the rapid activation of AMPK was observed after exposure of neurons to Zinc, one may speculate an interplay between zinc-induced TJ formation and the AMPK pathway. Interestingly, Zinc-induced AMPK activation was mediated by LKB1 in the absence of changes in intracellular AMP levels or CaMKKβ activation [130].

6. AMPK and Co-Culture Models

The involvement of AMPK in TJ formation has also been demonstrated in several models of direct cell-cell co-culture. Tang et al. investigated the role of lymphocytes in the modulation of the epithelial barrier since lymphocytes are recruited by epithelial cells during infection. To mimic an infection state, they used a direct co-culture of MDCK cells with lymphocytes and a Ca^{2+} switch model to measure the TJ formation. The time course of ZO-1 relocation after Ca^{2+} switch was accelerated in the co-culture compared to MDCK alone. Furthermore, lymphocytes drastically increased AMPK phosphorylation in comparison to MDCK alone after a Ca^{2+}-switch. To link the increased AMPK phosphorylation and TJ formation, the authors used Compound-C and MDCK expressing an shRNA directed against AMPKα1. In both cases, the beneficial effect of lymphocytes was abolished and a slower TJ assembly was observed, thereby confirming the requirement for AMPK in the TJ formation [131]. Similar experiments were performed in a co-culture model of mesenchymal stromal cells (MSC) and MDCK cells. Bone marrow-derived MSC can modulate epithelial TJ at the time of their Ca^{2+}-induced assembly. The relocation of ZO-1 to MDCK cell-cell contacts was indeed significantly accelerated in the presence of MSC compared to a MDCK cell culture alone. Furthermore, AMPK activation and activity were also enhanced in the co-culture model. The addition of Compound-C or STO-609 abolished this AMPK activation and ZO-1 relocation. On the other hand, the co-culture of MSC with MDCK expressing an shRNA directed against LKB1 did not suppress the AMPK activity and ZO-1 relocation. This work further supports a role for CaMKK in the activation of AMPK and ZO-1 protein relocation at TJ during a Ca^{2+}-switch, independently of LKB1 activity [118].

Patkee et al. worked on metformin and its role in airway epithelial TJ in a model of co-culture with *P. aeruginosa* (a respiratory pathogen) to mimic an infection and TJ disruption with higher glucose permeability. The addition of *P. aeruginosa* into the culture of airway epithelial cells produced a significant decrease in TER. Metformin treatment attenuated the fall in TER produced by *P. aeruginosa*. AICAR pre-treatment also attenuated the *P. aeuringosa*-induced reduction of TER. On the other hand, this increasing TER was prevented by pre-treatment with the AMPK inhibitor Compound-C. To explain this increased TER phenomenon, the authors investigated the effect of *P. aeruginosa* and metformin on the abundance of TJ proteins. They found a decline in claudin-1 and occludin abundance in the co-culture with *P. aeruginosa*. The addition of metformin enhanced the expression of these two TJ proteins. These data indicate a potential AMPK-dependence that may be responsible for metformin's ability to increase the airway epithelial barrier function [132].

7. Conclusions

TJ are key constituents of polarized epithelial cells. It is well established that the presence of TJ is indispensable for tissue compartmentalization and cellular homeostasis. Their disruption represents one of the earliest markers of epithelial injury and diseases. Accumulating evidence demonstrates that AMPK is a key factor in the formation of TJ via several signaling pathways (Figure 3). Further investigations concerning the impact of AMPK on epithelial maintenance in baseline conditions and in diseased conditions may lead to innovative therapies.

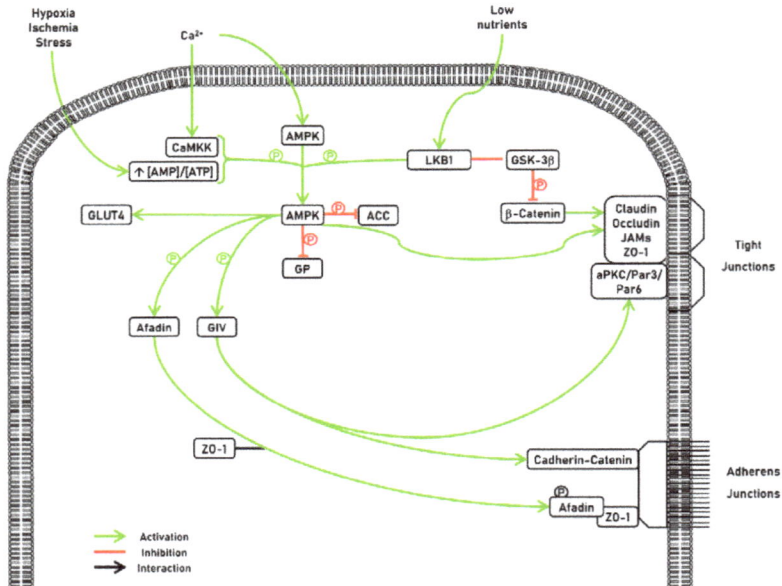

Figure 3. Representative schema of AMPK activators and substrates.

AMPK is involved in tight junctions (TJ) formation. AMPK can be activated by two mains upstream kinases: Ca^{2+}-Calmodulin Kinase Kinase (CaMKK) and Liver Kinase B1 (LKB1). Once activated, AMPK can have several effects. First, AMPK can modulate lipid metabolism by targeting the fatty acid synthesis pathway by the phosphorylation and inhibition of Acetyl CoA Carboxylase (ACC). Second, AMPK also exerts a potent effect on glucose metabolism trough the translocation of glucose transporter 4 (GLUT4) to the cell membrane. Third, AMPK also regulates glycogen metabolism. AMPK activation phosphorylates and decreases the activation of glycogen synthase (GS), thus reducing glycogen synthesis. AMPK is involved in TJ formation. Activated AMPK phosphorylates afadin and induces its association with ZO-1. AMPK also phosphorylates G-alpha interacting vesicle associated protein (GIV), which regulates cell polarity and morphogenesis, as well as cell-cell junction formation through its ability to bind Par3 and the Cadherin-catenin complex.

Author Contributions: Conceptualization, P.R., F.J. and M.J.C.; Writing-Original Draft Preparation, P.R.; Writing-Review & Editing, P.R., J.W., M.J.C. and F.J.; Supervision, F.J. and M.J.C.

funding: This research received no external funding.

Acknowledgments: F.J. is a Fellow of the Fonds National de la Recherche Scientifique (FNRS) and received support from the University of Liège (Fonds Spéciaux à la Recherche, Fonds Léon Fredericq) and the FNRS (Research Credits 2013 and 2016).

Conflicts of Interest: All the authors declared no competing interests.

References

1. Drubin, D.G.; Nelson, W.J. Origins of Cell Polarity. *Cell* **1996**, *84*, 335–344. [CrossRef]
2. Yeaman, C.; Grindstaff, K.K.; Nelson, W.J. New Perspectives on Mechanisms Involved in Generating Epithelial Cell Polarity. *Physiol. Rev.* **1999**, *79*, 73–98. [CrossRef] [PubMed]
3. Royer, C.; Lu, X. Epithelial cell polarity: A major gatekeeper against cancer? *Cell Death Differ.* **2011**, *18*, 1470–1477. [CrossRef] [PubMed]
4. Marchiando, A.M.; Graham, W.V.; Turner, J.R. Epithelial Barriers in Homeostasis and Disease. *Annu. Rev. Pathol. Mech. Dis.* **2010**, *5*, 119–144. [CrossRef] [PubMed]
5. Wang, Y.; Lu, X. *Cell Polarity: A Key Defence Mechanism Against Infection and Cancer Cell Invasion?* Springer: Cham, Switzerland, 2015.
6. Frisch, S.M.; Francis, H. Disruption of Epithelial Cell-Matrix Interactions Induces Apoptosis. *J. Cell Biol.* **1994**, *124*, 619–626. [CrossRef] [PubMed]
7. Klezovitch, O.; Fernandez, T.E.; Tapscott, S.J.; Vasioukhin, V. Loss of cell polarity causes severe brain dysplasia in Lgl1 knockout mice. *Genes Dev.* **2004**, *18*, 559–571. [CrossRef] [PubMed]
8. Basile, D.P.; Anderson, M.D.; Sutton, T.A. Pathophysiology of acute kidney injury. *Compr. Physiol.* **2012**, *2*, 1303–1353. [CrossRef] [PubMed]
9. Epstein, F.H.; Fish, E.M.; Molitoris, B.A. Alterations in Epithelial Polarity and the Pathogenesis of Disease States. *N. Engl. J. Med.* **1994**, *330*, 1580–1588. [CrossRef] [PubMed]
10. Fakhoury, M.; Negrulj, R.; Mooranian, A.; Al-Salami, H. Inflammatory bowel disease: Clinical aspects and treatments. *J. Inflamm. Res.* **2014**, *7*, 113–120. [CrossRef] [PubMed]
11. Wang, X.; Tully, O.; Ngo, B.; Zitin, M.; Mullin, J.M. Epithelial tight junctional changes in colorectal cancer tissues. *Sci. World J.* **2011**, *11*, 826–841. [CrossRef] [PubMed]
12. Förster, C. Tight junctions and the modulation of barrier function in disease. *Histochem. Cell Biol.* **2008**, *130*, 55–70. [CrossRef] [PubMed]
13. Caplan, M.J.; Seo-Mayer, P.; Zhang, L. Epithelial junctions and polarity: Complexes and kinases. *Curr. Opin. Nephrol. Hypertens.* **2008**, *17*, 506–512. [CrossRef] [PubMed]
14. Li, J.; Zhong, L.; Wang, F.; Zhu, H. Dissecting the role of AMP-activated protein kinase in human diseases. *Acta Pharm. Sin. B* **2017**, *7*, 249–259. [CrossRef] [PubMed]
15. Canfield, P.E.; Geerdes, A.M.; Molitoris, B.A. Effect of reversible ATP depletion on tight-junction integrity in LLC-PK1 cells. *Am. J. Physiol.* **1991**, *261*, F1038–F1045. [CrossRef] [PubMed]
16. Kwon, O.; Nelson, W.J.; Sibley, R.; Huie, P.; Scandling, J.D.; Dafoe, D.; Alfrey, E.; Myers, B.D. Backleak, Tight Junctions, and Cell–Cell Adhesion in Postischemic Injury to the Renal Allograft. *J. Clin. Investig.* **1998**, *101*, 2054–2064. [CrossRef] [PubMed]
17. Zhang, L.; Li, J.; Young, L.H.; Caplan, M.J. AMP-activated protein kinase regulates the assembly of epithelial tight junctions. *Proc. Natl. Acad. Sci. USA* **2006**, *103*, 17272–17277. [CrossRef] [PubMed]
18. Farquhar, M.G.; Palade, G.E. Junctional complexes in various epithelia. *J. Cell Biol.* **1963**, *17*, 375–412. [CrossRef] [PubMed]
19. Cereijido, M.; Valdés, J.; Shoshani, L.; Contreras, R.G. Role of tight junctions in establishing and maintaining cell polarity. *Annu. Rev. Physiol.* **1998**, *60*, 161–177. [CrossRef] [PubMed]
20. Campbell, H.K.; Maiers, J.L.; DeMali, K.A. Interplay between tight junctions & adherens junctions. *Exp. Cell Res.* **2017**, *358*, 39–44. [CrossRef] [PubMed]

21. Stevenson, B.R.; Siliciano, J.D.; Mooseker, M.S.; Goodenough, D.A. Identification of ZO-I: A High Molecular Weight Polypeptide Associated with the Tight Junction (Zonula Occludens) in a Variety of Epithelia. *J. Cell Biol.* **1986**, *103*, 755–766. [CrossRef] [PubMed]

22. Mandel, L.J.; Bacallao, R.; Zampighi, G. Uncoupling of the molecular "fence" and paracellular "gate" functions in epithelial tight junctions. *Nature* **1993**, *361*, 552–555. [CrossRef] [PubMed]

23. Dragsten, P.R.; Blumenthal, R.; Handler, J.S. Membrane asymmetry in epithelia: Is the tight junction a barrier to diffusion in the plasma membrane? *Nature* **1981**, *294*, 718–722. [CrossRef] [PubMed]

24. Van Meer, G.; Simons, K.; Simons, K. The function of tight junctions in maintaining differences in lipid composition between the apical and the basolateral cell surface domains of MDCK cells. *EMBO J.* **1986**, *5*, 1455–1464. [PubMed]

25. Van Itallie, C.M. The Molecular Physiology of Tight Junction Pores. *Physiology* **2004**. [CrossRef] [PubMed]

26. Anderson, J.M.; Van Itallie, C.M. Physiology and function of the tight junction. *Cold Spring Harb. Perspect. Biol.* **2009**, *1*, a002584. [CrossRef] [PubMed]

27. Fanning, S.; Mitic, L.L.; Anderson, J.M. Transmembrane proteins in the tight junction barrier. *J. Am. Soc. Nephrol.* **1999**, *10*, 1337–1345. [PubMed]

28. Tsukita, S.; Furuse, M.; Itoh, M. Multifunctional strands in tight junctions. *Nat. Rev. Mol. Cell Biol.* **2001**, *2*, 285–293. [CrossRef] [PubMed]

29. Matter, K.; Balda, M.S. Signalling to and from tight junctions. *Nat. Rev. Mol. Cell Biol.* **2003**, *4*, 225–236. [CrossRef] [PubMed]

30. Gonzalezmariscal, L. Tight junction proteins. *Prog. Biophys. Mol. Biol.* **2003**, *81*, 1–44. [CrossRef]

31. Furuse, M.; Hirase, T.; Itoh, M.; Nagafuchi, A.; Yonemura, S.; Tsukita, S.; Tsukita, S. Occludin: A novel integral membrane protein localizing at tight junctions. *J. Cell Biol.* **1993**, *123*, 1777–1788. [CrossRef] [PubMed]

32. Furuse, M.; Fujita, K.; Hiiragi, T.; Fujimoto, K.; Tsukita, S. Claudin-1 and -2: Novel integral membrane proteins localizing at tight junctions with no sequence similarity to occludin. *J. Cell Biol.* **1998**, *141*, 1539–1550. [CrossRef] [PubMed]

33. Martin-Padura, I.; Lostaglio, S.; Schneemann, M.; Williams, L.; Romano, M.; Fruscella, P.; Panzeri, C.; Stoppacciaro, A.; Ruco, L.; Villa, A.; et al. Junctional adhesion molecule, a novel member of the immunoglobulin superfamily that distributes at intercellular junctions and modulates monocyte transmigration. *J. Cell Biol.* **1998**, *142*, 117–127. [CrossRef] [PubMed]

34. Tsukita, S.; Furuse, M. Occludin and claudins in tight-junction strands: Leading or supporting players? *Trends Cell Biol.* **1999**, *9*, 268–273. [CrossRef]

35. Muresan, Z.; Paul, D.L.; Goodenough, D.A. Occludin 1B, a variant of the tight junction protein occludin. *Mol. Biol. Cell* **2000**, *11*, 627–634. [CrossRef] [PubMed]

36. Morita, K.; Furuse, M.; Fujimoto, K.; Tsukita, S. Claudin multigene family encoding four-transmembrane domain protein components of tight junction strands. *Proc. Natl. Acad. Sci. USA* **1999**, *96*, 511–516. [CrossRef] [PubMed]

37. Itoh, M.; Furuse, M.; Morita, K.; Kubota, K.; Saitou, M.; Tsukita, S. Direct binding of three tight junction-associated MAGUKs, ZO-1, ZO-2, and ZO-3, with the COOH termini of claudins. *J. Cell Biol.* **1999**, *147*, 1351–1363. [CrossRef] [PubMed]

38. Saitou, M.; Furuse, M.; Sasaki, H.; Schulzke, J.D.; Fromm, M.; Takano, H.; Noda, T.; Tsukita, S. Complex phenotype of mice lacking occludin, a component of tight junction strands. *Mol. Biol. Cell* **2000**, *11*, 4131–4142. [CrossRef] [PubMed]

39. Gow, A.; Southwood, C.M.; Li, J.S.; Pariali, M.; Riordan, G.P.; Brodie, S.E.; Danias, J.; Bronstein, J.M.; Kachar, B.; Lazzarini, R.A. CNS myelin and sertoli cell tight junction strands are absent in Osp/claudin-11 null mice. *Cell* **1999**, *99*, 649–659. [CrossRef]

40. Gow, A.; Davies, C.; Southwood, C.M.; Frolenkov, G.; Chrustowski, M.; Ng, L.; Yamauchi, D.; Marcus, D.C.; Kachar, B. Deafness in Claudin 11-null mice reveals the critical contribution of basal cell tight junctions to stria vascularis function. *J. Neurosci.* **2004**, *24*, 7051–7062. [CrossRef] [PubMed]

41. Aurrand-Lions, M.; Johnson-Leger, C.; Wong, C.; Du Pasquier, L.; Imhof, B.A. Heterogeneity of endothelial junctions is reflected by differential expression and specific subcellular localization of the three JAM family members. *Blood* **2001**, *98*, 3699–3707. [CrossRef] [PubMed]

42. Aurrand-Lions, M.A.; Duncan, L.; Du Pasquier, L.; Imhof, B.A. Cloning of JAM-2 and JAM-3: An emerging junctional adhesion molecular family? *Curr. Top. Microbiol. Immunol.* **2000**, *251*, 91–98. [PubMed]

43. Bazzoni, G.; Martinez-Estrada, O.M.; Orsenigo, F.; Cordenonsi, M.; Citi, S.; Dejana, E. Interaction of junctional adhesion molecule with the tight junction components ZO-1, cingulin, and occludin. *J. Biol. Chem.* **2000**, *275*, 20520–20526. [CrossRef] [PubMed]

44. Schneeberger, E.E.; Lynch, R.D. The tight junction: A multifunctional complex. *Am. J. Physiol. Cell Physiol.* **2004**, *286*, C1213–C1228. [CrossRef] [PubMed]

45. Guillemot, L.; Paschoud, S.; Pulimeno, P.; Foglia, A.; Citi, S. The cytoplasmic plaque of tight junctions: A scaffolding and signalling center. *Biochim. Biophys. Acta Biomembr.* **2008**, *1778*, 601–613. [CrossRef] [PubMed]

46. Furuse, M. Molecular basis of the core structure of tight junctions. *Cold Spring Harb. Perspect. Biol.* **2010**, *2*, a002907. [CrossRef] [PubMed]

47. Fanning, A.S.; Van Itallie, C.M.; Anderson, J.M. Zonula occludens-1 and -2 regulate apical cell structure and the zonula adherens cytoskeleton in polarized epithelia. *Mol. Biol. Cell* **2012**, *23*, 577–590. [CrossRef] [PubMed]

48. Ikenouchi, J.; Umeda, K.; Tsukita, S.; Furuse, M.; Tsukita, S. Requirement of ZO-1 for the formation of belt-like adherens junctions during epithelial cell polarization. *J. Cell Biol.* **2007**, *176*, 779–786. [CrossRef] [PubMed]

49. Van Itallie, C.M.; Anderson, J.M. Claudin interactions in and out of the tight junction. *Tissue Barriers* **2013**, *1*, e25247. [CrossRef] [PubMed]

50. Giepmans, B.N.; Moolenaar, W.H. The gap junction protein connexin43 interacts with the second PDZ domain of the zona occludens-1 protein. *Curr. Biol.* **1998**, *8*, 931–934. [CrossRef]

51. Umeda, K.; Ikenouchi, J.; Katahira-Tayama, S.; Furuse, K.; Sasaki, H.; Nakayama, M.; Matsui, T.; Tsukita, S.; Furuse, M.; Tsukita, S. ZO-1 and ZO-2 independently determine where claudins are polymerized in tight-junction strand formation. *Cell* **2006**, *126*, 741–754. [CrossRef] [PubMed]

52. Schmidt, A.; Utepbergenov, D.I.; Mueller, S.L.; Beyermann, M.; Schneider-Mergener, J.; Krause, G.; Blasig, I.E. Occludin binds to the SH3-hinge-GuK unit of zonula occludens protein 1: Potential mechanism of tight junction regulation. *Cell. Mol. Life Sci.* **2004**, *61*, 1354–1365. [CrossRef] [PubMed]

53. Saitou, M.; Fujimoto, K.; Doi, Y.; Itoh, M.; Fujimoto, T.; Furuse, M.; Takano, H.; Noda, T.; Tsukita, S. Occludin-deficient embryonic stem cells can differentiate into polarized epithelial cells bearing tight junctions. *J. Cell Biol.* **1998**, *141*, 397–408. [CrossRef] [PubMed]

54. Itoh, M.; Nagafuchi, A.; Moroi, S.; Tsukita, S. Involvement of ZO-1 in cadherin-based cell adhesion through its direct binding to alpha catenin and actin filaments. *J. Cell Biol.* **1997**, *138*, 181–192. [CrossRef] [PubMed]

55. Rajasekaran, A.K.; Hojo, M.; Huima, T.; Rodriguez-Boulan, E. Catenins and zonula occludens-1 form a complex during early stages in the assembly of tight junctions. *J. Cell Biol.* **1996**, *132*, 451–463. [CrossRef] [PubMed]

56. Yamamoto, T.; Harada, N.; Kano, K.; Taya, S.; Canaani, E.; Matsuura, Y.; Mizoguchi, A.; Ide, C.; Kaibuchi, K. The Ras target AF-6 interacts with ZO-1 and serves as a peripheral component of tight junctions in epithelial cells. *J. Cell Biol.* **1997**, *139*, 785–795. [CrossRef] [PubMed]

57. Fukuhara, A.; Irie, K.; Nakanishi, H.; Takekuni, K.; Kawakatsu, T.; Ikeda, W.; Yamada, A.; Katata, T.; Honda, T.; Sato, T.; et al. Involvement of nectin in the localization of junctional adhesion molecule at tight junctions. *Oncogene* **2002**, *21*, 7642–7655. [CrossRef] [PubMed]

58. Yokoyama, S.; Tachibana, K.; Nakanishi, H.; Yamamoto, Y.; Irie, K.; Mandai, K.; Nagafuchi, A.; Monden, M.; Takai, Y. alpha-catenin-independent recruitment of ZO-1 to nectin-based cell-cell adhesion sites through afadin. *Mol. Biol. Cell* **2001**, *12*, 1595–1609. [CrossRef] [PubMed]

59. Yamada, A.; Fujita, N.; Sato, T.; Okamoto, R.; Ooshio, T.; Hirota, T.; Morimoto, K.; Irie, K.; Takai, Y. Requirement of nectin, but not cadherin, for formation of claudin-based tight junctions in annexin II-knockdown MDCK cells. *Oncogene* **2006**, *25*, 5085–5102. [CrossRef] [PubMed]

60. Ooshio, T.; Kobayashi, R.; Ikeda, W.; Miyata, M.; Fukumoto, Y.; Matsuzawa, N.; Ogita, H.; Takai, Y. Involvement of the interaction of afadin with ZO-1 in the formation of tight junctions in Madin-Darby canine kidney cells. *J. Biol. Chem.* **2010**, *285*, 5003–5012. [CrossRef] [PubMed]

61. Fanning, A.S.; Jameson, B.J.; Jesaitis, L.A.; Anderson, J.M. The tight junction protein ZO-1 establishes a link between the transmembrane protein occludin and the actin cytoskeleton. *J. Biol. Chem.* **1998**, *273*, 29745–29753. [CrossRef] [PubMed]

62. Kemphues, K.J.; Priess, J.R.; Morton, D.G.; Cheng, N.S. Identification of genes required for cytoplasmic localization in early C. elegans embryos. *Cell* **1988**, *52*, 311–320. [CrossRef]

63. Hung, T.J.; Kemphues, K.J. PAR-6 is a conserved PDZ domain-containing protein that colocalizes with PAR-3 in Caenorhabditis elegans embryos. *Development* **1999**, *126*, 127–135. [PubMed]

64. Tabuse, Y.; Izumi, Y.; Piano, F.; Kemphues, K.J.; Miwa, J.; Ohno, S. Atypical protein kinase C cooperates with PAR-3 to establish embryonic polarity in Caenorhabditis elegans. *Development* **1998**, *125*, 3607–3614. [PubMed]

65. Chen, J.; Zhang, M. The Par3/Par6/aPKC complex and epithelial cell polarity. *Exp. Cell Res.* **2013**, *319*, 1357–1364. [CrossRef] [PubMed]

66. Assémat, E.; Bazellières, E.; Pallesi-Pocachard, E.; Le Bivic, A.; Massey-Harroche, D. Polarity complex proteins. *Biochim. Biophys. Acta Biomembr.* **2008**, *1778*, 614–630. [CrossRef] [PubMed]

67. Izumi, Y.; Hirose, T.; Tamai, Y.; Hirai, S.; Nagashima, Y.; Fujimoto, T.; Tabuse, Y.; Kemphues, K.J.; Ohno, S. An atypical PKC directly associates and colocalizes at the epithelial tight junction with ASIP, a mammalian homologue of Caenorhabditis elegans polarity protein PAR-3. *J. Cell Biol.* **1998**, *143*, 95–106. [CrossRef] [PubMed]

68. Carling, D. AMPK signalling in health and disease. *Curr. Opin. Cell Biol.* **2017**, *45*, 31–37. [CrossRef] [PubMed]

69. Carling, D.; Viollet, B. Beyond energy homeostasis: The expanding role of AMP-activated protein kinase in regulating metabolism. *Cell Metab.* **2015**, *21*, 799–804. [CrossRef] [PubMed]

70. Kurumbail, R.G.; Calabrese, M.F. Structure and Regulation of AMPK. In *AMP-activated Protein Kinase*; Springer: Cham, Switzerland, 2016; pp. 3–22.

71. Garcia, D.; Shaw, R.J. AMPK: Mechanisms of Cellular Energy Sensing and Restoration of Metabolic Balance. *Mol. Cell* **2017**, *66*, 789–800. [CrossRef] [PubMed]

72. Hardie, D.G. AMP-activated protein kinase: An energy sensor that regulates all aspects of cell function. *Genes Dev.* **2011**, *25*, 1895–1908. [CrossRef] [PubMed]

73. Hardie, D.; Scott, J.; Pan, D.; Hudson, E. Management of cellular energy by the AMP-activated protein kinase system. *FEBS Lett.* **2003**, *546*, 113–120. [CrossRef]

74. Ross, F.A.; MacKintosh, C.; Hardie, D.G. AMP-activated protein kinase: A cellular energy sensor that comes in 12 flavours. *FEBS J.* **2016**, *283*, 2987–3001. [CrossRef] [PubMed]

75. Steinberg, G.R.; Kemp, B.E. AMPK in Health and Disease. *Physiol. Rev.* **2009**, *89*, 1025–1078. [CrossRef] [PubMed]

76. Calabrese, M.F.; Rajamohan, F.; Harris, M.S.; Caspers, N.L.; Magyar, R.; Withka, J.M.; Wang, H.; Borzilleri, K.A.; Sahasrabudhe, P.V.; Hoth, L.R.; et al. Structural Basis for AMPK Activation: Natural and Synthetic Ligands Regulate Kinase Activity from Opposite Poles by Different Molecular Mechanisms. *Structure* **2014**, *22*, 1161–1172. [CrossRef] [PubMed]

77. Moffat, C.; Harper, M.E. Metabolic Functions of AMPK: Aspects of Structure and of Natural Mutations in the Regulatory Gamma Subunits. *IUBMB Life* **2010**, *62*, 739–745. [CrossRef] [PubMed]

78. Salt, I.; Celler, J.W.; Hawley, S.A.; Prescott, A.; Woods, A.; Carling, D.; Hardie, D.G. AMP-activated protein kinase: Greater AMP dependence, and preferential nuclear localization, of complexes containing the alpha2 isoform. *Biochem. J.* **1998**, *334*, 177–187. [CrossRef] [PubMed]

79. Hudson, E.R.; Pan, D.A.; James, J.; Lucocq, J.M.; Hawley, S.A.; Green, K.A.; Baba, O.; Terashima, T.; Hardie, D.G. A novel domain in AMP-activated protein kinase causes glycogen storage bodies similar to those seen in hereditary cardiac arrhythmias. *Curr. Biol.* **2003**, *13*, 861–866. [CrossRef]

80. Kemp, B.E. Bateman domains and adenosine derivatives form a binding contract. *J. Clin. Investig.* **2004**, *113*, 182–184. [CrossRef] [PubMed]

81. Hardie, D.; Hawley, S. AMP-activated protein kinase: The energy charge hypothesis revisited. *Bioessays* **2001**, *23*, 1112–1119. [CrossRef] [PubMed]

82. Xiao, B.; Sanders, M.J.; Underwood, E.; Heath, R.; Mayer, F.V.; Carmena, D.; Jing, C.; Walker, P.A.; Eccleston, J.F.; Haire, L.F.; Saiu, P.; et al. Structure of mammalian AMPK and its regulation by ADP. *Nature* **2011**, *472*, 230–233. [CrossRef] [PubMed]

83. Gowans, G.J.; Hawley, S.A.; Ross, F.A.; Hardie, D.G. AMP is a true physiological regulator of AMP-activated protein kinase by both allosteric activation and enhancing net phosphorylation. *Cell Metab.* **2013**, *18*, 556–566. [CrossRef] [PubMed]

84. Oligschlaeger, Y.; Miglianico, M.; Chanda, D.; Scholz, R.; Thali, R.F.; Tuerk, R.; Stapleton, D.I.; Gooley, P.R.; Neumann, D. The Recruitment of AMP-activated Protein Kinase to Glycogen Is Regulated by Autophosphorylation. *J. Biol. Chem.* **2015**, *290*. [CrossRef] [PubMed]

85. Mitchelhill, K.I.; Michell, B.J.; House, C.M.; Stapleton, D.; Dyck, J.; Gamble, J.; Ullrich, C.; Witters, L.A.; Kemp, B.E. Posttranslational Modifications of the 5'-AMP-activated Protein Kinase 1 Subunit. *J. Biol. Chem.* **1997**, *272*, 24475–24479. [CrossRef] [PubMed]

86. Hawley, S.A.; Boudeau, J.; Reid, J.L.; Mustard, K.J.; Udd, L.; Mäkelä, T.P.; Alessi, D.R.; Hardie, D.G. Complexes between the LKB1 tumor suppressor, STRAD alpha/beta and MO25 alpha/beta are upstream kinases in the AMP-activated protein kinase cascade. *J. Biol.* **2003**, *2*, 28. [CrossRef] [PubMed]

87. Lizcano, J.M.; Göransson, O.; Toth, R.; Deak, M.; Morrice, N.A.; Boudeau, J.; Hawley, S.A.; Udd, L.; Mäkelä, T.P.; Hardie, D.G.; Alessi, D.R. LKB1 is a master kinase that activates 13 kinases of the AMPK subfamily, including MARK/PAR-1. *EMBO J.* **2004**, *23*, 833–843. [CrossRef] [PubMed]

88. Shaw, R.J.; Kosmatka, M.; Bardeesy, N.; Hurley, R.L.; Witters, L.A.; DePinho, R.A.; Cantley, L.C. The tumor suppressor LKB1 kinase directly activates AMP-activated kinase and regulates apoptosis in response to energy stress. *Proc. Natl. Acad. Sci. USA* **2004**, *101*, 3329–3335. [CrossRef] [PubMed]

89. Hawley, S.A.; Selbert, M.A.; Goldstein, E.G.; Edelman, A.M.; Carling, D.; Hardie, D.G. 5'-AMP activates the AMP-activated protein kinase cascade, and Ca^{2+}/calmodulin activates the calmodulin-dependent protein kinase I cascade, via three independent mechanisms. *J. Biol. Chem.* **1995**, *270*, 27186–27191. [CrossRef] [PubMed]

90. Hurley, R.L.; Anderson, K.A.; Franzone, J.M.; Kemp, B.E.; Means, A.R.; Witters, L.A. The Ca^{2+}/calmodulin-dependent protein kinase kinases are AMP-activated protein kinase kinases. *J. Biol. Chem.* **2005**, *280*, 29060–29066. [CrossRef] [PubMed]

91. Hawley, S.A.; Pan, D.A.; Mustard, K.J.; Ross, L.; Bain, J.; Edelman, A.M.; Frenguelli, B.G.; Hardie, D.G. Calmodulin-dependent protein kinase kinase-β is an alternative upstream kinase for AMP-activated protein kinase. *Cell Metab.* **2005**, *2*, 9–19. [CrossRef] [PubMed]

92. Woods, A.; Dickerson, K.; Heath, R.; Hong, S.-P.; Momcilovic, M.; Johnstone, S.R.; Carlson, M.; Carling, D. Ca^{2+}/calmodulin-dependent protein kinase kinase-beta acts upstream of AMP-activated protein kinase in mammalian cells. *Cell Metab.* **2005**, *2*, 21–33. [CrossRef] [PubMed]

93. Lage, R.; Dié Guez, C.; Vidal-Puig, A.; Ló Pez, M. AMPK: A metabolic gauge regulating whole-body energy homeostasis. *Trends Mol. Med.* **2008**, *14*, 539–549. [CrossRef] [PubMed]

94. Brown, M.S.; Brunschede, G.Y.; Goldstein, J.L. Inactivation of 3-hydroxy-3-methylglutaryl coenzyme A reductase in vitro. An adenine nucleotide-dependent reaction catalyzed by a factor in human fibroblasts. *J. Biol. Chem.* **1975**, *250*, 2502–2509. [PubMed]

95. Marcinko, K.; Steinberg, G.R. The role of AMPK in controlling metabolism and mitochondrial biogenesis during exercise. *Exp. Physiol. Exp Physiol* **2014**, *9912*, 1581–1585. [CrossRef] [PubMed]

96. Beg, Z.H.; Allmann, D.W.; Gibson, D.M. Modulation of 3-hydroxy-3-methylglutaryl coenzyme A reductase activity with cAMP and wth protein fractions of rat liver cytosol. *Biochem. Biophys. Res. Commun.* **1973**, *54*, 1362–1369. [CrossRef]

97. Carlson, C.A.; Kim, K.H. Regulation of hepatic acetyl coenzyme A carboxylase by phosphorylation and dephosphorylation. *J. Biol. Chem.* **1973**, *248*, 378–380. [CrossRef]

98. Holmes, B.F.; Kurth-Kraczek, E.J.; Winder, W.W. Chronic activation of 5'-AMP-activated protein kinase increases GLUT-4, hexokinase, and glycogen in muscle. *J. Appl. Physiol.* **1999**, *87*, 1990–1995. [CrossRef] [PubMed]

99. Carling, D.; Grahame Hardie, D. The substrate and sequence specificity of the AMP-activated protein kinase. Phosphorylation of glycogen synthase and phosphorylase kinase. *Biochim. Biophys. Acta Mol. Cell Res.* **1989**, *1012*, 81–86. [CrossRef]

100. Kishton, R.J.; Barnes, C.E.; Nichols, A.G.; Cohen, S.; Gerriets, V.A.; Siska, P.J.; Macintyre, A.N.; Goraksha-Hicks, P.; de Cubas, A.A.; Liu, T.; et al. AMPK Is Essential to Balance Glycolysis and Mitochondrial Metabolism to Control T-ALL Cell Stress and Survival. *Cell Metab.* **2016**, *23*, 649–662. [CrossRef] [PubMed]

101. Liu, L.; Ulbrich, J.; Müller, J.; Wüstefeld, T.; Aeberhard, L.; Kress, T.R.; Muthalagu, N.; Rycak, L.; Rudalska, R.; Moll, R.; et al. Deregulated MYC expression induces dependence upon AMPK-related kinase 5. *Nature* **2012**, *483*, 608–612. [CrossRef] [PubMed]

102. Cereijido, M.; Robbins, E.S.; Dolan, W.J.; Rotunno, C.A.; Sabatini, D.D. Polarized monolayers formed by epithelial cells on a permeable and translucent support. *J. Cell Biol.* **1978**, *77*, 853–880. [CrossRef] [PubMed]

103. Matter, K.; Balda, M.S. Functional analysis of tight junctions. *Methods* **2003**, *30*, 228–234. [CrossRef]

104. Zheng, B.; Cantley, L.C. Regulation of epithelial tight junction assembly and disassembly by AMP-activated protein kinase. *Proc. Natl. Acad. Sci. USA* **2007**, *104*, 819–822. [CrossRef] [PubMed]

105. Mandai, K.; Nakanishi, H.; Satoh, A.; Obaishi, H.; Wada, M.; Nishioka, H.; Itoh, M.; Mizoguchi, A.; Aoki, T.; Fujimoto, T.; et al. Afadin: A novel actin filament-binding protein with one PDZ domain localized at cadherin-based cell-to-cell adherens junction. *J. Cell Biol.* **1997**, *139*, 517–528. [CrossRef] [PubMed]

106. Zhang, L.; Jouret, F.; Rinehart, J.; Sfakianos, J.; Mellman, I.; Lifton, R.P.; Young, L.H.; Caplan, M.J. AMP-activated protein kinase (AMPK) activation and glycogen synthase kinase-3β (GSK-3β) inhibition induce Ca^{2+}-independent deposition of tight junction components at the plasma membrane. *J. Biol. Chem.* **2011**, *286*, 16879–16890. [CrossRef] [PubMed]

107. Aznar, N.; Patel, A.; Rohena, C.C.; Dunkel, Y.; Joosen, L.P.; Taupin, V.; Kufareva, I.; Farquhar, M.G.; Ghosh, P. AMP-activated protein kinase fortifies epithelial tight junctions during energetic stress via its effector GIV/Girdin. *Elife* **2016**, *5*, e20795. [CrossRef] [PubMed]

108. Bhandari, D.; Lopez-Sanchez, I.; To, A.; Lo, I.-C.; Aznar, N.; Leyme, A.; Gupta, V.; Niesman, I.; Maddox, A.L.; Garcia-Marcos, M.; et al. Cyclin-dependent kinase 5 activates guanine nucleotide exchange factor GIV/Girdin to orchestrate migration–proliferation dichotomy. *Proc. Natl. Acad. Sci. USA* **2015**, *112*, E4874–E4883. [CrossRef] [PubMed]

109. Sasaki, K.; Kakuwa, T.; Akimoto, K.; Koga, H.; Ohno, S. Regulation of epithelial cell polarity by PAR-3 depends on Girdin transcription and Girdin-Gαi3 signaling. *J. Cell Sci.* **2015**, *128*, 2244–2258. [CrossRef] [PubMed]

110. Houssin, E.; Tepass, U.; Laprise, P. Girdin-mediated interactions between cadherin and the actin cytoskeleton are required for epithelial morphogenesis in Drosophila. *Development* **2015**, *142*, 1777–1784. [CrossRef] [PubMed]

111. Schumann, A.; Nutten, S.; Donnicola, D.; Comelli, E.M.; Mansourian, R.; Cherbut, C.; Corthesy-Theulaz, I.; Garcia-Rodenas, C. Neonatal antibiotic treatment alters gastrointestinal tract developmental gene expression and intestinal barrier transcriptome. *Physiol. Genom.* **2005**, *23*, 235–245. [CrossRef] [PubMed]

112. Peng, L.; Li, Z.-R.; Green, R.S.; Holzman, I.R.; Lin, J. Butyrate enhances the intestinal barrier by facilitating tight junction assembly via activation of AMP-activated protein kinase in Caco-2 cell monolayers. *J. Nutr.* **2009**, *139*, 1619–1625. [CrossRef] [PubMed]

113. 113. Mullin, J.M.; Laughlin, K.V.; Ginanni, N.; Marano, C.W.; Clarke, H.M.; Peralta Soler, A. Increased Tight Junction Permeability Can Result from Protein Kinase C Activation/Translocation and Act as a Tumor Promotional Event in Epithelial Cancers. *Ann. N. Y. Acad. Sci.* **2006**, *915*, 231–236. [CrossRef]

114. Miao, W.; Wu, X.; Wang, K.; Wang, W.; Wang, Y.; Li, Z.; Liu, J.; Li, L.; Peng, L. Sodium Butyrate Promotes Reassembly of Tight Junctions in Caco-2 Monolayers Involving Inhibition of MLCK/MLC2 Pathway and Phosphorylation of PKCβ2. *Int. J. Mol. Sci.* **2016**, *17*, 1696. [CrossRef] [PubMed]

115. Wu, G. Functional Amino Acids in Growth, Reproduction, and Health. *Adv. Nutr.* **2010**, *1*, 31–37. [CrossRef] [PubMed]

116. Li, N.; Lewis, P.; Samuelson, D.; Liboni, K.; Neu, J. Glutamine regulates Caco-2 cell tight junction proteins. *Am. J. Physiol. Liver Physiol.* **2004**, *287*, G726–G733. [CrossRef] [PubMed]

117. Wang, B.; Wu, Z.; Ji, Y.; Sun, K.; Dai, Z.; Wu, G. L-Glutamine Enhances Tight Junction Integrity by Activating CaMK Kinase 2–AMP-Activated Protein Kinase Signaling in Intestinal Porcine Epithelial Cells. *J. Nutr.* **2016**, *146*, 501–508. [CrossRef] [PubMed]

118. Rowart, P.; Erpicum, P.; Krzesinski, J.-M.; Sebbagh, M.; Jouret, F. Mesenchymal Stromal Cells Accelerate Epithelial Tight Junction Assembly via the AMP-Activated Protein Kinase Pathway, Independently of Liver Kinase B1. *Stem Cells Int.* **2017**, *2017*. [CrossRef] [PubMed]

119. Park, H.-Y.; Kunitake, Y.; Hirasaki, N.; Tanaka, M.; Matsui, T. Theaflavins enhance intestinal barrier of Caco-2 Cell monolayers through the expression of AMP-activated protein kinase-mediated Occludin, Claudin-1, and ZO-1. *Biosci. Biotechnol. Biochem.* **2015**, *79*. [CrossRef] [PubMed]

120. Egawa, M.; Kamata, H.; Kushiyama, A.; Sakoda, H.; Fujishiro, M.; Horike, N.; Yoneda, M.; Nakatsu, Y.; Ying, G.; Jun, Z.; et al. Long-term Forskolin Stimulation Induces AMPK Activation and Thereby Enhances

Tight Junction Formation in Human Placental Trophoblast BeWo Cells. *Placenta* **2008**, *29*, 1003–1008. [CrossRef] [PubMed]

121. Lee, J.H.; Koh, H.; Kim, M.; Kim, Y.; Lee, S.-H.; Karess, R.E.; Shong, M.; Kim, J.-M.J.; Kim, J.-M.J.; Chung, J. Energy-dependent regulation of cell structure by AMP-activated protein kinase. *Nature* **2007**, *447*, 1017–1020. [CrossRef] [PubMed]

122. Bilder, D. Epithelial polarity and proliferation control: Links from the Drosophila neoplastic tumor suppressors. *Genes Dev.* **2004**, *18*, 1909–1925. [CrossRef] [PubMed]

123. Faubert, B.; Boily, G.; Izreig, S.; Griss, T.; Samborska, B.; Dong, Z.; Dupuy, F.; Chambers, C.; Fuerth, B.J.; Viollet, B.; et al. AMPK Is a Negative Regulator of the Warburg Effect and Suppresses Tumor Growth In Vivo. *Cell Metab.* **2013**, *17*, 113–124. [CrossRef] [PubMed]

124. Sun, X.; Yang, Q.; Rogers, C.J.; Du, M.; Zhu, M.-J. AMPK improves gut epithelial differentiation and barrier function via regulating Cdx2 expression. *Cell Death Differ.* **2017**, *24*, 819–831. [CrossRef] [PubMed]

125. Mullin, J.M.; Skrovanek, S.M.; Ramalingam, A.; DiGuilio, K.M.; Valenzano, M.C. Methionine restriction fundamentally supports health by tightening epithelial barriers. *Ann. N. Y. Acad. Sci.* **2016**, *1363*, 59–67. [CrossRef] [PubMed]

126. Thomas, D.; Surdin-Kerjan, Y. Metabolism of sulfur amino acids in Saccharomyces cerevisiae. *Microbiol. Mol. Biol. Rev.* **1997**, *61*, 503–532. [PubMed]

127. Ogawa, T.; Tsubakiyama, R.; Kanai, M.; Koyama, T.; Fujii, T.; Iefuji, H.; Soga, T.; Kume, K.; Miyakawa, T.; Hirata, D.; et al. Stimulating S-adenosyl-l-methionine synthesis extends lifespan via activation of AMPK. *Proc. Natl. Acad. Sci. USA* **2016**, *113*, 11913–11918. [CrossRef] [PubMed]

128. Wang, X.; Valenzano, M.C.; Mercado, J.M.; Zurbach, E.P.; Mullin, J.M. Zinc Supplementation Modifies Tight Junctions and Alters Barrier Function of CACO-2 Human Intestinal Epithelial Layers. *Dig. Dis. Sci.* **2013**, *58*, 77–87. [CrossRef] [PubMed]

129. Miyoshi, Y.; Tanabe, S.; Suzuki, X.T. Cellular zinc is required for intestinal epithelial barrier maintenance via the regulation of claudin-3 and occludin expression. *Am. J. Physiol.* **2016**, *311*, G105–G116. [CrossRef] [PubMed]

130. Eom, J.-W.; Lee, J.-M.; Koh, J.-Y.; Kim, Y.-H. AMP-activated protein kinase contributes to zinc-induced neuronal death via activation by LKB1 and induction of Bim in mouse cortical cultures. *Mol. Brain* **2016**, *9*, 14. [CrossRef] [PubMed]

131. Tang, X.X.; Chen, H.; Yu, S.; Zhang, L.; Caplan, M.J.; Chan, H.C. Lymphocytes accelerate epithelial tight junction assembly: Role of AMP-activated protein kinase (AMPK). *PLoS ONE* **2010**, *5*, e12343. [CrossRef] [PubMed]

132. Patkee, W.R.A.; Carr, G.; Baker, E.H.; Baines, D.L.; Garnett, J.P. Metformin prevents the effects of Pseudomonas aeruginosa on airway epithelial tight junctions and restricts hyperglycaemia-induced bacterial growth. *J. Cell. Mol. Med.* **2016**, *20*, 758–764. [CrossRef] [PubMed]

International Journal of
Molecular Sciences

MDPI

Review

Hypothalamic AMPK as a Mediator of Hormonal Regulation of Energy Balance

Baile Wang [1,2] and Kenneth King-Yip Cheng [3,*]

1   State Key Laboratory of Pharmaceutical Biotechnology, The University of Hong Kong,
     Hong Kong, China; blwong@connect.hku.hk
2   Department of Medicine, The University of Hong Kong, Hong Kong, China
3   Department of Health Technology and Informatics, The Hong Kong Polytechnic University,
     Hong Kong, China
*   Correspondence: kenneth.ky.cheng@polyu.edu.hk; Tel.: +852-3400-8912

Received: 19 October 2018; Accepted: 7 November 2018; Published: 11 November 2018

Abstract: As a cellular energy sensor and regulator, adenosine monophosphate (AMP)-activated protein kinase (AMPK) plays a pivotal role in the regulation of energy homeostasis in both the central nervous system (CNS) and peripheral organs. Activation of hypothalamic AMPK maintains energy balance by inducing appetite to increase food intake and diminishing adaptive thermogenesis in adipose tissues to reduce energy expenditure in response to food deprivation. Numerous metabolic hormones, such as leptin, adiponectin, ghrelin and insulin, exert their energy regulatory effects through hypothalamic AMPK via integration with the neural circuits. Although activation of AMPK in peripheral tissues is able to promote fatty acid oxidation and insulin sensitivity, its chronic activation in the hypothalamus causes obesity by inducing hyperphagia in both humans and rodents. In this review, we discuss the role of hypothalamic AMPK in mediating hormonal regulation of feeding and adaptive thermogenesis, and summarize the diverse underlying mechanisms by which central AMPK maintains energy homeostasis.

Keywords: hypothalamus; adenosine monophosphate-activated protein kinase; adipose tissue; food intake; adaptive thermogenesis; beiging

---

1. The Hypothalamus and Energy Balance

The hypothalamus, a central integrator of the central nervous system (CNS), plays a critical role in the homeostatic regulation of appetite and energy expenditure by integrating hormonal, neuronal, and environmental signals [1]. It senses peripheral and central nutrients availability to modulate food intake and energy metabolism. Dysfunction of this highly-regulated system leads to energy imbalance, which initiates the development and progression of obesity and its related metabolic complications. There are different areas in the hypothalamus, which are believed to exert diverse functions in energy balance. Early in 1940, Hetherington and Ranson found that electrolytic lesions in the lateral hypothalamic area (LHA) cause inhibition of food intake, identifying LHA as a "feeding center" in the brain [2]. Subsequent studies showed that electrical stimulation of the LHA increases food intake [3], whereas lesions in the ventromedial hypothalamus (VMH) lead to the similar appetite-inducing effect [4]. Follow-up works demonstrated that not only lesions in the LHA and the VMH, but also disruption of other hypothalamic nuclei, including the arcuate nucleus (ARC), the dorsomedial hypothalamus (DMH), and the paraventricular nucleus (PVN), results in energy imbalance and obesity [5–8]. Among all of these regions, the ARC is critical in regulating feeding behavior and energy metabolism. The ARC is located near the median eminence (ME), which has abundant fenestrated capillaries that can lead to a 'penetrable' blood–brain barrier (BBB). The distinguished feature of the ME facilitates ARC neurons to sense hormonal and nutritional signals from the periphery [9], which is the reason why the ARC serves as the integration center of central and peripheral neural inputs.

The recent development of advanced techniques, including electrophysiology, optogenetics, and chemogenetics, enable us to identify distinct neuronal populations in the hypothalamus in rodents. The two best-studied neuronal populations in the ARC, which have opposite effects in appetite regulation, are the orexigenic Neuropeptide Y (NPY) and Agouti-related protein (AgRP) co-expressing neurons and the anorexigenic pro-opiomelanocortin (POMC) neurons. NPY/AgRP neurons are activated under a fasting condition, which drives hunger to promote food intake [10], whereas POMC neurons release alpha-melanocyte-stimulating hormone (α-MSH) to send satiety signals [11]. These two neuronal populations project to many second-order neurons in the PVN, VMH, DMH, and LHA [12–14]. The activity of these neurons is regulated by numerous neurotransmitters and/or hormones. For instance, the neurotransmitter serotonin exerts its anorexigenic effects by stimulating POMC neurons and suppressing NPY/AgRP neurons [15,16].

2. AMPK, an Energy Sensor and Regulator

AMP-activated protein kinase (AMPK), an evolutionarily conserved serine/threonine protein kinase, is a nutrient sensor that senses the ratio of AMP: adenosine triphosphate (ATP) or adenosine diphosphate (ADP): ATP to maintain energy balance in both peripheral tissues and the CNS. The heterotrimeric complex AMPK consists of three subunits, i.e., α catalytic subunit and β and γ regulatory subunits. Each subunit has several isoforms (α1, α2, β1, β2, γ1, γ2, γ3), suggesting 12 possibilities of heterotrimer combinations [17]. Some of these isoforms are tissue-specific and exert different functions under different physiological conditions [18,19]. For instance, heterotrimers containing the α1 isoform mainly exist in adipose tissues and the liver, whereas those containing α2 are predominantly expressed in skeletal muscles, the heart, and the brain [20,21]. The activity of AMPK can be regulated by both allosteric activation and phosphorylation at threonine 172 (Thr172) in the α-subunit. Specifically, allosteric activation is triggered by the increased intracellular AMP:ATP (or ADP:ATP) ratio, which facilitates the binding of AMP and/or ADP to the γ-subunit [22], while phosphorylation of AMPK is regulated by several upstream kinases, including liver kinase B1 (LKB1) [23,24], calcium-/calmodulin-dependent kinase kinase β (CaMKKβ) [25–27], TGFβ-activated kinase 1 (TAK1) [28,29], and the phosphatases including Mg^{2+}-/Mn^{2+}-dependent protein phosphatase 1E (PPM1E) [30], protein phosphatase 2A (PP2A), and protein phosphatase 2C (PP2C) [31]. The activated AMPK then shuts down ATP consumption and converts to ATP-producing pathways to stimulate carbohydrate and lipid metabolism by enhancing mitochondrial functions [21]. On the other hand, phosphorylation of AMPK at serine 485 (Ser485) in the α1 subunit or at serine 491 (Ser491) in the α2 subunit by protein kinase A (PKA) [32,33], autophosphorylation [32,34], or other protein kinases (such as Akt (also known as protein kinase B) [35,36] or the 70-kDa ribosomal protein S6 kinase (p70S6K) [37]) inhibits AMPK activity. The reduced AMPK activity in peripheral tissues, including liver, skeletal muscle, and adipose tissue causes glucose intolerance and lower exercise capacity, resulting in type 2 diabetes and obesity [21]. On the contrary, activation of AMPK by metabolic hormones, such as adiponectin and leptin, or a pharmacological compound, such as metformin, promotes insulin sensitivity and fatty acid oxidation in the peripheral tissues. Therefore, AMPK has been proposed as a promising drug target for obesity and type 2 diabetes [38,39].

3. Hypothalamic AMPK in the Regulation of Energy Balance

Apart from its crucial role in peripheral tissues, AMPK also plays a pivotal role in theCNS, especially in the hypothalamus. Activity of AMPK in the hypothalamus is induced by fasting but inhibited by feeding, hypothermia, and leptin, whereas high-fat diet (HFD) feeding blunts the leptin action and increases AMPK activity in the hypothalamus [40–43]. Specifically, AMPK activity is increased in AgRP-expressing neurons under fasting condition [44]. Hypothalamic AMPK modulates the functions of different neuronal populations (such as POMC and NPY/AgRP neurons), thereby controlling appetite and energy consumption to maintain energy homeostasis [45,46]. In addition, hypothalamic AMPK has been shown to control dietary selection, first- and second-phase insulin secretion, lipid metabolism, and hepatic gluconeogenesis, all of which are crucial for energy balance at the whole-body level [47–50]. Early studies revealed that pharmacological or adenovirus-mediated activation of AMPK in the medial hypothalamus significantly promotes food intake as a result of increased transcriptional levels of *NPY* and *AgRP* [51,52]. On the contrary, inhibition of AMPK by adenovirus-mediated overexpression of the dominant negative form of AMPK inhibits food intake [53]. Genetic-specific deletion of AMPKα2 in POMC neurons reduces energy expenditure and hence increases adiposity in mice, whereas deletion of this energy sensor in AgRP neurons prevents age-dependent obesity by promoting the anorexigenic effect of melanocortin [54]. Oh TS et al. recently demonstrated that AMPK regulates *NPY* and *POMC* transcription via autophagy in response to glucose deprivation in the mouse hypothalamic cell line [55]. A knockin mouse model with an activating mutation of AMPKγ2 (R302Q) gradually develops obesity due to elevated excitability of AgRP neurons and its associated hyperphagia [56]. Indeed, humans carrying this activating mutation have higher adiposity and dysregulated glucose balance [56]. Another protein-altering variant in AMPKγ1 has been recently shown to be associated with body mass index (BMI), which is identified by exome-targeted genotyping array [57]. Lentivirus-mediated overexpression of the constitutive active form of AMPK in corticotropin-releasing hormone (CRH) positive neurons in PVN leads to a food preference to a high carbohydrate diet over a HFD and obesity in mice [48]. In addition, AMPK activates the p21-activated kinase (PAK) signaling pathway in AgRP neurons, thereby mediating fasting-induced excitatory synaptic plasticity, neuronal activation, and feeding [44]. Apart from its direct action on POMC, NPY, and AgRP neurons, AMPK activity is also crucial to maintain excitatory synaptic input to AgRP neurons upon food deprivation [58]. In the following sections, we will discuss the key hormonal factors that positively or negatively regulate hypothalamic AMPK activity to control appetite and the underlying neuronal regulation.

4. Key Hormonal Factors That Regulate Food Intake via Hypothalamic AMPK

4.1. Leptin

Adipose tissue is an active and dynamic endocrine organ that secretes an array of hormones, bioactive peptides, and metabolites (collectively called adipokines), which control systemic energy, lipid and glucose homeostasis [59]. Leptin is the first identified adipokine that plays an indispensable role in controlling food intake and energy expenditure by mediating the crosstalk between adipose tissues and the hypothalamus [60]. The leptin receptor is abundantly expressed in POMC and NPY neurons in different regions of the hypothalamus [61–63]. Mutations in the *ob* gene (which encodes leptin) or the *leptin receptor* gene lead to severe obesity in humans and rodents mainly due to hyperphagia [64–66]. Leptin stimulates AMPK activation in skeletal muscle but reduces AMPK activity in the hypothalamus [38]. Noticeably, the inhibitory effect of leptin on AMPK activity is independent of the classic leptin signal transducer and activator of transcription 3 (STAT3) pathway [52]. The reduction of AMPK activity by leptin leads to an altered expression of neuropeptides, including NPY, AgRP, and α-MSH, in the ARC and the PVN [51,52]. Leptin selectively depolarizes POMC neurons and stimulates β-endorphin and α-MSH to inhibit AMPK activity [58,67,68]. AMPK also coordinates with other signaling networks, including mammalian target of rapamycin complex 1 (mTORC1) and phosphatidylinositol 3 kinase (PI3K), to fine-tune the hypothalamic actions of leptin [37,52,69]. For instance, PI3K-Akt-mTOR-p70S6K has been shown to phosphorylate AMPK at Ser485 and Ser491 in the hypothalamus upon leptin stimulation, which in turn reduces AMPK activity, leading to the inhibitory effect on food intake [37]. A more recent study also reports that leptin activates mTORC1 to repress AMPK activity via the PI3K-Akt axis [70]. Furthermore, the well-established downstream targets of AMPK in the peripheral tissues, such as acetyl-CoA carboxylase (ACC) and peroxisome proliferator-activated receptor gamma coactivator 1-alpha (PGC-1α), have been shown to mediate the hypothalamic function of AMPK [71,72]. Inhibition of AMPK by leptin increases the intracellular level of malonyl-CoA in the ARC and palmitoyl-CoA in the PVN through ACC [71]. Pharmacological blocking of the increase of these fatty acids attenuates leptin-induced suppression of food intake. A subsequent study demonstrated that inactivation of ACC by knocking in Serine 79 and Serine 212 with alanine in ACC impairs appetite in response to both fasting and cold in mice [73]. Genetic deletion of PGC-1α in AgRP neurons but not POMC neurons blunts the anorexigenic effect of leptin [72]. At the molecular level, knockdown of PGC-1α significantly reduces the mRNA level of *AgRP* in an AgRP-immortalized cell line under starvation but not fed state [72].

4.2. Adiponectin

Adiponectin is the most abundant adipokine that exerts multiple beneficial effects on the cardiometabolic system mainly via its insulin-sensitizing and anti-inflammatory properties [74,75]. In contrast to the increased level of leptin, the circulating level of adiponectin is reduced in humans with obesity and diabetes [76,77]. Adiponectin promotes glucose uptake and fatty acid oxidation in the skeletal muscle, suppresses glucose production in the liver, and induces vasorelaxation in the blood vessels [74,75]. These metabolic and vascular actions of adiponectin are largely mediated by AMPK [74,75]. Apart from its endocrine actions in the peripheral tissues, adiponectin also regulates feeding and energy expenditure via the hypothalamus [78]. Adiponectin can be detected in the cerebrospinal fluid (CSF) of mice after intravenous injection of recombinant full-length adiponectin, which promotes adaptive thermogenesis in brown adipose tissue (BAT) via the sympathetic nervous system (SNS)-uncoupling protein 1 (UCP1) axis [79]. Subsequent studies demonstrate that adiponectin is also detectable in human CSF, despite some studies having argued that adiponectin cannot pass through the blood–brain barrier [80–83]. In stark contrast to its abundant expression in circulation, only a trace amount of the trimeric and low-molecular-mass hexameric form (~0.1% of serum concentration), but no high-molecular-weight form, of adiponectin can be detected in CSF [80]. Importantly, the key signaling molecules (including the adiponectin receptors AdipoR1 and AdipoR2, the adaptor proteins containing an NH_2-terminal Bin/Amphiphiphysin/Rvs domain, a central pleckstrin homology domain, and a COOH-terminal phosphotyrosine binding domain (APPL)1 and APPL2) mediating adiponectin actions in peripheral tissues can also be detected in different regions of the hypothalamus [84–88]. With regard to feeding regulation, two early studies demonstrated opposite effects of adiponectin on food intake via distinct mechanisms in the hypothalamus [89,90]. The first study by Kubota et al. demonstrated that intravenous injection of full-length adiponectin increases AMPK activity in the hypothalamus, which in turn promotes food intake and decreases energy expenditure under a refeeding condition [89]. These adiponectin actions are abolished by siRNA-mediated knockdown expression of AdipoR1 or adenovirus-mediated overexpression of dominant negative AMPK. Genetic abrogation of adiponectin has a similar effect on hypothalamic AMPK activity and appetite. On the contrary, the study by Coope A et al. showed that intracerebroventricular (i.c.v.) injection of adiponectin reduces food intake via AdipoR1 in a fasted state [90]. Such change is accompanied by activations of insulin (increased phosphorylation of insulin receptor substrate 1[IRS1], Akt, and forkhead box protein O1 (FOXO1)) and leptin (STAT3 phosphorylation) signaling as well as an increase of AdipoR1-APPL1 interaction. Consistent with Coope A et al.'s study, a recent study demonstrated that i.c.v. injection of adiponectin decreases body weight as a consequence of reduced food consumption and increased adaptive thermogenesis in the BAT, and such effect of adiponectin is diminished in rats with a nutritional imbalance during their neonatal period [91]. On the other hand, the peroxisome proliferator-activated receptor (PPAR)γ agonist pioglitazone, a well-established insulin-sensitizing drug, boosts food intake and reduces energy expenditure by inducing adiponectin production in adipocytes, which in turn increases and decreases mRNA expression of *NPY* and *POMC* in the hypothalamus, respectively, via the AdipoR1-AMPK-dependent pathway [92]. Surprisingly, patch-clamp electrophysiology experiments reveal that adiponectin specifically depolarizes POMC neurons and inhibits NPY neurons in a

PI3K-dependent and AMPK-independent manner [93]. The discrepancy of adiponectin actions on food intake and hypothalamic neuronal activity may be due to the different nutritional states and concentrations of glucose used in the experiments. Indeed, two recent studies from Yada's research group show that adiponectin exerts opposite effects on feeding and POMC neuron activity under low and high glucose concentration, despite the fact that the suppressive effect of adiponectin on NPY neurons is independent of glucose [94,95]. In addition to its direct action on hypothalamic AMPK, several studies report that adiponectin is able to modulate the actions of insulin and leptin on the hypothalamus, thereby controlling energy homeostasis [89,90,93,96].

Recently, Okada-Iwabu et al. discovered an orally active synthetic small molecule of adiponectin receptor agonist (namely AdipoRon) [97]. Treatment with AdipoRon not only improves metabolic health but also prolongs the lifespan in obese and diabetic mouse models [97]. Intraperitoneal injection of AdipoRon attenuates corticosterone-induced body weight gain, depression, and neuroinflammation in mice, indicating that AdipoRon can penetrate and target the CNS [98]. Since AdipoRon has been proposed for the treatment of type 2 diabetes, it is, therefore, interesting to investigate whether AdipoRon has any effect on hypothalamic function, as adiponectin, in the regulation of feeding and energy expenditure.

4.3. Ghrelin

The stomach-derived hormone ghrelin, released during fasting, is the first circulating factor that has been reported to stimulate appetite in humans [99]. The orexigenic action of ghrelin is mediated by NPY and AgRP peptides [100]. Central or peripheral administration of ghrelin upregulates hypothalamic AMPK activity in both the ARC and the VMH in rats via growth hormone secretagogue receptor [51,101–103]. AMPK activation by ghrelin can be controlled at the transcriptional level by the transcriptional factor ALL1-fused gene from chromosome 4 (AF4), its upstream kinase CaMKKβ, the Sirtuin 1 (SIRT1)-p53 pathway, or glucose availability [104–107]. Inhibition of AMPK activity abolishes the orexigenic action of ghrelin [101,108–110]. On the contrary, knockin of an activating mutation in AMPKγ2 potentiates the orexigenic action of ghrelin under a refeeding condition [56]. There are multiple downstream targets of hypothalamic AMPK to mediate the orexigenic effect of ghrelin. First, ghrelin increases the release of intracellular Ca^{2+} to activate the CaMKKβ pathway, and, thus, facilitates AMPK phosphorylation in NPY neurons in the ARC [58,111,112]. Second, ghrelin activates AMPK and increases cytosolic Ca^{2+} in NPY neurons in the ARC [112]. Third, ghrelin triggers a hypothalamic mitochondrial function via uncoupling protein 2 (UCP2), which antagonizes the reactive oxidative species (ROS) production, allowing AMPK-mediated fatty acid oxidation for the support of synaptic plasticity and neuronal activation of NPY neurons [108]. Fourth, López et al. report that ghrelin inhibits fatty acid synthesis via the AMPK-ACC-dependent pathway, leading to reduced production of malonyl-CoA (the product of ACC), which in turn promotes carnitine palmitoyltransferase I (CPT1) activity in the mitochondria [101]. A subsequent study indicated that the regulation of ghrelin on fatty acid metabolism only occurs in the VMH but not the ARC [102]. Lastly, like leptin, ghrelin modulates the activity of presynaptic neurons that activate NPY neurons in AMPK-dependent and positive feedback loop manners [58].

4.4. Insulin

Insulin is exclusively produced by pancreatic β cells, and secreted in response to different nutrient stimuli, including glucose, fatty acid, and amino acids, after a meal. Apart from its glucose lowering and lipogenic actions, insulin also acts as an anorexigenic hormone. Insulin-deficient animals are hyperphagia, whereas their voracious appetite could be rectified by central administration of insulin [113,114]. Brain insulin resistance, a status in which neurons fail to respond to a physiological concentration of insulin, causes a dysregulation of energy homeostasis and cognitive functions [115,116]. The role of central insulin signaling in maintaining energy balance could be verified, at least in part, using neuron-specific insulin receptor knockout (NIRKO) mice. NIRKO mice have an elevated plasma insulin level, increased food consumption, and are susceptible to diet-induced obesity without alterations in brain development or neuronal survival [117]. In addition, i.c.v. injection of insulin [118–120] or insulin analogues [121] reduces both body weight and food intake, while intrahypothalamic infusion of an anti-insulin antibody results in opposite effects [122]. Insulin exerts a broad suppressive effect on AMPKα2 activity in different regions of the hypothalamus, including the PVN, the ARC and the LHA [52]. Indeed, the suppressive action of insulin on AMPKα2 activity is comparable to that of leptin [52]. Streptozotocin (STZ)-induced β cell loss and subsequent insulin deficiency lead to activation of AMPK and increase expression of NPY in the hypothalamus, resulting in hyperphagia in rats [53]. Insulin treatment reverses STZ-induced AMPK activation in the hypothalamus. Pharmacological or molecular inhibition of AMPK in the hypothalamus reverses STZ-induced hyperphagia [53]. Moreover, the inhibitory effect of insulin on AMPK activity and food intake can be further potentiated by i.c.v. injection of the amino acid taurine [123], and the effect of insulin on AMPK activation depends on the extracellular glucose concentration [124]. On the other hand, the anorexigenic action and the inhibitory effect of insulin on hypothalamic AMPK are largely abolished by cold exposure [125]. Apart from its direct action, Han et al. found that hypoglycemia triggered by insulin increases AMPKα2 activity in the hypothalamus [126]. This phenomenon was remarkable in hypothalamic ARC, VMH, and PVN [126]. Interestingly, insulin has been shown to inhibit AMPK activity by inducing phosphorylation of AMPK at Ser485 and Ser491 in skeletal muscle, ischemic heart, and hepatoma HepG2 cells in an Akt-dependent manner; however, whether insulin exerts a similar effect on hypothalamic AMPK phosphorylations in response to feeding remains elusive [35,127].

4.5. Glucagon-Like Peptide-1 (GLP-1)

GLP-1 is not only an incretin hormone secreted by intestinal L cells [128], but also a neuropeptide produced by preproglucagon neurons in the nucleus of the solitary tract (NTS) in the brainstem, which projects to hypothalamic nuclei to regulate appetite [129,130]. Hypothalamic GLP-1 level is reduced under fasting condition, while central administration of GLP-1 inhibits food intake in fasted rats [131,132]. This anorectic effect of GLP-1 is mediated by its inhibitory effect on fasting-induced hypothalamic AMPK activation [132,133]. HFD or central administration of fructose has been shown to inhibit the anorectic action of GLP-1 [134,135]. In addition, expression of the *proglucagon* gene (which encodes GLP-1) in the brain is regulated by transcription factor 7 like 2 (TCF7L2), which is associated with the risk of diabetes [136]. Transgenic overexpression of the dominant negative form of TCF7L2 driven by the proglucagon promoter represses the expression of GLP-1 in the brain, leading to a defective repression of AMPK activity in response to feeding. The defect can be reversed by treatment with the cyclic adenosine monophosphate (cAMP)-promoting agent forskolin, indicating that GLP-1 mediates its anorectic effect via the AMPK-PKA-cAMP axis [136]. Of note, this signaling axis also mediates the anorectic effect of the GLP-1 receptor in the hindbrain [137]. Similar to GLP-1, targeted injection of the GLP-1 receptor agonist liraglutide or exendin-4 into the VMH inhibits food intake in humans and rodents [138,139]. The anorexic effect of exendin-4 can be reversed by pre-injection with the AMPK activator AICAR in the VMH [139]. Further analysis reveals that mTOR but not ACC acts as a downstream mediator of AMPK for the hypophagic effect of exendin-4 [139].

5. The Role of Hypothalamic AMPK in the Regulation of Energy Expenditure

Total energy expenditure consists of basal metabolism, physical activity, and adaptive thermogenesis. Among these three components, adaptive thermogenesis in response to cold temperature or dietary intake is predominantly controlled by the hypothalamus. Adaptive thermogenesis is mainly mediated by BAT, which dissipates heat via the mitochondrial protein UCP1 in brown adipocytes. In the past few years, great advances have been made to broaden our knowledge on the inducible thermogenic adipocytes (beige adipocytes) in subcutaneous white adipose tissue (sWAT). Under certain circumstances (such as cold exposure, β-adrenergic stimulation, intermittent fasting, or exercise), beige adipocytes could be induced within WAT, especially in sWAT [140–144]. This process is called white fat beiging or browning, and is largely regulated by the crosstalk between the hypothalamus, the SNS, and adipose tissues. Beiging of WAT not only enhances energy expenditure, but also improves glucose metabolism, insulin sensitivity, and hyperlipidemia to ameliorate obesity and its related cardiometabolic complications [145–148]. The activation of BAT and beiging of WAT is, at least in part, controlled by the hypothalamus-SNS axis [149]. As the interscapular brown adipocytes only exist in human infants [150], and the gene expression profiles of the inducible UCP-1 positive cells in human adults share high similarities with mouse beige adipocytes rather than classical brown adipocytes [151,152], it is possible that induction of beige adipocytes in humans could be a potential therapeutic target for the prevention of obesity and its related metabolic syndromes.

An early study showed that whole-body depletion of AMPKα2 leads to elevated sympathetic activity and increased catecholamine secretion [153], suggesting the potential role of AMPK in beiging via the SNS. Indeed, emerging evidence suggests that numerous hormonal factors regulate adipose tissue beiging and adaptive thermogenesis via inhibition of hypothalamic AMPK activity, which will be further discussed in the following sections.

5.1. Leptin

As mentioned above, leptin is known to inhibit AMPK activity in the hypothalamus, which is accompanied by enhanced whole-body energy expenditure [52,60]. Mice with deletion of protein tyrosine phosphatase 1B (PTP1B), an inhibitor of leptin signaling, have diminished activation of hypothalamic AMPKα2, accompanied by an upregulation of UCP1 expression and mitochondrial density in BAT [154]. Central administration of leptin increases sympathetic outflow to adipose tissues via AMPKα2 [155]. Further studies demonstrated that sensitizing the leptin signaling in POMC neurons by deletion of PTP1B increases energy expenditure and promotes the conversion of WAT into BAT [156,157], although whether the inhibition of AMPK contributes to these changes is unknown. In addition, sympathetic denervation abolishes the potentiating effect of PTP1B deletion on WAT beiging [157]. These data collectively indicate that leptin regulates adaptive thermogenesis in adipose tissues via the SNS.

5.2. Thyroid Hormones

Thyroid hormones, including triiodothyronine (T3) and thyroxine (T4), have been found to raise energy expenditure via their peripheral actions in BAT or central action in the hypothalamus [158,159]. Stereotaxic injection of T3 (an active form of thyroid hormone) into the VMH (where AMPK and thyroid hormone receptors are highly co-expressed) stimulates SNS activity and BAT thermogenesis by inactivating hypothalamic AMPK via the thyroid hormone receptors [158]. Subsequent studies indicate that administration of T3 in the VMH but not the ARC is able to induce beiging in sWAT and thermogenesis via inhibition of hypothalamic AMPK activity [160,161]. Inactivation of the lipogenic pathway in the VMH attenuates the central action of T3 on BAT thermogenesis [158]. Genetic deletion of AMPKα1 in steroidogenic factor 1 (SF1) neurons in the VMH mimics the central effect of T3 on BAT metabolism [162]. Ablation of UCP1 completely abolishes the thermogenic action of central T3 administration [161]. At the molecular level, T3 relives endoplasmic reticulum (ER) stress and ceramide level in the VMH via an AMPK-dependent pathway, which has been shown to promote beiging in WAT and reduce obesity [162]. Substitution therapy with Levothroxin, a synthetic form of T4, promotes the basal metabolic rate and BAT activity in human subjects with a condition of hypothyroid state [163], but it remains unknown whether this is mediated by central or peripheral action of AMPK.

5.3. BMP8B

Bone morphogenetic protein 8B (BMP8B), a member of transforming growth factor β, acts both centrally and peripherally to increase BAT thermogenesis in female rodents [164]. mRNA of *BMP8B* can be detected in the brain and its receptors, including ALK4, ALK5, and ALK7, which are expressed in the VMH and the LHA. BMP8B-deficient mice display impaired thermogenesis and reduced AMPK activity in the VMH [164,165]. Acute i.c.v. injection of BMP8B in the VMH rather than the LHA enhances sympathetic outflow to BAT but not to the kidney, which can be abolished by expression of the constitutively active form of AMPK or potentiated by expression of the dominant negative form of AMPK in the VMH, respectively [164,165]. In addition, central administration of BMP8B exerts its thermogenic effect in BAT via upregulating orexin, a key modulator of BAT thermogenesis, in the LHA via glutamatergic signaling [165].

5.4. GLP-1

Activation of GLP-1 receptors in the hypothalamus not only inhibits appetite but also regulates BAT function. Stimulation of GLP-1 receptors by their agonist liraglutide in the VMH triggers both BAT thermogenesis and WAT beiging in mice, which are mediated by AMPK inhibition [138]. Central administration of the GLP-1 receptor agonist increases sympathetic outflow to BAT, leading to increased ability of glucose and lipid clearance and thermogenesis in BAT [166]. Although injection of liraglutide in DMH has no effect on BAT thermogenesis, injection of native GLP-1 in DMH increases the core body temperature and thermogenic program in BAT [138,167]. However, whether the thermogenic actions of native GLP-1 and exendin-4 are also mediated by AMPK signaling remains unclear, which warrants further investigation. With regard to human studies, the effect of GLP-1 receptor agonists and GLP-1 on energy expenditure remains inconclusive [168].

5.5. Estradiol

Estrogens are known to play a key role in the regulation of energy balance. The central action of estradiol on BAT thermogenesis has been recently identified and linked to the AMPK pathway [169,170]. Estradiol binds to its receptor (estrogen receptor α) in the VMH to diminish AMPK activity and enhances BAT thermogenesis without affecting feeding behavior [169]. Similar to the action of thyroid hormone, estradiol is able to relieve endoplasmic reticulum (ER) stress and reduce ceramide synthesis in the VMH, which in turn promotes BAT thermogenesis.

6. Conclusions and Future Perspectives

The diverse mechanisms driven by the hormonal factors convey on the hypothalamic AMPK signaling axis, supporting a critical role of AMPK in controlling feeding behavior and energy expenditure to maintain whole-body energy homeostasis (Table 1). AMPK in the VMH is crucial for BAT thermogenesis and beiging of sWAT, whereas AMPK in the ARC regulates food intake. Inhibition of AMPK activity by estradiol and thyroid hormone protects the hypothalamus from lipotoxicity and ER stress, which are the central pathogenic pathways that contribute to insulin and leptin resistance in obesity [171]. A recent study pinpointed that AMPK in SF1 neurons in the VMH regulates BAT thermogenesis via the SNS [172]; however, whether other hypothalamic neuronal population(s) mediates the inhibitory effect of AMPK activation on BAT functions remains unclear.

Considering the vital roles of hypothalamic AMPK, drugs that specifically target central AMPK are worth developing to prevent obesity and its related metabolic syndromes. As the regulatory effects of AMPK are differential in the periphery and centrally, the best therapeutic strategy is to specifically target hypothalamic AMPK without altering its functions in peripheral tissues. In this respect, the use of nanoparticles or exosomes [173], optogenetic neuromodulations [58], or chimeric proteins (targeting peptides associated with effective molecules or steroid hormones) [174,175], drawing from the implementations in other diseases, might be innovative strategies to achieve specific modulation of hypothalamic AMPK activity. However, despite the high specificity, we cannot exclude the possibility that these strategies may also affect other neuronal populations near the target hypothalamic region, which would result in limited efficacy and undesired side effects [176]. Another important issue is how to address the long-term influence of the altered hypothalamic AMPK activity. As AMPK is a canonical regulator of glucose and lipid metabolism, whether the sustained inhibition of hypothalamic AMPK may lead to lipotoxicity or other deleterious effects in neurons still needs further investigation. Taken together, great endeavors are required to advance our understanding of neuronal and hormonal regulation of hypothalamic AMPK, and AMPK in the hypothalamus will be a fascinating therapeutic target if we can address all of the above concerns properly.

Table 1. Actions of hormonal factors on hypothalamic AMPK activity, food intake, and energy expenditure.

Hormonal Factors	Hypothalamic AMPK Activity	Food Intake	Energy Expenditure
Adiponectin	↑	↑↓	↑↓
Ghrelin	↑	↑	-
Leptin	↓	↓	↑
Insulin	↓	↓	↑
GLP-1 and its analogues	↓	↓	↑
Thyroid hormones	↓	-	↑
BMP8B	↓	-	↑
Estradiol	↓	-	↑

funding: This research was funded by National Natural Science Foundation of China (NSFC) (grant number: 81270881) and PolyU Start-up funding (grant number: SUF-KC).

Conflicts of Interest: The authors declare no conflict of interest.

References

1. Dietrich, M.O.; Horvath, T.L. Hypothalamic control of energy balance: Insights into the role of synaptic plasticity. *Trends Neurosci.* **2013**, *36*, 65–73. [CrossRef] [PubMed]
2. Hetherington, A.W.; Ranson, S.W. Hypothalamic lesions and adiposity in the rat. *Anat. Rec.* **1940**, *78*, 149–172. [CrossRef]
3. Anand, B.K.; Brobeck, J.R. Hypothalamic control of food intake in rats and cats. *Yale J. Biol. Med.* **1951**, *24*, 123–140. [PubMed]
4. Brobeck, J.R.; Tepperman, J.; Long, C. Experimental hypothalamic hyperphagia in the albino rat. *Yale J. Biol. Med.* **1943**, *15*, 831–853. [PubMed]
5. Bernardis, L.L. Disruption of diurnal feeding and weight gain cycles in weanling rats by ventromedial and dorsomedial hypothalamic lesions. *Physiol. Behav.* **1973**, *10*, 855–861. [CrossRef]
6. Leibowitz, S.F.; Hammer, N.J.; Chang, K. Hypothalamic paraventricular nucleus lesions produce overeating and obesity in the rat. *Physiol. Behav.* **1981**, *27*, 1031–1040. [CrossRef]
7. Fukushima, M.; Tokunaga, K.; Lupien, J.; Kemnitz, J.; Bray, G. Dynamic and static phases of obesity following lesions in PVN and VMH. *Am. J. Physiol. Regul. Integr. Comp. Physiol.* **1987**, *253*, R523–R529. [CrossRef] [PubMed]
8. Choi, S.; Dallman, M.F. Hypothalamic Obesity: Multiple Routes Mediated by Loss of Function in Medial Cell Groups 1. *Endocrinology* **1999**, *140*, 4081–4088. [CrossRef] [PubMed]
9. Rodriguez, E.M.; Blazquez, J.L.; Guerra, M. The design of barriers in the hypothalamus allows the median eminence and the arcuate nucleus to enjoy private milieus: The former opens to the portal blood and the latter to the cerebrospinal fluid. *Peptides* **2010**, *31*, 757–776. [CrossRef] [PubMed]
10. Mizuno, T.M.; Makimura, H.; Silverstein, J.; Roberts, J.L.; Lopingco, T.; Mobbs, C.V. Fasting regulates hypothalamic neuropeptide Y., agouti-related peptide, and proopiomelanocortin in diabetic mice independent of changes in leptin or insulin. *Endocrinology* **1999**, *140*, 4551–4557. [CrossRef] [PubMed]
11. Biebermann, H.; Kühnen, P.; Kleinau, G.; Krude, H. *Appetite Control*; The Neuroendocrine Circuitry Controlled by POMC, MSH, and AGRP; Springer: Berlin, Germany, 2012; pp. 47–75.
12. Bagnol, D.; Lu, X.Y.; Kaelin, C.B.; Day, H.E.; Ollmann, M.; Gantz, I.; Akil, H.; Barsh, G.S.; Watson, S.J. Anatomy of an endogenous antagonist: Relationship between Agouti-related protein and proopiomelanocortin in brain. *J. Neurosci.* **1999**, *19*, RC26. [CrossRef] [PubMed]

13. Kleinridders, A.; Konner, A.C.; Bruning, J.C. CNS-targets in control of energy and glucose homeostasis. *Curr. Opin. Pharmacol.* **2009**, *9*, 794–804. [CrossRef] [PubMed]
14. Waterson, M.J.; Horvath, T.L. Neuronal Regulation of Energy Homeostasis: Beyond the Hypothalamus and Feeding. *Cell Metab.* **2015**, *22*, 962–970. [CrossRef] [PubMed]
15. Sohn, J.W.; Xu, Y.; Jones, J.E.; Wickman, K.; Williams, K.W.; Elmquist, J.K. Serotonin 2C receptor activates a distinct population of arcuate pro-opiomelanocortin neurons via TRPC channels. *Neuron* **2011**, *71*, 488–497. [CrossRef] [PubMed]
16. Heisler, L.K.; Jobst, E.E.; Sutton, G.M.; Zhou, L.; Borok, E.; Thornton-Jones, Z.; Liu, H.Y.; Zigman, J.M.; Balthasar, N.; Kishi, T.; et al. Serotonin reciprocally regulates melanocortin neurons to modulate food intake. *Neuron* **2006**, *51*, 239–249. [CrossRef] [PubMed]
17. Hardie, D.G. AMPK–sensing energy while talking to other signaling pathways. *Cell Metab.* **2014**, *20*, 939–952. [CrossRef] [PubMed]
18. Dasgupta, B.; Chhipa, R.R. Evolving Lessons on the Complex Role of AMPK in Normal Physiology and Cancer. *Trends Pharmacol. Sci.* **2016**, *37*, 192–206. [CrossRef] [PubMed]
19. Ross, F.A.; Jensen, T.E.; Hardie, D.G. Differential regulation by AMP and ADP of AMPK complexes containing different gamma subunit isoforms. *Biochem. J.* **2016**, *473*, 189–199. [CrossRef] [PubMed]
20. O'Neill, H.M. AMPK and Exercise: Glucose Uptake and Insulin Sensitivity. *Diabetes Metab. J.* **2013**, *37*, 1–21. [CrossRef] [PubMed]
21. Steinberg, G.R.; Kemp, B.E. AMPK in Health and Disease. *Physiol. Rev.* **2009**, *89*, 1025–1078. [CrossRef] [PubMed]
22. Hardie, D.G.; Ross, F.A.; Hawley, S.A. AMPK: A nutrient and energy sensor that maintains energy homeostasis. *Nat. Rev. Mol. Cell Biol.* **2012**, *13*, 251–262. [CrossRef] [PubMed]
23. Woods, A.; Johnstone, S.R.; Dickerson, K.; Leiper, F.C.; Fryer, L.G.; Neumann, D.; Schlattner, U.; Wallimann, T.; Carlson, M.; Carling, D. LKB1 is the upstream kinase in the AMP-activated protein kinase cascade. *Curr. Biol.* **2003**, *13*, 2004–2008. [CrossRef] [PubMed]
24. Shaw, R.J.; Kosmatka, M.; Bardeesy, N.; Hurley, R.L.; Witters, L.A.; DePinho, R.A.; Cantley, L.C. The tumor suppressor LKB1 kinase directly activates AMP-activated kinase and regulates apoptosis in response to energy stress. *Proc. Natl. Acad. Sci USA* **2004**, *101*, 3329–3335. [CrossRef] [PubMed]
25. Woods, A.; Dickerson, K.; Heath, R.; Hong, S.P.; Momcilovic, M.; Johnstone, S.R.; Carlson, M.; Carling, D. Ca^{2+}/calmodulin-dependent protein kinase kinase-beta acts upstream of AMP-activated protein kinase in mammalian cells. *Cell Metab.* **2005**, *2*, 21–33. [CrossRef] [PubMed]
26. Hawley, S.A.; Pan, D.A.; Mustard, K.J.; Ross, L.; Bain, J.; Edelman, A.M.; Frenguelli, B.G.; Hardie, D.G. Calmodulin-dependent protein kinase kinase-beta is an alternative upstream kinase for AMP-activated protein kinase. *Cell Metab.* **2005**, *2*, 9–19. [CrossRef] [PubMed]
27. Hurley, R.L.; Anderson, K.A.; Franzone, J.M.; Kemp, B.E.; Means, A.R.; Witters, L.A. The Ca^{2+}/calmodulin-dependent protein kinase kinases are AMP-activated protein kinase kinases. *J. Biol. Chem.* **2005**, *280*, 29060–29066. [CrossRef] [PubMed]
28. Momcilovic, M.; Hong, S.-P.; Carlson, M. Mammalian TAK1 activates Snf1 protein kinase in yeast and phosphorylates AMP-activated protein kinase in vitro. *J. Biol. Chem.* **2006**, *281*, 25336–25343. [CrossRef] [PubMed]
29. Xie, M.; Zhang, D.; Dyck, J.R.; Li, Y.; Zhang, H.; Morishima, M.; Mann, D.L.; Taffet, G.E.; Baldini, A.; Khoury, D.S. A pivotal role for endogenous TGF-β-activated kinase-1 in the LKB1/AMP-activated protein kinase energy-sensor pathway. *Proc. Natl. Acad. Sci. USA* **2006**, *103*, 17378–17383. [CrossRef] [PubMed]
30. Voss, M.; Paterson, J.; Kelsall, I.R.; Martin-Granados, C.; Hastie, C.J.; Peggie, M.W.; Cohen, P.T. Ppm1E is an in cellulo AMP-activated protein kinase phosphatase. *Cell Signal.* **2011**, *23*, 114–124. [CrossRef] [PubMed]

31. Davies, S.P.; Helps, N.R.; Cohen, P.T.; Hardie, D.G. 5′-AMP inhibits dephosphorylation, as well as promoting phosphorylation, of the AMP-activated protein kinase. Studies using bacterially expressed human protein phosphatase-2C alpha and native bovine protein phosphatase-2AC. *FEBS Lett.* **1995**, *377*, 421–425. [CrossRef] [PubMed]

32. Hurley, R.L.; Barre, L.K.; Wood, S.D.; Anderson, K.A.; Kemp, B.E.; Means, A.R.; Witters, L.A. Regulation of AMP-activated protein kinase by multisite phosphorylation in response to agents that elevate cellular cAMP. *J. Biol. Chem.* **2006**, *281*, 36662–36672. [CrossRef] [PubMed]

33. Pulinilkunnil, T.; He, H.; Kong, D.; Asakura, K.; Peroni, O.D.; Lee, A.; Kahn, B.B. Adrenergic regulation of AMP-activated protein kinase in brown adipose tissue in vivo. *J. Biol. Chem.* **2011**, *286*, 8798–8809. [CrossRef] [PubMed]

34. Hawley, S.A.; Ross, F.A.; Gowans, G.J.; Tibarewal, P.; Leslie, N.R.; Hardie, D.G. Phosphorylation by Akt within the ST loop of AMPK-alpha1 down-regulates its activation in tumour cells. *Biochem. J.* **2014**, *459*, 275–287. [CrossRef] [PubMed]

35. Horman, S.; Vertommen, D.; Heath, R.; Neumann, D.; Mouton, V.; Woods, A.; Schlattner, U.; Wallimann, T.; Carling, D.; Hue, L.; et al. Insulin antagonizes ischemia-induced Thr172 phosphorylation of AMP-activated protein kinase alpha-subunits in heart via hierarchical phosphorylation of Ser485/491. *J. Biol. Chem.* **2006**, *281*, 5335–5340. [CrossRef] [PubMed]

36. Ning, J.; Xi, G.; Clemmons, D.R. Suppression of AMPK activation via S485 phosphorylation by IGF-I during hyperglycemia is mediated by AKT activation in vascular smooth muscle cells. *Endocrinology* **2011**, *152*, 3143–3154. [CrossRef] [PubMed]

37. Dagon, Y.; Hur, E.; Zheng, B.; Wellenstein, K.; Cantley, L.C.; Kahn, B.B. p70S6 kinase phosphorylates AMPK on serine 491 to mediate leptin's effect on food intake. *Cell Metab.* **2012**, *16*, 104–112. [CrossRef] [PubMed]

38. Minokoshi, Y.; Kim, Y.B.; Peroni, O.D.; Fryer, L.G.; Muller, C.; Carling, D.; Kahn, B.B. Leptin stimulates fatty-acid oxidation by activating AMP-activated protein kinase. *Nature* **2002**, *415*, 339–343. [CrossRef] [PubMed]

39. Yamauchi, T.; Kamon, J.; Minokoshi, Y.; Ito, Y.; Waki, H.; Uchida, S.; Yamashita, S.; Noda, M.; Kita, S.; Ueki, K.; et al. Adiponectin stimulates glucose utilization and fatty-acid oxidation by activating AMP-activated protein kinase. *Nat. Med.* **2002**, *8*, 1288–1295. [CrossRef] [PubMed]

40. Martin, T.L.; Alquier, T.; Asakura, K.; Furukawa, N.; Preitner, F.; Kahn, B.B. Diet-induced obesity alters AMP kinase activity in hypothalamus and skeletal muscle. *J. Biol. Chem.* **2006**, *281*, 18933–18941. [CrossRef] [PubMed]

41. Cavaliere, G.; Viggiano, E.; Trinchese, G.; De Filippo, C.; Messina, A.; Monda, V.; Valenzano, A.; Cincione, R.I.; Zammit, C.; Cimmino, F.; et al. Long Feeding High-Fat Diet Induces Hypothalamic Oxidative Stress and Inflammation, and Prolonged Hypothalamic AMPK Activation in Rat Animal Model. *Front. Physiol.* **2018**, *9*, 818. [CrossRef] [PubMed]

42. McCrimmon, R.J.; Fan, X.; Ding, Y.; Zhu, W.; Jacob, R.J.; Sherwin, R.S. Potential role for AMP-activated protein kinase in hypoglycemia sensing in the ventromedial hypothalamus. *Diabetes* **2004**, *53*, 1953–1958. [CrossRef] [PubMed]

43. Cao, C.; Gao, T.; Cheng, M.; Xi, F.; Zhao, C.; Yu, W. Mild hypothermia ameliorates muscle wasting in septic rats associated with hypothalamic AMPK-induced autophagy and neuropeptides. *Biochem. Biophys. Res. Commun.* **2017**, *490*, 882–888. [CrossRef] [PubMed]

44. Kong, D.; Dagon, Y.; Campbell, J.N.; Guo, Y.; Yang, Z.; Yi, X.; Aryal, P.; Wellenstein, K.; Kahn, B.B.; Sabatini, B.L.; et al. A Postsynaptic AMPK → p21-Activated Kinase Pathway Drives Fasting-Induced Synaptic Plasticity in AgRP Neurons. *Neuron* **2016**, *91*, 25–33. [CrossRef] [PubMed]

45. Huynh, M.K.; Kinyua, A.W.; Yang, D.J.; Kim, K.W. Hypothalamic AMPK as a Regulator of Energy Homeostasis. *Neural Plast.* **2016**, *2016*, 2754078. [CrossRef] [PubMed]

46. Lopez, M.; Nogueiras, R.; Tena-Sempere, M.; Dieguez, C. Hypothalamic AMPK: A canonical regulator of whole-body energy balance. *Nat. Rev. Endocrinol.* **2016**, *12*, 421–432. [CrossRef] [PubMed]

47. Yang, C.S.; Lam, C.K.; Chari, M.; Cheung, G.W.; Kokorovic, A.; Gao, S.; Leclerc, I.; Rutter, G.A.; Lam, T.K. Hypothalamic AMP-activated protein kinase regulates glucose production. *Diabetes* **2010**, *59*, 2435–2443. [CrossRef] [PubMed]

48. Okamoto, S.; Sato, T.; Tateyama, M.; Kageyama, H.; Maejima, Y.; Nakata, M.; Hirako, S.; Matsuo, T.; Kyaw, S.; Shiuchi, T.; et al. Activation of AMPK-Regulated CRH Neurons in the PVH is Sufficient and Necessary to Induce Dietary Preference for Carbohydrate over Fat. *Cell Rep.* **2018**, *22*, 706–721. [CrossRef] [PubMed]

49. Kume, S.; Kondo, M.; Maeda, S.; Nishio, Y.; Yanagimachi, T.; Fujita, Y.; Haneda, M.; Kondo, K.; Sekine, A.; Araki, S.I.; et al. Hypothalamic AMP-Activated Protein Kinase Regulates Biphasic Insulin Secretion from Pancreatic beta Cells during Fasting and in Type 2 Diabetes. *eBioMedicine* **2016**, *13*, 168–180. [CrossRef] [PubMed]

50. Park, S.; Kim, D.S.; Kang, S.; Shin, B.K. Chronic activation of central AMPK attenuates glucose-stimulated insulin secretion and exacerbates hepatic insulin resistance in diabetic rats. *Brain Res. Bull.* **2014**, *108*, 18–26. [CrossRef] [PubMed]

51. Andersson, U.; Filipsson, K.; Abbott, C.R.; Woods, A.; Smith, K.; Bloom, S.R.; Carling, D.; Small, C.J. AMP-activated protein kinase plays a role in the control of food intake. *J. Biol. Chem.* **2004**, *279*, 12005–12008. [CrossRef] [PubMed]

52. Minokoshi, Y.; Alquier, T.; Furukawa, N.; Kim, Y.B.; Lee, A.; Xue, B.; Mu, J.; Foufelle, F.; Ferre, P.; Birnbaum, M.J.; et al. AMP-kinase regulates food intake by responding to hormonal and nutrient signals in the hypothalamus. *Nature* **2004**, *428*, 569–574. [CrossRef] [PubMed]

53. Namkoong, C.; Kim, M.S.; Jang, P.G.; Han, S.M.; Park, H.S.; Koh, E.H.; Lee, W.J.; Kim, J.Y.; Park, I.S.; Park, J.Y. Enhanced hypothalamic AMP-activated protein kinase activity contributes to hyperphagia in diabetic rats. *Diabetes* **2005**, *54*, 63–68. [CrossRef] [PubMed]

54. Claret, M.; Smith, M.A.; Batterham, R.L.; Selman, C.; Choudhury, A.I.; Fryer, L.G.; Clements, M.; Al-Qassab, H.; Heffron, H.; Xu, A.W.; et al. AMPK is essential for energy homeostasis regulation and glucose sensing by POMC and AgRP neurons. *J. Clin. Investig.* **2007**, *117*, 2325–2336. [CrossRef] [PubMed]

55. Oh, T.S.; Cho, H.; Cho, J.H.; Yu, S.W.; Kim, E.K. Hypothalamic AMPK-induced autophagy increases food intake by regulating NPY and POMC expression. *Autophagy* **2016**, *12*, 2009–2025. [CrossRef] [PubMed]

56. Yavari, A.; Stocker, C.J.; Ghaffari, S.; Wargent, E.T.; Steeples, V.; Czibik, G.; Pinter, K.; Bellahcene, M.; Woods, A.; Martinez de Morentin, P.B.; et al. Chronic Activation of gamma2 AMPK Induces Obesity and Reduces beta Cell Function. *Cell Metab.* **2016**, *23*, 821–836. [CrossRef] [PubMed]

57. Turcot, V.; Lu, Y.; Highland, H.M.; Schurmann, C.; Justice, A.E.; Fine, R.S.; Bradfield, J.P.; Esko, T.; Giri, A.; Graff, M.; et al. Protein-altering variants associated with body mass index implicate pathways that control energy intake and expenditure in obesity. *Nat. Genet.* **2018**, *50*, 26–41. [CrossRef] [PubMed]

58. Yang, Y.; Atasoy, D.; Su, H.H.; Sternson, S.M. Hunger states switch a flip-flop memory circuit via a synaptic AMPK-dependent positive feedback loop. *Cell* **2011**, *146*, 992–1003. [CrossRef] [PubMed]

59. Kwok, K.H.; Lam, K.S.; Xu, A. Heterogeneity of white adipose tissue: Molecular basis and clinical implications. *Exp. Mol. Med.* **2016**, *48*, e215. [CrossRef] [PubMed]

60. Friedman, J.M.; Halaas, J.L. Leptin and the regulation of body weight in mammals. *Nature* **1998**, *395*, 763–770. [CrossRef] [PubMed]

61. Cheung, C.C.; Clifton, D.K.; Steiner, R.A. Proopiomelanocortin neurons are direct targets for leptin in the hypothalamus. *Endocrinology* **1997**, *138*, 4489–4492. [CrossRef] [PubMed]

62. Hakansson, M.L.; Brown, H.; Ghilardi, N.; Skoda, R.C.; Meister, B. Leptin receptor immunoreactivity in chemically defined target neurons of the hypothalamus. *J. Neurosci.* **1998**, *18*, 559–572. [CrossRef] [PubMed]

63. Mercer, J.G.; Hoggard, N.; Williams, L.M.; Lawrence, C.B.; Hannah, L.T.; Morgan, P.J.; Trayhurn, P. Coexpression of leptin receptor and preproneuropeptide Y mRNA in arcuate nucleus of mouse hypothalamus. *J. Neuroendocrinol.* **1996**, *8*, 733–735. [CrossRef] [PubMed]

64. Pelleymounter, M.A.; Cullen, M.J.; Baker, M.B.; Hecht, R. Effects of the obese gene product on body weight regulation in ob/ob mice. *Science* **1995**, *269*, 540–543. [CrossRef] [PubMed]

65. Chen, H.; Charlat, O.; Tartaglia, L.A.; Woolf, E.A.; Weng, X.; Ellis, S.J.; Lakey, N.D.; Culpepper, J.; Moore, K.J.; Breitbart, R.E.; et al. Evidence that the diabetes gene encodes the leptin receptor: Identification of a mutation in the leptin receptor gene in db/db mice. *Cell* **1996**, *84*, 491–495. [CrossRef]

66. Clement, K.; Vaisse, C.; Lahlou, N.; Cabrol, S.; Pelloux, V.; Cassuto, D.; Gourmelen, M.; Dina, C.; Chambaz, J.; Lacorte, J.M.; et al. A mutation in the human leptin receptor gene causes obesity and pituitary dysfunction. *Nature* **1998**, *392*, 398–401. [CrossRef] [PubMed]

67. Cowley, M.A.; Smart, J.L.; Rubinstein, M.; Cerdan, M.G.; Diano, S.; Horvath, T.L.; Cone, R.D.; Low, M.J. Leptin activates anorexigenic POMC neurons through a neural network in the arcuate nucleus. *Nature* **2001**, *411*, 480–484. [CrossRef] [PubMed]

68. Poleni, P.E.; Akieda-Asai, S.; Koda, S.; Sakurai, M.; Bae, C.R.; Senba, K.; Cha, Y.S.; Furuya, M.; Date, Y. Possible involvement of melanocortin-4-receptor and AMP-activated protein kinase in the interaction of glucagon-like peptide-1 and leptin on feeding in rats. *Biochem. Biophys. Res. Commun.* **2012**, *420*, 36–41. [CrossRef] [PubMed]

69. Cota, D.; Matter, E.K.; Woods, S.C.; Seeley, R.J. The role of hypothalamic mammalian target of rapamycin complex 1 signaling in diet-induced obesity. *J. Neurosci.* **2008**, *28*, 7202–7208. [CrossRef] [PubMed]

70. Watterson, K.R.; Bestow, D.; Gallagher, J.; Hamilton, D.L.; Ashford, F.B.; Meakin, P.J.; Ashford, M.L. Anorexigenic and orexigenic hormone modulation of mammalian target of rapamycin complex 1 activity and the regulation of hypothalamic agouti-related protein mRNA expression. *Neurosignals* **2013**, *21*, 28–41. [CrossRef] [PubMed]

71. Gao, S.; Kinzig, K.P.; Aja, S.; Scott, K.A.; Keung, W.; Kelly, S.; Strynadka, K.; Chohnan, S.; Smith, W.W.; Tamashiro, K.L.; et al. Leptin activates hypothalamic acetyl-CoA carboxylase to inhibit food intake. *Proc. Natl. Acad. Sci. USA* **2007**, *104*, 17358–17363. [CrossRef] [PubMed]

72. Gill, J.F.; Delezie, J.; Santos, G.; Handschin, C. PGC-1alpha expression in murine AgRP neurons regulates food intake and energy balance. *Mol. Metab.* **2016**, *5*, 580–588. [CrossRef] [PubMed]

73. Galic, S.; Loh, K.; Murray-Segal, L.; Steinberg, G.R.; Andrews, Z.B.; Kemp, B.E. AMPK signaling to acetyl-CoA carboxylase is required for fasting- and cold-induced appetite but not thermogenesis. *eLife* **2018**, *7*. [CrossRef] [PubMed]

74. Wang, Z.V.; Scherer, P.E. Adiponectin, the past two decades. *J. Mol. Cell Biol.* **2016**, *8*, 93–100. [CrossRef] [PubMed]

75. Cheng, K.K.; Lam, K.S.; Wang, B.; Xu, A. Signaling mechanisms underlying the insulin-sensitizing effects of adiponectin. *Best Pract. Res. Clin. Endocrinol. Metab.* **2014**, *28*, 3–13. [CrossRef] [PubMed]

76. Hotta, K.; Funahashi, T.; Arita, Y.; Takahashi, M.; Matsuda, M.; Okamoto, Y.; Iwahashi, H.; Kuriyama, H.; Ouchi, N.; Maeda, K.; et al. Plasma concentrations of a novel, adipose-specific protein, adiponectin, in type 2 diabetic patients. *Arterioscler. Thromb. Vasc. Biol.* **2000**, *20*, 1595–1599. [CrossRef] [PubMed]

77. Weyer, C.; Funahashi, T.; Tanaka, S.; Hotta, K.; Matsuzawa, Y.; Pratley, R.E.; Tataranni, P.A. Hypoadiponectinemia in obesity and type 2 diabetes: Close association with insulin resistance and hyperinsulinemia. *J. Clin. Endocrinol. Metab.* **2001**, *86*, 1930–1935. [CrossRef] [PubMed]

78. Thundyil, J.; Pavlovski, D.; Sobey, C.G.; Arumugam, T.V. Adiponectin receptor signalling in the brain. *Br. J. Pharmacol.* **2012**, *165*, 313–327. [CrossRef] [PubMed]

79. Qi, Y.; Takahashi, N.; Hileman, S.M.; Patel, H.R.; Berg, A.H.; Pajvani, U.B.; Scherer, P.E.; Ahima, R.S. Adiponectin acts in the brain to decrease body weight. *Nat. Med.* **2004**, *10*, 524–529. [CrossRef] [PubMed]

80. Kusminski, C.M.; McTernan, P.G.; Schraw, T.; Kos, K.; O'Hare, J.P.; Ahima, R.; Kumar, S.; Scherer, P.E. Adiponectin complexes in human cerebrospinal fluid: Distinct complex distribution from serum. *Diabetologia* **2007**, *50*, 634–642. [CrossRef] [PubMed]

81. Neumeier, M.; Weigert, J.; Buettner, R.; Wanninger, J.; Schaffler, A.; Muller, A.M.; Killian, S.; Sauerbruch, S.; Schlachetzki, F.; Steinbrecher, A.; et al. Detection of adiponectin in cerebrospinal fluid in humans. *Am. J. Physiol. Endocrinol. Metab.* **2007**, *293*, E965–E969. [CrossRef] [PubMed]

82. Spranger, J.; Verma, S.; Gohring, I.; Bobbert, T.; Seifert, J.; Sindler, A.L.; Pfeiffer, A.; Hileman, S.M.; Tschop, M.; Banks, W.A. Adiponectin does not cross the blood-brain barrier but modifies cytokine expression of brain endothelial cells. *Diabetes* **2006**, *55*, 141–147. [CrossRef] [PubMed]

83. Pan, W.; Tu, H.; Kastin, A.J. Differential BBB interactions of three ingestive peptides: Obestatin, ghrelin, and adiponectin. *Peptides* **2006**, *27*, 911–916. [CrossRef] [PubMed]

84. Klein, I.; Sanchez-Alavez, M.; Tabarean, I.; Schaefer, J.; Holmberg, K.H.; Klaus, J.; Xia, F.; Marcondes, M.C.; Dubins, J.S.; Morrison, B.; et al. AdipoR1 and 2 are expressed on warm sensitive neurons of the hypothalamic preoptic area and contribute to central hyperthermic effects of adiponectin. *Brain Res.* **2011**, *1423*, 1–9. [CrossRef] [PubMed]

85. Guillod-Maximin, E.; Roy, A.F.; Vacher, C.; Aubourg, A.; Bailleux, V.; Lorsignol, A.; Pénicaud, L.; Parquet, M.; Taouis, M. Adiponectin receptors are expressed in hypothalamus and colocalized with proopiomelanocortin and neuropeptide Y in rodent arcuate neurons. *J. Endocrinol.* **2009**, *200*, 93–105. [CrossRef] [PubMed]

86. Benomar, Y.; Amine, H.; Crepin, D.; Al Rifai, S.; Riffault, L.; Gertler, A.; Taouis, M. Central Resistin/TLR4 Impairs Adiponectin Signaling, Contributing to Insulin and FGF21 Resistance. *Diabetes* **2016**, *65*, 913–926. [CrossRef] [PubMed]

87. Wang, B.; Li, A.; Li, X.; Ho, P.W.; Wu, D.; Wang, X.; Liu, Z.; Wu, K.K.; Yau, S.S.; Xu, A.; et al. Activation of hypothalamic RIP-Cre neurons promotes beiging of WAT via sympathetic nervous system. *EMBO Rep.* **2018**, *19*. [CrossRef] [PubMed]

88. Cheng, K.K.; Lam, K.S.; Wang, Y.; Huang, Y.; Carling, D.; Wu, D.; Wong, C.; Xu, A. Adiponectin-induced endothelial nitric oxide synthase activation and nitric oxide production are mediated by APPL1 in endothelial cells. *Diabetes* **2007**, *56*, 1387–1394. [CrossRef] [PubMed]

89. Kubota, N.; Yano, W.; Kubota, T.; Yamauchi, T.; Itoh, S.; Kumagai, H.; Kozono, H.; Takamoto, I.; Okamoto, S.; Shiuchi, T.; et al. Adiponectin stimulates AMP-activated protein kinase in the hypothalamus and increases food intake. *Cell Metab.* **2007**, *6*, 55–68. [CrossRef] [PubMed]

90. Coope, A.; Milanski, M.; Araujo, E.P.; Tambascia, M.; Saad, M.J.; Geloneze, B.; Velloso, L.A. AdipoR1 mediates the anorexigenic and insulin/leptin-like actions of adiponectin in the hypothalamus. *FEBS Lett.* **2008**, *582*, 1471–1476. [CrossRef] [PubMed]

91. Halah, M.P.; Marangon, P.B.; Antunes-Rodrigues, J.; Elias, L.L.K. Neonatal nutritional programming impairs adiponectin effects on energy homeostasis in adult life of male rats. *Am. J. Physiol. Endocrinol. Metab.* **2018**, *315*, E29–E37. [CrossRef] [PubMed]

92. Quaresma, P.G.; Reencober, N.; Zanotto, T.M.; Santos, A.C.; Weissmann, L.; de Matos, A.H.; Lopes-Cendes, I.; Folli, F.; Saad, M.J.; Prada, P.O. Pioglitazone treatment increases food intake and decreases energy expenditure partially via hypothalamic adiponectin/adipoR1/AMPK pathway. *Int. J. Obes.* **2016**, *40*, 138–146. [CrossRef] [PubMed]

93. Sun, J.; Gao, Y.; Yao, T.; Huang, Y.; He, Z.; Kong, X.; Yu, K.J.; Wang, R.T.; Guo, H.; Yan, J.; et al. Adiponectin potentiates the acute effects of leptin in arcuate Pomc neurons. *Mol. Metab.* **2016**, *5*, 882–891. [CrossRef] [PubMed]

94. Suyama, S.; Maekawa, F.; Maejima, Y.; Kubota, N.; Kadowaki, T.; Yada, T. Glucose level determines excitatory or inhibitory effects of adiponectin on arcuate POMC neuron activity and feeding. *Sci. Rep.* **2016**, *6*, 30796. [CrossRef] [PubMed]

95. Suyama, S.; Lei, W.; Kubota, N.; Kadowaki, T.; Yada, T. Adiponectin at physiological level glucose-independently enhances inhibitory postsynaptic current onto NPY neurons in the hypothalamic arcuate nucleus. *Neuropeptides* **2017**, *65*, 1–9. [CrossRef] [PubMed]

96. Wang, C.; Mao, X.; Wang, L.; Liu, M.; Wetzel, M.D.; Guan, K.L.; Dong, L.Q.; Liu, F. Adiponectin sensitizes insulin signaling by reducing p70 S6 kinase-mediated serine phosphorylation of IRS-1. *J. Biol. Chem.* **2007**, *282*, 7991–7996. [CrossRef] [PubMed]

97. Okada-Iwabu, M.; Yamauchi, T.; Iwabu, M.; Honma, T.; Hamagami, K.; Matsuda, K.; Yamaguchi, M.; Tanabe, H.; Kimura-Someya, T.; Shirouzu, M.; et al. A small-molecule AdipoR agonist for type 2 diabetes and short life in obesity. *Nature* **2013**, *503*, 493–499. [CrossRef] [PubMed]

98. Nicolas, S.; Debayle, D.; Bechade, C.; Maroteaux, L.; Gay, A.S.; Bayer, P.; Heurteaux, C.; Guyon, A.; Chabry, J. Adiporon, an adiponectin receptor agonist acts as an antidepressant and metabolic regulator in a mouse model of depression. *Transl. Psychiatry* **2018**, *8*, 159. [CrossRef] [PubMed]

99. Wren, A.; Seal, L.; Cohen, M.; Brynes, A.; Frost, G.; Murphy, K.; Dhillo, W.; Ghatei, M.; Bloom, S. Ghrelin enhances appetite and increases food intake in humans. *J. Clin. Endocrinol. Metab.* **2001**, *86*, 5992–5995. [CrossRef] [PubMed]

100. Chen, H.Y.; Trumbauer, M.E.; Chen, A.S.; Weingarth, D.T.; Adams, J.R.; Frazier, E.G.; Shen, Z.; Marsh, D.J.; Feighner, S.D.; Guan, X.M.; et al. Orexigenic action of peripheral ghrelin is mediated by neuropeptide Y and agouti-related protein. *Endocrinology* **2004**, *145*, 2607–2612. [CrossRef] [PubMed]

101. López, M.; Lage, R.; Saha, A.K.; Pérez-Tilve, D.; Vázquez, M.J.; Varela, L.; Sangiao-Alvarellos, S.; Tovar, S.; Raghay, K.; Rodríguez-Cuenca, S. Hypothalamic fatty acid metabolism mediates the orexigenic action of ghrelin. *Cell Metab.* **2008**, *7*, 389–399. [CrossRef] [PubMed]

102. Gao, S.; Casals, N.; Keung, W.; Moran, T.H.; Lopaschuk, G.D. Differential effects of central ghrelin on fatty acid metabolism in hypothalamic ventral medial and arcuate nuclei. *Physiol. Behav.* **2013**, *118*, 165–170. [CrossRef] [PubMed]

103. Lim, C.T.; Kola, B.; Feltrin, D.; Perez-Tilve, D.; Tschop, M.H.; Grossman, A.B.; Korbonits, M. Ghrelin and cannabinoids require the ghrelin receptor to affect cellular energy metabolism. *Mol. Cell. Endocrinol.* **2013**, *365*, 303–308. [CrossRef] [PubMed]

104. Anderson, K.A.; Ribar, T.J.; Lin, F.; Noeldner, P.K.; Green, M.F.; Muehlbauer, M.J.; Witters, L.A.; Kemp, B.E.; Means, A.R. Hypothalamic CaMKK2 contributes to the regulation of energy balance. *Cell Metab.* **2008**, *7*, 377–388. [CrossRef] [PubMed]

105. Komori, T.; Doi, A.; Nosaka, T.; Furuta, H.; Akamizu, T.; Kitamura, T.; Senba, E.; Morikawa, Y. Regulation of AMP-activated protein kinase signaling by AFF4 protein, member of AF4 (ALL1-fused gene from chromosome 4) family of transcription factors, in hypothalamic neurons. *J. Biol. Chem.* **2012**, *287*, 19985–19996. [CrossRef] [PubMed]

106. Velásquez, D.A.; Martínez, G.; Romero, A.; Vázquez, M.J.; Boit, K.D.; Dopeso-Reyes, I.G.; López, M.; Vidal, A.; Nogueiras, R.; Diéguez, C. The central sirtuin1/p53 pathway is essential for the orexigenic action of ghrelin. *Diabetes* **2011**, DB_100802.

107. Lockie, S.H.; Stark, R.; Mequinion, M.; Ch'ng, S.; Kong, D.; Spanswick, D.C.; Lawrence, A.J.; Andrews, Z.B. Glucose Availability Predicts the Feeding Response to Ghrelin in Male Mice, an Effect Dependent on AMPK in AgRP Neurons. *Endocrinology* **2018**, *159*, 3605–3614. [CrossRef] [PubMed]

108. Andrews, Z.B.; Liu, Z.W.; Walllingford, N.; Erion, D.M.; Borok, E.; Friedman, J.M.; Tschöp, M.H.; Shanabrough, M.; Cline, G.; Shulman, G.I. UCP2 mediates ghrelin's action on NPY/AgRP neurons by lowering free radicals. *Nature* **2008**, *454*, 846–851. [CrossRef] [PubMed]

109. Kola, B.; Hubina, E.; Tucci, S.A.; Kirkham, T.C.; Garcia, E.A.; Mitchell, S.E.; Williams, L.M.; Hawley, S.A.; Hardie, D.G.; Grossman, A.B. Cannabinoids and ghrelin have both central and peripheral metabolic and cardiac effects via AMP-activated protein kinase. *J. Biol. Chem.* **2005**, *280*, 25196–25201. [CrossRef] [PubMed]

110. Wren, A.; Small, C.; Ward, H.; Murphy, K.; Dakin, C.; Taheri, S.; Kennedy, A.; Roberts, G.; Morgan, D.; Ghatei, M. The novel hypothalamic peptide ghrelin stimulates food intake and growth hormone secretion. *Endocrinology* **2000**, *141*, 4325–4328. [CrossRef] [PubMed]

111. Andrews, Z.B. Central mechanisms involved in the orexigenic actions of ghrelin. *Peptides* **2011**, *32*, 2248–2255. [CrossRef] [PubMed]

112. Kohno, D.; Sone, H.; Minokoshi, Y.; Yada, T. Ghrelin raises $[Ca^{2+}]$ i via AMPK in hypothalamic arcuate nucleus NPY neurons. *Biochem. Biophys. Res. Commun.* **2008**, *366*, 388–392. [CrossRef] [PubMed]

113. Sipols, A.J.; Baskin, D.G.; Schwartz, M.W. Effect of intracerebroventricular insulin infusion on diabetic hyperphagia and hypothalamic neuropeptide gene expression. *Diabetes* **1995**, *44*, 147–151. [CrossRef] [PubMed]

114. Schwartz, M.W.; Figlewicz, D.P.; Baskin, D.G.; Woods, S.C.; Porte, D., Jr. Insulin in the brain: A hormonal regulator of energy balance. *Endocr. Rev.* **1992**, *13*, 387–414. [PubMed]

115. Arnold, S.E.; Arvanitakis, Z.; Macauley-Rambach, S.L.; Koenig, A.M.; Wang, H.Y.; Ahima, R.S.; Craft, S.; Gandy, S.; Buettner, C.; Stoeckel, L.E.; et al. Brain insulin resistance in type 2 diabetes and Alzheimer disease: Concepts and conundrums. *Nat. Rev. Neurol.* **2018**, *14*, 168–181. [CrossRef] [PubMed]

116. Cetinkalp, S.; Simsir, I.Y.; Ertek, S. Insulin resistance in brain and possible therapeutic approaches. *Curr. Vasc. Pharmacol.* **2014**, *12*, 553–564. [CrossRef] [PubMed]

117. Bruning, J.C.; Gautam, D.; Burks, D.J.; Gillette, J.; Schubert, M.; Orban, P.C.; Klein, R.; Krone, W.; Muller-Wieland, D.; Kahn, C.R. Role of brain insulin receptor in control of body weight and reproduction. *Science* **2000**, *289*, 2122–2125. [CrossRef] [PubMed]

118. Woods, S.C.; Lotter, E.C.; McKay, L.D.; Porte, D., Jr. Chronic intracerebroventricular infusion of insulin reduces food intake and body weight of baboons. *Nature* **1979**, *282*, 503–505. [CrossRef] [PubMed]

119. Woods, S.C.; Seeley, R.J.; Porte, D., Jr.; Schwartz, M.W. Signals that regulate food intake and energy homeostasis. *Science* **1998**, *280*, 1378–1383. [CrossRef] [PubMed]

120. Richardson, R.D.; Ramsay, D.S.; Lernmark, A.; Scheurink, A.J.; Baskin, D.G.; Woods, S.C. Weight loss in rats following intraventricular transplants of pancreatic islets. *Am. J. Physiol.* **1994**, *266*, R59–R64. [CrossRef] [PubMed]

121. Air, E.L.; Strowski, M.Z.; Benoit, S.C.; Conarello, S.L.; Salituro, G.M.; Guan, X.M.; Liu, K.; Woods, S.C.; Zhang, B.B. Small molecule insulin mimetics reduce food intake and body weight and prevent development of obesity. *Nat. Med.* **2002**, *8*, 179–183. [CrossRef] [PubMed]

122. McGowan, M.K.; Andrews, K.M.; Grossman, S.P. Chronic intrahypothalamic infusions of insulin or insulin antibodies alter body weight and food intake in the rat. *Physiol. Behav.* **1992**, *51*, 753–766. [CrossRef]

123. Solon, C.S.; Franci, D.; Ignacio-Souza, L.M.; Romanatto, T.; Roman, E.A.; Arruda, A.P.; Morari, J.; Torsoni, A.S.; Carneiro, E.M.; Velloso, L.A. Taurine enhances the anorexigenic effects of insulin in the hypothalamus of rats. *Amino Acids* **2012**, *42*, 2403–2410. [CrossRef] [PubMed]

124. Cai, F.; Gyulkhandanyan, A.V.; Wheeler, M.B.; Belsham, D.D. Glucose regulates AMP-activated protein kinase activity and gene expression in clonal, hypothalamic neurons expressing proopiomelanocortin: Additive effects of leptin or insulin. *J. Endocrinol.* **2007**, *192*, 605–614. [CrossRef] [PubMed]

125. Roman, E.A.; Cesquini, M.; Stoppa, G.R.; Carvalheira, J.B.; Torsoni, M.A.; Velloso, L.A. Activation of AMPK in rat hypothalamus participates in cold-induced resistance to nutrient-dependent anorexigenic signals. *J. Physiol.* **2005**, *568*, 993–1001. [CrossRef] [PubMed]

126. Han, S.M.; Namkoong, C.; Jang, P.G.; Park, I.S.; Hong, S.W.; Katakami, H.; Chun, S.; Kim, S.W.; Park, J.Y.; Lee, K.U.; et al. Hypothalamic AMP-activated protein kinase mediates counter-regulatory responses to hypoglycaemia in rats. *Diabetologia* **2005**, *48*, 2170–2178. [CrossRef] [PubMed]

127. Valentine, R.J.; Coughlan, K.A.; Ruderman, N.B.; Saha, A.K. Insulin inhibits AMPK activity and phosphorylates AMPK Ser(4)(8)(5)/(4)(9)(1) through Akt in hepatocytes, myotubes and incubated rat skeletal muscle. *Arch. Biochem. Biophys.* **2014**, *562*, 62–69. [CrossRef] [PubMed]

128. Lim, G.E.; Brubaker, P.L. Glucagon-like peptide 1 secretion by the L-cell: The view from within. *Diabetes* **2006**, *55*, S70–S77. [CrossRef]

129. Goldstone, A.P.; Morgan, I.; Mercer, J.G.; Morgan, D.G.; Moar, K.M.; Ghatei, M.A.; Bloom, S.R. Effect of leptin on hypothalamic GLP-1 peptide and brain-stem pre-proglucagon mRNA. *Biochem. Biophys. Res. Commun.* **2000**, *269*, 331–335. [CrossRef] [PubMed]

130. Trapp, S.; Richards, J.E. The gut hormone glucagon-like peptide-1 produced in brain: Is this physiologically relevant? *Curr. Opin. Pharmacol.* **2013**, *13*, 964–969. [CrossRef] [PubMed]

131. Turton, M.D.; O'Shea, D.; Gunn, I.; Beak, S.A.; Edwards, C.M.; Meeran, K.; Choi, S.J.; Taylor, G.M.; Heath, M.M.; Lambert, P.D.; et al. A role for glucagon-like peptide-1 in the central regulation of feeding. *Nature* **1996**, *379*, 69–72. [CrossRef] [PubMed]

132. Seo, S.; Ju, S.; Chung, H.; Lee, D.; Park, S. Acute effects of glucagon-like peptide-1 on hypothalamic neuropeptide and AMP activated kinase expression in fasted rats. *Endocr. J.* **2008**, *55*, 867–874. [CrossRef] [PubMed]

133. Hurtado-Carneiro, V.; Sanz, C.; Roncero, I.; Vazquez, P.; Blazquez, E.; Alvarez, E. Glucagon-like peptide 1 (GLP-1) can reverse AMP-activated protein kinase (AMPK) and S6 kinase (P70S6K) activities induced by fluctuations in glucose levels in hypothalamic areas involved in feeding behaviour. *Mol. Neurobiol.* **2012**, *45*, 348–361. [CrossRef] [PubMed]

134. Williams, D.L.; Hyvarinen, N.; Lilly, N.; Kay, K.; Dossat, A.; Parise, E.; Torregrossa, A.M. Maintenance on a high-fat diet impairs the anorexic response to glucagon-like-peptide-1 receptor activation. *Physiol. Behav.* **2011**, *103*, 557–564. [CrossRef] [PubMed]

135. Burmeister, M.A.; Ayala, J.; Drucker, D.J.; Ayala, J.E. Central glucagon-like peptide 1 receptor-induced anorexia requires glucose metabolism-mediated suppression of AMPK and is impaired by central fructose. *Am. J. Physiol. Endocrinol. Metab.* **2013**, *304*, E677–E685. [CrossRef] [PubMed]

136. Shao, W.; Wang, D.; Chiang, Y.T.; Ip, W.; Zhu, L.; Xu, F.; Columbus, J.; Belsham, D.D.; Irwin, D.M.; Zhang, H.; et al. The Wnt signaling pathway effector TCF7L2 controls gut and brain proglucagon gene expression and glucose homeostasis. *Diabetes* **2013**, *62*, 789–800. [CrossRef] [PubMed]

137. Hayes, M.R.; Leichner, T.M.; Zhao, S.; Lee, G.S.; Chowansky, A.; Zimmer, D.; De Jonghe, B.C.; Kanoski, S.E.; Grill, H.J.; Bence, K.K. Intracellular signals mediating the food intake-suppressive effects of hindbrain glucagon-like peptide-1 receptor activation. *Cell Metab.* **2011**, *13*, 320–330. [CrossRef] [PubMed]

138. Beiroa, D.; Imbernon, M.; Gallego, R.; Senra, A.; Herranz, D.; Villarroya, F.; Serrano, M.; Ferno, J.; Salvador, J.; Escalada, J.; et al. GLP-1 agonism stimulates brown adipose tissue thermogenesis and browning through hypothalamic AMPK. *Diabetes* **2014**, *63*, 3346–3358. [CrossRef] [PubMed]

139. Burmeister, M.A.; Brown, J.D.; Ayala, J.E.; Stoffers, D.A.; Sandoval, D.A.; Seeley, R.J.; Ayala, J.E. The glucagon-like peptide-1 receptor in the ventromedial hypothalamus reduces short-term food intake in male mice by regulating nutrient sensor activity. *Am. J. Physiol. Endocrinol. Metab.* **2017**, *313*, E651–E662. [CrossRef] [PubMed]

140. Barbatelli, G.; Murano, I.; Madsen, L.; Hao, Q.; Jimenez, M.; Kristiansen, K.; Giacobino, J.; De Matteis, R.; Cinti, S. The emergence of cold-induced brown adipocytes in mouse white fat depots is determined predominantly by white to brown adipocyte transdifferentiation. *Am. J. Physiol. Endocrinol. Metab.* **2010**, *298*, E1244–E1253. [CrossRef] [PubMed]

141. Rosen, E.D.; Spiegelman, B.M. What we talk about when we talk about fat. *Cell* **2014**, *156*, 20–44. [CrossRef] [PubMed]

142. Li, G.; Xie, C.; Lu, S.; Nichols, R.G.; Tian, Y.; Li, L.; Patel, D.; Ma, Y.; Brocker, C.N.; Yan, T.; et al. Intermittent Fasting Promotes White Adipose Browning and Decreases Obesity by Shaping the Gut Microbiota. *Cell Metab.* **2017**, *26*, 672–685. [CrossRef] [PubMed]

143. Stanford, K.I.; Middelbeek, R.J.; Goodyear, L.J. Exercise Effects on White Adipose Tissue: Beiging and Metabolic Adaptations. *Diabetes* **2015**, *64*, 2361–2368. [CrossRef] [PubMed]

144. Bartelt, A.; Bruns, O.T.; Reimer, R.; Hohenberg, H.; Ittrich, H.; Peldschus, K.; Kaul, M.G.; Tromsdorf, U.I.; Weller, H.; Waurisch, C.; et al. Brown adipose tissue activity controls triglyceride clearance. *Nat. Med.* **2011**, *17*, 200–205. [CrossRef] [PubMed]

145. Seale, P.; Conroe, H.M.; Estall, J.; Kajimura, S.; Frontini, A.; Ishibashi, J.; Cohen, P.; Cinti, S.; Spiegelman, B.M. Prdm16 determines the thermogenic program of subcutaneous white adipose tissue in mice. *J. Clin. Investig.* **2011**, *121*, 96–105. [CrossRef] [PubMed]

146. Liu, W.; Bi, P.; Shan, T.; Yang, X.; Yin, H.; Wang, Y.-X.; Liu, N.; Rudnicki, M.A.; Kuang, S. miR-133a regulates adipocyte browning in vivo. *PLoS Genet.* **2013**, *9*, e1003626. [CrossRef] [PubMed]

147. Castillo-Quan, J.I. From white to brown fat through the PGC-1α-dependent myokine irisin: Implications for diabetes and obesity. *Dis. Model. Mech.* **2012**, *5*, 293–295. [CrossRef] [PubMed]

148. Bi, P.; Shan, T.; Liu, W.; Yue, F.; Yang, X.; Liang, X.R.; Wang, J.; Li, J.; Carlesso, N.; Liu, X.; et al. Inhibition of Notch signaling promotes browning of white adipose tissue and ameliorates obesity. *Nat. Med.* **2014**, *20*, 911–918. [CrossRef] [PubMed]

149. Contreras, C.; Nogueiras, R.; Dieguez, C.; Medina-Gomez, G.; Lopez, M. Hypothalamus and thermogenesis: Heating the BAT, browning the WAT. *Mol. Cell Endocrinol.* **2016**, *438*, 107–115. [CrossRef] [PubMed]

150. Lidell, M.E.; Betz, M.J.; Dahlqvist Leinhard, O.; Heglind, M.; Elander, L.; Slawik, M.; Mussack, T.; Nilsson, D.; Romu, T.; Nuutila, P.; et al. Evidence for two types of brown adipose tissue in humans. *Nat. Med.* **2013**, *19*, 631–634. [CrossRef] [PubMed]

151. Wu, J.; Bostrom, P.; Sparks, L.M.; Ye, L.; Choi, J.H.; Giang, A.H.; Khandekar, M.; Virtanen, K.A.; Nuutila, P.; Schaart, G.; et al. Beige adipocytes are a distinct type of thermogenic fat cell in mouse and human. *Cell* **2012**, *150*, 366–376. [CrossRef] [PubMed]

152. Harms, M.; Seale, P. Brown and beige fat: Development, function and therapeutic potential. *Nat. Med.* **2013**, *19*, 1252–1263. [CrossRef] [PubMed]

153. Viollet, B.; Andreelli, F.; Jorgensen, S.B.; Perrin, C.; Geloen, A.; Flamez, D.; Mu, J.; Lenzner, C.; Baud, O.; Bennoun, M.; et al. The AMP-activated protein kinase alpha2 catalytic subunit controls whole-body insulin sensitivity. *J. Clin. Investig.* **2003**, *111*, 91–98. [CrossRef] [PubMed]

154. Xue, B.; Pulinilkunnil, T.; Murano, I.; Bence, K.K.; He, H.; Minokoshi, Y.; Asakura, K.; Lee, A.; Haj, F.; Furukawa, N.; et al. Neuronal protein tyrosine phosphatase 1B deficiency results in inhibition of hypothalamic AMPK and isoform-specific activation of AMPK in peripheral tissues. *Mol. Cell. Biol.* **2009**, *29*, 4563–4573. [CrossRef] [PubMed]

155. Tanida, M.; Yamamoto, N.; Shibamoto, T.; Rahmouni, K. Involvement of hypothalamic AMP-activated protein kinase in leptin-induced sympathetic nerve activation. *PLoS ONE* **2013**, *8*, e56660. [CrossRef] [PubMed]

156. Banno, R.; Zimmer, D.; De Jonghe, B.C.; Atienza, M.; Rak, K.; Yang, W.; Bence, K.K. PTP1B and SHP2 in POMC neurons reciprocally regulate energy balance in mice. *J. Clin. Investig.* **2010**, *120*, 720–734. [CrossRef] [PubMed]

157. Dodd, G.T.; Decherf, S.; Loh, K.; Simonds, S.E.; Wiede, F.; Balland, E.; Merry, T.L.; Munzberg, H.; Zhang, Z.Y.; Kahn, B.B.; et al. Leptin and insulin act on POMC neurons to promote the browning of white fat. *Cell* **2015**, *160*, 88–104. [CrossRef] [PubMed]

158. Lopez, M.; Varela, L.; Vazquez, M.J.; Rodriguez-Cuenca, S.; Gonzalez, C.R.; Velagapudi, V.R.; Morgan, D.A.; Schoenmakers, E.; Agassandian, K.; Lage, R.; et al. Hypothalamic AMPK and fatty acid metabolism mediate thyroid regulation of energy balance. *Nat. Med.* **2010**, *16*, 1001–1008. [CrossRef] [PubMed]

159. Sjogren, M.; Alkemade, A.; Mittag, J.; Nordstrom, K.; Katz, A.; Rozell, B.; Westerblad, H.; Arner, A.; Vennstrom, B. Hypermetabolism in mice caused by the central action of an unliganded thyroid hormone receptor alpha1. *EMBO J.* **2007**, *26*, 4535–4545. [CrossRef] [PubMed]

160. Martinez-Sanchez, N.; Moreno-Navarrete, J.M.; Contreras, C.; Rial-Pensado, E.; Ferno, J.; Nogueiras, R.; Dieguez, C.; Fernandez-Real, J.M.; Lopez, M. Thyroid hormones induce browning of white fat. *J. Endocrinol.* **2017**, *232*, 351–362. [CrossRef] [PubMed]

161. Alvarez-Crespo, M.; Csikasz, R.I.; Martinez-Sanchez, N.; Dieguez, C.; Cannon, B.; Nedergaard, J.; Lopez, M. Essential role of UCP1 modulating the central effects of thyroid hormones on energy balance. *Mol. Metab.* **2016**, *5*, 271–282. [CrossRef] [PubMed]

162. Martinez-Sanchez, N.; Seoane-Collazo, P.; Contreras, C.; Varela, L.; Villarroya, J.; Rial-Pensado, E.; Buque, X.; Aurrekoetxea, I.; Delgado, T.C.; Vazquez-Martinez, R.; et al. Hypothalamic AMPK-ER Stress-JNK1 Axis Mediates the Central Actions of Thyroid Hormones on Energy Balance. *Cell Metab.* **2017**, *26*, 212–229. [CrossRef] [PubMed]

163. Broeders, E.P.; Vijgen, G.H.; Havekes, B.; Bouvy, N.D.; Mottaghy, F.M.; Kars, M.; Schaper, N.C.; Schrauwen, P.; Brans, B.; van Marken Lichtenbelt, W.D. Thyroid Hormone Activates Brown Adipose Tissue and Increases Non-Shivering Thermogenesis–A Cohort Study in a Group of Thyroid Carcinoma Patients. *PLoS ONE* **2016**, *11*, e0145049. [CrossRef] [PubMed]

164. Whittle, A.J.; Carobbio, S.; Martins, L.; Slawik, M.; Hondares, E.; Vazquez, M.J.; Morgan, D.; Csikasz, R.I.; Gallego, R.; Rodriguez-Cuenca, S.; et al. BMP8B increases brown adipose tissue thermogenesis through both central and peripheral actions. *Cell* **2012**, *149*, 871–885. [CrossRef] [PubMed]

165. Martins, L.; Seoane-Collazo, P.; Contreras, C.; Gonzalez-Garcia, I.; Martinez-Sanchez, N.; Gonzalez, F.; Zalvide, J.; Gallego, R.; Dieguez, C.; Nogueiras, R.; et al. A Functional Link between AMPK and Orexin Mediates the Effect of BMP8B on Energy Balance. *Cell Rep.* **2016**, *16*, 2231–2242. [CrossRef] [PubMed]

166. Kooijman, S.; Wang, Y.; Parlevliet, E.T.; Boon, M.R.; Edelschaap, D.; Snaterse, G.; Pijl, H.; Romijn, J.A.; Rensen, P.C. Central GLP-1 receptor signalling accelerates plasma clearance of triacylglycerol and glucose by activating brown adipose tissue in mice. *Diabetologia* **2015**, *58*, 2637–2646. [CrossRef] [PubMed]

167. Lee, S.J.; Sanchez-Watts, G.; Krieger, J.P.; Pignalosa, A.; Norell, P.N.; Cortella, A.; Pettersen, K.G.; Vrdoljak, D.; Hayes, M.R.; Kanoski, S.E.; et al. Loss of dorsomedial hypothalamic GLP-1 signaling reduces BAT thermogenesis and increases adiposity. *Mol. Metab.* **2018**, *11*, 33–46. [CrossRef] [PubMed]

168. Maciel, M.G.; Beserra, B.T.S.; Oliveira, F.C.B.; Ribeiro, C.M.; Coelho, M.S.; Neves, F.A.R.; Amato, A.A. The effect of glucagon-like peptide 1 and glucagon-like peptide 1 receptor agonists on energy expenditure: A systematic review and meta-analysis. *Diabetes Res. Clin. Pract.* **2018**, *142*, 222–235. [CrossRef] [PubMed]

169. Martinez de Morentin, P.B.; Gonzalez-Garcia, I.; Martins, L.; Lage, R.; Fernandez-Mallo, D.; Martinez-Sanchez, N.; Ruiz-Pino, F.; Liu, J.; Morgan, D.A.; Pinilla, L.; et al. Estradiol regulates brown adipose tissue thermogenesis via hypothalamic AMPK. *Cell Metab.* **2014**, *20*, 41–53. [CrossRef] [PubMed]

170. Gonzalez-Garcia, I.; Contreras, C.; Estevez-Salguero, A.; Ruiz-Pino, F.; Colsh, B.; Pensado, I.; Linares-Pose, L.; Rial-Pensado, E.; Martinez de Morentin, P.B.; Ferno, J.; et al. Estradiol Regulates Energy Balance by Ameliorating Hypothalamic Ceramide-Induced ER Stress. *Cell Rep.* **2018**, *25*, 413–423. [CrossRef] [PubMed]

171. Konner, A.C.; Bruning, J.C. Selective insulin and leptin resistance in metabolic disorders. *Cell Metab.* **2012**, *16*, 144–152. [CrossRef] [PubMed]

172. Seoane-Collazo, P.; Roa, J.; Rial-Pensado, E.; Linares-Pose, L.; Beiroa, D.; Ruiz-Pino, F.; Lopez-Gonzalez, T.; Morgan, D.A.; Pardavila, J.A.; Sanchez-Tapia, M.J.; et al. SF1-Specific AMPKα1 deletion protects against diet-induced obesity. *Diabetes* **2018**, *67*, 2213–2226. [CrossRef] [PubMed]

173. Milbank, E.; Martinez, M.C.; Andriantsitohaina, R. Extracellular vesicles: Pharmacological modulators of the peripheral and central signals governing obesity. *Pharmacol. Ther.* **2016**, *157*, 65–83. [CrossRef] [PubMed]

174. Finan, B.; Yang, B.; Ottaway, N.; Stemmer, K.; Muller, T.D.; Yi, C.X.; Habegger, K.; Schriever, S.C.; Garcia-Caceres, C.; Kabra, D.G.; et al. Targeted estrogen delivery reverses the metabolic syndrome. *Nat. Med.* **2012**, *18*, 1847–1856. [CrossRef] [PubMed]

175. Finan, B.; Clemmensen, C.; Zhu, Z.; Stemmer, K.; Gauthier, K.; Muller, L.; De Angelis, M.; Moreth, K.; Neff, F.; Perez-Tilve, D.; et al. Chemical hybridization of glucagon and thyroid hormone optimizes therapeutic impact for metabolic disease. *Cell* **2016**, *167*, 843–857. [CrossRef] [PubMed]

176. Shughrue, P.J.; Lane, M.V.; Merchenthaler, I. Comparative distribution of estrogen receptor-alpha and -beta mRNA in the rat central nervous system. *J. Comp. Neurol.* **1997**, *388*, 507–525. [CrossRef]

International Journal of
Molecular Sciences

MDPI

Review

Nutritional Modulation of AMPK-Impact upon Metabolic-Inflammation

Claire L. Lyons [1,2] and Helen M. Roche [2,3,*]

1   Unit of Molecular Metabolism, Lund University Diabetes Center, Clinical Research Center, Lund University, 205 02 Malmö, Sweden; claire.lyons@med.lu.se
2   Nutrigenomics Research Group, UCD Institute of Food and Health, UCD Conway Institute of Biomolecular and Biomedical Research, University College Dublin, Belfield, 4 Dublin, Ireland
3   Institute of Global Food Security, Queen's University Belfast BT7 1NN, Northern Ireland, UK
*   Correspondence: helen.roche@ucd.ie; Tel.: +353-1-716-6845

Received: 7 September 2018; Accepted: 6 October 2018; Published: 9 October 2018

Abstract: Nutritional status provides metabolic substrates to activate AMP-Activated Protein Kinase (AMPK), the energy sensor that regulates metabolism. Recent evidence has demonstrated that AMPK has wider functions with respect to regulating immune cell metabolism and function. One such example is the regulatory role that AMPK has on NLRP3-inlflammasome and IL-1β biology. This in turn can result in subsequent negative downstream effects on glucose, lipid and insulin metabolism. Nutrient stress in the form of obesity can impact AMPK and whole-body metabolism, leading to complications such as type 2 diabetes and cancer risk. There is a lack of data regarding the nature and extent that nutrient status has on AMPK and metabolic-inflammation. However, emerging work elucidates to a direct role of individual nutrients on AMPK and metabolic-inflammation, as a possible means of modulating AMPK activity. The posit being to use such nutritional agents to re-configure metabolic-inflammation towards more oxidative phosphorylation and promote the resolution of inflammation. The complex paradigm will be discussed within the context of if/how dietary components, nutrients including fatty acids and non-nutrient food components, such as resveratrol, berberine, curcumin and the flavonoid genistein, modulate AMPK dependent processes relating to inflammation and metabolism.

Keywords: AMPK; IL-1β; NLRP3; nutrition; dietary fatty acids; metabolic-inflammation; nutrigenomics

1. Introduction

AMP-activated protein kinase (AMPK) is a serine/threonine kinase (once thought of as solely an energy sensor) and has since been assigned widespread roles in metabolism [1], inflammation [2] and Type 2 Diabetes (T2D) [3]. It is responsible for adapting cellular metabolism in response to nutritional and environmental variations. This involves activating pathways to produce energy whilst also inhibiting energy-consuming pathways. The body uses adenosine triphosphate (ATP) as an energy source, which is broken down to adenosine diphosphate (ADP) and then adenosine monophosphate (AMP). When ATP levels are low, the ADP:AMP ratio increases and in turn activates AMPK to activate pathways to replenish ATP levels and restore energy. AMPK is a heterotrimer composed of a catalytic α (α) (α1 and α2) subunit and regulatory beta (β) (β1 and β2) and γ (γ) (γ1, γ2 and γ3) subunits. AMPK can be activated through both allosteric and phosphorylation (p) means. The threonine-172 (Thr-172) residue within the activation loop of the kinase domain on the α-subunit is the main phosphorylation site for AMPK activation [4]. The upstream protein kinases, liver kinase B1 (LKB1), Ca^{2+}/calmodulin-dependent protein kinase kinases (CaMKK) [5] and AMPK kinase (AMPKK) [4], are responsible for AMPK activation by phosphorylation. The allosteric activation of AMPK occurs through AMP and helps to prevent the dephosphorylation of AMPK [6]. AMP has also been shown to directly increase the CaMKKβ- and LKB1-mediated α-Thr[172] phosphorylation [7]. AMPK activation, in response to low energy status, re-configures glucose, lipid and mitochondrial metabolism towards adenosine triphosphate (ATP) production and decreases anabolic pathways that would otherwise further deplete ATP levels (Figure 1). Insulin can inhibit AMPK activity through phosphorylation of the Ser[485/491] in the α1/α2 site in multiple tissues, including skeletal muscle and liver, without affecting Thr[172] phosphorylation, through the Akt pathway [8]. One of the main mechanisms of AMPK activation is the prevention of its dephosphorylation by AMP, especially within the context of metabolic disease [6]. The main pathways activated are those that are involved in growth and metabolism [9]. It activates fatty acid oxidation (FAO) to generate ATP and inhibits unnecessary pathways, such as fatty acid synthesis [1]. Alternatively, AMPK affects the cell cycle and neuronal membrane excitability, as a way to regulate ATP levels [10]. In this review, we will discuss how AMPK can be activated by nutrients, such as glucose and lipids, but is mainly determined by nutrient status. The widespread effects of AMPK on metabolism, ranging from nutrient to mitochondrial metabolism, will be covered. In times of stress, such as obesity or direct influence by fatty acids, AMPK activation becomes altered and leads to dysregulated metabolism. Furthermore, there is an increasing amount of research pointing to the effect of metabolic signals on immune function and inflammation, and how a reciprocal relationship exists. The literature to date has focused on the role of nutritional status on AMP activation and how this affects cellular processes and metabolism. However, whilst AMPK-mediated metabolic-inflammation is a critical biological interaction, there is a paucity of data in relation to the nature and the extent to which individual nutrients affects cellular process and metabolism, thus this will be a focus within the review.

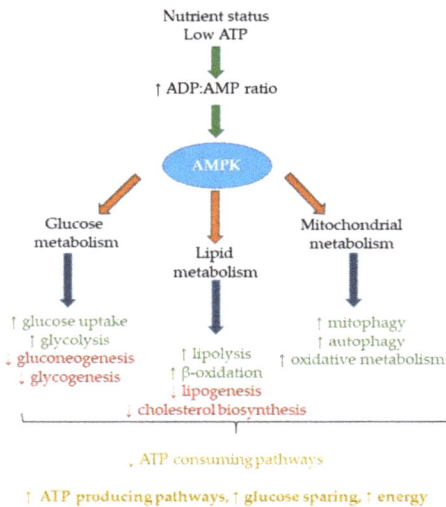

Figure 1. The role of AMP-Activated Protein Kinase (AMPK) on whole-body metabolism. AMPK is a nutrient sensor, which is activated in response to low adenosine triphosphate (ATP) levels, and an increased adenosine diphosphate: adenosine monophosphate (ADP:AMP) ratio. As a result, it activates pathways that produce ATP through glucose, lipid and mitochondrial metabolism pathways, thus increasing ATP levels. Conversely, pathways that deplete ATP are inhibited by AMPK. An ↑ arrow represents an upregulation of the process and ↓ represents a downregulation of the process. AMPK = AMP-Activated Protein Kinase, ATP = Adenosine triphosphate, ADP = Adenosine diphosphate, AMP = Adenosine monophosphate.

2. AMPK Activation, Metabolism and Nutrient Status

Metabolism is fueled by the nutrients we consume and is a tightly regulated process. Nutrient status therefore has a direct effect on the energy status of the organism. On the cellular level, AMPK activation is dependent upon energy status in the form of low ATP levels, and an increased ADP:AMP ratio. It is activated by energetic stress characterized by low levels of ATP. AMPK activation re-configures cellular metabolism, switching on catabolic pathways to generate ATP and switching off anabolic pathways that would otherwise deplete ATP. The metabolic impact of AMPK activation/de-activation has already been thoroughly reviewed by Herzig and Shaw [11]. Briefly, as illustrated in Figure 1, lipid and glucose metabolism are re-configured to supply energy; whilst protein metabolism, particularly protein synthesis is shut down. AMPK activation promotes glucose uptake and glycolysis, and activates lipolysis and oxidation, which is associated with significant upregulation of mitochondrial metabolism, mitophagy and autophagy. Conversely, gluconeogenesis and glycogenesis, as well as fatty acid synthesis/lipogenesis and cholesterol biosynthesis, are attenuated. In essence, AMPK activation promotes glucose sparing and oxidative metabolism to generate maximal ATP from cellular energy substrates. This process is used by most quiescent cells, as opposed to the glucose dependent, glycolytic metabolism relied upon by activated immune and proliferating cells. The current review will focus on the effects of AMPK on cellular metabolism in the context of inflammation and whole-body responses to obesity. For more detailed reviews about AMPK and the effects on lipid, glucose and mitochondrial metabolism, the readers are directed to the following reviews [9,12,13]. Recent work suggests that glucose metabolites mediate AMPK activation via non-canonical energy sensing mechanisms. Whilst it is well acknowledged that cellular glucose status activates AMPK as described above, more recent work suggests that glucose status affects AMPK activation, independent of increasing AMP:ATP and ADP ratios. It was initially presumed that low glucose status activated AMPK via reduced glucose catabolism leading to ATP depletion via the canonical energy sensing mechanism. But Zhang and colleagues recently showed that AMPK was activated when mouse embryonic fibroblasts were transferred from high to low (below 5 mM) glucose concentrations without any changes in cellular AMP/ATP or ADP/ATP ratios [14]. It was demonstrated that during glycolysis, glucose is converted to fructose-1, 6-bisphosphate (FBP), which is then processed or sensed by FBP aldolases. Such low glucose status reduces FBP-bound aldolase, which in turn activates AMPK via LKB1 phosphorylation. Thus, metabolites such as FBP indicative of poor glucose availability modulate FBP aldolases, which in turn sense low FBP and activate AMPK.

2.1. Nutrient Status and Impaired AMPK Action

On the whole-body level, energy status also affects AMPK activation. Obesity, in its simplest form, is caused by a chronic energy imbalance, wherein caloric intake exceeds caloric expenditure. This represents a metabolic stress event and hence affects AMPK activation, as illustrated in Figure 2. Obesity is associated with reduced AMPK activation, concomitant with alterations in glycolysis, insulin sensitivity, hepatic lipid metabolism and inflammation. The AMP metabolic stress signal results in β-myristoylation, allowing AMPK membrane association and facilitating the phosphorylation required for activation [7]. In man AMPK activity, as determined by its phosphorylation state, the P-AMPK/T-AMPK ratio, was reduced in visceral adipose tissue (VAT), rather than subcutaneous adipose tissue (SAT), of obese humans [15]. Obesity also affects immune cell AMPK status. Macrophage phosphorylated AMPK (pAMPK) expression was 33% lower in mouse models of genetic obesity compared to their lean counterparts [16]. Rats on a high fat diet (HFD) also show reduced renal AMPK activity [17]. Obesity arises as a result of excess energy intake, usually excess fat with or without surplus simple carbohydrates/sugars. Later in the review, we will deal with the impact of different dietary components within the context of obesity. It is important to note that high fat diets are usually preferentially enriched in saturated fatty acid (SFA), and initial data would suggest that SFA enriched high fat diets are particularly potent with respect to reducing pAMPK expression in adipose tissue and bone marrow derived macrophages (BMDM) [2]. Interestingly, recent data suggests that altered AMPK activation is not just a function of obesity but may play a role in energy homeostasis. Chronic activation of AMPK, through a mutation in the $\gamma2$ subunit, was involved in hyperphagia, obesity and impaired pancreatic function, which was observed in both mice and humans [18]. Therefore, long-term, non-discriminate activation of AMPK should also be taken with caution. It is possible that the adverse effect of obesity on pAMPK is mediated via adiponectin, the adipokine responsible for regulating glucose levels and fatty acid synthesis, which is decreased in obesity [19]. Adiponectin can activate AMPK via LKB1, but this is impaired in the skeletal muscle of *ob/ob* mice [20]. AMPK can be pharmacologically induced through treatment with the agonist, 5′-aminoimidazole-4-carboxamide ribonucleotide (AICAR). Mice given AICAR in conjunction with an HFD showed increased plasma adiponectin levels, reduced macrophage infiltration into the kidney, and reduced kidney hypertrophy, but the impact on adipose tissue biology was not investigated in this study [21].

2.2. The Involvement of AMPK in Insulin Resistance

AMPK expression is altered in insulin resistant obese individuals, with reduced expression compared to insulin sensitive body mass index-matched counterparts [15]. $AMPK\beta1^{-/-}$ contributes to insulin resistance with decreased phosphorylation of protein kinase B (Akt), increased adipose non-esterified fatty acids, hyperglycemia and hyperinsulinemia [16]. Serum leptin levels increased, and adiponectin reduced, indicating the negative effect of AMPK signaling disruption on adipose biology. Furthermore, in a co-culture system of $AMPK\alpha1^{-/-}$ macrophages and 3T3-L1 adipocytes, insulin stimulated phosphorylation of the insulin receptor and subsequent insulin stimulated glucose uptake, were both decreased with the deletion of the AMPK subunit. This provides confirmation that AMPKs anti-inflammatory effect in macrophages leads to a positive effect on adipose biology [22]. Adipocyte-specific deletion of $AMPK\beta$ subunits had a deleterious effect on glucose tolerance and insulin sensitivity, in response to an HFD. Further analysis demonstrated that reduced energy expenditure in brown adipose tissue and hepatic steatosis, but not white adipose tissue inflammation, was responsible for the insulin resistant phenotype [23].

Leclerc and colleagues demonstrated that glucose could regulate and reduce AMPK activity in both human and rodent islets, which resulted in decreased insulin secretion [24]. AMPK can regulate insulin-induced gene expression of L-type pyruvate kinase and pre-proinsulin promotor in islets [25]. A study by Mottillo and colleagues utilized an inducible model for deletion of the two AMPK β subunits in adipocytes (i$\beta1\beta2$AKO) [26], whereby deletion of AMPK exacerbated the insulin resistant phenotype in terms of hepatic steatosis and glucose tolerance in response to an HFD [23]. AMPK knockout models are associated with differing phenotypes, based on which subunit is targeted. Those with the $\alpha1/2$ subunit deficiency show increased lipolysis [27], whereas $\beta1/2$ deficiency show reduced oxidative metabolism [23]. AMPK activation has been shown to inhibit the first phase of adipogenesis [28] but also increase peroxisome proliferator-activated receptor γ (PPAR-γ) expression [29] in other studies. Despite these confounding results, AMPK$\alpha1$- and AMPK$\alpha2$-deficient mice display hypertrophic adiposity following an HFD [27,28,30], but these were conducted in whole body knock-out models. Global $AMPK\alpha1^{-/-}$ mice display increased adiposity, systemic insulin resistance and increased inflammation. When the bone marrow cells were deficient in AMPK$\alpha1$, the mice had insulin resistance but no obesity, whereas adipocyte AMPK$\alpha1$ deficiency had the same phenotype as the global knockout [31]. In genetically obese *ob/ob* mice, improved insulin sensitivity was a consequence of increased AMPK levels and glucose production in the liver, but no effects were observed in the skeletal muscle of these mice [20]. When AMPK levels were increased with metformin in HFD-rats switched to a chow diet, the metformin group had reduced weight gain and plasma glucose and triglyceride levels compared to a diet switch alone [17]. The authors speculated that metformin was increasing renal FAO by modulating AMPK/acetyl-CoA carboxylase (ACC) pathway and thereby reducing renal lipotoxicity. Metformin treatment in HFD mice also improves glucose metabolism and decreases the adipocyte size compared to HFD fed mice alone [32], thereby illustrating the multitude of effects that AMPK plays in whole body metabolism.

2.3. AMPK and Its Link to Cancer

As in obesity, AMPK is downregulated in cancer. AMPK has been implicated in cancer due to its effects on cellular growth and metabolism [33,34]. Tumor cells downregulate AMPK, and thus re-configure cellular metabolism towards glycolytic metabolism to enhance cell growth and proliferation. The mechanistic basis of which is complex and may be related to LKB1 dependent AMPK Thr^{172} phosphorylation to downregulate AMPK-mediated metabolism, over-signalling of the insulin/IGF1-regulated protein kinase Akt/PKB pathway, and/or other mechanisms [35]. LKB, the upstream kinase of AMPK activation, is a tumor suppressor but is mutated in cancer cells. Mammalian target of rapamycin (mTOR) controls cell growth and proliferation and is a target of AMPK. When LKB1 is deficient in cancer cells, mTOR activity inhibition is lost and thus cell proliferation is increased [36]. Cell growth is an ATP-consuming pathway and as such, AMPK directly inhibits mTOR activity. Fatty acid synthesis is increased in many cancers and through AMPKs ability to inhibit this process, AMPK is thought to have anti-cancer roles. Pharmacological AMPK activation, with metformin or salicylate, may protect against cancer initiation and development. Aspirin treated patients with CVD and metformin treated patients with type 2 diabetes have a lower incidence of cancer [37,38]. Whilst these studies seem promising, it is important to note that whether AMPK is a tumor suppressor or potential oncogene, at different cancer stages, is highly controversial and ripe for investigation [35,39,40]. Given the important regulatory role AMPK mediates within the context of metabolism, the next section will deal with the extent to which metabolism may alter inflammation. Also, the impact of nutrient-derived compounds is detailed in Section 4.

3. The Intersection between Metabolism and Inflammation

It is now appreciated that there is a complex and dynamic inter-relationship between metabolism and inflammation. Thus, rather than investigating simple pathways/biological processes, there is an intense interest in identifying potential hubs that co-regulate metabolism and inflammation, and in this context, AMPK is an interesting candidate. As stated previously, obesity can affect AMPK activity, but it is difficult to know if the HFD obesogenic effect reflects energy overload or a direct impact of dietary fatty acid (FA) composition. Interestingly, there is an increasing body of evidence that suggests that the nature of fatty acids may co-regulate inflammation and pAMPK expression. Attenuated pAMPK expression is associated with an inverse increase in inflammatory markers in both mouse models of obesity [16] and in obese humans [15]. In nutrient-rich conditions, and with inflammatory stimuli, AMPK phosphorylation and the activity of the α 1 subunit of AMPK is reduced. However, when AMPK is activated there is a subsequent reduction in both nuclear factor kappa B (NF-κB) and tumour necrosis factor α (TNF-α) secretion in macrophages [22].

3.1. AMPK and the NLRP3 Inflammasome

The nucleotide-binding domain, leucine-rich-containing family, pyrin domain-containing-3 (NLPR3) inflammasome, mediates interleukin-1β (IL-1β) production, but requires a metabolic product, reactive oxygen species (ROS), for its complete activation. NLRP3 has been shown to play a downstream role in metabolism, with Nlrp3$^{-/-}$ mice on an HFD displaying increased FAO [41]. Caspase-1 is required to process pro-IL-1β to IL-1β activation and *ob/ob* mice administered a caspase-1 inhibitor also display increased FAO [42]. AMPK can no longer reduce FAO when the regulatory β-1 subunit is deleted, thereby altering the tetrameric formation of AMPK. This occurs through changes in ACC phosphorylation and altered mitochondrial content [16]. AMPK can inhibit inducible nitric oxide synthase production as a way to reduce inflammation [43], while AMPK inhibition blocks autophagy increasing mitochondrial ROS production, an instrumental step in NLRP3 inflammasome activation [44]. ROS and nicotinamide adenine dinucleotide phosphate inhibition, as well as AMPK activation by AICAR, can all prevent the lipopolysaccharide (LPS)- and palmitic acid (PA)- induction of pro-IL-1β and mature IL-1β production [44]. NLRP3 inflammasome components and the activation of caspase-1 and IL-1β were significantly upregulated in monocyte-derived macrophages from T2D patients and were reduced following treatment with metformin, an insulin sensitizing drug that activates AMPK. The ability of metformin to reduce NLPR3 inflammasome activation was attributed to a reduction in mitochondrial ROS, a known activator of the inflammasome [45].

Metformin has been widely used in the treatment of T2D [3], wherein it induces not only metabolic but also positive anti-inflammatory effects. Metformin indirectly activates AMPK by inhibiting mitochondrial ATP synthesis [46], while decreasing the secretion of inflammatory markers, cyclooxygenase-2 and IL-1β, in the kidney of HFD-fed rats [17]. Citrate is involved in FA synthesis and has also been noted as being a potent inflammatory stimulus [47]. Accumulation of citrate can inhibit phosphofructokinase and back up the glycolytic pathway. The treatment of BMDM with citrate increased NLRP3-dependent IL-1β production and caspase-1 cleavage [48]. The isocitrate/pyruvate cycle has also been implicated in increasing glucose-stimulated insulin secretion, but this area of research warrants further investigation [47]. Treatment of BMDM with α-ketoglutarate has been shown to increase interleukin-10 (IL-10) but also significantly enhance IL-1β expression compared to LPS alone [49]. Conflicting reports however demonstrate that LPS-primed BMDM stimulated with α-ketoglutarate derivative attenuated the LPS-induced IL-1β mRNA and IL-1β protein, through reduced activity of prolyl hydroxylases, which is responsible for stabilising hypoxia-inducible factor 1-α (HIF-1α) protein [50].

3.2. Cellular Metabolism and Its Effect on Immune Cell Function

Early work suggested that sub-acute chronic inflammation that typified diet related diseases including obesity, T2D and atherosclerosis disrupted metabolism [51–53], wherein cytokines impeded a range of metabolic pathways such as insulin signaling, and reverse cholesterol transport [51–55]. However, it is now evident that there is a much more dynamic reciprocal regulatory relationship between metabolism and inflammation, wherein the nature of cellular metabolism determines immune cells functionality and response. At the simplest level, the cellular balance or switch between oxidative phosphorylation (OxPHOS) versus glucose dependent glycolytic metabolism, defines the functional nature of inactive versus activated immune and proliferating cells. Interestingly immune cells preferentially utilize different metabolic pathways depending on their pro-inflammatory, anti-inflammatory or resolving nature. The metabolic pathway selected reflects the functionality and requirements of the immune cell; whether it is a rapid increase in energy to carry out an innate immune response, or a prolonged availability of energy to enable adaptive responses, such as healing and repair [56]. An elegant combined transcriptomic and metabolomic analysis of BMDM functionality identified important transcriptional and metabolic rewiring during the macrophage polarization [57]. Inactive BMDM rely on the tricarboxylic acid (TCA) cycle and OxPHOS for cellular metabolism, wherein fatty acids and glutamine, but to a lesser extent glucose, are fully oxidized for maximal ATP generation. AMPK plays a key role in modulating metabolism and the balance between OxPHOS and glycolytic metabolism [11]. Upon activation, pro-inflammatory immune cells shift from utilizing the TCA cycle and generation of ATP via OxPHOS to anaerobic glycolysis [57]; this metabolic switch favors the use of glucose as the energy substrate [58]. The pro-inflammatory gram-negative bacterial product, LPS, is often used to induce a classically activated M1 macrophage M1 pro-inflammatory response, as it increases glycolysis and decreases oxygen consumption [50]. This results in a decrease in the AMP:ADP ratio, and a depletion of ATP. When this occurs, there is a shift in ATP production from OxPHOS to glycolysis, as glycolysis produces less ATP. As a result, AMPK would thus be activated to replenish the ATP levels [49]. Increased glycolysis is facilitated by a switch in the phosphofructokinase isoenzymes to its' ubiquitous isoform, leading to increased phosphorylation of fructose-6-phosphate, the rate limiting step in glycolysis [56].

3.3. Metabolic Reprogramming of T Cells

Metabolic reprogramming occurs in T cells, with regulatory T cells favoring OxPHOS and T effector cells switching to glycolysis upon activation. Treg cells have increased AMPK activation and in turn increased lipid oxidation, while T effector cells activate the opposing mTOR pathway [59]. T effector cells have significantly increased glucose transporter 1 (GLUT1) expression compared to T regulatory cells, to deal with the increased need for glycolysis. T regulatory cells require lipid metabolism for their differentiation which is mediated by AMPK where metformin administration to mice increases the number of T reg cells [60]. T effector cells can switch their metabolism under low glucose conditions by decreasing glycolysis and maintaining ATP levels by employing glutamine-dependent OxPHOS [61]. Specifically, LKB1 is critical for multiple T cell functions including development, viability, metabolism and activation. LKB1 was found to be involved in glucose metabolism as T cells lacking LKB1 had increased glycolysis and glucose uptake. A consequence of the disrupted metabolism in LKB1-deficient T cells was an increase in inflammatory CD4$^+$ and CD8$^+$ cells. In vivo studies in mice lacking T-cell specific AMPKα1 confirmed that AMPK activation was required for these processes [62]. AMPKα1 is required for T effector cells to maintain metabolic flexibility in times of nutrient stress and for T helper cell development in response to infection [61]. When T effector cells need to revert back to metabolically quiescent memory cells, AMPKα1null CD8 T cells were unable to make this transition, further confirming the role that AMPK plays in T cell metabolic switching [63]. Pozanski and colleagues have reviewed T cell and natural kill cell immune-metabolism, and recent emerging evidence suggests that AMPK may also play a role in the metabolic phenotype of natural killer cells [64]. Given the important inter-relationship between metabolism and inflammation, which is partly attributable to AMPK, we now address the extent to which this is sensitive to nutritional status and/or nutritional interventions. Thus, to explore the extent to which AMPK metabolic-inflammation may be either up- or down-regulated by different nutritional components, and how metabolic switching can also lead to beneficial changes in the cell [65]

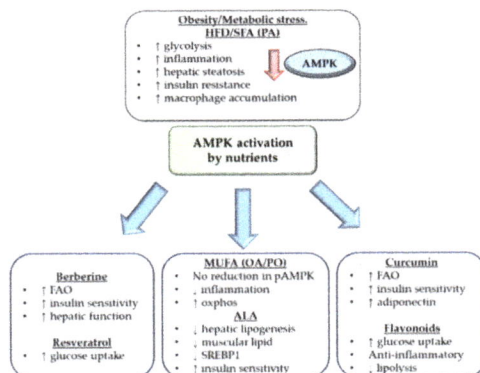

Figure 2. Whole body AMPK modulation and the impact of different nutrients on AMPK activation. Excess nutrient consumption through high fat diet (HFD) and obesity can downregulate AMP-Activated Protein Kinase (AMPK) expression and cause dysregulated metabolism, inflammation and insulin resistance. Nutrients, including monounsaturated fatty acids (MUFA), α linoleic acid (ALA), berberine, resveratrol, curcumin and flavonoids can all activate AMPK and downstream positive effects in relation to improved mitochondrial metabolism, improved liver function and reduced inflammation. Therefore, modulation of AMPK through nutrient intervention can improve whole body metabolism. An ↑ arrow represents an upregulation of the process or increased expression and ↓ represents a downregulation of the process or decreased expression. AMPK = AMP-Activated Protein Kinase, MUFA = Monounsaturated fatty acids, OA = Oleic acid, PO = Palmitoleic acid, ALA = α linoleic acid, SREBP1 = Sterol regulatory element-binding protein 1c, FAO = fatty acid oxidation.

4. Modulation of AMPK Activation by Nutrients

4.1. Fatty Acids Differentially Affect AMPK Function

One of the main drivers of obesity is the excessive consumption of dietary fat. Dietary fat can have widespread negative effects in relation to inflammation, insulin resistance and metabolism, depending on the nature of the fatty acids. Briefly, fatty acids can be either saturated or unsaturated, wherein the latter are sub-divided into monounsaturated (MUFA) or polyunsaturated (PUFA) fatty acids depending on the number of double bonds in the carbon chain. The inclusion of double bonds and the degree of unsaturation has very different effects on metabolism and inflammation. From the experimental point of view, rodent high fat diets are usually enriched with lard which is rich in the saturated fatty acid (SFA) lauric acid (C12:0) or palm oil which is enriched in another SFA palmitic acid (PA) (C16:0). Whilst it has long been acknowledged that obesity is associated with reduced pAMPK expression and a pro-inflammatory phenotype, the true nature of the apparent reciprocal regulation and putative impact of different dietary components is ill defined.

In vitro studies suggest that different fatty acids modulate AMPK mediated inflammation. AMPKβ1$^{-/-}$ BMDM display a predominant M1 profile, which was further exacerbated with two saturated fatty acids, PA and stearic acid (C18:0). Hematopoietic deletion of AMPKβ1$^{-/-}$ was sufficient to induce systemic inflammation in an HFD setting [16]. PA-treated macrophages display increased inflammatory markers, but this is reduced with metformin co-treatment, a phenomenon mimicked in vivo in HFD mice treated with metformin. When AMPK was inhibited with compound C (CC) and included with metformin in vitro, inflammatory markers were increased, thus confirming that metformin is mediating its anti-inflammatory effect through AMPK activation [32]. Interestingly the AMPK activator, AICAR impeded LPS- and FA-induced cytokine response, in part through NF-κB inhibition [16]. However, AICAR has its non-specific limitations, it can inhibit NF-κB DNA binding in human macrophages without AMPK activation [66].

Unsaturated fatty acid may have less of an adverse effect on AMPK mediated metabolism and inflammation (Figure 2). Two groups have compared the impact of palmitic acid and oleic acid (OA) on AMPK. In vitro, direct stimulation with PA reduces pAMPK expression in macrophages. In contrast two MUFA, both palmitoleic acid (PO) (C16:1) and OA (C18:1) do not [2,67]. In vivo, feeding SFA enriched HFD amplifies insulin resistance, inflammation and reduces pAMPK expression. In contrast, feeding an HFD derived from the MUFA, OA did not reduce adipose pAMPK, despite obesity, compared to an SFA enriched HFD [2]. In order to define if the apparent co-regulation of IL-1β inflammation and AMPK was fatty acid dependent, a series of in vitro experiments using CC and AICAR demonstrated that when AMPK was inhibited with CC, IL-1β was increased in BMDM. Conversely, AMPK activation with AICAR in conjunction with PA could inhibit IL-1β secretion [2]. When AMPK is activated pharmacologically through AICAR or by overexpression, it reduces TNF-α- and PA- induced increases in NF-κB expression in endothelial cells [68]. Alpha-linoleic acid (ALA) is n-3 polyunsaturated fatty acid (n-3 PUFA) (C18:3 n-3 *cis*-Δ^3) which acts as a naturally occurring antioxidant and is a cofactor for mitochondrial respiratory enzymes [69]. It is important to note that α- linoleic acid is not a fatty acid per se, but an organosulfur compound derived from the fatty acid caprylic acid (C8:0). ALA was shown to improve both insulin-stimulated glucose uptake and FAO in the skeletal muscle of diabetes prone, obese rats. This was attributed to AMPK activation and reduced lipid accumulation, as dominant-negative AMPK prevented the positive effects of ALA [70]. Subsequent studies found an improvement in whole body glucose tolerance in HFD + ALA rats but did not attribute this to an AMPK mediated lowering of intramuscular lipid [71]. Within the liver of HFD rats, ALA can reduce hepatic lipogenesis through reduced SREBP-1 expression. As a result, hepatic steatosis was attenuated, and this was mediated in part by AMPK activation [72]. ALA can directly increase AMPK in hepatocytes but its lipid lowering and beneficial effects on hepatic metabolism were not AMPK-dependent [73]. In terms of understanding the relative impact of fatty acids versus glucose, Kratz and colleagues identified a distinct population of metabolically activated macrophages (MMe), following a palmitate, glucose and insulin challenge within the adipose tissue. MMe macrophages display attributes of both alternatively activated M2 anti-inflammatory markers (lipid metabolism) but secreted M1-associated pro-inflammatory cytokines [74]. It was noteworthy that PA was more potent in this system compared to glucose and insulin metabolic challenges.

4.2. Reversing Metabolic Inflammation through AMPK

The concept of reversing HFD induced metabolic-inflammation is intriguing and involves AMPK. Bone marrow derived macrophages (BMDM) derived from HFD mice retained a "dietary memory" with increased mRNA levels of TNF-α, interleukin-6 (IL-6) and nitric oxide synthase 2, compared to those of a low-fat diet [75], which could be reversed with incubation of the n-6 MUFA, *cis*-PO (C16:1 *cis*-Δ^9). This PO mediated anti-inflammatory effect was AMPK-dependent. Both OA and PO maintain AMPK expression, and this is thought to mediate their anti-inflammatory effects [75]. Work in Thp-1 human macrophages containing a knockdown of AMPK showed the same inflammatory response by OA when AMPK could not function [2]. Co-incubation of PO with PA, restored pAMPK levels to that of PO alone in BMDM. AMPK inhibition results in PO-induced increases in nitric oxide expression [75]. PA, in a similar manner to the inflammatory stimulus LPS, can increase the cells use of glycolysis in BMDM, while exposure to PO increased OxPHOS. Furthermore, when PO was co-incubated with PA, they displayed similar oxygen consumption levels compared to PO alone. This was thought to reflect the increased M2 phenotype induced through PO incubation [75]. Anti-inflammatory M2 macrophages employ the OxPHOS pathway, which is capable of producing more ATP than glycolysis [57]. M2 macrophages rely on FA as an energy source [58]. The inhibition of various steps of the mitochondrial OxPHOS pathway were able to reduce the expression of arginine 1, a common M2 marker, and abrogate the anti-inflammatory potential of these cells, but had no effect on classically activated macrophages [58]. Activation of pyruvate kinase isozymes M2 enhanced IL-10 secretion in LPS-stimulated BMDM, favouring an M2 phenotype and reduced M1 polarisation [65]. Macrophage M1 polarisation was reduced through AMPK activation by fibronectin type III domain-containing 5 (FNDC5), and worsened when FNDC5 was inhibited, thus lowering AMPK levels in the adipose tissue of HFD-fed mice [76]. Palmitate-treated macrophages treated with AICAR was able to reduce the number of M1 macrophages while subsequently increasing the M2 macrophage population [32]. Microarray analysis of adipose tissue following an 8wk SFA diet showed upregulation of pathways involved in immune function and inflammation. Consumption of a MUFA diet showed no change or downregulation of the same pathways [77]. MUFA-HFD mice display increased energy expenditure compared to SFA-HFD mice, and given the role of AMPK as an energy sensor, the involvement of AMPK in regulating this process cannot be ruled out [2].

4.3. AMPK Regulation by Resveratrol, Berberine and Curcumin

Resveratrol, a natural phytochemical, increases glucose uptake in insulin-resistant 3T3-L1 adipocytes by increasing the phosphorylation of pAkt and downstream AMPK activation. A similar effect was noted in vivo in resveratrol treated KKAy mice who demonstrated increased pAkt and elevated AMPK expression [78]. Resveratrol can inhibit cell proliferation in the HT-29 cancer cell line, which was AMPK-dependent. The ability of resveratrol to increase ROS production was elucidated as the mechanism of AMPK activation [79]. Berberine is a naturally occurring plant-derived compound that can indirectly activate AMPK. Berberine can increase energy expenditure and improve liver function in obese mice by stimulating fatty acid oxidation and inhibiting lipogenic genes [80]. Berberine inhibits complex I of the mitochondria to reduce respiration in cultured myotubes and muscle mitochondria. Furthermore, a derivative of berberine demonstrated improved insulin sensitivity and reduced adiposity in vivo in HFD rats [81], owing to the possibility of nutrient intervention treatments. Berberine has positive effects on cancer by inhibiting their metastatic potential and reducing the inflammatory COX-2 pathway by increasing AMPK phosphorylation through increased ROS production [82]. Curcumin is a naturally occurring polyphenol and is the active constituent of turmeric. In a review by Shehzad and colleagues detailing curcumin studies over a 10-year period, this polyphenol was found to have multiple anti-inflammatory and insulin sensitizing effects, involving the suppression of anti-inflammatory transcription factors, upregulation of adiponectin and interactions with glucose and insulin signal transduction pathways [83]. HFD supplemented with 0.15% curcumin reduced body weight, adiposity, serum lipid and glucose levels, and insulin resistance in mice compared to HFD alone controls. Furthermore, beneficial effects on hepatic lipid accumulation and metabolism were noted with reductions in ACC and fatty acid synthase but increased FAO [84]. Curcumin can also increase FAO in muscle which was mediated in part by AMPK and improved palmitate-induced insulin resistance in L6-myotubes [85]. Curcumin demonstrates chemo preventative properties by inducing cell death through p53 phosphorylation, but this is prevented if AMPK action is inhibited with CC treatment in an ovarian cancer cell line [86].

4.4. AMPK Regulation by Flavonoids

Plant flavonoids are plant secondary metabolites known to have beneficial health effects [87]. Genistein, a soy isoflavone, can reduce blood glucose in a KK-Ay/Ta Jcl T2D mouse model. When used in vitro in L6 myotubes, genistein stimulates phosphorylation of AMPK and increases glucose uptake by increasing GLUT4 translocation to the plasma membrane, which was prevented when AMPK was inhibited with CC [88]. Genistein has anti-inflammatory properties in macrophages through inhibition of NF-κB mediated by AMPK [89]. In a diabetic rat model (Zucker *fa*/*fa*) fed soy protein, AMPK and FAO were increased in the skeletal muscle and associated with decreased weight gain and improved glucose and triglyceride levels, demonstrating the positive effect that flavonoids can have on overall metabolism [90]. Apigenin, another member of the flavonoid family, dose-dependently increases AMPK expression, while inhibiting adipogenic gene expression and lipolysis in 3T3-L1 adipocytes [91]. In vivo administration of apigenin to HFD-mice reduced the HFD-induced weight gain, increased serum glucose and muscle pro-inflammatory cytokine expression. The apigenin treated mice displayed increased OxPHOS and citrate synthase activity compared to the muscle HFD alone mice, through improvements in mitochondrial biogenesis and mitochondrial function [92]. Similar improvements were observed by Jung and colleagues, with additional benefits observed in relation to improved hepatic steatosis, reduced plasma FFA, and downregulation of lipogenic genes in the liver of apigenin + HFD mice [93]. Flavonoids have been ascribed cancer preventative roles through the induction of apoptotic pathways mediated through AMPK signaling pathways [94]. The ability of FA to undergo metabolic reconfiguration, combined with the anti-oxidant and anti-inflammatory properties of nutrients and how these are involved in the diet-induced modulation of inflammation therefore requires further investigation. FA are an important source of metabolites and no doubt their metabolism will feed more into the immuno-metabolism circuit, however to date the nature and extent to which dietary FA and/or nutritional status can modulate this dynamic two-way process requires definition.

5. Conclusions and Future Perspectives

AMPK is an exciting target given its co-regulatory role with respect to metabolism and inflammation. It is evident that Western dietary elements, which are obesogenic in nature, attenuate AMPK, the metabolic impact of which probably explains the sub-acute pro-inflammatory phenotype that typifies obesity, T2D and CVD. The challenge remains to discern if/how certain nutrients either up- or down-regulate cellular AMPK status. It is evident that feeding SFA are particularly deleterious, and therefore negatively impact upon metabolism and sub-acute chronic inflammation that typifies obesity T2D. Speculatively it is probable that high-fructose diets will also have deleterious effect on AMPK, given the fact that feeding fructose increases endogenous synthesis of SFA. Conversely, there is some evidence to suggest that feeding high fat diets rich in MUFA and PUFA, or supplemented with natural AMPK agonists, such as resveratrol and berberine, do not attenuate AMPK. Furthermore, sirtuin 1 (SIRT) is another nutrient sensor that has widespread effects on metabolism in response to caloric restriction and may also be involved in regulating metabolism and inflammation [95]. Whilst the paradigm of nutritional enhancement of AMPK activity is worthy of investigation in light of the obesity epidemic, to date the impact of these nutritional interventions have been observed in mice. Therefore, we need to define if/how this translates to humans.

Author Contributions: C.L.L. wrote the review. H.M.R. advised on content, wrote and critically evaluated the manuscript. All authors approved the final submission.

funding: This work was supported by Science Foundation Ireland (SFI) principal investigator award (11/PI/1119); Joint Programming Healthy Life for a Healthy Diet (JPI HDHL) funded EU Food Biomarkers Alliance "FOODBALL" (14/JP-HDHL/B3076); the Irish Department of Agriculture, Food and the Marine, "Healthy Beef" (13/F/514) and "ImmunoMet-dietary manipulation of microbiota diversity for controlling immune function" (14/F/828) programmes.

Conflicts of Interest: The authors declare no conflict of interest.

Abbreviations

ACC	Acetyl-CoA carboxylase
ADP	Adenosine diphosphate
AICAR	5′-aminoimidazole-4-carboxamide ribonucleotide
Akt	Protein kinase B
ALA	A-linoleic acid
AMP	Adenosine monophosphate
AMPK	AMP-Activated Protein Kinase
AMPKK	AMPK kinase
ATP	Adenosine triphosphate
BMDM	Bone marrow derived macrophages
CaMKK	Ca^{2+}/calmodulin-dependent protein kinase kinases
CC	Compound C
CVD	Cardiovascular disease
FA	Fatty acid
FAO	Fatty acid oxidation
FBP	Fructose-1,6-bisphosphate
FNDC5	Fibronectin type III domain-containing 5
Glut1	Glucose transporter 1
HIF-1α	Hypoxia-inducible factor 1-α
HFD	High fat diet
IL-1β	Interleukin-1β
IL-6	Interleukin 6
IL-10	Interleukin 10
LKB1	Liver kinase B1
LPS	Lipopolysaccharide
Mme	Metabolically activated macrophages
mTOR	Mammalian target of rapamycin
MUFA	Monounsaturated fatty acids
OA	Oleic acid
OxPHOS	Oxidative phosphorylation
P	Phosphorylation
PA	Palmitic acid
PO	Palmitoleic acid
PPAR-γ	Peroxisome proliferator-activated receptor γ
PUFA	Polyunsaturated fatty acid
ROS	Reactive oxygen species
SAT	Subcutaneous adipose tissue
SFA	Saturated fatty acid
SIRT1	Sirtuin
SREBP-1	Sterol regulatory element-binding protein 1
T2D	Type 2 diabetes
TCA	Tricarboxylic acid
Thr	Threonine
TNF-α	Tumour necrosis factor α
VAT	Visceral adipose tissue
α	Alpha
β	Beta
γ	Gamma

References

1. Hardie, D.G. The AMP-activated protein kinase pathway—New players upstream and downstream. *J. Cell Sci.* **2004**, *117*, 5479–5487. [CrossRef] [PubMed]
2. Finucane, O.M.; Lyons, C.L.; Murphy, A.M.; Reynolds, C.M.; Klinger, R.; Healy, N.P.; Cooke, A.A.; Coll, R.C.; McAllan, L.; Nilaweera, K.N.; et al. Monounsaturated fatty acid-enriched high-fat diets impede adipose NLRP3 inflammasome-mediated IL-1? Secretion and insulin resistance despite obesity. *Diabetes* **2015**, *64*, 2116–2128. [CrossRef] [PubMed]
3. Knowler, W.C.; Connor, E.B.; Fowler, S.E.; Hamman, R.F.; Lachin, J.M.; Walker, E.A.; Nathan, D.M. Reduction in the incidence of type 2 diabetes with lifestyle intervention or metformin. *N. Engl. J. Med.* **2002**, *346*, 393–403. [PubMed]
4. Stein, S.C.; Woods, A.; Jones, J.A.; Davison, M.D.; Carling, D. The regulation of AMP-activated protein kinase by phosphorylation. *Biochem. J.* **2000**, *345*, 437–443. [CrossRef] [PubMed]
5. Hawley, S.A.; Davison, M.; Woods, A.; Davies, S.P.; Beri, R.K.; Carling, D.; Hardie, D.G. Characterization of the AMP-activated protein kinase kinase from rat liver and identification of threonine 172 as the major site at which it phosphorylates AMP-activated protein kinase. *J. Biol. Chem.* **1996**, *271*, 27879–27887. [CrossRef]
6. Xiao, B.; Sanders, M.J.; Underwood, E.; Heath, R.; Mayer, F.V.; Carmena, D.; Jing, C.; Walker, P.A.; Eccleston, J.F.; Haire, L.F.; et al. Structure of mammalian AMPK and its regulation by ADP. *Nature* **2011**, *472*, 230–233. [CrossRef] [PubMed]
7. Oakhill, J.S.; Chen, Z.P.; Scott, J.W.; Steel, R.; Castelli, L.A.; Ling, N.; Macaulay, S.L.; Kemp, B.E. β-Subunit myristoylation is the gatekeeper for initiating metabolic stress sensing by AMP-activated protein kinase (AMPK). *Proc. Natl. Acad. Sci. USA* **2010**, *107*, 19237–19241. [CrossRef] [PubMed]
8. Valentine, R.J.; Coughlan, K.A.; Ruderman, N.B.; Saha, A.K. Insulin inhibits AMPK activity and phosphorylates AMPK Ser485/491 through Akt in hepatocytes, myotubes and incubated rat skeletal muscle. *Arch. Biochem. Biophys.* **2014**, *562*, 62–69. [CrossRef] [PubMed]
9. Mihaylova, M.M.; Shaw, R.J. The AMPK signalling pathway coordinates cell growth, autophagy and metabolism. *Nat. Cell Biol.* **2011**, *13*, 1016–1023. [CrossRef] [PubMed]
10. Hardie, D.G.; Ross, F.A.; Hawley, S.A. AMPK: A nutrient and energy sensor that maintains energy homeostasis. *Nat. Rev. Mol. Cell Biol.* **2012**, *13*, 251–262. [CrossRef] [PubMed]
11. Herzig, S.; Shaw, R.J. AMPK: Guardian of metabolism and mitochondrial homeostasis. *Nat. Rev. Mol. Cell Biol.* **2017**, *19*, 121–135. [CrossRef] [PubMed]
12. O'Neill, L.A.J.; Hardie, D.G. Metabolism of inflammation limited by AMPK and pseudo-starvation. *Nature* **2013**, *493*, 346–355. [CrossRef] [PubMed]
13. O'Neill, H.M.; Holloway, G.P. AMPK regulation of fatty acid metabolism and mitochondrial biogenesis: Implications for obesity. *Mol. Cell. Endocrinol.* **2013**, *366*, 135–151. [CrossRef] [PubMed]
14. Zhang, C.S.; Hawley, S.A.; Zong, Y.; Li, M.; Wang, Z.; Gray, A.; Ma, T.; Cui, J.; Feng, J.W.; Zhu, M.; et al. Fructose-1,6-bisphosphate and aldolase mediate glucose sensing by AMPK. *Nature* **2017**, *548*, 112–116. [CrossRef] [PubMed]

15. Gauthier, M.S.; O'Brien, E.L.; Bigornia, S.; Mott, M.; Cacicedo, J.M.; Xu, X.J.; Gokce, N.; Apovian, C.; Ruderman, N. Decreased AMP-activated protein kinase activity is associated with increased inflammation in visceral adipose tissue and with whole-body insulin resistance in morbidly obese humans. *Biochem. Biophys. Res. Commun.* **2011**, *404*, 382–387. [CrossRef] [PubMed]

16. Galic, S.; Fullerton, M.D.; Schertzer, J.D.; Sikkema, S.; Marcinko, K.; Walkley, C.R.; Izon, D.; Honeyman, J.; Chen, Z.P.; van Denderen, B.J.; et al. Hematopoietic AMPK β1 reduces mouse adipose tissue macrophage inflammation and insulin resistance in obesity. *J. Clin. Investig.* **2011**, *121*, 4903–4915. [CrossRef] [PubMed]

17. Tikoo, K.; Sharma, E.; Amara, V.R.; Pamulapati, H.; Dhawale, V.S. Metformin improves metabolic memory in high fat diet (HFD)-induced renal dysfunction. *J. Biol. Chem.* **2016**, *291*, 21848–21856. [CrossRef] [PubMed]

18. Yavari, A.; Stocker, C.J.; Ghaffari, S.; Wargent, E.T.; Steeples, V.; Czibik, G.; Pinter, K.; Bellahcene, M.; Woods, A.; Martínez de Morentin, P.B.; et al. Chronic Activation of γ2 AMPK Induces Obesity and Reduces β Cell Function. *Cell Metab.* **2016**, *23*, 821–836. [CrossRef] [PubMed]

19. Arita, Y.; Kihara, S.; Ouchi, N.; Takahashi, M.; Maeda, K.; Miyagawa, J.; Hotta, K.; Shimomura, I.; Nakamura, T.; Miyaoka, K.; et al. Paradoxical Decrease of an Adipose-Specific Protein, Adiponectin, in Obesity. *Biochem. Biophys. Res. Commun.* **1999**, *257*, 79–83. [CrossRef] [PubMed]

20. Kadowaki, T.; Yamauchi, T.; Kubota, N.; Hara, K.; Ueki, K.; Tobe, K. Adiponectin and adiponectin receptors in insulin resistance, diabetes, and the metabolic syndrome. *J. Clin. Investig.* **2006**, *116*, 1784–1792. [CrossRef] [PubMed]

21. Declèves, A.-E.; Mathew, A.V.; Cunard, R.; Sharma, K. AMPK mediates the initiation of kidney disease induced by a high-fat diet. *J. Am. Soc. Nephrol.* **2011**, *22*, 1846–1855. [CrossRef] [PubMed]

22. Yang, Z.; Kahn, B.B.; Shi, H.; Xue, B.Z. Macrophage α1 AMP-activated protein kinase (α1 AMPK) antagonizes fatty acid-induced inflammation through SIRT1. *J. Biol. Chem.* **2010**, *285*, 19051–19059. [CrossRef] [PubMed]

23. Mottillo, E.P.; Desjardins, E.M.; Crane, J.D.; Smith, B.K.; Green, A.E.; Ducommun, S.; Henriksen, T.I.; Rebalka, I.A.; Razi, A.; Sakamoto, K.; et al. Lack of Adipocyte AMPK Exacerbates Insulin Resistance and Hepatic Steatosis through Brown and Beige Adipose Tissue Function. *Cell Metab.* **2016**, *24*, 118–129. [CrossRef] [PubMed]

24. Leclerc, I.; Woltersdorf, W.W.; da Silva Xavier, G.; Rowe, R.L.; Cross, S.E.; Korbutt, G.S.; Rajotte, R.V.; Smith, R.; Rutter, G.A. Metformin, but not leptin, regulates AMP-activated protein kinase in pancreatic islets: Impact on glucose-stimulated insulin secretion. *Am. J. Physiol. Endocrinol. Metab.* **2004**, *286*, E1023–E1031. [CrossRef] [PubMed]

25. Da Silva Xavier, G.; Leclerc, I.; Varadi, A.; Tsuboi, T.; Moule, S.K.; Rutter, G.A. Role of AMP-activated protein kinase in the regulation by glucose of islet β cell gene expression. *Proc. Natl. Acad. Sci. USA* **2000**, *97*, 4023–4028. [CrossRef] [PubMed]

26. Mottillo, E.P.; Balasubramanian, P.; Lee, Y.-H.; Weng, C.; Kershaw, E.E.; Granneman, J.G. Coupling of lipolysis and de novo lipogenesis in brown, beige, and white adipose tissues during chronic β3-adrenergic receptor activation. *J. Lipid Res.* **2014**, *55*, 2276–2286. [CrossRef] [PubMed]

27. Kim, S.J.; Tang, T.; Abbott, M.; Viscarra, J.A.; Wang, Y.; Sul, H.S. AMPK phosphorylates desnutrin/ATGL and HSL to regulate lipolysis and fatty acid oxidation within adipose tissue. *Mol. Cell. Biol.* **2016**, *36*, 1961–1976. [CrossRef] [PubMed]

28. Habinowski, S.A.; Witters, L.A. The Effects of AICAR on Adipocyte Differentiation of 3T3-L1 Cells. *Biochem. Biophys. Res. Commun.* **2001**, *286*, 852–856. [CrossRef] [PubMed]

29. Ceddia, R.B. The role of AMP-activated protein kinase in regulating white adipose tissue metabolism. *Mol. Cell. Endocrinol.* **2013**, *366*, 194–203. [CrossRef] [PubMed]

30. Villena, J.A.; Viollet, B.; Andreelli, F.; Kahn, A.; Vaulont, S.; Sul, H.S. Induced Adiposity and Adipocyte Hypertrophy in Mice Lacking the AMP-Activated Protein Kinase-alpha2 Subunit. *Diabetes* **2004**, *53*, 2242–2249. [CrossRef] [PubMed]

31. Zhang, W.; Zhang, X.; Wang, H.; Guo, X.; Li, H.; Wang, Y.; Xu, X.; Tan, L.; Mashek, M.T.; Zhang, C.; et al. AMP-activated protein kinase α1 protects against diet-induced insulin resistance and obesity. *Diabetes* **2012**, *61*, 3114–3125. [PubMed]

32. Jing, Y.; Wu, F.; Li, D.; Yang, L.; Li, Q.; Li, R. Metformin improves obesity-associated inflammation by altering macrophages polarization. *Mol. Cell. Endocrinol.* **2018**, *461*, 256–264. [CrossRef] [PubMed]

33. Li, W.; Saud, S.M.; Young, M.R.; Chen, G.; Hua, B. Targeting AMPK for cancer prevention and treatment. *Oncotarget* **2015**, *6*, 7365–7378. [CrossRef] [PubMed]
34. Motoshima, H.; Goldstein, B.J.; Igata, M.; Araki, E. AMPK and cell proliferation—AMPK as a therapeutic target for atherosclerosis and cancer. *J. Physiol.* **2006**, *574*, 63–71. [CrossRef] [PubMed]
35. Hardie, D.G. Molecular Pathways: Is AMPK a Friend or a Foe in Cancer? *Clin. Cancer Res.* **2015**, *21*, 3836–3840. [CrossRef] [PubMed]
36. Carretero, J.; Medina, P.P.; Blanco, R.; Smit, L.; Tang, M.; Roncador, G.; Maestre, L.; Conde, E.; Lopez-Rios, F.; Clevers, H.C.; et al. Dysfunctional AMPK activity, signalling through mTOR and survival in response to energetic stress in LKB1-deficient lung cancer. *Oncogene* **2007**, *26*, 1616–1625. [CrossRef] [PubMed]
37. Rothwell, P.M.; Fowkes, F.G.; Belch, J.F.; Ogawa, H.; Warlow, C.P.; Meade, T.W. Effect of daily aspirin on long-term risk of death due to cancer: Analysis of individual patient data from randomised trials. *Lancet* **2011**, *377*, 31–41. [CrossRef]
38. Evans, J.M.M.; Donnelly, L.A.; Emslie-Smith, A.M.; Alessi, D.R.; Morris, A.D. Metformin and reduced risk of cancer in diabetic patients. *BMJ* **2005**, *330*, 1304–1305. [CrossRef] [PubMed]
39. Liang, J.; Mills, G.B. AMPK: A Contextual Oncogene or Tumor Suppressor? *Cancer Res.* **2013**, *73*, 2929–2935. [CrossRef] [PubMed]
40. Yang, Y. Metformin for cancer prevention. *Front. Med.* **2011**, *5*, 115–117. [CrossRef] [PubMed]
41. Vandanmagsar, B.; Youm, Y.H.; Ravussin, A.; Galgani, J.E.; Stadler, K.; Mynatt, R.L.; Ravussin, E.; Stephens, J.M.; Dixit, V.D. The NLRP3 inflammasome instigates obesity-induced inflammation and insulin resistance. *Nat. Med.* **2011**, *17*, 179–188. [CrossRef] [PubMed]
42. Stienstra, R.; Joosten, L.A.; Koenen, T.; van Tits, B.; van Diepen, J.A.; van den Berg, S.A.; Rensen, P.C.; Voshol, P.J.; Fantuzzi, G.; Hijmans, A.; et al. The inflammasome-mediated caspase-1 activation controls adipocyte differentiation and insulin sensitivity. *Cell Metab.* **2010**, *12*, 593–605. [CrossRef] [PubMed]
43. Pilon, G.; Dallaire, P.; Marette, A. Inhibition of Inducible Nitric-oxide Synthase by Activators of AMP-activated Protein Kinase. *J. Biol. Chem.* **2004**, *279*, 20767. [CrossRef] [PubMed]
44. Wen, H.; Gris, D.; Lei, Y.; Jha, S.; Zhang, L.; Huang, M.T.; Brickey, W.J.; Ting, J.P. Fatty acid-induced NLRP3-ASC inflammasome activation interferes with insulin signaling. *Nat. Immunol.* **2011**, *12*, 408–415. [CrossRef] [PubMed]
45. Lee, H.M.; Kim, J.J.; Kim, H.J.; Shong, M.; Ku, B.J.; Jo, E.K. Upregulated NLRP3 inflammasome activation in patients with type 2 diabetes. *Diabetes* **2013**, *62*, 194–204. [CrossRef] [PubMed]
46. Hawley, S.A.; Ross, F.A.; Chevtzoff, C.; Green, K.A.; Evans, A.; Fogarty, S.; Towler, M.C.; Brown, L.J.; Ogunbayo, O.A.; Evans, A.M.; et al. Use of cells expressing γ subunit variants to identify diverse mechanisms of AMPK activation. *Cell Metab.* **2010**, *11*, 554–565. [CrossRef] [PubMed]
47. Iacobazzi, V.; Infantino, V. Citrate—New functions for an old metabolite. *Biol. Chem.* **2014**, *395*, 387–399. [CrossRef] [PubMed]
48. Wolf, A.J.; Reyes, C.N.; Liang, W.; Becker, C.; Shimada, K.; Wheeler, M.L.; Cho, H.C.; Popescu, N.I.; Coggeshall, K.M.; Arditi, M.; et al. Hexokinase Is an Innate Immune Receptor for the Detection of Bacterial Peptidoglycan. *Cell* **2016**, *166*, 624–636. [CrossRef] [PubMed]
49. Mills, E.L.; Kelly, B.; Logan, A.; Costa, A.S.H.; Varma, M.; Bryant, C.E.; Tourlomousis, P.; Däbritz, J.H.M.; Gottlieb, E.; Latorre, I.; et al. Succinate Dehydrogenase Supports Metabolic Repurposing of Mitochondria to Drive Inflammatory Macrophages. *Cell* **2016**, *167*, 457–470. [CrossRef] [PubMed]
50. Tannahill, G.M.; Curtis, A.M.; Adamik, J.; Palsson-McDermott, E.M.; McGettrick, A.F.; Goel, G.; Frezza, C.; Bernard, N.J.; Kelly, B.; Foley, N.H.; et al. Succinate is an inflammatory signal that induces IL-1β through HIF-1α. *Nature* **2013**, *496*, 238–242. [CrossRef] [PubMed]
51. McArdle, M.A.; Finucane, O.M.; Connaughton, R.M.; McMorrow, A.M.; Roche, H.M. Mechanisms of obesity-induced inflammation and insulin resistance: Insights into the emerging role of nutritional strategies. *Front. Endocrinol.* **2013**, *4*, 52. [CrossRef] [PubMed]
52. Cooke, A.A.; Connaughton, R.M.; Lyons, C.L.; McMorrow, A.M.; Roche, H.M. Fatty acids and chronic low grade inflammation associated with obesity and the metabolic syndrome. *Eur. J. Pharmacol.* **2016**, *785*, 207–214. [CrossRef] [PubMed]
53. Murphy, A.M.; Lyons, C.L.; Finucane, O.M.; Roche, H.M. Interactions between differential fatty acids and inflammatory stressors-impact on metabolic health. *Prostaglandins Leukot. Essent. Fat. Acids* **2015**, *92*, 49–55. [CrossRef] [PubMed]

54. Reynolds, C.M.; McGillicuddy, F.C.; Harford, K.A.; Finucane, O.M.; Mills, K.H.; Roche, H.M. Dietary saturated fatty acids prime the NLRP3 inflammasome via TLR4 in dendritic cells-implications for diet-induced insulin resistance. *Mol. Nutr. Food Res.* **2012**, *6*, 1212–1222. [CrossRef] [PubMed]

55. O'Reilly, M.; Dillon, E.; Guo, W.; Finucane, O.; McMorrow, A.; Murphy, A.; Lyons, C.; Jones, D.; Ryan, M.; Gibney, M.; et al. High-Density Lipoprotein Proteomic Composition, and not Efflux Capacity, Reflects Differential Modulation of Reverse Cholesterol Transport by Saturated and Monounsaturated Fat Diets. *Circulation* **2016**, *133*, 1838–1850. [CrossRef] [PubMed]

56. Rodríguez-Prados, J.C.; Través, P.G.; Cuenca, J.; Rico, D.; Aragonés, J.; Martín-Sanz, P.; Cascante, M.; Boscá, L. Substrate fate in activated macrophages: A comparison between innate, classic, and alternative activation. *J. Immunol.* **2010**, *185*, 605–614. [CrossRef] [PubMed]

57. Jha, A.K.; Huang, S.C.; Sergushichev, A.; Lampropoulou, V.; Ivanova, Y.; Loginicheva, E.; Chmielewski, K.; Stewart, K.M.; Ashall, J.; Everts, B.; et al. Network Integration of Parallel Metabolic and Transcriptional Data Reveals Metabolic Modules that Regulate Macrophage Polarization. *Immunity* **2015**, *42*, 419–430. [CrossRef] [PubMed]

58. Vats, D.; Mukundan, L.; Odegaard, J.I.; Zhang, L.; Smith, K.L.; Morel, C.R.; Wagner, R.A.; Greaves, D.R.; Murray, P.J.; Chawla, A. Oxidative metabolism and PGC-1beta attenuate macrophage-mediated inflammation. *Cell Metab.* **2006**, *4*, 13–24. [CrossRef] [PubMed]

59. MacIver, N.J.; Michalek, R.D.; Rathmell, J.C. Metabolic regulation of T lymphocytes. *Annu. Rev. Immunol.* **2013**, *31*, 259–283. [CrossRef] [PubMed]

60. Michalek, R.D.; Gerriets, V.A.; Jacobs, S.R.; Macintyre, A.N.; MacIver, N.J.; Mason, E.F.; Sullivan, S.A.; Nichols, A.G.; Rathmell, J.C. Cutting edge: Distinct glycolytic and lipid oxidative metabolic programs are essential for effector and regulatory CD4+ T cell subsets. *J. Immunol.* **2011**, *186*, 3299–3303. [CrossRef] [PubMed]

61. Blagih, J.; Coulombe, F.; Vincent, E.E.; Dupuy, F.; Galicia-Vázquez, G.; Yurchenko, E.; Raissi, T.C.; van der Windt, G.J.; Viollet, B.; Pearce, E.L.; et al. The Energy Sensor AMPK Regulates T Cell Metabolic Adaptation and Effector Responses In Vivo. *Immunity* **2015**, *42*, 41–54. [CrossRef] [PubMed]

62. MacIver, N.J.; Blagih, J.; Saucillo, D.C.; Tonelli, L.; Griss, T.; Rathmell, J.C.; Jones, R.G. The Liver Kinase B1 Is a Central Regulator of T Cell Development, Activation, and Metabolism. *J. Immunol.* **2011**, *187*, 4187–4198. [CrossRef] [PubMed]

63. Rolf, J.; Zarrouk, M.; Finlay, D.K.; Foretz, M.; Viollet, B.; Cantrell, D.A. AMPKα1: A glucose sensor that controls CD8 T-cell memory. *Eur. J. Immunol.* **2013**, *43*, 889–896. [CrossRef] [PubMed]

64. Poznanski, S.M.; Barra, N.G.; Ashkar, A.A.; Schertzer, J.D. Immunometabolism of T cells and NK cells: Metabolic control of effector and regulatory function. *Inflamm. Res.* **2018**, *67*, 813–828. [CrossRef] [PubMed]

65. Palsson-McDermott, E.M.; Curtis, A.M.; Goel, G.; Lauterbach, M.A.R.; Sheedy, F.J.; Gleeson, L.E.; van den Bosch, M.W.M.; Quinn, S.R.; Domingo-Fernandez, R.; Johnston, D.G.W.; et al. Pyruvate kinase M2 regulates Hif-1α activity and IL-1β induction and is a critical determinant of the warburg effect in LPS-activated macrophages. *Cell Metab.* **2015**, *21*, 65–80. [CrossRef] [PubMed]

66. Kirchner, J.; Brüne, B.; Namgaladze, D. AICAR inhibits NFκB DNA binding independently of AMPK to attenuate LPS-triggered inflammatory responses in human macrophages. *Sci. Rep.* **2018**, *8*, 7081. [CrossRef] [PubMed]

67. Moon, J.S.; Nakahira, K.; Choi, A.M. Fatty acid synthesis and NLRP3-inflammasome. *Oncotarget* **2015**, *6*, 21765–21766. [CrossRef] [PubMed]

68. Cacicedo, J.M.; Yagihashi, N.; Keaney, J.F., Jr; Ruderman, N.B.; Ido, Y. AMPK inhibits fatty acid-induced increases in NF-kappaB transactivation in cultured human umbilical vein endothelial cells. *Biochem. Biophys. Res. Commun.* **2004**, *324*, 1204–1209. [CrossRef] [PubMed]

69. Dicter, N.; Madar, Z.; Tirosh, O. α-Lipoic Acid Inhibits Glycogen Synthesis in Rat Soleus Muscle via Its Oxidative Activity and the Uncoupling of Mitochondria. *J. Nutr.* **2002**, *132*, 3001–3006. [CrossRef] [PubMed]

70. Lee, W.J.; Song, K.H.; Koh, E.H.; Won, J.C.; Kim, H.S.; Park, H.S.; Kim, M.S.; Kim, S.W.; Lee, K.U.; Park, J.Y. α-Lipoic acid increases insulin sensitivity by activating AMPK in skeletal muscle. *Biochem. Biophys. Res. Commun.* **2005**, *332*, 885–891. [CrossRef] [PubMed]

71. Timmers, S.; de Vogel-van den Bosch, J.; Towler, M.C.; Schaart, G.; Moonen-Kornips, E.; Mensink, R.P.; Hesselink, M.K.; Hardie, D.G.; Schrauwen, P. Prevention of high-fat diet-induced muscular lipid accumulation in rats by α lipoic acid is not mediated by AMPK activation. *J. Lipid Res.* **2010**, *51*, 352–359. [CrossRef] [PubMed]

72. Park, K.G.; Min, A.K.; Koh, E.H.; Kim, H.S.; Kim, M.O.; Park, H.S.; Kim, Y.D.; Yoon, T.S.; Jang, B.K.; Hwang, J.S.; et al. Alpha-lipoic acid decreases hepatic lipogenesis through adenosine monophosphate-activated protein kinase (AMPK)-dependent and AMPK-independent pathways. *Hepatology* **2008**, *48*, 1477–1486. [CrossRef] [PubMed]

73. Yang, Y.; Li, W.; Liu, Y.; Sun, Y.; Li, Y.; Yao, Q.; Li, J.; Zhang, Q.; Gao, Y.; Gao, L.; et al. Alpha-lipoic acid improves high-fat diet-induced hepatic steatosis by modulating the transcription factors SREBP-1, FoxO1 and Nrf2 via the SIRT1/LKB1/AMPK pathway. *J. Nutr. Biochem.* **2014**, *25*, 1207–1217. [CrossRef] [PubMed]

74. Kratz, M.; Coats, B.R.; Hisert, K.B.; Hagman, D.; Mutskov, V.; Peris, E.; Schoenfelt, K.Q.; Kuzma, J.N.; Larson, I.; Billing, P.S.; et al. Metabolic dysfunction drives a mechanistically distinct proinflammatory phenotype in adipose tissue macrophages. *Cell Metab.* **2014**, *20*, 614–625. [CrossRef] [PubMed]

75. Chan, K.L.; Pillon, N.J.; Sivaloganathan, D.M.; Costford, S.R.; Liu, Z.; Théret, M.; Chazaud, B.; Klip, A. Palmitoleate Reverses High Fat-induced Proinflammatory Macrophage Polarization via AMP-activated Protein Kinase (AMPK). *J. Biol. Chem.* **2015**, *290*, 16979–16988. [CrossRef] [PubMed]

76. Xiong, X.Q.; Geng, Z.; Zhou, B.; Zhang, F.; Han, Y.; Zhou, Y.B.; Wang, J.J.; Gao, X.Y.; Chen, Q.; Li, Y.H.; et al. FNDC5 attenuates adipose tissue inflammation and insulin resistance via AMPK-mediated macrophage polarization in obesity. *Metabolism* **2018**, *83*, 31–41. [CrossRef] [PubMed]

77. Van Dijk, S.J.; Feskens, E.J.; Bos, M.B.; Hoelen, D.W.; Heijligenberg, R.; Bromhaar, M.G.; de Groot, L.C.; de Vries, J.H.; Müller, M.; Afman, L.A. A saturated fatty acid—Rich diet induces an obesity-linked proinflammatory gene expression profile in adipose tissue of subjects at risk of metabolic syndrome. *Am. J. Clin. Nutr.* **2009**, *90*, 1656–1664. [CrossRef] [PubMed]

78. Chen, S.; Zhao, Z.; Ke, L.; Li, Z.; Li, W.; Zhang, Z.; Zhou, Y.; Feng, X.; Zhu, W. Resveratrol improves glucose uptake in insulin-resistant adipocytes via Sirt1. *J. Nutr. Biochem.* **2018**, *55*, 209–218. [CrossRef] [PubMed]

79. Hwang, J.T.; Kwak, D.W.; Lin, S.K.; Kim, H.M.; Kim, Y.M.; Park, O.J. Resveratrol Induces Apoptosis in Chemoresistant Cancer Cells via Modulation of AMPK Signaling Pathway. *Ann. N. Y. Acad. Sci.* **2007**, *1095*, 441–448. [CrossRef] [PubMed]

80. Kim, W.S.; Lee, Y.S.; Cha, S.H.; Jeong, H.W.; Choe, S.S.; Lee, M.R.; Oh, G.T.; Park, H.S.; Lee, K.U.; Lane, M.D.; et al. Berberine improves lipid dysregulation in obesity by controlling central and peripheral AMPK activity. *Am. J. Physiol. Endocrinol. Metab.* **2009**, *296*, 812–819. [CrossRef] [PubMed]

81. Turner, N.; Li, J.Y.; Gosby, A.; To, S.W.; Cheng, Z.; Miyoshi, H.; Taketo, M.M.; Cooney, G.J.; Kraegen, E.W.; James, D.E.; et al. Berberine and its more biologically available derivative, dihydroberberine, inhibit mitochondrial respiratory complex I: A mechanism for the action of berberine to activate AMP-activated protein kinase and improve insulin action. *Diabetes* **2008**, *57*, 1414–1418. [CrossRef] [PubMed]

82. Kim, H.S.; Kim, M.J.; Kim, E.J.; Yang, Y.; Lee, M.S.; Lim, J.S. Berberine-induced AMPK activation inhibits the metastatic potential of melanoma cells via reduction of ERK activity and COX-2 protein expression. *Biochem. Pharmacol.* **2011**, *83*, 385–394. [CrossRef] [PubMed]

83. Shehzad, A.; Ha, T.; Subhan, F.; Lee, Y.S. New mechanisms and the anti-inflammatory role of curcumin in obesity and obesity-related metabolic diseases. *Eur. J. Nutr.* **2011**, *50*, 151–161. [CrossRef] [PubMed]

84. Um, M.Y.; Hwang, K.H.; Ahn, J.; Ha, T.Y. Curcumin Attenuates Diet-Induced Hepatic Steatosis by Activating AMP-Activated Protein Kinase. *Basic Clin. Pharmacol. Toxicol.* **2013**, *113*, 152–157. [CrossRef] [PubMed]

85. Na, L.X.; Zhang, Y.L.; Li, Y.; Liu, L.Y.; Li, R.; Kong, T.; Sun, C.H. Curcumin improves insulin resistance in skeletal muscle of rats. *Nutr. Metab. Cardiovasc. Dis.* **2011**, *21*, 526–533. [CrossRef] [PubMed]

86. Pan, W.; Yang, H.; Cao, C.; Song, X.; Wallin, B.; Kivlin, R.; Lu, S.; Hu, G.; Di, W.; Wan, Y. AMPK mediates curcumin-induced cell death in CaOV3 ovarian cancer cells. *Oncol. Rep.* **1994**, *20*, 1553–1559.

87. Panche, A.N.; Diwan, A.D.; Chandra, S.R. Flavonoids: An overview. *J. Nutr. Sci.* **2016**, *5*, e47. [CrossRef] [PubMed]

88. Ha, B.G.; Nagaoka, M.; Yonezawa, T.; Tanabe, R.; Woo, J.T.; Kato, H.; Chung, U.I.; Yagasaki, K. Regulatory mechanism for the stimulatory action of genistein on glucose uptake in vitro and in vivo. *J. Nutr. Biochem.* **2012**, *23*, 501–519. [CrossRef] [PubMed]

89. Ji, G.; Zhang, Y.; Yang, Q.; Cheng, S.; Hao, J.; Zhao, X.; Jiang, Z. Genistein Suppresses LPS-Induced Inflammatory Response through Inhibiting NF-κB following AMP Kinase Activation in RAW 264.7 Macrophages. *PLoS ONE* **2012**, *7*, e53101. [CrossRef] [PubMed]

90. Palacios-González, B.; Zarain-Herzberg, A.; Flores-Galicia, I.; Noriega, L.G.; Alemán-Escondrillas, G.; Zariñan, T.; Ulloa-Aguirre, A.; Torres, N.; Tovar, A.R. Genistein stimulates fatty acid oxidation in a leptin receptor-independent manner through the JAK2-mediated phosphorylation and activation of AMPK in skeletal muscle. *Biochim. Biophys. Acta Mol. Cell Biol. Lipids* **2014**, *1841*, 132–140. [CrossRef] [PubMed]

91. Ono, M.; Fujimori, K. Antiadipogenic Effect of Dietary Apigenin through Activation of AMPK in 3T3-L1 Cells. *J. Agric. Food Chem.* **2011**, *59*, 13346–13352. [CrossRef] [PubMed]

92. Choi, W.H.; Son, H.J.; Jang, Y.J.; Ahn, J.; Jung, C.H.; Ha, T.Y. Apigenin Ameliorates the Obesity-Induced Skeletal Muscle Atrophy by Attenuating Mitochondrial Dysfunction in the Muscle of Obese Mice. *Mol. Nutr. Food Res.* **2017**, *61*, 1700218. [CrossRef] [PubMed]

93. Jung, U.J.; Cho, Y.Y.; Choi, M.S. Apigenin Ameliorates Dyslipidemia, Hepatic Steatosis and Insulin Resistance by Modulating Metabolic and Transcriptional Profiles in the Liver of High-Fat Diet-Induced Obese Mice. *Nutrients* **2016**, *8*, 305. [CrossRef] [PubMed]

94. Sung, B.; Chung, H.Y.; Kim, N.D. Role of Apigenin in Cancer Prevention via the Induction of Apoptosis and Autophagy. *J. Cancer Prev.* **2016**, *21*, 216–226. [CrossRef] [PubMed]

95. Chang, H.C.; Guarente, L. SIRT1 and other sirtuins in metabolism. *Trends Endocrinol. Metab.* **2014**, *25*, 138–145. [CrossRef] [PubMed]

International Journal of
Molecular Sciences

MDPI

Review

Involvement of 5′AMP-Activated Protein Kinase (AMPK) in the Effects of Resveratrol on Liver Steatosis

Jenifer Trepiana [1,2,†], Iñaki Milton-Laskibar [1,2,3,†], Saioa Gómez-Zorita [1,2,3], Itziar Eseberri [1,2,3], Marcela González [4], Alfredo Fernández-Quintela [1,2,3,*] and María P. Portillo [1,2,3]

1   Department of Nutrition and Food Sciences, University of the Basque Country (UPV/EHU),
    01006 Vitoria, Spain; jenifer.trepiana@ehu.eus (J.T.); inaki.milton@ehu.eus (I.M.-L.);
    saio86@hotmail.com (S.G.-Z.); itziar.eseberri@ehu.eus (I.E.); mariapuy.portillo@ehu.eus (M.P.P.)
2   Nutrition and Obesity Group, Lucio Lascaray Research Institute, 01006 Vitoria, Spain
3   Biomedical Research Networking Centres, Physiopathology of Obesity and Nutrition (CIBERobn),
    Institute of Health Carlos III, 28029 Madrid, Spain
4   Nutrition and Food Science Department, Faculty of Biochemistry and Biological Sciences,
    National University of Litoral and National Scientific and Technical Research Council (CONICET),
    3000 Santa Fe, Argentina; maidagon@fbcb.unl.edu.ar
*   Correspondence: alfredo.fernandez@ehu.eus; Tel.: +34-945-013-066
†   These authors equally contributed to this manuscript.

Received: 16 October 2018; Accepted: 30 October 2018; Published: 5 November 2018

Abstract: This review focuses on the role of 5′AMP-activated protein kinase (AMPK) in the effects of resveratrol (RSV) and some RSV derivatives on hepatic steatosis. In vitro studies, performed in different hepatic cell models, have demonstrated that RSV is effective in preventing liver TG accumulation by activating AMPK, due to its phosphorylation. These preventive effects have been confirmed in studies conducted in animal models, such as mice and rats, by administering the phenolic compound at the same time as the diet which induces TG accumulation in liver. The literature also includes studies focused on other type of models, such as animals showing alcohol-induced steatosis or even steatosis induced by administering chemical products. In addition to the preventive effects of RSV on hepatic steatosis, other studies have demonstrated that it can alleviate previously developed liver steatosis, thus its role as a therapeutic tool has been proposed. The implication of AMPK in the delipidating effects of RSV in in vivo models has also been demonstrated.

Keywords: resveratrol; AMPK; hepatocyte; liver; steatosis

1. Introduction

The most benign form of non-alcoholic fatty liver disease (NAFLD) is known as simple hepatic steatosis, and it consists of excessive fat accumulation in the liver. This hepatic alteration is considered to be the main contributor to chronic liver disease development in western societies. Moreover, the prevalence of NAFLD is expected to increase with the greater incidence of metabolic syndrome and obesity, which are closely related to hepatic steatosis [1]. Indeed, hepatic steatosis has been proposed as the hepatic manifestation of metabolic syndrome. This health alteration is defined as an intrahepatic triglyceride (TG) accumulation greater than 5% of the liver weight, or by $\geq$5% of hepatocytes showing TG content [2].

5′AMP-activated protein kinase (AMPK) is a master regulator of energy homeostasis, activated by low cellular energy status [3,4]. It restores energy balance during metabolic stress both at cellular and physiological levels [5]. Under conditions of energy depletion, the activation of AMPK leads to the inhibition of ATP-consuming pathways (e.g., fatty acid synthesis, cholesterol synthesis, and gluconeogenesis), and the stimulation of ATP generating processes (e.g., fatty acid oxidation and glycolysis), thus restoring overall cellular energy homeostasis [6–8]. The activation of AMPK is regulated by its heterotrimeric structure, which consists of a catalytic subunit α (α1 and α2), a regulatory and structurally crucial β subunit (β1 and β2), and a regulatory subunit γ (γ1, γ2, and γ3), with unique tissue and species-specific expression profiles [3]. Given its key role in regulating energy balance, AMPK may have therapeutic interest for the treatment of diseases in humans, such as insulin resistance, type 2 diabetes, obesity, NAFLD, cardiovascular diseases, and cancer [3,4,9].

Several molecules have been identified as AMPK activators. Among these molecules, resveratrol (RSV; 3,5,4′-trihydroxy-*trans*-stilbene) (Figure 1), a non-flavonoid polyphenol, has generated great interest [10]. Numerous studies have been carried out using RSV and different models of obesity and liver steatosis in rodents [2,11]. The vast majority of preclinical studies have demonstrated that RSV is able to prevent liver TG accumulation, although more evidence is still needed with regard to human beings [12–14].

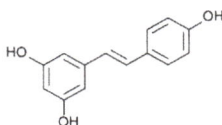

Figure 1. Chemical structure of *trans*-resveratrol (3,5,4′-trihydroxy-*trans*-stilbene).

As far as RSV-mediated AMPK activation is concerned, the role that is played in this process by sirtuin 1 (SIRT1) must be highlighted. Indeed, in both cases (AMPK and SIRT1), their regulation occurs in response to similar stimuli, namely nutrient availability and energy expenditure [15,16]. In this regard, while SIRT1 activation is enhanced by changes in the NAD^+/NADH ratio (Figure 2), the activation of AMPK is mediated by changes in the AMP/ATP ratio [16]. In fact, although it is widely accepted that both molecules regulate the activation of each other [17], no consensus has been reached regarding which is activated first. In this regard, while some authors have reported that SIRT1 activates AMPK through the deacetylation of liver kinase B1 (LKB1, Figure 2) [18,19], other authors have suggested that AMPK activation results in an increased NAD^+/NADH ratio, and thus, in greater SIRT1 activation [16,20]. In a study that was carried out in brains from Alzheimer's disease patients, the authors observed that RSV increased cytosolic Ca^{2+} levels and so increased AMPK activation via calcium/calmodulin-dependent protein kinase kinase-β (CaMKKβ, Figure 2) [21]. In line with this study, Park et al. analyzed the metabolic diseases that are associated with aging, reporting that RSV increased cyclic adenosine monophosphate (cAMP) through phosphodiesterase (PDE) inhibition, without activating adenylyl cyclase activity directly. In fact, PDE inhibition prevented diet-induced obesity by the CaMKKβ/AMPK/SIRT1 pathway (Figure 2) [22].

This review focuses on the role of AMPK in the effects of RSV and some RSV derivatives on hepatic steatosis (Figure 2). For this purpose, in vitro and in vivo studies have been included.

2. In Vitro Studies

A wide range of in vitro studies have been performed to date using different hepatic cell models, such as human hepatoma cells (HepG2), murine hepatocytes (Hepa 1–6), rat hepatoma cells (H4IIEC3) or even primary hepatocytes. In some cases, hepatocyte TG accumulation was induced by including large amounts of glucose, insulin, free fatty acid (FFA) and/or alcohol in the incubation medium. These simulate the induction of fatty liver by following an obesogenic dietary pattern, as in the case of studies addressed to animal models, and imitate the situation that very often takes place in humans from western societies. In addition other models, such as hepatocytes incubated with fatty acids in the incubation medium and hepatocytes treated with an agonist of the transcription factor liver X receptor (LXR), have also been used (Table 1).

Table 1. In vitro studies conducted to test the effect of resveratrol in liver cells.

Reference	Cell Line	Experimental Design	Effect of Resveratrol	Mechanism of Action
[23]	HepG2	24 h culture with RSV (10 μmol/L) 1 h culture with RSV (50 μmol/L) (AMPKα1 activity determination)	Prevention of high-glucose-induced lipid accumulation (in HepG2 cells)	↑ AMPK (Thr-172) and ACC (Ser-79) phosphorylation (10 μmol/L RSV) ↑ AMPKα1 activity
[18]	HepG2 HEK293	HepG2 24 h culture with RSV (1–100 μM) HEK293 cells pretreated with splitomicin (100 μM) for 24 h and incubated with RSV (50 μM) for an additional 1 h	↓ High glucose-induced TG accumulation: RSV (10–50 μM)	↑ SIRT1 activity (dose dependent (10–100 μM RSV) ↓ FAS protein expression ↑ AMPKα phosphorylation (Thr172) ↑ ACC phosphorylation (Ser79)
[24]	HepG2	6 or 24 h culture with 10, 25, and 50 μM of RSV	↓ TG accumulation	↑ AMPKα phosphorylation (Thr172) ↓ srebf1 and fasn gene expressions
[25]	Hepa 1–6 cell line (murine hepatocytes)	24 h culture with T0901317 (1 μM, LXR activator: ↑ liver fat accumulation) + RSV (40 μM), with or without compound C (10 μM, AMPK inhibitor)	↓ T0901317-induced fat accumulation	↓ T0901317-induced fat accumulation (via AMPK activation)
[26]	H4IIEC3 rat hepatoma cells	24 h culture with FFA [oleic acid and palmitic acid 2:1, 0.5 mM] or T0901317 (10 μM, LXR activator) + RSV or SY-102 (RSV derivative, 3.3–50 μM)	↓ FFA-induced lipid accumulation by RSV (50 mM) and SY-102 (30 μM) ↓ T0901317-induced SREBP-1 maturation by RSV (30 μM) and SY-102 (10 μM)	↓ srebf1 and fasn gene expression by SY-102 (50 μM) via AMPK/LXR pathway
[27]	HepG2	48 h treatment Control, oleic acid + alcohol (O + A), O + A-RSV (5, 15, 45, 135 μM), O + A-AICAR-RSV 45 μM, O + A-Compound C-RSV 45 μM	↓ Lipid accumulation (15, 45, 135 μM) ↓ Hepatocyte TG content (45 and 135 μM) Attenuated hepatic steatosis	↑ AMPKα phosphorylation * ↑ ACC phosphorylation * ↓ SREBP1c and lipin protein expression
[28]	HepG2	24 h culture with 0.2 mM palmitate Additional 24 h treated with RSV (20–80 μM) Also exposed to 3-MA autophagy inhibitor for 1 h or siRNAs before the addition of RSV	↓ Lipid content	↑ LC3-II protein expression and SQSTM1 protein degradation (3-MA pre-treatment inhibited this effect) ↑ SIRT1 protein expression/activity and cyclic AMP levels ↑ AMPK (Thr-172) and PRKA (Ser-96) phosphorylation
[29]	Primary hepatocytes from C57BL/6 mice	Treatment with NEFA, NEFA + RSV (50 and 100 μM), NEFA + Nicotinamide, NEFA + Compound C, NEFA + RSV + Nicotinamide, NEFA + RSV + Compound C Treatment length not specified	↓ NEFA increased expression of several inflammatory markers	↑ AMPK phosphorylation * ↑ sirt1 gene and SIRT1 protein expressions ↓ phosphorylation IκBα and NF-κB p65 ↓ il-1β, il-6, and tnf-α gene expression

3-MA: 3-Methyladenine, ACC: acetyl CoA carboxylase, AICAR: 5-Aminoimidazole-4-carboxamide ribonucleotide, AMPK: 5′AMP-activated protein kinase, FAS: fatty acid synthase, FFA: free fatty acid, IκBα: inhibitor of kappa B, IL-1β: interleukin 1 beta, IL-6: interleukin 6, LC3-II: microtubule-associated protein1 light chain 3, LXR: liver x receptor, NEFA: non-esterified fatty acid, NF-κB: nuclear factor-κB, PRKA: monophosphate-activated protein kinase, RSV: resveratrol, SIRT1: silent information regulator 1, SREBF1: sterol regulatory element binding transcription factor 1, SREBP1: sterol regulatory element binding transcription factor 1, SQSTM1: sequestosome 1, TG: triglyceride, TNF-α: tumor necrosis factor α. ↑: increased, ↓: decreased. * The measured phosphorylation residue is not mentioned in the article.

Zang et al. [23] examined the effect of several polyphenols, including RSV, on AMPK activity and lipid levels in human hepatoma HepG2 cells. In a previous in vitro study, these authors had reported that exposing HepG2 to elevated glucose (30 mmol/L) for 24 h induced decreased AMPK and acetyl CoA carboxylase (ACC) phosphorylation, as well as hepatocellular lipid accumulation [30]. Then, in the following study [23], when HepG2 were incubated with RSV (10 μmol/L) for 24 h, this treatment significantly stimulated ACC phosphorylation and the activity of AMPKα1, which is the predominant isoform in hepatocytes. By contrast, AMPKα2 activity was not modified. Consequently, RSV induced the phosphorylation of ACC by AMPK, rendering ACC inactive, which translates into a decrease in lipid synthesis rates (Figure 2). This fact inhibited high-glucose–induced accumulation of TG in HepG2 cells [23].

Later, Hou et al. [18] studied the mechanisms underlying the effects of RSV on lipid accumulation in HepG2 hepatocytes also exposed to a high concentration of glucose. These authors previously had demonstrated that the dysfunction of hepatic AMPK induced by hyperglycemia represents a key mechanism for hepatic lipid accumulation [23]. The results obtained indicated that RSV prevented the impairment of AMPK phosphorylation and that of its downstream target ACC, the elevation in fatty acid synthase (FAS) expression, and, therefore, the lipid accumulation in human HepG2 hepatocytes. All of these effects were mediated by increased SIRT1 activity, LKB1 phosphorylation, and AMPK activity. In order to confirm these results, the authors performed further experimental procedures. After several experimental approaches, they observed that the positive effects of RSV on AMPK phosphorylation were largely abolished by pharmacological and genetic inhibition of SIRT1, suggesting that the stimulation of AMPK and lipid-lowering effect of the phenolic compound depended on SIRT1 activity. On the other hand, Hou and coworkers, after an adenoviral overexpression of SIRT1, detected a stimulation of basal AMPK signaling in HepG2 cells. Moreover, LKB1 is required for the activation of AMPK induced by RSV and SIRT1. The authors put forward that SIRT1 functions as a novel upstream regulator for LKB1/AMPK signaling and it plays an essential role in the regulation of hepatocyte lipid metabolism (Figure 2).

In another study reported by Shang et al. [24], fat accumulation was also induced in HepG2 hepatocytes by incubating these cells with large amounts of glucose and insulin in the medium. Moreover, RSV was included in the incubation medium at a final concentration of 0.1%. This phenolic compound prevented TG accumulation. This effect seemed to be due to the activation of AMPK and the down-regulation of *srebf1* and *fasn* gene expression, two important AMPK targets and two key genes that are involved in lipogenesis, a metabolic process that allows hepatocytes to accumulate TG.

In other studies, the accumulation of TG in hepatocytes was induced by incubating cells with fatty acids. Thus, Zhang et al. [28] used HepG2 cells incubated with palmitate, a common approach to inducing steatosis in cell culture, to study the role of autophagy in the beneficial effect of RSV on hepatocyte TG accumulation. For this purpose, they incubated these cells with palmitate at 0.1–0.5 mM for 24 h, which were then treated with or without a series of RSV concentrations (10–100 μM) for a further 24 h, or with or without RSV (40 μM) for different time intervals (6, 2, and 48 h). The optimal condition for inducing steatosis was HepG2 incubation with palmitate (0.2 mM) for 24 h. RSV reduced intracellular lipid content induced by palmitate in a dose-dependent manner. RSV also induced autophagy through the up-regulation of LC3-II and the degradation of SQSTM1, as compared with cells treated with palmitate alone, in which autophagy was inhibited by 3-MA (an inhibitor of the early stages) pre-treatment.

The decreased hepatic lipid content stimulated by RSV was attenuated in the presence of autophagy inhibitors (3-MA, BafA1, or CQ). In addition, RSV increased SIRT1 expression in palmitic-stimulated-HepG2 cells in dose- and time-dependent manners. They also validated that the intracellular lipid content reduction induced by RSV was reversed by EX-527 (an inhibitor of SIRT1) or SIRT1 siRNA in palmitic-stimulated HepG2 cells.

By using primary mice hepatocytes, instead of a cell line culture, Tian et al. [29] carried out an experiment where cells were incubated with different fatty acid concentrations, RSV (50 and 100 μM), nicotinamide (a SIRT1 inhibitor, 10 mM), and compound C (an AMPKα inhibitor, 10 mM).

Hepatocyte TG content, phosphorylation levels of IκBα and NF-κB p65, as well as gene expression of *tnf-α*, *il-6*, and *il-1b* were higher in fatty acid-treated hepatocytes than in control cells. Phosphorylation of IκBα and NF-κB p65 and the inflammatory markers levels were greater in cells that were treated with fatty acids, RSV and Compound C than in those treated with fatty acids and RSV, indicating AMPK and SIRT1 inhibition can avoid RSV induced improvements. In addition, RSV reversed the reduction in AMPK phosphorylation and SIRT1 protein expression (Figure 2) induced by fatty acids. Taken as a whole, these results show that RSV inhibits the NF-κB inflammatory pathway via AMPK-SIRT1 pathway.

Tang et al. [27] combined incubation with fatty acid and the addition of ethanol to the medium to induce TG accumulation. Their experimental model consisted of HepG2 cells subjected to 48 h culture with 100 μM oleic acid + 87 mM alcohol (O + A) plus the respective treatments with RSV and/or the AMPK activator AICAR or the AMPK inhibitor compound C. The experimental groups used were control, O + A, O + A-RSV (RSV 5, 15, 45, 135 μM), O + A-AICAR (0.5 mM AICAR), O + A-Compound C-RSV (3 μM Compound C and RSV 45 μM). After 48 h treatment, Oil Red O staining showed that the combination of oleic acid and alcohol increased lipid accumulation. That effect was partially reversed by RSV 15, 45, and 135 μM, but not by RSV 5 μM. Similar results were appreciated when TG contents were measured. All RSV doses, except 5 and 15 μM, reversed the effect induced by fat and alcohol. To elucidate the possible mechanisms implicated in the delipidating effect of the polyphenol, 45 μM of RSV were used in combination with AICAR or Compound C. The AMPK phosphorylation ratio was lower in cells that were supplemented by oleic acid and alcohol as compared to control group. The decreased AMPK phosphorylation ratio induced by O + A was reversed when AICAR was added to cultured cells. Finally, RSV stimulated AMPK in a similar manner to AICAR. By the same token, pACC levels were lower in the O + A group than in the control group. At the same time, these levels were normalized by AICAR and RSV, while in the case of the AMPK inhibitor intermediate values were observed. Greater protein expression of sterol regulatory element binding protein 1 (SREBP-1c) and lipin1 were also found in the O + A group when compared with the control group. Again, RSV and AICAR reversed that effect, while the AMPK inhibitor increased both protein expressions, inhibiting the effects that were induced by the polyphenol treatment. All of these results indicated that RSV was preventing hepatocyte lipid accumulation via AMPK-Lipin1.

Finally, a different model was used by Gao et al. [25]. These authors examined the activity of RSV on the suppression of fat accumulation induced by a liver X receptor (LXR) activator. Due to the fact that *srebf1* and *chrebp*, the two master genes that are responsible for de novo lipogenesis are LXR target genes, the activation of the latter leads to hepatic fat accumulation [31,32]. For this purpose, murine Hepa 1–6 hepatocytes (CRL-1830) were incubated with T0901317 (an agonist and thus activator of LXR, 1 μM) with or without RSV (40 μM), and compound C (10 μM) for 24 h. The authors observed that RSV suppressed the hepatocyte lipid droplet increase induced by the LXR agonist (Nile Red staining), while this effect was greatly repressed by Compound C. These results demonstrated that the effect of RSV was indeed mediated by AMPK (Figure 2). RSV prevented the hepatocyte lipid droplet increase induced by T0901317, without reversing the increase in mRNA levels of *srebf1*, *chrebp*, *acc*, and *fasn* induced by the LXR activator.

The effects of RSV derivatives have also been addressed in the literature. Choi et al. [26] worked with SY-102, a synthesized derivative of RSV, which has lower cytotoxicity than the parent compound, but similar potency. In this study, H4IIEC3 rat hepatoma cells were treated with a FFA mixture (0.5 mM, 2:1 oleic acid and palmitic acid) with or without RSV or its derivative (0, 3.3, 10, 30, or 50 μM) for 24 h. Moreover, they also used T0901317 instead of the FFA mixture, in order to activate SREBP-1 expression in cells. By these means, the involvement of the inhibition of this transcriptional factor in the therapeutic effects of RSV and its derivatives could be investigated. The authors observed that T0901317 treatment induced SREBP-1 maturation, and that RSV (30 μM) and SY-102 (10 μM) addition inhibited this process in a dose-dependent manner. Consistent with these results, they also observed that SY-102 reduced *fasn* gene expression induced by T0901317.

Additionally, they examined whether inhibition of AMPK blocked the effect of SY-102 on the T0901317-mediated SREBP-1 induction. Compound C reversed the inhibitory effect of SY-102 on *srebf1* and *fasn* expression, confirming that SY-102 inhibits the maturation and transcriptional activation of SREBP-1 via AMPK/LXR. Based on these data, the authors conclude that RSV and SY-102 are effective in reducing cell lipid accumulation by inhibiting LXR agonist induced SREBP-1 activation, and thus, reducing the expression of key genes in de novo lipogenesis (*fasn*). Moreover, while using AMPK inhibitors, the authors demonstrated the involvement of this kinase in the effects mentioned.

As far as in vitro experiments are concerned, primary mice hepatocytes were cultured with different non-esterified fatty acid (NEFA) concentrations, RSV (50 and 100 µM), nicotinamide (SIRT1 inhibitor, 10 mM), and compound C (10 mM). TG content, phosphorylation levels of IκBα and NF-κB p65 as well as gene expressions of *tnf-a*, *il-6*, and *il-1b* were higher in NEFA-treated hepatocytes than in the controls. RSV supplementation reversed NEFAs effect in a dose-dependent manner. NEFA reduced AMPK phosphorylation and SIRT1 protein expression, whereas RSV increased it. The phosphorylation of IκBα and NF-κB p65 and the levels of inflammatory markers were greater in the cells treated with NEFA+RSV+Compound C than in the cells that were treated with NEFA+RSV, indicating that the inhibition of AMPK and SIRT1 can avoid RSV mediated improvements (Figure 2). Based on these results, it could be considered that the inhibition of the NF-κB inflammatory pathway produced by RSV occurs through the AMPK-SIRT1 pathway.

All of these studies demonstrate that RSV is effective in preventing TG accumulation in different hepatocyte models by activating AMPK, due to its phosphorylation (Figure 2). According to the experimental designs used, this ability takes place under different metabolic conditions. Furthermore, depending on the used experimental conditions, the beneficial effects of RSV occurs in a dose range of 15 and 135 µM. Moreover, this steatosis preventive effect has also been demonstrated by some RSV derivatives, such as SY-102 and Z-TMS.

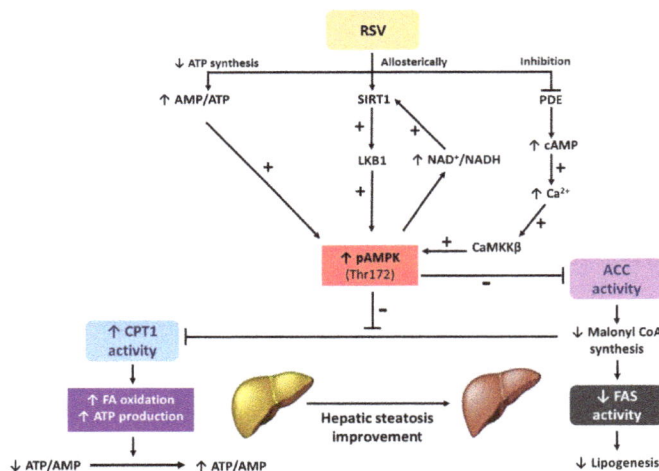

Figure 2. Effects of resveratrol (RSV) on hepatic steatosis improvement through 5′ AMP-activated protein kinase (AMPK) activation. ACC: acetyl-CoA-carboxylase, AMPK: 5′ AMP-activated protein kinase, AMP: adenosine monophosphate, ATP: adenosine triphosphate, cAMP: cyclic adenosine monophosphate, CaMKKβ: Ca^{2+}/calmodulin-dependent protein kinase kinase β, CPT1: carnitine palmitoyltransferase 1, FA: fatty acid, FAS: fatty acid synthase, LKB1: liver kinase B1, NAD$^+$: oxidized nicotinamide adenine dinucleotide, NADH: reduced nicotinamide adenine dinucleotide, PDE: phosphodiesterases, RSV: resveratrol, SIRT1: silent information regulator 1. (+: activation, −: inhibition, ➔: regulation direction; ⊣: regulation inhibition).

Int. J. Mol. Sci. **2018**, *19*, 3473

3. In Vivo Studies on Hepatic Steatosis Prevention

Studies that were conducted to analyze the preventive effect of RSV on hepatic steatosis development have been carried out in mice or rats, by administering the phenolic compound at the same time as the diet that induces TG accumulation in liver (Table 2). There are also studies in the literature focused on other types of models, such as animals showing alcohol steatohepatitis or even steatosis induced by administering chemical products.

The first study devoted to analyzing the effects of RSV on liver steatosis induced by diet, as well as the involvement of AMPK, was carried out by Baur et al. [33]. In that study, the authors used middle-aged (one-year-old) male C57BL/6NIA mice fed either a standard diet (SD) or an equivalent high-calorie diet (60% of calories from fat, HC) supplemented or not with RSV at a dose of 22.4 mg/kg/day (0.04%; HCR) for six months. At 18 months of age, the high-calorie diet greatly increased the size and weight of livers, which was associated with fatty liver, whereas RSV prevented these alterations. Indeed, in the histological examination of liver sections, they observed a loss of cellular integrity and the accumulation of large lipid droplets in the livers of the HC but not the HCR group. Searching for a mechanistic explanation to this effect, the authors demonstrated that RSV-treated mice showed a strong tendency towards inducing phosphorylation of AMPK, as well as phosphorylation of ACC at Ser79, thereby decreasing its activity, and decreased expression of *fasn* gene [33].

Moreover, the livers of the RSV-treated mice exhibited considerably more mitochondria than those of mice from HC controls and were not significantly different from those of the SD group. Due to the lack of tissue availability, the authors performed a parallel experiment in the same cohort of animals, but supplemented for six weeks with 186 mg/kg/day. In this case, the authors also observed increased mitochondrial biogenesis in liver through the deacetylation of PGC-1α (peroxisome proliferator activated receptor gamma, coactivator 1α), a master regulator of this process. It has been reported that AMPK promotes PGC-1α activation with the involvement of SIRT1 [34], and in turn, the acetylation status of PGC-1α is considered a marker of SIRT1 activity in vivo. Importantly, since Baur et al. did not find changes in SIRT1 protein levels in RSV-treated mice. As a result of the increase in mitochondrial number and PGC-1α activation, these authors suggested that SIRT1 enzymatic activity was enhanced by RSV [33].

Table 2. Preclinical studies conducted in vivo to test the preventive effect of resveratrol on hepatic steatosis.

Reference	Animal Model	Experimental Design	Effect of Resveratrol	Mechanism of Action
[33]	One-year-old male C57BL/6NIA mice	HFD (60% of calories as fat) RSV dose: 22.4 mg/kg bw/day Length: 6 months	Fatty liver development prevention (organ size) Prevention of cellular integrity loss and large lipid droplet accumulation	↑AMPK phosphorylation (Thr172) ↑ACC phosphorylation (Ser79)
[35]	6–8 week-old male C57BL/6J mice	LFD (10% of calories as fat) RSV dose: 200 and 400 mg/kg bw/day 3 groups: LF diet+ethanol, LF diet + ethanol + RSV200, LF diet + ethanol + RSV400 Length: 2 weeks	Prevention of liver weight, liver lipid droplets, hepatic TG content and serum ALT level increase	↑*sirt1* gene and SIRT1 protein expressions ↑ AMPKα and β phosphorylation * ↑ total AMPK levels ↑ ACC phosphorylation * ↓ SREBP1c protein expression ↓*fasn, gpat1, scd1, acca, me* ↑*acox1, mcad* and *cpt1a* gene expression ↓ *pparγ* gene expression
[36]	Male Sprague Dawley rats	Obesogenic diet (45% of calories as fat) RSV dose: 30 mg/kg bw/day Length: 6 weeks	↓ Hepatic fat content	↑ AMPK phosphorylation (Thr172) ↑ ACC phosphorylation (Ser79)
[25]	Male C57BL/6 mice induced by LXR receptor	Groups: Control, T0901317 (LXR activator) and T0901317+ RSV RSV dose: 200 mg/kg bw/day Length: 5 days	Prevention of the increase in liver size, fat accumulation and TG content (induced by LXR activator)	↑ AMPK phosphorylation (Thr172) ↑ACC phosphorylation (Ser79) ↓ *srebf1, chrebp* and *acc* expression (RSV alone)
[29]	4 week-old C57BL/6 mice	HFD (60% of calories as fat) RSV dose: 30 mg/kg bw/day Length: 60 day	↓ Liver weight ↓ GGT, AST, ALT, ALP, LDH plasma levels ↓ IL-1β, IL-6, and TNF-α plasma levels	↑ AMPK phosphorylation (Thr172) ↑ SIRT1 protein expression ↑ IkBα and NF-kB p65 phosphorylation * ↓ *il-1β, il-6,* and *tnf-α* gene expression
[37]	5 week-old male C57BL/6 mice	HFD (45% of energy as fat) RSV dose: 0.1% resveratrol (w/w) Length: 18 weeks	No changes in liver weight and serum AST and ALT levels	↑ AMPK phosphorylation (Thr172) ↑ACC phosphorylation (Ser79) ↓ FAS protein expression ↓ Hepatic adipogenic protein expression
[38]	Pups from female Wistar rats	Control diet RSV dose: 20 mg/kg bw/day Length: 3 weeks (lactation period)	↓ Hepatic lipid accumulation	↑ AMPK phosphorylation (Ser403) ↑ SIRT1 protein expression ↑ Active/precursor SREBP-1c protein ratio ↓ ACC protein expression ↓ FAS protein expression ↓Hepatic adipogenic protein expression

ACC: acetyl CoA carboxylase, ACOX1: acyl-Coenzyme A oxidase 1, ALP: alkaline phosphatase, ALT: alanine aminotransferase, AMPK: 5′AMP-activated protein kinase, AST: aspartate aminotransferase, bw: body weight, chrebp: carbohydrate response element binding protein, CPT1a: carnitine palmitoyltransferase 1a, FAS: fatty acid synthase, GGT: gamma glutamil transpeptidase, GPAT1: glycerol-3-phosphate acyltransferase 1, HFD: High-fat diet, IL-1β: interleukine-1β, IL-6: interleukine-1β, LDH: lactate dehydrogenase, LFD: low fat diet, LXR: liver x receptor, MCAD: mitochondrial medium-chain acyl-CoA dehydrogenase, ME: malic enzyme, pIkBα: phospho-inhibitory subunit of NF-KBα, NF-KBβ p65: nuclear factor kappa-light-chain-enhancer of activated B cells subunit p65, PPARγ: peroxisome proliferator-activated receptor γ, SCD1: stearoyl-Coenzyme A desaturase 1, RSV: resveratrol, SIRT1: silent information regulator 1, SREBF1: sterol regulatory element binding transcription factor 1, SREBP-1c: sterol regulatory element binding protein-1c, TG: triglyceride, TNF-α: tumor necrosis factor-α. ↑: increased, ↓: decreased. * The measured phosphorylation residue is not mentioned in the article.

Int. J. Mol. Sci. **2018**, *19*, 3473

In the work reported by Tian et al. [29], the authors not only evaluated the effect of RSV on TG accumulation, but they also extended the study to inflammatory markers. For this purpose, they fed C57BL/6 mice a high-fat (HF) diet (60% fat) for 60 days, treated or not with RSV (30 mg/kg body weight/day). Resveratrol-treated mice showed decreased liver weight and reduced hepatic TG content to an intermediate level between that found in mice fed the HF diet and those observed in a control group fed a standard diet. Regarding plasmatic levels, reduced levels of transaminases were also observed. Moreover, RSV supplementation avoided the inactivation of AMPKα induced by the HF diet, as shown by the increased pAMPKα/AMPKα ratio. Moreover, SIRT1 expression followed the same trend. As far as inflammation markers are concerned, phosphorylated IkBα and NF-kB p65, IL-1β, Il-6, and TNF-α were overexpressed in HF diet-fed mice. However, this effect was avoided by RSV.

Other studies reported in the literature have been performed in rats. In a study that was carried out by our group we fed Sprague-Dawley rats a high-fat high-sucrose diet supplemented or not with the amount of RSV needed to achieve a dose of were 30 mg/kg body weight/day for six weeks. RSV did not reduce liver weight or serum ALT and AST concentrations, but did prevent increased hepatic fat infiltration. To elucidate the mechanisms involved in the delipidating effect we analyzed some the activity of enzymes involved in lipogenesis and fatty acid oxidation, as well as other mitochondrial markers. The polyphenol limited hepatic lipogenesis as shown by the reduction in ACC activity, although no changes were observed in FAS, malic enzyme (ME), and glucose-6P-dehydrogenase (G6PDH) activities, or in gene expression of *srebf1*, a transcriptional factor that regulates FAS. Moreover, RSV increased carnitine-palmitoyl-1a (CPT1a) and acyl-CoA oxidase (ACO) activities and activated PGC-1α, indicating an increase in mitochondrial and peroxisomal fatty acid oxidation, even though the gene expression of *pparγ* and *pgc1α* was not modified. *Tfam* and *cox2* expression, two indicators of mitochondrial genesis and oxidative phosphorylation respectively, remained unaltered by RSV treatment. In view of these results, we proposed that increased CPT1a activity was not due to increased mitochondria number. As far as AMPK was concerned, in this study RSV increased its activity since phosphorylated AMPK/total AMPK protein ratio was augmented. For that reason we proposed the potential involvement of the AMPK in RSV reducing fatty liver infiltration effect [36].

The influence of RSV on imprinting has also been analyzed. This term refers to the fact that nutritional and other environmental factors in early life have a profound influence on lifelong health. Tanaka et al. [38] carried out a study devoted to investigating whether maternal RSV administration affected lipogenesis in male offspring, by means of lactation, once they reached adult age. If this was the case, the authors also wanted to determine the mechanisms that are involved in the effects observed. For this purpose, pregnant Wistar rats were divided into two groups fed a control diet with RSV or not. The group supplemented with the polyphenol received a RSV dose of 20 mg/kg body weight/day by gavage during lactation (three weeks), while the control group received the vehicle. Once the lactation period was completed, six male pups for each group (control and RSV-treated group) were fed a standard diet. Animals were sacrificed when they were 36 week-old. When liver sections of the animals in both groups were stained (hematoxylin-eosin), the lipid droplet amount found in the samples of the RSV group was lower than that found in the control group. No information was reported regarding liver weight or liver TG content. Using the western-blot technique, a significant increase in the pAMPK/AMPK ratio was found in the RSV-treated group. In the same way, greater SIRT1 protein expression was also found in the animals of this group when compared with the control group. Moreover, when analyzing ACC and FAS, the two key enzymes of lipogenesis, lower activities, and protein expressions were found. Finally, the authors also measured the proteolytic processing of SREBP-1c, which is the transcriptional factor that regulates *acc* and *fasn* gene expression. In this case, a significantly lower SREBP-1c active form/precursor form ratio was found in the RSV group, suggesting that, in this case, the activation of this transcriptional factor was lower. Based on the results obtained, the authors concluded that maternal RSV intake during lactation effectively down-regulated hepatic lipogenesis in adult male rats. Moreover, the authors suggested that such effects could be due to epigenetic modifications induced by the polyphenol. The lipid lowering effects that were observed in the animals from the RSV group seemed to be mediated, at least in part, by the greater AMPK-SIRT1 axis activation induced by the compound (Figure 2).

In addition to non-alcoholic liver steatosis, hepatic TG accumulation can also be induced in humans by chronic high intake of ethanol. Thus, animal models imitating this situation have been developed, and the effectiveness of RSV on ethanol-induced steatosis has been addressed. In this context, Ajmo et al. [35] performed a study in mice that were fed a low-fat diet supplemented with ethanol (29% of total calories). Mice were divided into three experimental groups: (1) low-fat diet plus ethanol, (2) low-fat diet plus ethanol and 200 mg/kg body weight RSV (E+R200), and (3) low-fat diet plus ethanol and 400 mg/kg body weight/day RSV (E+R400). RSV was added to the diets during the last two weeks of the study.

RSV treatment increased SIRT1 expression levels and stimulated AMPK activity in livers of ethanol-fed mice. The protective action of RSV was in whole or in part mediated through the up-regulation of a SIRT1-AMPK signaling system (Figure 2) in the livers of ethanol-fed mice. The final conclusion was that RSV treatment led to reduced lipid synthesis and increased rates of fatty acid oxidation, thus preventing alcoholic liver steatosis and so it might represent a promising agent for the prevention and treatment of human alcoholic fatty liver disease.

Finally, Gao et al. [25] examined the effect of RSV on the suppression of fat accumulation induced by the activation of LXR, a transcriptional factor that facilitates TG accumulation through a compound known as T0901317, as they carried out with murine Hepa 1–6 hepatocytes. They divided C57BL/6 mice into three groups: the carrier solution-treated control group, a group treated intraperitoneally with T0901317 (5 mg/kg body weight/day), a group treated intraperitoneally with T0901317 (5 mg/kg body weight/day) and orally with RSV (200 mg/kg body weight/day) for a short period (five days). In the T0901317-treated group the liver became bigger, with higher lipid and number of bright red spots in tissue sections, analyzed by staining in tissue sections, and increased serum TG and cholesterol levels. RSV prevented liver size increase and reduced hepatic TG amounts. Moreover, it completely blocked serum TG and cholesterol levels elevation. By contrast, transaminase levels were not affected by treatments.

When gene expression analysis was carried out, it was observed that T0901317 markedly increased the expression of *acc*, *srebf1* and *chreb*, but RSV did not have any impact on these genes. However, treating with RSV alone, *acc*, *srebf1*, and *chreb* gene expression decreased. Immunohistochemistry and western analysis revealed that mice treated at the same time with T0901317 and RSV showed a marked increase in pAMPK (Thr-172) and pACC (Ser-79, inactivation) as compared to mice treated with T0901317 alone. The conclusion of this study is that RSV activated AMPK by increasing its phosphorylation at the post-translational level, and consequently reduced carboxylase activity of ACC, thus suppressing lipogenesis and fat accumulation in the liver (Figure 2).

As in the case of in vitro studies, the effects of RSV derivatives have also been analyzed in animal models. Tung et al. carried out a study aimed at analyzing the inhibitory effect of piceatannol, a natural stilbene that is an analog and a metabolite of RSV, on HF diet-induced obesity in C57BL/6 mice [37]. The authors also included in the experimental design a group supplemented with RSV. For this purpose, five week-old male C57BL/6 mice were fed a HFD (45% of the energy as fat) for 18 weeks, alone or supplemented with RSV (0.1% *w/w*) or piceatannol (0.1% or 0.25% *w/w*). At the end of the experimental period, lower liver weights were found in the two groups supplemented with piceatannol when compared with the HF group, while this parameter remained unchanged in the group receiving RSV. As far as serum transaminases is concerned, aspartate aminotransferase (AST) and alanine aminotransferase (ALT) levels remained unchanged in all of the treated groups when compared with the HF group. The activation of AMPK in the liver, along with the expression of different adipogenic proteins, was assessed by means of western-blot. Greater AMPK phosphorylation (in threonine 172 residue) was found in all of the groups treated (with no difference among them) in comparison with the group fed the HFD alone. This means that RSV and piceatannol significantly enhanced the activation of this kinase. Moreover, a similar phosphorylation pattern was also observed in the case of acetyl CoA carboxylase (ACC) (in serine 79 residue) (Figure 2). In this case, the greater phosphorylation resulted in a lower activation of this lipogenic enzyme. Finally, the authors also found a significant reduction in the protein expression of FAS in the groups in which RSV and piceatannol were administered. Based on those results, it could be stated that, under these experimental conditions, both RSV and piceatannol are effective in inducing the activation of AMPK in the liver of mice that were fed an obesogenic diet. Moreover, a concomitant reduction in the activation or protein expression of key lipogenic proteins (ACC and FAS) was also described. Nevertheless, in these conditions piceatannol seemed to be more effective than RSV in reducing liver lipid content. Although the activation of AMPK may be involved in the effects observed, due to the fact that the activity of enzymes that are involved in fatty acid oxidation (such as CPT1a) was not measured by the authors, no conclusion can be drawn in this respect.

Altogether, these studies confirm the results obtained in in vitro studies using isolated hepatocytes. In fact, when RSV administration takes place at the same time as the cause inducing TG accumulation, this phenolic compound prevents partial or totally this alteration. This beneficial effect has been reported in a very huge range of doses, from 22.4 to 400 mg/kg body weight/day and using experimental periods, from two weeks to six months. As in the case of in vitro studies, it seemed that the activation of AMPK was mediating, at least in part, the hepatic delipidating effect. Moreover, piceatannol, a RSV derivative that shows two hydroxyl groups in the benzene ring, instead of three as in the case of RSV, shows the same effect but it was even more effective that its parent compound.

4. In Vivo Studies on Hepatic Steatosis Treatment

There are also studies devoted to analyzing the effect of RSV on previously developed liver steatosis in the literature, in other words, studies that address its therapeutic effects on this hepatic metabolic alteration (Table 3). Shang et al. [24] carried out an experiment by feeding rats HF-diet (59% calorie from lard fat, 21% from protein, 20% from carbohydrate) for six weeks. At the end of this period, half of the HF group was treated orally with RSV (100 mg/kg body weight/day; HR group), and the other half was treated with saline. At the end of the 16th week, all of the animals were sacrificed. Liver histology showed that rats fed the HF diet developed liver steatosis and insulin resistance, which were markedly improved by 10 weeks of RSV administration. In addition, this compound promoted the phosphorylation of AMPK, which in this study suppressed the expression of genes that are related to lipogenesis, thus contributing to the improvement of liver steatosis and insulin resistance. The authors concluded that, by reducing TG accumulation and improving insulin resistance through AMPK activation, RSV could protect the liver from NAFLD.

Table 3. Preclinical studies conducted in vivo to test the therapeutic effect of resveratrol on hepatic steatosis.

Reference	Animal Model	Experimental Design	Effect of Resveratrol	Mechanism of Action
[24]	Male Wistar rats (180–200 g)	Acute treatment: Fed stated rats RSV dose: 100 mg/kg bw/day Length: 4 h. Chronic treatment: High-fat diet (59% of calories as fat) RSV dose: 100 mg/kg bw/day Length: 10 weeks	↓ Hepatic fat content	↑ AMPK phosphorylation (Thr172) ↓ srebf1 and fasn gene expressions
[39]	Obese male Zucker rats and lean heterozygous littermates	STD RSV dose: 10 mg/kg bw/day Length: 8 weeks	No change in liver weight ↓ Liver TG and cholesterol content	↑ AMPK phosphorylation (Thr172) ↓ ACC phosphorylation *
[40]	4 week-old male C57BL/KsJ-db/db mice	RSV dose: 0.005% and 0.02% (w/w) Length: 6 weeks	↓ Hepatic fat content (only in 0.02% RSV group)	↓ ACC phosphorylation * ↑ srebf1 gene expression (0.02% RSV) ↑ PPARα protein expression (0.02% RSV) ↑ UCP2 protein expression ↑ AMPK phosphorylation *
[41]	5 week-old male C57BL/6N mice	HFD RSV dose: 30 mg/kg bw/day Length: 2 weeks	↓ Hepatic fat content	↑ AKT phosphorylation (Ser473 and Thr308) ↓ AMPKα phosphorylation (Thr172)
[42]	4 week-old male C57BL/6 mice expressing HBV X protein	RSV dose: 30 mg/kg bw/day Length: 2, 3, 7, and 14 days	Histopatology alteration reversion ↓ Serum ALT levels	↓ srebf1 and lxrα gene expressions (from day 2 in advance) ↑ AMPK phosphorylation (Thr172) (from day 3 in advance) ↓ pparγ and acc gene expressions. ↑ SIRT1 protein expression and activity (from day 7 in advance) ↓ fasn gene expression
[26]	Male ICR mice (20–25 g)	HFD. RSV dose: 15 or 45 mg/kg bw/day. (same doses of SY-102, a RSV derivative). Length: 2 days	↓ Hepatic TG levels (by SY-102 and RSV)	↑ AMPK phosphorylation (Thr172) (by SY-102) ↓ srebf1 and fasn mRNA levels (by SY-102 and RSV)
[43]	8 week-old male KKAy mice (genetic model of obesity) 8 week-old male C57BL/6J mice (control)	Chow diet (AIN93G) RSV dose: 2 or 4 g/kg diet Length: 12 weeks	↓ Hepatic fat content (Oil Red) and TG levels Hepatic steatosis attenuation (histological study) ↓ MDA levels	↑ AMPK phosphorylation (Thr172) ↑ SIRT1 protein expression ↑ FOXO1 phosphorylation (Thr24) ↓ ROS levels ↑ GSH levels, GPx and SOD activities ↑ hsl gene expression and HSL phosphorylation (Ser660) ↑ atgl gene expression and ATGL protein expression
[44]	6 week-old male C57BL/6J mice	HFD RSV dose: 8 mg/kg bw/day Length: 4 weeks	↓ Liver weight ↓ Plasma levels of Fetuin-A and ALT ↓ Hepatic index	↑ ampk gene expression ↓ fetuin-A gene expression ↓ Fetuin-A protein expression ↓ nfκβ gene expression

Table 3. *Cont.*

Reference	Animal Model	Experimental Design	Effect of Resveratrol	Mechanism of Action
[28]	8 week-old 129/SvJ mice (male)	HFD (60% of energy as fat) RSV dose: 0.4% (*w/w*) Length: 8 weeks	↓ Hepatic fat content	↑ cyclic AMP levels ↑ PRKA phosphorylation (Ser96) ↑ AMPK phosphorylation (Thr172) ↑ SIRT1 protein expression
[1]	6 week-old male Wistar rats	HFHS RSV dose: 30 mg/kg bw/day Length: 6 weeks	↓ Hepatic TG content ↑ Plasma TG release (from liver) ↓ Liver fatty acid uptake	↑ AMPK phosphorylation (Thr172) ↑ CPT1a activity ↑ CS activity ↑ MTP activity ↓ FATP5 protein expression

ACC: acetyl CoA carboxylase, ALT: alanine aminotransferase, AMP: adenosine monophosphate, AMPK: 5′AMP-activated protein kinase, ATGL: adipose triglyceride lipase, ATP: adenosine triphosphate, bw: body weight, CPT1a: carnitine palmitoyltransferase 1a, CS: citrate synthase, FAS: fatty acid synthase, FATP5: fatty acid transport protein 5, FOXO1: forkhead box protein O1, GPx: glutathione peroxidase, GSH: reduced glutathione, HFD: High-fat diet, HFHS: High-fat high-sucrose diet, HSL: hormone sensitive lipase, LXR: liver x receptor, MDA: malonaldehide, MTP: microsomal triglyceride transfer protein, NF-KBβ p65: nuclear factor kappa-light-chain-enhancer of activated B cells, PPARγ: peroxisome proliferator-activated receptor γ, PPARα: peroxisome proliferator-activated receptor α, PRKA: protein kinase A, ROS: reactive oxygen species, RSV: resveratrol, SIRT1: silent information regulator 1, SOD: superoxide dismutase, SREBF-1c: sterol regulatory element binding factor-1c, SREBP: sterol regulatory element-binding protein, STD: standard diet, TG: triglyceride, UCP2: uncoupling Protein 2. ↑: increased, ↓: decreased. * The measured phosphorylation residue is not mentioned in the article.

In another study, Lee et al. [44] fed C57BL/6J mice a standard or a HF diet (45% fat), and then continued the experiment for four additional weeks treating mice with RSV (8 mg/kg body weight/day) or vehicle (control group), administered by using an osmotic pump. The aim of this study was to analyze both the relationship between adiponectin and fetuin-A and whether RSV alters both cytokines and several related factors.

At the end of the treatment, liver weight and hepatic index were reduced by RSV supplementation. Serum TG and ALT were also normalized to control levels, but serum AST, increased by the HF diet, was not re-established by the polyphenol. These data indicate that this polyphenol improved liver function, which is apparently mediated by fetuin-A, a hepatokine that mediates fatty liver-induced cardiometabolic diseases. As occurred in serum, hepatic *fetuin-A* gene expression and protein levels were significantly increased by the HF diet, and RSV re-established control values. Gene expression of *ampk* in liver was reduced by the HF diet and highly increased in RSV-treated mice, to a level even higher than that observed in the controls. In addition, *nf-κb* gene expression, a downstream target of fetuin-A, was higher in HF animals. Taking these results into account, the authors concluded that under their experimental conditions RSV probably improved the steatosis by the AMPK-NF-κB-fetuin-A axis.

Choi et al. [26] who evaluated the effect of SY-102, a synthesized derivative of RSV, on cell cultures also carried out an experiment in ICR mice with this compound. Animals were randomly assigned to six groups: a control group fed a purified diet for five days and the other groups were fed a HF diet (27% safflower oil; 59% fat-derived calories) for five days. During the last two days of the experimental treatment, mice were treated daily with the RSV or the RSV derivative (SY-102) by oral injection (15 and 45 mg/kg body weight/day, respectively). Liver TG content increased by two-fold in the HF group, which decreased back to the control level by treatment with SY-102. This compound also recovered the impaired AMPK phosphorylation caused by the HF, in a dose-dependent manner and prevented the increase *srebf1* and *fasn* mRNA levels induced by HF diet. The authors concluded that SY-102 improved HF-induced fatty liver in mice through the AMPK/LXR pathway.

Zhang et al. [28] determined the role of autophagy in the beneficial effect of RSV on hepatic steatosis. The authors used 129/SvJ mice (eight-week-old) that were fed a HF diet (containing 60% fat) or chow diet (10% fat) for four weeks to induce hepatic steatosis. After this period, HF-fed mice were further divided into two subgroups, which were fed chow diet (HF+chow group) or chow diet containing RSV (0.4%) (HF+RSV group) for four additional weeks. The authors observed that after four weeks of treatment, RSV reduced the weight of HF-fed mice, without significant changes in food intake. The HFD+chow group showed increases in lipid content in the liver compared with the control group, which were markedly attenuated by the addition of RSV (0.4%) (HFD+RSV group) to the diet for a further four weeks. RSV significantly increased SIRT1 activity, the pAMPK and pPRKA (protein kinase A) levels in liver tissues of mice fed with HFD (HFD+RSV group). RSV also increased adenylate cyclase expression and cAMP levels in liver tissues of HFD-fed mice (HFD+RSV group). In conclusion, RSV-induced autophagy in response to hepatic steatosis through the cAMP-PRKA-AMPK-SIRT1 signaling pathway (Figure 2).

Recently, in a study carried out in our group [1], we fed rats a high-fat high-sucrose diet for six weeks. At the end of this period, nine animals were sacrificed in order to determine whether hepatic steatosis had been induced. The liver lipid content of these rats was compared with that of a matched group of rats fed a standard diet for six weeks. Once the development of hepatic steatosis was confirmed, the remaining rats were shifted to a standard diet (STD) supplemented with RSV (30 mg/kg body weight/day) or not for six additional weeks. At the end of the experimental period (12 weeks), significantly lower hepatic TG levels were found in the group treated with RSV. In order to know the mechanisms that are involved in this hepatic lipid lowering effects of RSV, the activities of several enzymes involved in lipid metabolism were assessed. CPT1a and citrate synthase (CS) activities were observed in the group receiving the compound, which points toward increased mitochondrial fatty acid oxidation and density (respectively). Moreover, when measuring the effects of RSV on microsomal triacylglycerol transfer protein (MTP) activity, a greater activation of this enzyme was appreciated when the polyphenol was administered to the animals, suggesting that, in this group, the plasma TG release from the liver was enhanced. The activations of AMPK and ACC, as well as the protein levels of fatty acid transport protein 5 (FATP5) were studied by means of western-blot. RSV significantly increased AMPK phosphorylation (in threonine 172 residue) and thus activation, while decreasing the protein expression of FATP5. In the case of ACC, although no significant differences were found between both groups, a trend towards an increased phosphorylation status (the serine 79 residue) was observed in the animals that were supplemented with RSV. Based on the data obtained, we concluded that RSV was effective in reducing hepatic steatosis that was previously induced by an obesogenic feeding pattern. It could be suggested that the increased AMPK activation found in the group receiving RSV could, to some extent, influence the greater fatty acid oxidation through the reduction of ACC activity, and thus malonyl-CoA production (an inhibitor of CPT1a).

Kang et al. [41] hypothesized that RSV action on insulin signaling could depend on the metabolic state of cells and that it is tissue specific. Thus, they evaluated the effect of RSV on insulin action in insulin-sensitive tissues in mice fed a HF diet. For that purpose, they treated diet-induced obese C57BL/6N mice with RSV, at a dose of 30 mg/kg body weight/day for two weeks. RSV did not affect body weight in HF-fed mice, but reduced both fasting glucose and insulin levels, suggesting an insulin-sensitizing effect. In order to explain this effect, they assessed insulin signaling pathway and the activation state of AMPK in liver. RSV restored diminished insulin signaling induced by phosphorylation of AKT in Ser473 and Thr308 residues. By contrast with the most common studies in the literature, AMPK was more active in HF mice when compared to control mice and this activation was partially reverted after RSV treatment (not to the level of control mice).

After a histological examination of the liver, fat increase induced by the HF diet was reversed by the short-term RSV treatment. It is important to point out that this effect was independent of body weight change that, as previously indicated, was not significantly reduced after RSV treatment.

In addition to works addressing models of liver steatosis that are associated to diet, other experiments have been performed by using genetically models of obesity or diabetes. Rivera et al. [39] treated obese Zucker rats (Zucker *fa/fa* rats), which show a strong liver steatosis early in life, and their heterozygous lean littermates (Zucker *Fa/Fa*) with RSV at a dose of 10 mg/kg body weight/d for eight weeks. Food intake, body weight, liver weight, as well as hepatic TG and cholesterol content were notably greater in obese rats than in their lean littermates. RSV treatment attenuated the increase in hepatic TG content, by reducing it by half and cholesterol content, which reached lean levels. These reductions seemed to be due to AMPK activation by phosphorylation of its Thr172 residue and the consequent ACC inactivation by phosphorylation (Figure 2).

Do et al. [40] analyzed the effects of RSV on steatosis in a model of diabetic mice (*db/db* mice). In this study, they distributed animals into four experimental groups: control group, a group treated with 0.001% (*w/w*) of the hypoglycemic agent rosiglitazone, a group treated with 0.005% (*w/w*) RSV, and a group treated with 0.02% (*w/w*) RSV for six weeks. This review will focus only on the effects of RSV. Regarding hepatic lipid content, only the highest dose of this phenolic compound reduced TG when mice were compared with the controls, but it did not affect hepatic cholesterol levels. With regard to protein levels, no quantification figures appear in the publication and only protein bands can be observed; consequently, only data commented by authors can be discussed. Phosphorylated ACC, measured as an indicator of de novo lipogenesis, was reduced by RSV and the protein expression was lower with the highest dose of RSV compared with the lowest one. Moreover, 0.02% RSV increased *srebf1* gene and PPARα protein expression, although no differences were observed in *ppara* gene expression. As far as fatty acid β-oxidation is concerned, RSV increased UCP2 protein expression, which is regulated by PPARα. When they determined AMPK activation, they observed higher phosphorylated AMPK protein levels in RSV-treated mice compared with control mice; this activation was higher at a dose of 0.005% of RSV. Authors concluded that RSV was able to reduce liver steatosis in their experimental model of type 2 diabetes by activating AMPK signaling.

Lin et al. [42] studied the effects of RSV on hepatitis B virus (HBV) associated fatty livers at an early stage of pathogenesis. The work was carried out in C57BL/6 transgenic mice, which spontaneously develop hepatocellular carcinoma at the age of 13–16 months because they express the HBV X protein, specifically in hepatocytes, and in wild type mice. Animals received orally RSV at a dose of 30 mg/kg body weight/day and they were sacrificed at 2, 3, 7, and 14 days. In the transgenic animals, microsteatosis, pleomorphic and bizarre nuclei, ballooning and abnormal arrangements of the sinusoids were observed at 4–6 weeks. The polyphenol mitigated liver impairment and reverted histopathological alterations. After 14 days of treatment, RSV-treated mice had no fatty liver and serum ALT was significantly reduced. Regarding the mechanism of action, RSV reduced *srebf1* and *lxrα* but no *lxrβ* gene expressions on day 2. The ratio of phosphorylated AMPK/total AMPK, as representative of its activation, was increased on day 3, and the gene expression of *pparγ* and *acc* was decreased. Moreover, SIRT1 was activated, its protein expression increased and *fasn* gene expression decreased on day 7 of treatment with RSV. Finally, no changes were observed in phosphorylated AKT/total AKT ratio, as representative of AKT activation, and *scd1* gene expression but no changes were observed. When considering these results, it can be observed that the reduction in hepatic lipid content was, at least in part, due to a reduction in the lipogenic pathway; probably the down-regulation of srebf1 decreased *acc* and *fasn* gene expression, but the reduction in *srebf1* was not due to the AMPK pathway.

Finally, Zhu et al. [43] investigated the potential benefits of RSV on the amelioration of oxidative stress and hepatic steatosis in a model of genetic obesity, the KKAy mouse. In this study, C57BL/6J mice were used as controls. The KKAy mice were randomly divided into three groups: a standard group that was fed a chow diet (KKAy group), a group that was treated with a low dose of RSV (KKAy + Low RSV), and a group treated with a high dose of RSV (KKAy + High RSV). The control and KKAy groups were fed a standard AIN93G diet, while the Low RSV and High RSV groups were fed a standard AIN93G diet supplemented with RSV at doses of 2 and 4 g/kg diet, respectively, for 12 weeks. They observed that in the KKAy group body weight and hepatic index were higher in comparison to values observed in C57BL/6J mice, but they were reversed in the KKAy+high RSV group. Serum levels of FFA and malonaldehyde (MDA) in KKAy mice were higher than those in C57BL/6J mice, while the superoxide dismutase (SOD) level was decreased. MDA levels in the RSV treatment group were significantly decreased when compared with the KKAy group. ROS level was decreased by RSV, while levels of glutathione (GSH), glutathione peroxidase (GPx), and SOD were increased.

The authors also carried out a histological study of liver, where they noticed fresh bright red liver tissue in C57BL/6J mice, but in non-treated KKAy mice. The degree of hepatic steatosis evaluated by Oil Red was significantly alleviated by RSV. Similarly, the liver TG level was reduced in RSV-treated mice as compared with non-treated KKAy mice, whereas there was an insignificant change in TC level. *hsl* and *atgl* mRNA levels were decreased in KKAy mice compared with C57BL/6J, and this decrease was reversed by high RSV treatment. Moreover, pHSL protein was highly expressed in both the low and high RSV treatment groups and SIRT1, pAMPK α, pFOXO1 and FOXO1 (cytoplasm) in KKAy mice was reversed by RSV treatment, when compared with C57BL/6J mice. The authors concluded that RSV ameliorated hepatic steatosis inducing up-regulation of SIRT1 and AMPK.

The studies described demonstrate that, in addition to its preventive effect, RSV is also able to reduce hepatic TG accumulation, independently of the steatosis model used, thus representing a potential interesting approach to treat liver lipid alteration. As well as in the case of steatosis prevention, the improvement that was observed in steatosis previously developed is also due, at least in part, to increased AMPK activation.

Author Contributions: All the authors contributed equally to the manuscript writing and original draft preparation. A.F.-Q. reviewed and edited the final version of the manuscript. M.P.P supervised the final version of the manuscript.

funding: This study was supported by grants from the Instituto de Salud Carlos III (CIBERObn), Government of the Basque Country (IT-572-13).

Int. J. Mol. Sci. **2018**, *19*, 3473

Acknowledgments: The authors acknowledge Esperanza Irles for her collaboration.

Conflicts of Interest: The authors declare no conflict of interest.

Abbreviations

ACC	acetyl-CoA-carboxylase
ACOX1	Acyl-Coenzyme A oxidase 1
AICAR	5-Aminoimidazole-4-carboxamide ribonucleotide
AKT	Protein kinase B
ALP	Alkaline phosphatase,
ALT	Alanine aminotransferase
AMP	Adenosine monophosphate
AMPK	5'AMP-activated protein kinase
AST	Aspartate aminotransferase
ATGL	Adipose triglyceride lipase
ATP	Adenosine triphosphate
BW	Body weight
CaMKKβ	Ca^{2+}/calmodulin-dependent protein kinase kinase β
cAMP	Cyclic adenosine monophosphate
CHREBP	Carbohydrate response element binding protein
CPT	Carnitine palmitoyl transferase
CS	Citrate synthase
FAS	Fatty acid synthase
FATP5	Fatty acid transport protein 5
FFA	Free fatty acid
FOXO 1	Forkhead box protein O1
GGT	Gamma glutamil transpeptidase
GPAT1	Glycerol-3-phosphate acyltransferase 1
G6PDH	Glucose-6P-dehydrogenase
GSH	Glutathione
GPx	Glutathione peroxidase
HBV	Hepatitis B virus
HFD	High-fat diet
HFHS	High-fat high-sucrose diet
HSL	Hormone sensitive lipase
IkBα	Inhibitor of nuclear factor *kappa* subunit α
IL	Interleukin
LDH	Lactate dehydrogenase
LFD	Low fat diet
LKB1	Liver kinase B1
LXR	Liver X receptor
MCAD	Mitochondrial medium-chain acyl-CoA dehydrogenase
MDA	Malonaldehyde
ME	Malic enzyme
MTP	Microsomal triglyceride transfer protein
NAD^+	Oxidized nicotinamide adenine dinucleotide
NADH	Reduced nicotinamide adenine dinucleotide
NAFLD	Non-alcoholic fatty liver disease
NF-kB p65	Nuclear factor kappa-light-chain-enhancer of activated B cells
O+A	Oleic acid *plus* alcohol
PDE	Phosphodiesterase
pIkBα	Phospho-inhibitory subunit of NF-KBα
PPAR	Peroxisome proliferator-activated receptor

Int. J. Mol. Sci. **2018**, *19*, 3473

PRKA	Protein kinase A
ROS	Reactive oxygen species
RSV	Resveratrol
si RNA	Small interfering RNA
SIRT1	Silent information regulator 1; Surtuin-1
SCD	Stearoyl-Coenzyme A desaturase 1
SQSTM1	Sequestosome 1
SOD	Superoxide dismutase
SREBF	Sterol regulatory element-binding transcription factor
SREBP	Sterol regulatory element-binding protein
STD	Standard diet
TG	Triglyceride
TNF-α	Tumor necrosis factor- α
UCP	Uncoupling protein

References

1. Milton-Laskibar, I.; Aguirre, L.; Fernández-Quintela, A.; Rolo, A.P.; Soeiro Teodoro, J.; Palmeira, C.M.; Portillo, M.P. Lack of additive effects of resveratrol and energy restriction in the treatment of hepatic steatosis in rats. *Nutrients* **2017**, *9*, 737. [CrossRef] [PubMed]
2. Aguirre, L.; Portillo, M.P.; Hijona, E.; Bujanda, L. Effects of resveratrol and other polyphenols in hepatic steatosis. *World J. Gastroenterol.* **2014**, *20*, 7366–7380. [CrossRef] [PubMed]
3. Day, E.A.; Ford, R.J.; Steinberg, G.R. Ampk as a therapeutic target for treating metabolic diseases. *Trends Endocrinol. Metab.* **2017**, *28*, 545–560. [CrossRef] [PubMed]
4. Carling, D. Ampk signalling in health and disease. *Curr. Opin. Cell Biol.* **2017**, *45*, 31–37. [CrossRef] [PubMed]
5. Garcia, D.; Shaw, R.J. Ampk: Mechanisms of cellular energy sensing and restoration of metabolic balance. *Mol. Cell* **2017**, *66*, 789–800. [CrossRef] [PubMed]
6. Carling, D. The amp-activated protein kinase cascade—A unifying system for energy control. *Trends Biochem. Sci.* **2004**, *29*, 18–24. [CrossRef] [PubMed]
7. Hardie, D.G. Minireview: The amp-activated protein kinase cascade: The key sensor of cellular energy status. *Endocrinology* **2003**, *144*, 5179–5183. [CrossRef] [PubMed]
8. Cool, B.; Zinker, B.; Chiou, W.; Kifle, L.; Cao, N.; Perham, M.; Dickinson, R.; Adler, A.; Gagne, G.; Iyengar, R.; et al. Identification and characterization of a small molecule ampk activator that treats key components of type 2 diabetes and the metabolic syndrome. *Cell Metab.* **2006**, *3*, 403–416. [CrossRef] [PubMed]
9. Woods, A.; Williams, J.R.; Muckett, P.J.; Mayer, F.V.; Liljevald, M.; Bohlooly, Y.M.; Carling, D. Liver-specific activation of ampk prevents steatosis on a high-fructose diet. *Cell Rep.* **2017**, *18*, 3043–3051. [CrossRef] [PubMed]
10. Hardie, D.G. Ampk: A target for drugs and natural products with effects on both diabetes and cancer. *Diabetes* **2013**, *62*, 2164–2172. [CrossRef] [PubMed]
11. Fernández-Quintela, A.; Milton-Laskibar, I.; González, M.; Portillo, M.P. Antiobesity effects of resveratrol: Which tissues are involved? *Ann. N. Y. Acad. Sci.* **2017**, *1403*, 118–131. [CrossRef] [PubMed]
12. Heebøll, S.; Thomsen, K.L.; Pedersen, S.B.; Vilstrup, H.; George, J.; Grønbæk, H. Effects of resveratrol in experimental and clinical non-alcoholic fatty liver disease. *World J. Hepatol.* **2014**, *6*, 188–198. [CrossRef] [PubMed]
13. Heebøll, S.; Thomsen, K.L.; Clouston, A.; Sundelin, E.I.; Radko, Y.; Christensen, L.P.; Ramezani-Moghadam, M.; Kreutzfeldt, M.; Pedersen, S.B.; Jessen, N.; et al. Effect of resveratrol on experimental non-alcoholic steatohepatitis. *Pharmacol. Res.* **2015**, *95–96*, 34–41. [CrossRef] [PubMed]
14. Heebøll, S.; El-Houri, R.B.; Hellberg, Y.E.; Haldrup, D.; Pedersen, S.B.; Jessen, N.; Christensen, L.P.; Grønbaek, H. Effect of resveratrol on experimental non-alcoholic fatty liver disease depends on severity of pathology and timing of treatment. *J. Gastroenterol. Hepatol.* **2016**, *31*, 668–675. [CrossRef] [PubMed]
15. Chen, D.; Bruno, J.; Easlon, E.; Lin, S.J.; Cheng, H.L.; Alt, F.W.; Guarente, L. Tissue-specific regulation of sirt1 by calorie restriction. *Genes Dev.* **2008**, *22*, 1753–1757. [CrossRef] [PubMed]

16. Cantó, C.; Auwerx, J. Pgc-1alpha, sirt1 and ampk, an energy sensing network that controls energy expenditure. *Curr. Opin. Lipidol.* **2009**, *20*, 98–105. [CrossRef] [PubMed]

17. Ruderman, N.B.; Xu, X.J.; Nelson, L.; Cacicedo, J.M.; Saha, A.K.; Lan, F.; Ido, Y. Ampk and sirt1: A long-standing partnership? *Am. J. Physiol. Endocrinol. Metab.* **2010**, *298*, E751–E760. [CrossRef] [PubMed]

18. Hou, X.; Xu, S.; Maitland-Toolan, K.A.; Sato, K.; Jiang, B.; Ido, Y.; Lan, F.; Walsh, K.; Wierzbicki, M.; Verbeuren, T.J.; et al. Sirt1 regulates hepatocyte lipid metabolism through activating amp-activated protein kinase. *J. Biol. Chem.* **2008**, *283*, 20015–20026. [CrossRef] [PubMed]

19. Lan, F.; Cacicedo, J.M.; Ruderman, N.; Ido, Y. Sirt1 modulation of the acetylation status, cytosolic localization, and activity of lkb1. Possible role in amp-activated protein kinase activation. *J. Biol. Chem.* **2008**, *283*, 27628–27635. [CrossRef] [PubMed]

20. Fulco, M.; Cen, Y.; Zhao, P.; Hoffman, E.P.; McBurney, M.W.; Sauve, A.A.; Sartorelli, V. Glucose restriction inhibits skeletal myoblast differentiation by activating SIRT1 through AMPK-mediated regulation of Nampt. *Dev. Cell* **2008**, *14*, 661–673. [CrossRef] [PubMed]

21. Vingtdeux, V.; Giliberto, L.; Zhao, H.; Chandakkar, P.; Wu, Q.; Simon, J.E.; Janle, E.M.; Lobo, J.; Ferruzzi, M.G.; Davies, P.; et al. Amp-activated protein kinase signaling activation by resveratrol modulates amyloid-beta peptide metabolism. *J. Biol. Chem.* **2010**, *285*, 9100–9113. [CrossRef] [PubMed]

22. Park, S.J.; Ahmad, F.; Philp, A.; Baar, K.; Williams, T.; Luo, H.; Ke, H.; Rehmann, H.; Taussig, R.; Brown, A.L.; et al. Resveratrol ameliorates aging-related metabolic phenotypes by inhibiting camp phosphodiesterases. *Cell* **2012**, *148*, 421–433. [CrossRef] [PubMed]

23. Zang, M.; Xu, S.; Maitland-Toolan, K.A.; Zuccollo, A.; Hou, X.; Jiang, B.; Wierzbicki, M.; Verbeuren, T.J.; Cohen, R.A. Polyphenols stimulate amp-activated protein kinase, lower lipids, and inhibit accelerated atherosclerosis in diabetic LDL receptor-deficient mice. *Diabetes* **2006**, *55*, 2180–2191. [CrossRef] [PubMed]

24. Shang, J.; Chen, L.L.; Xiao, F.X.; Sun, H.; Ding, H.C.; Xiao, H. Resveratrol improves non-alcoholic fatty liver disease by activating amp-activated protein kinase. *Acta Pharmacol. Sin.* **2008**, *29*, 698–706. [CrossRef] [PubMed]

25. Gao, M.; Liu, D. Resveratrol suppresses t0901317-induced hepatic fat accumulation in mice. *AAPS J.* **2013**, *15*, 744–752. [CrossRef] [PubMed]

26. Choi, Y.J.; Suh, H.R.; Yoon, Y.; Lee, K.J.; Kim, D.G.; Kim, S.; Lee, B.H. Protective effect of resveratrol derivatives on high-fat diet induced fatty liver by activating amp-activated protein kinase. *Arch. Pharm. Res.* **2014**, *37*, 1169–1176. [CrossRef] [PubMed]

27. Tang, L.Y.; Chen, Y.; Rui, B.B.; Hu, C.M. Resveratrol ameliorates lipid accumulation in HepG2 cells, associated with down-regulation of lipin1 expression. *Can. J. Physiol. Pharmacol.* **2016**, *94*, 185–189. [CrossRef] [PubMed]

28. Zhang, Y.; Chen, M.L.; Zhou, Y.; Yi, L.; Gao, Y.X.; Ran, L.; Chen, S.H.; Zhang, T.; Zhou, X.; Zou, D.; et al. Resveratrol improves hepatic steatosis by inducing autophagy through the camp signaling pathway. *Mol. Nutr. Food Res.* **2015**, *59*, 1443–1457. [CrossRef] [PubMed]

29. Tian, Y.; Ma, J.; Wang, W.; Zhang, L.; Xu, J.; Wang, K.; Li, D. Resveratrol supplement inhibited the NF-κB inflammation pathway through activating AMPKα-sirt1 pathway in mice with fatty liver. *Mol. Cell. Biochem.* **2016**, *422*, 75–84. [CrossRef] [PubMed]

30. Zang, M.; Zuccollo, A.; Hou, X.; Nagata, D.; Walsh, K.; Herscovitz, H.; Brecher, P.; Ruderman, N.B.; Cohen, R.A. AMP-activated protein kinase is required for the lipid-lowering effect of metformin in insulin-resistant human HepG2 cells. *J. Biol. Chem.* **2004**, *279*, 47898–47905. [CrossRef] [PubMed]

31. Cha, J.Y.; Repa, J.J. The liver x receptor (LXR) and hepatic lipogenesis. The carbohydrate-response element-binding protein is a target gene of LXR. *J. Biol. Chem.* **2007**, *282*, 743–751. [CrossRef] [PubMed]

32. Repa, J.J.; Liang, G.; Ou, J.; Bashmakov, Y.; Lobaccaro, J.M.; Shimomura, I.; Shan, B.; Brown, M.S.; Goldstein, J.L.; Mangelsdorf, D.J. Regulation of mouse sterol regulatory element-binding protein-1c gene (SREBP-1c) by oxysterol receptors, LXRalpha and LXRbeta. *Genes Dev.* **2000**, *14*, 2819–2830. [CrossRef] [PubMed]

33. Baur, J.A.; Pearson, K.J.; Price, N.L.; Jamieson, H.A.; Lerin, C.; Kalra, A.; Prabhu, V.V.; Allard, J.S.; Lopez-Lluch, G.; Lewis, K.; et al. Resveratrol improves health and survival of mice on a high-calorie diet. *Nature* **2006**, *444*, 337–342. [CrossRef] [PubMed]

34. Cantó, C.; Gerhart-Hines, Z.; Feige, J.N.; Lagouge, M.; Noriega, L.; Milne, J.C.; Elliott, P.J.; Puigserver, P.; Auwerx, J. AMPK regulates energy expenditure by modulating NAD+ metabolism and SIRT1 activity. *Nature* **2009**, *458*, 1056–1060. [CrossRef] [PubMed]

35. Ajmo, J.M.; Liang, X.; Rogers, C.Q.; Pennock, B.; You, M. Resveratrol alleviates alcoholic fatty liver in mice. *Am. J. Physiol. Gastrointest. Liver Physiol.* **2008**, *295*, G833–G842. [CrossRef] [PubMed]
36. Alberdi, G.; Rodríguez, V.M.; Macarulla, M.T.; Miranda, J.; Churruca, I.; Portillo, M.P. Hepatic lipid metabolic pathways modified by resveratrol in rats fed an obesogenic diet. *Nutrition* **2013**, *29*, 562–567. [CrossRef] [PubMed]
37. Tung, Y.C.; Lin, Y.H.; Chen, H.J.; Chou, S.C.; Cheng, A.C.; Kalyanam, N.; Ho, C.T.; Pan, M.H. Piceatannol exerts anti-obesity effects in c57bl/6 mice through modulating adipogenic proteins and gut microbiota. *Molecules* **2016**, *21*, 1419. [CrossRef] [PubMed]
38. Tanaka, M.; Kita, T.; Yamasaki, S.; Kawahara, T.; Ueno, Y.; Yamada, M.; Mukai, Y.; Sato, S.; Kurasaki, M.; Saito, T. Maternal resveratrol intake during lactation attenuates hepatic triglyceride and fatty acid synthesis in adult male rat offspring. *Biochem. Biophys. Rep.* **2017**, *9*, 173–179. [CrossRef] [PubMed]
39. Rivera, L.; Morón, R.; Zarzuelo, A.; Galisteo, M. Long-term resveratrol administration reduces metabolic disturbances and lowers blood pressure in obese Zucker rats. *Biochem. Pharmacol.* **2009**, *77*, 1053–1063. [CrossRef] [PubMed]
40. Do, G.M.; Jung, U.J.; Park, H.J.; Kwon, E.Y.; Jeon, S.M.; McGregor, R.A.; Choi, M.S. Resveratrol ameliorates diabetes-related metabolic changes via activation of amp-activated protein kinase and its downstream targets in db/db mice. *Mol. Nutr. Food Res.* **2012**, *56*, 1282–1291. [CrossRef] [PubMed]
41. Kang, W.; Hong, H.J.; Guan, J.; Kim, D.G.; Yang, E.J.; Koh, G.; Park, D.; Han, C.H.; Lee, Y.J.; Lee, D.H. Resveratrol improves insulin signaling in a tissue-specific manner under insulin-resistant conditions only: In vitro and in vivo experiments in rodents. *Metabolism* **2012**, *61*, 424–433. [CrossRef] [PubMed]
42. Lin, H.C.; Chen, Y.F.; Hsu, W.H.; Yang, C.W.; Kao, C.H.; Tsai, T.F. Resveratrol helps recovery from fatty liver and protects against hepatocellular carcinoma induced by hepatitis b virus × protein in a mouse model. *Cancer Prev. Res.* **2012**, *5*, 952–962. [CrossRef] [PubMed]
43. Zhu, W.; Chen, S.; Li, Z.; Zhao, X.; Li, W.; Sun, Y.; Zhang, Z.; Ling, W.; Feng, X. Effects and mechanisms of resveratrol on the amelioration of oxidative stress and hepatic steatosis in KKAy mice. *Nutr. Metab. (Lond.)* **2014**, *11*, 35. [CrossRef] [PubMed]
44. Lee, H.J.; Lim, Y.; Yang, S.J. Involvement of resveratrol in crosstalk between adipokine adiponectin and hepatokine fetuin-A in vivo and in vitro. *J. Nutr. Biochem.* **2015**, *26*, 1254–1260. [CrossRef] [PubMed]

International Journal of
Molecular Sciences

MDPI

Article

AMPK Activation Reduces Hepatic Lipid Content by Increasing Fat Oxidation In Vivo

Marc Foretz [1,2,3,]*, Patrick C. Even [4] and Benoit Viollet [1,2,3,]*

[1] INSERM, U1016, Institut Cochin, Département d'Endocrinologie Métabolisme et Diabète, 24, rue du Faubourg Saint Jacques, 75014 Paris, France
[2] CNRS, UMR8104, 75014 Paris, France
[3] Université Paris Descartes, Sorbonne Paris Cité, 75014 Paris, France
[4] UMR PNCA, AgroParisTech, INRA, Université Paris-Saclay, 75005 Paris, France;
patrick.even@agroparistech.fr
* Correspondence: marc.foretz@inserm.fr (M.F.); benoit.viollet@inserm.fr (B.V.);
Tel.: +33-1-4441-2438 (M.F.); +33-1-4441-2401 (B.V.); Fax: +33-1-4441-2421 (M.F.); +33-1-4441-2401 (B.V.)

Received: 1 August 2018; Accepted: 15 September 2018; Published: 19 September 2018

Abstract: The energy sensor AMP-activated protein kinase (AMPK) is a key player in the control of energy metabolism. AMPK regulates hepatic lipid metabolism through the phosphorylation of its well-recognized downstream target acetyl CoA carboxylase (ACC). Although AMPK activation is proposed to lower hepatic triglyceride (TG) content via the inhibition of ACC to cause inhibition of de novo lipogenesis and stimulation of fatty acid oxidation (FAO), its contribution to the inhibition of FAO in vivo has been recently questioned. We generated a mouse model of AMPK activation specifically in the liver, achieved by expression of a constitutively active AMPK using adenoviral delivery. Indirect calorimetry studies revealed that liver-specific AMPK activation is sufficient to induce a reduction in the respiratory exchange ratio and an increase in FAO rates in vivo. This led to a more rapid metabolic switch from carbohydrate to lipid oxidation during the transition from fed to fasting. Finally, mice with chronic AMPK activation in the liver display high fat oxidation capacity evidenced by increased [C^{14}]-palmitate oxidation and ketone body production leading to reduced hepatic TG content and body adiposity. Our findings suggest a role for hepatic AMPK in the remodeling of lipid metabolism between the liver and adipose tissue.

Keywords: AMPK; liver; lipid metabolism; fatty acid oxidation; indirect calorimetry

---

1. Introduction

AMP-activated protein kinase (AMPK) is a phylogenetically conserved serine/threonine protein kinase viewed as a fuel gauge monitoring systemic and cellular energy status which plays a crucial role in protecting cellular function under energy-restricted conditions [1]. AMPK is a heterotrimeric protein consisting of a catalytic α-subunit and two regulatory subunits, β and γ, with each subunit existing as at least two isoforms. AMPK is activated in response to a variety of metabolic stresses that typically change the cellular AMP:ATP ratio caused by increasing ATP consumption or reducing ATP production, as seen following glucose deprivation and inhibition of mitochondrial oxidative phosphorylation as well as exercise and muscle contraction. Activation of AMPK initiates metabolic changes to reprogram metabolism by switching cells from an anabolic to a catabolic state, shutting down the ATP-consuming synthetic pathways and restoring energy balance. This regulation involves AMPK-dependent phosphorylation of key regulators of many important pathways [2,3].

One of the first identified AMPK targets is acetyl CoA carboxylase (ACC), playing a role in the control of fatty acid metabolism via the regulation of malonyl-CoA synthesis [4]. Malonyl-CoA is both a critical precursor of biosynthesis of fatty acids and an inhibitor of fatty acid uptake into mitochondria via the transport system involving carnitine palmitoyltransferase-1. By inhibiting ACC and lowering the concentration of its reaction product malonyl-CoA, AMPK activation is expected to coordinate the partitioning of fatty acids between oxidative and biosynthetic pathways by increasing fatty acid oxidation (FAO) capacity and inhibiting de novo lipogenesis (DNL), respectively. For these reasons, AMPK has emerged as a promising therapeutic target to treat metabolic disorders that occur in conditions such as nonalcoholic fatty liver disease (NAFLD). There is now literature precedence demonstrating the impact of hepatic AMPK activation in the setting of NAFLD [5]. In addition, recent advances in the development of allosteric and isoform-biased small-molecule AMPK activators have reinforced the potential for the pharmacological activation of AMPK as a treatment modality for hepatic steatosis [6–9]. Recent evidences showed that regulation of hepatic lipogenesis by AMPK activation mainly resides in the phosphorylation and inactivation of ACC, but not in the control of lipogenic gene expression [7,8,10]. Accordingly, genetic mouse models of hepatic AMPK deficiency and ACC with knock-in phosphorylation mutations confirmed the importance of the activation of AMPK and phosphorylation of ACC for the improvement of fatty liver disease induced by AMPK-activating drugs [7,8,11]. These studies also provided in vitro and in vivo evidence for the contribution of both hepatic FAO and DNL in the reduction of hepatic triglyceride (TG) accumulation mediated through pharmacological AMPK activation. However, one study recently questioned the effect of liver-specific activation of AMPK on FAO rates in vivo [10]. In that study, by using a genetic mouse model expressing in the liver a gain-of-function AMPKγ1 mutant, Woods et al. demonstrated that the effect of hepatic AMPK activation in the protection against hepatic steatosis is largely dependent on the suppression of de novo lipogenesis, but not on the stimulation of hepatic fatty acid oxidation [10]. Therefore, in the present study, we examined the impact of AMPK activation in the liver on hepatic lipid metabolism and determined its effect on FAO rates in vivo, measured by indirect calorimetry.

2. Results

As a first step to elucidating the impact of AMPK activation in the liver on hepatic lipid metabolism in vivo, we generated a mouse model in which AMPK activation specifically in the liver is achieved by expression of a constitutively active AMPK. Mice were injected intravenously with an adenovirus expressing a constitutively active form of AMPKα2 (Ad AMPK-CA) or GFP as a control (Ad GFP). This resulted in AMPK-CA expression restricted to the liver and undetectable in all other tissues (Figure 1A and data not shown). High levels of hepatic AMPK-CA expression were maintained until day 8, with no change in endogenous AMPKα expression (Figure 1A). ACC protein levels were low in Ad AMPK-CA livers, but the phospho-ACC/total ACC ratio was twice that in control livers, demonstrating an increase in ACC phosphorylation and therefore AMPK activation following AMPK-CA expression (Figure 1B). Decreased hepatic malonyl CoA levels in Ad AMPK-CA compared to Ad GFP livers also confirmed inhibition of ACC activity (Figure 1C). The levels of carnitine palmitoyltransferase (CPT)-1a and -2 mRNA expression were similar in the liver of Ad GFP and AMPK-CA mice (Figure 1D). There were no significant changes in body weight and food intake during the week following the injection of AMPK-CA or GFP adenoviruses (Figure 1E,F).

We studied the metabolic consequences of AMPK activation in the liver by monitoring energy expenditure and respiratory exchange ratio (RER) during a 22-h fasting period, determined by indirect calorimetry. The values of RER provide an approximation of carbohydrate and lipid oxidation to generate energy, ranging from 1.0 to approximately 0.7, respectively. In fed mice, the RER associated with the total and resting metabolism rates was lower in Ad AMPK-CA than in Ad GFP mice. During fasting, Ad AMPK-CA mice reached maximal rates of lipid oxidation after only 3 h of fasting, whereas such rates were not achieved until 12 h in Ad GFP mice (Figure 2, upper panel). Thus, AMPK activation in the liver enhances lipid oxidation, leading to a more rapid metabolic switch from carbohydrate to lipid oxidation during the transition from fed to fasting. Thereafter, RER stabilized at the same values in Ad GFP and Ad AMPK-CA mice, suggesting that the rate of lipid oxidation reached the same maximum intensity in both groups. Total and resting metabolic rates and spontaneous activity were similar in Ad GFP and Ad AMPK-CA mice (Figure 2, middle and lower panels). All mice exhibited a period of intense activity during the night period between 00:00 and 06:00 h. According to previous observations of fed mice, this hyperactivity was probably related to the fact that the mice were fasted and seeking for food. Analysis of the changes in total metabolism and RER induced by bursts of spontaneous activity that occurred during the light period (i.e., when RER was lower in Ad AMPK-CA than in control Ad GFP mice) showed that the utilization of glucose and lipids by the working muscles was very similar in both groups (same changes in RER). These observations agree with the conclusion that the rapid mobilization and utilization of lipids in Ad AMPK-CA mice in response to fasting is probably specific to the constitutive activation of AMPK in the liver.

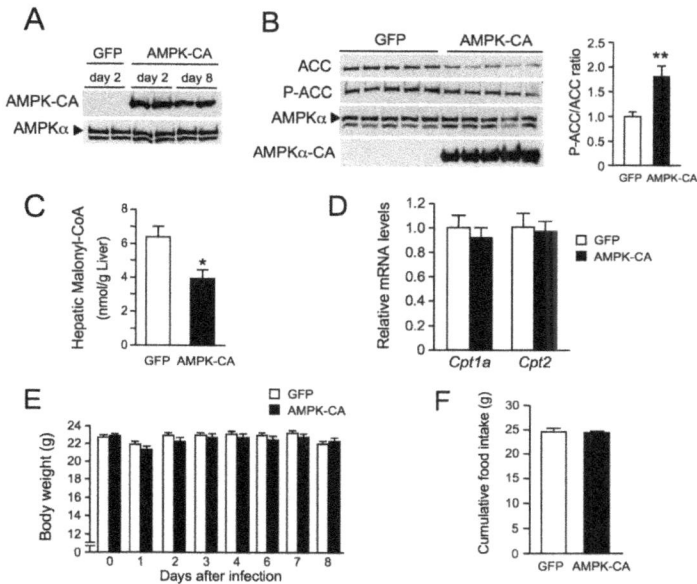

Figure 1. Effects of the expression of an active form of AMP-activated protein kinase (AMPK) in the liver on body weight and food intake. Ten-week-old male C57BL/6J mice received injections of adenovirus (Ad) expressing the green fluorescent protein (GFP) or a constitutively active form of AMPKα2 (AMPK-CA) and were studied for the indicated times after adenovirus injection and in the indicated nutritional state. (**A**) Western blot analysis of liver lysates with antibodies raised against pan-AMPKα and myc-tagged AMPK-CA was performed on days 2 and day 8 after adenovirus administration; (**B**) Western blot analysis of liver lysates from fed mice 48 h after the injection of Ad GFP or Ad AMPK-CA, with the antibodies indicated. Each lane represents a liver sample from an individual mouse. The panel on the right shows Ser79 phosphorylated acetyl CoA carboxylase/ total acetyl CoA carboxylase (P-ACC/ACC) ratios from the quantification of immunoblot images ($n = 5$); (**C**) Hepatic malonyl-CoA levels in 8 h-fasted mice 48 h after the injection of Ad GFP or Ad AMPK-CA ($n = 5$); (**D**) Effect of AMPK activation in the liver on the expression of *Cpt1a* and *Cpt2* genes. Total RNA was isolated from the liver of 24 h-fasted mice 48 h after the injection of Ad GFP or Ad AMPK-CA ($n = 6$). The expression of *Cpt1a* and *Cpt2* genes was assessed by real-time quantitative RT-PCR. Relative mRNA levels are expressed as fold-activation relative to levels in Ad GFP livers; (**E**) Body weight changes and (**F**) cumulative food intake measured for 8 days after adenovirus administration ($n = 11$–12 per group). Data are means ± standard error of mean (SEM). * $p < 0.05$, ** $p < 0.01$ versus Ad GFP mice by unpaired two-tailed Student's *t*-test (**B–D,F**) or by one-way ANOVA with Bonferroni post-hoc test (**E**).

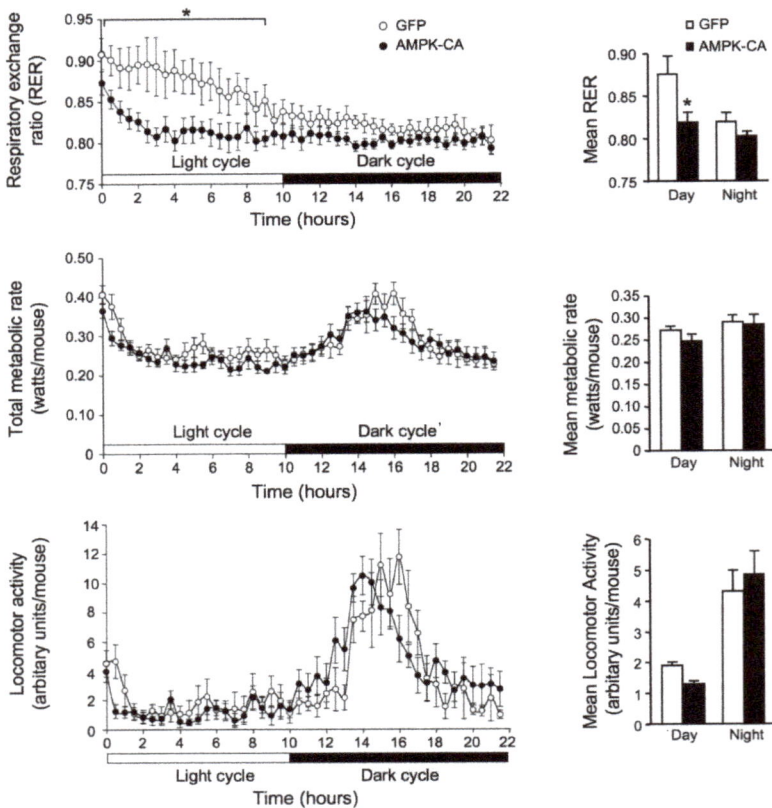

Figure 2. Effects of the expression of an active form of AMPK in the liver on respiratory exchange ratio. Whole-animal indirect calorimetry was used to assess oxygen consumption (VO_2) and carbon dioxide production (VCO_2) in mice infected with Ad GFP or Ad AMPK-CA for 48 h. Fed adenovirus-infected mice were placed in a metabolic chamber at 10:00 h. They were kept in the cage for 22 h, with free access to water but no food. Upper panel: The respiratory exchange ratio (RER = VO_2/VCO_2) was calculated from VO_2 and VCO_2 data and plotted at 15-min intervals. An RER of 1.0 is expected for glucose oxidation and an RER of 0.7 corresponds to lipid oxidation. The right panel shows mean RER results for Ad GFP and Ad AMPK-CA mice (*n* = 6 per group) during light or dark periods. Middle panel: Total metabolic rate. The right panel shows mean metabolic rates for Ad GFP and Ad AMPK-CA mice (*n* = 6 per group) during light and dark periods. Lower panel: Locomotor activity. The right panel shows the mean locomotor activity results for Ad GFP- and Ad AMPK-CA mice (*n* = 6 per group) during light and dark periods. Data are means ± SEM. * $p < 0.05$ versus Ad GFP mice by one-way ANOVA with Bonferroni post-hoc test.

We then investigated whether AMPK activation mediated increased hepatic fatty acid oxidation, by measuring the rate of β-oxidation through assays of [^{14}C]-palmitoyl-CoA oxidation in the liver (Figure 3A). Rates of palmitoyl-CoA oxidation were ~25% higher in Ad AMPK-CA mice than in control Ad GFP mice. Indirect support for the increase in FAO is provided by the increase in plasma ketone bodies in Ad AMPK-CA mice (Figure 3B) and a corresponding decrease in plasma triglyceride (TG) and free fatty acid (FFA) concentrations (Figure 3C,D). To determine whether AMPK activation increased fatty acid utilization, [C^{14}]-palmitate was injected into Ad AMPK-CA and Ad GFP mice and its incorporation into lipids measured. AMPK activation was associated with an increase in hepatic fatty acid uptake of ~25% (Figure 3E). These findings were correlated with increased expression of the fatty acid transporters *Slc27a4* (fatty acid transport protein 4, *Fatp4*), *Cd36* (fatty acid translocase, *Fat*), and *Fabp4* (fatty acid binding protein 4) (Figure 3F).

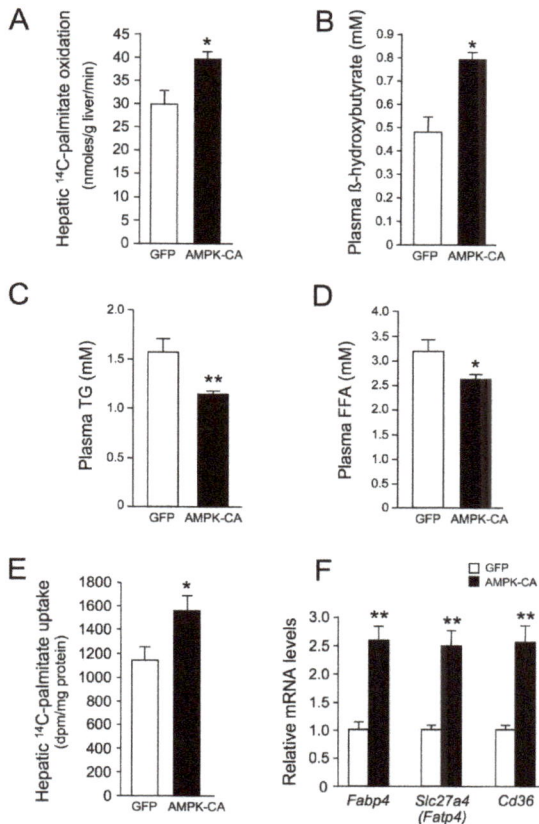

Figure 3. Long-term adenovirus-mediated expression of an active form of AMPK in the liver increases hepatic lipid oxidation and fatty acid uptake. Ten-week-old male C57BL/6J mice received injections of Ad GFP or Ad AMPK CA and were studied at the indicated times after adenovirus injection and in the indicated nutritional state. (**A**) Hepatic [1-^{14}C]-palmitate oxidation in fed mice 48 h after the injection of Ad GFP or Ad AMPK-CA ($n = 4$); (**B**) Plasma β-hydroxybutyrate levels in 24 h-fasted mice 48 h after the injection of Ad GFP or Ad AMPK-CA ($n = 6$); (**C**) Plasma triglyceride (TG) and (**D**) plasma free fatty acid (FFA) levels in overnight-fasted mice 8 days after the injection of Ad GFP or Ad AMPK-CA ($n = 12$); (**E**) Hepatic [1-^{14}C]-palmitate uptake in 24 h-fasted mice 48 h after the injection of Ad GFP or Ad AMPK-CA ($n = 5$); (**F**) Effect of AMPK activation in the liver on the expression of the fatty acid transporters. Total RNA was isolated from the liver of 24 h-fasted mice 48 h after the injection of Ad GFP or Ad AMPK-CA ($n = 5$). The expression of *Slc27a4* (*Fatp4*), *Cd36*, and *Fabp4* genes was assessed by real-time quantitative RT-PCR. Relative mRNA levels are expressed as fold-activation relative to levels in Ad GFP livers. Data are means ± SEM. * $p < 0.05$, ** $p < 0.01$ versus Ad GFP-infected mice by unpaired two-tailed Student's *t*-test.

Long-term (8 days) expression of AMPK-CA in liver was sufficient to modify hepatic lipid content and lowered TG levels by ~45% and cholesterol levels by ~10% (Figure 4A,B). This is in line with low abundance of lipid droplets in hepatocytes from Ad AMPK-CA compared to Ad GFP mice revealed by liver ultrastructure changes using transmission electron microscopy (Figure 4C). The increase in hepatic β-oxidation was related to systemic changes in adiposity and resulted in a significant decrease in body fat mass (Figure 5A). This decrease was confirmed by the careful weighing of adipose tissue from epidydimal and inguinal fat pads (Figure 5B). The epidydimal fat pads were much smaller, as was adipocyte diameter (Figure 5C–E). As a result, plasma leptin concentration, a marker of adiposity, was halved in Ad AMPK-CA mice (Figure 5F).

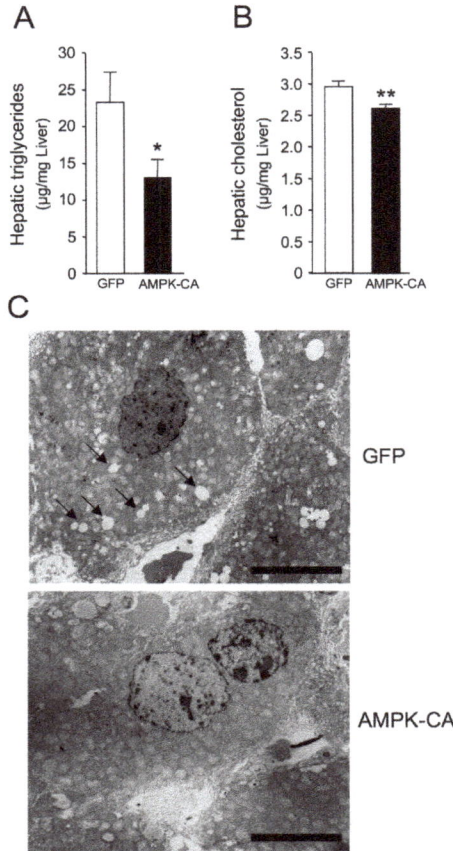

Figure 4. Long-term adenovirus-mediated expression of an active form of AMPK in the liver reduces hepatic lipid accumulation. Ten-week-old male C57BL/6J mice received injections of Ad GFP or Ad AMPK-CA. Fed mice were studied on day 8 after adenovirus administration. (**A**) Liver triglyceride content and (**B**) liver cholesterol content (n = 9–10). Data are means ± SEM. * $p < 0.05$, ** $p < 0.01$ versus Ad GFP-infected mice by unpaired two-tailed Student's t-test; (**C**) Representative images of transmission electron microscopy showing the ultrastructure change in Ad GFP and Ad AMPK-CA livers. Scale bar: 10 μm. Black arrowheads in insets depict lipid droplets.

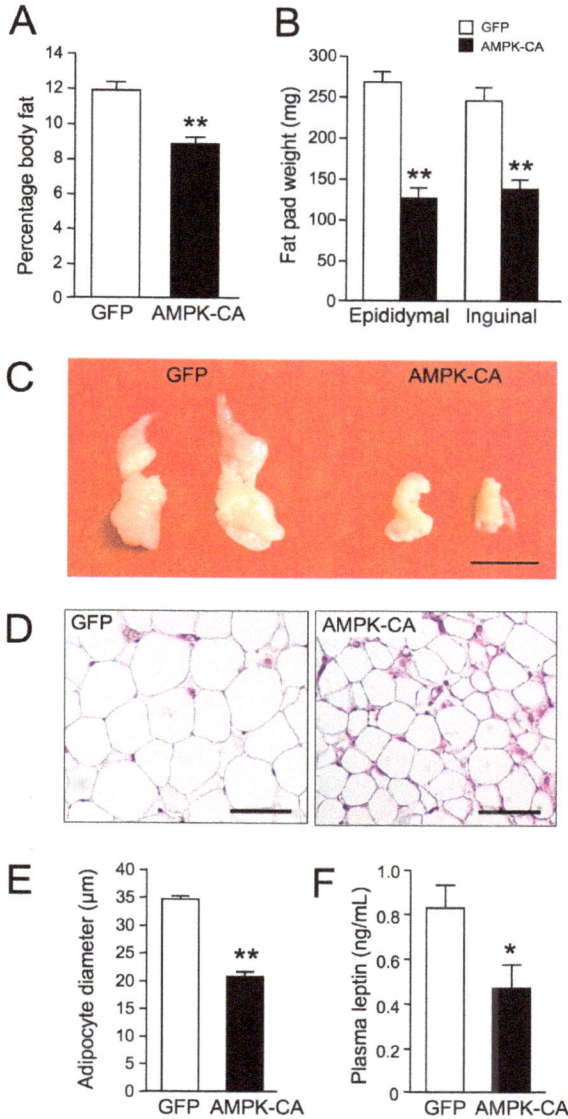

Figure 5. Long-term adenovirus-mediated expression of an active form of AMPK in the liver diminishes peripheral adiposity. Ten-week-old male C57BL/6J mice received injections of Ad GFP or Ad AMPK-CA. Fed mice were studied on day 8 after adenovirus administration. (**A**) Body fat content was measured by dual X-ray absorptiometry ($n = 10$ per group); (**B**) Epididymal and inguinal subcutaneous fat-pad weight ($n = 10$ per group); (**C**) Representative epididymal white fat pads fixed in formalin. Scale bar: 1 cm; (**D**) Representative hematoxylin-and-eosin-stained sections of epididymal adipose tissues. Scale bars: 50 μm; (**E**) Mean adipocyte size in epididymal white adipose tissues. The diameter of at least 200 cells per sample was determined ($n = 4$ mice per group); (**F**) Plasma leptin levels in fed mice ($n = 10$ per group). Data are means ± SEM. * $p < 0.05$, ** $p < 0.001$ versus Ad GFP mice by unpaired two-tailed Student's *t*-test.

3. Discussion

Activation of AMPK has been reported to reduce hepatic lipid content in many preclinical studies, yet the importance of hepatic FAO and DNL for its TG-lowering effect has been unclear [6–10]. Here, we report that AMPK activation in the liver is capable of significant reduction in liver TG through the stimulation of fatty acid utilization, as evidenced by a reduction of RER and increased palmitate oxidation and ketone body production. These results are reminiscent of the acute effect of the direct AMPK activator A-769662, showing a concurrent drop in RER in fed rats and leading to the reduction in liver TGs after chronic treatment of obese mice [9]. Importantly, it has been demonstrated that A-769662 acts in an AMPK-dependent manner to induce fat utilization [12]. We also recently confirmed that A-769662 was capable to restore hepatic fatty acid oxidation after chronic treatment of a fatty liver mouse model [8]. Further support for a significant role of lipid oxidation following hepatic AMPK activation recently came from a study investigating the therapeutic beneficial of the β1-biased activator PF-06409577 in a high-fat-fed mouse model, where the contribution of de novo lipogenesis is essentially negligible for hepatic TG accumulation [7]. Acute or 42 days' dosing with PF-06409577 resulted in a large increase in circulating β-hydroxybutyrate and lower hepatic TG, an effect that was lost in mice lacking AMPK specifically in the liver [7]. In addition to the impact on FAO, activation of AMPK in the liver has also been largely documented as a source of inhibition of DNL [6–10]. Thus, our results substantially contribute to the current view that following hepatic AMPK activation, lowering of hepatic TGs may arise through the capacity of AMPK to combine between both inhibition of TG synthesis and stimulation of lipid utilization [5,7,8]. Definitive evidence for a dual effect of hepatic AMPK activation on lipid synthesis and utilization is provided by in vitro and in vivo studies with AMPK-deficient mouse models and primary culture of hepatocytes treated with various pharmacological activators of AMPK [7,8]. The balance and contribution between inhibition of DNL and stimulation of FAO may depend on the source of hepatic TGs at the origin of the development of hepatic steatosis. Consistent with this notion, pharmacological AMPK-induced inhibition of DNL has been suggested to play a significant role in the improvement of hepatic steatosis of animal models where DNL mainly contributes to hepatic TG accumulation [7,8]. Similarly, transgenic mice expressing specifically in the liver a naturally occurring gain-of-function AMPKγ1 mutant were completely protected against hepatic steatosis when fed a high-fructose diet, known to increase hepatic lipogenesis [10]. In that study, the effect of AMPK activation was relying exclusively on the inhibition of DNL, because no difference in FAO and RER was detected, despite hepatic AMPK activation. However, it is possible that mice fed a high-sucrose diet preferentially oxidize carbohydrates as their primary source of energy and this obscures the effect of AMPK activation on fat oxidation due to the competition between glucose and fat for substrate oxidation [13]. Intriguingly, in the same study, these mice expressing a gain-of-function AMPKγ1 mutant in the liver failed to stimulate FAO and reduce hepatic lipids when fed a high-fat diet [10]. These results contrast with the effectiveness of the AMPKα2-CA mutant used in the present study and various direct AMPK activators in stimulating hepatic FAO and reducing hepatic TG accumulation in vivo [7,8]. What causes this discrepancy is unclear, but we speculate that basal AMPK activity increased by mutation of the AMPKγ1 subunit is insufficient to fully phosphorylate and inactivate ACC and therefore presumably to stimulate FAO in vivo. Given the observation of the lowering in AMPK activity in the liver of high-fat-fed mice and fatty liver mouse models [8,14–16], AMPK activation probably needs to reach a higher threshold before the stimulation of FAO can be effective [8], providing an alternative explanation for the absence of a significant effect on hepatic lipid content in mice expressing the gain-of-function AMPKγ1 mutant on a high-fat diet.

AMPK has been proposed as a potential pharmacological target for the treatment of NAFLD due to its capacity to increase FAO and inhibit DNL in the liver [5]. One mechanism by which AMPK regulates the partitioning of fatty acids between oxidative and biosynthetic pathways is accomplished by the phosphorylation and inactivation of ACC, the rate-controlling enzyme for the synthesis of malonyl-CoA, which is both a critical precursor for biosynthesis of fatty acids and a potent inhibitor of long-chain fatty acyl-CoA transport into mitochondria for β-oxidation. This is supported by the observation of increased fatty acid synthesis and reduced FAO in the liver of mice lacking AMPK phosphorylation sites on ACC1/ACC2 [11]. In addition, these mice are resistant to the inhibition of lipogenesis in vivo induced by the AMPK-activating drugs metformin and A-769662 [11]. The effects of metformin and the direct AMPK activator PF-06409577 on lipid synthesis are abolished in hepatocytes isolated from these mice as well as liver AMPK-deficient mice [7,8,17]. The direct AMPK activator A-769662 also failed to increase fatty acid oxidation in these hepatocytes with mutation at the AMPK phosphorylation sites on ACC isoforms [11]. Thus, the action of AMPK in the improvement of hepatic steatosis is likely mediated through the phosphorylation of ACC to increase FAO and suppress DNL [11]. Recent studies performed in mice and humans treated with pharmacological inhibitors of ACC support the concept that direct inhibition of ACC is a promising therapeutic option for the management of fatty liver disease [18,19].

We have previously shown that short-term (48 h) expression of AMPK-CA in the liver paradoxically induced a concomitant hepatic lipid accumulation and increase in fatty acid oxidation [20]. Interestingly, a similar phenotype is observed during the physiological response to fasting, where hepatic TG contents rise significantly [21]. We hypothesized that in response to short-term AMPK activation, the hepatic lipid oxidation capacity is overloaded by the uptake of mobilized fatty acids from adipose tissue, which are stored temporarily as TGs in the liver until they are oxidized [20]. As anticipated, we report here that long-term (8 days) expression of AMPK-CA finally leads to a decrease in hepatic lipid content, but also in a reduction of body adiposity. However, body weight of Ad AMPK-CA mice was not significantly altered due to the small amount of fat mass loss compared to total body weight. Our data are corroborated with the effect of chronic treatment with the AMPK activators metformin, AICAR, or A-769662 in mice, which are associated with reduced fatty liver and fat pad weight [9,22–24], although no change in body composition was reported in diet-induced obese (DIO) mice treated with the AMPK β1-biased activator PF-06409577 [7]. Overall, these observations suggest a role for hepatic AMPK in the remodeling of lipid metabolism through crosstalk between liver and adipose tissue. However, the nature of the hepatic signal triggering the mobilization of fatty acids from adipose tissue to the liver remains to be elucidated. One possibility is the secretion of liver-derived proteins known as hepatokines, which could act on adipose tissue to stimulate lipolysis. FGF21 and Angptl3 are reasonable candidates playing important roles in the regulation of lipid metabolism [25,26]. Interestingly, FGF21 expression is induced by metformin and AICAR in hepatocytes [27].

In conclusion, chronic AMPK activation in the liver increases lipid oxidation, thereby decreasing hepatic lipid content and body adiposity, suggesting a role for hepatic AMPK in the remodeling of lipid metabolism between the liver and adipose tissue. Overall, our data emphasizes the potential therapeutic implications for hepatic AMPK activation in vivo.

4. Material and Methods

4.1. Reagents and Antibodies

Adenovirus expressing GFP and a myc epitope-tagged constitutively active form of AMPKα2 (AMPK-CA) were generated as previously described [20]. Primary antibodies directed against total AMPKα (#2532), total acetyl-CoA carboxylase (ACC) (#3676), and ACC phosphorylated at Ser79 (#3661) were purchased from Cell Signaling Technology (Danvers, MA, USA) and myc epitope tag (clone 9E10) from Sigma (Saint-Quentin-Fallavier, France). HRP-conjugated secondary antibodies were purchased from Calbiochem (Burlington, MA, USA).

4.2. Animals

Animal studies were approved by the Paris Descartes University ethics committees (no. CEEA34.BV.157.12) and performed under a French authorization to experiment on vertebrates (no. 75-886) in accordance with the European guidelines. C57BL/6J mice were obtained from Harlan France (Gannat, France). All mice were maintained in a barrier facility under a 12-h light/12-h dark cycle with free access to water and standard mouse diet (in terms of energy: 65% carbohydrate, 11% fat, 24% protein).

4.3. Metabolic Parameters

Blood was collected into heparin-containing tubes, and centrifuged to obtain plasma. Plasma leptin levels were assessed using mouse ELISA kit (Crystal Chem, Elk Grove Village, IL, USA). Plasma triglyceride, free fatty acid, and β-hydroxybutyrate levels were determined enzymatically (Dyasis, Grabels, France).

4.4. Liver Triglyceride, Cholesterol, and Malonyl-CoA Contents

For the extraction of total lipids from the liver, a portion of frozen tissue was homogenized in acetone (500 μL/50 mg tissue) and incubated on a rotating wheel overnight at 4 °C. Samples were centrifuged at 4 °C for 10 min at $5000 \times g$, and the triglyceride and cholesterol concentrations of the supernatants were determined with enzymatic colorimetric assays (Diasys, Grabels, France). Hepatic malonyl CoA ester content was measured using a modified high-performance liquid chromatography method [28].

4.5. Indirect Calorimetry

Mice were placed in a metabolic cage from 10:00 h until 08:00 h the next day (22 h). The metabolic cage was continuously connected to an open-circuit, indirect calorimetry system controlled by a computer running a data acquisition and analysis program, as previously described [29]. Mice were housed with free access to water but no food. Air flow through the chamber was regulated at 0.5 L/min by a mass flow-meter, and temperature was maintained close to thermoneutrality (30 °C ± 1 °C). Oxygen consumption (VO_2) and carbon dioxide production (VCO_2) were recorded at one-second intervals. Spontaneous activity was measured by means of 3 piezo-electric force transducers positioned in triangle under the metabolic cage, with sampling of the electrical signal at 100 Hz. Data were averaged every 10 s and stored on a hard disk for further processing. Computer-assisted processing of respiratory exchanges and spontaneous activity signals was performed to extract the respiratory exchanges specifically associated with spontaneous activity (Kalman filtering method) [29]. This separation provided information about total, resting, and activity-related O_2 consumption and CO_2 production. The respiratory exchange ratio (RER) was calculated as the ratio of VCO_2 produced/VO_2 consumed.

4.6. Assessment of Fatty Acid Oxidation in Liver Homogenates

The rate of mitochondrial palmitate oxidation was measured in fresh liver homogenate from fed mice anesthetized with a xylamine/ketamine mixture via intraperitoneal injection according to a modified version of the method described by Yu et al. [30]. The rate of palmitate oxidation was assessed by collecting and counting the radiolabeled acid-soluble metabolites (ASMs) produced from the oxidation of [1-^{14}C]-palmitate. Briefly, a portion of liver (200 mg) was homogenized in 19 volumes of ice-cold buffer containing 250 mM sucrose, 1 mM EDTA, and 10 mM Tris–HCl pH 7.4. For the assessment of palmitate oxidation, 75 μL of liver homogenate was incubated with 425 μL of reaction mixture (pH 7.4) in a 25 mL flask. The reaction mixture contained 100 mM sucrose, 10 mM Tris–HCl, 80 mM KCl, 5 mM K_2HPO_4, 1 mM $MgCl_2$, 0.2 mM EDTA, 1 mM dithiothreitol, 5.5 mM ATP, 1 mM NAD, 0.03 mM cytochrome C, 2 mM L-carnitine, 0.5 mM malate, and 0.1 mM coenzyme A. The reaction was started by adding 120 μM palmitate plus 1.7 μCi [1-^{14}C]-palmitate (56 mCi/mmol) complexed with fatty acid-free bovine serum albumin in a 5:1 molar ratio. Each homogenate was incubated in triplicate in the presence or absence of 75 μM antimycin A plus 10 μM rotenone to inhibit mitochondrial β-oxidation. After 30 min of incubation at 37 °C in a shaking water bath, the reaction was stopped by adding 200 µL of ice-cold 3 M perchloric acid. The radiolabeled ASMs produced from the oxidation of [1-^{14}C]-palmitate were assayed in the supernatants of the acid precipitate. ASM radioactivity was determined by liquid scintillation counting. Mitochondrial β-oxidation was calculated as the difference between the total β-oxidation rate and the peroxisomal β-oxidation rate, which was determined following incubation of the homogenate with antimycin A and rotenone. Data are expressed in nanomoles of radiolabeled ASM produced per gram of liver per hour.

4.7. Palmitate Uptake by the Liver

The in vivo uptake of palmitate by the liver was assessed by injection of 10 µCi [1-^{14}C]-palmitate (56 mCi/mmol) complexed with 1% fatty acid-free bovine serum albumin in a final volume of 200 µL PBS via the inferior vena cava in anesthetized 24 h-fasted mice by the intraperitoneal injection of a xylamine/ketamine mixture. Four minutes after injection, the superior vena cava was clamped and the hepatic portal vein was sectioned. A needle was inserted into the inferior vena cava toward the liver, and 10 mL of ice-cold PBS was injected under pressure with a syringe. At the end of this procedure, the liver was pale and the fluid emerging from the portal vein was clear. The liver was removed and used for lipid extraction and for the measurement of radioactivity by scintillation counting. Rates of palmitate uptake are expressed as disintegrations per minute (dpm) per gram of protein per hour.

4.8. Fat Mass and Histomorphometry

The total body fat content of mice was determined by dual energy X-ray absorptiometry (Lunar PIXImus2 mouse densitometer; GE Healthcare, Chicago, IL, USA), in accordance with the manufacturer's instructions. Body weight was determined and the left and right epididymal and inguinal white fat pads were harvested and weighed. Epididymal fat pads were then fixed in 10% neutral buffered formalin and embedded in paraffin. Tissues were cut into 4-µm sections and stained with hematoxylin and eosin. For the determination of adipocyte size, photomicrographs of the stained sections were obtained at ×100 magnification. Mean adipocyte diameter was calculated from measurements of at least 200 cells per sample.

4.9. Injection of Recombinant Adenovirus

Male C57BL/6J mice were anesthetized with isoflurane before the injection (between 9:00 and 10:00 h) into the penis vein of 1 × 10^9 pfu of either Ad GFP or Ad AMPK-CA in a final volume of 200 µL of sterile 0.9% NaCl. Mice were sacrificed 48 h or 8 days after adenovirus injection, as indicated in the figure legends. For the eight-day studies, mouse weight and food intake were measured daily.

4.10. Isolation of Total mRNA and Quantitative RT-PCR Analysis

Total RNA from mouse liver tissue was extracted using Trizol (Invitrogen, Carlsbad, CA, USA), and single-strand cDNA was synthesized from 5 µg of total RNA with random hexamer primers (Applied Biosystems, Foster City, CA, USA) and Superscript II (Life Technologies, Carlsbad, CA, USA). Real-time RT-PCR reactions were carried out in a final volume of 20 µL containing 125 ng of reverse-transcribed total RNA, 500 nM of primers, and 10 µL of 2× PCR mix containing Sybr Green (Roche, Meylan, France). The reactions were performed in 96-well plates in a LightCycler 480 instrument (Roche) with 40 cycles. The relative amounts of the mRNAs studied were determined by means of the second-derivative maximum method, with LightCycler 480 analysis software and 18S RNA as the invariant control for all studies. The sense and antisense PCR primers used, respectively, were as follows: for *Cpt1a*, 5′-AGATCAATCGGACCCTAGACAC-3′, 5′-CAGCGAGTAGCGCATAGTCA-3′; for *Cpt2*, 5′-CAGCACAGCATCGTACCCA-3′, 5′-TCCCAATG CCGTTCTCAAAAT-3′; for *Cd36*, 5′-TGGCTAAATGAGACTGGGACC-3′, 5′-ACATCACCACTCCAATCCCAAG-3′; for *Slc27a4* (*Fatp4*), 5′-GCACACTCAGCCGCCTGCTTCA-3′, 5′-TCACAGCTTCTCCTCGCCTGCCTG-3′; for *Fabp4*, 5′-GT

GATGCCTTTGTGGGAACCT-3′, 5′-ACTCTTGTGGAAGTCGCCT-3′; for 18S, 5′-GTAACCCGTTGAA CCCCATT-3′, 5′-CCATCCAATCGGTAGTAGCG-3′.

4.11. Western Blot Analysis

After the indicated incubation time in the figure legends, cultured hepatocytes were lysed in ice-cold lysis buffer containing 50 mM Tris, pH 7.4, 1% Triton X-100, 150 mM NaCl, 1 mM EDTA, 1 mM EGTA, 10% glycerol, 50 mM NaF, 5 mM sodium pyrophosphate, 1 mM Na_3VO_4, 25 mM sodium-β-glycerophosphate, 1 mM DTT, 0.5 mM PMSF and protease inhibitors (Complete Protease Inhibitor Cocktail; Roche). Lysates were sonicated on ice for 15 s to shear DNA and reduce viscosity. The tissues were homogenized in ice-cold lysis buffer using a ball-bearing homogenizer (Retsch, Eragny, France). The homogenate was centrifuged for 10 min at $10,000 \times g$ at 4 °C, and the supernatants were removed for determination of total protein content with a BCA protein assay kit (Thermo Fisher Scientific, Waltham, MA, USA). Fifty micrograms of protein from the supernatant were separated on 7.5% or 10% SDS-PAGE gels and transferred to nitrocellulose membranes. The membranes were blocked for 30 min at 37 °C with Tris-buffered saline supplemented with 0.05% NP40 and 5% nonfat dry milk. Immunoblotting was performed following standard procedures, and the signals were detected by chemiluminescence reagents (Thermo). X-ray films were scanned, and band intensities were quantified by Image J (NIH) densitometry analysis.

4.12. Transmission Electron Microscopy

Livers were fixed in 3% glutaraldehyde, 0.1 M sodium phosphate buffer (pH 7.4) for 24 h at 4 °C, postfixed with 1% osmium tetroxide, dehydrated with 100% ethanol, and embedded in epoxy resin. For ultrastructure analysis, ultrathin slices (70–100 nm thick) were cut from the resin blocks with a Reichert Ultracut S ultramicrotome (Reichert Technologies, Depew, NY, USA), stained with lead citrate and uranyl acetate, and examined in a transmission electron microscope (model 1011; JEOL, Tokyo, Japan) at the Cochin Institute electron microscopy facility.

4.13. Statistical Analysis

Results are expressed as means ± standard error of mean (SEM). Comparisons between groups were made by unpaired two-tailed Student's *t*-test or ANOVA for multiple comparisons where appropriate. Differences between groups were considered statistically significant when $p < 0.05$.

Author Contributions: M.F. designed and performed experiments, interpreted data, and wrote the manuscript. P.C.E. performed and analyzed indirect calorimetry experiments. B.V. interpreted the data and wrote the manuscript.

funding: This work was supported by grants from Inserm, CNRS, Université Paris Descartes, Agence Nationale de la Recherche (2010 BLAN 1123 01), Région Ile-de-France (CORDDIM), and Société Francophone du Diabète (SFD).

Acknowledgments: The authors thank Alain Schmitt (Cellular imaging: Electron microscopy facility at Institut Cochin, Paris, France) for transmission electron microscopy pictures and Jason Dyck (University of Alberta, Edmonton, Canada) for malonyl-CoA assays.

Conflicts of Interest: The authors declare that they have no conflicts of interest with the contents of this article.

References

1. Hardie, D.G. AMP-activated protein kinase: Maintaining energy homeostasis at the cellular and whole-body levels. *Annu. Rev. Nutr.* **2014**, *34*, 31–55. [CrossRef] [PubMed]
2. Hardie, D.G.; Lin, S.C. AMP-activated protein kinase—Not just an energy sensor. *F1000Res* **2017**, *6*, 1724. [CrossRef] [PubMed]
3. Day, E.A.; Ford, R.J.; Steinberg, G.R. AMPK as a Therapeutic Target for Treating Metabolic Diseases. *Trends Endocrinol. Metab.* **2017**, *28*, 545–560. [CrossRef] [PubMed]

4.   Viollet, B.; Foretz, M.; Guigas, B.; Horman, S.; Dentin, R.; Bertrand, L.; Hue, L.; Andreelli, F. Activation of AMP-activated protein kinase in the liver: A new strategy for the management of metabolic hepatic disorders. *J. Physiol.* **2006**, *574*, 41–53. [CrossRef] [PubMed]

5.   Smith, B.K.; Marcinko, K.; Desjardins, E.M.; Lally, J.S.; Ford, R.J.; Steinberg, G.R. Treatment of nonalcoholic fatty liver disease: Role of AMPK. *Am. J. Physiol. Endocrinol. Metab.* **2016**, *311*, E730–E740. [CrossRef] [PubMed]

6.   Gomez-Galeno, J.E.; Dang, Q.; Nguyen, T.H.; Boyer, S.H.; Grote, M.P.; Sun, Z.; Chen, M.; Craigo, W.A.; van Poelje, P.D.; MacKenna, D.A.; et al. A Potent and Selective AMPK Activator That Inhibits de Novo Lipogenesis. *ACS Med. Chem. Lett.* **2010**, *1*, 478–482. [CrossRef] [PubMed]

7.   Esquejo, R.M.; Salatto, C.T.; Delmore, J.; Albuquerque, B.; Reyes, A.; Shi, Y.; Moccia, R.; Cokorinos, E.; Peloquin, M.; Monetti, M.; et al. Activation of Liver AMPK with PF-06409577 Corrects NAFLD and Lowers Cholesterol in Rodent and Primate Preclinical Models. *EBioMedicine* **2018**, *31*, 122–132. [CrossRef] [PubMed]

8.   Boudaba, N.; Marion, A.; Huet, C.; Pierre, R.; Viollet, B.; Foretz, M. AMPK Re-Activation Suppresses Hepatic Steatosis but its Downregulation Does Not Promote Fatty Liver Development. *EBioMedicine* **2018**, *28*, 194–209. [CrossRef] [PubMed]

9.   Cool, B.; Zinker, B.; Chiou, W.; Kifle, L.; Cao, N.; Perham, M.; Dickinson, R.; Adler, A.; Gagne, G.; Iyengar, R.; et al. Identification and characterization of a small molecule AMPK activator that treats key components of type 2 diabetes and the metabolic syndrome. *Cell Metab.* **2006**, *3*, 403–416. [CrossRef] [PubMed]

10. Woods, A.; Williams, J.R.; Muckett, P.J.; Mayer, F.V.; Liljevald, M.; Bohlooly, Y.M.; Carling, D. Liver-Specific Activation of AMPK Prevents Steatosis on a High-Fructose Diet. *Cell Rep.* **2017**, *18*, 3043–3051. [CrossRef] [PubMed]

11. Fullerton, M.D.; Galic, S.; Marcinko, K.; Sikkema, S.; Pulinilkunnil, T.; Chen, Z.P.; O'Neill, H.M.; Ford, R.J.; Palanivel, R.; O'Brien, M.; et al. Single phosphorylation sites in Acc1 and Acc2 regulate lipid homeostasis and the insulin-sensitizing effects of metformin. *Nat. Med.* **2013**, *19*, 1649–1654. [CrossRef] [PubMed]

12. Hawley, S.A.; Fullerton, M.D.; Ross, F.A.; Schertzer, J.D.; Chevtzoff, C.; Walker, K.J.; Peggie, M.W.; Zibrova, D.; Green, K.A.; Mustard, K.J.; et al. The ancient drug salicylate directly activates AMP-activated protein kinase. *Science* **2012**, *336*, 918–922. [CrossRef] [PubMed]

13. Randle, P.J.; Garland, P.B.; Hales, C.N.; Newsholme, E.A. The glucose fatty-acid cycle. Its role in insulin sensitivity and the metabolic disturbances of diabetes mellitus. *Lancet* **1963**, *1*, 785–789. [CrossRef]

14. Lindholm, C.R.; Ertel, R.L.; Bauwens, J.D.; Schmuck, E.G.; Mulligan, J.D.; Saupe, K.W. A high-fat diet decreases AMPK activity in multiple tissues in the absence of hyperglycemia or systemic inflammation in rats. *J. Physiol. Biochem.* **2013**, *69*, 165–175. [CrossRef] [PubMed]

15. Muse, E.D.; Obici, S.; Bhanot, S.; Monia, B.P.; McKay, R.A.; Rajala, M.W.; Scherer, P.E.; Rossetti, L. Role of resistin in diet-induced hepatic insulin resistance. *J. Clin. Investig.* **2004**, *114*, 232–239. [CrossRef] [PubMed]

16. Yu, X.; McCorkle, S.; Wang, M.; Lee, Y.; Li, J.; Saha, A.K.; Unger, R.H.; Ruderman, N.B. Leptinomimetic effects of the AMP kinase activator AICAR in leptin-resistant rats: Prevention of diabetes and ectopic lipid deposition. *Diabetologia* **2004**, *47*, 2012–2021. [CrossRef] [PubMed]

17. Hawley, S.A.; Ford, R.J.; Smith, B.K.; Gowans, G.J.; Mancini, S.J.; Pitt, R.D.; Day, E.A.; Salt, I.P.; Steinberg, G.R.; Hardie, D.G. The Na$^+$/Glucose Cotransporter Inhibitor Canagliflozin Activates AMPK by Inhibiting Mitochondrial Function and Increasing Cellular AMP Levels. *Diabetes* **2016**, *65*, 2784–2794. [CrossRef] [PubMed]

18. Harriman, G.; Greenwood, J.; Bhat, S.; Huang, X.; Wang, R.; Paul, D.; Tong, L.; Saha, A.K.; Westlin, W.F.; Kapeller, R.; et al. Acetyl-CoA carboxylase inhibition by ND-630 reduces hepatic steatosis, improves insulin sensitivity, and modulates dyslipidemia in rats. *Proc. Natl. Acad. Sci. USA* **2016**, *113*, E1796–E1805. [CrossRef] [PubMed]

19. Kim, C.W.; Addy, C.; Kusunoki, J.; Anderson, N.N.; Deja, S.; Fu, X.; Burgess, S.C.; Li, C.; Ruddy, M.; Chakravarthy, M.; et al. Acetyl CoA Carboxylase Inhibition Reduces Hepatic Steatosis but Elevates Plasma Triglycerides in Mice and Humans: A Bedside to Bench Investigation. *Cell Metab.* **2017**, *26*, 576. [CrossRef] [PubMed]

20. Foretz, M.; Ancellin, N.; Andreelli, F.; Saintillan, Y.; Grondin, P.; Kahn, A.; Thorens, B.; Vaulont, S.; Viollet, B. Short-term overexpression of a constitutively active form of AMP-activated protein kinase in the liver leads to mild hypoglycemia and fatty liver. *Diabetes* **2005**, *54*, 1331–1339. [CrossRef] [PubMed]

21. Lin, X.; Yue, P.; Chen, Z.; Schonfeld, G. Hepatic triglyceride contents are genetically determined in mice: Results of a strain survey. *Am. J. Physiol. Gastrointest. Liver Physiol.* **2005**, *288*, G1179–G1189. [CrossRef] [PubMed]

22. Borgeson, E.; Wallenius, V.; Syed, G.H.; Darshi, M.; Lantero Rodriguez, J.; Biorserud, C.; Ragnmark Ek, M.; Bjorklund, P.; Quiding-Jarbrink, M.; Fandriks, L.; et al. AICAR ameliorates high-fat diet-associated pathophysiology in mouse and ex vivo models, independent of adiponectin. *Diabetologia* **2017**, *60*, 729–739. [CrossRef] [PubMed]

23. Lin, H.Z.; Yang, S.Q.; Chuckaree, C.; Kuhajda, F.; Ronnet, G.; Diehl, A.M. Metformin reverses fatty liver disease in obese, leptin-deficient mice. *Nat. Med.* **2000**, *6*, 998–1003. [CrossRef] [PubMed]

24. Henriksen, B.S.; Curtis, M.E.; Fillmore, N.; Cardon, B.R.; Thomson, D.M.; Hancock, C.R. The effects of chronic AMPK activation on hepatic triglyceride accumulation and glycerol 3-phosphate acyltransferase activity with high fat feeding. *Diabetol. Metab. Syndr.* **2013**, *5*, 29. [CrossRef] [PubMed]

25. Hotta, Y.; Nakamura, H.; Konishi, M.; Murata, Y.; Takagi, H.; Matsumura, S.; Inoue, K.; Fushiki, T.; Itoh, N. Fibroblast growth factor 21 regulates lipolysis in white adipose tissue but is not required for ketogenesis and triglyceride clearance in liver. *Endocrinology* **2009**, *150*, 4625–4633. [CrossRef] [PubMed]

26. Shimamura, M.; Matsuda, M.; Kobayashi, S.; Ando, Y.; Ono, M.; Koishi, R.; Furukawa, H.; Makishima, M.; Shimomura, I. Angiopoietin-like protein 3, a hepatic secretory factor, activates lipolysis in adipocytes. *Biochem. Biophys. Res. Commun.* **2003**, *301*, 604–609. [CrossRef]

27. Nygaard, E.B.; Vienberg, S.G.; Orskov, C.; Hansen, H.S.; Andersen, B. Metformin stimulates FGF21 expression in primary hepatocytes. *Exp. Diabetes Res.* **2012**, *2012*, 465282. [CrossRef] [PubMed]

28. Dyck, J.R.; Barr, A.J.; Barr, R.L.; Kolattukudy, P.E.; Lopaschuk, G.D. Characterization of cardiac malonyl-CoA decarboxylase and its putative role in regulating fatty acid oxidation. *Am. J. Physiol.* **1998**, *275*, H2122–H2129. [CrossRef] [PubMed]

29. Even, P.C.; Mokhtarian, A.; Pele, A. Practical aspects of indirect calorimetry in laboratory animals. *Neurosci. Biobehav. Rev.* **1994**, *18*, 435–447. [CrossRef]

30. Yu, X.X.; Drackley, J.K.; Odle, J. Rates of mitochondrial and peroxisomal beta-oxidation of palmitate change during postnatal development and food deprivation in liver, kidney and heart of pigs. *J. Nutr.* **1997**, *127*, 1814–1821. [CrossRef] [PubMed]

International Journal of
Molecular Sciences

MDPI

Review

AMPK: An Epigenetic Landscape Modulator

Brendan Gongol [1,2], Indah Sari [2,†], Tiffany Bryant [2,†], Geraldine Rosete [2,†] and Traci Marin [1,3,*]

[1]   Department of Medicine, University of California, San Diego, CA 92093, USA; brengong@gmail.com
[2]   Department of Cardiopulmonary Sciences, School of Allied Health Professions, Loma Linda University,
      Loma Linda, CA 92350, USA; isari@llu.edu (I.S.); tprescott@llu.edu (T.B.); grosete@llu.edu (G.R.)
[3]   Department of Health Sciences, Victor Valley College, Victorville, CA 92395, USA
*   Correspondence: traci.marin@vvc.edu; Tel.: +1-760-887-2294
†   These authors contributed equally to this work.

Received: 26 September 2018; Accepted: 17 October 2018; Published: 19 October 2018

Abstract: Activated by AMP-dependent and -independent mechanisms, AMP-activated protein kinase (AMPK) plays a central role in the regulation of cellular bioenergetics and cellular survival. AMPK regulates a diverse set of signaling networks that converge to epigenetically mediate transcriptional events. Reversible histone and DNA modifications, such as acetylation and methylation, result in structural chromatin alterations that influence transcriptional machinery access to genomic regulatory elements. The orchestration of these epigenetic events differentiates physiological from pathophysiological phenotypes. AMPK phosphorylation of histones, DNA methyltransferases and histone post-translational modifiers establish AMPK as a key player in epigenetic regulation. This review focuses on the role of AMPK as a mediator of cellular survival through its regulation of chromatin remodeling and the implications this has for health and disease.

Keywords: AMPK; epigenetics; chromatin remodeling; histone modification; DNA methylation

---

1. Introduction

Epigenetic regulation gives rise to a spectrum of cellular phenotypes observed in a single organism independent of primary DNA sequence. Such regulation is hereditable and stable, as occurs in the determination of cell type, but also transient, producing a particular phenotypic outcome to ensure survival [1–3]. Influencing gene expression, epigenetics promotes organismal adaption by offering substantial functional variability in response to environmental stimuli [1–3]. This regulation occurs, in part, through nucleosomal remodeling as a result of histone, DNA, and DNA-binding protein modifications that include: phosphorylation, acetylation, O-GlcNAcylation, ribosylation, and methylation. Such signature modifications or marks characterize nucleosome remodeling and determine the degree of gene activation or silencing. At a fundamental level, stressors such as nutrient deprivation or heightened physical activity trigger dynamic epigenetic markings that orchestrate adaptive gene regulation to improve survivability [4–7].

AMP-activated protein kinase (AMPK), a master regulator of energy homeostasis and a key mediator of adaptation and cell survival, is activated by conditions that produce energy deprivation such as hypoxia, exercise, nutrient starvation, and infection [8]. AMPK functions as a heterotrimeric serine/threonine protein kinase composed of a catalytic α-subunit, a scaffolding β-subunit, and a regulatory γ-subunit. Both α and β-subunits exist in two isoforms (α1, α2, and β1, β2), while the regulatory γ-subunit exists in three isoforms (γ1, γ2, γ3). The combination of α, β, and γ isoforms can form 12 different heterotrimeric complexes, each demonstrating unique functions and subcellular or tissue-specific distributions [9–11]. Once activated, AMPK phosphorylates targets with well-defined consensus sequences to regulate bioenergetics by modulating metabolic pathways that promote ATP production and limit energy expenditure [12]. Such roles of AMPK in cellular homeostasis are facilitated by gene regulation as a result of chromatin conformational dynamics. To that end, AMPK orchestrates these changes by phosphorylating several histones and proteins involved in nucleosome remodeling that enhance mitochondrial biogenesis and function [13]. Additionally, during acute metabolic stressors, such as fasting or exercise, AMPK associates with chromatin at promoters of genes involved in lipid and glucose metabolism [14].

Chronic AMPK activation plays a role in cellular and organismal inheritance, evidenced by its multi-isoform 2R-ohnologue characteristics, which are often evolutionarily conserved in gene coding regions to support basic survival functions and to increase the possibility for complex tissue diversity and adaptation [15–17]. Such adaptation for survival early in life initiates epigenetic programming that correlates with AMPK activation and determines predisposition to disease. For example, insults during development, such as placental insufficiency or maternal metabolic disorders, increase the probability of adverse metabolic disorders in later stages of life, such as diabetes mellitus, metabolic syndrome, insulin resistance, hypertension, vascular disease, and cancer [18]. Furthermore, metabolic disorders, such as maternal obesity, are often associated with reduced AMPK expression or activity and concomitant loss of its epigenetic mechanisms associated with adaptation and survival [19,20]. Although the mechanistic basis for this regulation, as well as its implications for inheritance, is still under exploration, these collective observations point to a fundamentally important role of AMPK as an epigenetic regulator.

2. Histone Modification

Histones are nuclear-localized, primarily positively charged proteins. However, a number of histone splice variants have been described with broad cellular functions and localizations [21,22]. Within the nucleus, histones are packaged into an octamer consisting of two H2A, H2B, H3, and H4 [23]. These octameric histone complexes associate via charge–charge interactions with DNA and make up a nucleosome, which is linked to other nucleosomes by histone H1. Given their tight association with DNA, histones serve as mediators between stress response signaling cascades and the regulation of nucleosomal structure that ultimately influence gene expression and cellular survival [24]. AMPK, both directly and indirectly, regulates the post-translational modification (PTM) status of histones that play a major role in the regulation of nucleosome structure.

2.1. Histone Phosphorylation

Bioinformatic analysis has identified several histones that contain an AMPK phosphorylation consensus sequence. These include: H1FX, H2AFX, H2AFY, H2AFY2, H2AFZ, H2BFM, and H3F3B [25], suggesting that AMPK plays an important role in epigenetic regulation through histone phosphorylation. Ultimately, such histone phosphorylation promotes cross-talk between epigenetic regulators to facilitate spatio-temporal nucleosome structural changes that influence transcriptional machinery [24]. For example, activation by glucose deprivation or UV radiation results in the colocalization of AMPK and phosphorylating serine 36 in H2B (H2B^{S36}) throughout the *carnitine palmitoyltransferase 1C* (*CPT1C*) and *cyclin-dependent kinase inhibitor* (*p21*) promoter regions and gene bodies. This exposes TP53 or tumor protein (p53) DNA binding sites and subsequent association with RNA polymerase II (Figure 1) [26,27]. These events enhance CPT1C and p21 expression, which play a role in cellular survival via activating autophagy and transporting long-chain fatty acids to the mitochondria, collectively enhancing β-oxidation and energy production [28–30]. Related to cellular survival, charged multivesicular body protein 1B (CHMP1b) is phosphorylated by AMPK, forms a "shell" around nucleosomes enriched with H3 phosphorylation and acetylation, and influences gene transcription [25]. This might play a role in the transition of active and inactive nucleosome regions and potentially heritable epigenetic marks [31].

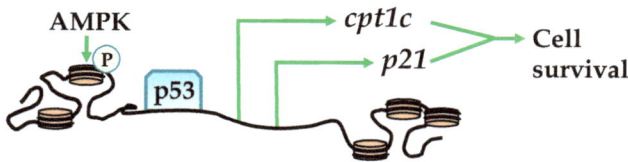

Figure 1. AMPK (AMP-activated protein kinase) promotes cell survival through histone phosphorylation. AMPK phosphorylates H2B to promote chromatin relaxation at tumor protein p53 (p53) recognized promoters and transcription of *carnitine palmitoyltransferase 1C* (*cpt1c*) and *cyclin-dependent kinase inhibitor* (*p21*) to enhance cell survival.

2.2. Histone Acetylation

Lysine acetylation of the N-terminal tails of histones disrupts the charge–charge interactions between DNA and histone tails, producing a euchromatin, a relaxed and active, chromatin state. This reaction changes histone–DNA associations, histone–histone associations between adjacent nucleosomes, and histone–regulatory protein interactions [32]. Histone acetylation occurs via transfer of an acetyl group from acetyl coenzyme-A (acetyl-CoA) to the to the ε-ammonium (NH_3^+) group of lysine catalyzed by histone acetyltransferases (HATs). HAT families are diverse and promote a spectrum of interactions and functions. Histone deacetylation restores the DNA-histone interaction promoting a heterochromatic, a condensed and silent chromatin state [33]. Antagonistically, histone deacetylation, catalyzed by histone deacetylases (HDACs), restores the DNA-histone interaction promoting a euchromatic chromatin state [33]. AMPK regulates the activity of both HATs and HDACs by influencing cofactor or substrate availability through direct phosphorylation.

Indirectly, 5-Aminoimidazole-4-carboxamide ribonucleotide (AICAR) activation of AMPK affects global HAT activity by increasing available acetyl-CoA levels through several mechanisms. AMPK phosphorylates acetyl-CoA carboxylase (ACC) to prevent the conversion of acetyl-CoA to malonyl-CoA increasing available acetyl donating groups for HATs (Figure 2A). AMPK also increases the formation of acetyl-CoA by phosphorylating acetyl-CoA synthetase short-chain family member 2 (ACSS2), causing its nuclear translocation for the conversion of acetate, the byproduct of HDAC histone deacetylation, to acetyl-CoA (Figure 2A). This mechanism has been shown to increase HAT acetylation of H3 at transcription factor EB (TFEB)-responsive promoters to activate genes important for autophagy and lysosomal function [34,35]. Yet, in general, increased acetyl-CoA levels in response to AICAR results in increased H3 acetylation at lysine 14 ($H3^{K14}$) and H4 acetylation at lysines 5,8,12,16 ($H4^{K5,8,12,16}$) [36]. AMPK activated by metformin regulates histone acetylation in a metabolic state-dependent manner. For example, metformin reduces H3 acetylation by decreasing bioavailability of mitochondrial acetyl-CoA in breast cancer, bringing acetylation status to normal in promoters of cancer-specific genes [37].

AMPK also regulates substrate availability of HDACs. The class III HDACs, sirtuins (SIRTs), couple deacetylation with nicotinamide adenine dinucleotide (NAD^+) hydrolysis to produce *O*-acetyl-ADP-ribose [38]. AMPK indirectly activates SIRT1, in part, by increasing the $NAD^+/NADH$ ratio [38,39]. Once activated, SIRT1 plays a fundamental role in chromatin organization by interacting with and deacetylating a variety of transcription factors and coregulators [40]. However, in contrary to activating the SIRTs, AMPK indirectly inhibits class I and II HDACs by increasing β-hydroxybutyrate (βOHB) during fatty acid oxidation. βOHB acts similarly to the HDAC inhibitor butyrate, increasing global histone acetylation (Figure 2B) [41]. However, in addition to globally inhibiting class I and class II HDACs, AMPK activation also promotes HDACs 4, 5, and 7 hyperphosphorylation and translocation from the nucleus (Figure 2B) [42]. Both of these events increase global histone acetylation. However, despite these global effects, the regulation of gene expression is often promoter- and gene cluster-specific, suggesting the importance of direct phosphorylation events of AMPK on specific HATs and HDACs.

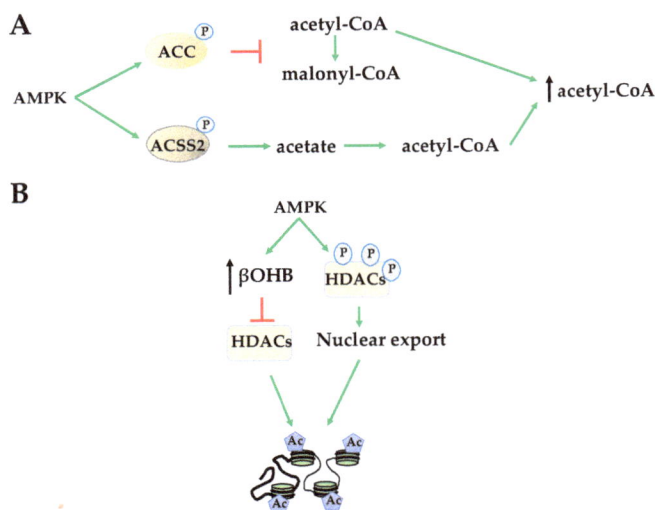

Figure 2. AMPK activation increases acetyl-CoA and promotes histone acetylation. (**A**) Through phosphorylation, AMPK inhibits acetyl-CoA carboxylase (ACC) while activating acetyl-CoA synthetase short-chain family member 2 (ACSS2) to increase acetyl-CoA availability for acetylation; (**B**) AMPK increases β-hydroxybutyrate (βOHB) to inhibit histone deacetylases (HDACs) and promotes HDAC nuclear export via hyperphosphorylation. Both of these events increase histone acetylation.

Investigations into epigenetically regulated networks identified HAT1 and retinoblastoma binding protein 7 (RBBP7) as direct targets of AMPK that dimerize following phosphorylation. This results in enhanced euchromatin structure at the promoters of *peroxisome proliferator–activated receptor gamma coactivator–1α* (*PGC-1α*), *transcription factor A* (*Tfam*), *nuclear respiratory factors 1* and *2* (*NRF1* and *NRF2*), and *uncoupling proteins 2* and *3* (*UCP2* and *UCP3*). The corresponding induction of these genes produces enhanced mitochondrial function (Figure 3A) [13]. In addition to HAT1 and RBBP7, AMPK phosphorylates HDAC5, promoting its dissociation from the promoters releasing its suppressive effects to increase expression of an array of metabolic genes, including glucose transporter type 4 (GLUT-4) (Figure 3B) [43].

Although less characterized, it is likely that AMPK phosphorylates an array of HATs including transcription regulator family member A (SIN3), CREB binding protein (CREBBP), elongator acetyltransferase complex subunits (ELP) 2, 3, and 4, and K(lysine) acetyltransferases (KAT) 2A, 2B, 6A, 6B, 7, and 8, and HDACs 1–5, 8–9, and 10, 11, as well as SIRTs 2, 3, 4, 5, 6, and 7; all of which contain AMPK consensus sequences and are opportune for future study [13].

2.3. Histone Methylation

Histone methylation occurs on lysine and arginine, and its effects on gene expression are dependent upon the site and degree of methylation. For example, methylation of H3^{K4}, H3^{K36}, and H3^{K79} promotes an active euchromatin state; while methylation of H3^{K9} and H3^{K27} promotes a silent heterochromatic state [44–46]. Both lysine-specific and arginine-specific histone methyltransferases use S-adenosyl methionine (SAM) as cofactors and methyl donor catalyzing the transfer of one, two, or three methyl groups. The demethylases, however, serve a variety of functions and include two primary classes: flavin adenine dinucleotide (FAD)-dependent amine oxidase and Fe(II) and α-ketoglutarate-dependent hydroxylase. The regulation of demethylases via α-ketoglutarate, a tricarboxylic acid (TCA) cycle intermediate, suggests that AMPK may play a role in regulating histone methylation and demethylation because AMPK is activated by conditions where TCA cycle intermediates are depleted and, once activated, these intermediates are restored. For example, enhanced amino acid, fatty acid, or glycogen catabolism by AMPK increases α-ketoglutarate, which, in turn, may increase histone demethylase activity [47]. In addition to restoring α-ketoglutarate levels, AMPK phosphorylates and inhibits fumarase a TCA cycle enzyme that converts fumarate to malate. Subsequently, elevated fumarate levels inhibit lysine-specific demethylase 2A (KDM2A), restoring H3^{K36} dimethylation (H3^{K36me2}) at promoters of genes mediating cell growth (Figure 4A) [48]. In addition to these indirect effects, AMPK activates the demethyltransferase lysine demethylase 5 (KDM5) and lysine-specific histone demethylase-1 (LSD1) that remove H3^{K4} trimethylation (H3^{K4me3}) [49]. Although AMPK activates histone demethyltransferase activity, it inhibits several complexes that regulate methyltransferase activity including the methylation complex polycomb repressive complex 2 (PRC2) and the histone methyltransferase-containing COMPASS complex (complex proteins associated with Set1) [20,50]. While the inhibition of PRC2 occurs through the phosphorylation of histone methyl transferase enhancer of zeste homolog 2 (EZH2), which decreases H2^{K27} monomethylation (H2^{K27me1}), inhibition of the COMPASS complex by AMPK decreases H3^{K4me3}, a marker of transcriptional activation (Figure 4B) [50–52]. Globally, inhibition of PRC2 results in the upregulation of tumor suppressor PRC2 target genes and suppression of tumor growth, while COMPASS complex inhibition is a protective mechanism that halts chromatin marking in the presence of nutrient deficiency to promote stress tolerance (Figure 4B) [20,50–52].

However, the orchestration between the methylation and demethylation activities of AMPK is promoter-specific. For example, AMPK recruits PRC2 and LSD1 to the promoter of *caudal type homeobox2* (*Cdx2*), increasing H3^{K4me3} but not H3^{K27me3} [53]. This increases CDX2 expression, which transactivates solute carrier family 5 member 8 (SLC5A8) to increase immune function and cell survival under several stress conditions (Figure 4C) [54,55].

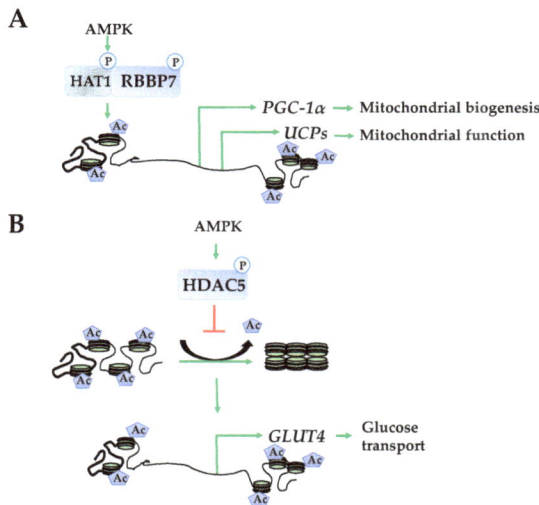

Figure 3. AMPK inhibits histone deacetylase activity to promote histone acetylation. (**A**) AMPK phosphorylates histone acetylase1 and retinoblastoma binding protein 7 (RBBP7) to form a complex that acetylates histones at promoters of *peroxisome proliferator-activated receptor gamma coactivator-1α* (*PGC-1α*), *uncoupling proteins* (*UCPs*) to increase mitochondrial biogenesis and function; (**B**) AMPK phosphorylates and inhibits HDAC5 enhancing acetylation at the *glucose transporter type 4* (*GLUT-4*) promoter increasing its transcription.

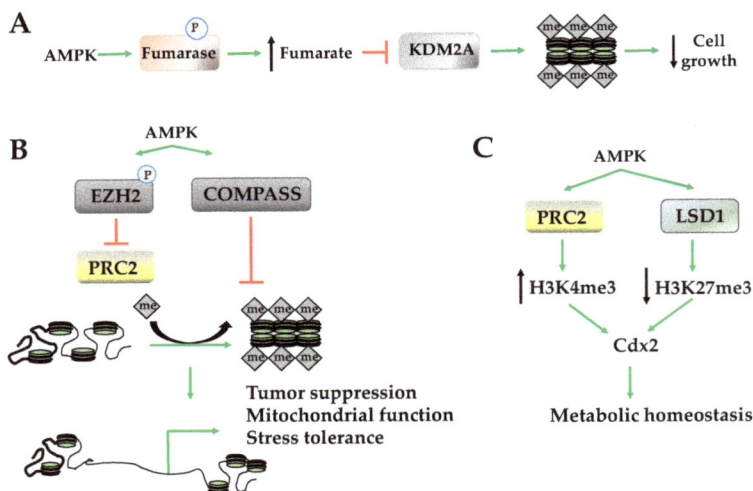

Figure 4. AMPK regulates histone methylation. (**A**) AMPK phosphorylates fumarase increase fumarate, which inhibits lysine-specific demethylase 2A (KDM2A), increasing histone methylation at cell growth; (**B**) AMPK decreases histone methylation through inhibition of polycomb repressive complex 2 (PRC2) and complex proteins associated with set1 (COMPASS complex); (**C**) AMPK modulates histone 3 methylations status to increase caudal type homeobox2 [cell differentiation]) (Cdx2) expression and metabolic homeostasis. LSD1: lysine-specific histone demethylase-1; EZH2: zeste homolog 2.

2.4. Histone O-GlcNAcylation

O-GlcNAcylation is the addition of an O-linked N-acetylglucosamine (O-GlcNAc) group to a serine or threonine by O-GlcNAc transferase (OGT). The donor substrate for O-GlcNAcylation, uridine diphosphate N-acetylglucosamine (UDP-GlcNAc), is a product of the hexamine biosynthetic pathway (HBP) in which glucose, fatty acid, amino acid, and ATP metabolism converge implicating its role as a nutrient and stress sensor [56]. O-GlcNAcase (OGA) removes O-GlcNAc through hydrolysis. The dynamics of histone O-GlcNAcylation regulate a variety of activities including other histone PTMs such as acetylation, methylation, and phosphorylation, and is highly sensitive to a spectrum of cellular stressors such as hypoxia, heat shock, and starvation (Figure 5A) [57]. AMPK phosphorylates O-GlcNAc transferase (OGT) promoting its dissociation from chromatin. This inhibits its O-GlcNAcylation of H2B^{S112} in response to extracellular glucose through the hexosamine biosynthesis pathway (HBP) and promotes H2B^{K120} mono-ubiquitination and transcriptional activation (Figure 5B) [56].

2.5. Histone Ribosylation

Histone ADP-ribosylation can occur as addition of mono- or poly-ADP-ribosylation units by mono-ADP ribosyltransferases (ARTs) or poly-(ADP-ribose) polymerases (PARPs), respectively. NAD$^+$ serves as the main source for histone ADP-ribosylation, which can occur on many amino acids restructuring chromatin into a euchromatin state [58,59]. AMPK phosphorylates and activates PARP1 [60,61]. Substrates of PARP1 include PARP1 itself and the tail of histones H1, H2A, H2B, H3, and H4, creating a poly(ADP-ribose)ylated mark at transcription start sites in transcriptionally necessary genes during metaphase [62]. Contrarily, when PARP1 is not active as a polymerase, it binds to DNA and promotes a heterochromatin state [59]. This is underscored by AMPK's phosphorylation of PARP1, causing its dissociation from the B-cell lymphoma protein 6 (Bcl-6) to increase its expression [61].

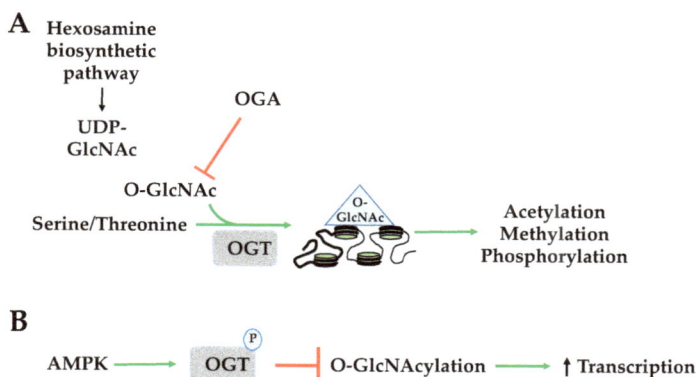

Figure 5. AMPK regulates histone *O*-GlcNAcylation and ADP-ribosylation. (**A**) *O*-GlcNAc transferase adds *O*-GlcNAc to a serine or threonine to produce *O*-GlcNAcylation. Nutrient and stress sensor stimulates hexosamine biosynthetic pathway (HBP) to create UDP-GlcNAc, which is a donor substrate for *O*-GlcNAcylation. Histone *O*-GlcNAcylation regulates acetylation, methylation, and phosphorylation; (**B**) AMPK phosphorylates *O*-GlcNAc transferase and inhibits *O*-GlcNAcylation to increase transcription. OGA: *O*-GlcNAcase; OGT: *O*-GlcNAc transferase.

3. DNA Modification

More abundantly enriched in promoter regions, cytosine-guanine dinucleotides (CpG) are susceptible to methylation at the 5′ cytosine position by a family of DNA methyltransferases (DNMTs). Each DNA methyltransferase member has a slightly different functional role in the global regulation of global DNA methylation. While DNMT 3a and 3b create the methyl "marks" that are carried through mitosis, DNMT1 maintains these methylation marks and regulates the dynamics of DNA modification and nucleosomal remodeling [63]. Following methylation, a number of functional changes occur. Due to its electron donating effects, methylation weakens Watson–Crick base pairing and recruits methyl-CpG binding domain proteins (MBDs), HDACs, and transcriptional repressors that collectively organize chromatin into a heterochromatic, a transcriptionally inactive conformation. These CpG methylations are highly dynamic- and stimulation-dependent. Removal of the CpG methylation requires oxidation of 5-methycytosine (5-mC) to 5-hydroxymethylcytosine (5-hmC), which is then converted to 5-formylcytosine (5-fC), and 5-carboxylcytosine (5-caC) followed by complete removal of the functional group by ten-eleven translocation hydroxylases (TETs) [64,65]. Although TETs are required to actively remove CpG methyl groups, the process occurs passively and semi-conservatively during DNA replication [66].

DNA Methylation

AMPK regulates global methylation by changing the substrates required for DNMT and TET activity. For example, SAM, the methyl donor for DNMT3a and DNMT3b, is converted to S-adenosylhomocysteine (SAH) upon CpG methylation [67]. Therefore, while SAM is required for DNMT3a and DNMT3b activity, SAH has an opposing inhibitory effect [67]. Supporting DNMT3a and DNMT3b activation, AMPK increases mitochondrial function and serine hydroxymethyltransferase 2 (SHMT2). Once activated, SHMT2 facilitates the one carbon transfer from folate to SAH, making SAM, and increasing the SAM/SAH ratio [67]. AMPK also increases the SAM/SAH ratio by transactivating let-7 micro RNA that subsequently degrades H19, relieving a direct inhibitory effect on S-adenosylhomocysteine hydrolase (SAHH). Subsequent activation of SAHH enables DNMT3b activation [68]. Taken together, these studies indicate that AMPK increases the SAM/SAH ratio that has an activating effect on DNMT3a and DNMT3b. These AMPK-dependent shifts in methylation also correlate to metabolic adaptive situations. For example, promoter methylation of *cytochrome C oxidase subunit 4I1* (*COX4I*) and *fatty acid binding protein 3* (*FABP3*) increases during exercise, while *peroxisome proliferator-activated receptor δ* (*PPARδ*) promoter methylation increases after fasting (Figure 6A) [69]. However, the overall effect AMPK has on DNA methylation is likely to be stimulation-, tissue-, promoter-, and DNMT isoform-specific. Paralleling its role as an epigenetic metabolic regulator in histone acetylation through HAT1 and RBBP7, AMPK also phosphorylates DNMT1[S730], inhibiting methyl CpG in *PGC-1α*, *Tfam*, *NRF1*, *NRF2*, *UCP2* and *UCP3* promoters [13]. This results in improved mitochondrial function, a requirement for cell survival [13,70].

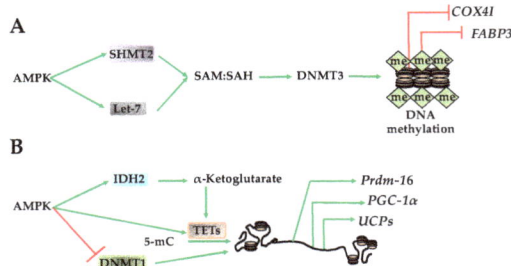

Figure 6. AMPK regulates DNA methylation. (**A**) AMPK activates serine hydroxymethyltransferase 2 (SHMT2) and Let-7 to increase SAM:SAH ratio providing substrate for DNA methyltransferase 3 (DNMT3) methylation of promoters such as *COX4I* and *FABP3*; (**B**) AMPK directly regulates isocitrase dehydrogenase 2 (IDH2) to yield α-ketoglutarates to promote ten-eleven translocation hydroxylases (TETs) activation. AMPK likely also directly regulates TETs catalyzing the conversion of 5-mc to unmethylated cytosine while inhibiting DNA methyltransferase 1 (DNMT1). These actions result in decreased promoter methylation of *PR domain containing 16* (*Prdm16*), *PGC-1α*, and the *uncoupler proteins*.

In addition to the DNMTs, AMPK also regulates TETs, which, in turn, govern loci-specific CpG methylation. The effect of AMPK on TETs occurs through its regulation of TCA cycle intermediates. For example, AMPK regulates isocitrase dehydrogenase 2 (IDH2) to increase the levels of α-ketoglutarate, an activator of TET1–TET4 resulting in CpG demethylation [47,71]. Activation of TET1–TET4 results in transactivation of the PR domain containing 16 (Prdm16) gene in progenitor cells that promotes brown adipogenesis, therefore supporting AMPK's role as an epigenetic regulator of metabolic function (Figure 6B) [47].

4. Approaches to Elucidating the AMPK-Modulated Epigenetic Landscape

Although biochemical studies have resolved a number of mechanistic insights into the role AMPK plays in the epigenetic regulation of chromatin structure, the study of AMPK on epigenetic function is still in its infancy. A number of predictive algorithms and studies have identified a potential role of AMPK on a number of epigenetic regulators. However, the role they may play under different physiological stimulations and their tissue specificity are still unknown. Because of the interconnected nature between regulatory cascades, the study of AMPK on epigenetic function in a disease-relevant capacity requires systems biological and bioinformatics approaches.

Scientific advances in the computational and Big Data arenas have resulted in novel experimental approaches, databases, and bioinformatics techniques that can be used in the exploration of AMPK epigenetic-regulated signaling pathways. Examples of tools used to explore chromatin characteristics include H3[K27] acetylation H3[K4] and monomethylation (H3[K4me1]) immunoprecipitation sequencing (IP-Seq); assay for transposase-accessible chromatin using sequencing (ATAC-seq); assay for transposase-accessible chromatin using sequencing (ATAC-seq); and Hi-C [72]. The relationship between DNA, histone modifications, histone remodeling, followed by Hi-C structural mapping, provides an integrative understanding of how AMPK can influence chromatin architecture. The identification of AMPK-regulated networks can be explored using consensus sequence mapping and machine learning in computing environments such as R/Bioconductor computing [25,73–75]. The expression of effected genes and loci and their application of these loci to disease-relevant stimuli can be further explored via cross-referencing with expression profiles housed in the Gene Expression Omnibus (GEO) database, Sequence Read Archive (SRA), Single Nucleotide Polymorphism database (dbSNP), 3D Genome database C (3DGD), and CR2Cancer [76–79]. Integration of these datasets can provide a comprehensive picture of the influence AMPK has on epigenetic signaling cascades in addition to their genetic loci and disease-specific regulation (Figure 7).

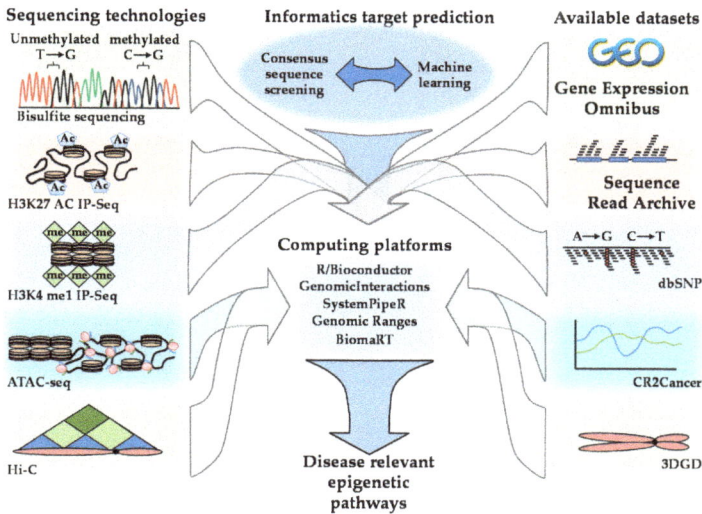

Figure 7. Methods for resolving disease-relevant AMPK-regulated epigenetic networks. Developed sequencing technologies are listed on the left-hand figure panels. Publicly available data repositories are listed on the right. Computing platforms R and relevant Bioconductor packages including Genomic interactions, SystemPipeR, Genomic Ranges, and BioMart can be used to integrate sequencing technologies, publicly available datasets, and AMPK target predictions as illustrated in the center panels.

5. Conclusions

AMPK's role as an epigenetic landscape modulator is underscored by its multifunctional kinase effects that regulate histones and epigenetic enzymes to mediate histone and DNA modifications. AMPK is crucial for cell survival and adaptation as evident by its activation upon nutrient depletion and stressors resulting in histone phosphorylation, acetylation, methylation, O-GlcNAcylation, and ribosylation. Additional, AMPK regulates DNA methylation mediated by influencing the SAM/SAH ratio, TET regulation and inhibition of DNMT1. Technological advances in data analysis continue to reveal the remarkable and diverse roles of AMPK in signaling pathways and epigenetic regulation, providing opportunities to exploit novel therapies relating to health and disease.

funding: This research received no external funding.

Acknowledgments: We would like to acknowledge David A. Johnson for his thoughtful review of the manuscript and comment to aid in clarity. We would also like to acknowledge John Shyy for his continued support and guidance.

Conflicts of Interest: The authors declare no conflict of interest.

Int. J. Mol. Sci. **2018**, *19*, 3238

Abbreviations

AMPK	AMP-activated protein kinase
CPT1C	carnitine palmitoyltransferase 1C
p21	cyclin-dependent kinase inhibitor
p53	TP53 or tumor protein
HAT	histone deacetylases
acetyl-CoA	acetyl coenzyme-A
HDAC	histone deacetylases
AICAR	5-Aminoimidazole-4-carboxamide ribonucleotide
ACC	acetyl-CoA carboxylase
ACSS2	acetyl-CoA synthetase short-chain family member 2
TFEB	transcription factor EB
SIRT	sirtuin
NAD+	nicotinamide adenine dinucleotide
βOHB	β-hydroxybutyrate
RBBP7	retinoblastoma binding protein 7
PGC-1α	peroxisome proliferator-activated receptor gamma coactivator-1α
Tfam	transcription factor A
NRF	nuclear respiratory factors
UCP	uncoupling proteins
SAM	S-adenosyl methionine
GLUT-4	glucose transporter type 4
KDM2A	lysine-specific demethylase 2A
KDM5	demethyltransferase lysine demethylase 5
LSD1	lysine-specific histone demethylase-1
PRC2	polycomb repressive complex 2
COMPASS	complex proteins associated with Set1
EZH2	enhancer of zeste homolog 2
Cdx2	caudal type homeobox2
SLC5A8	solute carrier family 5 member 8
O-GlcNAc	O-linked N-acetylglucosamine
OGT	O-GlcNAc transferase
UDP-GlcNAc	uridine diphosphate N-acetylglucosamine
HBP	hexosamine biosynthesis pathway
ART	mono-ADP ribosyltransferase
PTM	Post translational modification
PARP	poly-(ADP-ribose) polymerase
Bcl-6	B-cell Lymphoma protein 6
CpG	cytosine-guanine dinucleotides
DNMT	DNA methyltransferase
MBD	methyl-CpG binding domain proteins
5-mC	5-methycytosine
5-hmC	5-hydroxymethylcytosine
5-fC	5-formylcytosine
5-caC	5-carboxylcytosine
TETs	ten-eleven translocation hydroxylases
SAH	S-adenosylhomocysteine
SHMT2	serine hydroxymethyltransferase 2
SAHH	S-adenosylhomocysteine hydrolase
COX4I	cytochrome C oxidase subunit 4I1
FABP3	fatty acid binding protein 3
PPARδ	peroxisome proliferator-activated receptor δ
TCA	tricarboxylic acid
IDH2	isocitrase dehydrogenase 2

IP-Seq	immunoprecipitation sequencing
ATAC-seq	assay for transposase-accessible chromatin using sequencing
GEO	gene expression omnibus
SRA	sequence read archive
dbSNP	single nucleotide polymorphism database
3DGD	3D genome database

References

1. Trerotola, M.; Relli, V.; Simeone, P.; Alberti, S. Epigenetic inheritance and the missing heritability. *Hum. Genom.* **2015**, *28*, 9–17. [CrossRef] [PubMed]
2. Bunkar, N.; Pathak, N.; Lohiya, N.K.; Mishra, P.K. Epigenetics: A key paradigm in reproductive health. *Clin. Exp. Reprod. Med.* **2016**, *43*, 59–81. [CrossRef] [PubMed]
3. Boland, M.J.; Nazor, K.L.; Loring, J.F. Epigenetic regulation of pluripotency and differentiation. *Circ. Res.* **2014**, *115*, 311–324. [CrossRef] [PubMed]
4. Barrès, R.; Osler, M.E.; Yan, J.; Rune, A.; Fritz, T.; Caidahl, K.; Krook, A.; Zierath, J.R. Non-CpG methylation of the *PGC-1alpha* promoter through DNMT3B controls mitochondrial density. *Cell Metab.* **2009**, *10*, 189–198. [CrossRef] [PubMed]
5. Barrès, R.; Yan, J.; Egan, B.; Treebak, J.T.; Rasmussen, M.; Fritz, T.; Caidahl, K.; Krook, A.; O'Gorman, D.J.; Zierath, J.R. Acute exercise remodels promoter methylation in human skeletal muscle. *Cell Metab.* **2012**, *15*, 405–411. [CrossRef] [PubMed]
6. Nitert, M.D.; Dayeh, T.; Volkov, P.; Elgzyri, T.; Hall, E.; Nilsson, E.; Yang, B.T.; Lang, S.; Parikh, H.; Wessman, Y.; et al. Impact of an exercise intervention on DNA methylation in skeletal muscle from first-degree relatives of patients with type 2 diabetes. *Diabetes* **2012**, *61*, 3322–3332. [CrossRef] [PubMed]
7. Ferraro, E.; Giammarioli, A.M.; Chiandotto, S.; Spoletini, I.; Rosano, G. Exercise-induced skeletal muscle remodeling and metabolic adaptation: Redox signaling and role of autophagy. *Antioxid. Redox Signal.* **2014**, *21*, 154–176. [CrossRef] [PubMed]
8. Kim, J.; Yang, G.; Kim, Y.; Ha, J. AMPK activators: Mechanisms of action and physiological activities. *Exp. Mol. Med.* **2016**, *48*, e224. [CrossRef] [PubMed]
9. Rabinovitch, R.C.; Samborska, B.; Faubert, B.; Ma, E.H.; Gravel, S.P.; Andrzejewski, S.; Raissi, T.C.; Pause, A.; St-Pierre, J.; Jones, R.G. AMPK Maintains Cellular Metabolic Homeostasis through Regulation of Mitochondrial Reactive Oxygen Species. *Cell Rep.* **2017**, *21*, 1–9. [CrossRef] [PubMed]
10. Ross, F.A.; Jensen, T.E.; Hardie, D.G. Differential regulation by AMP and ADP of AMPK complexes containing different gamma subunit isoforms. *Biochem. J.* **2016**, *473*, 189–199. [CrossRef] [PubMed]
11. Vara-Ciruelos, D.; Dandapani, M.; Gray, A.; Egbani, E.O.; Evans, A.M.; Hardie, D.G. Genotoxic Damage Activates the AMPK-α1 Isoform in the Nucleus via Ca^{2+}/CaMKK2 Signaling to Enhance Tumor Cell Survival. *Mol. Cancer Res.* **2018**, *16*, 345–357. [CrossRef] [PubMed]
12. Langendorf, C.G.; Ngoei, K.R.; Scott, J.W.; Ling, N.X.; Issa, S.M.; Gorman, M.A.; Parker, M.W.; Sakamoto, K.; Oakhill, J.S.; Kemp, B.E. Structural basis of allosteric and synergistic activation of AMPK by furan-2-phosphonic derivative C2 binding. *Nat. Commun.* **2016**, *7*, 10912. [CrossRef] [PubMed]
13. Marin, T.L.; Gongol, B.; Zhang, F.; Martin, M.; Johnson, D.A.; Xiao, H.; Wang, Y.; Subramaniam, S.; Chien, S.; Shyy, J.Y. AMPK promotes mitochondrial biogenesis and function by phosphorylating the epigenetic factors DNMT1, RBBP7, and HAT1. *Sci. Signal.* **2017**, *10*, eaaf7478. [CrossRef] [PubMed]
14. Ratman, D.; Mylka, V.; Bougarne, N.; Pawlak, M.; Caron, S.; Hennuyer, N.; Paumelle, R.; De Cauwer, L.; Thommis, J.; Rider, M.H.; et al. Chromatin recruitment of activated AMPK drives fasting response genes co-controlled by GR and PPARα. *Nucleic Acids Res.* **2016**, *44*, 10539–10553. [CrossRef] [PubMed]
15. Satake, M.; Kawata, M.; McLysaght, A.; Makino, T. Evolution of vertebrate tissues driven by differential modes of gene duplication. *DNA Res.* **2012**, *19*, 305–316. [CrossRef] [PubMed]
16. Ross, F.A.; MacKintosh, C.; Hardie, D.G. AMP-activated protein kinase: A cellular energy sensor that comes in 12 flavours. *FEBS J.* **2016**, *283*, 2987–3001. [CrossRef] [PubMed]
17. Hardie, D.G. Keeping the home fires burning: AMP-activated protein kinase. *J. R. Soc. Interface* **2018**, *15*, 20170774. [CrossRef] [PubMed]

18. Laker, R.C.; Wlodek, M.E.; Connelly, J.J.; Yan, Z. Epigenetic origins of metabolic disease: The impact of the maternal condition to the offspring epigenome and later health consequences. *Food Sci. Hum. Wellness* **2013**, *2*, 1–11. [CrossRef]

19. Zhu, M.J.; Han, B.; Tong, J.; Ma, C.; Kimzey, J.M.; Underwood, K.R.; Xiao, Y.; Hess, B.W.; Ford, S.P.; Nathanielsz, P.W.; et al. AMP-activated protein kinase signaling pathways are down regulated and skeletal muscle development impaired in fetuses of obese, over-nourished sheep. *J. Physiol.* **2008**, *586*, 2651–2664. [CrossRef] [PubMed]

20. Demoinet, E.; Li, S.; Roy, R. AMPK blocks starvation-inducible transgenerational defects in *Caenorhabditis elegans*. *Proc. Natl. Acad. Sci. USA* **2017**, *114*, E2689–E2698. [CrossRef] [PubMed]

21. Bönisch, C.; Hake, S.B. Histone H2A variants in nucleosomes and chromatin: More or less stable? *Nucleic Acids Res.* **2012**, *40*, 10719–10741. [CrossRef] [PubMed]

22. Zlatanova, J.S.; Srebreva, L.N.; Banchev, T.B.; Tasheva, B.T.; Tsanev, R.G. Cytoplasmic pool of histone H1 in mammalian cells. *J. Cell Sci.* **1990**, *96*, 461–468. [PubMed]

23. Mariño-Ramírez, L.; Kann, M.G.; Shoemaker, B.A.; Landsman, D. Histone structure and nucleosome stability. *Expert Rev. Proteom.* **2005**, *2*, 719–729. [CrossRef] [PubMed]

24. Sawicka, A.; Seiser, C. Sensing core histone phosphorylation- a matter of perfect timing. *Biochim. Biophys. Acta* **2014**, *1839*, 711–718. [CrossRef] [PubMed]

25. Marin, T.L.; Gongol, B.; Martin, M.; King, S.J.; Smith, L.; Johnson, D.A.; Subramaniam, S.; Chien, S.; Shyy, J.Y. Identification of AMP-activated protein kinase targets by a consensus sequence search of the proteome. *BMC Syst. Biol.* **2015**, *9*, 13. [CrossRef] [PubMed]

26. Jones, R.G.; Plas, D.R.; Kubek, S.; Buzzai, M.; Mu, J.; Xu, Y.; Birnbaum, M.J.; Thompson, C.B. AMP-activated protein kinase induces a p53-dependent metabolic checkpoint. *Mol. Cell* **2005**, *18*, 283–293. [CrossRef] [PubMed]

27. Bungard, D.; Fuerth, B.J.; Zeng, P.Y.; Faubert, B.; Maas, N.L.; Viollet, B.; Carling, D.; Thompson, C.B.; Jones, R.G.; Berger, S.L. Signaling kinase AMPK activates stress-promoted transcription via histone H2B phosphorylation. *Science* **2010**, *329*, 1201–1205. [CrossRef] [PubMed]

28. Roa-Mansergas, X.; Fadó, R.; Atari, M.; Mir, J.F.; Muley, H.; Serra, D.; Casals, N. CPT1C promotes human mesenchymal stem cells survival under glucose deprivation through the modulation of autophagy. *Sci. Rep.* **2018**, *8*, 6997. [CrossRef] [PubMed]

29. Lee, I.H.; Kawai, Y.; Fergusson, M.M.; Rovira, I.I.; Bishop, A.J.; Motoyama, N.; Cao, L.; Finkel, T. Atg7 modulates p53 activity to regulate cell cycle and survival during metabolic stress. *Science* **2012**, *336*, 225–228. [CrossRef] [PubMed]

30. Itahana, Y.; Itahana, K. Emerging Roles of p53 Family Members in Glucose Metabolism. *Int. J. Mol. Sci.* **2018**, *19*, 776. [CrossRef] [PubMed]

31. Stauffer, D.R.; Howard, T.L.; Nyun, T.; Hollenberg, S.M. CHMP1 is a novel nuclear matrix protein affecting chromatin structure and cell-cycle progression. *J. Cell Sci.* **2001**, *114*, 2383–2393. [PubMed]

32. Javaid, N.; Choi, S. Acetylation- and Methylation-Related Epigenetic Proteins in the Context of Their Targets. *Genes* **2017**, *8*, 196. [CrossRef] [PubMed]

33. Marmorstein, R.; Zhou, M.M. Writers and readers of histone acetylation: Structure, mechanism, and inhibition. *Cold Spring Harb. Perspect. Biol.* **2014**, *6*, a018762. [CrossRef] [PubMed]

34. Bulusu, V.; Tumanov, S.; Michalopoulou, E.; van den Broek, N.J.; MacKay, G.; Nixon, C.; Dhayade, S.; Schug, Z.T.; Vande Voorde, J.; Blyth, K.; et al. Acetate recapturing by nuclear acetyl-CoA synthetase 2 prevents loss of histone acetylation during oxygen and serum limitation. *Cell Rep.* **2017**, *18*, 647–658. [CrossRef] [PubMed]

35. Li, X.; Yu, W.; Qian, X.; Xia, Y.; Zheng, Y.; Lee, J.H.; Li, W.; Lyu, J.; Rao, G.; Zhang, X.; et al. Nucleus-translocated ACSS2 promotes gene transcription for lysosomal biogenesis and autophagy. *Mol. Cell* **2017**, *66*, 684–697. [CrossRef] [PubMed]

36. Galdieri, L.; Gatla, H.; Vancurova, I.; Vancura, A. Activation of AMP-activated Protein Kinase by Metformin Induces Protein Acetylation in Prostate and Ovarian Cancer Cells. *J. Biol. Chem.* **2016**, *291*, 25154–25166. [CrossRef] [PubMed]

37. Cuyàs, E.; Fernández-Arroyo, S.; Joven, J.; Menendez, J.A. Metformin targets histone acetylation in cancer prone epithelial cells. *Cell Cycle* **2016**, *15*, 3355–3361. [CrossRef] [PubMed]

38. Ruderman, N.B.; Xu, X.J.; Nelson, L.; Cacicedo, J.M.; Saha, A.K.; Lan, F.; Ido, Y. AMPK and SIRT1: A long-standing partnership? *Am. J. Physiol. Endocrinol. Metab.* **2010**, *298*, E751–E760. [CrossRef] [PubMed]

39. Salminen, A.; Kauppinen, A.; Kaarniranta, K. AMPK/Snf1 signaling regulates histone acetylation: Impact on gene expression and epigenetic functions. *Cell. Signal.* **2016**, *28*, 887–895. [CrossRef] [PubMed]

40. Zhang, T.; Kraus, W.L. SIRT1-dependent regulation of chromatin and transcription: Linking NAD(+) metabolism and signaling to the control of cellular functions. *Biochim. Biophys. Acta* **2010**, *1804*, 1666–1675. [CrossRef] [PubMed]

41. Shimazu, T.; Hirschey, M.D.; Newman, J.; He, W.; Shirakawa, K.; Le Moan, N.; Grueter, C.A.; Lim, H.; Saunders, L.R.; Stevens, R.D.; et al. Suppression of oxidative stress by β-hydroxybutyrate, an endogenous histone deacetylase inhibitor. *Science* **2013**, *339*, 211–214. [CrossRef] [PubMed]

42. Mihaylova, M.M.; Vasquez, D.S.; Ravnskjaer, K.; Denechaud, P.D.; Yu, R.T.; Alvarez, J.G.; Downes, M.; Evans, R.M.; Montminy, M.; Shaw, R.J. Class IIa histone deacetylases are hormone-activated regulators of FOXO and mammalian glucose homeostasis. *Cell* **2011**, *145*, 607–621. [CrossRef] [PubMed]

43. McGee, S.L.; van Denderen, B.J.; Howlett, K.F.; Mollica, J.; Schertzer, J.D.; Kemp, B.E.; Hargreaves, M. AMP-activated protein kinase regulates *GLUT4* transcription by phosphorylating histone deacetylase 5. *Diabetes* **2008**, *57*, 860–867. [CrossRef] [PubMed]

44. Ng, S.; Yue, W.; Oppermann, U.; Klose, R. Dynamic protein methylation in chromatin biology. *Cell. Mol. Life Sci.* **2009**, *66*, 407–422. [CrossRef] [PubMed]

45. Feitag, M. Histone Methylation by SET Domain Proteins in Fungi. *Annu. Rev. Microbiol.* **2017**, *71*, 413–439. [CrossRef] [PubMed]

46. Wei, S.; Li, C.; Yin, Z.; Wen, J.; Meng, H.; Xue, L.; Wang, J. Histone methylation in DNA repair and clinical practice: New findings during the past 5-years. *J. Cancer* **2018**, *9*, 2072–2081. [CrossRef] [PubMed]

47. Yang, Q.; Liang, X.; Sun, X.; Zhang, L.; Fu, X.; Rogers, C.; Berim, A.; Zhang, S.; Wang, S.; Wang, B.; et al. AMPK/α-ketoglutarate axis dynamically mediates DNA demethylation in the *Prdm16* promoter and brown adipogenesis. *Cell Metab.* **2016**, *24*, 542–554. [CrossRef] [PubMed]

48. Wang, T.; Yu, Q.; Li, J.; Hu, B.; Zhao, Q.; Ma, C.; Huang, W.; Zhuo, L.; Fang, H.; Liao, L.; et al. O-GlcNAcylation of fumarase maintains tumour growth under glucose deficiency. *Nat. Cell Biol.* **2017**, *19*, 833–843. [CrossRef] [PubMed]

49. Eissenberg, J.C.; Shilatifard, A. Histone H3 lysine 4 (H3K4) methylation in development and differentiation. *Dev. Biol.* **2010**, *339*, 240–249. [CrossRef] [PubMed]

50. Wan, L.; Xu, K.; Wei, Y.; Zhang, J.; Han, T.; Fry, C.; Zhang, Z.; Wang, Y.V.; Huang, L.; Yuan, M.; et al. Phosphorylation of EZH2 by AMPK Suppresses PRC2 Methyltransferase Activity and Oncogenic Function. *Mol. Cell* **2018**, *69*, 279–291. [CrossRef] [PubMed]

51. Tang, G.; Guo, J.; Zhu, Y.; Huang, Z.; Liu, T.; Cai, J.; Yu, L.; Wang, Z. Metformin inhibits ovarian cancer via decreasing H3K27 trimethylation. *Int. J. Oncol.* **2018**, *52*, 1899–1911. [CrossRef] [PubMed]

52. Aloia, L.; Di Stefano, B.; Di Croce, L. Polycomb complexes in stem cells and embryonic development. *Development* **2013**, *140*, 2525–2534. [CrossRef] [PubMed]

53. Sun, X.; Yang, Q.; Rogers, C.J.; Du, M.; Zhu, M.J. AMPK improves gut epithelial differentiation and barrier function via regulating Cdx2 expression. *Cell Death Differ.* **2017**, *24*, 819–831. [CrossRef] [PubMed]

54. Kakizaki, F.; Aoki, K.; Miyoshi, H.; Carrasco, N.; Aoki, M.; Taketo, M.M. CDX transcription factors positively regulate expression of solute carrier family 5, member 8 in the colonic epithelium. *Gastroenterology* **2010**, *138*, 627–635. [CrossRef] [PubMed]

55. Gurav, A.; Sivaprakasam, S.; Bhutia, Y.D.; Boettger, T.; Singh, N.; Ganapathy, V. Slc5a8, a Na+-coupled high-affinity transporter for short-chain fatty acids, is a conditional tumour suppressor in colon that protects against colitis and colon cancer under low-fibre dietary conditions. *Biochem. J.* **2015**, *469*, 267–278. [CrossRef] [PubMed]

56. Yang, X.; Qian, K. Protein O-GlcNAcylation: Emerging mechanisms and functions. *Nat. Rev. Mol. Cell Biol.* **2017**, *18*, 452–465. [CrossRef] [PubMed]

57. Dehennaut, V.; Leprince, D.; Lefebvre, T. O-GlcNAcylation, an Epigenetic Mark. Focus on the Histone Code, TETFamily Proteins, and Polycomb Group Proteins. *Front. Endocrinol.* **2014**, *5*, 155. [CrossRef] [PubMed]

58. Poirier, G.G.; de Murcia, G.; Jongstra-Bilen, J.; Niedergang, C.; Mandel, P. Poly(ADP-ribosyl)ation of polynucleosomes causes relaxation of chromatin structure. *Proc. Natl. Acad. Sci. USA* **1982**, *79*, 3423–3427. [CrossRef] [PubMed]

59. Ciccarone, F.; Zampieri, M.; Caiafa, P. PARP1 orchestrates epigenetic events setting up chromatin domains. *Semin. Cell Dev. Biol.* **2017**, *63*, 123–134. [CrossRef] [PubMed]

60. Walker, J.W.; Jijon, H.B.; Madsen, K.L. AMP-activated protein kinase is a positive regulator of poly(ADP-ribose) polymerase. *Biochem. Biophys. Res. Commun.* **2006**, *342*, 336–341. [CrossRef] [PubMed]

61. Gongol, B.; Marin, T.; Peng, I.C.; Woo, B.; Martin, M.; King, S.; Sun, W.; Johnson, D.A.; Chien, S.; Shyy, J.Y. AMPKα2 exerts its anti-inflammatory effects through PARP-1 and Bcl-6. *Proc. Natl. Acad. Sci. USA* **2013**, *110*, 3161–3166. [CrossRef] [PubMed]

62. Lodhi, N.; Kossenkov, A.V.; Tulin, A.V. Bookmarking promoters in mitotic chromatin: Poly (ADP-ribose) polymerase-1 as an epigenetic mark. *Nucleic Acids Res.* **2014**, *42*, 7028–7038. [CrossRef] [PubMed]

63. Dunn, J.; Qiu, H.; Kim, S.; Jjingo, D.; Hoffman, R.; Kim, C.W.; Jang, I.; Son, D.J.; Kim, D.; Pan, C.; et al. Flow-dependent epigenetic DNA methylation regulates endothelial gene expression and atherosclerosis. *J. Clin. Investig.* **2014**, *124*, 3187–3199. [CrossRef] [PubMed]

64. Hill, P.W.; Amouroux, R.; Hajkova, P. DNA methylation, Tet proteins and 5-hydroxymethylcytosine in epigenetic reprogramming: An emerging complex story. *Genomics* **2014**, *104*, 324–333. [CrossRef] [PubMed]

65. Ficz, G.; Branco, M.R.; Seisenberger, S.; Santos, F.; Krueger, F.; Hore, T.A.; Marques, C.J.; Andrews, S.; Reik, W. Dynamic regulation of 5-hydroxymethylcytosine in mouse ES cells and during differentiation. *Nature* **2011**, *473*, 398–402. [CrossRef] [PubMed]

66. Chen, Z.X.; Riggs, A.D. DNA methylation and demethylation in mammals. *J. Biol. Chem.* **2011**, *286*, 18347–18353. [CrossRef] [PubMed]

67. Cuyàs, E.; Fernández-Arroyo, S.; Verdura, S.; García, R.Á.; Stursa, J.; Werner, L.; Blanco-González, E.; Montes-Bayón, M.; Joven, J.; Viollet, B.; et al. Metformin regulates global DNA methylation via mitochondrial one-carbon metabolism. *Oncogene* **2018**, *37*, 963–970. [CrossRef] [PubMed]

68. Zhong, T.; Men, Y.; Lu, L.; Geng, T.; Zhou, J.; Mitsuhashi, A.; Shozu, M.; Maihle, N.J.; Carmichael, G.G.; Taylor, H.S.; et al. Metformin alters DNA methylation genome-wide via the H19/SAHH axis. *Oncogene* **2017**, *36*, 2345–2354. [CrossRef] [PubMed]

69. Lane, S.C.; Camera, D.M.; Lassiter, D.G.; Areta, J.L.; Bird, S.R.; Yeo, W.K.; Jeacocke, N.A.; Krook, A.; Zierath, J.R.; Burke, L.M.; et al. Effects of sleeping with reduced carbohydrate availability on acute training responses. *J. Appl. Physiol.* **2015**, *119*, 643–655. [CrossRef] [PubMed]

70. Snyder, C.M.; Chandel, N.S. Mitochondrial regulation of cell survival and death during low-oxygen conditions. *Antioxid. Redox Signal.* **2009**, *11*, 2673–2683. [CrossRef] [PubMed]

71. Wu, H.; Zhang, Y. Reversing DNA methylation: Mechanisms, genomics, and biological functions. *Cell* **2014**, *156*, 45–68. [CrossRef] [PubMed]

72. Jiang, S.; Mortazavi, A. Integrating ChIP-seq with other functional genomics data. *Brief. Funct. Genom.* **2018**, *17*, 104–115. [CrossRef] [PubMed]

73. Xue, L.; Tao, W.A. Current technologies to identify protein kinase substrates in high throughput. *Front. Biol.* **2013**, *8*, 216–227. [CrossRef] [PubMed]

74. Harmston, N.; Ing-Simmons, E.; Perry, M.; Barešić, A.; Lenhard, B. GenomicInteractions: An R/Bioconductor package for manipulating and investigating chromatin interaction data. *BMC Genom.* **2015**, *16*, 963. [CrossRef] [PubMed]

75. Backman, T.W.H.; Girke, T. systemPipeR: NGS workflow and report generation environment. *BMC Bioinform.* **2016**, *20*, 388. [CrossRef] [PubMed]

76. Deininger, P.; Morales, M.E.; White, T.B.; Baddoo, M.; Hedges, D.J.; Servant, G.; Srivastav, S.; Smither, M.E.; Concha, M.; DeHaro, D.L.; et al. A comprehensive approach to expressionof L1 loci. *Nucleic Acids Res.* **2017**, *45*, e31. [CrossRef] [PubMed]

77. Bhagwat, M. Searching NCBI's dbSNP database. *Curr. Protoc. Bioinform.* **2010**. [CrossRef]

78. Li, C.; Dong, X.; Fan, H.; Wang, C.; Ding, G.; Li, Y. The 3DGD: A database of genome 3D structure. *Bioinformatics* **2014**, *30*, 1640–1642. [CrossRef] [PubMed]

79. Ru, B.; Sun, J.; Tong, Y.; Wong, C.N.; Chandra, A.; Tang, A.T.S.; Chow, L.K.Y.; Wun, W.L.; Levitskaya, Z.; Zhang, J. CR2Cancer: A database for chromatin regulators in human cancer. *Nucleic Acids Res.* **2018**, *46*, D918–D924. [CrossRef] [PubMed]

International Journal of
Molecular Sciences

MDPI

Review

Reciprocal Regulation of AMPK/SNF1 and Protein Acetylation

Ales Vancura *, Shreya Nagar, Pritpal Kaur, Pengli Bu, Madhura Bhagwat and Ivana Vancurova

Department of Biological Sciences, St. John's University, New York, NY 11439, USA;
shreya.nagar17@my.stjohns.edu (S.N.); pritpal.kaur17@my.stjohns.edu (P.K.); bup@stjohns.edu (P.B.);
madhura.bhagwat16@my.stjohns.edu (M.B.); vancuroi@stjohns.edu (I.V.)
* Correspondence: vancuraa@stjohns.edu

Received: 26 September 2018; Accepted: 24 October 2018; Published: 25 October 2018

Abstract: Adenosine monophosphate (AMP)-activated protein kinase (AMPK) serves as an energy sensor and master regulator of metabolism. In general, AMPK inhibits anabolism to minimize energy consumption and activates catabolism to increase ATP production. One of the mechanisms employed by AMPK to regulate metabolism is protein acetylation. AMPK regulates protein acetylation by at least five distinct mechanisms. First, AMPK phosphorylates and inhibits acetyl-CoA carboxylase (ACC) and thus regulates acetyl-CoA homeostasis. Since acetyl-CoA is a substrate for all lysine acetyltransferases (KATs), AMPK affects the activity of KATs by regulating the cellular level of acetyl-CoA. Second, AMPK activates histone deacetylases (HDACs) sirtuins by increasing the cellular concentration of NAD^+, a cofactor of sirtuins. Third, AMPK inhibits class I and II HDACs by upregulating hepatic synthesis of α-hydroxybutyrate, a natural inhibitor of HDACs. Fourth, AMPK induces translocation of HDACs 4 and 5 from the nucleus to the cytoplasm and thus increases histone acetylation in the nucleus. Fifth, AMPK directly phosphorylates and downregulates p300 KAT. On the other hand, protein acetylation regulates AMPK activity. Sirtuin SIRT1-mediated deacetylation of liver kinase B1 (LKB1), an upstream kinase of AMPK, activates LKB1 and AMPK. AMPK phosphorylates and inactivates ACC, thus increasing acetyl-CoA level and promoting LKB1 acetylation and inhibition. In yeast cells, acetylation of Sip2p, one of the regulatory β-subunits of the SNF1 complex, results in inhibition of SNF1. This results in activation of ACC and reduced cellular level of acetyl-CoA, which promotes deacetylation of Sip2p and activation of SNF1. Thus, in both yeast and mammalian cells, AMPK/SNF1 regulate protein acetylation and are themselves regulated by protein acetylation.

Keywords: AMP-activated protein kinase; epigenetics; protein acetylation; KATs; HDACs; acetyl-CoA; NAD^+

1. AMPK Links Metabolism and Signaling with Protein Acetylation, Epigenetics, and Transcriptional Regulation

AMP-activated protein kinase (AMPK) is highly conserved across eukaryotes and serves as an energy sensor and master regulator of metabolism, functioning as a fuel gauge monitoring systemic and cellular energy status [1–3]. Activation of AMPK occurs when the intracellular AMP/ATP ratio increases. In general, AMPK inhibits anabolism to minimize energy consumption and activates catabolism to increase ATP production.

AMPK is a heterotrimeric complex composed of subunit and two regulatory subunits, α and γ. The human genome contains two genes encoding two distinct subunits, 1 and 2, two α subunits, $\beta1$ and $\beta2$, and three γ subunits, $\gamma1$, $\gamma2$, and $\gamma3$ [1–3]. Different combinations of α, β, and γ subunits can produce 12 distinct AMPK complexes; however, it is not known whether these complexes differ in substrate specificities, subcellular localization, or other aspects of regulation. The subunit features the catalytic protein kinase domain, the α subunit contains a carbohydrate-binding domain that allows AMPK to interact with glycogen [4], and the γ subunit contains domains that bind AMP and thus impart AMPK regulation by cellular energy state [5–7].

The AMPK complex is activated more than 100-fold by phosphorylation on Thr172 of the catalytic α subunit [8]. The major upstream kinase targeting this site is the tumor suppressor liver kinase B1 (LKB1) [9–11]. LKB1 is responsible for most of AMPK activation under low energy conditions in the majority of tissues, including liver and muscle [2,12–14]. LKB1 is also responsible for AMPK activation in response to mitochondrial insults [15].

AMPK targets a number of metabolic enzymes and transporters, such as glucose transporter (GLUT) 1 and GLUT4, glycogen synthase (GS), acetyl-CoA carboxylase (ACC), and hydroxymethylglutaryl-CoA reductase (HMGCR) [16–18]. AMPK also regulates metabolism at the transcriptional level by phosphorylating sterol regulatory element-binding protein 1 (SREBP1), carbohydrate-responsive element-binding protein (ChREBP), transcriptional coactivator peroxisome proliferator-activated receptor gamma coactivator 1-alpha (PGC1) and transcriptional factor forkhead box O3 (FOXO3) [19–22].

One of the most important targets of AMPK is mechanistic target of rapamycin complex 1 (mTORC1). mTOR is a conserved serine/threonine protein kinase from the phosphatidylinositol-3-kinase (PI3K) family. mTOR is found in all eukaryotes and forms the catalytic subunit of mTORC1 and mTORC2. mTORC1 is regulated by nutrients and growth factors, and functions as a master regulator of cell growth and metabolism by phosphorylating a host of targets [23]. The AMPK-dependent mechanisms of mTORC1 inhibition are mediated by phosphorylation of the tuberous sclerosis complex (TSC) and raptor subunit of mTORC1 [24,25]. TSC functions as a GTPase activating protein (GAP) for the small GTPase Rheb, which directly binds and activates mTORC1. Thus, by downregulating Rheb, TSC inhibits mTORC1 and downregulation of TSC, which therefore leads to activation of mTORC1. In addition to integrating signals from several growth factor pathways, TSC is also regulated by AMPK. Activated AMPK directly phosphorylates TSC2 on serine residues that are distinct from those regulated by growth factor pathways, resulting in TSC activation and mTORC1 inhibition. In addition to TSC2, AMPK also phosphorylates mTORC1 subunit Raptor, leading again to mTORC1 inhibition [23,25]. Under low energy conditions, AMPK assembles in a complex with v-ATPase, Ragulator, scaffold protein Axin, and LKB1 on the lysosome surface, resulting in AMPK activation. At the same time, mTORC1 dissociates from the Ragulator and lysosome, resulting in mTORC1 inhibition [26,27]. These results further illustrate that AMPK and mTORC1 are inversely regulated and represent a molecular switch between catabolism and anabolism.

This review focuses on the previously little explored role of AMPK in regulation of acetyl-CoA and NAD+ homeostasis and on reciprocal regulation of AMPK and protein acetylation, which places AMPK at the interface between metabolism and other essential cellular functions, including transcription, replication, DNA repair, and aging [14,28–30].

2. Protein Acetylation

Protein acetylation is a posttranslational protein modification in which the acetyl group from acetyl-CoA is transferred onto ε-amino group of lysine residues. Histones were the first proteins known to be acetylated. More recently, genomic and proteomic approaches in bacteria, yeast, and higher eukaryotes identified many non-histone proteins that are acetylated, suggesting that acetylation extends beyond histones. A number of proteomic studies show that acetylation occurs at thousands of sites throughout eukaryotic cells and that the human proteome contains at least ~2500 acetylated proteins [31–34]. In comparison, similar analyses of human and mouse proteins identified ~2200 phosphoproteins [35,36]. Thus, it appears that protein acetylation is as widespread as phosphorylation [37]. In human cells, acetylated proteins are involved in the regulation of diverse cellular processes, including chromatin remodeling, the cell cycle, RNA metabolism, cytoskeleton dynamics, membrane trafficking, and key metabolic pathways, such as glycolysis, gluconeogenesis, and the citric acid cycle [32,34]. In general, protein acetylation can both activate and inhibit enzymatic activity of proteins as well as interactions between proteins [28,29,38,39]. Acetylation of histones affects the chromatin structure and transcriptional regulation by two mechanisms. It neutralizes positive charges of lysines and thus diminishes interaction of histone tails with DNA. By forming acetyllysines, histone acetylation creates sites that are recognized and bound by proteins and protein complexes that contain bromodomains. Many of these bromodomain-containing protein complexes covalently or noncovalently modify chromatin structure and thus regulate transcription [40,41].

2.1. KATs and HDACs

The enzymes that catalyze protein acetylation were originally called histone acetyltransferases (HATs) [40]. With the realization that histones are not the only substrates, these enzymes are now more commonly referred to as lysine acetyltransferases (KATs) [41–45]. The human genome contains 22 genes that encode proteins currently known to possess protein acetyltransferase activity [39]. The KATs can be classified into three major groups: The GNAT, MYST, and p300/CBP families. Most KATs are catalytic subunits of multiprotein complexes; the noncatalytic subunits of these complexes are typically responsible for substrate recognition, regulation, and subcellular localization.

Protein acetylation is a dynamic modification. The acetyl groups are removed from proteins by histone deacetylases (HDACs), sometimes also called lysine deacetylases (KDACs) to indicate that acetylated histones are not the only substrates [39]. However, the name HDACs is still more commonly used. HDACs hydrolyze the amide linkage between the acetyl group and amino group of lysine residues, yielding acetate. HDACs are grouped into four classes. Class I, II, and IV are Zn^{2+}-dependent amidohydrolases, while class III uses NAD^+ as a cosubstrate [46]. Class III HDACs are known as the sirtuins [47].

The dynamic balance between protein acetylation and deacetylation, mediated by the activities of KATs and HDACs, is well regulated in healthy cells, but is often dysregulated in cancer and other pathologic conditions. For example, change in the acetylation status of chromatin histones alters the structure of chromatin and expression pattern of genes in cancer cells [43].

2.2. Nonenzymatic Acetylation of Mitochondrial Proteins

The reactivity of metabolites depends on the presence of nucleophilic or electrophilic groups. The carbonyl group is electrophilic and can be further enzymatically activated by adding electronegative groups, such as thiols or phosphates. Compounds containing a reactive thioester group in the form of CoA are common in many metabolic pathways and reactions, involving fatty acid synthesis, tricarboxylic acid cycle, amino acid metabolism, and protein acetylation [48]. Protein acetylation by KATs employs a common catalytic mechanism which involves the formation of a ternary complex of KAT-acetyl-CoA-histone and the deprotonation of the ε-amino group of lysine by a glutamate or aspartate residue within the active site of a KAT, followed by a nucleophilic attack on the carbonyl group of acetyl-CoA [42].

High concentration of acetyl-CoA coupled with high pH, conditions that exist in the mitochondrial matrix, create a permissive environment for non-enzymatic acetylation of proteins [49,50]. In *Saccharomyces cerevisiae*, about 4000 lysine acetylation sites were identified, many of them on mitochondrial proteins [49,51]. The acetylation of mitochondrial proteins correlates with acetyl-CoA levels in mitochondria, as demonstrated by the fact that acetylation of mitochondrial proteins is dependent on *PDA1*, encoding a subunit of the pyruvate dehydrogenase (PDH) complex [49]. The acetylation of mitochondrial proteins was also elevated by introducing the *cit1Δ* mutation. *CIT1* encodes mitochondrial citrate synthase; *cit1Δ* mutants are not able to utilize acetyl-CoA for citrate synthesis and probably have an elevated level of mitochondrial acetyl-CoA. These results suggest that most of the mitochondrial acetyl-CoA in exponentially growing cells is derived from glycolytically-produced pyruvate that was translocated into mitochondria and converted to acetyl-CoA by the PDH complex. Inactivation of the PDH complex results in about a 30% decrease in cellular acetyl-CoA; this indicates that mitochondrial acetyl-CoA represents about 30% of the cellular pool. However, since mitochondria occupy only 1–2% of the cellular volume in *S. cerevisiae* [52], the mitochondrial acetyl-CoA concentration is about 20–30-fold higher than the concentration in the nucleocytosolic compartment and is probably within the millimolar range [49,50,53,54]. Due to the extrusion of protons across the inner mitochondrial membrane, the pH of the mitochondrial matrix is higher than the pH in the cytosol or nucleus, about 8.0 [50,55]. The high pH coupled with the high concentration of acetyl-CoA in the mitochondrial matrix create a permissive environment for non-enzymatic acetylation of mitochondrial proteins [49,50]. However, these considerations do not exclude the possibility that at least some protein acetylation in the mitochondria is catalyzed by KATs. In addition, some acyl-CoAs, such as 3-hydroxy-3-methylglutaryl-CoA, and glutaryl-CoA, are sufficiently reactive under the in vivo conditions and are able to non-enzymatically modify proteins [56].

3. AMPK Regulation of Protein Acetylation

AMPK regulates protein acetylation by at least five distinct mechanisms (Figure 1). First, AMPK phosphorylates and inhibits ACC and thus regulates acetyl-CoA homeostasis. Second, AMPK activates sirtuin SIRT1 by increasing the cellular concentration of NAD$^+$, a cofactor of sirtuins. Third, AMPK inhibits class I and II histone deacetylases (HDACs) by upregulating hepatic synthesis of α-hydroxybutyrate, a natural inhibitor of HDACs. Fourth, AMPK induces translocation of HDACs 4 and 5 from the nucleus to the cytoplasm and thus increases histone acetylation in the nucleus. Fifth, AMPK directly phosphorylates and downregulates p300 KAT.

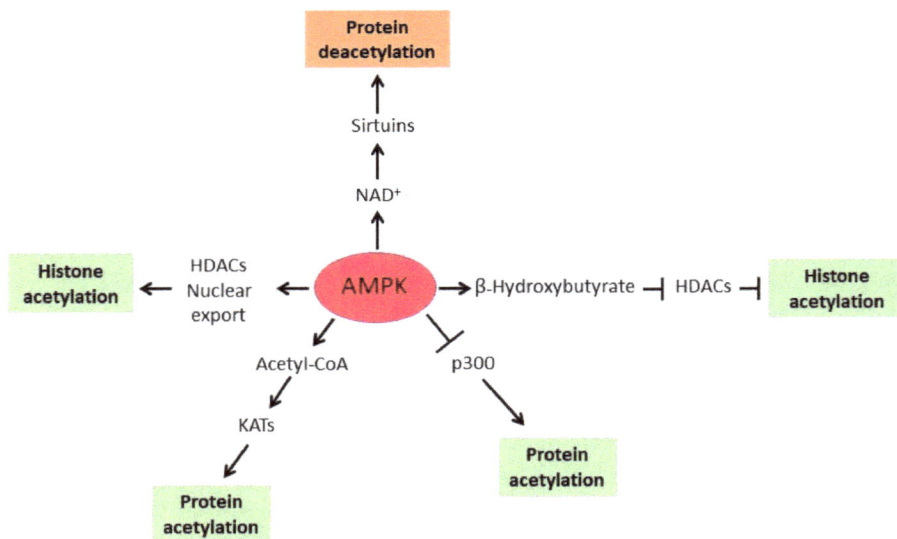

Figure 1. AMP-activated protein kinase (AMPK) regulates protein acetylation by several different mechanisms: (i) AMPK phosphorylates and inhibits acetyl-CoA carboxylase (ACC) and thus elevates acetyl-CoA level and activity of lysine acetyltransferases (KATs); (ii) AMPK increases the cellular concentration of NAD$^+$ and thus activates sirtuins; (iii) AMPK upregulates hepatic synthesis of α-hydroxybutyrate, and thus inhibits histone deacetylases (HDACs) and promotes histone acetylation; (iv) AMPK increases histone acetylation in the nucleus by inducing nuclear export of HDACs; (v) AMPK directly phosphorylates and downregulates p300 KAT. Arrows denote activation and t-bars denote inhibition.

3.1. Acetyl-CoA Level Regulates Protein Acetylation

Acetyl-CoA is the donor of acetyl groups for protein acetylation and KATs depend on intermediary metabolism for supplying acetyl-CoA in the nucleocytosolic compartment (Figure 1). Acetyl-CoA is thus a key metabolite that links metabolism with signaling, chromatin structure, and transcription [29,30,45,53,57–59]. Changing metabolic conditions drive fluctuations of the cellular level of acetyl-CoA to the extent that the activity of KATs is regulated by the availability of acetyl-CoA, resulting in dynamic protein acetylations that regulate a variety of cell functions, including transcription, replication, DNA repair, cell cycle progression, and aging. Acetyl-CoA can freely diffuse through the nuclear pore complex and changes in the pool of available acetyl-CoA in the cytoplasm cause changes in protein acetylation in both the nucleus and cytoplasm. However, the mitochondrial pool of acetyl-CoA is biochemically isolated and cannot be used for histone acetylation in the nucleocytosolic compartment [60]. In mammalian cells, glycolytically produced pyruvate is translocated from the cytosol into mitochondria, where pyruvate dehydrogenase converts it into acetyl-CoA. Acetyl-CoA then enters the tricarboxylic acid (TCA) cycle and condenses with oxaloacetate, producing citrate. Citrate can be subsequently exported from the mitochondrial matrix into the cytosol, where ATP-citrate lyase (ACL) converts it into acetyl-CoA and oxaloacetate. This acetyl-CoA is then used by KATs for protein acetylation in the nucleocytosolic compartment [61], in addition to being a precursor of several anabolic pathways, including de novo synthesis of fatty acids. Since ACL generates acetyl-CoA from glucose-derived citrate, glucose availability affects histone acetylation in an ACL-dependent manner, and when the synthesis of acetyl-CoA is compromised, rapid histone deacetylation ensues [60,61].

Translocation of pyruvate dehydrogenase complex (PDH) from the mitochondria to the nucleus provides an alternative mechanism for synthesis of acetyl-CoA in the nucleus. PDH translocated to the nucleus in a cell-cycle-dependent manner and in response to serum, epidermal growth factor, or mitochondrial stress. Inhibition of nuclear PDH decreased acetylation of specific lysine residues in histones and transcription of genes important for G1-S phase progression [62]. In addition to ACL and nuclear PDH, direct de novo synthesis of acetate from pyruvate for acetyl-CoA production occurs under conditions of nutritional excess [63]. The conversion of pyruvate to acetate takes place either by coupling to reactive oxygen species (ROS) or by the activity of keto acid dehydrogenases, which function under certain conditions as pyruvate decarboxylase [63].

Since nucleocytosolic acetyl-CoA is also used for de novo synthesis of fatty acids, histone acetylation and synthesis of fatty acids compete for the same acetyl-CoA pool. ACC catalyzes the carboxylation of acetyl-CoA to malonyl-CoA, the first and rate-limiting reaction in the de novo synthesis of fatty acids. The ACC activity affects the concentration of nucleocytosolic acetyl-CoA. Attenuated expression of yeast ACC encoded by the *ACC1* gene, increases global acetylation of chromatin histones as well as non-histone proteins, and alters transcriptional regulation [64]. Direct pharmacological inhibition of ACC in human cancer cells also induces histone acetylation [65,66]. ACC is phosphorylated and inhibited by AMPK. In yeast, inactivation of the SNF1 complex, the budding yeast ortholog of mammalian AMPK [67–69], results in increased Acc1p activity, reduced pool of cellular acetyl-CoA, and globally decreased histone acetylation [70]. Activation of AMPK with metformin or with the AMP mimetic 5-aminoimidazole-4-carboxamide ribonucleotide (AICAR) increases the inhibitory phosphorylation of ACC, and decreases the conversion of acetyl-CoA to malonyl-CoA, leading to increased protein acetylation and altered gene expression in prostate and ovarian cancer cells [65].

3.2. NAD$^+$ Synthesis Regulates Protein Acetylation

NAD$^+$ is a cofactor used by many oxidoreductases to carry electrons in redox reactions. In addition, NAD$^+$ is used as a cosubstrate by a group of HDACs called sirtuins, named after the budding yeast protein Sir2. In mammals, there are seven NAD$^+$-dependent sirtuins. Similar to the dependence of KATs on acetyl-CoA level, the activity of sirtuins is regulated by metabolically driven changes in the cellular level of NAD+ [71,72]. AMPK activation induces expression of nicotinamide phosphoribosyltransferase (NAMPT), the rate-limiting enzyme in the NAD+ salvage pathway that converts nicotinamide to nicotinamide mononucleotide to enable NAD+ biosynthesis. AMPK activation thus increases NAD+ level, elevating SIRT1 activity [73,74]. SIRT1-mediated protein deacetylation subsequently activates downstream targets, including peroxisome proliferator-activated receptor gamma coactivator 1-α (PGC-1α) and forkhead box protein O1 (FOXO1) [74]. In addition, AMPK directly phosphorylates SIRT1 at T344, which results in dissociation of SIRT1 from the inhibitory bladder cancer protein 1 (DBC1), and SIRT1-mediated deacetylation of p53 and inhibition of its transcriptional activity [75]. Yet another mechanism for SIRT1 activation by AMPK involves phosphorylation of glyceraldehyde 3-phosphate dehydrogenase (GAPDH) by AMPK, leading to nuclear translocation of GAPDH and GAPDH-dependent dissociation of SIRT1 from DBC1 [76].

3.3. α-Hydroxybutyrate Synthesis Regulates Protein Acetylation

The ketone body α-hydroxybutyrate is structurally similar to butyrate, an effective inhibitor of class I and II HDACs [77]. Ketone bodies, including α-hydroxybutyrate, are produced during starvation or prolonged exercise, when liver switches the metabolic mode from catabolism of glucose to catabolism of triacylglycerols and fatty acids. This metabolic switch is partly orchestrated by AMPK and activation of AMPK increases fatty acid oxidation, leading to production of ketone bodies, including α-hydroxybutyrate [77]. Administration of exogenous α-hydroxybutyrate or inducing catabolism of fatty acids by calorie restriction increased global histone acetylation in mouse tissues [77]. Acetylation of histones in the promoters of genes required for protection against oxidative stress was also increased, leading to increased expression of the corresponding genes and elevated protection against oxidative stress [77]. These results indicate that AMPK activation induced by calorie restriction or prolonged exercise leads to α-hydroxybutyrate-mediated inhibition of HDACs and globally increased histone acetylation, and may represent one of the health-promoting mechanisms of calorie restriction.

3.4. AMPK Induces Nuclear Export of HDACs

Type II HDACs belong into two subgroups, IIa and IIb. HDACs of the IIa subgroup, HDAC4, HDAC5, HDAC7, and HDAC9, are able to shuttle between the nucleus and cytoplasm [78–81]. AMPK phosphorylates HDAC5 at Ser259 and Ser498, which promotes export of HDAC5 from the nucleus to the cytoplasm. This removal of HDAC5 from the nucleus results in decreased occupancy of HDAC5 and increased histone acetylation at promoters of glucose transporter member 4 (GLUT4), myogenin, and α-catenin genes, leading to increased transcription of the corresponding genes [82–86]. It appears that AMPK regulates expression of host of genes involved in differentiation or development by phosphorylating HDAC5 and HDAC4 and promoting their export from the nucleus. AMPK also mediates nuclear accumulation of transcription factor hypoxia-inducible factor 1α (HIF-1α) by a mechanism that involves HDAC5. Activation of nuclear AMPK promotes nuclear export of HDAC5, presumably by directly phosphorylating HDAC5. Cytosolic HDAC then deacetylates heat shock protein 70 (HSP70), triggering dissociation of HSP70 from HIF-1 and nuclear transport of HIF-1α [84]. Depending on their phosphorylation level, also HDAC7 and HDAC9 shuttle between nucleus and cytoplasm. However, it is not known whether they are AMPK substrates and whether AMPK regulates their nucleocytosolic shuttling.

3.5. AMPK Phosphorylates p300 KAT and Histone H3

AMPK directly phosphorylates transcriptional coactivator p300 KAT on Ser89, which inhibits the interaction of p300 with peroxisome proliferator-activated receptor γ (PPAR-γ) and retinoid acid receptor [19]. The AMPK-mediated phosphorylation of p300 also results in decreased acetylation and reduced activity of transcription factors nuclear factor kappa B (NFκB) and SMAD3 [87,88]. AMPK also promotes histone acetylation indirectly through phosphorylation of histone H2B at Ser36, particularly at promoters occupied by p53. During metabolic or genotoxic stress, AMPK translocates to the nucleus, binds to chromatin, and phosphorylates histone H2B at Ser36. This phosphorylation leads to increased assembly and recruitment of KATs to specific promoters, associated with increased transcription [89]. An analogous situation was described also in yeast. SNF1, the yeast AMPK ortholog, phosphorylates histone H3 at Ser10, which results in increased acetylation of Lys14 of histone H3 by KAT Gcn5 in the promoter of the *INO1* gene [90]. It appears that SNF1-mediated phosphorylation of Ser10 of histone H3 does not represent a general mechanism of histone acetylation. Rather, inactivation of SNF1 results in decreased nucleocytosolic level of acetyl-CoA by increasing conversion of acetyl-CoA into malonyl-CoA, which results in globally reduced acetylation of histone and non-histone proteins [70].

4. AMPK Is Regulated by Protein Acetylation

AMPK activity is regulated by an upstream kinase LKB1, which activates AMPK by phosphorylating it on Thr172. LKB1 is a low energy sensor that regulates tumorigenesis and apoptosis by regulating AMPK and mechanistic target of rapamycin (mTOR) pathways [10]. LKB1 is acetylated and the acetylation reduces its ability to activate AMPK (Figure 2). Deacetylation of LKB1 by SIRT1 activates LKB1 and AMPK and increases inhibitory phosphorylation of ACC [91]. Inhibition of ACC increases acetyl-CoA level and protein acetylation [65] and presumably should also promote LKB1 acetylation and reduced activation of AMPK (Figure 2). It is tempting to speculate that LKB1 acetylation and diminished activation of AMPK form a regulatory loop with ACC, which contributes to regulation of acetyl-CoA homeostasis and protein acetylation [70]. Increased acetyl-CoA level would promote LKB1 acetylation and diminished activation of AMPK. Decreased activity of AMPK would result in lower AMPK-mediated phosphorylation and inhibition of ACC, increased conversion of acetyl-CoA to malonyl-CoA, decreased acetyl-CoA level and decreased protein acetylation. Decreased acetyl-CoA level would also result in hypoacetylation of LKB1 and increased activation of AMPK. This, in turn, would lead to increased phosphorylation and inhibition of ACC, decreased conversion of acetyl-CoA to malonyl-CoA, increased acetyl-CoA level, and increased protein acetylation. In addition, activation of AMPK would promote NAD^+ synthesis, leading to elevated SIRT1 activity (Figure 2). This homeostatic mechanism would contribute to the regulation of the nucleocytosolic level of acetyl-CoA and NAD^+ within certain limits and would prevent gross hypoacetylation or hyperacetylation of proteins, a condition that might alter regulation of many essential processes. It also appears that SIRT1 and AMPK mediate the positive effect of some dietary compounds, such as resveratrol and other polyphenols. Resveratrol increases SIRT1 activity, which activates AMPK signaling, presumably by deacetylating LKB1. Activated AMPK then suppresses lipid accumulation in hepatocytes [92].

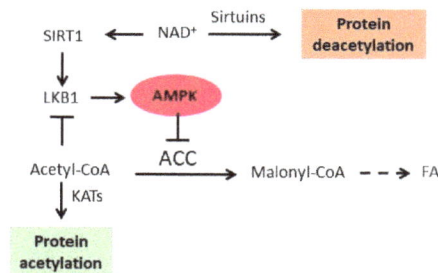

Figure 2. Model of the feedback regulation of AMPK by protein acetylation. AMPK phosphorylates and inhibits ACC, thus increasing acetyl-CoA cellular level and promoting KAT-mediated protein acetylation. Acetylation of liver kinase B1 (LKB1) inhibits the ability of LKB1 to activate AMPK. AMPK also promotes synthesis of NAD^+, thus activating SIRT1 and other sirtuins and promoting protein deacetylation. SIRT1 deacetylates and activates LKB1, resulting in AMPK activation. Arrows denote activation, t-bars denote inhibition, and dashed arrow indicates multistep pathway.

Int. J. Mol. Sci. **2018**, *19*, 3314

An analogous regulatory loop seems to operate in yeast. The yeast SNF1 complex consists of the catalytic α subunit Snf1p, one of three different regulatory α subunits, Sip1p, Sip2p, or Gal83p, and the stimulatory γ subunit Snf4p [93]. Acetylation of Sip2p, one of the regulatory α-subunits of the SNF1 complex, results in inhibition of SNF1. The level of Sip2p acetylation depends on the nucleocytosolic level of acetyl-CoA and is increased when *ACC1* transcription is repressed [64] and decreased in *snf1*Δ cells [70]. The acetylation of Sip2p increases its interaction and inhibition of Snf1p [94]. This results in activation of the *ACC1* gene and reduced cellular level of acetyl-CoA, which promotes deacetylation of Sip2p and activation of SNF1. Thus, in both yeast and mammalian cells, AMPK/SNF1 regulate protein acetylation and are themselves regulated by protein acetylation. Since SNF1 also phosphorylates Sch9, a yeast ortholog of the Akt kinase, inhibition of SNF1 by Sip2p results in reduced phosphorylation of Sch9, ultimately leading to extended life span. Acetylation of Sip2p thus promotes life span extension and Sip2p acetylation mimetics are more resistant to oxidative stress [94].

5. Conclusions

AMPK is an energy sensor and master regulator of metabolism, functioning as a fuel gauge. AMPK phosphorylates and regulates a number of metabolic enzymes and transporters, as well as transcription factors. In this review article, we have focused on the role of AMPK in regulation of protein acetylation and on regulation of AMPK by protein acetylation. Taken together, AMPK regulates protein acetylation by regulating synthesis of acetyl-CoA, NAD^+, and α-hydroxybutyrate, as well as by directly phosphorylating and regulating KATs and HDACs. Using these two general mechanisms, AMPK contributes to the global regulation of protein acetylation and connects epigenetic chromatin modifications with the cellular metabolic state. Since protein acetylation appears to be as widespread as protein phosphorylation, it endows AMPK with yet another mechanism of regulation of cellular and organismal physiology.

funding: The work was supported by NIH GM120710 grant (to A.V.).

Conflicts of Interest: The authors declare no conflict of interest.

References

1. Herzig, S.; Shaw, R. AMPK: Guardian of metabolism and mitochondrial homeostasis. *Nat. Rev. Mol. Cell. Biol.* **2018**, *19*, 121–135. [CrossRef] [PubMed]
2. Hardie, D.G. Keeping the home fires burning: AMP-activated protein kinase. *J. R. Soc. Interface* **2018**, *15*, 20170774. [CrossRef] [PubMed]
3. Kjobsted, R. AMPK in skeletal muscle function and metabolism. *FASEB J.* **2018**, *32*, 1741–1777. [CrossRef] [PubMed]
4. Hudson, E.R.; Pan, D.A.; James, J.; Lucocq, J.M.; Hawley, S.A.; Green, K.A.; Baba, O.; Terashima, T.; Hardie, D.G. A novel domain in AMP-activated protein kinase causes glycogen storage bodies similar to those seen in hereditary cardiac arrhythmias. *Curr. Biol.* **2003**, *13*, 861–866. [CrossRef]
5. Xiao, B.; Heath, R.; Saiu, P.; Leiper, F.C.; Leone, P.; Jing, C.; Walker, P.A.; Haire, L.; Eccleston, J.F.; Davis, C.T.; et al. Structural basis for AMP binding to mammalian AMP-activated protein kinase. *Nature* **2007**, *449*, 496–500. [CrossRef] [PubMed]
6. Hardie, D.G.; Carling, D.; Gamblin, S.J. AMP-activated protein kinase: Also regulated by ADP? *Trends Biochem Sci.* **2011**, *36*, 470–477. [CrossRef] [PubMed]
7. Gowans, G.J.; Hawley, S.A.; Ross, F.A.; Hardie, D.G. AMP is a true physiological regulator of AMP-activated protein kinase by both allosteric activation and enhancing net phosphorylation. *Cell Metab.* **2013**, *18*, 556–566. [CrossRef] [PubMed]
8. Hawley, S.A.; Davison, M.; Woods, A.; Davies, S.P.; Beri, R.K.; Carling, D.; Hardie, D.G. Characterization of the AMP-activated protein kinase kinase from rat liver and identification of threonine 172 as the major site at which it phosphorylates AMP-activated protein kinase. *J. Biol. Chem.* **1996**, *271*, 27879–27887. [CrossRef] [PubMed]

9. Hawley, S.A.; Boudeau, J.; Reid, J.L.; Mustard, K.J.; Udd, L.; Mäkelä, T.P.; Alessi, D.R.; Hardie, D.G. Complexes between the LKB1 tumor suppressor, STRAD α/β and MO25 α/β are upstream kinases in the AMP-activated protein kinase cascade. *J. Biol.* **2003**, *2*, 28. [CrossRef] [PubMed]

10. Shaw, R.J.; Kosmatka, M.; Bardeesy, N.; Hurley, R.L.; Witters, L.A.; de Pinho, R.A.; Cantley, L.C. The tumor suppressor LKB1 kinase directly activates AMP-activated kinase and regulates apoptosis in response to energy stress. *Proc. Natl. Acad. Sci. USA* **2004**, *101*, 3329–3335. [CrossRef] [PubMed]

11. Woods, A.; Johnstone, S.R.; Dickerson, K.; Leiper, F.C.; Fryer, L.G.; Neumann, D.; Schlattner, U.; Wallimann, T.; Carlson, M.; Carling, D. LKB1 is the upstream kinase in the AMP-activated protein kinase cascade. *Curr. Biol.* **2003**, *13*, 2004–2008. [CrossRef] [PubMed]

12. Shaw, R.J.; Lamia, K.A.; Vasquez, D.; Koo, S.H.; Bardeesy, N.; Depinho, R.A.; Montminy, M.; Cantley, L.C. The kinase LKB1 mediates glucose homeostasis in liver and therapeutic effects of metformin. *Science* **2005**, *310*, 1642–1646. [CrossRef] [PubMed]

13. Hardie, D.G. AMPK-Sensing energy while talking to other signaling pathways. *Cell Metab.* **2014**, *20*, 939–952. [CrossRef] [PubMed]

14. Burkewitz, K.; Zhang, Y.; Mair, W.B. AMPK at the nexus of energetics and aging. *Cell Metab.* **2014**, *20*, 10–25. [CrossRef] [PubMed]

15. Shackelford, D.B.; Shaw, R.J. The LKB1-AMPK pathway: Metabolism and growth control in tumour suppression. *Nat Rev Cancer.* **2009**, *9*, 563–575. [CrossRef] [PubMed]

16. Carling, D.; Zammit, V.A.; Hardie, D.G. A common bicyclic protein kinase cascade inactivates the regulatory enzymes of fatty acid and cholesterol biosynthesis. *FEBS Lett.* **1987**, *223*, 217–222. [CrossRef]

17. Munday, M.R.; Campbell, D.G.; Carling, D.; Hardie, D.G. Identification by amino acid sequencing of three major regulatory phosphorylation sites on rat acetyl-CoA carboxylase. *Eur. J. Biochem.* **1988**, *175*, 331–338. [CrossRef] [PubMed]

18. Wu, N.; Zheng, B.; Shaywitz, A.; Dagon, Y.; Tower, C.; Bellinger, G.; Shen, C.H.; Wen, J.; Asara, J.; McGraw, T.E.; et al. AMPK-dependent degradation of TXNIP upon energy stress leads to enhanced glucose uptake via GLUT1. *Mol. Cell.* **2013**, *49*, 1167–1175. [CrossRef] [PubMed]

19. Yang, W.; Hong, Y.H.; Shen, X.Q.; Frankowski, C.; Camp, H.S.; Leff, T. Regulation of transcription by AMP-activated protein kinase: Phosphorylation of p300 blocks its interaction with nuclear receptors. *J. Biol. Chem.* **2001**, *276*, 38341–38344. [CrossRef] [PubMed]

20. Koo, S.H.; Flechner, L.; Qi, L.; Zhang, X.; Screaton, R.A.; Jeffries, S.; Hedrick, S.; Xu, W.; Boussouar, F.; Brindle, P.; et al. The CREB coactivator TORC2 is a key regulator of fasting glucose metabolism. *Nature* **2005**, *437*, 1109–1111. [CrossRef] [PubMed]

21. Li, Y.; Xu, S.; Mihaylova, M.M.; Zheng, B.; Hou, X.; Jiang, B.; Park, O.; Luo, Z.; Lefai, E.; Shyy, J.Y.; et al. AMPK phosphorylates and inhibits SREBP activity to attenuate hepatic steatosis and atherosclerosis in diet-induced insulin-resistant mice. *Cell Metab.* **2011**, *13*, 376–388. [CrossRef] [PubMed]

22. Mihaylova, M.M.; Vasquez, D.S.; Ravnskjaer, K.; Denechaud, P.D.; Yu, R.T.; Alvarez, J.G.; Downes, M.; Evans, R.M.; Montminy, M.; Shaw, R.J. Class IIa histone deacetylases are hormone-activated regulators of FOXO and mammalian glucose homeostasis. *Cell* **2011**, *145*, 607–621. [CrossRef] [PubMed]

23. Saxton, R.A.; Sabatini, D.M. mTOR Signaling in Growth, Metabolism, and Disease. *Cell* **2017**, *169*, 361–371. [CrossRef] [PubMed]

24. Inoki, K.; Zhu, T.; Guan, K.L. TSC2 mediates cellular energy response to control cell growth and survival. *Cell* **2003**, *115*, 577–590. [CrossRef]

25. Gwinn, D.M.; Shackelford, D.; Egan, D.F.; Mihaylova, M.M.; Mery, A.; Vasquez, D.S.; Turk, B.E.; Shaw, R.J. AMPK phosphorylation of raptor mediates a metabolic checkpoint. *Mol. Cell* **2008**, *30*, 214–226. [CrossRef] [PubMed]

26. Zhang, Y.L.; Guo, H.; Zhang, C.S.; Lin, S.Y.; Yin, Z.; Peng, Y.; Luo, H.; Shi, Y.; Lian, G.; Zhang, C.; et al. AMP as a low-energy charge signal autonomously initiates assembly of AXIN-AMPK-LKB1 complex for AMPK activation. *Cell Metab.* **2013**, *18*, 546–555. [CrossRef] [PubMed]

27. Zhang, C.S.; Jiang, B.; Li, M.; Zhu, M.; Peng, Y.; Zhang, Y.L.; Wu, Y.Q.; Li, T.Y.; Liang, Y.; Lu, Z.; et al. The lysosomal v-ATPase-Ragulator complex is a common activator for AMPK and mTORC1, acting as a switch between catabolism and anabolism. *Cell Metab.* **2014**, *20*, 526–540. [CrossRef] [PubMed]

28. Salminen, A.; Kauppinen, A.; Kaarniranta, K. AMPK/Snf1 signaling regulates histone acetylation: Impact on gene expression and epigenetic functions. *Cell Signal.* **2016**, *28*, 887–895. [CrossRef] [PubMed]

29. Guarente, L. The logic linking protein acetylation and metabolism. *Cell Metab.* **2011**, *14*, 151–153. [CrossRef] [PubMed]

30. Pietrocola, F.; Galluzzi, L.; Bravo-San Pedro, J.M.; Madeo, F.; Kroemer, G. Acetyl coenzyme A: A central metabolite and second messenger. *Cell Metab.* **2015**, *21*, 805–821. [CrossRef] [PubMed]

31. Kim, S.C.; Sprung, R.; Chen, Y.; Xu, Y.; Ball, H.; Pei, J.; Cheng, T.; Kho, Y.; Xiao, H.; Xiao, L.; et al. Substrate and functional diversity of lysine acetylation revealed by a proteomics survey. *Mol. Cell* **2006**, *23*, 607–618. [CrossRef] [PubMed]

32. Choudhary, C.; Kumar, C.; Nielsen, M.L.; Rehman, M.; Walther, T.C.; Olsen, J.V.; Mann, M. Lysine acetylation targets protein complexes and co-regulates major cellular functions. *Science* **2009**, *325*, 834–840. [CrossRef] [PubMed]

33. Wang, Q.; Zhang, Y.; Yang, C.; Xiong, H.; Lin, Y.; Yao, J.; Li, H.; Xie, L.; Zhao, W.; Yao, Y.; et al. Acetylation of metabolic enzymes coordinates carbon source utilization and metabolic flux. *Science* **2010**, *327*, 1004–1007. [CrossRef] [PubMed]

34. Zhao, S.; Xu, W.; Jiang, W.; Yu, W.; Lin, Y.; Zhang, T.; Yao, J.; Zhou, L.; Zeng, Y.; Li, H.; et al. Regulation of cellular metabolism by protein lysine acetylation. *Science* **2010**, *327*, 1000–1004. [CrossRef] [PubMed]

35. Olsen, J.V.; Blagoev, B.; Gnad, F.; Macek, B.; Mortensen, P.; Mann, M. Global, in vivo, and site-specific phosphorylation dynamics in signaling networks. *Cell* **2006**, *127*, 635–648. [CrossRef] [PubMed]

36. Villen, J.; Beausoleil, S.A.; Gerber, S.A.; Gygi, S.P. Large scale phosphorylation analysis of mouse liver. *Proc Natl. Acad. Sci. USA* **2007**, *104*, 1488–1493. [CrossRef] [PubMed]

37. Kim, G.W.; Yang, X.J. Comprehensive lysine acetylomes emerging from bacteria to humans. *Trens. Biochem. Sci.* **2011**, *36*, 211–220. [CrossRef] [PubMed]

38. Galdieri, L.; Zhang, T.; Rogerson, D.; Lleshi, R.; Vancura, A. Protein acetylation and acetyl coenzyme A metabolism in budding yeast. *Eukaryot. Cell* **2014**, *13*, 1472–1483. [CrossRef] [PubMed]

39. Drazic, A.; Myklebust, L.M.; Ree, R.; Arnesen, T. The world of protein acetylation. *Biochim. Biophys. Acta* **2016**, *1864*, 1372–1401. [CrossRef] [PubMed]

40. Roth, S.Y.; Denu, J.M.; Allis, C.D. Histone acetyltransferases. *Annu. Rev. Biochem.* **2001**, *70*, 81–120. [CrossRef] [PubMed]

41. Kouzarides, T. Chromatin modifications and their function. *Cell* **2007**, *128*, 693–705. [CrossRef] [PubMed]

42. Albaugh, B.N.; Arnold, K.M.; Denu, J.M. KAT(ching) metabolism by the tail: Insight into the links between lysine acetyltransferases and metabolism. *ChemBioChem* **2011**, *12*, 290–298. [CrossRef] [PubMed]

43. Farria, A.; Li, W.; Dent, S.Y.R. KATs in cancer: Function and therapies. *Oncogene* **2015**, *34*, 4901–4913. [CrossRef] [PubMed]

44. Fan, J.; Krautkramer, K.A.; Feldman, J.L.; Denu, J.M. Metabolic regulation of histone post-translational modifications. *ACS Chem. Biol.* **2015**, *10*, 95–108. [CrossRef] [PubMed]

45. Janke, R.; Dodson, A.E.; Rine, J. Metabolism and epigenetics. *Annu. Rev. Cell Dev. Biol.* **2015**, *31*, 473–496. [CrossRef] [PubMed]

46. Haberland, M.; Montgomery, R.L.; Olson, E.N. The many roles of histone deacetylases in development and physiology: Implications for disease and therapy. *Nat. Rev. Genet.* **2009**, *10*, 32–42. [CrossRef] [PubMed]

47. Chalkiadaki, A.; Guarente, L. The multifaceted functions of sirtuins in cancer. *Nat. Rev. Cancer* **2015**, *15*, 608–624. [CrossRef] [PubMed]

48. Wagner, G.R.; Hirschey, M.D. Nonezymatic protein acylation as a carbon stress regulated by sirtuin deacylases. *Mol. Cell* **2014**, *54*, 5–16. [CrossRef] [PubMed]

49. Weinert, B.T.; Iesmantavicius, V.; Moustafa, T.; Scholz, C.; Wagner, S.A.; Magnes, C.; Zechner, R.; Choudhary, C. Acetylation dynamics and stoichiometry in *Saccharomyces cerevisiae*. *Mol. Systems Biol.* **2014**, *10*, 716. [CrossRef] [PubMed]

50. Wagner, G.R.; Payne, R.M. Widespread and enzyme-independent *N*-acetylation and *N*-succinylation of proteins in the chemical conditions of the mitochondrial matrix. *J. Biol. Chem.* **2013**, *288*, 29036–29045. [CrossRef] [PubMed]

51. Henriksen, P.; Wagner, S.A.; Weinert, B.T.; Sharma, S.; Bacinskaja, G.; Rehman, M.; Juffer, A.H.; Walther, T.C.; Lisby, M.; Choudhary, C. Proteome-wide analysis of lysine acetylation suggests its broad regulatory scope in Saccharomyces cerevisiae. *Mol. Cell. Proteomics* **2012**, *11*, 1510–1522. [CrossRef] [PubMed]

52. Uchida, M.; Sun, Y.; Mcdermott, G.; Knoechel, C.; le Gros, M.A.; Parkinson, D.; Drubin, D.G.; Larabell, C.A. Quantitative analysis of yeast internal architecture using soft X-ray tomography. *Yeast* **2011**, *28*, 227–236. [CrossRef] [PubMed]

53. Cai, L.; Tu, B.P. Acetyl-CoA drives the transcriptional growth program in yeast. *Cell Cycle* **2011**, *10*, 3045–3046. [CrossRef] [PubMed]

54. Cai, L.; Sutter, B.M.; Li, B.; Tu, B.P. Acetyl-CoA induces cell growth and proliferation by promoting the acetylation of histones at growth genes. *Mol. Cell* **2011**, *42*, 426–437. [CrossRef] [PubMed]

55. Casey, J.R.; Grinstein, S.; Orlowski, J. Sensors and regulators of intracellular pH. *Nat. Rev. Mol. Cell Biol.* **2010**, *11*, 50–61. [CrossRef] [PubMed]

56. Wagner, G.R.; Bhatt, D.P.; O'Connell, T.M.; Thompson, J.W.; Dubois, L.G.; Backos, D.S.; Yang, H.; Mitchell, G.A.; Ilkayeva, O.R.; Stevens, R.D.; et al. A Class of Reactive Acyl-CoA Species Reveals the Non-enzymatic Origins of Protein Acylation. *Cell Metab.* **2017**, *25*, 823–837. [CrossRef] [PubMed]

57. Cai, L.; Tu, B.P. On acetyl-CoA as a gauge of cellular metabolic state. *Cold Spring Harbor Symp. Quant. Biol.* **2011**, *76*, 195–202. [CrossRef] [PubMed]

58. Shi, L.; Tu, B.P. Acetyl-CoA and the regulation of metabolism: Mechanisms and consequences. *Curr. Opin. Cell Biol.* **2015**, *33*, 125–131. [CrossRef] [PubMed]

59. Sivanand, S.; Viney, I.; Wellen, K.E. Spatiotemporal Control of Acetyl-CoA Metabolism in Chromatin Regulation. *Trends Biochem Sci.* **2018**, *43*, 61–74. [CrossRef] [PubMed]

60. Takahashi, H.; MacCaffery, J.M.; Irizarry, R.A.; Boeke, J.D. Nucleocytosolic acetyl-coenzyme A synthetase is required for histone acetylation and global transcription. *Mol. Cell* **2006**, *23*, 207–217. [CrossRef] [PubMed]

61. Wellen, K.E.; Hatzivassiliou, G.; Sachdeva, U.M.; Bui, T.V.; Cross, J.R.; Thompson, C.B. ATP-citrate lyase links cellular metabolism to histone acetylation. *Science* **2009**, *324*, 1076–1080. [CrossRef] [PubMed]

62. Sutendra, G.; Kinnaird, A.; Dromparis, P.; Paulin, R.; Stenson, T.H.; Haromy, A.; Hashimoto, K.; Zhang, N.; Flaim, E.; Michelakis, E.D. A nuclear pyruvate dehydrogenase complex is important for the generation of acetyl-CoA and histone acetylation. *Cell* **2014**, *158*, 84–97. [CrossRef] [PubMed]

63. Liu, X.; Cooper, D.E.; Cluntun, A.A.; Warmoes, M.O.; Zhao, S.; Reid, M.A.; Liu, J.; Lund, P.J.; Lopes, M.; Garcia, B.A.; et al. Acetate production from glucose and coupling to mitochondrial metabolism in mammals. *Cell* **2018**, *175*, 502–513. [CrossRef] [PubMed]

64. Galdieri, L.; Vancura, A. Acetyl-CoA carboxylase regulates global histone acetylation. *J. Biol. Chem.* **2012**, *28*, 23865–23876. [CrossRef] [PubMed]

65. Galdieri, L.; Gatla, H.; Vancurova, I.; Vancura, A. Activation of AMP-activated protein kinase by metformin induces protein acetylation in prostate and ovarian cancer cells. *J. Biol. Chem.* **2016**, *291*, 25154–25166. [CrossRef] [PubMed]

66. Vancura, A.; Vancurova, I. Metformin induces protein acetylation in cancer cells. *Oncotarget* **2017**, *8*, 39939–39940. [CrossRef] [PubMed]

67. Hardie, D.G.; Carling, D.; Carlson, M. The AMP-activated/SNF1 protein kinase subfamily: Metabolic sensors of the eukaryotic cell? *Annu. Rev. Biochem.* **1998**, *67*, 821–855. [CrossRef] [PubMed]

68. Hardie, D.G. AMP-activated/SNF1 protein kinases: Conserved guardians of cellular energy. *Nat. Rev. Mol. Cell Biol.* **2007**, *8*, 774–785. [CrossRef] [PubMed]

69. Hedbacker, K.; Carlson, M. SNF1/AMPK pathways in yeast. *Front. Biosci.* **2009**, *13*, 2408–2420. [CrossRef]

70. Zhang, M.; Galdieri, L.; Vancura, A. The yeast AMPK homolog SNF1 regulates acetyl coenzyme A homeostasis and histone acetylation. *Mol. Cell. Biol.* **2013**, *33*, 4701–4717. [CrossRef] [PubMed]

71. Haigis, M.C.; Guarente, L.P. Mammalian sirtuins-emerging roles in physipology, aging, and calorie restriction. *Genes Dev.* **2006**, *20*, 2913–2921. [CrossRef] [PubMed]

72. Houtkooper, R.H.; Pirinen, E.; Auwerx, J. Sirtuins as regulators of metabolism and healthspan. *Nat. Rev. Mol. Cell Biol.* **2012**, *13*, 225–238. [CrossRef] [PubMed]

73. Fulco, M.; Ce, Y.; Zhao, P.; Hoffman, E.P.; McBurney, M.W.; Sauve, A.A.; DSartorelli, V. Glucose restriction inhibits skeletal myoblast differentiation by activating SIRT1 through AMPK-mediated regulation of Nampt. *Dev. Cell* **2008**, *14*, 661–673. [CrossRef] [PubMed]

74. Canto, C.; Gerhart-Hines, Z.; Feige, J.N.; Lagouge, M.; Noriega, L.; Milne, J.C.; Elliot, P.J.; Puigserver, P.; Auwerx, J. AMPK regulates energy expenditure by modulating NAD$^+$ metabolism and SIRT1 activity. *Nature* **2009**, *458*, 1056–1060. [CrossRef] [PubMed]

75. Lau, A.W.; Liu, P.; Inuzuka, H.; Gao, D. SIRT1 phosphorylation by AMP-activated protein kinase regulates p53 acetylation. *Am. J. Cancer Res.* **2014**, *4*, 245–255. [PubMed]

76. Chang, C.; Su, H.; Zhang, D.; Wang, Y.; Shen, Q.; Liu, B.; Huang, R.; Zhou, T.; Peng, C.; Wong, C.C.; et al. AMPK-dependent phosphorylation of GAPDH triggers Sirt1 activation and is necessary for autophagy upon glucose starvation. *Mol. Cell* **2015**, *60*, 930–940. [CrossRef] [PubMed]

77. Shimazu, T.; Hirschey, M.D.; Newman, J.; He, W.; Shirakawa, K.; Le Moan, N.; Grueter, C.A.; Lim, H.; Saunders, L.R.; Stevens, R.D.; et al. Suppression of oxidative stress by -hydroxybutyrate, an endogenous histone deacetylase inhibitor. *Science* **2013**, *339*, 211–214. [CrossRef] [PubMed]

78. Verdin, E.; Dequiedt, F.; Kasler, H.G. Class II histone deacetylase: Versatile regulators. *Trends Genet.* **2003**, *19*, 286–293. [CrossRef]

79. Yang, X.J.; Gregoire, S. Class II histone deacetylases: From sequence to function, regulation, and clinical implication. *Mol. Cell. Biol.* **2005**, *25*, 2873–2884. [CrossRef] [PubMed]

80. Martin, M.; Kettmann, R.; Dequiedt, F. Class IIa histone deacetylases: Regulating the regulators. *Oncogene* **2007**, *26*, 5450–5467. [CrossRef] [PubMed]

81. Parra, M. Class IIa HDACs-new insights into their function in physiology and pathology. *FEBS J.* **2015**, *282*, 1736–1744. [CrossRef] [PubMed]

82. McGee, S.L.; van Denderen, B.J.; Howlett, K.F.; Mollica, J.; Schertzer, J.D.; Kemp, B.E.; Hargreaves, M. AMP-actoivated protein kinase regulates GLUT4 transcription by phosphorylating histone deacetylase 5. *Diabetes* **2008**, *57*, 860–867. [CrossRef] [PubMed]

83. Zhao, J.X.; Yue, W.F.; Zhu, M.J.; Du, M. AMP-activated protein kinase regulates -catenin transcription via histone deacetylase 5. *J. Biol. Chem.* **2011**, *286*, 16426–16434. [CrossRef] [PubMed]

84. Chen, S.; Yin, C.; Lao, T.; Liang, D.; He, D.; Wang, C.; Sang, N. AMPK-HDAC5 pathway facilitates nuclear accumulation of HIF-1 and functional activation of HIF-1 by deacetylating HSP70 in the cytosol. *Cell Cycle* **2015**, *14*, 2520–2536. [CrossRef] [PubMed]

85. Fu, X.; Zhao, J.X.; Liang, J.; Zhu, M.J.; Foretz, M.; Viollet, B.; Du, M. AMP-activated protein kinase mediates myogenin expression and myogenesis via histone deacetylase 5. *Am. J. Phys. Cell Physiol.* **2013**, *305*, C887–C895. [CrossRef] [PubMed]

86. Fu, X.; Zhao, J.X.; Zhu, M.J.; Foretz, M.; Viollet, B.; Dodson, M.V.; Du, M. AMP-activated protein kinase a1 but not a2 catalytic subunit potentiates myogenin expression and myogenesis. *Mol. Cell. Biol.* **2013**, *33*, 4517–4525. [CrossRef] [PubMed]

87. Zhang, Y.; Qiu, J.; Wang, X.; Zhang, Y.; Xia, M. AMP-activated protein kinase suppresses endothelial cell inflammation through phosphorylation of transcriptional coactivator p300. *Arterioscler. Thromb. Vasc. Biol.* **2011**, *31*, 2897–2908. [CrossRef] [PubMed]

88. Lim, J.Y.; Oh, M.A.; Kim, W.H.; Sohn, H.Y.; Park, S.I. AMP-activated protein kinase inhibits TGF-B-induced fibrogenic responses of hepatic stellate cells by targeting transcriptional coactivator p300. *J. Cell. Physiol.* **2012**, *227*, 1081–1089. [CrossRef] [PubMed]

89. Bungard, D.; Fuerth, B.J.; Zeng, P.Y.; Faubert, B.; Maas, N.L.; Viollet, B.; Carling, D.; Thompson, C.B.; Jones, R.G.; Berger, S.L. Signaling kinase AMPK activates stress-promoted transcription via histone H2B phosphorylation. *Science* **2010**, *329*, 1201–1205. [CrossRef] [PubMed]

90. Lo, W.S.; Duggan, L.; Emree, N.C.; Belotserkovskaya, R.; Lane, W.S.; Shiekhattar, R.; Berger, S.L. Snf1-a histone kinase that works in concert with the histone acetyltransferase Gcn5 to regulate transcription. *Science* **2001**, *293*, 1142–1146. [CrossRef] [PubMed]

91. Lan, F.; Cacicedo, J.M.; Ruderman, N.; Ido, Y. SIRT1 modulation of the acetylation status, cytosolic localization, and activity of LKB1. Possible role in AMP-activated protein kinase activation. *J. Biol. Chem.* **2008**, *283*, 27628–27635. [CrossRef] [PubMed]

92. Hou, X.; Xu, S.; Maitland-Toolan, K.A.; Sato, K.; Jiang, B.; Ido, Y.; Lan, F.; Walsh, K.; Wierzbicki, M.; Verbeuren, T.J.; et al. SIRT1 regulates hepatocyte lipid metabolism through activating AMP-activated protein kinase. *J. Biol. Chem.* **2008**, *283*, 20015–20026. [CrossRef] [PubMed]

93.   Jiang, R.; Carlson, M. The Snf1 protein kinase and its activating subunit, Snf4, interact with distinct domains of the Sip1/Sip2/Gal83 component in the kinase complex. *Mol. Cell. Biol.* **1997**, *17*, 2099–2106. [CrossRef] [PubMed]

94.   Lu, J.-Y.; Lin, Y.-Y.; Sheu, J.-C.; Wu, J.-T.; Lee, F.-J.; Chen, Y.; Lin, M.-I.; Chiang, F.-T.; Tai, T.-Y.; Berger, S.L.; et al. Acetylation of yeast AMPK controls intrinsic aging idependently of caloric restriction. *Cell* **2011**, *146*, 969–979. [CrossRef] [PubMed]

International Journal of
Molecular Sciences

MDPI

Article

Co-Expression Network Analysis of AMPK and Autophagy Gene Products during Adipocyte Differentiation

Mahmoud Ahmed [1], Jin Seok Hwang [1], Trang Huyen Lai [1], Sahib Zada [1], Huynh Quoc Nguyen [1], Trang Min Pham [1], Miyong Yun [2] and Deok Ryong Kim [1,*]

[1] Department of Biochemistry and Convergence Medical Sciences and Institute of Health Sciences,
     Gyeongsang National University School of Medicine, Jinju 527-27, Korea;
     ma7moud_sha3ban@hotmail.com (M.A.); cloud8104@naver.com (J.S.H.);
     tranghuyen20493@gmail.com (T.H.L.); s.zada.qau@gmail.com (S.Z.); nqh2412@gmail.com (H.Q.N.);
     phamminhtrang010895@gmail.com (T.M.P.)
[2] Department of Bioindustry and Bioresource Engineering, College of Life Sciences, Sejong University,
     Seoul 05006, Korea; myyun91@gmail.com
*   Correspondence: drkim@gnu.ac.kr; Tel.: +82-10-4190-7190

Received: 12 April 2018; Accepted: 12 June 2018; Published: 19 June 2018

Abstract: Autophagy is involved in the development and differentiation of many cell types. It is essential for the pre-adipocytes to respond to the differentiation stimuli and may contribute to reorganizing the intracellulum to adapt the morphological and metabolic demands. Although AMPK, an energy sensor, has been associated with autophagy in several cellular processes, how it connects to autophagy during the adipocyte differentiation remains to be investigated. Here, we studied the interaction between AMPK and autophagy gene products at the mRNA level during adipocyte differentiation using public-access datasets. We used the weighted-gene co-expression analysis to detect and validate multiple interconnected modules of co-expressed genes in a dataset of MDI-induced 3T3-L1 pre-adipocytes. These modules were found to be highly correlated with the differentiation course of the adipocytes. Several novel interactions between AMPK and autophagy gene products were identified. Together, it is possible that AMPK-autophagy interaction is temporally and locally modulated in response to the differentiation stimuli.

Keywords: AMPK; autophagy; co-expression; microarrays; 3T3-L1; adipocyte; differentiation

---

1. Introduction

Autophagy is essential for the white adipocyte differentiation. The knockdown of *Atg5* and/or *Atg7* gene in the 3T3-L1 pre-adipocyte prevents its maturation upon the chemical induction [1,2]. Secondary to that, the pre-adipocyte fails to accumulate triglycerides and to form the fat droplets, which is a characteristic of mature white adipocytes [3]. This is yet to be reconciled with another observation in cells known to contain large quantities of lipids (e.g., hepatocytes) where autophagy takes part in lipid degradation [4]. AMP-activated protein kinase (AMPK), which can be activated at the low level of energy such as starvation, stimulates autophagy through the inhibition of the mTOR activity [5,6] and/or the direct phosphorylation of ULK1 [7,8]. Autophagy and AMPK regulate several aspects of the lipid metabolism and the cell response to changing energy levels. Therefore, the AMPK-autophagy interaction could be consequential in the context of adipocyte differentiation.

3T3-L1 pre-adipocyte is a mouse fibroblast with the potential to differentiate into a mature adipocyte when treated with the MDI differentiation induction medium (160 nM insulin, 250 nM dexamethasone, and 0.5 mM 1-methyl-3-isobutylxanthine) [9]. Upon induction, the pre-adipocyte undergoes multiple metabolic and morphological changes to reach maturation. Evidently, several of these changes can be observed at the transcription level of multiple adipogenic and lipogenic markers [10,11].

The aim of this work is to identify the potential AMPK-autophagy connections, in the broad sense of the pathways, that are both novel and consequential in the adipocyte differentiation, mainly using the weighted-gene co-expression network analysis (WGCNA) [12]. This approach focuses on identifying co-expressed pairs of genes across the differentiation stages, and enables the direct use of similar datasets to test and validate the findings even though they might be performed on different platforms.

In this study, we applied the WGCNA approach to a microarrays dataset of MDI-induced adipocyte at eight different time points corresponding to three differentiation stages. We identified two networks/modules among autophagy and AMPK gene products that were correlated with the differentiation course. By analyzing these modules in one of the datasets, we were able to specify several potential novel AMPK-autophagy interactions and connect these to candidate functions through the known annotations. Finally, we checked these findings in three independent datasets of similar design and found the networks/modules to be well preserved.

2. Results

2.1. Preparing Data and Annotations

First, we retrieved several microarray datasets from the Gene expression omnibus (GEO) [13]. We sought arrays of MDI-induced 3T3-L1 pre-adipocytes at three or more time points. After excluding the ones with varying designs and limited annotations, four datasets were included in the analysis (Table 1); one dataset (GSE34150) was chosen for the main analysis, and the rest were reserved for testing and validation. In GSE34150, the total RNA from 24 samples of MDI-induced pre-adipocytes were collected at eight different time points corresponding to three adipocyte differentiation stages (0 day, undifferentiated; two and four days, differentiating; 6–18 days, maturing). The initial quality assessment included checking the distribution of the intensities from all probes at the log scale, hierarchal clustering and multi-dimensional scaling analysis (MDS). Groups of samples from different stage of differentiation showed similar distributions, appropriate clustering and separation across the two different dimensions of MDS. Furthermore, to ensure the reliability of the analysis, we examined the expression of a number of differentiation and lipogenesis markers (Appendix A). The *Pparg*, *Cebpa* and *Lpl* genes, essential factors for adipocyte differentiation, were highly expressed in differentiating and maturing cells compared to the undifferentiated cells. Expression of most lipogenic genes (*Pparg*, *Cebpa*, *Lpl*, *Scd1*, *Scd2*, *Dgat1*, *Dgat2* and *Fasn*) was correlated with the development of 3T3-L1 pre-adipocytes into mature adipocytes.

Table 1. MDI-induced 3T3-L1 microarrays' datasets.

Series ID	Platform ID	Samples	Included	(Contact, Year)	Reference
GSE15018	GPL6845	54	18	(Chin, 2009)	[14]
GSE20696	GPL1261	8	8	(Mikkelsen, 2010)	[15]
GSE34150	GPL6885	24	24	(Irmler, 2011)	[16]
GSE69313	GPL6246	48	12	(Renbin, 2015)	[17]

The gene ontology (GO) terms: AMP-activated protein kinase activity (AMPK) and autophagy were used to identify 14 and 167 genes of known functions in the corresponding biological processes (BP), respectively. A total of 181 genes was used in the downstream analysis to limit the input to WGCNA, over-representation and defining novel interactions between and among AMPK and autophagy pathways. GO was also used to define the terms involving these interacting gene products and link them to known molecular functions (MF) and cellular components (CC). Appendix A contains a detailed discussion for the data inclusion criteria, quality assessment and obtaining the GO annotations.

2.2. Detecting Co-Expression Modules of AMPK and Autophagy Genes in Differentiating Adipocytes

Constructing co-expression networks is a multi-step process. First, the Pearson's correlation coefficient was calculated between each pair of the genes of interest ($n = 181$) across all samples ($n = 24$). Second, these correlations were raised to the power 5 to obtain an adjacency matrix of all possible pairs. Third, the adjacency matrix was used to calculate the Topological Overlap Matrix (TOM) as a reliable similarity measure. Finally, TOM similarity between pairs of genes were used to calculate the weight of their connection in a network of all possible pairs and a distance (1 - TOM) to cluster the pairs into highly interconnected modules/colors (Figure 1). Appendix B provides a detailed discussion of the previous steps and the rationale for the different choices that were made in this analysis.

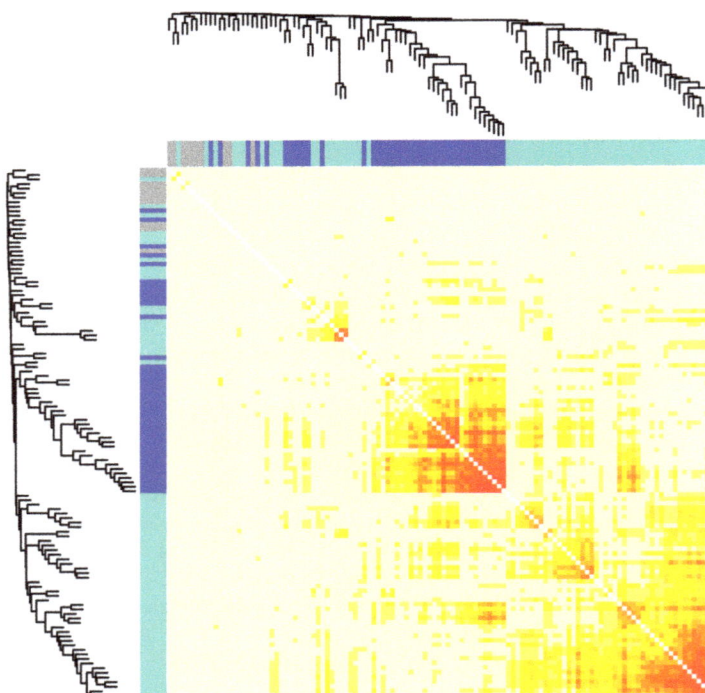

Figure 1. Clustering of AMPK and autophagy genes by their pairwise distances. Pairwise topological overlap matrix (TOM) similarities of AMPK and autophagy genes ($n = 181$) were calculated from their expression values in the GSE34150 dataset. Distances between each pair of genes were derived as 1 - TOM and shown as color values (small, red or large, yellow). A hierarchal tree and colored segments of the clusters were shown on the top and side.

Among all possible pairwise correlations between the 181 genes of interest, two groups/modules of highly co-expressed gene products were formed (blue, 42; turquoise, 66), and the rest were unassigned (gray, 10). Genes that code for the subunits of the AMPK complex fell into different modules; four AMPK genes, *Prkaa2, Prkab2, Prkag2* and *Prkag3*, in the blue module along with 38 of the genes involved in autophagy; and two AMPK genes, *Prkab2 and Prkag1*, in the turquoise module together with 63 of the autophagy genes (Table 2). In the following sections, we describe the significance and the interactions of the individual members of these modules.

Table 2. AMPK and autophagy genes in different modules/colors.

Module/Color	AMPK	Autophagy
blue	*Prkaa2, Prkab1, Prkag2, Prkag3, Smok3b*	*Acbd5, Atg5, Atm, Bmf, Bok, Casp1, Cln3, Dapk1, Dhrsx, Fbxl2, Fbxo7, Fis1, Hif1a, Lep, Map1lc3a, Mcl1, Mid2, Optn, Pik3c2a, Pink1, Prkaa2, Rab39b, Rab8a, Rragc, Sh3bp4, Sh3glb1, Sqstm1, Tbc1d5, Tnfaip3, Tpcn1, Tpcn2, Trim8, Trp53inp2, Vcp, Wdr45, Yod1, Zc3h12a, Zfyve1*
turquoise	*4921509C19Rik, Prkab2, Prkag1*	*Ager, Akt1, Bcl2, Becn1, Capn10, Cdkn2a, D17Wsu92e, Dap, Dcn, Depdc5, Ei24, Eif4g1, Eif4g2, Foxo1, Fundc1, Fundc2, Hmgb1, Hspa8, Htr2b, Ifng, Lamp2, Lars, Lmx1b, Lrrk2, Map1lc3b, Map2k1, Mapt, Mt3, Nbr1, Nlrp6, Pik3c3, Pik3r2, Pik3r4, Pim2, Plaa, Plekhf1, Plk2, Pycard, Rasip1, Rnf5, Rraga, Rragb, Sirt2, Smcr8, Smurf1, Stk11, Tcirg1, Tmem74, Trim21, Trp53inp1, Tsc2, Ubqln1, Ulk1, Usp10, Usp13, Usp30, Usp33, Vps4a, Vps4b, Wdr6, Wipi1, Wipi2, Xbp1*

2.3. Correlating the Detected Modules to the Stage of Differentiation

To establish the biological significance of these modules, we used the expression values of their individual members to calculate a representative summary—the first principal component (PC)—for each module. Then, we calculated the Pearson's correlation coefficient for the first PC with the stage of differentiation (undifferentiated, differentiating or maturing) of all 24 samples. Both modules showed a reasonable correlation with sample stages (>0.8 for the blue and >0.3 for the turquoise module) (Figure 2A). In other words, the expression values of the members of the blue module, and to a less extent the turquoise, capture a lot of the observed differences between the cells as they progress from a differentiation stage to the next.

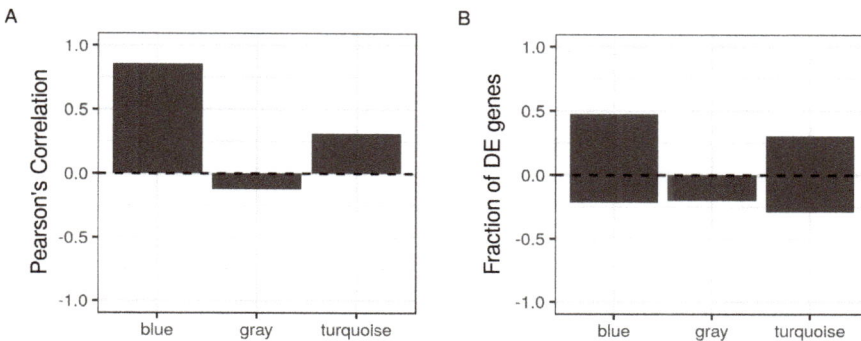

Figure 2. Correlations and over-representation of the detected modules in differentiation stages. The expression values of the members of the detected modules in the GSE34150 dataset (42, blue; 10, gray; and 66, turquoise) were used to calculate two representative summary statistics. (**A**) the first principal component (PC) across samples were correlated to the sample stages using Pearson's correlation (bars); (**B**) the fraction of differentially expressed (DE) genes across differentiation stages (bars).

2.4. Testing the Over-Representation of the Modules over the Differentiation Course

Again, we considered the expression values of the individual members of each module to calculate the fraction of the differentially expressed genes (DE) across differentiation stages. Both blue and turquoise modules had a significant fraction of their member genes (>0.5) either up or downregulated at the differentiating or maturing stage compared to the control undifferentiated cell stage (Figure 2B). These fractions were significantly higher than the expected fractions of DE genes in randomly selected modules of the corresponding sizes. The calculated p-values were adjusted for multiple testing using the False Discovery Rate (FDR). Adjusted p-values less than 0.1 were considered significant.

2.5. Visualizing Modules and Identifying Novel AMPK-Autophagy Interactions

To visually explore the detected modules, we treated each of their members as a node in a network graph. Nodes were divided into two networks based on the module to which they belong. Each pair of nodes was connected by an edge that has a weight calculated from the TOM similarity measure between the corresponding pair of genes. Edges with weights less than a minimum threshold (0.1) were excluded to obtain a less condense network (Figure 3). Evidently, some nodes did not share edges that passes this threshold and were not included in the network graph. In addition, nodes were labeled with the corresponding official gene symbol and colored as AMPK or autophagy genes; and the edges were colored by the novelty of the connection. The latter was determined mainly based on previous reports in the STRING database (textmining evidence), which is evidence extracted from abstracts of scientific literature.

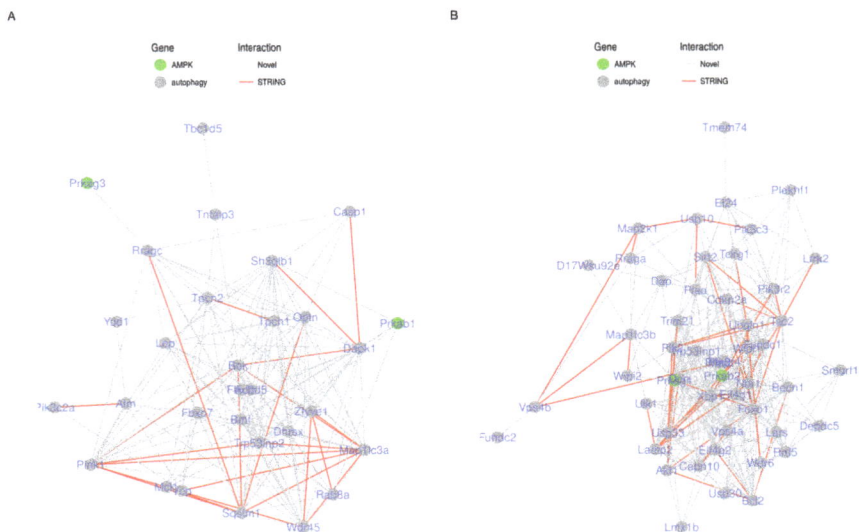

Figure 3. Network representation of the AMPK and autophagy modules. Members of the blue (**A**) and turquoise (**B**) modules are shown as a nodes. Each pair of nodes is connected by an edge if the corresponding pairwise topological overlap matrix (TOM) similarity/weight is above the threshold 0.1. Nodes are colored by gene category (AMPK, green or autophagy, gray). Edges are colored by type of interaction (STRING, red or Novel, gray).

By representing the modules in graphs, we were able to calculate different statistics to identify influential genes/nodes and important interactions/edges. Considering various centrality measures, we ranked the genes in each module by their influence on the modules (Table 3). *Trp53inp2, Map1lc3a, Wadr45, Pink1* and *Dapk1* genes were the most influential nodes in the graph of the blue module with a hub score more than 0.95, while *Foxo1, Dcn* and *Xbp1* genes had the highest scores in the turquoise module. Edges between AMPK and autophagy nodes that were not previously reported in the STRING database (text-mining evidence) were considered novel potential interactions (Table 4). The protein kinase AMP-activated non-catalytic subunit beta 1 (Prkab1) showed a potential interaction with several autophagy gene products including Bcl2-modifying factor (Bmf), death associated protein kinase 1 (Dapk1), Ras-associated protein Rab8a (Rab8a), SH3 domain, GRB2-like, endophilin B1 (Sh3glb1) and transformation related protein 53 inducible nuclear protein 2 (Trp53inp2) as part of the blue module. Similarly, the gamma subunit 1 (Prkag1) in the turquoise module revealed a novel binding ability to some well-known autophagy-related gene products such as Becn1, Fundc1, Lamp2 or Map1lc3b and also showed a novel interaction with some other gene products including calpain-like cysteine protease (Capn10), cyclin-dependent kinase inhibitor 2a (Cdkn2a) for p16INK4a and p14ARF, and ubiquitin-associated proteins (Trp53inp1, Nbr1, Usp33). In addition, there were some novel interactions between AMPK and autophagy gene products across the two modules, indicated as "inbetween" in Table 4.

Table 3. Top five hubs in the different module networks.

Module/Color	Gene	Degree	Betweenness	Closeness	Hub Score
blue	Trp53inp2	19	14.67	0.16	1
	Map1lc3a	19	22.71	0.16	0.99
	Wdr45	18	11.23	0.16	0.98
	Pink1	18	12.91	0.16	0.97
	Dapk1	18	14.48	0.16	0.96
turquoise	Foxo1	28	70.89	0.28	1
	Dcn	25	36.23	0.28	0.96
	Xbp1	24	29.59	0.28	0.93
	Plk2	24	60.28	0.27	0.9
	Eif4g1	24	41.01	0.27	0.9

Table 4. Summary of reported and novel AMPK-autophagy interactions.

Module/Color	AMPK	Autophagy
blue	Prkab1 Prkag3	Bmf, Dapk1, Rab8a, Sh3glb1, Trp53inp2 Rragc [4]
inbetween	Prkab1 Prkab2 Prkag3	Tsc2 [2,4], Ubqln1, Wipi1 Zc3h12a Usp33
turquoise	Prkag1	Akt1 [1,3,4], Bcl2, Becn1, Capn10, Cdkn2a, Dcn [3,4], Eif4g1, Foxo1 [2], Fundc1, Lamp2, Lars, Map1lc3b, Nbr1, Plk2, Sirt2 [3,4], Trim21, Trp53inp1, Usp33, Vps4a, Wipi2 [3,4], Xbp1

[1] Coexpression in the same or in other species (transferred by homology). [2] Database gathered from curated databases. [3] Experiments gathered from other protein–protein interaction databases. [4] Textmining extracted from the abstracts of scientific literature.

2.6. Testing for Molecular Functions and Cellular Components Enrichment by the Detected Modules

We used a list-based enrichment to specify the contributions of the modules to the differentiation process. The mouse gene ontology Molecular Function (MF) and Cellular Components (CC) terms were submitted to an enrichment analysis by the gene members of the detected modules (42 for blue and 66 for turquoise). The significant terms (FDR < 0.1) are shown in Figure 4 stratified by the category (MF/CC) and the module (blue/turquoise). As expected, the two modules share a number of MF terms, namely; ubiquitin-like protein binding, ubiquitinyl hydrolase activity and phospholipid binding. At the same time, several terms had significance by mutually exclusive enrichment by the modules. This includes the nucleoside binding and the ubiquitin-like protein transferase activity by the blue module; and a few protein kinase terms by the turquoise module. Similarly, two of the CC terms; extrinsic component of membrane and outer membrane were enriched by both modules, while others related to only one of the two modules.

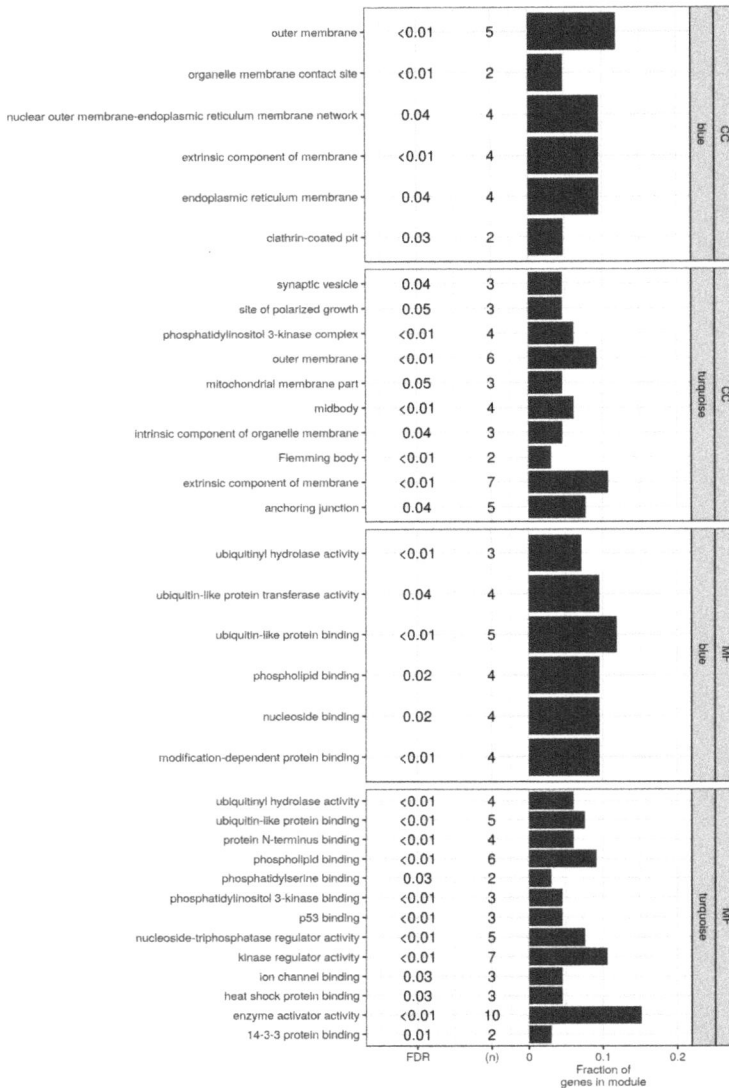

Figure 4. Enrichment of the gene ontology terms by the detected modules. The list of genes in the two detected modules (42, blue and 66, turquoise) were used to test for gene ontology terms enrichment. All terms in the Molecular Function (MF) and the Cellular component (CC) categories of the gene ontology were considered. Only significant terms at a false discovery rate (FDR) less than 0.1 are shown. For each term, the count (*n*) and the fractions of hits (bars) in the module are shown.

Building on these two pieces of the analysis, the suggested novel AMPK interactions and the gene ontology enrichment by the members of the modules, we set out to specify the kind of functions that the AMPK-autophagy interactions were likely to be involved in. Table 5 shows how AMPK is functionally connected to autophagy gene products through several gene ontology terms such as membrane components, and regulation of kinases, enzymatic and ubiquitin activity. For examples, Prkab1 interacts with Sh3glb1, a Bax-interacting protein at the mitochondrial outer membrane and also with Trp53Inp2 for the uniquitin-like activity. In addition, Prkag1 associates with many autophagy gene products such as Usp33 for the uniquitin activity at the membrane anchoring junction, Cdkn2a for the regulation of kinase activity, and other cellular components for their molecular functions.

Table 5. AMPK and autophagy interactions by gene ontology term.

Module/Color	Ontology	AMPK	Term	Autophgy
	CC	*Prkab1*	outer membrane	*Sh3glb1*
blue	MF		nucleoside binding	*Dapk1, Rab8a*
			ubiquitin-like protein binding	*Trp53inp2*
		Prkag3	nucleoside binding	*Rragc*
	CC	*Prkag1*	anchoring junction	*Usp33*
			extrinsic component of membrane	*Becn1, Wipi2*
			Flemming body	*Vps4a*
			intrinsic component of organelle membrane	*Fundc1, Lamp2*
			midbody	*Sirt2, Vps4a*
			mitochondrial membrane part	*Fundc1*
			outer membrane	*Bcl2, Capn10, Fundc1*
			phosphatidylinositol 3-kinase complex	*Becn1*
turquoise	MF		14-3-3 protein binding	*Akt1*
			enzyme activator activity	*Lars*
			kinase regulator activity	*Cdkn2a, Dcn*
			nucleoside-triphosphatase regulator activity	*Lars*
			p53 binding	*Cdkn2a*
			phosphatidylinositol 3-kinase binding	*Becn1, Xbp1*
			phospholipid binding	*Akt1, Wipi2*
			protein N-terminus binding	*Cdkn2a, Dcn*
			ubiquitin-like protein binding	*Nbr1, Sirt2*
			ubiquitinyl hydrolase activity	*Usp33*

2.7. Preservation of AMPK-Autophagy Networks across Independent Datasets

Finally, we validated these findings in three independent datasets of similar MDI-induced 3T3-L1 cells at different time points or differentiation stages. Three GEO microarray datasets (GSE15018, GSE20696 and GSE69313) were used to perform this step of the analysis (Table 1). The average log expression of the 181 genes of interest from the three datasets were first compared to these in the main dataset (Figure 5). As expected, the averages are highly correlated between the datasets (>0.74), a pre-requisite for the following module preservation analysis. A moderate to high preservation of the modules was observed in the three independent datasets (Figure 6). Generally, modules with a Z summary values between 5 and 10 are considered moderately preserved and these above 10 are considered highly preserved. In fact, the two modules: blue and turquoise, showed a summary statistics in that first category with at least 6 and 7, respectively, indicating that the interaction modules of the main dataset are well preserved in other independent datasets.

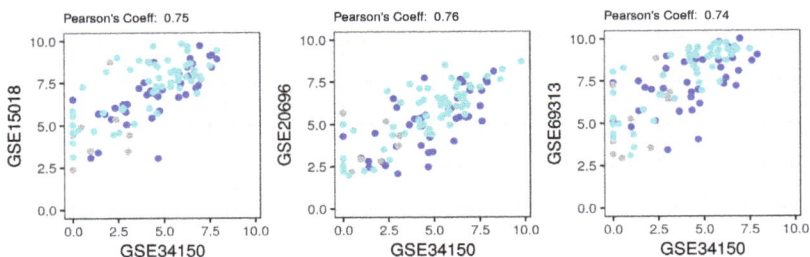

Figure 5. Average expression of AMPK and autophagy in multiple MDI-induced 3T3-L1 microarrays datasets. The log average expression values of AMPK and autophagy genes (n = 181) in the MDI-induced 3T3-L1 datasets (GSE15018, GSE20696 and GSE69313) are compared to the corresponding averages in the main dataset (GSE34150). Individual values are shown as colored points by their assigned modules. The Pearson's correlation coefficient of the corresponding values is shown on top.

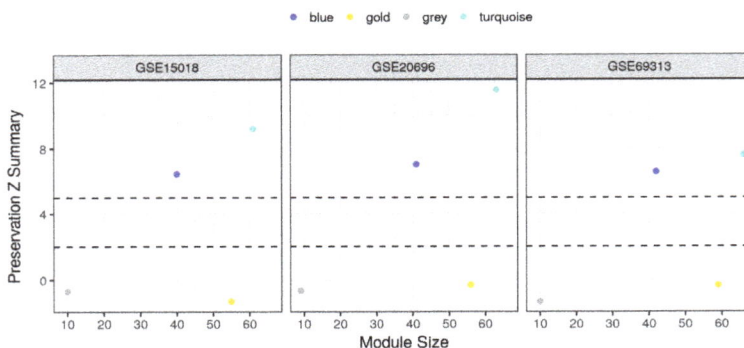

Figure 6. Module preservation Z summary across multiple MDI-induced 3T3-L1 microarrays datasets. The GSE34150 dataset was used to detect the highly co-expressed modules among AMPK and autophagy genes (42, blue; 66, turquoise; 10, gray, unassigned; and 55, gold , randomly assigned). The detected modules were used as a reference to calculate several preservation statistics in three independent datasets of similar design (GSE15018, GSE20696 and GSE69313). Z summary statistics and sizes of four modules are shown as colored points.

2.8. Validation of Selected Gene Products Correlations with Prkab1 and Prkag1

We selected several autophagy gene products that are highly correlated with the AMPK subunits for experimental validation using RT-qPCR (Figure 7A,B). The relative mRNA level of each group of genes were used to calculate the Pearson's correlation coefficients with two AMPK subunits: Prkab1 and Prkag1. Although the resulting coefficient may vary, these calculated earlier due to the different sensitivities between microarrays and RT-qPCR, the directions of the correlation were the same as ones that we observed in the dataset (Figure 7C,D). In agreement with the suggested potential interaction of Prkab1 with Wipi1, Rab8a and Trp53inp2, strong correlations were validated. Prkag1 showed strong to moderate correlations with Becn1, Sirt2 and Trim21 as previously predicted by WGCNA.

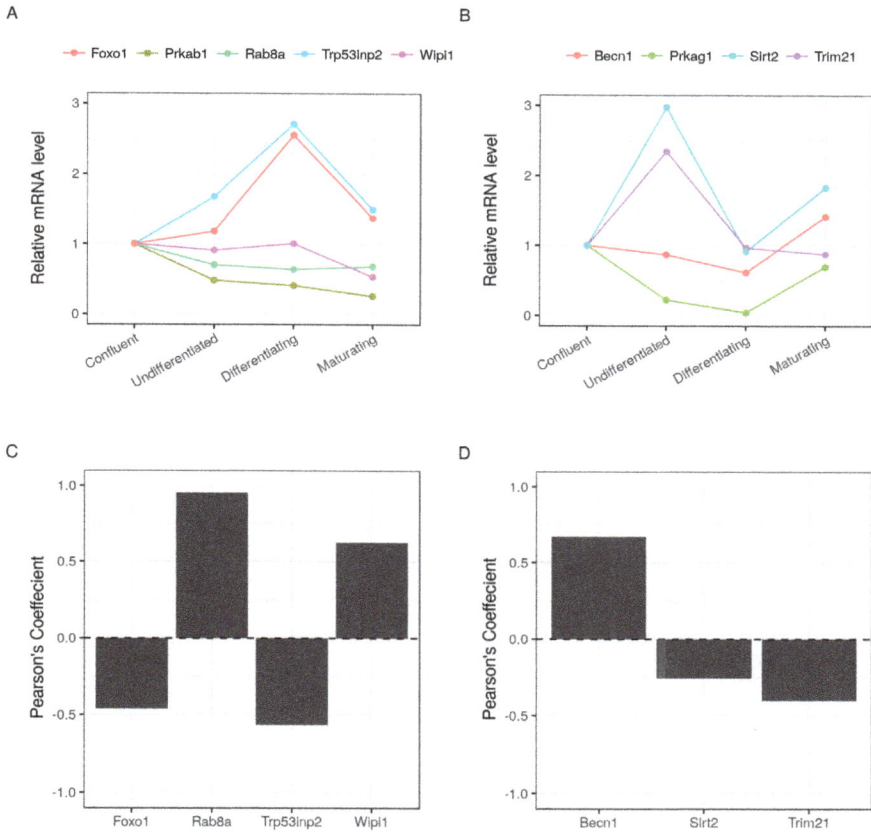

Figure 7. Validation of selected gene products expression and co-expression with AMPK subunits. Three independent samples of MDI-induced 3T3-L1 cells at four different time points corresponding to confluent, undifferentiated, differentiating and maturating stages were used to check the mRNA level of several gene products. (**A,B**) the $\Delta\Delta C_t$ values of five and four gene products, respectively, normalized by 18S and relative to the confluent cell stage are shown as points; (**C,D**) the Pearson's coefficient of four and three gene products with Prkab1 and Prkag1, respectively, are shown as bars.

3. Discussion

The adipocyte differentiation is a well regulated complex process. On one hand, this complexity allows for flexibility in response to different stimuli. For example, the over-expression of LC3 in 3T3-L1 pre-adipocytes produced a downstream activation of key regulators of adipogenesis and resulted in a differentiation pattern similar to that of the MDI induction [18]. On the other hand, this process likely involves a wide range of changes in transcription, translation and protein modification. In a previous study from our laboratory, we suggested that many autophagy genes were functionally associated with adipocyte differentiation using the RNA-Seq expression data [19]. We also showed that the mRNA level of key autophagy genes is specifically regulated at different time points, and clusters of these genes respond to the differentiation stimulus in a time-dependent manner. In particular, the subsets of organelle specific autophagy (e.g., mitophagy, reticulophagy, etc.) are highly regulated, suggesting a role in reorganizing the interacellulum and removing parts of the cell to adapt the morphological and metabolic changes of the mature adipocyte. Here, we explore the connection between autophagy and AMPK, which was established in conditions such as starvation, as it applies to the pre-adipocytes response to differentiation stimuli.

One aspect of this connection can be deduced from the observed positive correlation of the detected modules with the differentiation course (Figure 2A). In addition, the blue and the turquoise modules scores a significant (p-value < 0.001) protein–protein interaction (PPI) enrichment with an average clustering co-efficient of 0.4 and 0.5, respectively (STRING web interface). Although a detailed molecular link would be less clearer, AMPK gene products were evenly split among these modules, and had multiple edges with highly influential nodes (hubs) of well studied autophagy genes (Tables 3 and 4). Thus, AMPK gene products are part of biologically connected autophagy modules, which seem to be fairly consequential in the process of adipocyte differentiation.

The AMPK complex is formed of one catalytic subunit (α) and two non-catalytic regulatory subunits (β and γ), each has more than one isoform encoded by a separate gene [20]. The different subunits contribute to the stability and activity of the complex, whereas the combinations of the different isoforms give rise to complexes that behave differently and/or are specific to certain tissues [21,22]. Probes corresponding to the genes that code the different isoforms of the subunits were consistently expressed at different levels. Moreover, they showed varying correlations with the cell differentiation stage (data not shown). We considered the subunits and the isoforms of the AMPK complex individually. *Prkab1* and *Prkag1* were expressed at higher levels, and they, therefore, are the main AMPK side of the reported interaction with the autophagy pathway (Tables 4 and 5).

The adipocyte differentiation is characterized by events of increased lipogenesis and intracellular remodeling. AMPK is known for inhibiting the former and stimulating the latter. The absence of a reliable signal from the catalytic subunits of AMPK in our analysis may only enable a partial view. Nevertheless, we observe enrichment of ubiquitin activity and certain cellular component terms, probably akin to a form of localization, by autophagy genes co-expressed with the regulatory (β and γ) subunits of AMPK (Figure 4). One of these terms, ubiquitin-like protein binding, includes two gene products, *Trp53inp2* and *Nbr1*. Both are known to help the formation of autophagosome and the selective removal of ubiquitinated proteins through binding to LC3 [23–25]. Few other terms related to the organelle membranes appear interesting. Perhaps, the localization of AMPK at certain intracellular locations mediates the autophagy selectivity, as suggested before [26]. This is consistent with the description of two emerging mechanisms of AMPK regulation, namely by ubiquitination and sub-cellular distribution [27].

Together, it is possible that AMPK-autophagy connection is determined by the energy supply and demand of the differentiating cells. It is more likely that the two pathways interact dexterously with some temporospatial agility. For example, AMPK activates autophagy early in the differentiation course in response to the differentiation stimulus. In a later stage, autophagy might remove ubiquitinated AMPK to allow the accelerated fat accumulation. Finally, the localization of AMPK to certain intracellular organelles could guide their recycling or removal by selective autophagy. The implications of these features do not escape us, both AMPK and autophagy are involved in disorders such as obesity and diabetes [28,29]. The manipulation of one pathway could affect the outcomes controlled by the other. In addition, the timely intervention during adipocyte differentiation could preferentially favor certain consequences of the pathways' interplay.

Nassiri and colleagues provided a system view of the adipogenesis and ranked the involved biological processes to identify the coordinated activity among them using the NASFinder method of publicly available omics data [30]. According to their analyses, the translational machinery, mitochondrial associated pathways, PPAR signaling, insulin and leptin signaling, and some membrane associated complexes are coordinately upregulated after adipocyte induction. Although they might be associated with AMPK or autophagy at the broad concept, they do not show any specific coordination between AMPK pathway and autophagy process. Here, we used the widely known WGCNA method to detect the conserved genetic networks of AMPK-autophagy gene products that might contribute to the process of differentiation [31,32]. Typically, it needs to input the list of differentially expressed genes among three or more experimental conditions [12]. In this study, we limited the analysis to the probes that mapped uniquely to AMPK and autophagy genes as defined in their gene ontology terms. The downside of limiting the analysis to a predefined set of genes is that the prospective findings would be limited to the available annotation, potential loss of signals from probes that map to genes not in the predefined gene set and the inclusion of probes that map to genes that are not actively changing among the conditions. On the other hand, this approach allows for simplifying the analysis steps and the interpretation of the results. The detected networks are more likely to have biologically meaningful consequences since they are formed of nodes that are known for certain functions in their pathways/gene sets. In addition, this allows for including genes that are highly correlated even though they don't show the highest degree of differentiation among conditions. Certainly, some of these genes are involved in the biology of adipocyte differentiation either by maintaining essential cellular processes or they show subtle changes that wouldn't be typically picked by the differential expression approach.

4. Materials and Methods

4.1. Data and Annotation Sources

4.1.1. Gene Ontology

The Gene Ontology (GO) terms AMP-activated protein kinase (AMPK) (GO:0004679) and autophagy (GO:0006914) were used to identify the gene products (14 and 167, respectively) with known functions in the corresponding biological processes [33]. Similarly, GO was used to identify the molecular function (MF) and cellular component (CC) terms containing these gene products. GO was accessed through the GO.db and the mouse organism package org.Mm.eg.db [34,35].

4.1.2. Microarrays Expression Data

To identify the relevant datasets, we queried the NCBI Gene Expression Omnibus (GEO) metadata by GEOmetadb [36]. The term '3T3-L1' was used to search the titles of all entries, the query results were then searched manually and datasets of similar induction time-course design were included. The expression and the annotation data were then obtained using a GEOquery [37]. Table 1 summaries the four datasets that were used in this analysis. GSE34150 consists of 24 samples of MDI-induced 3T3-L1 pre-adipocytes at eight different time points corresponding to three differentiation stages (0 day, undifferentiated; two and four days, differentiating; 6–18 days, maturating).

4.1.3. Protein–Protein Interactions

The STRING database was used to query all possible AMPK-autophagy protein–protein interactions that are reported with different evidence types [38]. The HUGO symbols of 181 genes were mapped to the ENSEMBL IDs before querying the database. STRINGdb was used to do the mapping, construct the query and obtain the results. The interactions were matched against the edges of the co-expression networks of the detected modules to label the edges with the type of evidence when they were previously reported.

4.2. Weighted-Gene Co-Expression Network Analysis

The package WGCNA was used to apply most of the necessary steps for weighted-gene co-expression network analysis on the GSE34150 dataset as described in the original publications [39]. Briefly, a co-expression measure (Pearson's correlation coefficient) was calculated between each pair of genes. The coefficients were raised to the power of 5 to form an adjacency matrix. The adjacency matrix was then used to calculate the topological overlap similarity matrix (TOM). To detect modules and assign genes to them, a dissimilarity matrix is obtained $(1 - \text{TOM})$ and used as distances between genes. A hierarchical clustering was then performed and a gene tree is built. Upon cutting the tree at a certain height, genes nearby are assigned to modules, referred to as colors (names are arbitrarily assigned). The detected modules were then used to find the correlation with the phenotype and the preservation in independent datasets. To correlate the modules to the sample phenotypes or to each other, an eigengen or the principal components (PC) were calculated from the expression of their respective members and used as a representative summary. Finally, a module preservation analysis was performed by calculating various summary statistics on the detected modules in the test datasets [40].

4.3. Network Visualization and Analysis

The igraph package was used to visualize and analyze the detected modules [41]. The genes of interest were treated as nodes in a network graph and were connected by an edge if its weight—calculated from the TOM similarity between each pair of genes—passed a minimum threshold. Several graph statistics were used to determine the importance/centrality of genes and their interactions.

4.4. Gene Modules Over-Representation

The limma package was used to test for the over-representation of the detected modules in the GSE34150 dataset [42]. An index of the modules as gene sets, the expression data and comparison matrix based on the differentiation stage were used as input. A gene set is considered *over-represented* when it has a significantly higher fraction of differentially expressed genes than a randomly selected module of the same size. The clusterProfiler package was used to apply a similar list-based enrichment of GO terms by the detected modules [43]. Tests were adjusted for multiple testing using the False Discovery Rate (FDR) and a cutoff (0.1) was applied.

4.5. Cell Culture and RT-qPCR

3T3-L1 pre-adipocytes were cultured and induced for differentiation using MDI protocol as described before [19]. Total RNA was collected at four different time points corresponding the the major differentiation stages of the adipocytes (-2 day, full confluence; 0 day, undifferentiated; 10 h differentiating; and -6 day, maturating). The list of the primers that were used in the reaction are provided in (Appendix A). The C_t values from the RT-qPCR reaction were normalized by a reference gene 18S and calibrated by the confluent samples ($\Delta\Delta C_t$) using the pcr R package [44].

4.6. Software Environment and Reproducibility

The data were obtained, processed and analyzed in an R environment and using multiple Bioconductor packages [45,46]. The full analysis was done and reproduced in an isolated environment based on docker (bioconductor/release_base2) [47]. The scripts for reproducing the analysis, figures and tables are available at https://github.com/MahShaaban/aacna. The instructions for reproducing the analysis are described in Appendix C.

5. Conclusions

In summary, we used the WGCNA to investigate the interactions of AMPK and autophagy gene products in the context of adipocyte differentiation. Two co-expression networks were found to be highly correlated with the time course of differentiation. We were able to validate the case of these networks in other independent datasets of similar experimental designs. These networks appear to be consequential in the response of the pre-adipocyte to the differentiation stimulus. Finally, we present several novel potential interactions between AMPK and autophagy gene products and link them to potential functions and cellular sites.

Author Contributions: M.A. conceived, performed and reported the analysis; J.S.H. performed the RT-qPCR experiment; H.Q.N., S.Z., T.H.L., T.M.P. performed the cell culture and the differentiation induction assay; M.Y. consulted on the design and the execution of the study; D.R.K. supervised the study and contributed to writing the manuscript

Funding: This research received no external funding. **Acknowledgments:** This study was supported by the Basic

Science Research Program through the National Research Foundation of Korea (NRF) funded by the Ministry of Education Science and Technology (2018R1D1A1B07043715) and by the Ministry of Science, ICT and Future Planning (NRF-2015R1A5A2008833).

Conflicts of Interest: The authors declare no conflict of interest.

Abbreviations

AMPK	AMP-Activated Protein Kinase
CC	Cellular Components
FDR	False Discovery Rate
GEO	Gene Expression Omnibus
GO	Gene Ontology
MDS	Multi-Dimensional Scaling
mTOR	Mechanistic Target Of Rapamycin Kinase
MF	Molecular Function
PC	Principal Component
PPI	Protein–Protein Interaction
RT-qPCR	Real-Time Quantitative Polymerase Chain Reaction
TOM	Topological Overlap Matrix
ULK1	Unc-51 Like Autophagy Activating Kinase 1
WGCNA	Weighted Gene Co-expression Network Analysis

Appendix A. Datasets and Annotations

This appendix contains details of the datasets and the gene annotation used in the study.

Appendix A.1. Time Point of Samples in the Microarrays Datasets

We queried the metadata of the Gene Expression omnibus (GEO) for microarrays datasets of MDI-induced 3T3-L1 pre-adipocytes at different time points that covers the various differentiation stages. The datasets were then manually checked for containing sufficient phenotype and annotation data. A few datasets were generated using custom microarrays chips of a few thousand probes and were excluded for not containing a sufficient number of the probes of interest. In Table A1, we listed four datasets included in this study and the time points (in hours) of their samples.

Table A1. Sample time points in the different datasets.

Time Point (hours)	GSE15018	GSE20696	GSE34150	GSE69313
−48		2		
0		2	3	3
1	1			
2	1			
3	1			
4	1			
5	1			
6	1			3
7	1			

Table A1. *Cont.*

Time Point (hours)	GSE15018	GSE20696	GSE34150	GSE69313
8	1			
9	1			
10	1			
11	1			
12	1			
13	1			
14	1			
15	1			
16	1			
17	1			
18	1			
24				3
48		2	3	
72				3
96			3	
144			3	
168		2		
192			3	
240			3	
336			3	
432			3	

Appendix A.2. Comparing the Average Expression in Four Datasets

To ensure that the different datasets are exhibiting comparable probe expression, we compared the average expression of all common probes in three datasets (GSE15018, GSE20696 and GSE69313) with the main (GSE34150) dataset (Figure A1).

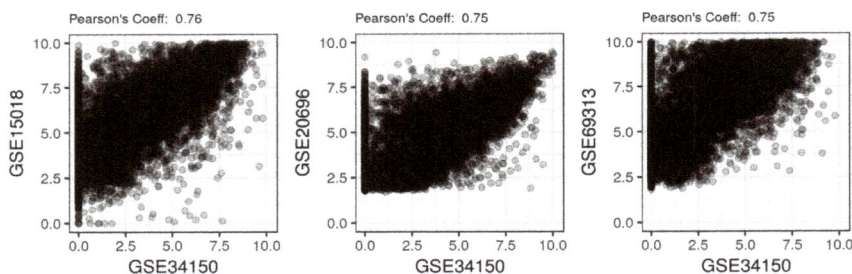

Figure A1. Average expression of all probes in multiple MDI-induced 3T3-L1 microarrays datasets. The log average expression values of all probes in the MDI-induced 3T3-L1 datasets (GSE15018, GSE20696 and GSE69313) are compared to the corresponding averages in the main dataset (GSE34150). Individual values are shown as points. The Pearson's correlation coefficient of the corresponding values is shown on top.

Appendix A.3. Data Quality Assessment

The accession number (GSE34150) was used to obtain the expression matrix and the metadata of the dataset. Several quality assessment measures were applied to ensure the suitability of the data for the downstream analysis (Figure A2).

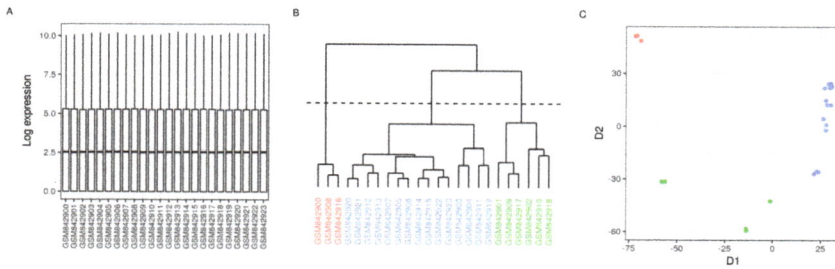

Figure A2. Quality assessment and exploration of the main microarrays dataset. Twenty-four samples of MDI-treated 3T3-L1 cells of the microarray series (GSE34150) were obtained from GEO along with the corresponding annotation data (GPL6885); (**A**) the distribution of the log expression of (n = 25,697) probes from all samples as box plots; (**B**) hierarchical clustering based on the euclidean distances of all samples; (**C**) multi-dimensional scaling (MDS) of all samples. Colors represent the cell stage/time point (green, undifferentiated; dark green, differentiating and red, maturing).

Appendix A.4. Confirming Differentiation and Lipogenesis

Several adipogenic and lipogenic of markers transcriptional changes are expected during the course of the 3T3-L1 cell differentiation. Figure A3 show the log expression level of some of these markers at different stages in the GSE34150 dataset.

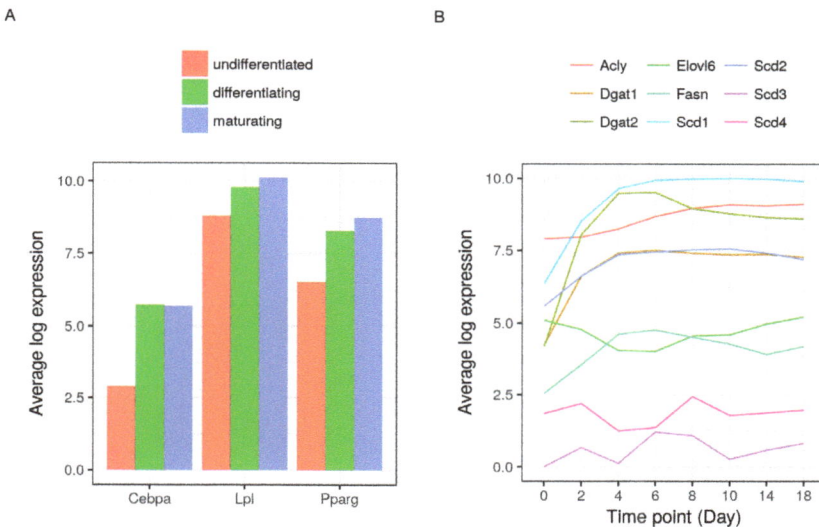

Figure A3. Differentiation and lipogensis markers in differentiating adipocytes. Average log expression values from 24 samples of MDI-induced 3T3-L1 cells (GSE34150) at 3 differentiation stages and 8 time points (0 day, undifferentiated (red); 2 and 4 days, differentiating (green); 6–18 days, maturing (blue)) from (**A**) differentiation markers and (**B**) lipogenesis markers are shown as bars and lines, respectively. *Cebpa*, CCAAT/enhancer binding protein (C/EBP), alpha; *Lpl*, lipoprotein lipase; *Pparg*, peroxisome proliferator activated receptor gamma; *Acly*, ATP Citrate Lyase; *Dgat*, Diacylglycerol O-Acyltransferase; *Elov6*, Fatty Acid Elongase 6; *Fasn*, Fatty Acid Synthase; *Scd*, Stearoyl-CoA Desaturase.

Appendix A.5. RT-qPCR Primer Sequences

The following nine primers were used to validate the expression and correlations of the corresponding genes products with the subunits of AMPK during the time course of 3T3-L1 differentiation (Table A2).

Table A2. RT-qPCR primer sequences.

Name	Forward (5′ to 3′)	Reverse (3′ to 5′)
18S	ACCGCAGCTAGGAATAATGGA	GCCTCAGTTCCGAAAACCA
Becn1	CAGGAACTCACAGCTCCATTAC	CCATCCTGGCGAGTTTCAATA
Prkab1	GAGATCAAGGCTCCAGAGAAAG	GTTGAAGGACCCAGACAAGTAG
Prkag1	GAACTGGAGGAGCACAAGATAG	GGGAGCCTGTGGATCTTATTT
Rab8a	GCTCGATGGCAAGAGGATTA	CTGTAGTAGGCTGTCGTGATTG
Sirt2	CATAGCCTCTAACCACCATAGC	GTAGCCTGTTGTCTGGGAATAA
Trim21	GATAGCCCAGAATACCAAGAAGAG	GCCCATCTTCCTCACAGAATAG
Trp53inp2	GGTGAAGCGCTGGAACAT	CACAACTACCTCAGCGCAGC
Wipi1	GTGTGTCTAGACGACGAGAATG	GACTTCTGAGGTAGGCTTCTTG

Appendix A.6. Gene Ontology Annotation

To define the sets of genes involved in the AMPK and autophagy pathways, we turned to the gene ontology (GO) annotations. GO identifies the AMP-dependent protein kinase activity (GO:0004679) as catalysis of the reaction: ATP + a protein = ADP + a phosphoprotein. In total, 14 genes were identified to be involved in the pathway and its regulation and are referred to as AMPK genes in the manuscript. Similarly, GO defines the term autophagy (GO:0006914) as the catabolic process in which the cells digest parts of their own cytoplasm. It contains 10 children/subcategory terms and 167 genes. The gene symbols of AMPK and autophagy gene sets are listed in Table A3.

Table A3. Gene members of the AMPK and autopahgy gene ontology terms.

Category	Term	Genes
AMPK	AMP-activated protein kinase activity	Prkab1, Prkag1, Smok1, Smok2a, Prkaa1, Prkaa2, Prkab2, Prkag2, Smok2b, Prkag3, 4921509C19Rik, Smok3a, Smok3b, Smok3c
autophagy	autophagy of mitochondrion	Atg5, Rb1cc1, Cdkn2a, Capn10, Park2, Wipi1, Wdr45, Becn1, Fis1, Atg4b, Map1lc3a, Wdr45b, Cisd2, Fundc2, Map1lc3b, Atg12, Atg3, Pink1, Fbxo7, Fundc1, Atg7, Wipi2, Atg2b, Usp30, Atg9b, Ambra1, Atg4d, Atg4c, Atg9a, Atg2a, Atg4a
	autophagy of nucleus	Atg5, Wipi1, Wdr45, Becn1, Atg4b, Wdr45b, Atg12, Atg3, Wipi2, Trappc8, Atg2b, Becn2, Atg4d, Atg4c, Atg2a, Atg4a
	autophagy of peroxisome	Rb1cc1, Acbd5, Pik3r4, Trappc8, Pik3c3
	chaperone-mediated autophagy	Hspa8, Lamp2
	late endosomal microautophagy	Hspa8, Vps4b, Vps4a
	macroautophagy	Atg5, Cln3, Ei24, Nbr1, Sqstm1, Pik3c2a, Plaa, Ulk1, Tcirg1, Ubqln1, Becn1, Ubxn6, Map1lc3a, Map1lc3b, Atg3, Trp53inp2, Tbc1d5, Atg7, Pik3r4, Zfyve1, D17Wsu92e, Pik3c3, Yod1, Tmem74, Pik3c2b, Vcp, Atg14
	negative regulation of autophagy	Akt1, Bcl2, Eif4g2, Htr2b, Il3, Lep, Lepr, Mcl1, Mt3, Ptpn22, Tnfaip3, Rnf5, Mtor, Sirt2, Rraga, Washc1, Rasip1, Zkscan3, Dapl1, Wdr6, Tbc1d14, Lars, Bmf, Eif4g1, Dap, Kdm4a, Herc1, Rubcn
	positive regulation of autophagy	Ager, Dcn, Hif1a, Ifng, Pim2, Prkd1, Plk2, Trim21, Stk11, Tfeb, Tsc2, Xbp1, Mid2, Map2k1, Irgm2, Mefv, Sh3glb1, Nprl2, Becn1, Tmem59, Foxo1, Trp53inp1, Lrrk2, Mtdh, Trp53inp2, Dapk1, Optn, Plekhf1, Atg7, Scin, Uvrag, Tlr9, Trim8, Sh3bp4, Prkaa1, Ticam1, Prkaa2, Flcn, Ambra1, Zc3h12a, Dhrsx, Tpcn1, Rnf152, Trim65, Atg14
	protein targeting to vacuole involved in autophagy	Smurf1
	regulation of autophagy	Atm, Bcl2, Casp1, Hmgb1, Lmx1b, Rab8a, Mapt, Pik3r2, Ppp4k2a, Usp10, Xbp1, Park2, Bok, Rragd, Rragc, Iigp1, Trp53inp1, Lrrk2, Pycard, Cisd2, Dram2, Soga3, Kat8, Rab39b, Rraga, Mtcl1, Dram1, Mfsd8, Fbxl2, Usp13, 3110043O21Rik, Rptor, Chmp4b, Nlrp6, Ppp4k2b, Ppp4k2c, Usp33, Wdr41, Tpcn2, Smcr8, Lamp3, Rragb, Wdr24, Depdc5, Soga1, Fbxw7as1

Appendix B. Rational of Analysis Directive

This appendix contains a brief discussion of some of the decision that were made at different steps of the analysis and the rationals behind them. Specifically, we describe the steps of constructing the co-expression networks and the centrality measures that were applied to them.

Appendix B.1. Choosing the Network Power Threshold

A critical choice in constructing the co-expression networks is setting the soft threshold (power) to which the adjacency matrix is raised. A Scale Free Topology (SFT) measure is calculated by multiplying the a slope and a fitted R square and a mean connectivity of the networks at different power values (Figure A4). A power value of 5 was chosen to satisfy the SFT the most.

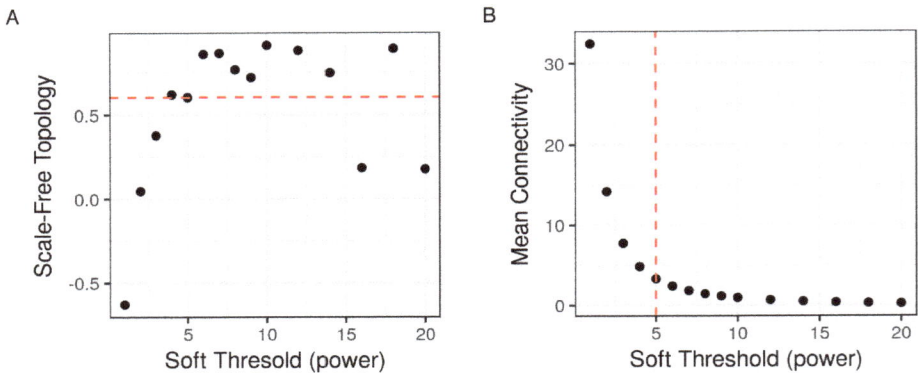

Figure A4. Scale free topology for multiple power values. Expression data from 24 samples (GSE34150) MDI-induced 3T3-L1 at different time points were used to calculate an $(n \times n)$ similarity matrix ($n = 181$ genes), which were used to obtain the weighed networks by raising them to multiple power values. For each value, a scale free topology index was calculated. (**A**) the fit indices, the slopes multiplied by the R squared R^2 values, are shown for each power value as points; (**B**) the mean connectivity, average edges shared by a node, for the resultant network at each power value are shown as points. Red lines represent the choice of power that satisfy both high R squared R^2 values and high connectivity.

Appendix B.2. Steps of Constructing the Weighed Co-Expression Networks

In this section, we discuss the different steps for calculating the similarity measures that was used in constructing the networks as well as comparing to the intermediary forms in representing the notation of co-expression. Particularly, the issue of penalizing the low correlation values. Three main steps are necessary:

- The absolute values of either Pearson's (default) or Spearman's coefficient can be used to provide an initial similarity measure s_{ij} between each pair of nodes (ij) as in: $s_{ij} = |\, cor(ij) \,|$.
- The similarity matrix is then transformed to and adjacency matrix by elevating it to a selected power β as in: $a_{ij} = s_{ij}^{\beta}$.
- This matrix is then used to calculate the connectivity/weight of each pair of nodes as follows:

$$w_{ij} = \frac{l_{ij} + a_{ij}}{min\{k_i + k_j\} + 1 - a_{ij}},$$

where $l_{ij} = \sum_u a_{iu} a_{uj}$ and $k_i = \sum_u a_{iu}$. These weights (also not shown) are finally used to calculate a dissimilarity measure for clustering and detecting the gene modules as in: $d_{ij}^w = 1 - w_{ij}$.

Figure A5A shows the cumulative distribution functions of the final TOM similarity measure and the two intermediaries; Pearson's correlation and the adjacency (after being raised to the power of 5).

A

B

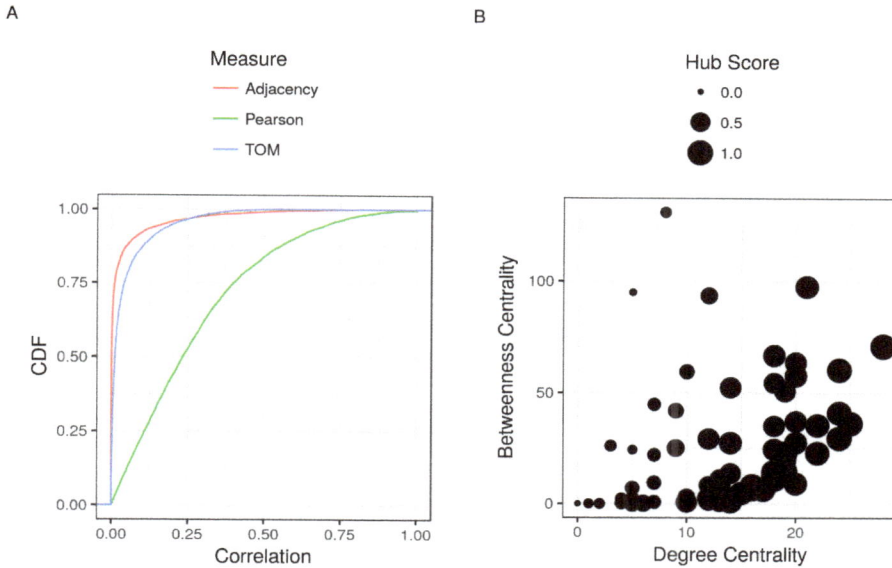

Figure A5. Gene similarity and node centrality measures. (**A**) the cumulative distribution function (CDF) of three correlation/similarity measures of AMPK and autophagy genes (*n* = 181) are shown as colored lines (green, Pearson's correlation coefficients; red, adjacency; and blue, TOM); (**B**) three centrality measures for all nodes in the two detected modules are as points. The degree centrality on the *x*-axis, the betweenness centrality on the *y*-axis and the hub score as the point size.

Appendix B.3. Node Centrality Measures

To determine the importance of each node/gene in the networks, we relied on multiple measures of centrality or node influence. These measures are calculated as follows:

- Degree Centrality: the number of edges/connections shared by a node. The Degree of a node v is given by:

$$Degree(v) = \sum_j a_{v,j},$$

where $a_{v,j}$ is the adjacency matrix of the network or the sum of the corresponding row.
- Betweenness Centrality: The number of the occasion of a node falls on the shortest between two other nodes. The Betweenness of a node v is given by:

$$Betweenness(v) = \sum_{s \neq v \neq t \in V} \frac{\sigma_{st}(v)}{\sigma_{st}}$$

or the fraction of the shortest paths between each pair of nodes (s, t) that passes through v.
- Hub Score/Eigenvector: a measure of the influence of a given node in a network. The Eigenvector of a node v is given by:

$$Eigenvector(v) = \frac{1}{\lambda} \sum_{t \in M(v)} x_t = \frac{1}{\lambda} \sum_{t \in G} a_{vt} x_t$$

or the sum of the scores x_i of its neighboring nodes $M(v)$ in the network G.

Figure A5B shows the different centrality measures and the correlations between them. Specifically, we found that the degree centrality is highly correlated with the hub scores (Pearson's coefficient about 0.73) and less so with the betweenness centrality (Pearson's coefficient about 0.44).

Appendix B.4. Network Preservation

The authors of the WGCNA method suggests using composite preservation summaries to evaluate the evidence of the preservation of the detected modules in the test dataset/s as opposed to using individual statistics as they measure different aspect of the preservation. In the main text, we showed the $Z_{summary}$ for the preservation of the two detected modules in three test datasets. Here, we briefly expand on what this statistics composed of and show additional statistics supporting the preservation of the modules.

The $Z_{summary}$ statistics is given by:

$$Z_{summary} = \frac{Z_{connectivity} + Z_{density}}{2}.$$

Each of the two Z summaries are further composed of several statistics, the details of which are provided in the references.

As opposed to the $Z_{summary}$ as evidence for module preservation, the median rank of the preservation is more robust to the module sizes. However, it is more informative in comparing preservation of modules relative to each other, as it is based on the ranks of the observed preservation statistics of the module. The lower the median rank of the module, the more preservation it exhibits in the test set. Figure A6 shows the median ranks/relative preservation of the modules with their sizes.

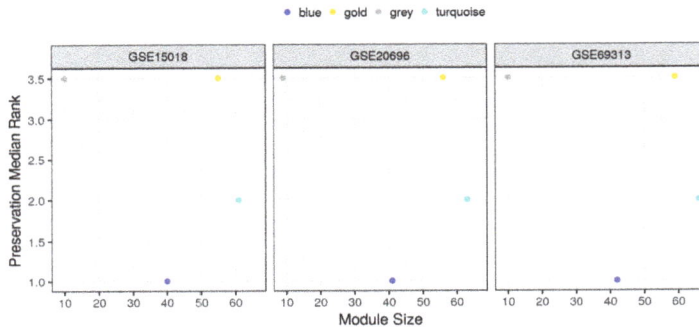

Figure A6. Module preservation ranks across multiple MDI-induced 3T3-L1 microarrays datasets. The GSE34150 dataset was used to detect the highly co-expressed modules among AMPK and autophagy genes (42, blue; 66, turquoise; 10, gray, unassigned; and 55, gold, randomly assigned). The detected modules were used as a reference to calculate several preservation statistics in three independent datasets of similar design (GSE15018, GSE20696 and GSE69313). The median ranks of the preservation statistics and the sizes of four modules are shown as colored points.

Appendix C. A Note on Reproducing the Analysis

This is a detailed description of the details required to reproduce the analysis from the source code. We first introduce the way to obtain the source code, the software environment and the commands to run the analysis script and generate the figures and tables that appear in this manuscript. In Figure A7, we show the workflow of the study listing the steps, datasets and software packages that were used in the analysis.

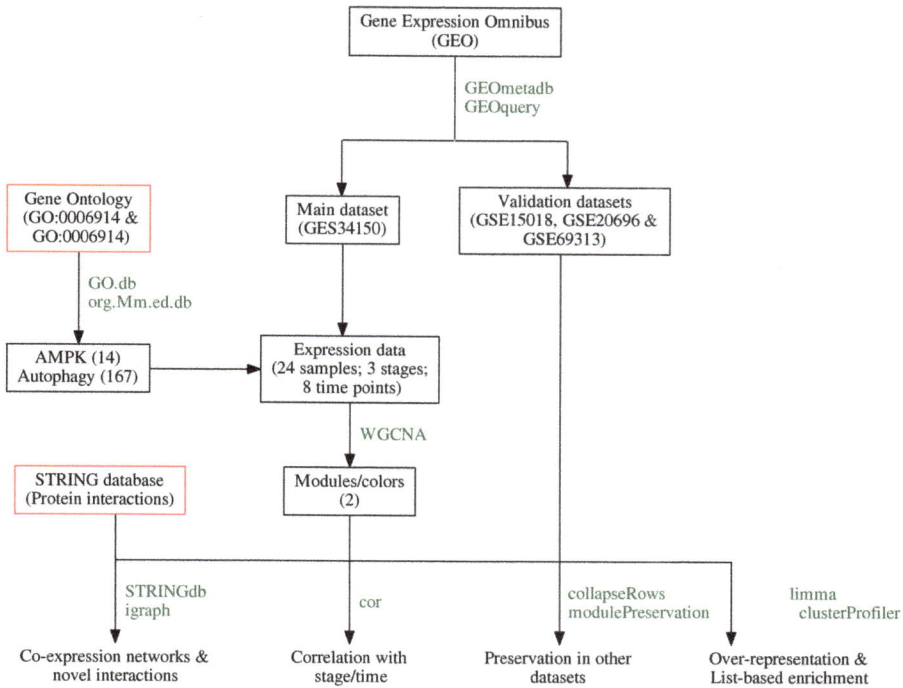

Figure A7. Workflow of the study.

Appendix C.1. Setting up the Docker Environment

The analysis was run on a docker image based on the the latest bioconductor/release_base2. Other R packages were added to the image and were made available as an image that can be obtained and launched on any local machine running docker:

```
$ docker pull mahshaaban/analysis_containers:bioc_wgcna,
$ docker run −it mahshaaban/analysis_containers:bioc_wgcna bash.
```

Appendix C.2. Obtaining the Source Code

The source code is hosted publicly on a repository on github in a form of research compendium. This includes the functions used throughout the analysis as an R package, the scripts to run the analysis and finally the scripts to reproduce the figures and tables in this manuscript. From within the container, git can be used to cloned the source code. The cloned repository contains a sub-folder called 'analysis/scripts', which can be used to reproduce the analysis from scratch:

- 01.analysis.R This script loads the required libraries, download the data and run all the steps of the analysis described in the manuscript,
- figures/A sub-folder with a separate file for each graph in the manuscript,
- tables/A sub-folder with a sepearte file for each table in the manuscript.

The following code clones the repository containing the source code:

```
$ git clone http://github.com/MahShaaban/aacna.
```

Appendix C.3. Running the Analysis

The analysis scripts is organized to be ran using a single 'make' command. This will first load the necessary functions and run the main analysis and save the data in an R object 'wgcna.rda'. This will be used to generate the figures and graphs. In addition, a log file is generated in the sub-folder 'log/' for each script that can be used for troubleshooting.

To do that, the 'make' command should be invoked from withing the 'analysis/' sub-folder.

```
$ cd aacna/analysis/
$ make
```

Appendix C.4. Details of the R Environment

The version of R that was used to perform this analysis is the 3.4.2 (28 September 2017) on x86_64-pc-linux-gnu. The 'DESCRIPTION' file in the main repository contains further details about the dependencies and the license of this work.

References

1. Zhang, Y.; Goldman, S.; Baerga, R.; Zhao, Y.; Komatsu, M.; Jin, S. Adipose-specific deletion of autophagy-related gene 7 (*atg7*) in mice reveals a role in adipogenesis. *Proc. Natl. Acad. Sci. USA* **2009**, *106*, 19860–19865. [CrossRef] [PubMed]
2. Baerga, R.; Zhang, Y.; Chen, P.H.; Goldman, S.; Jin, S. Targeted deletion of autophagy-related 5 (*atg5*) impairs adipogenesis in a cellular model and in mice. *Autophagy* **2009**, *5*, 1118–1130. [CrossRef] [PubMed]
3. Singh, R.; Xiang, Y.; Wang, Y.; Baikati, K.; Cuervo, A.M.; Luu, Y.K.; Tang, Y.; Pessin, J.E.; Schwartz, G.J.; Czaja, M.J. Autophagy regulates adipose mass and differentiation in mice. *J. Clin. Investig.* **2009**, *119*, 3329–3339. [CrossRef] [PubMed]
4. Singh, R.; Kaushik, S.; Wang, Y.; Xiang, Y.; Novak, I.; Komatsu, M.; Tanaka, K.; Cuervo, A.M.; Czaja, M.J. Autophagy regulates lipid metabolism. *Nature* **2009**, *458*, 1131–1135. [CrossRef] [PubMed]
5. Gwinn, D.M.; Shackelford, D.B.; Egan, D.F.; Mihaylova, M.M.; Mery, A.; Vasquez, D.S.; Turk, B.E.; Shaw, R.J. AMPK Phosphorylation of raptor mediates a metabolic checkpoint. *Mol. Cell* **2008**, *30*, 214–226. [CrossRef] [PubMed]
6. Inoki, K.; Zhu, T.; Guan, K.L. TSC2 Mediates Cellular Energy Response to Control Cell Growth and Survival. *Cell* **2003**, *115*, 577–590. [CrossRef]
7. Kim, J.; Kundu, M.; Viollet, B.; Guan, K.L. AMPK and mTOR regulate autophagy through direct phosphorylation of Ulk1. *Nat. Cell Biol.* **2011**, *13*, 132–141. [CrossRef] [PubMed]
8. Egan, D.F.; Shackelford, D.B.; Mihaylova, M.M.; Gelino, S.; Kohnz, R.A.; Mair, W.; Vasquez, D.S.; Joshi, A.; Gwinn, D.M.; Taylor, R.; et al. Phosphorylation of ULK1 (hATG1) by AMP-activated protein kinase connects energy sensing to mitophagy. *Science* **2011**, *331*, 456–461. [CrossRef] [PubMed]
9. Green, H.; Kehinde, O. An established preadipose cell line and its differentiation in culture II. Factors affecting the adipose conversion. *Cell* **1975**, *5*, 19–27. [CrossRef]
10. Ntambi, J.M.; Young-Cheul, K. Adipocyte differentiation and gene expression. *J. Nutr.* **2000**, *130*, 3122S–3126S. [CrossRef] [PubMed]
11. Roberts, R.; Hodson, L.; Dennis, A.L.; Neville, M.J.; Humphreys, S.M.; Harnden, K.E.; Micklem, K.J.; Frayn, K.N. Markers of de novo lipogenesis in adipose tissue: Associations with small adipocytes and insulin sensitivity in humans. *Diabetologia* **2009**, *52*, 882–890. [CrossRef] [PubMed]
12. Zhang, B.; Horvath, S. A General Framework for Weighted Gene Co-Expression Network Analysis. *Stat. Appl. Genet. Mol. Biol.* **2005**, *4*, Article17. [CrossRef] [PubMed]
13. Edgar, R.; Domrachev, M.; Lash, A.E. Gene Expression Omnibus: NCBI gene expression and hybridization array data repository. *Nucleic Acids Res.* **2002**, *30*, 207–210. [CrossRef] [PubMed]
14. Chin K. *Dataset: A Time Course Analysis of the Effects of Prieurianin in the Mouse Preadipocytes 3T3-L1 Cells*; The University of Toledo College of Medicine: Toledo, OH, USA, 2010.
15. Mikkelsen, T.S.; Xu, Z.; Zhang, X.; Wang, L.; Gimble, J.M.; Lander, E.S.; Rosen, E.D. Comparative epigenomic analysis of murine and human adipogenesis. *Cell* **2010**, *143*, 156–169. [CrossRef] [PubMed]

16. Horsch, M.; Beckers, J.; Adamski, J.; Halama, A. *Dataset: Genome-Wide Expression Profiling Analysis of a Time Course of Differentiating Adipocytes*; Helmholtz Zentrum Munchen GmbH: Neuherberg, Germany, 2015.

17. Zhang, M.; Zhang, Y.; Ma, J.; Guo, F.; Cao, Q.; Zhang, Y.; Zhou, B.; Chai, J.; Zhao, W.; Zhao, R. *Dataset: Effect of siRNA Knock-Down of FTO on 3T3-L1 Cell Differentiation*; China Academy of Space Technology: Beijing, China, 2015.

18. Hahm, J.R.; Ahmed, M.; Kim, D.R. RKIP phosphorylation–dependent ERK1 activation stimulates adipogenic lipid accumulation in 3T3-L1 preadipocytes overexpressing LC3. *Biochem. Biophys. Res. Commun.* **2016**, *478*, 12–17. [CrossRef] [PubMed]

19. Ahmed, M.; Quoc Nguyen, H.; Seok Hwang, J.; Zada, S.; Huyen Lai, T.; Soo Kang, S.; Ryong Kim, D. Systematic characterization of autophagy-related genes during the adipocyte differentiation using public-access data. *Oncotarget* **2018**, *9*, 15526–15541. [CrossRef] [PubMed]

20. Davies, S.P.; Hawley, S.A.; Woods, A.; Carling, D.; Haystead, T.A.; Hardie, D.G. Purification of the AMP-activated protein kinase on ATP-γ-sepharose and analysis of its subunit structure. *Eur. J. Biochem.* **1994**, *223*, 351–357. [CrossRef] [PubMed]

21. Dasgupta, B.; Chhipa, R.R. Evolving lessons on the complex role of AMPK in normal physiology and cancer. *Trends Pharmacol. Sci.* **2016**, *37*, 192–206. [CrossRef] [PubMed]

22. Ross, F.A.; Jensen, T.E.; Hardie, D.G. Differential regulation by AMP and ADP of AMPK complexes containing different subunit isoforms. *Biochem. J.* **2016**, *473*, 189–199. [CrossRef] [PubMed]

23. Mauvezin, C.; Orpinell, M.; Francis, V.A.; Mansilla, F.; Duran, J.; Ribas, V.; Palacín, M.; Boya, P.; Teleman, A.A.; Zorzano, A. *The Nuclear Cofactor DOR Regulates Autophagy in Mammalian and Drosophila Cells*; EMBO Reports; Nature Publishing Group: London, UK, 2010.

24. Sala, D.; Ivanova, S.; Plana, N.; Ribas, V.; Duran, J.; Bach, D.; Turkseven, S.; Laville, M.; Vidal, H.; Karczewska-Kupczewska, M.; et al. Autophagy-regulating TP53INP2 mediates muscle wasting and is repressed in diabetes. *J. Clin. Investig.* **2014**, *124*, 1914–1927. [CrossRef] [PubMed]

25. Kirkin, V.; Lamark, T.; Sou, Y.S.; Bjørkøy, G.; Nunn, J.L.; Bruun, J.A.; Shvets, E.; McEwan, D.G.; Clausen, T.H.; Wild, P.; et al. A Role for NBR1 in Autophagosomal Degradation of Ubiquitinated Substrates. *Mol. Cell* **2009**, *33*, 505–516. [CrossRef] [PubMed]

26. Liang, J.; Xu, Z.X.; Ding, Z.; Lu, Y.; Yu, Q.; Werle, K.D.; Zhou, G.; Park, Y.Y.; Peng, G.; Gambello, M.J.; et al. Myristoylation confers noncanonical AMPK functions in autophagy selectivity and mitochondrial surveillance. *Nat. Commun.* **2015**, *6*, 7926. [CrossRef] [PubMed]

27. Jeon, S.M. Regulation and function of AMPK in physiology and diseases. *Exp. Mol. Med.* **2016**, *48*, e245. [CrossRef] [PubMed]

28. Kola, B.; Grossman, A.; Korbonits, M. The role of AMP-activated protein kinase in obesity. *Front. Horm Res.* **2008**, *36*, 198–211. [PubMed]

29. Goldman, S.; Zhang, Y.; Jin, S. Autophagy and adipogenesis: implications in obesity and type II diabetes. *Autophagy* **2010**, *6*, 179–181. [CrossRef] [PubMed]

30. Nassiri, I.; Lombardo, R.; Lauria, M.; Morine, M.J.; Moyseos, P.; Varma, V.; Nolen, G.T.; Knox, B.; Sloper, D.; Kaput, J.; et al. Systems view of adipogenesis via novel omics-driven and tissue-specific activity scoring of network functional modules. *Sci. Rep.* **2016**, *6*, 28851. [CrossRef] [PubMed]

31. Stuart, J.M.; Segal, E.; Koller, D.; Kim, S.K. A gene-coexpression network for global discovery of conserved genetic modules. *Science* **2003**, *302*, 249–255. [CrossRef] [PubMed]

32. Carter, S.L.; Brechbühler, C.M.; Griffin, M.; Bond, A.T. Gene co-expression network topology provides a framework for molecular characterization of cellular state. *Bioinformatics (Oxford England)* **2004**, *20*, 2242–2250. [CrossRef] [PubMed]

33. Ashburner, M.; Ball, C.A.; Blake, J.A.; Botstein, D.; Butler, H.; Cherry, J.M.; Davis, A.P.; Dolinski, K.; Dwight, S.S.; Eppig, J.T.; et al. Gene Ontology: Tool for the unification of biology. *Nat. Genet.* **2000**, *25*, 25–29. [CrossRef] [PubMed]

34. Carlson, M. *org.Mm.eg.db: Genome Wide Annotation for Mouse*; Bioconductor: Seattle, WA, USA, 2016.

35. Carlson, M. *GO.db: A Set of Annotation Maps Describing the Entire Gene Ontology*; R Package Version 3.2.2; Bioconductor: Seattle, WA, USA, 2015

36. Zhu, Y.; Davis, S.; Stephens, R.; Meltzer, P.S.; Chen, Y. GEOmetadb: Powerful alternative search engine for the Gene Expression Omnibus. *Bioinformatics* **2008**, *24*, 2798–2800. [CrossRef] [PubMed]

37. Sean, D.; Meltzer, P.S. GEOquery: A bridge between the Gene Expression Omnibus (GEO) and BioConductor. *Bioinformatics* **2007**, *23*, 1846–1847.

38. Szklarczyk, D.; Morris, J.H.; Cook, H.; Kuhn, M.; Wyder, S.; Simonovic, M.; Santos, A.; Doncheva, N.T.; Roth, A.; Bork, P.; et al. The STRING database in 2017: Quality-controlled protein–protein association networks, made broadly accessible. *Nucleic Acids Res.* **2017**, *45*, D362–D368. [CrossRef] [PubMed]

39. Langfelder, P.; Horvath, S. WGCNA: An R package for weighted correlation network analysis. *BMC Bioinform.* **2008**, *9*, 559. [CrossRef] [PubMed]

40. Langfelder, P.; Luo, R.; Oldham, M.C.; Horvath, S. Is my network module preserved and reproducible? *PLoS Comput. Biol.* **2011**. [CrossRef] [PubMed]

41. Xu, K.; Tang, C.; Tang, R.; Ali, G.; Zhu, J. A Comparative Study of Six Software Packages for Complex Network Research. In Proceedings of the 2010 Second International Conference on Communication Software and Networks, Singapore, 26–28 February 2010; pp. 350–354.

42. Ritchie, M.E.; Phipson, B.; Wu, D.; Hu, Y.; Law, C.W.; Shi, W.; Smyth, G.K. limma powers differential expression analyses for RNA-sequencing and microarray studies. *Nucleic Acids Res.* **2015**, *43*, e47. [CrossRef] [PubMed]

43. Yu, G.; Wang, L.G.; Han, Y.; He, Q.Y. clusterProfiler: An R Package for Comparing Biological Themes among Gene Clusters. *OMICS J. Integr.e Biol.* **2012**, *16*, 284–287. [CrossRef] [PubMed]

44. Ahmed, M.; Kim, D.R. pcr: An R package for quality assessment, analysis and testing of qPCR data. *PeerJ* **2018**, *6*, e4473. [CrossRef] [PubMed]

45. R Core Team. *R: A Language and Environment for Statistical Computing*; R Foundation for Statistical Computing: Vienna, Austria, 2017.

46. Huber, W.; Carey, V.J.; Gentleman, R.; Anders, S.; Carlson, M.; Carvalho, B.S.; Bravo, H.C.; Davis, S.; Gatto, L.; Girke, T.; et al. Orchestrating high-throughput genomic analysis with Bioconductor. *Nat. Methods* **2015**, *12*, 115–121. [CrossRef] [PubMed]

47. Merkel, D. Docker: Lightweight Linux containers for consistent development and deployment. *Linux J.* **2014**, *2014*, 2.

International Journal of
Molecular Sciences

MDPI

Review

AMPK: Regulation of Metabolic Dynamics in the Context of Autophagy

Isaac Tamargo-Gómez [1,2] **and Guillermo Mariño** [1,2,*]

[1]   Instituto de Investigación Sanitaria del Principado de Asturias, 33011 Oviedo, Spain;
      isaactamargo13@gmail.com
[2]   Departamento de Biología Funcional, Universidad de Oviedo, 33011 Oviedo, Spain
*   Correspondence: marinoguillermo@uniovi.es; Tel.: +34-9-856-524-16; Fax: +34-9-856-524-19

Received: 26 October 2018; Accepted: 24 November 2018; Published: 29 November 2018

Abstract: Eukaryotic cells have developed mechanisms that allow them to link growth and proliferation to the availability of energy and biomolecules. AMPK (adenosine monophosphate-activated protein kinase) is one of the most important molecular energy sensors in eukaryotic cells. AMPK activity is able to control a wide variety of metabolic processes connecting cellular metabolism with energy availability. Autophagy is an evolutionarily conserved catabolic pathway whose activity provides energy and basic building blocks for the synthesis of new biomolecules. Given the importance of autophagic degradation for energy production in situations of nutrient scarcity, it seems logical that eukaryotic cells have developed multiple molecular links between AMPK signaling and autophagy regulation. In this review, we will discuss the importance of AMPK activity for diverse aspects of cellular metabolism, and how AMPK modulates autophagic degradation and adapts it to cellular energetic status. We will explain how AMPK-mediated signaling is mechanistically involved in autophagy regulation both through specific phosphorylation of autophagy-relevant proteins or by indirectly impacting in the activity of additional autophagy regulators.

Keywords: AMPK; autophagy; metabolism; mTOR; ULK

1. Introduction

Eukaryotic cells are able to adapt to adverse fluctuations in the cellular environment. In order to do so, cells have developed molecular sensors, which react to circumstances that may perturb cell homeostasis. AMPK (adenosine monophosphate-activated protein kinase) is one of the main cellular sensors able to link a variety of cellular functions and processes to energy availability. AMPK is an evolutionarily conserved protein kinase, present both in unicellular organisms, such as baker's yeast, and also in more complex multicellular eukaryotes, as mammals. In 1981, the role for AMPK yeast ortholog, SNF1 (Sucrose Non-fermenting 1) as the main energy sensor in this organism was described [1]. This kinase is responsible for activating alternate metabolic pathways when the main carbon or nitrogen sources change, thus adapting cellular metabolism to oscillations in the cellular environment. In higher eukaryotes, AMPK is activated when AMP (adenosine monophosphate):ATP (adenosine triphosphate) and/or ADP (adenosine diphosphate):ATP ratios increase [1,2]. Once activated, AMPK maintains energy homeostasis by two complementary actions, an inhibition of ATP-consuming anabolic processes coupled with an activation of ATP-generating catabolic processes [3]. Due to its importance for cellular energy homeostasis, the activity of AMPK is tightly controlled by multiple upstream regulators, which contributes to link cellular metabolism with oscillating parameters in the cellular environment and also with changes in cellular nutritional and energetic requirements [4]. Moreover, AMPK signaling pathways are involved in numerous physiological processes apart from their main metabolic functions, such as cytoskeleton remodeling and transcriptional control or regulation of essential cellular processes, such as apoptosis or autophagy [5].

Autophagy is the cellular process by which organelles, proteins, and different macromolecules are delivered to the lysosomes for degradation [6]. This process can be classified into at least three different pathways: Macroautophagy (which we will refer to as autophagy), microautophagy, and chaperone-mediated autophagy, which mainly differ in the way in which autophagic cargo is transferred to the lysosome [7–9]. Autophagy plays essential roles in all eukaryotic cells and has been implicated in multiple processes, such as cell differentiation, cell death, and regulation of innate and adaptive immune responses or antigen presentation, among many other processes in high eukaryotes [10,11]. Despite all these functions, autophagy's most evolutionarily conserved role, from yeast to mammals, is to sustain energy balance in the cell by providing ATP and building blocks (lipids, amino acids, nucleotides, etc.) out of the degradation of non-essential or damaged cellular structures. Although autophagy regulation is complex and a variety of signaling cascades and regulatory mechanisms modulate autophagic activity, AMPK is probably the most conserved autophagy inducer through evolution. AMPK activity is linked to autophagic degradation in almost all eukaryotic cells.

In this review, we describe the main metabolic regulatory functions of AMPK, including its prominent role in autophagy regulation. We will discuss how AMPK activity is key to the coordination of catabolic pathways which produce energy and micro-molecules with anabolic processes, which use energy and micro-molecules to synthesize new macro-molecules which may be essential to sustain cell viability when cells face significant alterations in the intracellular/extracellular environment.

2. AMPK: Structure and Activation Mechanism

Evolutionarily, AMPK is a highly conserved serine/threonine protein kinase, and a member of the AMPK-related kinase family, which is comprised of thirteen kinases in the human genome. In mammalian cells, AMPK exists as a heterotrimeric complex formed by a catalytic α subunit and regulatory β and γ subunits [12]. There are multiple isoforms for each subunit encoded by different genes, *PRKAA* (5′-AMP-activated protein kinase catalytic subunit alpha), *PRKAB* (5′-AMP-activated protein kinase subunit beta) and *PRKAG* (5′-AMP-activated protein kinase subunit gamma). In humans, there are seven genes encoding AMPK subunits: Two isoforms for the α subunit (α1 and α2), encoded by the genes *PRKAA1* and *PRKAA2*, two isoforms of the β subunit (β1 and β2), encoded by *PRKAB1*, and three isoforms of the γ subunit (γ1, γ2 and γ3), encoded by *PRKAG1, PRKAG2* and *PRKAG3*, respectively [13]. Each AMPK complex is composed by one α-subunit, one β-subunit, and one γ-subunit (Figure 1). The fact that all combinations are possible leads to twelve different conformations in which α, β and γ subunits may constitute a functional AMPK complex, which are normally associated with a specific tissue or a determined cell type, or determined subcellular localizations inside cells [14].

The α-subunit presents a serine/threonine kinase domain at the N-terminal region and a critical residue, Thr172. This conserved residue can be phosphorylated by several different upstream kinases, which constitutes the main mechanism by which AMPK activity is regulated at the short term. Different groups have shown that the LKB1 (Liver kinase B1) kinase is able to phosphorylate Thr172 in response to a variety of signals [15,16]. Furthermore, Thr172 can be phosphorylated by CAMKK2 (Calcium/calmodulin-dependent protein kinase kinase 2) kinase in response to calcium flux, independently of LKB1 [17]. Several other studies have suggested that the MAPKKK family member TAK1/MAP3K7 (Transforming growth factor beta-activated kinase 1)/(Mitogen-activated protein kinase 7) might also phosphorylate Thr172 [18,19]. Apart from the direct regulation of AMPK activity by phosphorylation, its γ-subunit acts a sensor that enables AMPK to respond to changes in AMP/ATP or ADP/ATP levels [20]. Moreover, AMP alone is able to directly modulate AMPK activity through three distinct mechanisms. First, AMP may stimulate the phosphorylation of Thr172 by acting directly on upstream kinases [21]. Second, through allosteric modulation, AMP allows AMPK to be a more attractive substrate for its upstream kinases [22]. Third, AMP could inhibit Thr172 dephosphorylation by protecting it from phosphatases activity, or also increase AMPK activity once Thr172 is phosphorylated, also in an allosteric fashion [23,24].

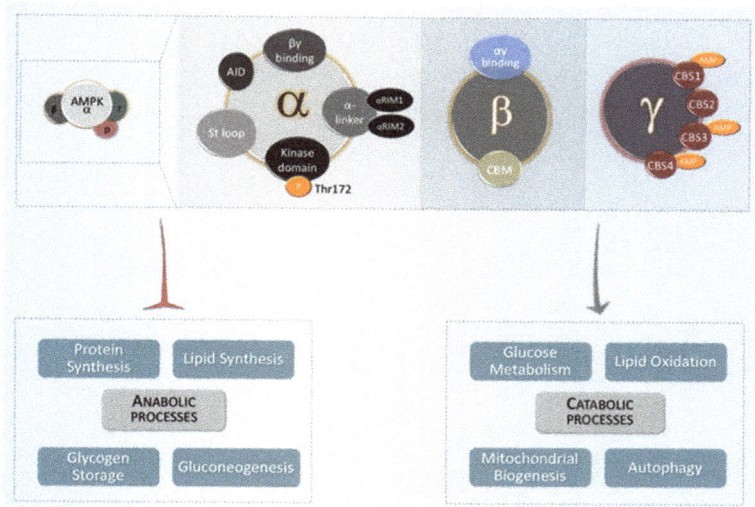

Figure 1. The domain structure of AMPK (adenosine monophosphate-activated protein kinase) heterotrimer. Functional AMPK complexes consist of one catalytic and two regulatory subunits. When activated, AMPK acts by decreasing energy-consuming anabolic processes (lipid synthesis, glycogen storage, gluconeogenesis, and protein synthesis) and increasing energy-providing catabolic processes that provide ATP (glucose metabolism, lipid oxidation, mitochondrial biogenesis and autophagy).

3. AMPK as an Energy-Sensing Kinase for Metabolic Regulation

AMPK is one of the main energy-sensing kinases in eukaryotic cells, able to regulate a wide variety of metabolic processes either by directly acting on metabolically-relevant proteins or by indirectly influencing gene expression [4,25]. During energy stress, AMPK directly activates metabolic enzymes and regulates different energy-consuming/producing pathways, (i.e., lipid or glucose metabolism, or mitochondrial biogenesis) in order to maintain an adequate energy balance. Moreover, AMPK can inhibit the activity of several transcription factors involved in anabolic routes, such as lipid, protein, and carbohydrate biosynthesis, to minimize ATP consumption. By contrast, AMPK promotes the activity of numerous transcription factors implicated in catabolism pathways to stimulate ATP production such as glucose uptake and metabolism (Figure 2) [26].

3.1. Lipid Metabolism

AMPK is a major regulator of cellular lipid metabolism. Once activated, AMPK is able to reduce the activity of essential enzymes for lipid synthesis and related processes, in order to couple their activity to cell energy levels. One of the best-studied AMPK functions in this context is the inhibitory phosphorylation of ACC1 (acetyl-CoA carboxylase 1) and ACC2 (acetyl-CoA carboxylase 2), which catalyze the first step in lipid synthesis. Several studies using animal models have shown that AMPK-mediated phosphorylation of Ser79 in ACC1 and Ser221 in ACC2 are mechanistically involved in the regulation of lipid homeostasis by AMPK [27]. Similarly, AMPK inhibits HMGCR (3-hydroxy-3-methylglutaryl-coenzyme A reductase), which catalyzes an essential step in cholesterol synthesis [28]. Conversely, AMPK positively promotes triglyceride conversion to fatty acids by stimulating lipases such as ATGL (adipocyte triglyceride lipase) and HSL (hormone-sensitive lipase) [29]. When cellular energy is low, free fatty acids can be imported into mitochondria for β-oxidation. This process requires the activity of the acyl-transferases form the CPT1 (Carnitine palmitoyl-transferase 1) family [30]. AMPK indirectly participates in the regulation of CPT1 because malonyl-CoA generated by ACC1 and ACC2 is a potent inhibitor of CPT1 activity. Therefore, AMPK activity directly decreases lipid synthesis and indirectly increases fatty acid import into mitochondria for β-oxidation [31,32].

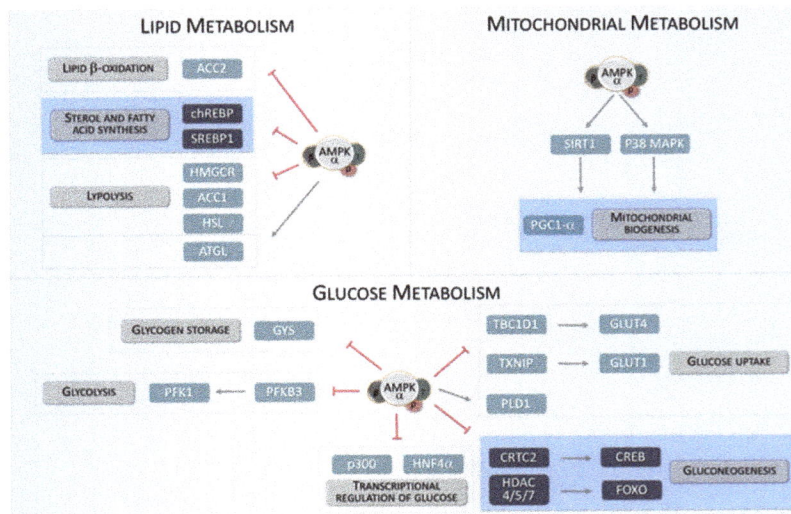

Figure 2. AMPK regulates different metabolic key targets. The metabolic pathways modulated by AMPK can be classified into three general categories: Lipid metabolism, mitochondrial metabolism, and glucose metabolism. The arrow indicates key targets for AMPK involved in these three metabolic categories. Transcriptional regulators are shown in dark squares.

In addition to its direct activity, AMPK is able to modulate lipid metabolism by regulating the activity of several transcription factors involved in lipid synthesis and associated processes. Specifically, AMPK phosphorylation inhibits the transcriptional activity of SREBP1 (sterol regulatory element binding protein 1), ChREBP (carbohydrate-responsive element binding protein) or HNF4α (hepatocyte nuclear factor 4α) [33–36]. Thus, by inhibiting phosphorylation of these transcription factors, AMPK negatively regulates lipogenic processes through the modulation of lipid-specific transcriptional programs.

3.2. Glucose Metabolism

In parallel to its inhibitory role for lipid synthesis, AMPK activity contributes to ATP generation through the modulation of a variety catabolic and anabolic pathways involved in glucose metabolism [37]. AMPK promotes glucose uptake by its inhibitory phosphorylation on TBC1D1 (TBC domain family member 1) and TXNIP (thioredoxin-interacting protein). These two factors respectively inhibit translocation of glucose transporters GLUT1 (Glucose transporter 1) and GLUT4 (Glucose transporter 4) to the plasma membrane. Thus, a high AMPK activity is associated to an increased presence of GLUT1 and GLUT4 glucose transporters in the plasma membrane [38,39]. Consistently, AMPK positively regulates glycolysis by phosphorylating PFKFB3 (6-phosphofructo-2-kinase/fructose-2,6-biphosphatase) [40]. Moreover, AMPK also inhibits glucose conversion into glycogen by inhibitory phosphorylation on different isoforms of GYS (glycogen synthase) [41,42]. Paradoxically, AMPK is involved in the regulation of glycogen supercompensation in skeletal muscle. Under conditions of prolonged physical exercise, sustained activation of AMPK boosts glycogen synthesis, especially in skeletal muscle. This is produced as a result of an increase in glucose uptake, which leads to an accumulation of intracellular G6P (glucose 6 phosphate) that allosterically activates the GYS, thus bypassing the inhibitory action of AMPK on this enzyme [43,44]. In parallel to its direct action on specific enzymes, AMPK also modulates glucose metabolism in a transcriptional fashion. This is the case for gluconeogenesis, which is activated during fasting or reduced glucose intake to maintain blood glucose levels. After re-feeding, a surge in insulin levels leads to phosphorylation of liver AMPK by either AKT (RAC-alpha serine/threonine-protein kinase) or LKB1 kinases, which in turn leads to transcriptional inhibition of gluconeogenesis key genes. This effect is partially achieved by AMPK-dependent phosphorylation and nuclear exclusion of CRTC2 (cyclic-AMP-regulated transcriptional co-activator 2) and HDACs (class IIA histone deacetylases), which are all essential co-factors for the transcription of gluconeogenic genes [45,46]. Thus, AMPK can influence glucose metabolism either by direct regulation of specific proteins or by transcriptional regulation of key genes involved in glucose metabolism [47].

3.3. Mitochondrial Biogenesis

AMPK activity is also important for maintaining mitochondrial function. In situations of energy imbalance, AMPK contributes to mitochondrial biogenesis in order to increase ATP production. During this process, cells increase their individual mitochondrial mass, which requires an increase in the expression of mitochondrial protein genes [48]. One of the main regulators of this process is PGC1-α (peroxisome proliferator-activated receptor-gamma coactivator), which transcriptionally controls the expression of a wide variety of mitochondrial genes. Overexpression of PGC1-α in muscle contributes to the conversion of type IIb fibers into type II and type I fibers, which are rich in mitochondria [49]. PGC1-α interacts with PPAR-γ (peroxisome proliferator-activated receptor-γ) and ERRs (estrogen-related receptors) and is regulated by numerous post-translational mechanisms [50]. These mechanisms include methylation, acetylation, and phosphorylation by upstream kinases. Different in vitro studies indicate that PGC1-α harbors two sites susceptible of being phosphorylated by AMPK, specifically Thr177 and Ser538. In addition, AMPK can indirectly regulate PGC1-α through phosphorylation of additional targets, such as HDAC5, SIRT1 and p38 MAPK. Moreover, AMPK promotes the activity of TFEB (transcription factor EB), which activates the gene encoding PGC1-α, *PPARGC1A*, as well as different genes involved in autophagy [51–54].

AMPK-dependent mitochondrial biogenesis has been specifically studied in skeletal muscle in response to exercise. Exercise activates AMPK in myocytes, leading to mitochondrial biogenesis upregulation [55]. Several studies have shown that overexpression of a constitutively active AMPK γ3-subunit induces mitochondrial biogenesis in mice [56]. In addition, AMPK activation improves muscle regeneration and protects muscle from age-related pathologies, in part by increasing autophagic activity [57].

4. AMPK: Regulation of Autophagy

4.1. Autophagy Regulation

Autophagy is an essential catabolic pathway conserved in all known nucleated cells [7,58]. Although autophagic degradation is constitutively active at basal levels, the main physiological autophagy inducer is nutrient deprivation and/or energy scarcity. This intrinsic characteristic has remained unaltered in organisms ranging from yeast to humans, which makes the involvement of AMPK in the regulation of this process logical.

An autophagy pathway starts with the formation of double-membrane vesicles called autophagosomes. These autophagosomes sequester cytoplasmic cargo (both through specific and non-specific mechanisms) and move along the cellular microtubule network until they eventually fuse with lysosomes [59]. Autophagosome-lysosome fusion allows for the degradation of autophagosome cargo and the autophagosomal inner membrane. Once degradation has occurred, the resulting biomolecules, such as amino acids, lipids, or nucleotides are recycled back to the cytoplasm and will be

Int. J. Mol. Sci. **2018**, *19*, 3812

reused by the cell to synthesize new biomolecules [60]. From a molecular perspective, the autophagy pathway requires the involvement of a group of evolutionarily conserved genes/proteins called ATG (AuTophaGy-related) proteins [61]. These proteins were originally described in yeast and are required for autophagosome formation, maturation, transport, or degradation, being involved in the different steps of the autophagic pathway in a hierarchical and temporally-coordinated fashion (Figure 3) [61,62].

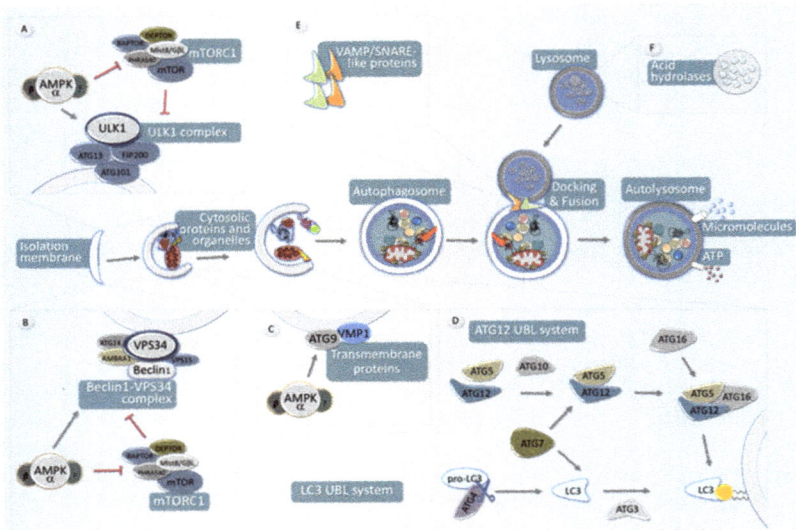

Figure 3. Scheme of the main steps for autophagy and their regulation by AMPK. An autophagy pathway starts with the formation of the isolation membrane, also known as phagophore. The autophagy implicates the coordinated temporal and spatial activation of numerous molecular components. (**A**): The ULK1-FIP200-ATG13-ATG101 complex is responsible for initiating the autophagic process. The activity of this protein complex is antagonistically regulated by mTORC1 (inhibitory phosphorylation) and by AMPK, which both activates the ULK1 complex as well as inhibits the activity of the mTORC1 complex. (**B**): The ClassIII PI3K complex formed by VPS34, Beclin1, ATG14, AMBRA1, and other subunits creates a membrane domain enriched in PtsIns3P, which drives the nucleation of ATG (AuTophaGy-related) proteins in the phagophore, either directly or indirectly. AMPK is able to increase the pro-autophagic function of this complex and to enhance its formation, whereas mTORC1 activity negatively regulates its function. (**C**): Two different transmembrane proteins, the vacuole membrane protein 1 (VMP1) and ATG9 participate in the recruitment of membranes to the phagophore. AMPK is able to phosphorylate ATG9, which increases its recruitment towards autophagosome formation sites. (**D**): Two ubiquitin-like (UBL) protein conjugation systems (ATG12- and LC3- UBLs) involving the participation of ATG4 cysteine proteinases (which activate LC3 by cleaving its carboxyl terminus), the E1-like enzyme ATG7 (common to both conjugation systems), and the E2-like enzymes ATG10 (ATG12 system), and ATG3 (LC3 system). In coordination, the activity of both systems is required to conjugate LC3 (and other members of this protein family homologous to yeast ATG8) to a phosphatidyl-ethanolamine lipid at the nascent pre-autophagosomal membrane. (**E**): Upon completion, fully-formed autophagosomes move along the microtubule network, eventually fusing with a lysosome, thus acquiring hydrolytic activity, and thus becoming autolysosomes. Several SNARE-like proteins (i.e., Syntaxin17 and VAMP8, among others) are required for efficient fusion between lysosomes and autophagosomes. Once content and inner membrane are degraded by acidic hydrolases, the resultant molecules (amino acids, nucleotides, lipids, etc.) are recycled back to the cytoplasm by membrane permeases.

In mammalian cells, autophagosome biogenesis requires the combined activity of two protein complexes, namely the Class III PI3-Kinase protein complex and ULK1/2 (unc-51-like kinase1/2)-containing complexes, which are recruited to autophagosome forming sites during autophagy initiation [63–65]. Mammalian ULK proteins are a family of serine/threonine kinases which are the orthologues of yeast ATG1 and whose activity is essential for the recruitment of autophagy-relevant proteins involved in autophagosome biogenesis [66,67]. There are four members of the ULK family in mammalian cells (ULK1-4). ULK1 is the main ATG1 functional orthologue in mammalian cells [68], although ULK2 has been shown to compensate for ULK1 loss. By contrast, ULK3 and ULK4 seem to have evolved to perform biological functions unrelated to autophagy [68]. ULK1 is part of a protein complex containing ATG13, FIP200, also known as RB1CC1 (RB1-inducible coiled-coil protein 1), and ATG101. As we will discuss later, the autophagy-promoting activity of this protein complex can be modulated through specific phosphorylation of its subunits.

The activity of ULK1 is not enough to promote efficient autophagosome biogenesis. In fact, the activity of the Class III PI3-Kinase protein complex, which contains the catalytic subunit VPS34 (Vacuolar protein sorting 34) and other variable regulatory subunits, is also required for autophagosome formation. This protein complex, which acts as a lipid kinase, generates PIP3P-enriched membrane domains at the site of autophagosome formation, which are required to recruit essential factors for autophagosome formation. Apart from the autophagy-relevant Class III PI3K complex, which is formed by VPS34, VPS15, Beclin1 (Coiled-coil, myosin-like BCL2 interacting protein), ATG14, and AMBRA1 (Activating molecule in Beclin1-regulated autophagy protein 1), VPS34 can form part of other Class III PI3K complexes. These alternative VPS34 protein complexes are comprised of different subunits and regulate vesicular trafficking in processes such as endocytosis or Golgi-mediated protein secretion [69,70]. The relative abundance of these different VPS43-containing complexes is variable and will be regulated according to cellular needs. Thus, in conditions of autophagy induction, most VPS34-containing complexes will consist of autophagy-relevant subunits. Thus, autophagy regulation at this level is achieved both by direct modulation of VPS34 kinase activity and by increasing the formation of autophagy-relevant VPS34-containing complexes.

In addition to the activity of these protein complexes and other ATG proteins, which are specifically involved in autophagy execution, multiple signaling cascades are able to regulate autophagic activity. The relative importance of the different regulatory inputs for autophagy execution is, in many cases, cell type-specific and tissue-dependent. However, there are some major autophagy regulatory circuits which have remained conserved through evolution and are present in most tissues/cell types from most multicellular organisms, including mammals, such as AMPK or mTORC1 [59,71].

4.2. AMPK Antagonizes mTORC1 to Regulate ULK Complex Activity

In mammalian cells, although many different pathways or signaling events may influence autophagy, the two main regulators for autophagic degradation are mTOR (mechanistic/mammalian Target Of Rapamycin) and AMPK kinases [72,73]. mTOR can be found forming two protein complexes, mTORC1 and mTORC2. Although mTORC2 involvement in autophagy dynamics is not totally neglectable, its contribution to autophagy regulation is marginal. By contrast, mTORC1 can be considered as the main autophagy suppressor in mammalian cells. mTORC1 is normally active in situations of high energy levels, high amino acid cellular content, or growth factors stimulation, all of which have a negative impact in autophagic degradation [74]. By contrast, AMPK activation positively regulates autophagic activity, as one may expect due to its pro-catabolic functions. Due to their antagonistic roles, mTORC1 and AMPK activities are molecularly connected, inhibition of mTORC1 activity being one of the main mechanisms by which AMPK increases autophagic degradation and vice versa.

When energy/growth factors or amino acids are abundant, mTORC1 represses autophagy through inhibitory phosphorylation of ATG13, which reduces the activity of the ULK1 complex, thus decreasing the rate of autophagosome formation [75,76]. In the same sense, ULK1 itself is a direct target of mTORC1, so mTORC1 can inhibit the autophagic process by acting both on ULK1 and ATG13 [75]. AMPK plays an opposite role to mTORC1 regarding ULK1 complex activity, thus positively regulating the first steps of autophagosome formation in response to a variety of pro-autophagic stimuli [77,78]. In fact, AMPK increases ULK1 activity by directly phosphorylating Ser467, Ser555, Thr574, and Ser637, which increases the recruitment of autophagy-relevant proteins (ATG proteins) to the membrane domains in which autophagosome formation takes place [79].

In addition, AMPK negatively regulates mTORC1 activity, which blocks its inhibitory effect on ULK1 by two complementary actions [74]. First, AMPK activates TSC2 (Tuberous sclerosis complex 2) by phosphorylating Thr1227 and Ser1345 residues, thus favoring the assembly of TSC1/TSC2 heterodimer, which negatively impacts mTORC1 activity [80]. Second, AMPK can inhibit mTORC1 by direct phosphorylation of RAPTOR (regulatory-associated protein of mTOR) Ser722 and Ser792 residues [81]. In addition, AMPK is able to promote autophagy by acting differentially at different levels of autophagy regulation, through specific phosphorylation in components of autophagy-initiating protein complexes. Again, the activities of AMPK and mTORC1 are antagonistic in relation to autophagy regulation, and they act together to couple autophagy regulation with multiple signaling pathways, with the ULK1 complex being one of the main checkpoints for the regulation of autophagy initiation (Table 1).

Table 1. Regulation of autophagy relevant proteins by AMPK. H, human; M, mouse; R, rat.

Protein	Phosphorylation Site(s)	Stage of Autophagy	Autophagy Function	Ref.
ATG9	Ser761(H, M, R)	Autophagosome elongation	Participates in the recruitment of lipids to the isolation membrane	[82]
BECN1	Ser91(M, R) Ser94(M, R)	Autophagosome biogenesis	Part of the III PI3KC3 complex	[83]
mTOR (RAPTOR)	Ser722(H, M) Ser792(H, M)	Regulation of Autophagy	Negative regulator of Autophagy	[81]
mTOR	Thr2446(H)	Regulation of Autophagy	Negative regulator of Autophagy	[84]
PAQR3	Thr32(H, M)	Autophagosome biogenesis	Facilitates the formation of pro-autophagic PI3KC3 III complex	[85]
RACK1	Thr50(H, M, R)	Autophagosome biogenesis	Promoting the assembly of the III PI3KC3 complex	[86]
TSC2	Ser1387(H, M, R) Thr1271(H, R)	Regulation of Autophagy	Negative regulator of Mtor	[80,87]
ULK1	Ser555(M, R) Ser467(H, M, R) Thr574(M, R) Ser637(M, R)	Autophagy Initiation	Part of the ULK1-complex/early steps of autophagosome biogenesis	[79]
VPS34	Thr163(H, M, R) Ser165(H, M, R)	Autophagosome biogenesis	Part of the III PI3KC3 complex	[83]

4.3. AMPK Regulates Class III PI3K Complex Activity

Apart from its role in regulating ULK1 activity, AMPK is also involved in the regulation of Class III PI3K complex activity. ULK1 itself exerts its pro-autophagic activity by phosphorylating several components of this complex, including Beclin1, AMBRA1, or the catalytic subunit VPS34 [65]. Interestingly, mTORC1 also inhibits autophagosome biogenesis through phosphorylation of ATG14L, an essential component of the pro-autophagic VPS34 complex [64].

AMPK regulation of autophagy also operates through phosphorylation in different subunits of the different Class III PI3K complexes, including VPS34 itself, which modifies their affinity for other components of the complex. Thus, AMPK regulates the relative abundance of the different VPS34-containing complexes, thus connecting the activity of processes involving vesicle trafficking to cellular energy status. In this regard, different biochemical studies have shown how AMPK regulates the composition of the Class III PI3K complex. For example, AMPK phosphorylation of Beclin1 at Thr388 increases Beclin1 binding to VPS34 and ATG14, which promotes higher autophagy activity upon glucose withdrawal than the wild-type control [83,88]. Similarly, AMPK phosphorylation of mouse Beclin1 at Ser-91 and Ser-94 increases the rate of autophagosome formation under nutrient stress conditions [83]. Apart from its activity towards components of the different Class III PI3K complexes, AMPK can also influence their composition by phosphorylating other proteins, which are relevant for the formation/stability of VPS34-containing complexes. Thus, AMPK-mediated phosphorylation of Thr32 on PAQR3 (progestin and adipo-Q receptors member 3), an ATG14L/VPS34 scaffolding protein, or that of Thr50 on the VPS34 associated protein RACK1 (Receptor for activated C kinase 1) has also been shown to enhance stability and pro-autophagic activity of Class III PI3K complexes [85,86].

In parallel to its activating phosphorylation in diverse components of the pro-autophagic VPS34 complexes, AMPK inhibits VPS34 complexes that do not contain pro-autophagic factors and are thus involved in different cellular vesicle trafficking processes, by direct phosphorylation of VPS34 on Thr163 and Ser165 [83]. Hence, in autophagy-promoting conditions, AMPK activation both enhances the activity of pro-autophagic VPS34 complexes and inhibits the formation of other different Class III PI3K complexes involved in autophagy-independent processes.

4.4. Additional AMPK Regulation of Autophagy

Apart from its direct activity towards autophagy-initiating complexes, AMPK can also influence autophagic activity by specific phosphorylation of ATG9, a transmembrane protein involved in autophagosome biogenesis by supplying vesicles which contribute to autophagosome elongation. In fact, AMPK-mediated phosphorylation in Ser761 of ATG9 increases recruitment of ATG9A (and ATG9-containing vesicles) to LC3-positive autophagosomes, thus enhancing autophagosome biogenesis [82].

Additionally, AMPK is able to influence autophagic activity in a transcriptional fashion (Table 2). In fact, under stress situations, AMPK directly phosphorylates the FOXO3 (Forkhead box O3) transcription factor, which regulates genes implicated in autophagy execution [89]. This activity antagonizes that of the mTOR, which on the other hand phosphorylates other members of the FOX (Forkhead box) family, such as FOXK2 (Forkhead box protein K2) and FOXK1 (Forkhead box protein K1), which compete with FOXO3 to repress genes implicated in autophagy [90]. A similar situation in which AMPK and mTOR activities antagonize each other can be found in relation with TFEB/TFE transcription factors, which control the expression of a variety of genes involved in lysosomal biogenesis and autophagy [91].

In situations of high energy, mTOR phosphorylates these factors of transcription and inhibits their function. By contrast, recent reports have shown that TFEB/TFE nuclear translocation is highly reduced either in cells deficient for AMPK or treated with AMPK inhibitors [92,93]. Consistently, it has been recently reported that AMPK activity is required for efficient dissociation of the transcriptional repressor BRD4 (Bromodomain-containing protein 4) from autophagy gene promoters in response to starvation [94].

Apart from its ability to regulate autophagy-relevant transcription factors through direct phosphorylation, AMPK also regulates different transcriptional regulators, such as EP300 [95] or Class IIa HDACs [47], which are involved in metabolism, autophagy and lysosomal functions (Table 2).

Table 2. Transcriptional regulation of autophagy through AMPK phosphorylation. H, human; M, mouse; R, rat.

Transcription Factor	Phosphorylation Site(s)	Target Gene (s)	Ref.
CHOP	Ser30(H, M, R)	*ATG5, MAP1LC3B*	[96]
FOXO3	Thr179(H) Ser399(H) Ser413(H) Ser555(H) Ser588(H) Ser626(H)	*ATG4B, GABARAPL1, ATG12, ATG14, GLUL, MAP1LC3, BECN1, PIK3CA, PIK3C3, ULK1, BNIP3, FBXO32*	[89]
HSF1	Ser121(H, M, R)	*ATG7*	[97]
Nrf2	Ser558(H, M)	*SQSTM1*	[98]
p53	Ser15(H, R)	*AEN, DRAM1, BAX, IGFBP3, BBC3, C12orf5, PRKAB1, PRKAB2, CDKN2A, SESN1, SESN2, DAPK1, BCL2, MCL1*	[99]
p73	Ser426(H)	*ATG5, DRAM1, ATG7, UVRAG*	[100]

4.5. Selective Degradation of Mitochondria by Autophagy

Apart from its general role in the regulation of bulk autophagic degradation, AMPK specifically participates in the regulation of mitophagy, the selective elimination of defective mitochondria through autophagy [101]. Mitochondria are organized in a dynamic network that changes its morphology through the combined actions of fission and fusion. Thus, mitochondria can be found in different distributions ranging from a single closed network to large numbers of small fragments. Recent studies have shown that an increase in mitochondrial fission is required in order to facilitate mitophagy [102]. This renders mitochondria susceptible to being engulfed by pre-autophagosomal isolation membranes, thus allowing mitophagy to take place. Consistently, mitochondrial stressors, such as electron transport chain poisons, or other stressors that damage mitochondria (and thus would increase mitophagy) have been shown to increase mitochondrial fission [103]. AMPK activation by energy imbalance and also a variety of other cellular and mitochondrial stressors promotes mitochondrial fission, thus coupling mitochondrial dynamics with mitophagic degradation [104]. This effect mainly relies on the ability of AMPK to phosphorylate and to activate the MFF (mitochondrial fission factor). MFF is a mitochondrial outer-membrane protein, which recruits cytoplasmic DRP1 (dynamin 1 like protein) to the mitochondrial outer membrane [105]. DRP recruitment to the mitochondrial outer-membrane increases mitochondrial fission, enabling the resulting fragmented mitochondria to undergo mitophagy [106].

In this context, AMPK phosphorylation of ULK1 on Ser555 has been shown to be critical for the development of exercise-induced mitophagy [107]. Thus, through its combined actions on key factors for autophagy regulation (by acting on ULK1 and other major autophagy regulators) and mitochondrial network dynamics (by specifically activating MFF), AMPK is able to coordinate autophagosome formation and mitochondrial size in order to enable efficient autophagic degradation of mitochondria.

5. Conclusions and Future Perspectives

Autophagy regulation has become increasingly complex in high eukaryotes, in which multiple signaling cascades are connected to autophagy key factors. However, AMPK probably remains as the major molecular autophagy inducer, counteracting the activity of mTORC1, which has also evolutionarily remained as the main molecular autophagy inhibitor. In parallel to its function in autophagy regulation, AMPK's role of adapting cellular metabolism to energetic availability has also been conserved in a diversity of organisms, from yeast to mammals. Thus, it is not surprising that substantial efforts have been made to identify new pharmacological AMPK activators. Recently, a variety of compounds able to increase AMPK activity, such as AICAR, Compound-13, PT-1, A769662 or benzimidazole have been identified [108]. Many of these drugs have shown great potential as research tools to modulate AMPK activity, and some of them have been successfully tested in animal models. However, and despite the substantial advances in this field, metformin (clinically developed in the late 1950s) is still the only AMPK-activating drug widely used in human patients. Interestingly, therapeutic AMPK modulation by metformin has shown promising results in the context of diverse metabolic conditions such as type II diabetes, fatty liver diseases, Alzheimer's, and in diverse types of cancers [109]. In fact, the beneficial effects of metformin are sometimes beyond the scope of its a priori potential. The fact that AMPK plays a pivotal role in autophagy regulation, together with the wide variety of processes for which autophagic activity is beneficial, suggests a potential mechanistic involvement of autophagy for some of the positive effects of AMPK activation. Future studies aimed at dissecting the precise molecular mechanisms by which AMPK exerts its wide variety of beneficial effects for human health will shed more light into these questions.

funding: This research was funded by Spain's Ministerio de Economía y Competitividad, grant number RYC-2013-12751, Spain's Ministerio de Economía y Competitividad, grant number BFU2015-68539, and the BBVA foundation, grant number BBM_BIO_3105.

Conflicts of Interest: The authors declare no conflict of interest.

Abbreviation

AID	autoinhibitory domain
AMPK	AMP-activated protein kinase
CBM	carbohydrate-binding module
CBS	cystathionine β-synthase repeats
ChREBP	carbohydrate-responsive element-binding protein
CREB	cAMP response element-binding protein
CTD	C-terminus domain
DEPTOR	DEP domain-containing mTOR-interacting protein
FOXO	forkhead box protein O
HDAC	histone deacetylase
HMGCR	HMG-CoA reductase
HNF4α	hepatocyte nuclear factor 4α
MCL1	myeloid cell leukaemia sequence 1
mLST8	mammalian lethal with SEC13 protein
NTD	N-terminus domain
PGC1α	peroxisome proliferator-activated receptor-γ co-activator 1α
PLD1	phospholipase D1
PRAS40	40 kDa Pro-rich AKT substrate
RAPTOR	regulatory-associated protein of mTOR
RIM	regulatory-subunit-interacting motif
SREBP1	sterol regulatory element-binding protein 1
ST-loop	serine/threonine enriched loop

References

1. Carlson, M.; Osmond, B.; Botstein, D. Mutants of yeast defective in sucrose utilization. *Genetics* **1981**, *98*, 25–40. [PubMed]
2. Hedbacker, K.; Carlson, M. SNF1/AMPK pathways in yeast. *Front. Biosci.* **2008**, *13*, 2408–2420. [CrossRef] [PubMed]
3. Hardie, D.G. The AMP-activated protein kinase pathway—New players upstream and downstream. *J. Cell Sci.* **2004**, *117*, 5479–5487. [CrossRef] [PubMed]
4. Herzig, S.; Shaw, R.J. AMPK: Guardian of metabolism and mitochondrial homeostasis. *Nat. Rev. Mol. Cell Biol.* **2018**, *19*, 121–135. [CrossRef] [PubMed]
5. Ke, R.; Xu, Q.; Li, C.; Luo, L.; Huang, D. Mechanisms of AMPK in the maintenance of atp balance during energy metabolism. *Cell Biol. Int.* **2018**, *42*, 384–392. [CrossRef] [PubMed]

6. Kroemer, G.; Marino, G.; Levine, B. Autophagy and the integrated stress response. *Mol. Cell* **2010**, *40*, 280–293. [CrossRef] [PubMed]

7. Marino, G.; Lopez-Otin, C. Autophagy: Molecular mechanisms, physiological functions and relevance in human pathology. *Cell. Mol. Life Sci.* **2004**, *61*, 1439–1454. [CrossRef] [PubMed]

8. Kaushik, S.; Cuervo, A.M. Chaperone-mediated autophagy: A unique way to enter the lysosome world. *Trends Cell Biol.* **2012**, *22*, 407–417. [CrossRef] [PubMed]

9. Sahu, R.; Kaushik, S.; Clement, C.C.; Cannizzo, E.S.; Scharf, B.; Follenzi, A.; Potolicchio, I.; Nieves, E.; Cuervo, A.M.; Santambrogio, L. Microautophagy of cytosolic proteins by late endosomes. *Dev. Cell* **2011**, *20*, 131–139. [CrossRef] [PubMed]

10. Levine, B.; Kroemer, G. Autophagy in the pathogenesis of disease. *Cell* **2008**, *132*, 27–42. [CrossRef] [PubMed]

11. Levine, B.; Mizushima, N.; Virgin, H.W. Autophagy in immunity and inflammation. *Nature* **2011**, *469*, 323–335. [CrossRef] [PubMed]

12. Hardie, D.G. AMPK: A key regulator of energy balance in the single cell and the whole organism. *Int. J. Obes. (Lond.)* **2008**, *32* (Suppl. 4), S7–S12. [CrossRef]

13. Willows, R.; Navaratnam, N.; Lima, A.; Read, J.; Carling, D. Effect of different gamma-subunit isoforms on the regulation of AMPK. *Biochem. J.* **2017**, *474*, 1741–1754. [CrossRef] [PubMed]

14. Ross, F.A.; Jensen, T.E.; Hardie, D.G. Differential regulation by AMP and ADP of AMPK complexes containing different gamma subunit isoforms. *Biochem. J.* **2016**, *473*, 189–199. [CrossRef] [PubMed]

15. Stein, S.C.; Woods, A.; Jones, N.A.; Davison, M.D.; Carling, D. The regulation of AMP-activated protein kinase by phosphorylation. *Biochem. J.* **2000**, *345 Pt 3*, 437–443. [CrossRef]

16. Hardie, D.G. AMPK: Positive and negative regulation, and its role in whole-body energy homeostasis. *Curr. Opin. Cell Biol.* **2015**, *33*, 1–7. [CrossRef] [PubMed]

17. Fogarty, S.; Hawley, S.A.; Green, K.A.; Saner, N.; Mustard, K.J.; Hardie, D.G. Calmodulin-dependent protein kinase kinase-β activates AMPK without forming a stable complex: Synergistic effects of Ca^{2+} and AMP. *Biochem. J.* **2010**, *426*, 109–118. [CrossRef] [PubMed]

18. Xie, M.; Zhang, D.; Dyck, J.R.; Li, Y.; Zhang, H.; Morishima, M.; Mann, D.L.; Taffet, G.E.; Baldini, A.; Khoury, D.S.; et al. A pivotal role for endogenous TGF-β-activated kinase-1 in the LKB1/AMP-activated protein kinase energy-sensor pathway. *Proc. Natl. Acad. Sci. USA* **2006**, *103*, 17378–17383. [CrossRef] [PubMed]

19. Herrero-Martin, G.; Hoyer-Hansen, M.; Garcia-Garcia, C.; Fumarola, C.; Farkas, T.; Lopez-Rivas, A.; Jaattela, M. Tak1 activates AMPK-dependent cytoprotective autophagy in trail-treated epithelial cells. *EMBO J.* **2009**, *28*, 677–685. [CrossRef] [PubMed]

20. Xiao, B.; Heath, R.; Saiu, P.; Leiper, F.C.; Leone, P.; Jing, C.; Walker, P.A.; Haire, L.; Eccleston, J.F.; Davis, C.T.; et al. Structural basis for AMP binding to mammalian AMP-activated protein kinase. *Nature* **2007**, *449*, 496–500. [CrossRef] [PubMed]

21. Oakhill, J.S.; Steel, R.; Chen, Z.P.; Scott, J.W.; Ling, N.; Tam, S.; Kemp, B.E. AMPK is a direct adenylate charge-regulated protein kinase. *Science* **2011**, *332*, 1433–1435. [CrossRef] [PubMed]

22. Hawley, S.A.; Boudeau, J.; Reid, J.L.; Mustard, K.J.; Udd, L.; Makela, T.P.; Alessi, D.R.; Hardie, D.G. Complexes between the LKB1 tumor suppressor, STRADα/β and MO25α/β are upstream kinases in the AMP-activated protein kinase cascade. *J. Biol.* **2003**, *2*, 28. [CrossRef] [PubMed]

23. Lin, S.C.; Hardie, D.G. AMPK: Sensing glucose as well as cellular energy status. *Cell Metab.* **2018**, *27*, 299–313. [CrossRef] [PubMed]

24. Gowans, G.J.; Hawley, S.A.; Ross, F.A.; Hardie, D.G. AMP is a true physiological regulator of AMP-activated protein kinase by both allosteric activation and enhancing net phosphorylation. *Cell Metab.* **2013**, *18*, 556–566. [CrossRef] [PubMed]

25. Mihaylova, M.M.; Shaw, R.J. The AMPK signalling pathway coordinates cell growth, autophagy and metabolism. *Nat. Cell Biol.* **2011**, *13*, 1016–1023. [CrossRef] [PubMed]

26. Garcia, D.; Shaw, R.J. AMPK: Mechanisms of cellular energy sensing and restoration of metabolic balance. *Mol. Cell* **2017**, *66*, 789–800. [CrossRef] [PubMed]

27. Fullerton, M.D.; Galic, S.; Marcinko, K.; Sikkema, S.; Pulinilkunnil, T.; Chen, Z.P.; O'Neill, H.M.; Ford, R.J.; Palanivel, R.; O'Brien, M.; et al. Single phosphorylation sites in Acc1 and Acc2 regulate lipid homeostasis and the insulin-sensitizing effects of metformin. *Nat. Med.* **2013**, *19*, 1649–1654. [CrossRef] [PubMed]

28. Willows, R.; Sanders, M.J.; Xiao, B.; Patel, B.R.; Martin, S.R.; Read, J.; Wilson, J.R.; Hubbard, J.; Gamblin, S.J.; Carling, D. Phosphorylation of AMPK by upstream kinases is required for activity in mammalian cells. *Biochem. J.* **2017**, *474*, 3059–3073. [CrossRef] [PubMed]

29. Ahmadian, M.; Abbott, M.J.; Tang, T.; Hudak, C.S.; Kim, Y.; Bruss, M.; Hellerstein, M.K.; Lee, H.Y.; Samuel, V.T.; Shulman, G.I.; et al. Desnutrin/ATGL is regulated by AMPK and is required for a brown adipose phenotype. *Cell Metab.* **2011**, *13*, 739–748. [CrossRef] [PubMed]

30. Kerner, J.; Hoppel, C. Fatty acid import into mitochondria. *Biochim. Biophys. Acta* **2000**, *1486*, 1–17. [CrossRef]

31. Saggerson, D. Malonyl-coa, a key signaling molecule in mammalian cells. *Annu. Rev. Nutr.* **2008**, *28*, 253–272. [CrossRef] [PubMed]

32. Fantino, M. Role of lipids in the control of food intake. *Curr. Opin. Clin. Nutr. Metab. Care* **2011**, *14*, 138–144. [CrossRef] [PubMed]

33. Li, Y.; Xu, S.; Mihaylova, M.M.; Zheng, B.; Hou, X.; Jiang, B.; Park, O.; Luo, Z.; Lefai, E.; Shyy, J.Y.; et al. AMPK phosphorylates and inhibits SREBP activity to attenuate hepatic steatosis and atherosclerosis in diet-induced insulin-resistant mice. *Cell Metab.* **2011**, *13*, 376–388. [CrossRef] [PubMed]

34. Sato, S.; Jung, H.; Nakagawa, T.; Pawlosky, R.; Takeshima, T.; Lee, W.R.; Sakiyama, H.; Laxman, S.; Wynn, R.M.; Tu, B.P.; et al. Metabolite regulation of nuclear localization of carbohydrate-response element-binding protein (ChREBP): Role of AMP as an allosteric inhibitor. *J. Biol. Chem.* **2016**, *291*, 10515–10527. [CrossRef] [PubMed]

35. Sato, Y.; Tsuyama, T.; Sato, C.; Karim, M.F.; Yoshizawa, T.; Inoue, M.; Yamagata, K. Hypoxia reduces HNF4α/MODY1 protein expression in pancreatic β-cells by activating AMP-activated protein kinase. *J. Biol. Chem.* **2017**, *292*, 8716–8728. [CrossRef] [PubMed]

36. Elhanati, S.; Kanfi, Y.; Varvak, A.; Roichman, A.; Carmel-Gross, I.; Barth, S.; Gibor, G.; Cohen, H.Y. Multiple regulatory layers of SREBP1/2 by SIRT6. *Cell Rep.* **2013**, *4*, 905–912. [CrossRef] [PubMed]

37. Hardie, D.G. AMPK: A target for drugs and natural products with effects on both diabetes and cancer. *Diabetes* **2013**, *62*, 2164–2172. [CrossRef] [PubMed]

38. Wu, N.; Zheng, B.; Shaywitz, A.; Dagon, Y.; Tower, C.; Bellinger, G.; Shen, C.H.; Wen, J.; Asara, J.; McGraw, T.E.; et al. AMPK-dependent degradation of TXNIP upon energy stress leads to enhanced glucose uptake via GLUT1. *Mol. Cell* **2013**, *49*, 1167–1175. [CrossRef] [PubMed]

39. Chavez, J.A.; Roach, W.G.; Keller, S.R.; Lane, W.S.; Lienhard, G.E. Inhibition of glut4 translocation by tbc1d1, a rab gtpase-activating protein abundant in skeletal muscle, is partially relieved by AMP-activated protein kinase activation. *J. Biol. Chem.* **2008**, *283*, 9187–9195. [CrossRef] [PubMed]

40. Domenech, E.; Maestre, C.; Esteban-Martinez, L.; Partida, D.; Pascual, R.; Fernandez-Miranda, G.; Seco, E.; Campos-Olivas, R.; Perez, M.; Megias, D.; et al. AMPK and PFKFB3 mediate glycolysis and survival in response to mitophagy during mitotic arrest. *Nat. Cell Biol.* **2015**, *17*, 1304–1316. [CrossRef] [PubMed]

41. Hardie, D.G. Ampk—Sensing energy while talking to other signaling pathways. *Cell Metab.* **2014**, *20*, 939–952. [CrossRef] [PubMed]

42. Bultot, L.; Guigas, B.; Von Wilamowitz-Moellendorff, A.; Maisin, L.; Vertommen, D.; Hussain, N.; Beullens, M.; Guinovart, J.J.; Foretz, M.; Viollet, B.; et al. AMP-activated protein kinase phosphorylates and inactivates liver glycogen synthase. *Biochem. J.* **2012**, *443*, 193–203. [CrossRef] [PubMed]

43. Hingst, J.R.; Bruhn, L.; Hansen, M.B.; Rosschou, M.F.; Birk, J.B.; Fentz, J.; Foretz, M.; Viollet, B.; Sakamoto, K.; Faergeman, N.J.; et al. Exercise-induced molecular mechanisms promoting glycogen supercompensation in human skeletal muscle. *Mol. Metab.* **2018**, *16*, 24–34. [CrossRef] [PubMed]

44. Janzen, N.R.; Whitfield, J.; Hoffman, N.J. Interactive roles for AMPK and glycogen from cellular energy sensing to exercise metabolism. *Int. J. Mol. Sci.* **2018**, *19*, 3344. [CrossRef] [PubMed]

45. Lee, J.M.; Seo, W.Y.; Song, K.H.; Chanda, D.; Kim, Y.D.; Kim, D.K.; Lee, M.W.; Ryu, D.; Kim, Y.H.; Noh, J.R.; et al. AMPK-dependent repression of hepatic gluconeogenesis via disruption of CREB/CRTC2 complex by orphan nuclear receptor small heterodimer partner. *J. Biol. Chem.* **2010**, *285*, 32182–32191. [CrossRef] [PubMed]

46. Di Giorgio, E.; Brancolini, C. Regulation of class iia hdac activities: It is not only matter of subcellular localization. *Epigenomics* **2016**, *8*, 251–269. [CrossRef] [PubMed]

47. Mihaylova, M.M.; Vasquez, D.S.; Ravnskjaer, K.; Denechaud, P.D.; Yu, R.T.; Alvarez, J.G.; Downes, M.; Evans, R.M.; Montminy, M.; Shaw, R.J. Class iia histone deacetylases are hormone-activated regulators of FOXO and mammalian glucose homeostasis. *Cell* **2011**, *145*, 607–621. [CrossRef] [PubMed]

48. Kiriyama, Y.; Nochi, H. Intra- and intercellular quality control mechanisms of mitochondria. *Cells* **2017**, *7*, 1. [CrossRef] [PubMed]

49. Spiegelman, B.M. Transcriptional control of mitochondrial energy metabolism through the PGC1 coactivators. *Novartis Found. Symp.* **2007**, *287*, 60–63; Discussion 63–69. [PubMed]

50. Eichner, L.J.; Giguere, V. Estrogen related receptors (errs): A new dawn in transcriptional control of mitochondrial gene networks. *Mitochondrion* **2011**, *11*, 544–552. [CrossRef] [PubMed]

51. Puigserver, P.; Rhee, J.; Lin, J.; Wu, Z.; Yoon, J.C.; Zhang, C.Y.; Krauss, S.; Mootha, V.K.; Lowell, B.B.; Spiegelman, B.M. Cytokine stimulation of energy expenditure through p38 map kinase activation of ppargamma coactivator-1. *Mol. Cell* **2001**, *8*, 971–982. [CrossRef]

52. Li, X.; Monks, B.; Ge, Q.; Birnbaum, M.J. Akt/pkb regulates hepatic metabolism by directly inhibiting PGC-1α transcription coactivator. *Nature* **2007**, *447*, 1012–1016. [CrossRef] [PubMed]

53. Teyssier, C.; Ma, H.; Emter, R.; Kralli, A.; Stallcup, M.R. Activation of nuclear receptor coactivator PGC-1α by arginine methylation. *Genes Dev.* **2005**, *19*, 1466–1473. [CrossRef] [PubMed]

54. Rodgers, J.T.; Lerin, C.; Haas, W.; Gygi, S.P.; Spiegelman, B.M.; Puigserver, P. Nutrient control of glucose homeostasis through a complex of PGC-1α and sirt1. *Nature* **2005**, *434*, 113–118. [CrossRef] [PubMed]

55. Kim, Y.; Triolo, M.; Hood, D.A. Impact of aging and exercise on mitochondrial quality control in skeletal muscle. *Oxid. Med. Cell. Longev.* **2017**, *2017*, 3165396. [CrossRef] [PubMed]

56. Pinter, K.; Grignani, R.T.; Watkins, H.; Redwood, C. Localisation of AMPK gamma subunits in cardiac and skeletal muscles. *J. Muscle Res. Cell Motil.* **2013**, *34*, 369–378. [CrossRef] [PubMed]

57. Burkewitz, K.; Zhang, Y.; Mair, W.B. AMPK at the nexus of energetics and aging. *Cell Metab.* **2014**, *20*, 10–25. [CrossRef] [PubMed]

58. Yin, Z.; Pascual, C.; Klionsky, D.J. Autophagy: Machinery and regulation. *Microb. Cell* **2016**, *3*, 588–596. [CrossRef] [PubMed]

59. Yang, Z.; Klionsky, D.J. Mammalian autophagy: Core molecular machinery and signaling regulation. *Curr. Opin. Cell Biol.* **2010**, *22*, 124–131. [CrossRef] [PubMed]

60. Singh, R.; Cuervo, A.M. Autophagy in the cellular energetic balance. *Cell Metab.* **2011**, *13*, 495–504. [CrossRef] [PubMed]

61. Itakura, E.; Mizushima, N. Characterization of autophagosome formation site by a hierarchical analysis of mammalian Atg proteins. *Autophagy* **2010**, *6*, 764–776. [CrossRef] [PubMed]

62. Koyama-Honda, I.; Itakura, E.; Fujiwara, T.K.; Mizushima, N. Temporal analysis of recruitment of mammalian Atg proteins to the autophagosome formation site. *Autophagy* **2013**, *9*, 1491–1499. [CrossRef] [PubMed]

63. Kaizuka, T.; Mizushima, N. Atg13 is essential for autophagy and cardiac development in mice. *Mol. Cell. Biol.* **2016**, *36*, 585–595. [CrossRef] [PubMed]

64. Park, J.M.; Jung, C.H.; Seo, M.; Otto, N.M.; Grunwald, D.; Kim, K.H.; Moriarity, B.; Kim, Y.M.; Starker, C.; Nho, R.S.; et al. The ULK1 complex mediates mtorc1 signaling to the autophagy initiation machinery via binding and phosphorylating ATG14. *Autophagy* **2016**, *12*, 547–564. [CrossRef] [PubMed]

65. Russell, R.C.; Tian, Y.; Yuan, H.; Park, H.W.; Chang, Y.Y.; Kim, J.; Kim, H.; Neufeld, T.P.; Dillin, A.; Guan, K.L. ULK1 induces autophagy by phosphorylating beclin-1 and activating VPS34 lipid kinase. *Nat. Cell Biol.* **2013**, *15*, 741–750. [CrossRef] [PubMed]

66. Hara, T.; Takamura, A.; Kishi, C.; Iemura, S.; Natsume, T.; Guan, J.L.; Mizushima, N. Fip200, a ulk-interacting protein, is required for autophagosome formation in mammalian cells. *J. Cell Biol.* **2008**, *181*, 497–510. [CrossRef] [PubMed]

67. Chan, E.Y.; Kir, S.; Tooze, S.A. Sirna screening of the kinome identifies ULK1 as a multidomain modulator of autophagy. *J. Biol. Chem.* **2007**, *282*, 25464–25474. [CrossRef] [PubMed]

68. Mizushima, N. The role of the Atg1/ULK1 complex in autophagy regulation. *Curr. Opin. Cell Biol.* **2010**, *22*, 132–139. [CrossRef] [PubMed]

69. Stjepanovic, G.; Baskaran, S.; Lin, M.G.; Hurley, J.H. Vps34 kinase domain dynamics regulate the autophagic pi 3-kinase complex. *Mol. Cell* **2017**, *67*, 528–534. [CrossRef] [PubMed]

70. Stjepanovic, G.; Baskaran, S.; Lin, M.G.; Hurley, J.H. Unveiling the role of vps34 kinase domain dynamics in regulation of the autophagic pi3k complex. *Mol. Cell Oncol.* **2017**, *4*, e1367873. [CrossRef] [PubMed]

71. Russell, R.C.; Yuan, H.X.; Guan, K.L. Autophagy regulation by nutrient signaling. *Cell Res.* **2014**, *24*, 42–57. [CrossRef] [PubMed]

72. Inoki, K.; Kim, J.; Guan, K.L. AMPK and mtor in cellular energy homeostasis and drug targets. *Annu. Rev. Pharmacol. Toxicol.* **2012**, *52*, 381–400. [CrossRef] [PubMed]

73. Kim, J.; Kundu, M.; Viollet, B.; Guan, K.L. AMPK and mtor regulate autophagy through direct phosphorylation of ULK1. *Nat. Cell Biol.* **2011**, *13*, 132–141. [CrossRef] [PubMed]

74. Hardie, D.G. Cell biology. Why starving cells eat themselves. *Science* **2011**, *331*, 410–411. [CrossRef] [PubMed]

75. Puente, C.; Hendrickson, R.C.; Jiang, X. Nutrient-regulated phosphorylation of ATG13 inhibits starvation-induced autophagy. *J. Biol. Chem.* **2016**, *291*, 6026–6035. [CrossRef] [PubMed]

76. Kamada, Y.; Yoshino, K.; Kondo, C.; Kawamata, T.; Oshiro, N.; Yonezawa, K.; Ohsumi, Y. Tor directly controls the Atg1 kinase complex to regulate autophagy. *Mol. Cell. Biol.* **2010**, *30*, 1049–1058. [CrossRef] [PubMed]

77. Dite, T.A.; Ling, N.X.Y.; Scott, J.W.; Hoque, A.; Galic, S.; Parker, B.L.; Ngoei, K.R.W.; Langendorf, C.G.; O'Brien, M.T.; Kundu, M.; et al. The autophagy initiator ulk1 sensitizes AMPK to allosteric drugs. *Nat. Commun.* **2017**, *8*, 571. [CrossRef] [PubMed]

78. Alers, S.; Loffler, A.S.; Wesselborg, S.; Stork, B. Role of AMPK-mtor-ULK1/2 in the regulation of autophagy: Cross talk, shortcuts, and feedbacks. *Mol. Cell. Biol.* **2012**, *32*, 2–11. [CrossRef] [PubMed]

79. Egan, D.F.; Shackelford, D.B.; Mihaylova, M.M.; Gelino, S.; Kohnz, R.A.; Mair, W.; Vasquez, D.S.; Joshi, A.; Gwinn, D.M.; Taylor, R.; et al. Phosphorylation of ULK1 (hATG1) by AMP-activated protein kinase connects energy sensing to mitophagy. *Science* **2011**, *331*, 456–461. [CrossRef] [PubMed]

80. Inoki, K.; Zhu, T.; Guan, K.L. Tsc2 mediates cellular energy response to control cell growth and survival. *Cell* **2003**, *115*, 577–590. [CrossRef]

81. Gwinn, D.M.; Shackelford, D.B.; Egan, D.F.; Mihaylova, M.M.; Mery, A.; Vasquez, D.S.; Turk, B.E.; Shaw, R.J. AMPK phosphorylation of raptor mediates a metabolic checkpoint. *Mol. Cell* **2008**, *30*, 214–226. [CrossRef] [PubMed]

82. Weerasekara, V.K.; Panek, D.J.; Broadbent, D.G.; Mortenson, J.B.; Mathis, A.D.; Logan, G.N.; Prince, J.T.; Thomson, D.M.; Thompson, J.W.; Andersen, J.L. Metabolic-stress-induced rearrangement of the 14-3-3ζ interactome promotes autophagy via a ULK1- and AMPK-regulated 14-3-3ζ interaction with phosphorylated Atg9. *Mol. Cell. Biol.* **2014**, *34*, 4379–4388. [CrossRef] [PubMed]

83. Kim, J.; Kim, Y.C.; Fang, C.; Russell, R.C.; Kim, J.H.; Fan, W.; Liu, R.; Zhong, Q.; Guan, K.L. Differential regulation of distinct vps34 complexes by AMPK in nutrient stress and autophagy. *Cell* **2013**, *152*, 290–303. [CrossRef] [PubMed]

84. Cheng, S.W.; Fryer, L.G.; Carling, D.; Shepherd, P.R. Thr2446 is a novel mammalian target of rapamycin (mtor) phosphorylation site regulated by nutrient status. *J. Biol. Chem.* **2004**, *279*, 15719–15722. [CrossRef] [PubMed]

85. Xu, D.Q.; Wang, Z.; Wang, C.Y.; Zhang, D.Y.; Wan, H.D.; Zhao, Z.L.; Gu, J.; Zhang, Y.X.; Li, Z.G.; Man, K.Y.; et al. PAQR3 controls autophagy by integrating AMPK signaling to enhance ATG14L-associated PI3K activity. *EMBO J.* **2016**, *35*, 496–514. [CrossRef] [PubMed]

86. Zhao, Y.; Wang, Q.; Qiu, G.; Zhou, S.; Jing, Z.; Wang, J.; Wang, W.; Cao, J.; Han, K.; Cheng, Q.; et al. RACK1 promotes autophagy by enhancing the Atg14l-beclin 1-Vps34-Vps15 complex formation upon phosphorylation by AMPK. *Cell Rep.* **2015**, *13*, 1407–1417. [CrossRef] [PubMed]

87. Huang, J.; Manning, B.D. The tsc1-tsc2 complex: A molecular switchboard controlling cell growth. *Biochem. J.* **2008**, *412*, 179–190. [CrossRef] [PubMed]

88. Kim, J.; Guan, K.L. AMPK connects energy stress to pik3c3/vps34 regulation. *Autophagy* **2013**, *9*, 1110–1111. [CrossRef] [PubMed]

89. Greer, E.L.; Oskoui, P.R.; Banko, M.R.; Maniar, J.M.; Gygi, M.P.; Gygi, S.P.; Brunet, A. The energy sensor AMP-activated protein kinase directly regulates the mammalian foxo3 transcription factor. *J. Biol. Chem.* **2007**, *282*, 30107–30119. [CrossRef] [PubMed]

90. Bowman, C.J.; Ayer, D.E.; Dynlacht, B.D. Foxk proteins repress the initiation of starvation-induced atrophy and autophagy programs. *Nat. Cell Biol.* **2014**, *16*, 1202–1214. [CrossRef] [PubMed]

91. Lapierre, L.R.; Kumsta, C.; Sandri, M.; Ballabio, A.; Hansen, M. Transcriptional and epigenetic regulation of autophagy in aging. *Autophagy* **2015**, *11*, 867–880. [CrossRef] [PubMed]

92. Young, N.P.; Kamireddy, A.; Van Nostrand, J.L.; Eichner, L.J.; Shokhirev, M.N.; Dayn, Y.; Shaw, R.J. AMPK governs lineage specification through tfeb-dependent regulation of lysosomes. *Genes Dev.* **2016**, *30*, 535–552. [CrossRef] [PubMed]

93. Kim, S.H.; Kim, G.; Han, D.H.; Lee, M.; Kim, I.; Kim, B.; Kim, K.H.; Song, Y.M.; Yoo, J.E.; Wang, H.J.; et al. Ezetimibe ameliorates steatohepatitis via AMP activated protein kinase-tfeb-mediated activation of autophagy and nlrp3 inflammasome inhibition. *Autophagy* **2017**, *13*, 1767–1781. [CrossRef] [PubMed]

94. Sakamaki, J.I.; Wilkinson, S.; Hahn, M.; Tasdemir, N.; O'Prey, J.; Clark, W.; Hedley, A.; Nixon, C.; Long, J.S.; New, M.; et al. Bromodomain protein brd4 is a transcriptional repressor of autophagy and lysosomal function. *Mol. Cell* **2017**, *66*, 517–532. [CrossRef] [PubMed]

95. Lin, Y.Y.; Kiihl, S.; Suhail, Y.; Liu, S.Y.; Chou, Y.H.; Kuang, Z.; Lu, J.Y.; Khor, C.N.; Lin, C.L.; Bader, J.S.; et al. Functional dissection of lysine deacetylases reveals that hdac1 and p300 regulate AMPK. *Nature* **2012**, *482*, 251–255. [CrossRef] [PubMed]

96. Dai, X.; Ding, Y.; Liu, Z.; Zhang, W.; Zou, M.H. Phosphorylation of chop (c/ebp homologous protein) by the AMP-activated protein kinase Alpha 1 in macrophages promotes chop degradation and reduces injury-induced neointimal disruption in vivo. *Circ. Res.* **2016**, *119*, 1089–1100. [CrossRef] [PubMed]

97. Dai, S.; Tang, Z.; Cao, J.; Zhou, W.; Li, H.; Sampson, S.; Dai, C. Suppression of the hsf1-mediated proteotoxic stress response by the metabolic stress sensor AMPK. *EMBO J.* **2015**, *34*, 275–293. [CrossRef] [PubMed]

98. Joo, M.S.; Kim, W.D.; Lee, K.Y.; Kim, J.H.; Koo, J.H.; Kim, S.G. AMPK facilitates nuclear accumulation of Nrf2 by phosphorylating at serine 550. *Mol. Cell. Biol.* **2016**, *36*, 1931–1942. [CrossRef] [PubMed]

99. Jones, R.G.; Plas, D.R.; Kubek, S.; Buzzai, M.; Mu, J.; Xu, Y.; Birnbaum, M.J.; Thompson, C.B. Amp-activated protein kinase induces a p53-dependent metabolic checkpoint. *Mol. Cell* **2005**, *18*, 283–293. [CrossRef] [PubMed]

100. Adamovich, Y.; Adler, J.; Meltser, V.; Reuven, N.; Shaul, Y. AMPK couples p73 with p53 in cell fate decision. *Cell Death Differ.* **2014**, *21*, 1451–1459. [CrossRef] [PubMed]

101. Zhang, C.S.; Lin, S.C. AMPK promotes autophagy by facilitating mitochondrial fission. *Cell Metab.* **2016**, *23*, 399–401. [CrossRef] [PubMed]

102. Shirihai, O.S.; Song, M.; Dorn, G.W. How mitochondrial dynamism orchestrates mitophagy. *Circ. Res.* **2015**, *116*, 1835–1849. [CrossRef] [PubMed]

103. Youle, R.J.; van der Bliek, A.M. Mitochondrial fission, fusion, and stress. *Science* **2012**, *337*, 1062–1065. [CrossRef] [PubMed]

104. Toyama, E.Q.; Herzig, S.; Courchet, J.; Lewis, T.L., Jr.; Loson, O.C.; Hellberg, K.; Young, N.P.; Chen, H.; Polleux, F.; Chan, D.C.; et al. Metabolism. AMP-activated protein kinase mediates mitochondrial fission in response to energy stress. *Science* **2016**, *351*, 275–281. [CrossRef] [PubMed]

105. Otera, H.; Wang, C.; Cleland, M.M.; Setoguchi, K.; Yokota, S.; Youle, R.J.; Mihara, K. Mff is an essential factor for mitochondrial recruitment of Drp1 during mitochondrial fission in mammalian cells. *J. Cell Biol.* **2010**, *191*, 1141–1158. [CrossRef] [PubMed]

106. Wang, C.; Youle, R. Cell biology: Form follows function for mitochondria. *Nature* **2016**, *530*, 288–289. [CrossRef] [PubMed]

107. Laker, R.C.; Drake, J.C.; Wilson, R.J.; Lira, V.A.; Lewellen, B.M.; Ryall, K.A.; Fisher, C.C.; Zhang, M.; Saucerman, J.J.; Goodyear, L.J.; et al. AMPK phosphorylation of ulk1 is required for targeting of mitochondria to lysosomes in exercise-induced mitophagy. *Nat. Commun.* **2017**, *8*, 548. [CrossRef] [PubMed]

108. Kim, J.; Yang, G.; Kim, Y.; Kim, J.; Ha, J. AMPK activators: Mechanisms of action and physiological activities. *Exp. Mol. Med.* **2016**, *48*, e224. [CrossRef] [PubMed]

109. Schulten, H.J. Pleiotropic effects of metformin on cancer. *Int. J. Mol. Sci.* **2018**, *19*, 2850. [CrossRef] [PubMed]

International Journal of
Molecular Sciences

MDPI

Review

Implication and Regulation of AMPK during Physiological and Pathological Myeloid Differentiation

Arnaud Jacquel [1,2,†], Frederic Luciano [1,2,†], Guillaume Robert [1,2,†] and Patrick Auberger [1,2,*,†]

[1] Université Côte d'Azur, C3M Inserm U1065, 06204 Nice, France; jacquel@unice.fr (A.J.); fluciano@unice.fr (F.L.); robertg@unice.fr (G.R.)
[2] Equipe Labellisée par la Fondation ARC, 94803 Villejuif, France
[*] Correspondence: auberger@unice.fr; Tel.: +33-4-89-06-43-06
[†] These authors contributed equally to this work.

Received: 6 September 2018; Accepted: 28 September 2018; Published: 30 September 2018

Abstract: AMP-activated protein kinase (AMPK) is a heterotrimeric serine/threonine kinase consisting of the arrangement of various α β, and γ isoforms that are expressed differently depending on the tissue or the cell lineage. AMPK is one of the major sensors of energy status in mammalian cells and as such plays essential roles in the regulation of cellular homeostasis, metabolism, cell growth, differentiation, apoptosis, and autophagy. AMPK is activated by two upstream kinases, the tumor suppressor liver kinase B1 (LKB1) and the calcium/calmodulin-dependent protein kinase kinase 2 (CAMKK2) through phosphorylation of the kinase on Thr172, leading to its activation. In addition, AMPK inhibits the mTOR pathway through phosphorylation and activation of tuberous sclerosis protein 2 (TSC2) and causes direct activation of unc-51-like autophagy activating kinase 1 (ULK1) via phosphorylation of Ser555, thus promoting initiation of autophagy. Although it is well established that AMPK can control the differentiation of different cell lineages, including hematopoietic stem cells (HSCs), progenitors, and mature hematopoietic cells, the role of AMPK regarding myeloid cell differentiation is less documented. The differentiation of monocytes into macrophages triggered by colony stimulating factor 1 (CSF-1), a process during which both caspase activation (independently of apoptosis induction) and AMPK-dependent stimulation of autophagy are necessary, is one noticeable example of the involvement of AMPK in the physiological differentiation of myeloid cells. The present review focuses on the role of AMPK in the regulation of the physiological and pathological differentiation of myeloid cells. The mechanisms of autophagy induction by AMPK will also be addressed, as autophagy has been shown to be important for differentiation of hematopoietic cells. In addition, myeloid malignancies (myeloid leukemia or dysplasia) are characterized by profound defects in the establishment of proper differentiation programs. Reinduction of a normal differentiation process in myeloid malignancies has thus emerged as a valuable and promising therapeutic strategy. As AMPK seems to exert a key role in the differentiation of myeloid cells, notably through induction of autophagy, we will also discuss the potential to target this pathway as a pro-differentiating and anti-leukemic strategy in myeloid malignancies.

Keywords: AMPK; monocytes; macrophages; differentiation; autophagy; AML; MDS; CML; CMML

1. Introduction

AMPK is an essential regulator and sensor of cellular energy status in mammalian cells. This kinase integrates changes in the AMP/ATP and ADP/ATP ratios, fine-tuning the balance between ATP consumption and synthesis [1,2]. AMPK acts to increase the rate of catabolic process and to decrease the rate of anabolic process to sustain intracellular energy homeostasis [3]. AMPK is activated by a wide range of stimuli, such as nutrient deprivation [4], cellular stresses [5], fasting or caloric restriction [6], caloric restriction mimetics [7–10], and nucleoside analogues such as (5-aminoimidazole-4-carboxamide-1-β-D-ribofuranoside; AICAR) [11], and has been evaluated in clinical trials for the treatment of metabolic diseases and cancers, including both hematopoietic malignancies and solid tumors [12–15]. In addition to its well-established role in maintaining energy homeostasis, AMPK is also extensively involved in promoting autophagy [16–18], longevity, and tumor suppression [19–22].

AMPK is a highly conserved heterotrimeric serine/threonine kinase that phosphorylates a plethora of cellular substrates implicated in numerous metabolic pathways through rapid or long-term adaptive responses to different metabolites [23]. The first thoroughly characterized AMPK substrate was acetyl-CoA-carboxylase, which catalyzes the carboxylation of acetyl-CoA to produce malonyl-CoA, highlighting an essential role for this kinase in the induction of fatty acid oxydation [24]. More generally, AMPK intervenes at the crossroad of different key cellular signaling pathways, monitoring cellular energy status, cell proliferation, differentiation and autophagy, notably through its negative control of the PI3K/AKT/mTOR pathway and its stimulatory effect on autophagy through direct phosphorylation of ULK1 (ATG1, a serine threonine kinase involved in the initiation of this catabolic process) [25–27]. AMPK negatively regulates energy-consuming cell proliferation/growth, whereas it induces an energy compensating program, such as autophagy. It is well established that AKT signaling favors glucose uptake and glycolysis, therefore increasing ATP production and reducing AMPK activation. To that end, AKT directly phosphorylates AMPK on ser487, hindering the activation of AMPK by LKB1 [28]. AKT signaling through mTORC1 stimulates ATP-consuming anabolic processes, whereas AMPK blocks anabolic metabolism to favor catabolic processes. In addition, AKT and AMPK exert opposing effects on mTORC1 signaling as well as nutrient and glycogen synthesis. In addition to regulating cell proliferation, AKT, through mTORC1, inhibits autophagy. Indeed, ULK1 activity is repressed by mTORC1 complex leading to autophagy inhibition. By contrast, phosphorylation of TSC2 by AMPK increases its activity, leading to mTORC1 inhibition and indirectly to autophagy induction. Finally, AMP directly phosphorylates ULK1 on ser555 to promote induction of autophagy. In conclusion, cross-talk between these two kinases is critical to switch from the catabolic to the anabolic states (and vice versa) and to balance autophagy [29].

Recent findings also indicate that AMPK can sense glucose availability independently of adenine nucleotide binding through the formation of a complex with axin at the lysosomal membrane [30,31]. Finally, in addition to its essential role in the regulation of cellular metabolism and autophagy, AMPK also plays a critical role in the regulation of anti-inflammatory responses [32,33] and cell growth and differentiation [4,16,34]. In this context, modulation of the AMPK pathway has emerged as a promising strategy in a wide range of human pathologies, including metabolic and inflammatory diseases and cancer.

There is increasing evidence that AMPK is also required for many types of cellular differentiation processes by providing through different mechanisms (increased metabolism and activation of autophagy, for instance) the fuel and nutrients (glucose, amino acids, lipids, and sugars) that sustain the modification of the transcription programs necessary for the completion of cellular differentiation. The present review emphasizes the role of AMPK in the differentiation of hematopoietic cells, with a special focus on the physiological and pathological differentiation of myeloid cells. The opportunity to use recently developed AMPK activators for the treatment of myeloid malignancies, including myelodysplastic syndrome (MDS), acute myeloid leukemia (AML), chronic myelogenous leukemia (CML), and chronic myelomonocytic leukemia (CMML) will also be addressed.

2. AMPK Structure and Activation

The structure and activation modalities of AMPK have been excellently reviewed recently [34,35]. Briefly, the AMPK catalytic subunits (α1 and α2) are encoded by two different genes called PRKAA1 and PRKAA2. The α subunits consist, from their N- to C-termini, of N- and C-lobes, which together form the kinase domain (α-KD) typical of serine/threonine kinases, an auto-inhibitory domain (α-AID), an α-linker domain, and finally a globular C-terminal domain (α-CTD). The scaffolding beta subunits (β1 or β2) are encoded by two different genes (PRKAB1 and PRKAB2). They display a myristoylated N-terminal region involved in the recruitment of AMPK to the membrane of mitochondria or phagophores, followed by a central carbohydrate-binding module domain (β-CBM) and a C-terminal subunit interaction domain (β-SID). The γ regulatory subunits (γ1, γ2, or γ3) are encoded by three different genes: PRKAG1, PRKAG2, and PRKAG3. The γ subunits are characterized by an N-terminal region of variable length and the presence of four tandem cystathionine-beta-synthase repeats (CBS1 to CBS4), that are involved in the binding of the regulatory nucleotides AMP, ADP, and ATP (Figure 1). The α, β, and γ AMPK subunits assemble into different heterotrimeric serine/threonine kinases depending on their tissue distribution and on the representativeness of the various isoforms (Figure 1). In myeloid cells, the α1β1γ1 complex is predominantly expressed. Of note, AMPK complexes composed of different gamma subunit isoforms (γ1 to γ3) exhibit exquisite variations in their sensitivity to increased AMP and ADP levels that enable subtle variations in the regulation of the kinase activity. This suggests that AMPK could respond differently and selectively to changes in adenine nucleotide depending on the nature of the γ subunit present in the complex [36]. AMPK is activated by phosphorylation at Thr172 in the C-lobe domain, an event that is achieved by one of the three main upstream kinases, the tumor suppressor LKB1, CaMKK2, or TAK1 [37]. Other regulatory sites, such as Ser487 in the α1 subunit and Ser491 in the α2 subunit, that are phosphorylated by either protein kinase A, protein kinase B (AKT), or ribosomal S6 kinase are implicated in the negative regulation of AMPK and thus favor activation of the mTOR pathway. Once activated, AMPK phosphorylates a plethora of protein substrates that are involved in the regulation of cell energetic metabolism, autophagy, modulation of proliferation, regulation of inflammatory responses, and differentiation [38]. Thus far, more than 50 AMPK substrates have been identified in mammalian cells that are involved in energy homeostasis but also in other functions not directly linked to metabolism regulation [39]. However, the phosphorylome of AMPK is far from completely known, as recently attested by the identification of TET2 (Tet-eleven translocation 2) phosphorylation by AMPK linking diabetes to cancer [40].

335

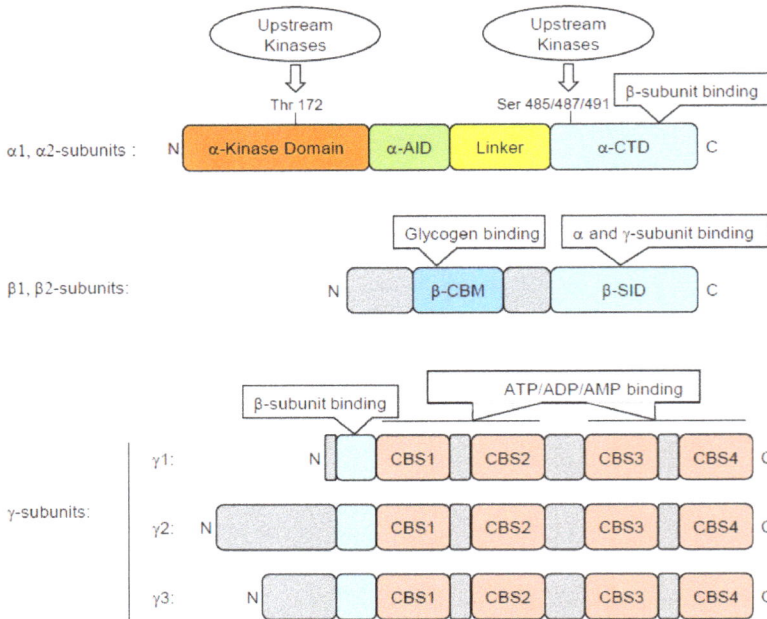

Figure 1. Structure of mammalian AMPK subunits. AMPK is a heterotrimeric protein consisting of 1 catalytic subunit (α subunit) and 2 regulatory subunits (β and γ subunits). The α subunit contains a kinase domain (α-KD), the activity of which relies on the phosphorylation of Thr172 by upstream AMPK kinases. The kinase domain is followed by an autoinhibitory domain (α-AID) that is joined to the COOH-terminal domains (α-CTD) by a less well conserved linker. The α-CTD domain binds to the C-terminal domain of the β subunit. The β-subunit contains two conserved regions: (1) a carbohydrate-binding module (β-CBM) that causes the mammalian complex to bind to glycogen particles and (2) a COOH- terminal subunit interaction domain (β-SID) that provides the bridge between the α- and γ-subunits. The γ-subunit contains variable NH2-terminal regions followed by a short sequence involved in binding to the β-subunit and by four tandem repeats of a cystathionine-β-synthase (CBS) motif that act in pairs to form the binding sites for adenine nucleotides (ATP, ADP and AMP).

3. AMPK and Differentiation of Hematopoietic Stem Cells

Hematopoietic stem cells (HSCs) are multipotent, self-renewing progenitor cells that can replenish all blood cell types in a process called "hematopoiesis." HSCs differentiate into two lineage-restricted, lymphoid and myelo-erythroid, oligopotent progenitor cells that will ultimately give rise to lymphocytes, granulocytes, macrophages, erythrocytes, and platelets. An alternative, "myeloid-based" model for blood lineage differentiation also implies a common myelo-lymphoid progenitor cell that generates progeny from both lineages. The mechanisms controlling HSC self-renewal and differentiation are influenced by a diverse set of cytokines, chemokines, receptors, and intracellular signaling molecules. Producing 500 billion cells a day from a limited number of HSCs is a highly energy-consuming process. How hematopoietic stem cells (HSCs) accommodate their energy requirement during growth and differentiation stages remains poorly understood. Recent studies in the literature have established that the LKB1 tumor suppressor, one of the AMPK upstream kinases, is critical for the maintenance of energy homeostasis in HSCs. In different studies, *Lkb1* invalidation in mice has been shown to cause loss of HSC quiescence, division, rapid depletion, and pancytopenia [41]. In addition, *Lkb1*-deficient bone marrow cells display drastic mitochondrial defects, defaults in lipid and nucleotide metabolism, and, as a consequence, decreased ATP consumption. Nevertheless, AMPK seems to exert only a marginal role in HSC depletion. While HSC exhaustion appears to occur largely independently of AMPK, the defects in mitochondrial function are also observed in *Ampk*-deficient mice [41], suggesting that AMPK can act in concert with LKB1 to regulate mitochondrial fitness but not HSC depletion. These data highlighted an essential role for LKB1 as an inhibitor of HSC proliferation but a likely more minor role of AMPK and mTOR in this process [42]. As mentioned previously, the LKB1/AMPK pathway is one of the main activators of autophagy [25,43]. It is therefore possible that the changes in mitochondrial mass and decreased ATP levels observed in the absence of LKB1 and AMPK could reflect alteration in the rates of autophagy in deficient mice.

Differentiation of HSCs into their different progeny, including myeloid precursors, requires high energy status levels. Recent studies have highlighted the crucial role of mitochondrial oxidative phosphorylation for the differentiation of HSCs. Indeed, specific invalidation of the PTEN-like mitochondrial phosphatase PTPMT-1 in HSCs resulted in an increase of the HSC pool and a block in differentiation [44]. Importantly, reintroduction of catalytically active PTPMT1, but not catalytically inactive PTPMT1 or PTPMP1 lacking mitochondrial localization signal, restores the differentiation capabilities of PTPMT1-deficient HSCs. This blockade in differentiation potential of HSCs is linked to altered mitochondrial metabolism in the absence of PTPMT1. While PTPMT1 seems to be essential for HSC commitment, depletion of PTPMT1 in myeloid or lymphoid progenitors failed to cause any defect in lineage-specific knock-out mice. It is likely that the altered mitochondrial metabolism caused by *PTPMT1* deficiency is sensed by AMPK and then relayed to the p53-p21/p57 pathway, as p21 and p57 expression was significantly upregulated in PTPMT1 depleted stem cells and early progenitors. AMPK on its own could exert its effect on mitochondria through phosphorylation and activation of PGC-1α and β, which are master transcriptional regulators of mitochondrial biogenesis [45].

4. AMPK and Physiological Monocyte Differentiation

Beyond their originally described role as conveyors of programmed cell death and inflammation, caspases are involved in a number of other cellular functions, the most prominent being cell differentiation, a conserved property across a plethora of cell types in divergent metazoan organisms (Figure 2). Caspases mediate DNA damage and morphological changes that are common to different cell fates. The specificity of their action may be controlled by a diversity of mechanisms, including protein–protein interactions, post-translational modifications, subcellular localization, and interaction with other fundamental cell processes such as autophagy and proteostasis. Macrophages, which are mainly derived from monocytes, are essential components of mammal tissue homeostasis. Some are seeded into tissues before birth, while others are continually replenished from blood monocytes. Monocytes are circulating blood leukocytes [46] that migrate into tissues where they differentiate into morphological and functionally heterogeneous cells, including macrophages, myeloid dendritic cells, and osteoclasts [47]. Their differentiation can be recapitulated ex vivo by incubation with cytokines, e.g., they differentiate into macrophages upon exposure to colony stimulating factor-1 (CSF-1) [48]. The biologic effects of CSF-1 are typically mediated by plasma membrane associated CSF-1R [49]. Downstream signaling pathways include PI3K-AKT and AMPK, which mediate caspase activation and autophagy, respectively [50,51]. We have depicted these two pathways in more detail. Monocyte differentiation triggered by CSF-1 receptor (CSF-1R) engagement is critically dependent on the oscillatory activation of the kinase AKT, leading within 2–3 days to the formation of a multi-molecular platform. This molecular complex includes the adaptor molecule Fas-associated death domain (FADD), the serine-threonine kinase receptor-interaction protein kinase1 (RIP1), the long and short isoforms of flice inhibitory protein (FLIP), and procaspase-8 [52,53]. In turn, active caspase-8 provokes a spatially restricted activation of caspase-3 and -7 that cleaves selected intracellular proteins to generate a resting macrophage phenotype. This proteolytic machinery is inhibited as soon as macrophages are activated with lipopolysaccharides, a condition that also impairs autophagy induction. Engaged CSF-1R also promotes autophagy [50] through increasing the expression of the purinergic receptor P2RY6 that activates the CAMKK2-AMPK-ULK1 pathway [51]. Caspase activation and autophagy induction are both required for proper generation of macrophages upon CSF-1 stimulation (Figure 3 provides a schematic view of signaling network involved). The molecular links between caspase activation and autophagy require further investigation. One possible hypothesis is that post-translational modifications of caspases and differentiation-specific cleavage sites in cellular protein targets may be instrumental in connecting caspases to autophagy in cells undergoing differentiation. Thus, the CAMKK2/AMPK/ULK1 axis appears essential for proper differentiation of monocytes into macrophages.

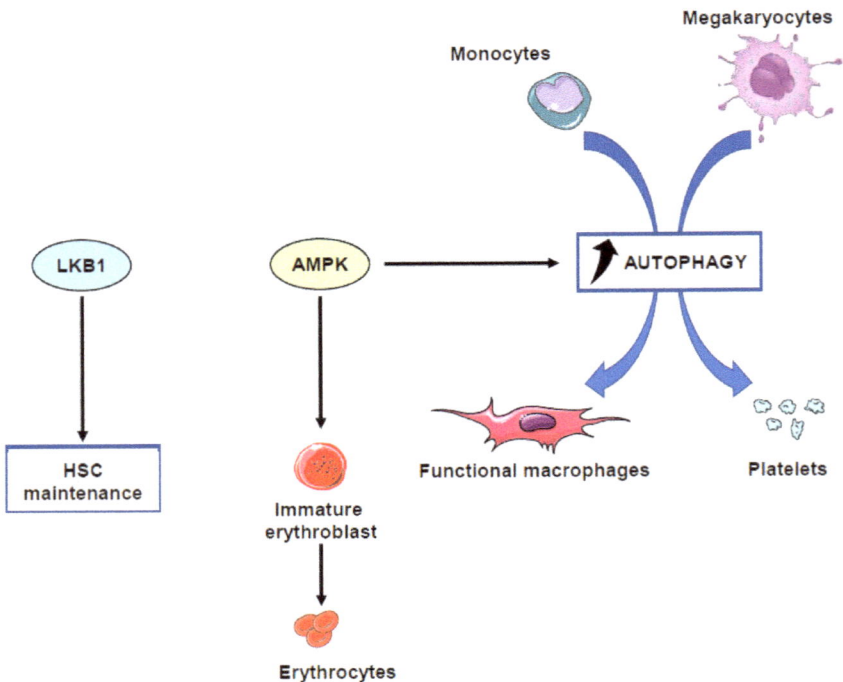

Figure 2. Role of LKB1 and AMPK in HSC maintenance and hematopoietic cell differentiation. There is genetic evidence that LKB1 is required for HSC maintenance since *Lkb1* depletion in mice results in loss of HSC quiescence, division and rapid depletion that contributes to pancytopenia. AMPK for its own is required for the early steps of erythroid differentiation, the production of functional macrophages from monocytes and the differentiation of HSCs into megakaryocytes and ultimately functional platelets.

Of note, in freshly isolated monocytes there is limited if any expression of AMPK subunit proteins. During CSF-1-mediated differentiation of monocytes into macrophages, there is a time-dependent increase in AMPKα1 expression and phosphorylation on Thr172 and a concomitant increase in ULK1 level and phosphorylation on Ser555 [51]. Increase in AMPKα1 and ULK1 expression at the protein but not the mRNA level, by a still unknown mechanism, does not rely on inhibition of the proteasome activity. This increase in AMPKα1 protein expression is further accompanied by elevated phosphorylation and activation of AMPK and ULK1 and correlates with the induction of autophagy and myeloid differentiation.

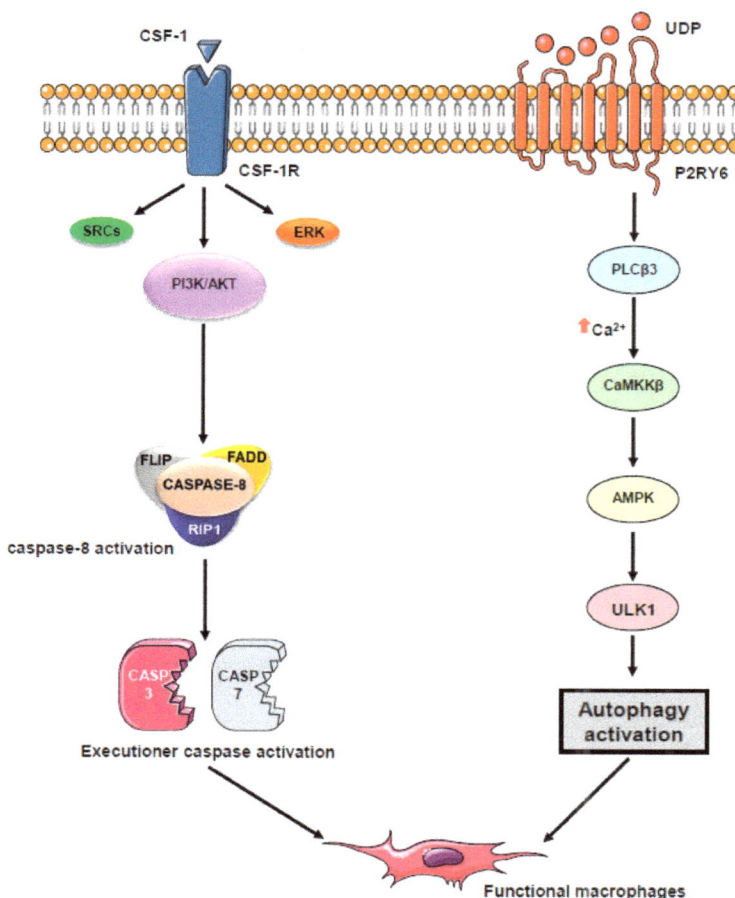

Figure 3. A schematic molecular view of the pathway involved in macrophagic differentiation. Engagement of the CSF-1 tyrosine kinase receptor by CSF-1 activates the PI3K/AKT pathway and induces caspase-8 activation within a FADD/RIP/FLIP multimolecular complex. In turn, active caspase-8 triggers a spatially restricted activation of caspase-3 and -7 that cleave selected intracellular proteins to generate a resting macrophage phenotype. Engaged CSF-1R also promotes autophagy through increasing the expression of the purinergic receptor P2RY6 that activates the CAMKK2-AMPK-ULK1 pathway. Caspase activation and autophagy induction are both required for proper generation of macrophages upon CSF-1 stimulation.

Other evidence for AMPK implication in myeloid cell differentiation comes from additional experiments performed in U937 and HEL acute myeloid leukemia (AML) cell lines, in which low doses of cytarabine were found to trigger the autophagy necessary for myeloid differentiation, as shown by increased expression of CD11b and morphologic features of differentiation. Inhibition of the mTOR pathway, likely independent of AMPK, was associated with cytarabine-mediated myeloid differentiation [54]. A last example of a role for AMPK in autophagy-mediated differentiation stems from in vivo experiments conducted with AMPKα1 mice. Indeed, AMPKα1 deficiency was shown to impair autophagy-mediated differentiation and decrease monocyte macrophage survival [55]. Finally, α1 and α2 AMPK isoforms appeared to display functional differences in the regulation of osteogenesis and osteoblast-associated induction of osteoclastogenesis [56].

5. The Role of AMPK in Hematological Cancers

AMPK has been reported to inhibit the growth of various hematological cancers. Indeed, in acute lymphoblastic leukemia (ALL) cell lines, AICAR triggered dose- and time-dependent inhibition of cell growth [57,58]. The pro-apoptotic effect of AMPK is mediated by activation of the p38 MAPK pathway, increased expression of cell-cycle inhibitors such as p27 and p53, and the downstream effects of the mTOR pathway. In B-cell chronic lymphocytic leukemia (B-CLL) cells, AMPK triggered apoptosis in a p53-independent manner [13]. In mantle cell lymphoma (MCL), AICAR-mediated stimulation of AMPK activity dampened phosphorylation of critical downstream effectors of mTOR signaling, such as 4E-BP1 and ribosomal protein S6, leading to cell growth inhibition [59]. In chronic myeloid leukemia (CML) and Philadelphia chromosome positive ALL, metformin and AICAR suppressed the mTOR activity and cell growth [58,60]. In myelodysplastic syndromes (MDS) and some acute myeloid leukemia (AML) cell lines, AICAR induced suppression of cell growth independently of apoptosis induction and AMPK activation [12]. Globally, indirect activators of AMPK induced suppression of cell growth and activation of cell death through both AMPK dependent and independent mechanisms in different hematological malignancies.

6. AMPK in the Regulation of Pathological Myeloid Differentiation

6.1. Myelodysplastic Syndromes

Myelodysplastic syndromes (MDS) constitute a set of myeloid progenitor cell diseases characterized by an ineffective production of mature and differentiated myeloid cells. This pathology strongly impacts the elderly, with an incidence of 20/100,000 people affected at 60 years and up to 50/100,000 at 80 years. This heterogeneous pathology is a multistep disease, the first stage being characterized by the appearance of abnormal cells in the bone marrow that are unable to differentiate and therefore to produce mature and functional cells in peripheral blood. At this stage, the disease is considered low-risk MDS and characterized by excessive apoptotic cell death of CD34-positive cells. In the next stage, CD34-positive progenitors accumulate mutations and hypermethylation of their DNA [61]. These genetic and epigenetic modifications render MDS cells less sensitive to apoptotic signals and contribute to abnormal cell growth. At this stage, the disease is defined as high-risk MDS and associated with an increase in medullar blast count, a higher probability for transformation to AML and a worse prognosis. Each stage of the disease is correlated with survival expectancy and probability for AML transformation by the Revised—International Prognostic Scoring System (IPSS-R). For classification of patients, IPSS-R takes into account the percentage of medullar blasts, the number of cytopenias and the cytogenetic status [62]. The recommended treatment in lower-risk MDS is the management of transfusion needs, whereas in high-risk MDS and AML, the only curative treatment remains bone marrow allograft, but few patients are eligible for this therapeutic option due to their advanced age. For non-eligible MDS and AML patients, the conventional chemotherapy consists of demethylating agents, including azacytidine or decitabine [63–65], or high-doses of cytarabine. Unfortunately, most patients fail to respond durably to these drugs, and there is currently no effective second-line option.

To adapt to the oxygen, nutrient, and glucose deprivation found during the different steps of tumor development, cancer cells modify their energetic needs through the fine-tuning of AMPK activity, a key metabolic sensor. As previously mentioned, AMPK is the master gene for the regulation of energy homeostasis. Its involvement in the regulation of apoptosis, autophagy and cell growth make this serine/threonine protein kinase a target of choice for the treatment of cancers in general and leukemia in particular [16,66,67]. Azacytidine-resistant MDS cell lines exhibited an altered response to apoptotic stimuli but displayed functional autophagy. Hence, triggering autophagy in resistant MDS cells induced autophagic cell death and bypassed the insensitivity to apoptosis [68,69]. Of note, it has recently been reported that the knockdown of *sf3b1*, a splicing factor mutated in 16% of MDS cases, deregulates the AMPK pathway [70]. This observation is of great interest since mutations of *sf3b1* are associated with good prognosis in MDS patients. Finally, cells issued from high-risk MDS patients exhibited a strong decrease of AMPK mRNA levels compared to those coming from low-risk MDS or healthy patients. This observation could explain the lower sensitivity of highly transformed MDS cells to variations in their energy environment.

6.2. Acute Myeloid Leukemia

The involvement of AMPK in the genesis and development of AML is well documented. Indeed, Saito et al. reported that leukemia initiating cells (LICs) responsible for the genesis of myeloid leukemia are protected from metabolic stress in the bone marrow through an AMPK-dependent mechanism [71]. This study clearly shows that dietary restriction induces metabolic cell survival. In contrast, depletion of AMPK expression in LIC reduces this cell population in the hypoxic bone marrow environment. Interestingly, human bone marrow stromal cells appear independent of the AMPK pathway, as knockout of AMPKα fails to sensitize these cells under dietary restriction or metformin treatment. Targeting AMPK depletes myeloid leukemia cells by disruption of glucose metabolism. In AML cell lines, exhibiting a high level of MAP kinase activation, AMPK activation was impaired. Indeed, Kawashima et al. demonstrated that glucose deprivation or metformin treatment could activate the AMPK pathway only in cells where ERK is weakly activated [72]. Moreover, in AML cells in which the MAP kinase pathway is overactivated, U0126, a MEK-specific inhibitor, was able to restore sustained AMPK activation under metformin treatment. The combination of an inhibitor of the MAPK pathway and an activator of the AMPK pathway leads to a significant decrease in cell growth and an increase in cell death in AML cells. This study suggests that both inhibition of an oncogenic pathway and activation of AMPK are necessary to trigger a strong anti-leukemic effect. The potent tumor suppressor role of the LKB1/AMPK pathway has been established in AML cells [73]. In this study, the authors showed that metformin activates the LKB1/AMPK pathway in AML and reduces tumor size in a mice xenograft model through the repression of mTOR-dependent oncogenic mRNA translation (c-MYC, CYCLIN-D, BCL-XL). It was recently demonstrated that co-activation of AMPK and mTORC1 could represent a good therapeutic strategy for AML [74]. Indeed, Sujobert et al. showed that GSK621, an AMPK direct activator, was highly cytotoxic for AML cells exhibiting a constitutive activation of mTORC1. As expected, this autophagy-dependent cell death induced by two concomitant signals was inhibited by rapamycin, an mTORC1 inhibitor.

Finally, although the strategy consisting of the use of AMPK agonists mimicking caloric restriction seems very promising, the development of an optimized and personalized therapy will need to block the predominant oncogenic pathway (overactivation due to mutations or translocation) and to activate AMPK in a concomitant manner. These complementary signals that will better impair cell growth and cell viability in AML will undoubtedly be a valuable option for patients in therapeutic failure.

6.3. Chronic Myelogenous Leukemia

Chronic myelogenous leukemia (CML) is a myeloproliferative syndrome linked to a hematopoietic stem cell disorder leading to increased production of granulocytes at all stages of differentiation. CML accounts for 15–20% of all cases of leukemia in adults [75] and is due to the t(9;22)(q34;q11) translocation, which encodes for the chimeric protein p210 BCR-ABL, a constitutively activated tyrosine kinase [76]. BCR-ABL expression leads to the engagement and activation of multiple pro-proliferative and anti-apoptotic cascades in transformed cells, including PI3K/AKT/mTOR and MAPK pathways [77–79]. Before the advent of targeted therapy, the gold standard for pharmacologic treatment of CML was α-interferon, but this treatment was associated with not-negligible toxicity and a median survival time of approximately five years [80]. In 2001, the identification of imatinib (Gleevec) as a small molecule ATP-pocket inhibitor of BCR-ABL dramatically re-defined the treatment of CML and had a major impact on the survival of patients with CML [81–83]. Imatinib mesylate along with second- (nilotinib, dasatinib, and bosutinib) and third-generation (ponatinib) tyrosine inhibitors (TKIs) have revolutionized the natural history of CML and have provided important treatment options for this leukemia that in the past was uniformly fatal [84]. Unfortunately, mutations rendering CML patients non-responsive to TKIs have been identified, including the threonine 315 to isoleucine (T315I) mutation and several others, which prevent binding of different TKIs to the active site of the ABL kinase, thereby avoiding its inhibition [85,86]. At present, more than 50 different mutation hotspots have been identified. BCR-ABL-independent mechanisms of resistance have also been reported to occur in tyrosine kinase inhibitor (TKI)-treated patients. In this regard, overexpression or hyperactivation of some members of the SRC family of kinases (LYN and HCK) have been described in cell lines and in some imatinib- and nilotinib-resistant patients [87–89]. Knowing that the PI3K/AKT/mTOR pathway is hyperactive in CML, indirect suppression of mTOR function by modulation of the AMPK pathway was proposed as an alternative therapeutic approach to overcome TKI resistances. Indeed, AMPK activation leads to mTOR inhibition through the phosphorylation and activation of the TSC1/2 complex [73,90] and/or a direct phosphorylation of the Raptor subunit on serines 722 and 792 [91], resulting in inactivation of the TORC1 complex. Most of the tested compounds that activate the AMPK pathway in the CML context are indirect activators. Among them, resveratrol, a naturally occurring substance found in grapes, triggers both apoptosis and autophagy in CML cells and is therefore able to overcome imatinib resistance [92,93]. Resveratrol-mediated autophagy is independent of BECLIN1 but mediated by a conjoint activation of AMPK and transcriptional upregulation of p62/SQSTM1 [93]. Interestingly, among the resistant cells that were sensitive to resveratrol, there were also cells expressing the T315I BCR-ABL mutant. Other studies have also pinpointed the anti-leukemic effects of AICAR or metformin on BCR-ABL transformed cells [12,60,94]. Unexpectedly, the effects of these compounds are often independent of AMPK activation. It has been established that the anti-leukemic effect of AICAR is dependent on protein kinase C-mediated autophagic cell death but independent of AMPK and apoptosis [12]. In conclusion, approaches to target cellular effectors of BCR-ABL, such as the PI3K/mTOR pathway, may provide alternative strategies to the use of TKIs to overcome resistance in refractory CML. The effect of direct AMPK activators on both TKI-sensitive and resistant CML cells has not been yet investigated.

6.4. Chronic Myelomonocytic Leukemia

Chronic myelomonocytic leukemia (CMML) is a paradigmatic chronic myeloid malignancy that associates features of myelodysplastic syndromes (MDS) and myeloproliferative neoplasms (MPN) [95]. CMML is a clonal disease of the hematopoietic stem cell characterized by a persistent monocytosis (>1×10^9/L), due to an accumulation of classical monocytes [96], and the aleatory presence of immature dysplastic granulocytes (PolyMorphoNuclear-MDSC or PMN-MDSC) in the peripheral blood of CMML patients [97]. These PMN-MDSCs that belong to the same clone as the leukemic monocytes appear to have immunosuppressive properties resembling those of the myeloid-derived suppressor cells (MDSCs), widely described in solid tumors. Whether these immature granulocytes contribute to autoimmune manifestations or immune-escape and progression of CMML is a conundrum and remains to be investigated. In recent years, large numbers of gene mutations have been discovered in CMML, none of which are specific to this condition, as they can be encountered with different frequencies in other myeloid neoplasms. These mutated genes encode signaling proteins (NRAS, KRAS, CBL, JAK2, FLT3, and several members of the Notch pathway), epigenetic regulators (TET2, ASXL1, EZH2, IDH1, and IDH2), and splicing factors (SF3B1, SRSF2, and ZRSF2) [98]. Mutations in the transcription regulators RUNX1, NPM1, and TP53 have also been reported in CMML. However, the role of these mutations in leukemogenesis is still unclear. Allogeneic stem cell transplantation is the only potentially curative option in patients suffering CMML, but it is associated with significant morbidity and mortality. Consequently, CMML patients are often treated like MDS patients with supportive care and hypomethylating agents, such as 5-azacitidine and decitabine, with overall response rates of 30–40% and complete remission rates of 7–17%, although with no impact on mutational allele burdens [99–101]. To improve their efficacy, azacytidine and decitabine have been combined with another drug, typically in "pick the winner" clinical trials with DNA-damaging drugs, immune-modulating drugs and histone deacetylase inhibitors, but none of these strategies has proven effective, and several were even found to be toxic [102,103]. There is therefore a strong need for alternative strategies aiming at increasing CMML patient overall survival. In a recent study, we have shown that CMML is characterized by defects in monocyte to macrophage differentiation [97]. These differentiation defects can be partly attributed to the presence of PMN-MDSCs that secrete high levels of alpha-defensins HNP1-3, which antagonize the purinergic receptor P2RY6 and inhibit AMPK-mediated autophagy in CMML patients [51]. Interestingly, we demonstrated that the physiological P2RY6 ligand UDP and the specific P2RY6 agonist MRS2693 can restore normal monocyte differentiation through re-induction of AMPK-dependent autophagy in primary myeloid cells from some, but not all, CMML patients (Figure 4). These results highlight an essential role for P2RY6-mediated autophagy through AMPK activation during the differentiation of human monocytes and pave the way for future therapeutic interventions for CMML. The use of direct AMPK activator in the CMML context could therefore be a promising therapeutic strategy.

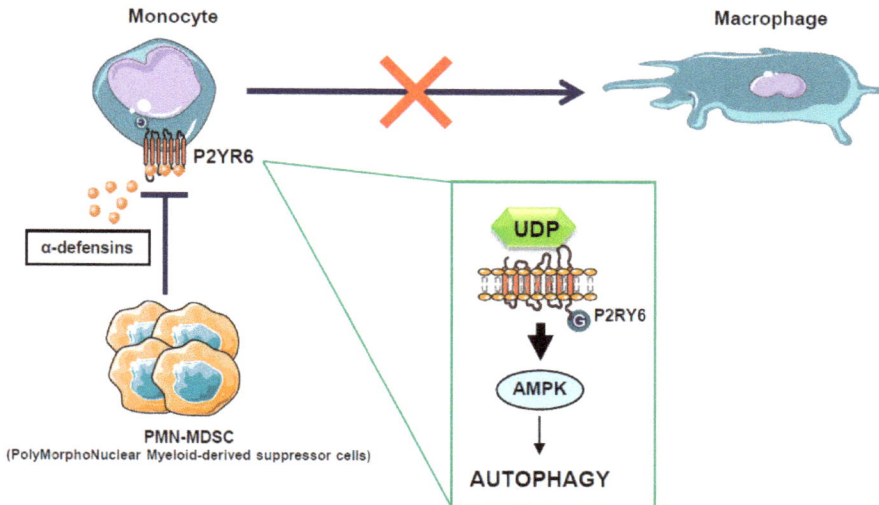

Figure 4. Mode of action for PMN-MDSC in the CMML context. In some CMML patients, PMN-MDSC secretes high levels of alpha-defensins, which antagonize the P2Y6 receptor, block autophagy activation, and inhibit the macrophagic differentiation of monocytes isolated from CMML patients. Alpha-defensin acts as a competitive inhibitor of UDP, the natural ligand of the P2Y6 receptor.

7. AMPK Modulators for the Treatment of Hematopoietic Malignancies

Due to its implication at the crossroad of cellular metabolism and proliferation, AMPK has emerged as an attractive and promising target for a great number of human pathological situations, including metabolic diseases and cancer. Pharmacological compounds, such as 5-aminoImidazole-4-carboxamide-1-β-D-ribofuranoside (AICAR), metformin, and natural occurring compounds such as resveratrol and spermidine, activate AMPK by direct or indirect mechanisms. It has been reported that metformin, AICAR, or resveratrol exert potent antitumoral effects in both solid tumors and hematopoietic malignancies either sensitive or resistant to their treatment of reference [12,60,93,104–106]. However, the anti-leukemic effects of some but not all of these molecules were linked to AMPK activation, making some natural AMPK activators promising molecules for the treatment of hematopoietic malignancies. Due to the potential therapeutic benefit of AMPK modulators in numerous diseases, direct and more specific modulators of AMPK have been developed that mainly target the β subunits, and less frequently the α and γ subunits (please refer to [35,107] for detailed reviews). These inhibitors included β1-selective compounds such as A-769662, PF-06409577, and GSK621. A-769662 has been evaluated in different settings and although efficient, appeared to exhibit off-target effects and cannot be considered per se as a specific AMPK activator. MT47-100 is an allosteric modulator of AMPK that activates the AMPK complexes containing the β1 isoform and inhibits those comprising the β2 subunits (Table 1). All these inhibitors have been tested in different settings, including cancers, with variable efficacy. Among the direct activators of AMPK, only GSK621, which selectively binds to the β1 subunit, has been studied in hematopoietic malignancies. Indeed, it was reported that GSK621 selectively kills AML cell lines and AML primary cells, sparing normal hematopoietic progenitors [74]. The lethality of GSK621 was abrogated by chemical inhibition or genetic ablation of mTORC1, suggesting a preferential effect in AML cells with over-activation of the mTORC1 pathway. Finally, the GSK621 cytotoxicity in AML cells was strictly dependent on the eIF2α/ATF4 signaling pathway activated through mTORC1. Although it is premature to speculate that activation of AMPK could represent a new strategic therapy for all myeloid malignancies, these promising findings strongly suggest that specific activators of AMPKα1, such as GSK621, may represent a therapeutic opportunity in AML and likely more globally in cancers in which mTORC1 is over-activated.

Table 1. AMPK activators include well-known pharmacological compounds, such as AICAR and metformin that act by increasing the AMP/ATP ratio, small molecules including A-769662, PF-06409577 and GSK621 that behave as AMPKβ1 subunit activator, and MT47-100, an AMPKβ1 activator and AMPKβ2 inhibitor. AMPK is also activated by a set of natural compounds including resveratrol and spermidine that indirectly increase the AMP/ATP ratio.

AMPK Activator	
Pharmacological Compounds	
AICAR Metformin	AMP:ATP ratio (up)
A-769662 PF-06409577 GSK621	AMPKβ1 subunit activators
MT47-100	AMPKβ1 subunit activator AMPKβ2 subunit inhibitor
Natural Compounds	
resveratrol spermidine	AMP:ATP ratio (up)

8. Conclusions and Outlook

In this review, we have discussed how AMPK is involved and regulated during myeloid lineage differentiation and how it can impact myeloid cell pathophysiology. As autophagy is known to play an important role in the process of hematopoietic cell differentiation [108], it was expected that AMPK would be a regulator of this process. However, the role of the AMPK pathway as an important actor of the metabolic modifications necessary for myeloid differentiation is just emerging. As an energy-consuming process, physiological myeloid differentiation requires adaptation of AMPK expression and activity in part to sustain autophagy that is necessary for this process. Hematopoietic malignancies are systematically characterized by profound defects in cell differentiation. Restoration of an effective differentiation process in myeloid malignancies has thus emerged as a pertinent therapeutic strategy. This notion is particularly well exemplified by the successful use of arsenic trioxide to promote acute promyelomonocytic leukemia redifferentiation, even leading to cure in a majority of patients. Alteration of AMPK expression and/or activity is found in hematological malignancies such as CMML, and activation of the AMPK pathway appears to restore normal differentiation in some CMML patients [51]. Beside myeloid differentiation, recent evidence in the literature indicates that AMPK could also play a key role during normal erythroid cell differentiation (Ladli, M et al., 2018, Haematologica, in press). While using small molecule AMPK activators to treat a panel of human metabolic and neurodegenerative diseases now appears achievable, further investigations should be carried out before such activators, direct or indirect, reach the clinic for the treatment of myeloid malignancies.

funding: This work was supported by the Fondation ARC pour la Recherche sur le Cancer (Equipe Labellisée 2017–2019), The Fondation de France, the Association Laurette Fugain, and the Institut National des Canceropole (INCA) (PRTK # 2017–2019). This work was also funded by the French government (National Research Agency, ANR) through the "Investments for the Future" Labex Signalife: Program reference # ANR-11-LABX-0028-01.

Acknowledgments: All authors participate equally to this work.

Conflicts of Interest: The authors have declared that no competing interests exist.

References

1. Lin, S.C.; Hardie, D.G. AMPK: Sensing Glucose as well as Cellular Energy Status. *Cell Metab.* **2018**, *27*, 299–313. [CrossRef] [PubMed]
2. Hardie, D.G. AMPK: Positive and negative regulation, and its role in whole-body energy homeostasis. *Curr. Opin. Cell Biol.* **2015**, *33*, 1–7. [CrossRef] [PubMed]
3. Hardie, D.G.; Schaffer, B.E.; Brunet, A. AMPK: An Energy-Sensing Pathway with Multiple Inputs and Outputs. *Trends Cell Biol.* **2016**, *26*, 190–201. [CrossRef] [PubMed]
4. Hardie, D.G. AMP-activated protein kinase: An energy sensor that regulates all aspects of cell function. *Genes Dev.* **2011**, *25*, 1895–1908. [CrossRef] [PubMed]
5. Steinberg, G.R.; Kemp, B.E. AMPK in Health and Disease. *Physiol. Rev.* **2009**, *89*, 1025–1078. [CrossRef] [PubMed]
6. Lopez-Lluch, G.; Navas, P. Calorie restriction as an intervention in ageing. *J. Physiol.* **2016**, *594*, 2043–2060. [CrossRef] [PubMed]
7. Pietrocola, F.; Castoldi, F.; Markaki, M.; Lachkar, S.; Chen, G.; Enot, D.P.; Durand, S.; Bossut, N.; Tong, M.; Malik, S.A.; et al. Aspirin Recapitulates Features of Caloric Restriction. *Cell Rep.* **2018**, *22*, 2395–2407. [CrossRef] [PubMed]
8. Pietrocola, F.; Pol, J.; Vacchelli, E.; Rao, S.; Enot, D.P.; Baracco, E.E.; Levesque, S.; Castoldi, F.; Jacquelot, N.; Yamazaki, T.; et al. Caloric Restriction Mimetics Enhance Anticancer Immunosurveillance. *Cancer Cell* **2016**, *30*, 147–160. [CrossRef] [PubMed]
9. Marino, G.; Pietrocola, F.; Madeo, F.; Kroemer, G. Caloric restriction mimetics: Natural/physiological pharmacological autophagy inducers. *Autophagy* **2014**, *10*, 1879–1882. [CrossRef] [PubMed]
10. Ben Sahra, I.; Le Marchand-Brustel, Y.; Tanti, J.F.; Bost, F. Metformin in cancer therapy: A new perspective for an old antidiabetic drug? *Mol. Cancer Ther.* **2010**, *9*, 1092–1099. [CrossRef] [PubMed]
11. Narkar, V.A.; Downes, M.; Yu, R.T.; Embler, E.; Wang, Y.X.; Banayo, E.; Mihaylova, M.M.; Nelson, M.C.; Zou, Y.; Juguilon, H.; et al. AMPK and PPARdelta agonists are exercise mimetics. *Cell* **2008**, *134*, 405–415. [CrossRef] [PubMed]
12. Robert, G.; Ben Sahra, I.; Puissant, A.; Colosetti, P.; Belhacene, N.; Gounon, P.; Hofman, P.; Bost, F.; Cassuto, J.P.; Auberger, P. Acadesine kills chronic myelogenous leukemia (CML) cells through PKC-dependent induction of autophagic cell death. *PLoS ONE* **2009**, *4*, e7889. [CrossRef] [PubMed]
13. Campas, C.; Lopez, J.M.; Santidrian, A.F.; Barragan, M.; Bellosillo, B.; Colomer, D.; Gil, J. Acadesine activates AMPK and induces apoptosis in B-cell chronic lymphocytic leukemia cells but not in T lymphocytes. *Blood* **2003**, *101*, 3674–3680. [CrossRef] [PubMed]
14. Guo, D.; Hildebrandt, I.J.; Prins, R.M.; Soto, H.; Mazzotta, M.M.; Dang, J.; Czernin, J.; Shyy, J.Y.; Watson, A.D.; Phelps, M.; et al. The AMPK agonist AICAR inhibits the growth of EGFRvIII-expressing glioblastomas by inhibiting lipogenesis. *Proc. Natl. Acad. Sci. USA* **2009**, *106*, 12932–12937. [CrossRef] [PubMed]

15. Su, R.Y.; Chao, Y.; Chen, T.Y.; Huang, D.Y.; Lin, W.W. 5-Aminoimidazole-4-carboxamide riboside sensitizes TRAIL- and TNFα-induced cytotoxicity in colon cancer cells through AMP-activated protein kinase signaling. *Mol. Cancer Ther.* **2007**, *6*, 1562–1571. [CrossRef] [PubMed]

16. Mihaylova, M.M.; Shaw, R.J. The AMPK signalling pathway coordinates cell growth, autophagy and metabolism. *Nat. Cell Biol.* **2011**, *13*, 1016–1023. [CrossRef] [PubMed]

17. Zhang, C.S.; Lin, S.C. AMPK Promotes Autophagy by Facilitating Mitochondrial Fission. *Cell Metab.* **2016**, *23*, 399–401. [CrossRef] [PubMed]

18. Herzig, S.; Shaw, R.J. AMPK: Guardian of metabolism and mitochondrial homeostasis. *Nat. Rev. Mol. Cell Biol.* **2018**, *19*, 121–135. [CrossRef] [PubMed]

19. Greer, E.L.; Dowlatshahi, D.; Banko, M.R.; Villen, J.; Hoang, K.; Blanchard, D.; Gygi, S.P.; Brunet, A. An AMPK-FOXO pathway mediates longevity induced by a novel method of dietary restriction in *C. elegans*. *Curr. Biol.* **2007**, *17*, 1646–1656. [CrossRef] [PubMed]

20. Mair, W.; Morantte, I.; Rodrigues, A.P.; Manning, G.; Montminy, M.; Shaw, R.J.; Dillin, A. Lifespan extension induced by AMPK and calcineurin is mediated by CRTC-1 and CREB. *Nature* **2011**, *470*, 404–408. [CrossRef] [PubMed]

21. Weir, H.J.; Yao, P.; Huynh, F.K.; Escoubas, C.C.; Goncalves, R.L.; Burkewitz, K.; Laboy, R.; Hirschey, M.D.; Mair, W.B. Dietary Restriction and AMPK Increase Lifespan via Mitochondrial Network and Peroxisome Remodeling. *Cell Metab.* **2017**, *26*, 884–896. [CrossRef] [PubMed]

22. Templeman, N.M.; Murphy, C.T. Regulation of reproduction and longevity by nutrient-sensing pathways. *J. Cell Biol.* **2018**, *217*, 93–106. [CrossRef] [PubMed]

23. Banko, M.R.; Allen, J.J.; Schaffer, B.E.; Wilker, E.W.; Tsou, P.; White, J.L.; Villen, J.; Wang, B.; Kim, S.R.; Sakamoto, K.; et al. Chemical genetic screen for AMPKalpha2 substrates uncovers a network of proteins involved in mitosis. *Mol. Cell* **2011**, *44*, 878–892. [CrossRef] [PubMed]

24. Winder, W.W.; Hardie, D.G. Inactivation of acetyl-CoA carboxylase and activation of AMP-activated protein kinase in muscle during exercise. *Am. J. Physiol.* **1996**, *270*, 299–304. [CrossRef] [PubMed]

25. Kim, J.; Kundu, M.; Viollet, B.; Guan, K.L. AMPK and mTOR regulate autophagy through direct phosphorylation of Ulk1. *Nat. Cell Biol.* **2011**, *13*, 132–141. [CrossRef] [PubMed]

26. Zhao, M.; Klionsky, D.J. AMPK-dependent phosphorylation of ULK1 induces autophagy. *Cell Metab.* **2011**, *13*, 119–120. [CrossRef] [PubMed]

27. Laker, R.C.; Drake, J.C.; Wilson, R.J.; Lira, V.A.; Lewellen, B.M.; Ryall, K.A.; Fisher, C.C.; Zhang, M.; Saucerman, J.J.; Goodyear, L.J.; et al. Ampk phosphorylation of Ulk1 is required for targeting of mitochondria to lysosomes in exercise-induced mitophagy. *Nat. Commun.* **2017**, *8*, 548. [CrossRef] [PubMed]

28. Hawley, S.A.; Ross, F.A.; Gowans, G.J.; Tibarewal, P.; Leslie, N.R.; Hardie, D.G. Phosphorylation by Akt within the ST loop of AMPK-alpha1 down-regulates its activation in tumour cells. *Biochem. J.* **2014**, *459*, 275–287. [CrossRef] [PubMed]

29. Manning, B.D.; Toker, A. AKT/PKB Signaling: Navigating the Network. *Cell* **2017**, *169*, 381–405. [CrossRef] [PubMed]

30. Zhang, Y.L.; Guo, H.; Zhang, C.S.; Lin, S.Y.; Yin, Z.; Peng, Y.; Luo, H.; Shi, Y.; Lian, G.; Zhang, C.; et al. AMP as a low-energy charge signal autonomously initiates assembly of AXIN-AMPK-LKB1 complex for AMPK activation. *Cell Metab.* **2013**, *18*, 546–555. [CrossRef] [PubMed]

31. Zhang, C.S.; Jiang, B.; Li, M.; Zhu, M.; Peng, Y.; Zhang, Y.L.; Wu, Y.Q.; Li, T.Y.; Liang, Y.; Lu, Z.; et al. The lysosomal v-ATPase-Regulator complex is a common activator for AMPK and mTORC1, acting as a switch between catabolism and anabolism. *Cell Metab.* **2014**, *20*, 526–540. [CrossRef] [PubMed]

32. O'Neill, L.A.; Hardie, D.G. Metabolism of inflammation limited by AMPK and pseudo-starvation. *Nature* **2013**, *493*, 346–355. [CrossRef] [PubMed]

33. He, C.; Li, H.; Viollet, B.; Zou, M.H.; Xie, Z. AMPK Suppresses Vascular Inflammation In Vivo by Inhibiting Signal Transducer and Activator of Transcription-1. *Diabetes* **2015**, *64*, 4285–4297. [CrossRef] [PubMed]

34. Hardie, D.G.; Lin, S.C. AMP-activated protein kinase—Not just an energy sensor. *F1000Reseach* **2017**, *6*, 1724. [CrossRef] [PubMed]

35. Olivier, S.; Foretz, M.; Viollet, B. Promise and challenges for direct small molecule AMPK activators. *Biochem. Pharmacol.* **2018**, *153*, 147–158. [CrossRef] [PubMed]

36. Ross, F.A.; Jensen, T.E.; Hardie, D.G. Differential regulation by AMP and ADP of AMPK complexes containing different gamma subunit isoforms. *Biochem. J.* **2016**, *473*, 189–199. [CrossRef] [PubMed]

37. Neumann, D. Is TAK1 a Direct Upstream Kinase of AMPK? *Int. J. Mol. Sci.* **2018**, *19*, 2412. [CrossRef] [PubMed]

38. Carling, D.; Thornton, C.; Woods, A.; Sanders, M.J. AMP-activated protein kinase: New regulation, new roles? *Biochem. J.* **2012**, *445*, 11–27. [CrossRef] [PubMed]

39. Schaffer, B.E.; Levin, R.S.; Hertz, N.T.; Maures, T.J.; Schoof, M.L.; Hollstein, P.E.; Benayoun, B.A.; Banko, M.R.; Shaw, R.J.; Shokat, K.M.; et al. Identification of AMPK Phosphorylation Sites Reveals a Network of Proteins Involved in Cell Invasion and Facilitates Large-Scale Substrate Prediction. *Cell Metab.* **2015**, *22*, 907–921. [CrossRef] [PubMed]

40. Wu, D.; Hu, D.; Chen, H.; Shi, G.; Fetahu, I.S.; Wu, F.; Rabidou, K.; Fang, R.; Tan, L.; Xu, S.; et al. Glucose-regulated phosphorylation of TET2 by AMPK reveals a pathway linking diabetes to cancer. *Nature* **2018**, *559*, 637–641. [CrossRef] [PubMed]

41. Nakada, D.; Saunders, T.L.; Morrison, S.J. Lkb1 regulates cell cycle and energy metabolism in haematopoietic stem cells. *Nature* **2010**, *468*, 653–658. [CrossRef] [PubMed]

42. Gan, B.; Hu, J.; Jiang, S.; Liu, Y.; Sahin, E.; Zhuang, L.; Fletcher-Sananikone, E.; Colla, S.; Wang, Y.A.; Chin, L.; et al. Lkb1 regulates quiescence and metabolic homeostasis of haematopoietic stem cells. *Nature* **2010**, *468*, 701–704. [CrossRef] [PubMed]

43. Egan, D.; Kim, J.; Shaw, R.J.; Guan, K.L. The autophagy initiating kinase ULK1 is regulated via opposing phosphorylation by AMPK and mTOR. *Autophagy* **2011**, *7*, 643–644. [CrossRef] [PubMed]

44. Yu, W.M.; Liu, X.; Shen, J.; Jovanovic, O.; Pohl, E.E.; Gerson, S.L.; Finkel, T.; Broxmeyer, H.E.; Qu, C.K. Metabolic regulation by the mitochondrial phosphatase PTPMT1 is required for hematopoietic stem cell differentiation. *Cell Stem Cell* **2013**, *12*, 62–74. [CrossRef] [PubMed]

45. Jager, S.; Handschin, C.; St-Pierre, J.; Spiegelman, B.M. AMP-activated protein kinase (AMPK) action in skeletal muscle via direct phosphorylation of PGC-1alpha. *Proc. Natl. Acad. Sci. USA* **2007**, *104*, 12017–12022. [CrossRef] [PubMed]

46. Ginhoux, F.; Jung, S. Monocytes and macrophages: Developmental pathways and tissue homeostasis. *Nat. Rev. Immunol.* **2014**, *14*, 392–404. [CrossRef] [PubMed]

47. Jakubzick, C.V.; Randolph, G.J.; Henson, P.M. Monocyte differentiation and antigen-presenting functions. *Nat. Rev. Immunol.* **2017**, *17*, 349–362. [CrossRef] [PubMed]

48. Sordet, O.; Rebe, C.; Plenchette, S.; Zermati, Y.; Hermine, O.; Vainchenker, W.; Garrido, C.; Solary, E.; Dubrez-Daloz, L. Specific involvement of caspases in the differentiation of monocytes into macrophages. *Blood* **2002**, *100*, 4446–4453. [CrossRef] [PubMed]

49. Stanley, E.R.; Chitu, V. CSF-1 receptor signaling in myeloid cells. *Cold Spring Harb. Perspect. Biol.* **2014**, *6*. [CrossRef] [PubMed]

50. Jacquel, A.; Obba, S.; Boyer, L.; Dufies, M.; Robert, G.; Gounon, P.; Lemichez, E.; Luciano, F.; Solary, E.; Auberger, P. Autophagy is required for CSF-1-induced macrophagic differentiation and acquisition of phagocytic functions. *Blood* **2012**, *119*, 4527–4531. [CrossRef] [PubMed]

51. Obba, S.; Hizir, Z.; Boyer, L.; Selimoglu-Buet, D.; Pfeifer, A.; Michel, G.; Hamouda, M.A.; Goncalves, D.; Cerezo, M.; Marchetti, S.; et al. The PRKAA1/AMPKalpha1 pathway triggers autophagy during CSF1-induced human monocyte differentiation and is a potential target in CMML. *Autophagy* **2015**, *11*, 1114–1129. [CrossRef] [PubMed]

52. Rebe, C.; Cathelin, S.; Launay, S.; Filomenko, R.; Prevotat, L.; L'Ollivier, C.; Gyan, E.; Micheau, O.; Grant, S.; Dubart-Kupperschmitt, A.; et al. Caspase-8 prevents sustained activation of NF-kappaB in monocytes undergoing macrophagic differentiation. *Blood* **2007**, *109*, 1442–1450. [CrossRef] [PubMed]

53. Jacquel, A.; Benikhlef, N.; Paggetti, J.; Lalaoui, N.; Guery, L.; Dufour, E.K.; Ciudad, M.; Racoeur, C.; Micheau, O.; Delva, L.; et al. Colony-stimulating factor-1-induced oscillations in phosphatidylinositol-3 kinase/AKT are required for caspase activation in monocytes undergoing differentiation into macrophages. *Blood* **2009**, *114*, 3633–3641. [CrossRef] [PubMed]

54. Chen, L.; Guo, P.; Zhang, Y.; Li, X.; Jia, P.; Tong, J.; Li, J. Autophagy is an important event for low-dose cytarabine treatment in acute myeloid leukemia cells. *Leuk. Res.* **2017**, *60*, 44–52. [CrossRef] [PubMed]

55. Zhang, M.; Zhu, H.; Ding, Y.; Liu, Z.; Cai, Z.; Zou, M.H. AMP-activated protein kinase alpha1 promotes atherogenesis by increasing monocyte-to-macrophage differentiation. *J. Biol. Chem.* **2017**, *292*, 7888–7903. [CrossRef] [PubMed]

56. Wang, Y.G.; Han, X.G.; Yang, Y.; Qiao, H.; Dai, K.R.; Fan, Q.M.; Tang, T.T. Functional differences between AMPK alpha1 and alpha2 subunits in osteogenesis, osteoblast-associated induction of osteoclastogenesis, and adipogenesis. *Sci. Rep.* **2016**, *6*, 32771. [CrossRef] [PubMed]

57. Sengupta, T.K.; Leclerc, G.M.; Hsieh-Kinser, T.T.; Leclerc, G.J.; Singh, I.; Barredo, J.C. Cytotoxic effect of 5-aminoimidazole-4-carboxamide-1-beta-4-ribofuranoside (AICAR) on childhood acute lymphoblastic leukemia (ALL) cells: Implication for targeted therapy. *Mol. Cancer* **2007**, *6*, 46. [CrossRef] [PubMed]

58. Vakana, E.; Platanias, L.C. AMPK in BCR-ABL expressing leukemias. Regulatory effects and therapeutic implications. *Oncotarget* **2011**, *2*, 1322–1328. [PubMed]

59. Drakos, E.; Atsaves, V.; Li, J.; Leventaki, V.; Andreeff, M.; Medeiros, L.J.; Rassidakis, G.Z. Stabilization and activation of p53 downregulates mTOR signaling through AMPK in mantle cell lymphoma. *Leukemia* **2009**, *23*, 784–790. [CrossRef] [PubMed]

60. Vakana, E.; Altman, J.K.; Glaser, H.; Donato, N.J.; Platanias, L.C. Antileukemic effects of AMPK activators on BCR-ABL-expressing cells. *Blood* **2011**, *118*, 6399–6402. [CrossRef] [PubMed]

61. Shallis, R.M.; Ahmad, R.; Zeidan, A.M. The genetic and molecular pathogenesis of myelodysplastic syndromes. *Eur. J. Haematol.* **2018**, *101*, 260–271. [CrossRef] [PubMed]

62. Greenberg, P.L.; Tuechler, H.; Schanz, J.; Sanz, G.; Garcia-Manero, G.; Sole, F.; Bennett, J.M.; Bowen, D.; Fenaux, P.; Dreyfus, F.; et al. Revised international prognostic scoring system for myelodysplastic syndromes. *Blood* **2012**, *120*, 2454–2465. [CrossRef] [PubMed]

63. Fenaux, P.; Mufti, G.J.; Hellstrom-Lindberg, E.; Santini, V.; Finelli, C.; Giagounidis, A.; Schoch, R.; Gattermann, N.; Sanz, G.; List, A.; et al. Efficacy of azacitidine compared with that of conventional care regimens in the treatment of higher-risk myelodysplastic syndromes: A randomised, open-label, phase III study. *Lancet Oncol.* **2009**, *10*, 223–232. [CrossRef]

64. Kantarjian, H.; Oki, Y.; Garcia-Manero, G.; Huang, X.; O'Brien, S.; Cortes, J.; Faderl, S.; Bueso-Ramos, C.; Ravandi, F.; Estrov, Z.; et al. Results of a randomized study of 3 schedules of low-dose decitabine in higher-risk myelodysplastic syndrome and chronic myelomonocytic leukemia. *Blood* **2007**, *109*, 52–57. [CrossRef] [PubMed]

65. Steensma, D.P.; Baer, M.R.; Slack, J.L.; Buckstein, R.; Godley, L.A.; Garcia-Manero, G.; Albitar, M.; Larsen, J.S.; Arora, S.; Cullen, M.T.; et al. Multicenter study of decitabine administered daily for 5 days every 4 weeks to adults with myelodysplastic syndromes: The alternative dosing for outpatient treatment (ADOPT) trial. *J. Clin. Oncol.* **2009**, *27*, 3842–3848. [CrossRef] [PubMed]

66. Hardie, D.G.; Ross, F.A.; Hawley, S.A. AMPK: A nutrient and energy sensor that maintains energy homeostasis. *Nat. Rev. Mol. Cell Biol.* **2012**, *13*, 251–262. [CrossRef] [PubMed]

67. Laplante, M.; Sabatini, D.M. mTOR signaling in growth control and disease. *Cell* **2012**, *149*, 274–293. [CrossRef] [PubMed]

68. Cluzeau, T.; Robert, G.; Puissant, A.; Jean-Michel, K.; Cassuto, J.P.; Raynaud, S.; Auberger, P. Azacitidine-resistant SKM1 myeloid cells are defective for AZA-induced mitochondrial apoptosis and autophagy. *Cell Cycle* **2011**, *10*, 2339–2343. [CrossRef] [PubMed]

69. Cluzeau, T.; Robert, G.; Mounier, N.; Karsenti, J.M.; Dufies, M.; Puissant, A.; Jacquel, A.; Renneville, A.; Preudhomme, C.; Cassuto, J.P.; et al. BCL2L10 is a predictive factor for resistance to azacitidine in MDS and AML patients. *Oncotarget* **2012**, *3*, 490–501. [CrossRef] [PubMed]

70. Dolatshad, H.; Pellagatti, A.; Fernandez-Mercado, M.; Yip, B.H.; Malcovati, L.; Attwood, M.; Przychodzen, B.; Sahgal, N.; Kanapin, A.A.; Lockstone, H.; et al. Disruption of SF3B1 results in deregulated expression and splicing of key genes and pathways in myelodysplastic syndrome hematopoietic stem and progenitor cells. *Leukemia* **2015**, *29*, 1092–1103. [CrossRef] [PubMed]

71. Saito, Y.; Chapple, R.H.; Lin, A.; Kitano, A.; Nakada, D. AMPK Protects Leukemia-Initiating Cells in Myeloid Leukemias from Metabolic Stress in the Bone Marrow. *Cell Stem Cell* **2015**, *17*, 585–596. [CrossRef] [PubMed]

72. Kawashima, I.; Mitsumori, T.; Nozaki, Y.; Yamamoto, T.; Shobu-Sueki, Y.; Nakajima, K.; Kirito, K. Negative regulation of the LKB1/AMPK pathway by ERK in human acute myeloid leukemia cells. *Exp. Hematol.* **2015**, *43*, 524–533. [CrossRef] [PubMed]

73. Green, A.S.; Chapuis, N.; Maciel, T.T.; Willems, L.; Lambert, M.; Arnoult, C.; Boyer, O.; Bardet, V.; Park, S.; Foretz, M.; et al. The LKB1/AMPK signaling pathway has tumor suppressor activity in acute myeloid leukemia through the repression of mTOR-dependent oncogenic mRNA translation. *Blood* **2010**, *116*, 4262–4273. [CrossRef] [PubMed]

74. Sujobert, P.; Poulain, L.; Paubelle, E.; Zylbersztejn, F.; Grenier, A.; Lambert, M.; Townsend, E.C.; Brusq, J.M.; Nicodeme, E.; Decrooqc, J.; et al. Co-activation of AMPK and mTORC1 Induces Cytotoxicity in Acute Myeloid Leukemia. *Cell Rep.* **2015**, *11*, 1446–1457. [CrossRef] [PubMed]
75. Sawyers, C.L. Chronic myeloid leukemia. *N. Engl. J. Med.* **1999**, *340*, 1330–1340. [CrossRef] [PubMed]
76. Ren, R. Mechanisms of BCR-ABL in the pathogenesis of chronic myelogenous leukaemia. *Nat. Rev. Cancer* **2005**, *5*, 172–183. [CrossRef] [PubMed]
77. Steelman, L.S.; Pohnert, S.C.; Shelton, J.G.; Franklin, R.A.; Bertrand, F.E.; McCubrey, J.A. JAK/STAT, Raf/MEK/ERK, PI3K/Akt and BCR-ABL in cell cycle progression and leukemogenesis. *Leukemia* **2004**, *18*, 189–218. [CrossRef] [PubMed]
78. Kim, J.H.; Chu, S.C.; Gramlich, J.L.; Pride, Y.B.; Babendreier, E.; Chauhan, D.; Salgia, R.; Podar, K.; Griffin, J.D.; Sattler, M. Activation of the PI3K/mTOR pathway by BCR-ABL contributes to increased production of reactive oxygen species. *Blood* **2005**, *105*, 1717–1723. [CrossRef] [PubMed]
79. Redig, A.J.; Vakana, E.; Platanias, L.C. Regulation of mammalian target of rapamycin and mitogen activated protein kinase pathways by BCR-ABL. *Leuk. Lymphoma* **2011**, *52* (Suppl. 1), 45–53. [CrossRef] [PubMed]
80. Hehlmann, R.; Berger, U.; Pfirrmann, M.; Hochhaus, A.; Metzgeroth, G.; Maywald, O.; Hasford, J.; Reiter, A.; Hossfeld, D.K.; Kolb, H.J.; et al. Randomized comparison of interferon alpha and hydroxyurea with hydroxyurea monotherapy in chronic myeloid leukemia (CML-study II): Prolongation of survival by the combination of interferon alpha and hydroxyurea. *Leukemia* **2003**, *17*, 1529–1537. [CrossRef] [PubMed]
81. Druker, B.J.; Talpaz, M.; Resta, D.J.; Peng, B.; Buchdunger, E.; Ford, J.M.; Lydon, N.B.; Kantarjian, H.; Capdeville, R.; Ohno-Jones, S.; et al. Efficacy and safety of a specific inhibitor of the BCR-ABL tyrosine kinase in chronic myeloid leukemia. *N. Engl. J. Med.* **2001**, *344*, 1031–1037. [CrossRef] [PubMed]
82. Fang, G.; Kim, C.N.; Perkins, C.L.; Ramadevi, N.; Winton, E.; Wittmann, S.; Bhalla, K.N. CGP57148B (STI-571) induces differentiation and apoptosis and sensitizes Bcr-Abl-positive human leukemia cells to apoptosis due to antileukemic drugs. *Blood* **2000**, *96*, 2246–2253. [PubMed]
83. Dan, S.; Naito, M.; Tsuruo, T. Selective induction of apoptosis in Philadelphia chromosome-positive chronic myelogenous leukemia cells by an inhibitor of BCR—ABL tyrosine kinase, CGP 57148. *Cell Death Differ.* **1998**, *5*, 710–715. [CrossRef] [PubMed]
84. Pasic, I.; Lipton, J.H. Current approach to the treatment of chronic myeloid leukaemia. *Leuk. Res.* **2017**, *55*, 65–78. [CrossRef] [PubMed]
85. Patel, A.B.; O'Hare, T.; Deininger, M.W. Mechanisms of Resistance to ABL Kinase Inhibition in Chronic Myeloid Leukemia and the Development of Next Generation ABL Kinase Inhibitors. *Hematol. Oncol. Clin. N. Am.* **2017**, *31*, 589–612. [CrossRef] [PubMed]
86. Soverini, S.; Mancini, M.; Bavaro, L.; Cavo, M.; Martinelli, G. Chronic myeloid leukemia: The paradigm of targeting oncogenic tyrosine kinase signaling and counteracting resistance for successful cancer therapy. *Mol. Cancer* **2018**, *17*, 49. [CrossRef] [PubMed]
87. Donato, N.J.; Wu, J.Y.; Stapley, J.; Gallick, G.; Lin, H.; Arlinghaus, R.; Talpaz, M. BCR-ABL independence and LYN kinase overexpression in chronic myelogenous leukemia cells selected for resistance to STI571. *Blood* **2003**, *101*, 690–698. [CrossRef] [PubMed]
88. Ferri, C.; Bianchini, M.; Bengio, R.; Larripa, I. Expression of LYN and PTEN genes in chronic myeloid leukemia and their importance in therapeutic strategy. *Blood Cells Mol. Dis.* **2014**, *52*, 121–125. [CrossRef] [PubMed]
89. Grosso, S.; Puissant, A.; Dufies, M.; Colosetti, P.; Jacquel, A.; Lebrigand, K.; Barbry, P.; Deckert, M.; Cassuto, J.P.; Mari, B.; et al. Gene expression profiling of imatinib and PD166326-resistant CML cell lines identifies Fyn as a gene associated with resistance to BCR-ABL inhibitors. *Mol. Cancer Ther.* **2009**, *8*, 1924–1933. [CrossRef] [PubMed]
90. Inoki, K.; Ouyang, H.; Zhu, T.; Lindvall, C.; Wang, Y.; Zhang, X.; Yang, Q.; Bennett, C.; Harada, Y.; Stankunas, K.; et al. TSC2 integrates Wnt and energy signals via a coordinated phosphorylation by AMPK and GSK3 to regulate cell growth. *Cell* **2006**, *126*, 955–968. [CrossRef] [PubMed]
91. Gwinn, D.M.; Shackelford, D.B.; Egan, D.F.; Mihaylova, M.M.; Mery, A.; Vasquez, D.S.; Turk, B.E.; Shaw, R.J. AMPK phosphorylation of raptor mediates a metabolic checkpoint. *Mol. Cell* **2008**, *30*, 214–226. [CrossRef] [PubMed]

Int. J. Mol. Sci. **2018**, *19*, 2991

92. Puissant, A.; Grosso, S.; Jacquel, A.; Belhacene, N.; Colosetti, P.; Cassuto, J.P.; Auberger, P. Imatinib mesylate-resistant human chronic myelogenous leukemia cell lines exhibit high sensitivity to the phytoalexin resveratrol. *FASEB J.* **2008**, *22*, 1894–1904. [CrossRef] [PubMed]

93. Puissant, A.; Robert, G.; Fenouille, N.; Luciano, F.; Cassuto, J.P.; Raynaud, S.; Auberger, P. Resveratrol promotes autophagic cell death in chronic myelogenous leukemia cells via JNK-mediated p62/SQSTM1 expression and AMPK activation. *Cancer Res.* **2010**, *70*, 1042–1052. [CrossRef] [PubMed]

94. Shi, R.; Lin, J.; Gong, Y.; Yan, T.; Shi, F.; Yang, X.; Liu, X.; Naren, D. The antileukemia effect of metformin in the Philadelphia chromosome-positive leukemia cell line and patient primary leukemia cell. *Anticancer Drugs* **2015**, *26*, 913–922. [CrossRef] [PubMed]

95. Patnaik, M.M.; Tefferi, A. Chronic myelomonocytic leukemia: 2018 update on diagnosis, risk stratification and management. *Am. J. Hematol.* **2018**, *93*, 824–840. [CrossRef] [PubMed]

96. Selimoglu-Buet, D.; Badaoui, B.; Benayoun, E.; Toma, A.; Fenaux, P.; Quesnel, B.; Etienne, G.; Braun, T.; Abermil, N.; Morabito, M.; et al. Accumulation of classical monocytes defines a subgroup of MDS that frequently evolves into CMML. *Blood* **2017**, *130*, 832–835. [CrossRef] [PubMed]

97. Droin, N.; Jacquel, A.; Hendra, J.B.; Racoeur, C.; Truntzer, C.; Pecqueur, D.; Benikhlef, N.; Ciudad, M.; Guery, L.; Jooste, V.; et al. Alpha-defensins secreted by dysplastic granulocytes inhibit the differentiation of monocytes in chronic myelomonocytic leukemia. *Blood* **2010**, *115*, 78–88. [CrossRef] [PubMed]

98. Itzykson, R.; Duchmann, M.; Lucas, N.; Solary, E. CMML: Clinical and molecular aspects. *Int. J. Hematol.* **2017**, *105*, 711–719. [CrossRef] [PubMed]

99. Ades, L.; Sekeres, M.A.; Wolfromm, A.; Teichman, M.L.; Tiu, R.V.; Itzykson, R.; Maciejewski, J.P.; Dreyfus, F.; List, A.F.; Fenaux, P.; et al. Predictive factors of response and survival among chronic myelomonocytic leukemia patients treated with azacitidine. *Leuk. Res.* **2013**, *37*, 609–613. [CrossRef] [PubMed]

100. Fianchi, L.; Criscuolo, M.; Breccia, M.; Maurillo, L.; Salvi, F.; Musto, P.; Mansueto, G.; Gaidano, G.; Finelli, C.; Aloe-Spiriti, A.; et al. High rate of remissions in chronic myelomonocytic leukemia treated with 5-azacytidine: Results of an Italian retrospective study. *Leuk. Lymphoma* **2013**, *54*, 658–661. [CrossRef] [PubMed]

101. Wijermans, P.W.; Ruter, B.; Baer, M.R.; Slack, J.L.; Saba, H.I.; Lubbert, M. Efficacy of decitabine in the treatment of patients with chronic myelomonocytic leukemia (CMML). *Leuk. Res.* **2008**, *32*, 587–591. [CrossRef] [PubMed]

102. Garcia-Manero, G.; Sekeres, M.A.; Egyed, M.; Breccia, M.; Graux, C.; Cavenagh, J.D.; Salman, H.; Illes, A.; Fenaux, P.; DeAngelo, D.J.; et al. A phase 1b/2b multicenter study of oral panobinostat plus azacitidine in adults with MDS, CMML or AML with 30% blasts. *Leukemia* **2017**, *31*, 2799–2806. [CrossRef] [PubMed]

103. Kobayashi, Y.; Munakata, W.; Ogura, M.; Uchida, T.; Taniwaki, M.; Kobayashi, T.; Shimada, F.; Yonemura, M.; Matsuoka, F.; Tajima, T.; et al. Phase I study of panobinostat and 5-azacitidine in Japanese patients with myelodysplastic syndrome or chronic myelomonocytic leukemia. *Int. J. Hematol.* **2018**, *107*, 83–91. [CrossRef] [PubMed]

104. Puissant, A.; Robert, G.; Auberger, P. Targeting autophagy to fight hematopoietic malignancies. *Cell Cycle* **2010**, *9*, 3470–3478. [CrossRef] [PubMed]

105. Santidrian, A.F.; Gonzalez-Girones, D.M.; Iglesias-Serret, D.; Coll-Mulet, L.; Cosialls, A.M.; de Frias, M.; Campas, C.; Gonzalez-Barca, E.; Alonso, E.; Labi, V.; et al. AICAR induces apoptosis independently of AMPK and p53 through up-regulation of the BH3-only proteins BIM and NOXA in chronic lymphocytic leukemia cells. *Blood* **2010**, *116*, 3023–3032. [CrossRef] [PubMed]

106. Van Den Neste, E.; Cazin, B.; Janssens, A.; Gonzalez-Barca, E.; Terol, M.J.; Levy, V.; Perez de Oteyza, J.; Zachee, P.; Saunders, A.; de Frias, M.; et al. Acadesine for patients with relapsed/refractory chronic lymphocytic leukemia (CLL): A multicenter phase I/II study. *Cancer Chemother. Pharmacol.* **2013**, *71*, 581–591. [CrossRef] [PubMed]

107. Kim, J.; Yang, G.; Kim, Y.; Kim, J.; Ha, J. AMPK activators: Mechanisms of action and physiological activities. *Exp. Mol. Med.* **2016**, *48*, e224. [CrossRef] [PubMed]

108. Riffelmacher, T.; Simon, A.K. Mechanistic roles of autophagy in hematopoietic differentiation. *FEBS J.* **2017**, *284*, 1008–1020. [CrossRef] [PubMed]

International Journal of
Molecular Sciences

MDPI

Article

AMP-activated Protein Kinase Controls Immediate Early Genes Expression Following Synaptic Activation Through the PKA/CREB Pathway

Sébastien Didier [†], Florent Sauvé [†], Manon Domise, Luc Buée, Claudia Marinangeli and Valérie Vingtdeux *

Université de Lille, Inserm, Centre Hospitalo-Universitaire de Lille, UMR-S1172—JPArc—Centre de Recherche Jean-Pierre AUBERT, F-59000 Lille, France; sebastien.didier1@gmail.com (S.D.); florent.sauve@inserm.fr (F.S.); manon.domise@inserm.fr (M.D.); luc.buee@inserm.fr (L.B.); c.marinangeli83@gmail.com (C.M.)
* Correspondence: valerie.vingtdeux@inserm.fr; Tel.: +33-320-298893
† These authors contributed equally to this work.

Received: 31 October 2018; Accepted: 20 November 2018; Published: 22 November 2018

Abstract: Long-term memory formation depends on the expression of immediate early genes (IEGs). Their expression, which is induced by synaptic activation, is mainly regulated by the $3',5'$-cyclic AMP (cAMP)-dependent protein kinase/cAMP response element binding protein (cAMP-dependent protein kinase (PKA)/ cAMP response element binding (CREB)) signaling pathway. Synaptic activation being highly energy demanding, neurons must maintain their energetic homeostasis in order to successfully induce long-term memory formation. In this context, we previously demonstrated that the expression of IEGs required the activation of AMP-activated protein kinase (AMPK) to sustain the energetic requirements linked to synaptic transmission. Here, we sought to determine the molecular mechanisms by which AMPK regulates the expression of IEGs. To this end, we assessed the involvement of AMPK in the regulation of pathways involved in the expression of IEGs upon synaptic activation in differentiated primary neurons. Our data demonstrated that AMPK regulated IEGs transcription via the PKA/CREB pathway, which relied on the activity of the soluble adenylyl cyclase. Our data highlight the interplay between AMPK and PKA/CREB signaling pathways that allows synaptic activation to be transduced into the expression of IEGs, thus exemplifying how learning and memory mechanisms are under metabolic control.

Keywords: AMPK; synaptic activation; PKA; CREB; soluble Adenylyl cyclase; Immediate early genes; transcription

1. Introduction

Long-term memory formation as well as long lasting forms of synaptic plasticity depend on the expression of new genes and proteins. These activity-regulated genes, referred to as immediate early genes (IEGs), encode for transcription factors and proteins that have the potential to transduce synaptic activity directly into immediate changes of neural function. They include, for example, *Arc/Arg3.1*, *Egr1/Zif268*, and *c-Fos*. These genes are indirect markers of neuronal activity and are used to map neuronal networks and circuits engaged in information processing and plasticity [1]. For instance, Arc (activity-regulated cytoskeleton-associated protein) is a cytosolic protein found in post-synaptic densities that regulates the endocytosis of AMPA receptors [2], Notch signaling, spine density, and morphology [3] through actin remodeling [4]. *Arc* knock-out (KO) mice display impairments in the formation of long-term memories while short-term memory is not affected [5]. Egr1/Zif268 and c-Fos interact with an array of other transcription factors to regulate gene expression. *Egr1* and *c-Fos* KO animals display deficits in complex behavioral tasks and memories [6,7].

Signaling pathways involved in activity-driven regulation of transcription and translation have been the object of many studies, however, not all the components have been elucidated. One of the most studied mediators of these transcriptional changes is the transcription factor 3′,5′-cyclic AMP (cAMP) response element-binding (CREB) protein [8]. Indeed, many of the IEGs contain cAMP response elements (CRE) and thus are regulated by the transcription factor CREB. CREB signaling is regulated by phosphorylation on its Ser^{133}, a key regulatory site where phosphorylation ensures the transcriptional function of CREB [9,10]. While several signaling pathways and kinases are known to induce CREB phosphorylation, the most important CREB kinase is the 3′,5′-cyclic AMP (cAMP)-dependent protein kinase (PKA). PKA activity, in turn, is known to be regulated upstream by signaling pathways leading to the increase of intracellular cAMP levels, and thus by the activity of adenylyl cyclases (ACs), the best characterized of which being the G protein-coupled receptors (GPCRs) [11].

Altogether, these processes are induced by synaptic activation and in particular by glutamatergic neurotransmission. Importantly, glutamatergic transmission is a highly energy-consuming process [12,13]. Within neurons, energy levels are regulated by the AMP-activated protein kinase (AMPK). AMPK is a Ser/Thr protein kinase, which is an important intracellular energy sensor and regulator. AMPK is composed of a catalytic subunit α and two regulatory subunits β and γ [14]. AMPK activity is regulated by the intracellular levels of adenine nucleotides AMP and ATP [15,16] and by the phosphorylation of its α subunit on Thr^{172} [17–19]. Interestingly, we recently reported that AMPK was necessary to maintain energy levels in neurons during synaptic activation [20]. Indeed, following glutamatergic synaptic stimulation we showed that AMPK activity was necessary to up-regulate glycolysis and mitochondrial respiration in order to maintain ATP levels within neurons. Failure to maintain energy homeostasis, through AMPK inhibition, prevented IEGs protein expression, synaptic plasticity, and hence long-term memory formation. This evidence strongly suggested that AMPK might act as a gatekeeper inside the neurons to allow signal transduction only in conditions where energy supplies are sufficient.

The goal of the present study was to determine the signaling pathway regulated by AMPK that allows the expression of IEGs. To this end, synaptic activation was induced in primary neurons, and AMPK and PKA signaling pathways were studied in these conditions. Our results showed that both signaling pathways were required for the expression of IEGs to occur. Interestingly, we also showed that the soluble adenylyl cyclase (AC) was responsible for PKA activation. Finally, inhibition of AMPK led to a downregulation of PKA pathway activation. Altogether, these data show how AMPK and PKA pathways interplay to regulate the expression of IEGs following synaptic activation.

2. Results

2.1. AMPK Activity is Required for Synaptic Activity-Induced IEGs Transcriptional Regulation

In order to determine the signaling pathway regulated by AMPK that allows for the expression of IEGs, we used primary neuronal cultures at 15 days in vitro (DIV) in which glutamatergic synaptic activation was induced using bicuculline and 4-aminopyridine (Bic/4-AP) as previously described [20–22]. As we recently showed, synaptic activation (SA) in this model led to the rapid activation of AMPK, as indicated by the increased phosphorylation of AMPK at Thr^{172}, and of its direct target, the Acetyl-CoA carboxylase (ACC), at Ser^{79}. Additionally, after 2 h of SA, a significant increase of the IEGs Arc, EgrI, and c-Fos expression was observed (Figure 1a, b). Further, AMPK inhibition using Compound C (Cc) prevented the expression of IEGs following SA (Figure 1c). These data demonstrated, as we previously reported [20], that proper AMPK activation is necessary for the expression of IEGs.

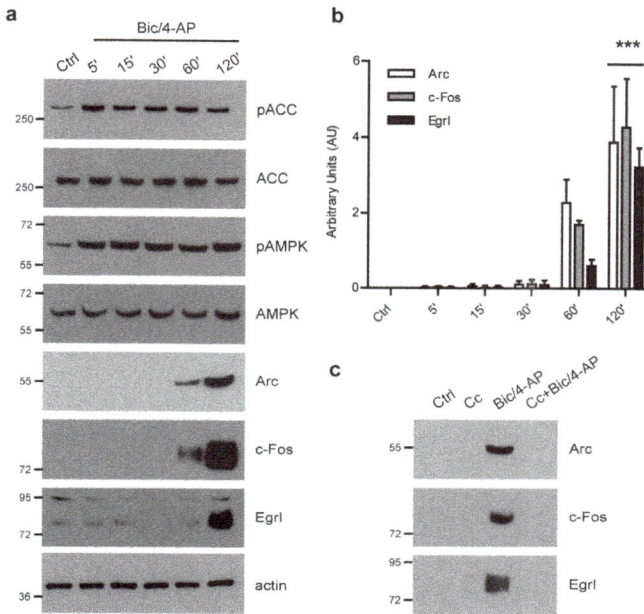

Figure 1. AMP-activated protein kinase (AMPK) is required for the expression of immediate early genes (IEGs) following glutamatergic activation. (**a**) Primary neurons at 15 days in vitro (DIV) treated with bicuculline and 4-aminopyridine (Bic/4-AP) (50μM/2.5mM) for the indicated time were subjected to immunoblotting with anti- phospho-AMPK (pAMPK), phospho-acetyl-CoA carboxylase (pACC), acetyl-CoA carboxylase (ACC), total AMPK, Arc, c-Fos, EgrI, and actin antibodies. Results are representative of at least four experiments. (**b**) Quantification of Western blot (WB) as in (**a**) showing Arc, c-Fos, and EgrI expression. Results show mean $\pm$ SD ($n = 4$). One-way ANOVA followed by Bonferroni's post-hoc test were used for evaluation of statistical significance, * $p < 0.05$, *** $p < 0.001$ compared to control condition. (**c**) Primary neurons at 15 DIV were pre-treated for 20 min in presence or absence of the AMPK inhibitor Compound C (Cc, 10 μM) prior to being treated with Bic/4-AP (50μM/2.5mM, 2 h). Cell lysates were subjected to immunoblotting with anti-Arc, c-Fos, EgrI, and actin antibodies. Results are representative of at least four experiments.

IEGs protein expression relies on the transcription of new genes, however, it was also proposed that it could result from the translation of a pre-existing pool of messenger RNA (mRNA) that is dendritically localized [23]. Therefore, we next thought to determine whether the expression of IEGs in our system was dependent on new mRNA expression or whether a pre-existing pool of mRNA could be sufficient to allow for the expression of IEGs following SA. To this end, translation was inhibited using anisomycin A and transcription inhibited using the RNA polymerase inhibitor actinomycin D. Both anisomycin A and actinomycin D repressed the expression of IEGs' proteins induced by SA (Figure 2a), showing that both *de novo* translation and transcription were necessary for the expression of IEGs. Altogether, these data implied that the expression of IEGs required new mRNA synthesis following SA. Indeed, Bic/4-AP stimulation led to a significant up-regulation of *Arc, c-Fos*, and *EgrI* mRNA (Figure 2b–d). We next assessed whether AMPK was required for this increased transcription to occur. Pre-treatment with the AMPK inhibitor Cc prevented the expression of IEGs, demonstrating that AMPK repression led to an inhibition of the activity-mediated IEG's mRNA levels of induction (Figure 2b–d). Altogether, these results showed that AMPK activity is involved in the transcriptional regulation of IEGs.

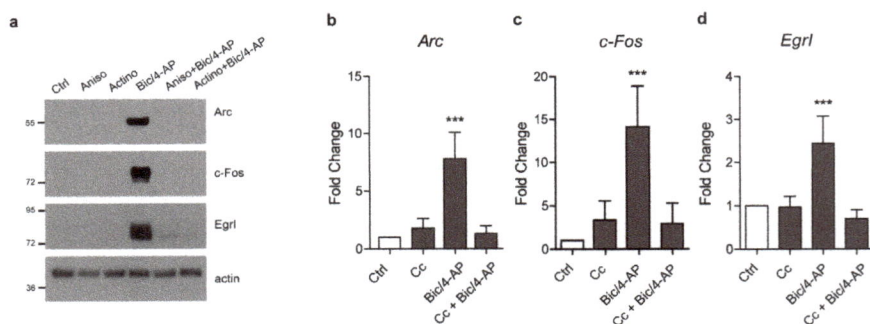

Figure 2. Expression of IEGs required *de novo* messenger RNA (mRNA) transcription and translation. (a) Primary neurons at 15 DIV co-treated with Bic/4-AP (50μM/2.5 mM) and the translation inhibitor anisomycin A (Aniso, 25 μM) or the transcription inhibitor actinomycin D (Actino, 10 μM) for 2 h were subjected to immunoblotting with anti-Arc, c-Fos, EgrI, and actin antibodies. Results are representative of at least four experiments. Results demonstrate that both translation and transcription are required for the expression of IEGs following synaptic activation (SA). (b–d) mRNA levels of *Arc*, *c-Fos*, and *EgrI* were determined by quantitative PCR in primary neurons at 15 DIV treated with Bic/4-AP (50μM/2.5mM, 30 min) after 20 min with or without pre-treatment with the AMPK inhibitor Compound C (Cc, 10 μM). Results show mean ± SD (n = 4–6). One-way ANOVA followed by Bonferroni's post-hoc test were used for evaluation of statistical significance, *** $p < 0.001$.

2.2. PKA Pathway Is Activated Following SA and Is Required for The Expression of IEGs

As the main pathway involved in the regulation of IEGs transcription is the PKA/CREB pathway, we questioned whether AMPK could cross-talk with this signaling pathway. We first assessed the activation of the PKA/CREB pathway following SA. To this end, we used an anti-phospho-PKA substrate antibody that detects proteins containing a phosphorylated Ser/Thr residue within the consensus sequence for PKA, thus giving an indirect readout of PKA activation status. Bic/4-AP stimulation led to a rapid and sustained activation of PKA, as observed using the anti-phospho-PKA substrate antibody as well as to the phosphorylation of CREB at Ser[133], a direct target of PKA (Figure 3a–c). Altogether, these data demonstrated that the PKA pathway was rapidly activated following SA and led to the activation of CREB.

We next determined whether the PKA pathway was required for the expression of IEGs following SA. To this end, primary neurons were pre-treated with the pharmacological PKA inhibitor H89, prior to being treated with Bic/4-AP (Figure 3d–f). Additionally, as H89 was reported to display off-target effects, to further validate the implication of PKA, we also used PKI 14–22 amide, a specific PKA peptide inhibitor (PKI) (Figure 3g–i). Results showed that both H89 and PKI prevented the PKA-substrate and CREB phosphorylation induced by SA (Figure 3d–i). Furthermore, our results showed that PKA inhibition by H89 or PKI led to a significant reduction of the expression of IEGs (Figure 3j,k). Consistent with previous reports, the present results show that PKA activation during SA is required for the expression of IEGs.

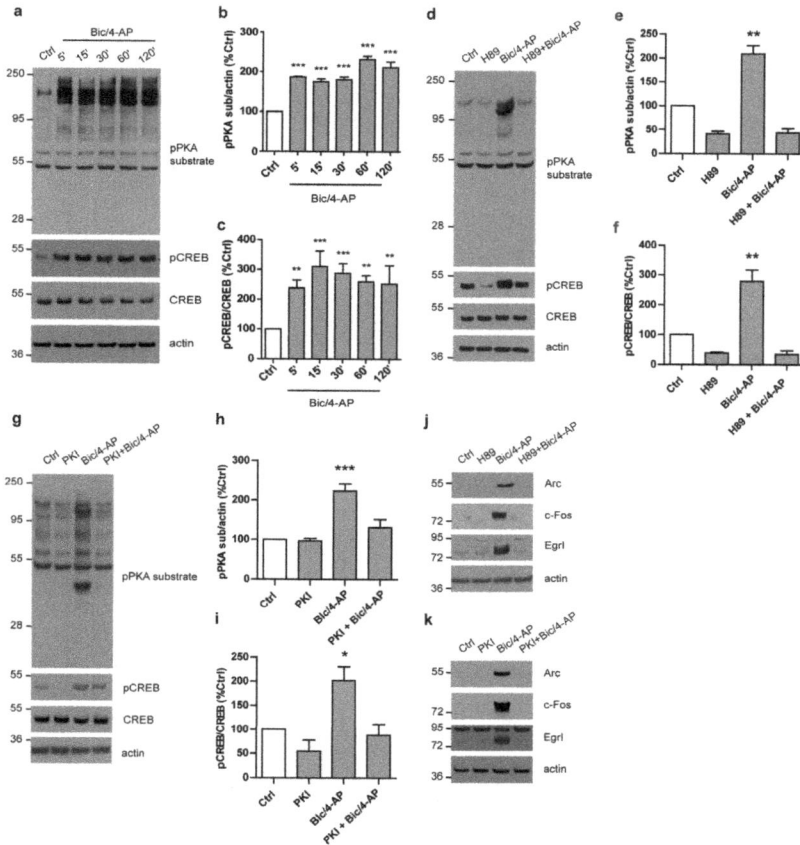

Figure 3. The cAMP-dependent protein kinase (PKA) pathway is rapidly activated following synaptic activation and is required for the expression of IEGs. (3′,5′-cyclic AMP = cAMP) (**a**) Primary neurons at 15 DIV treated with Bic/4-AP (50μM/2.5mM) for the indicated times were subjected to immunoblotting with anti- phospho-PKA substrate (pPKA sub), phospho-CREB (pCREB), total CREB, and actin antibodies. (cAMP response element binding = CREB) (**b**, **c**) Quantification of WB as in (**a**) showing the ratios pPKA sub/actin (**b**) and pCREB/CREB (**c**) expressed as a percentage of control ($n = 3$). (**d–g**) Primary neurons at 15 DIV treated with Bic/4-AP (50μM/2.5mM, 10 min) with or without 20 min pre-treatment with the PKA inhibitors H89 (20 μM, **d–f**) or PKA peptide inhibitor (PKI) (50 μM, **g–i**) were subjected to immunoblotting with anti- pPKA sub, pCREB, total CREB, and actin antibodies (**d,i**). Quantification of WB as in (**d**) and (**g**) showing the ratios pPKA sub/actin (**e,h**) and pCREB/CREB (**f,i**) expressed as a percentage of control ($n = 4$). (**j,k**) Primary neurons at 15 DIV treated with Bic/4-AP (50μM/2.5mM, 2 h) with or without 20 min pre-treatment with the PKA inhibitors H89 (20 μM, **j**) or PKI (50 μM, **k**) were subjected to immunoblotting with anti-Arc, cFos, EgrI, and actin antibodies. Results are representative of at least four experiments. Results show mean ± SD. One-way ANOVA followed by Bonferroni's post-hoc test were used for evaluation of statistical significance. * $p < 0.05$, ** $p < 0.01$, *** $p < 0.001$.

2.3. PKA Activation Following Synaptic Activation is Mediated by the Soluble AC

PKA is activated by the second messenger cAMP that is produced from ATP by AC. We next determined which of the ACs were responsible for PKA activation. To this end, neurons were pre-treated with various inhibitors of ACs, including inhibitors directed against the membrane bound

ACs, (SQ22536 or NKY80), or the specific inhibitor of the soluble AC (sAC) KH7 before Bic/4-AP stimulation [24]. Results showed that only KH7 inhibited PKA-substrate and CREB phosphorylation following SA (Figure 4a,c,d) and hence inhibited the expression of IEGs (Figure 4b). Thus, PKA was activated following SA-regulated expression of IEGs via the sAC activity, since only KH7 pre-treatment repressed PKA activation and the expression of IEGs.

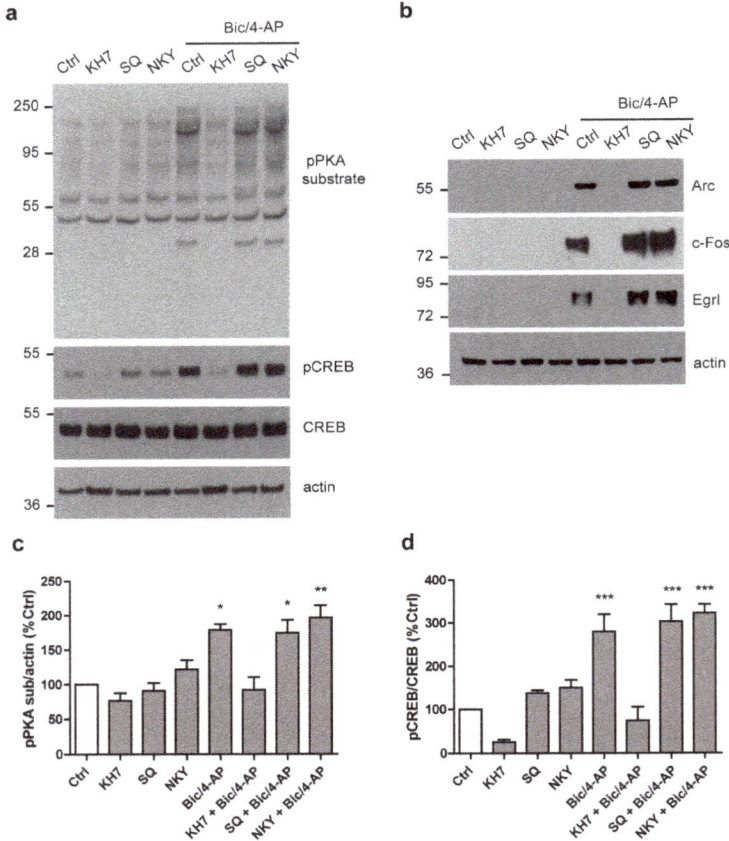

Figure 4. PKA activation following SA is dependent on soluble adenylyl cyclase (sAC). (**a**) Primary neurons at 15 DIV treated with Bic/4-AP (50μM/2.5mM, 10 min) with or without 20 min pre-treatment with the adenylyl cyclase (AC) inhibitors KH7 (20 μM), SQ22536 (SQ, 20 μM), and NKY80 (NKY, 20 μM) were subjected to immunoblotting with anti- phospho-PKA substrate (pPKA sub), phospho-CREB (pCREB), total CREB, and actin antibodies. (Ctrl was without pre-treatment) (**b**) Quantification of WB as in (a) showing the ratios pPKA sub/actin (c) and pCREB/CREB (d) expressed as a percentage of control (n = 4). (**c**) Primary neurons at 15 DIV treated with Bic/4-AP (2 h) with or without 20 min pre-treatment with the AC inhibitors KH7 (20 μM), SQ (20 μM), and NKY (20 μM) were subjected to immunoblotting with anti- Arc, c-Fos, EgrI, and actin antibodies. Results are representative of at least four experiments. Results show mean ± SD. One-way ANOVA followed by Bonferroni's post-hoc test were used for evaluation of statistical significance. * $p < 0.05$, ** $p < 0.01$, *** $p < 0.001$.

2.4. AMPK Regulates PKA Activation Following SA

To determine whether AMPK could be involved in the regulation of the PKA pathway, neurons were pre-treated with Cc as described above. Interestingly, Cc-pre-treatment prohibited PKA activation mediated by Bic/4-AP, as both PKA substrate and CREB were no longer phosphorylated (Figure 5a–c). Further experiments using short hairpin RNA (shRNA) directed against AMPK were performed. AMPK expression was down-regulated in primary neurons using shRNA directed against the $\alpha1$ and $\alpha2$ AMPK catalytic subunits (shAMPK) (Figure 5d,e). In these conditions, shAMPK reduced the phosphorylation of ACC following SA, confirming its inhibitory effect on AMPK signaling (Figure 5d). Moreover, following SA, shAMPK led to a significant reduction of PKA-substrate and CREB phosphorylation as compared to the control non-targeting shRNA (shNT) (Figure 5d,f,g), thus validating the results obtained with Cc.

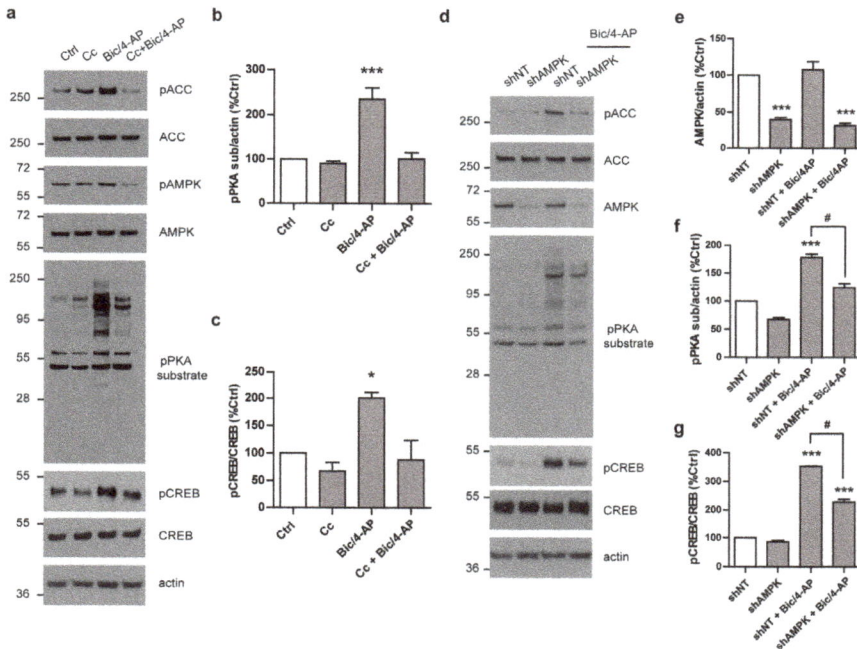

Figure 5. AMPK regulates PKA activation following SA. (**a**) Primary neurons at 15 DIV treated with Bic/4-AP (50μM/2.5mM, 10 min) with or without 20 min pre-treatment with the AMPK inhibitor Compound C (Cc, 10 μM) were subjected to immunoblotting with anti- phospho-AMPK (pAMPK), phospho-ACC (pACC), phospho-PKA substrate (pPKA sub), phospho-CREB (pCREB), total AMPK, ACC, CREB, and actin antibodies. (**b,c**) Quantification of WB as in (a) showing the ratios pPKA sub/actin (**b**) and pCREB/CREB (**c**) expressed as a percentage of control (*n* = 6). (**d**) 15 DIV primary neurons transduced for seven days with control non-targeting short hairpin RNA (shRNA) non-targeting shRNA (shNT) or with AMPK shRNA (shAMPK) were stimulated with Bic/4-AP (10 min) and subjected to immunoblotting with anti- pACC, pPKA sub, pCREB,total AMPK, ACC, CREB, and actin antibodies. (**e,f,g**) Quantification of WB as in (d) showing the ratios AMPK/actin (**e**), pPKA sub/actin (**f**), and pCREB/CREB (**g**) expressed as percentage of control (*n* = 3). Results show mean ± SD. One-way ANOVA followed by Bonferroni's post-hoc test were used for evaluation of statistical significance. * $p < 0.05$, ** $p < 0.01$, *** $p < 0.001$ as compared to Ctrl (**b,c**) or shNT (**e,f,g**), # $p < 0.05$ as compared to shNT + Bic/4-AP condition (**e,f**).

Altogether, these results show for the first time that AMPK activation cross-talks with the PKA pathway to regulate the expression of IEGs following SA.

3. Discussion

Changes in the expression of IEGs is an important process mediated by synaptic activity that is necessary for the conversion of short-term memory to long-term memory. With the present study, we extended on our previous data to determine the mechanism by which AMPK activity following synaptic activation led to the expression of IEGs. Here, we showed that SA led to the activation of the PKA/CREB pathway in an AMPK-dependent manner.

Whether AMPK directly or indirectly regulated the PKA/CREB pathway remains to be explored. However, in a previous report, we demonstrated that AMPK was required during SA to maintain intracellular ATP levels [20], therefore, it is possible that AMPK indirectly regulated the PKA pathway via controlling ATP levels. ATP, indeed, is converted by AC into cAMP, the second messenger that regulates PKA. Therefore, it is possible that the drop of ATP levels, due to AMPK inhibition, could lead to a parallel decrease of cAMP production, and eventually to a decrease of the signaling systems dependent on PKA.

Interestingly, our data showed that membrane-bound AC are not responsible for PKA activation following SA. Rather, it is the unconventional sAC (ADCY10) that is involved. sAC is distributed throughout the cytoplasm and in cellular organelles including the nucleus and mitochondria. Its functions are distinct from those of the transmembrane AC. For instance, it is insensible to G-proteins and forskolin regulation. However, in neuronal cells sAC activity can also be activated by intracellular Ca^{2+} elevations that increase its affinity for ATP [23] but also by bicarbonate anions (HCO_3^-) that increase the enzyme's V_{max}. Importantly, HCO_3^- can be metabolically generated within the cells under the action of carbonic anhydrases (CA), hence sAC activity can be modulated by metabolically generated HCO_3^- within the mitochondria [25–27]. Thus, mitochondrial metabolism regulated by AMPK could be another level of regulation of sAC, and hence cAMP production. Finally, we cannot exclude the possibility that AMPK could also regulate in a more direct fashion the sAC, through phosphorylation for instance. Finally, recent results have reported that activation of the mitochondrial cannabinoid receptor (mtCB1) caused inhibition of mitochondrial sAC, which resulted in reduction of PKA-dependent regulation of mitochondrial respiration, and eventually amnesic effects [28].

It is also interesting to note that AMPK was reported to be regulated by phosphorylation on $Ser^{485/491}$ on its catalytic subunits, respectively to α1 and α2. This phosphorylation occurs in response to agents that elevate intracellular cAMP, such as forskolin and isobutylmethylxanthine, and is likely to be mediated by PKA. These agents, however, act via membrane bound AC. Therefore, further investigations would be required to determine whether the sAC could also regulate PKA-mediated phosphorylation of AMPK. Interestingly, this phosphorylation of AMPK could be implicated in attenuating its activity given that it is associated with a down-regulation of its phosphorylation on Thr^{172} [29]. Further, in adipocytes, PKA was found to phosphorylate Ser^{173} on the AMPK α subunit to regulate lipolysis in response to PKA-activating signals [30]. Altogether these studies show that PKA can negatively regulate AMPK activity. It is therefore possible that PKA activation could in return repress AMPK activity, which could be an interesting mechanism to recover a basal AMPK activity state following SA.

Importantly, our data ([20], this study) suggest that neuronal energetic status may influence the formation of long-term memory. The hypothesis that AMPK influences these processes by maintaining ATP levels raises the question of long-term memory formation in an energetic stress environment. Metabolic disorders such as obesity and diabetes are characterized by peripheral metabolic dysfunction, but also cognitive deficits [31], elevated neurodegerenative disease risk, especially for Alzheimer's disease [32], and have recently been associated with central metabolic perturbations [33]. Interestingly, other studies have shown that several neurodegenerative diseases, including Alzheimer's disease, are not only associated with hypometabolism, but also to an activation of AMPK [34].

In conclusion, our study adds a player in the induction of signaling pathways involved in the regulation of the expression of IEGs, and hence memory formation. Altogether, our data suggest that through energy levels regulation, AMPK might indirectly control the activity of other signaling pathways, including those regulated by the second messenger cAMP.

4. Materials and Methods

4.1. Chemicals and Reagents/Antibodies

Antibodies directed against AMPKα (1/1000, Rabbit), ACC (1/1000, Rabbit), phospho-Ser^{79}ACC (1/1000, Rabbit), phospho-PKA substrate (RRXS*/T*) (1/2000, Rabbit), phospho-AMPK substrate (1/1000, Rabbit), and phospho-Ser^{133}CREB (1/1000, Rabbit) were obtained from Cell Signaling technology (Danvers, MA, USA). Anti phospho-Thr^{172}AMPKα (1/1000, Rabbit), CREB (1/500, Rabbit), Arc (1/500, Mouse), c-Fos (1/500, Mouse), and EgrI (1/500, Rabbit) antibodies were from Santa-Cruz (Dallas, TX, USA). Anti-actin (1/15 000, Mouse) antibody was from BD Bioscience (Franklin Lakes, NJ, USA). HRP-coupled secondary antibodies directed against the primary antibodies' hosts were obtained from Cell Signaling technology. Bicuculline (Bic), H 89, PKI 14-22 amide, NKY 80, SQ 22536, and KH 7 were purchased from Tocris (Bristol, UK), 4-aminopyridine (4-AP) was purchased from Sigma (St Louis, MO, USA), and Compound C (Cc) was from Santa Cruz (Dallas, TX, USA).

4.2. Primary Neuronal Cell Culture and Treatments

All animal experiments were performed according to procedures approved by the local Animal Ethical Committee following European standards for the care and use of laboratory animals (agreement APAFIS#4689-2016032315498524 v5 from CEEA75, Lille, France; approved on Oct 11, 2016). Primary neurons were prepared as previously described [35]. Briefly, fetuses at stage E18.5 were obtained from pregnant C57BL/6J wild-type female mice (The Jackson Laboratory, Bar Harbor, ME, USA). Forebrains were dissected in ice-cold dissection medium composed of Hanks' balanced salt solution (HBSS) (Invitrogen, Carlsbad, CA, USA) supplemented with 0.5 % *w/v* D-glucose (Sigma, St Louis, MO, USA) and 25 mM Hepes (Invitrogen, Carlsbad, CA, USA). Neurons were dissociated and isolated in ice-cold dissection medium containing 0.01 % *w/v* papain (Sigma, St Louis, MO, USA), 0.1 % *w/v* dispase (Sigma, St Louis, MO, USA), and 0.01 % *w/v* DNaseI (Roche, Rotkreuz, Switzerland), and by incubation at 37 °C for 15 min. Cells were spun down at 220 x g for 5 min at 4 °C, resuspended in Neurobasal medium supplemented with 2% B27, 1 mM NaPyr, 100 units/mL penicillin, 100 μg/ml streptomycin, and 2 mM Glutamax (Invitrogen, Carlsbad, CA, USA). For Western blots experiments, 12-well plates were seeded with 500,000 neurons per well and for RT-qPCR experiments, 6-well plates were seeded with 1,000,000 neurons per well. Fresh medium was added every 3 days (1:3 of starting volume). Cells were then treated and collected between DIV 14 to 17. For shRNA transduction, shRNA vectors from the TRC-Mm1.0 (Mouse) library, shAMPK α1 (CloneID:TRCN0000024000) shAMPK α2 (CloneID:TRCN0000024046), and non-targeting control shRNA (RHS6848) were obtained from Dharmacon, Lafayette, CO, USA. For the lentiviral production, HEK 293T cells were transfected for 72 h before collecting the supernatant as previously described [20]. Supernatant was concentrated using Amicon® Ultra 15-mL Centrifugal Filters (EMD Millipore, Burlington, MA, USA). Primary neuronal cultures were transduced with both AMPK α1 and AMPK α2 shRNA or the non-targeting shRNA at DIV 7, 7 days before performing experimentation.

4.3. Immunoblotting

For Western blot (WB) analysis, 15 μg of proteins from total cell lysates were separated in 8–16% Tris-Glycine gradient gels and transferred to nitrocellulose membranes. Membranes were then blocked in 5% fat-free milk in Tris Buffer Saline-0.01% Tween-20, and incubated with specific primary antibodies overnight at 4 °C. Proteins were thereafter detected via the use of Horseradish Peroxidase-conjugated secondary antibodies and electrochemiluminescence detection system (ThermoFisher Scientific,

Waltham, MA, USA). The Western blot bands corresponding to proteins of interest, or smears for phosphorylated-PKA substrate, were analyzed using the FIJI software v1.51n [36].

4.4. Quantitative Real-Time RT-PCR for the Expression of IEGs

Total RNA was isolated using the NucleoSpin® RNA kit (Macherey-Nagel, Düren, Germany) according to the manufacturer's instructions. One microgram of total RNA was reverse-transcribed using the Applied Biosystems High-Capacity cDNA reverse transcription kit (ThermoFisher Scientific, Watham, MA, USA). Real-time quantitative reverse transcription polymerase chain reaction (qRT-PCR) analyses were performed using Power SYBR Green PCR Master Mix (ThermoFisher Scientific, Watham, MA, USA) on a StepOneTM Real-Time PCR System (ThermoFisher Scientific, Watham, MA, USA) using the following primers: *β-actin* forward: 5′-CTAAGGCCAACCGTGAAAAG-3′, reverse: 5′-ACCAGAGGCATACAGGGACA-3′; *Arc* forward: 5′-GGTGAGCTGAAGCCACAAAT-3′, reverse: 5′-TTCACTGGTATGAATCACTGCTG-3′; *Egr1* forward: 5′-AAGACACCCCCCCATGAA-C-3′, reverse: 5′-CTCATCCGAGCGAGAAAAGC-3′; and *c-Fos* forward: 5′-CGAAGGGAACGGAATAAG-3′, reverse: 5′-CTCTGGGAAGCCAAGGTC-3′. The thermal cycler conditions were as follows: hold for 10 min at 95 °C, followed by 45 cycles of a two-step PCR consisting of a 95 °C step for 15 s followed by a 60 °C step for 25 s. Amplifications were carried out in triplicate, and the relative expression of target genes was determined by the ΔΔCT method using β-actin for normalization.

4.5. Statistical Analyses

All statistical analyses were performed using GraphPad Prism (Prism 5.0d, GraphPad Software Inc, La Jolla, CA, USA).

Author Contributions: Conceptualization, V.V. and C.M.; Investigation, S.D., F.S., M.D., C.M., and V.V.; Validation, S.D. and C.M.; Formal Analysis, S.D., F.S., and C.M.; Resources, L.B; Visualization, F.S.; Supervision, V.V.; Writing–Original Draft, C.M. and V.V.; Writing – Review and Editing, All authors; Funding Acquisition, L.B. and V.V.

funding: This research was funded by the French Fondation pour la cooperation Scientifique—Plan Alzheimer 2008–2012 (Senior Innovative Grant 2013) to VV, by the Fondation Vaincre Alzheimer (n°FR-16071p to VV), and in part through the Labex DISTALZ (Development of Innovative Strategies for a Transdisciplinary Approach to Alzheimer's disease). FS holds a doctoral scholarship from Lille 2 University.

Acknowledgments: We thank the animal core facility (animal facilities of Université de Lille-Inserm) of "Plateformes en Biologie Santé de Lille" as well as C. Degraeve, M. Besegher-Dumoulin, J. Devassine, R. Dehaynin, and D. Taillieu for animal care.

Conflicts of Interest: The authors declare no conflict of interest.

Abbreviations

AC	Adenylyl cyclase
ACC	Acetyl-CoA carboxylase
AMPK	AMP-activated protein kinase
Arc	Activity-regulated cytoskeleton-associated protein
Bic/4-AP	Bicuculline/4-aminopyridine
cAMP	3′,5′-cyclic AMP
Cc	Compound C
CREB	cAMP response element binding
DIV	Days in vitro
GPCRs	G protein-coupled receptors
IEGs	Immediate Early Genes
KO	Knock-out
PKA	cAMP-dependent protein kinase
SA	Synaptic activation
sAC	Soluble adenylyl cyclase
shRNA	Short hairpin RNA

References

1. Guzowski, J.F.; Timlin, J.A.; Roysam, B.; McNaughton, B.L.; Worley, P.F.; Barnes, C.A. Mapping behaviorally relevant neural circuits with immediate-early gene expression. *Curr. Opin. Neurobiol.* **2005**, *15*, 599–606. [CrossRef] [PubMed]

2. Shepherd, J.D.; Rumbaugh, G.; Wu, J.; Chowdhury, S.; Plath, N.; Kuhl, D.; Huganir, R.L.; Worley, P.F. *Arc/Arg3.1* mediates homeostatic synaptic scaling of AMPA receptors. *Neuron* **2006**, *52*, 475–484. [CrossRef] [PubMed]

3. Peebles, C.L.; Yoo, J.; Thwin, M.T.; Palop, J.J.; Noebels, J.L.; Finkbeiner, S. Arc regulates spine morphology and maintains network stability in vivo. *Proc. Natl. Acad. Sci. USA* **2010**, *107*, 18173–18178. [CrossRef] [PubMed]

4. Fukazawa, Y.; Saitoh, Y.; Ozawa, F.; Ohta, Y.; Mizuno, K.; Inokuchi, K. Hippocampal LTP is accompanied by enhanced F-actin content within the dendritic spine that is essential for late LTP maintenance in vivo. *Neuron* **2003**, *38*, 447–460. [CrossRef]

5. Plath, N.; Ohana, O.; Dammermann, B.; Errington, M.L.; Schmitz, D.; Gross, C.; Mao, X.; Engelsberg, A.; Mahlke, C.; Welzl, H.; et al. *Arc/Arg3.1* is essential for the consolidation of synaptic plasticity and memories. *Neuron* **2006**, *52*, 437–444. [CrossRef] [PubMed]

6. Jones, M.W.; Errington, M.L.; French, P.J.; Fine, A.; Bliss, T.V.; Garel, S.; Charnay, P.; Bozon, B.; Laroche, S.; Davis, S. A requirement for the immediate early gene Zif268 in the expression of late LTP and long-term memories. *Nat. Neurosci.* **2001**, *4*, 289–296. [CrossRef] [PubMed]

7. Paylor, R.; Johnson, R.S.; Papaioannou, V.; Spiegelman, B.M.; Wehner, J.M. Behavioral assessment of c-fos mutant mice. *Brain Res.* **1994**, *651*, 275–282. [CrossRef]

8. Alberini, C.M. Transcription factors in long-term memory and synaptic plasticity. *Physiol. Rev.* **2009**, *89*, 121–145. [CrossRef] [PubMed]

9. Gonzalez, G.A.; Montminy, M.R. Cyclic AMP stimulates somatostatin gene transcription by phosphorylation of CREB at serine 133. *Cell* **1989**, *59*, 675–680. [CrossRef]

10. Naqvi, S.; Martin, K.J.; Arthur, J.S. CREB phosphorylation at Ser133 regulates transcription via distinct mechanisms downstream of cAMP and MAPK signalling. *Biochem. J.* **2014**, *458*, 469–479. [CrossRef] [PubMed]

11. Pierce, K.L.; Premont, R.T.; Lefkowitz, R.J. Seven-transmembrane receptors. *Nat. Rev. Mol. Cell. Biol.* **2002**, *3*, 639–650. [CrossRef] [PubMed]

12. Attwell, D.; Laughlin, S.B. An energy budget for signaling in the grey matter of the brain. *J. Cereb. Blood Flow MeTable* **2001**, *21*, 1133–1145. [CrossRef] [PubMed]

13. Harris, J.J.; Jolivet, R.; Attwell, D. Synaptic energy use and supply. *Neuron* **2012**, *75*, 762–777. [CrossRef] [PubMed]

14. Kahn, B.B.; Alquier, T.; Carling, D.; Hardie, D.G. AMP-activated protein kinase: Ancient energy gauge provides clues to modern understanding of metabolism. *Cell. MeTable* **2005**, *1*, 15–25. [CrossRef] [PubMed]

15. Hardie, D.G.; Salt, I.P.; Hawley, S.A.; Davies, S.P. AMP-activated protein kinase: An ultrasensitive system for monitoring cellular energy charge. *Biochem. J.* **1999**, *338*, 717–722. [CrossRef] [PubMed]

16. Scott, J.W.; Hawley, S.A.; Green, K.A.; Anis, M.; Stewart, G.; Scullion, G.A.; Norman, D.G.; Hardie, D.G. CBS domains form energy-sensing modules whose binding of adenosine ligands is disrupted by disease mutations. *J. Clin. Invest.* **2004**, *113*, 274–284. [CrossRef] [PubMed]

17. Stapleton, D.; Mitchelhill, K.I.; Gao, G.; Widmer, J.; Michell, B.J.; Teh, T.; House, C.M.; Fernandez, C.S.; Cox, T.; Witters, L.A.; Kemp, B.E. Mammalian AMP-activated protein kinase subfamily. *J. Biol. Chem.* **1996**, *271*, 611–614. [CrossRef] [PubMed]

18. Hawley, S.A.; Davison, M.; Woods, A.; Davies, S.P.; Beri, R.K.; Carling, D.; Hardie, D.G. Characterization of the AMP-activated protein kinase kinase from rat liver and identification of threonine 172 as the major site at which it phosphorylates AMP-activated protein kinase. *J. Biol. Chem.* **1996**, *271*, 27879–27887. [CrossRef] [PubMed]

19. Stein, S.C.; Woods, A.; Jones, N.A.; Davison, M.D.; Carling, D. The regulation of AMP-activated protein kinase by phosphorylation. *Biochem. J.* **2000**, *345*, 437–443. [CrossRef] [PubMed]

20. Marinangeli, C.; Didier, S.; Ahmed, T.; Caillerez, R.; Domise, M.; Laloux, C.; Bégard, S.; Carrier, S.; Colin, M.; Marchetti, P.; et al. AMP-Activated Protein Kinase Is Essential for the Maintenance of Energy Levels during Synaptic Activation. *iScience* **2018**, *9*, 1–13. [CrossRef] [PubMed]

21. Hardingham, G.E.; Fukunaga, Y.; Bading, H. Extrasynaptic NMDARs oppose synaptic NMDARs by triggering CREB shut-off and cell death pathways. *Nat. Neurosci.* **2002**, *5*, 405–414. [CrossRef] [PubMed]

22. Hoey, S.E.; Williams, R.J.; Perkinton, M.S. Synaptic NMDA receptor activation stimulates alpha-secretase amyloid precursor protein processing and inhibits amyloid-beta production. *J. Neurosci.* **2009**, *29*, 4442–4460. [CrossRef] [PubMed]

23. Steward, O.; Farris, S.; Pirbhoy, P.S.; Darnell, J.; Driesche, S.J. Localization and local translation of Arc/Arg3.1 mRNA at synapses: Some observations and paradoxes. *Front. Mol. Neurosci.* **2014**, *7*. [CrossRef] [PubMed]

24. Bitterman, J.L.; Ramos-Espiritu, L.; Diaz, A.; Levin, L.R.; Buck, J. Pharmacological distinction between soluble and transmembrane adenylyl cyclases. *J. Pharmacol. Exp. Ther.* **2013**, *347*, 589–598. [CrossRef] [PubMed]

25. Tresguerres, M.; Buck, J.; Levin, L.R. Physiological carbon dioxide, bicarbonate, and pH sensing. *Pflugers. Arch.* **2010**, *460*, 953–964. [CrossRef] [PubMed]

26. Acin-Perez, R.; Salazar, E.; Brosel, S.; Yang, H.; Schon, E.A.; Manfredi, G. Modulation of mitochondrial protein phosphorylation by soluble adenylyl cyclase ameliorates cytochrome oxidase defects. *EMBO Mol. Med.* **2009**, *1*, 392–406. [CrossRef] [PubMed]

27. Acin-Perez, R.; Salazar, E.; Kamenetsky, M.; Buck, J.; Levin, L.R.; Manfredi, G. Cyclic AMP produced inside mitochondria regulates oxidative phosphorylation. *Cell. MeTable* **2009**, *9*, 265–276. [CrossRef] [PubMed]

28. Hebert-Chatelain, E.; Desprez, T.; Serrat, R.; Bellocchio, L.; Soria-Gomez, E.; Busquets-Garcia, A.; Pagano Zottola, A.C.; Delamarre, A.; Cannich, A.; Vincent, P.; et al. A cannabinoid link between mitochondria and memory. *Nature* **2016**, *539*, 555–559. [CrossRef] [PubMed]

29. Hurley, R.L.; Barré, L.K.; Wood, S.D.; Anderson, K.A.; Kemp, B.E.; Means, A.R.; Witters, L.A. Regulation of AMP-activated protein kinase by multisite phosphorylation in response to agents that elevate cellular cAMP. *J. Biol. Chem.* **2006**, *281*, 36662–36672. [CrossRef] [PubMed]

30. Djouder, N.; Tuerk, R.D.; Suter, M.; Salvioni, P.; Thali, R.F.; Scholz, R.; Vaahtomeri, K.; Auchli, Y.; Rechsteiner, H.; Brunisholz, R.A.; et al. PKA phosphorylates and inactivates AMPKalpha to promote efficient lipolysis. *EMBO J.* **2010**, *29*, 469–481. [CrossRef] [PubMed]

31. Bischof, G.N.; Park, D.C. Obesity and Aging: Consequences for Cognition, Brain Structure, and Brain Function. *Psychosom. Med.* **2015**, *77*, 697–709. [CrossRef] [PubMed]

32. Kivipelto, M.; Ngandu, T.; Fratiglioni, L.; Viitanen, M.; Kåreholt, I.; Winblad, B.; Helkala, E.L.; Tuomilehto, J.; Soininen, H.; Nissinen, A. Obesity and vascular risk factors at midlife and the risk of dementia and Alzheimer disease. *Arch. Neurol.* **2005**, *62*, 1556–1560. [CrossRef] [PubMed]

33. Hwang, J.J.; Jiang, L.; Hamza, M.; Sanchez Rangel, E.; Dai, F.; Belfort-DeAguiar, R.; Parikh, L.; Koo, B.B.; Rothman, D.L.; Mason, G.; Sherwin, R.S. Blunted rise in brain glucose levels during hyperglycemia in adults with obesity and T2DM. *JCI Insight* **2017**, *2*. [CrossRef] [PubMed]

34. Vingtdeux, V.; Davies, P.; Dickson, D.W.; Marambaud, P. AMPK is abnormally activated in tangle- and pre-tangle-bearing neurons in Alzheimer's disease and other tauopathies. *Acta Neuropathol.* **2011**, *121*, 337–349. [CrossRef] [PubMed]

35. Domise, M.; Didier, S.; Marinangeli, C.; Zhao, H.; Chandakkar, P.; Buée, L.; Viollet, B.; Davies, P.; Marambaud, P.; Vingtdeux, V. AMP-activated protein kinase modulates tau phosphorylation and tau pathology in vivo. *Sci. Rep.* **2016**, *6*. [CrossRef] [PubMed]

36. Schindelin, J.; Arganda-Carreras, I.; Frise, E.; Kaynig, V.; Longair, M.; Pietzsch, T.; Preibisch, S.; Rueden, C.; Saalfeld, S.; Schmid, B.; et al. Fiji: an open-source platform for biological-image analysis. *Nat Methods.* **2012**, *9*, 676–682. [CrossRef] [PubMed]

International Journal of
Molecular Sciences

MDPI

Article

Loss of *AMPKα2* Impairs Hedgehog-Driven Medulloblastoma Tumorigenesis

Honglai Zhang [1], Rork Kuick [2], Sung-Soo Park [1], Claire Peabody [1], Justin Yoon [1], Ester Calvo Fernández [1], Junying Wang [3], Dafydd Thomas [1], Benoit Viollet [4,5,6], Ken Inoki [3], Sandra Camelo-Piragua [1,* and Jean-François Rual [1,*]

[1]   Department of Pathology, University of Michigan Medical School, Ann Arbor, MI 48109, USA; honglaiz@med.umich.edu (H.Z.); parksung@med.umich.edu (S.-S.P.); peabodyc@umich.edu (C.P.); jxy673@case.edu (J.Y.); esfernan@med.umich.edu (E.C.F.); thomasda@med.umich.edu (D.T.)
[2]   Department of Biostatistics, School of Public Health, University of Michigan, Ann Arbor, MI 48109, USA; rork@umich.edu
[3]   Life Sciences Institute, University of Michigan, Ann Arbor, MI 48109, USA; wangjoy03@gmail.com (J.W.); inokik@umich.edu (K.I.)
[4]   Inserm, U1016, Institut Cochin, 75014 Paris, France; benoit.viollet@inserm.fr
[5]   CNRS, UMR8104, 75014 Paris, France
[6]   Université Paris Descartes, Sorbonne Paris cité, 75014 Paris, France
*   Correspondence: sandraca@med.umich.edu (S.C.-P.); jrual@med.umich.edu (J.-F.R.)

Received: 8 September 2018; Accepted: 19 October 2018; Published: 23 October 2018

Abstract: The AMP-activated protein kinase (AMPK) is a sensor of cellular energy status that has a dual role in cancer, i.e., pro- or anti-tumorigenic, depending on the context. In medulloblastoma, the most frequent malignant pediatric brain tumor, several in vitro studies previously showed that AMPK suppresses tumor cell growth. The role of AMPK in this disease context remains to be tested in vivo. Here, we investigate loss of *AMPKα2* in a genetically engineered mouse model of sonic hedgehog (SHH)-medulloblastoma. In contrast to previous reports, our study reveals that *AMPKα2* KO impairs SHH medulloblastoma tumorigenesis. Moreover, we performed complementary molecular and genomic analyses that support the hypothesis of a pro-tumorigenic SHH/AMPK/CNBP axis in medulloblastoma. In conclusion, our observations further underline the context-dependent role of AMPK in cancer, and caution is warranted for the previously proposed hypothesis that AMPK agonists may have therapeutic benefits in medulloblastoma patients. Note: an abstract describing the project was previously submitted to the American Society for Investigative Pathology PISA 2018 conference and appears in *The American Journal of Pathology* (Volume 188, Issue 10, October 2018, Page 2433).

Keywords: medulloblastoma; sonic hedgehog; AMPK

1. Introduction

Medulloblastoma is an embryonal tumor of the cerebellum and is the most common malignant brain tumor of childhood [1,2]. Genomics applied to medulloblastoma defined four medulloblastoma subgroups, each characterized by a distinct molecular/genetic signature, distinct patient demographics, and a distinct clinical profile (WNT, Sonic Hedgehog or SHH, Groups #3 and #4) [3,4]. More recently, integrative genomic analyses further underscored the highly heterogeneous and complex nature of medulloblastoma with a large spectrum of molecularly distinct consensus subgroups and subtypes within them [5–7]. Current therapeutic approaches to medulloblastoma are based on surgery, radiation, and non-targeted chemotherapy and are indistinguishably applied to all medulloblastoma subgroups. These therapies have led to significant improvements, with a 73% survival rate [1], but these results are achieved at a high cost to quality of life, e.g., neurocognitive or hormonal deficiencies [8,9]. Alternative therapeutic approaches are needed.

The AMP-activated protein kinase (AMPK) is a crucial energy sensor that controls cell metabolism and growth in response to low energy levels by phosphorylating a variety of substrates [10]. AMPK forms a heterotrimeric protein complex composed of three subunits that exist as multiple isoforms: a catalytic subunit (α1 or α2), a scaffolding subunit (β1 or β2), and an AMP-sensing subunit (γ1, γ2, or γ3) [10]. AMPK is allosterically activated by an increased AMP/ATP ratio when cells are metabolically starved. Upon metabolic stress, AMPK activation results in both the inhibition of anabolic, energy-consuming mechanisms (e.g., biosynthetic pathways and the cell cycle) and the activation of catabolic processes (e.g., the promotion of mitochondrial biogenesis and glycolysis) [10]. As the master metabolic guardian, AMPK modulates numerous targets, including both oncogenes and tumor suppressors [11,12]. Thus, AMPK has a dual role in cancer, tumor-suppressive, or pro-oncogenic, depending on the cellular or tissue context [13–15]. For example, as an inhibitor of the cell cycle and as a target of the LKB1 kinase [16–18], AMPK has long been considered an essential mediator of LKB1's tumor-suppressive effect in cancer [13,19]. The tumor suppressor role of AMPK is also exemplified by the observation that inactivation of *AMPKα1* in murine B-cell lineages promotes *Myc*-driven lymphomagenesis [20] and that *AMPKα2* suppresses murine embryonic fibroblast transformation and tumorigenesis [21]. Yet, loss of *AMPKα* impairs tumor growth in vivo in various carcinogenic contexts [22–27]. Indeed, maintenance of the metabolic balance by AMPK is likely a critical process for survival during the metabolic stress that can occur in the tumor microenvironment [13,14,26].

In SHH-driven medulloblastoma tumorigenesis, several studies support a tumor-suppressive role for AMPK [28–33]. First, in medulloblastoma cells, AMPK phosphorylates the SHH pathway transcription factor GLI1, promoting its proteasomal degradation, thus inhibiting SHH signaling and SHH-driven medulloblastoma [28–30,33]. Second, a potential tumor-suppressive role for AMPK in SHH-driven medulloblastoma was also inferred by the fact that the increased survival observed upon *Hk2* KO in a mouse model of medulloblastoma is accompanied by a gain of AMPK activity [31,32]. In light of these observations, AMPK agonists could have therapeutic value for the treatment of SHH medulloblastoma patients. As noted above, though, the multifaceted role of AMPK in cancer warrants caution [13–15]. As a matter of fact, D'amico et al. recently proposed that a non-canonical SHH/AMPK axis promotes medulloblastoma via activation of CCHC type nucleic acid binding protein (CNBP), ornithine decarboxylase 1 (ODC1), and polyamine metabolism [34]. Though indirectly inferred, this study supports a putative pro-tumorigenic role for AMPK in SHH medulloblastoma [34]. Importantly, in all of the aforementioned studies, the role of AMPK in medulloblastoma was only investigated in vitro in medulloblastoma cells.

The potential role of AMPK as a promoter or as a suppressor of medulloblastoma tumorigenesis warrants further investigation and remains to be tested in vivo. Here, we describe the analysis of loss of *AMPKα2* in a genetically engineered mouse model of SHH-driven medulloblastoma. Remarkably, in disagreement with the previous studies supporting a tumor-suppressive role for AMPK in medulloblastoma tumorigenesis [28–33], our analysis reveals that loss of *AMPKα2* impairs SHH-driven medulloblastoma tumorigenesis.

2. Results

2.1. AMPKα2 Is Required for SHH-Driven Medulloblastoma In Vivo

We investigated the role of *AMPKα2* in vivo in medulloblastoma using the [GFAP-tTA;TRE-SmoA1] genetically engineered mouse model of SHH-driven medulloblastoma [35,36] in combination with the *AMPKα2* KO mouse [37]. We note that, as previously described [37], *AMPKα2* KO mice are born with the expected Mendelian ratio, are fertile, appear indistinguishable from their WT littermates, and have a normal lifespan. Accordingly, *AMPKα2* KO mice exhibit normal cerebellar development with properly tri-laminated cerebellar cortex (molecular, Purkinje cell and internal granular cell layers).

Three groups of C57BL/6 mice were analyzed: (a) *TRE-SmoA1* mice: the oncogenic gain-of-function allele of *Smo*, i.e., *SmoA1*, is not expressed in the absence of *GFAP-tTA* in these negative control group mice; (b) [*GFAP-tTA;TRE-SmoA1*] mice: all mice develop medulloblastoma in this positive control group; (c) [*GFAP-tTA;TRE-SmoA1;AMPKα2*$^{-/-}$] mice: this test group allows us to assess the extent to which *AMPKα2* KO modulates tumor incidence. Genotyping data for each group is shown in Figure S1. For each study group, we performed histological (Figure 1) and survival (Figure 2) analyses.

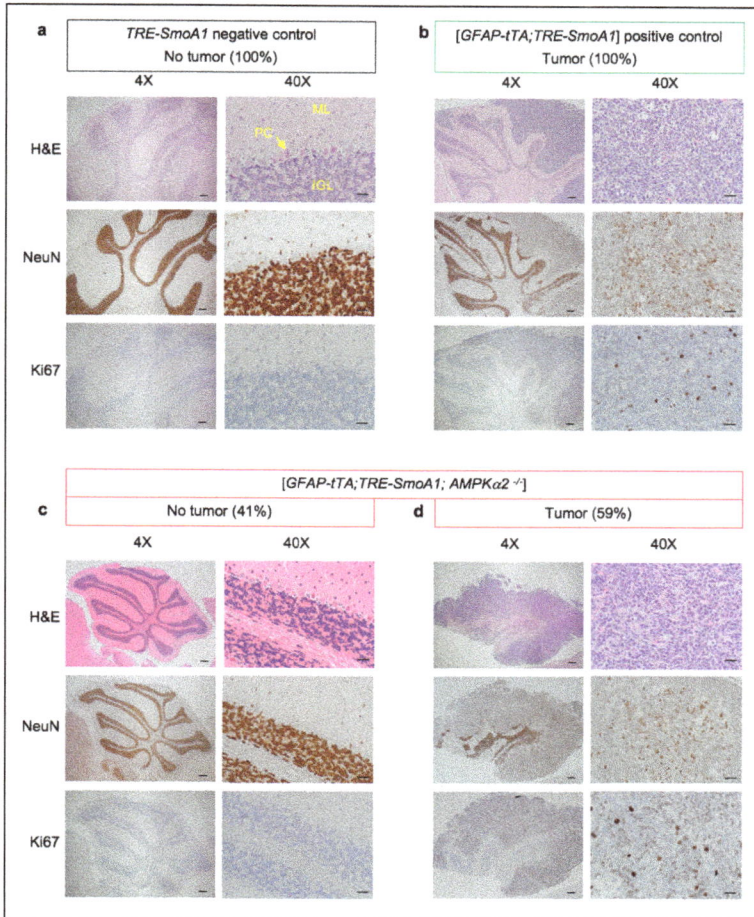

Figure 1. Loss of *AMPKα* impairs SHH medulloblastoma tumorigenesis in vivo. Histopathological examination of the cerebellums of (**a**) a Postnatal Day 63 (P63) *TRE-SmoA1* mouse, (**b**) a P56 [*GFAP-tTA;TRE-SmoA1*] mouse, (**c**) a one-year-old [*GFAP-tTA;TRE-SmoA1;AMPKα2*$^{-/-}$] mouse representative of the 7/17 (41%) [*GFAP-tTA;TRE-SmoA1;AMPKα2*$^{-/-}$] mice that do not show any evidence of tumor upon histological analysis, and (**d**) a P30 [*GFAP-tTA;TRE-SmoA1;AMPKα2*$^{-/-}$] mouse representative of the 10/17 (59%) [*GFAP-tTA;TRE-SmoA1;AMPKα2*$^{-/-}$] mice that show a tumor. The histological analysis includes H&E staining as well as Ki67 (marker of proliferation) and NeuN (marker of neuronal differentiation) immunohistochemistry (magnification: 4× and 40×). The different cell layers of the cerebellum, i.e., the molecular (ML), Purkinje cell (PC), and internal granular cell (IGL) layers are labeled in the 40× *TRE-SmoA1* control H&E picture. Scale bars: 200 μm (4×) or 25 μm (40×).

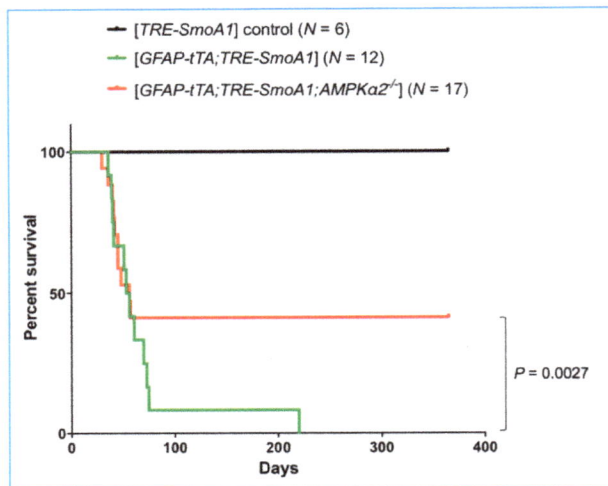

Figure 2. Loss of *AMPKα* increases survival in *SmoA1* mouse model of SHH medulloblastoma. Survival analysis of *AMPKα2* KO in *SmoA1*-driven medulloblastoma. Three groups of C57BL/6 mice were assessed for survival: (i) TRE-SmoA1, (ii) [GFAP-tTA;TRE-SmoA1], and (iii) [GFAP-tTA;TRE-SmoA1;AMPKα2$^{-/-}$]. The *AMPKα2* KO mice have a significantly lower chance than the *AMPKα2* WT mice of developing medulloblastoma [59% (10/17) for *AMPKα2* KO versus 100% (12/12) for *AMPKα2* WT; $p = 0.0027$, likelihood ratio chi-square test]. For the log-rank test, which is designed to be most powerful when hazards are proportional, we obtained $p = 0.108$. Indeed, the *AMPKα2* KO mice that develop medulloblastoma succumb to the tumor at the same pace as WT mice in the first 60 days.

Cerebellar development is normal in the *TRE-SmoA1* control mice. Indeed, as expected for adult mice, we observe properly tri-laminated molecular, Purkinje cell, and internal granular cell layers, but no external granular cell layer. The strong staining for the neuronal marker NeuN indicates the presence of normal cerebellar neurons in the internal granular layer. The absence of staining for Ki-67 indicates that cells are in a quiescent, non-proliferative stage (Figure 1a). As previously described [35,36], all [GFAP-tTA;TRE-SMOA1] mice develop large medulloblastoma tumors that can extend along the entire rostral-caudal length of the cerebellum, account for more than a quarter of the total cerebellar volume, express NeuN, indicative of neuronal origin, and show marked proliferative activity (positive for Ki67) (Figure 1b). The vast majority (11/12) of the [GFAP-tTA;TRE-SMOA1] mice succumb to the tumor within 80 days (Figure 2), as previously described [35,36]. One [GFAP-tTA;TRE-SMOA1] mouse survived beyond 200 days with a very large tumor mass identified post-mortem (Figure S2). We note that, apart from medulloblastoma formation, no other types of tumors can be detected in the brain or the other organs of [GFAP-tTA;TRE-SMOA1] mice.

In contrast, 41% (7/17) of the [GFAP-tTA;TRE-SmoA1;AMPKα2$^{-/-}$] mice survived beyond one year (Figure 2) and do not show any evidence of tumor upon histological analysis (Figure 1c). Two of these seven mice show focal, microscopic remnants of external granular cell (EGC) neurons in groups no larger than 10 cells, which were non-proliferative and non-neoplastic (Figure S3). These cells likely correspond to granular cell neurons that have failed to properly migrate into the internal granular layer, a process that is dependent on SHH signaling inhibition [38]. Thus, the *AMPKα2* KO mice have a significantly lower chance of developing medulloblastoma compared to the *AMPKα2* WT mice [59% (10/17) for *AMPKα2* KO versus 100% (12/12) for *AMPKα2* WT; $p = 0.0027$]. Our results demonstrate the pro-tumorigenic role of *AMPKα2* in medulloblastoma. We note that the number of pyknotic cells undergoing apoptosis in either the normal cerebellar tissues (very rare occurrence, <1%)

or the tumorigenic tissues (~10–20%) remains the same in the *AMPKα2* WT and KO mice. Thus, the effect of *AMPKα2* on apoptosis is not a likely to be a major factor underlying the pro-tumorigenic role of *AMPKα2* in medulloblastoma. Below, we explore the potential effect of *AMPKα2* KO on CNBP.

2.2. Loss of AMPKα2 Results in the Decreased Expression of the CNBP Protein

In response to SHH signaling, AMPK phosphorylates CNBP, thus promoting its interaction with SUFU and its stabilization [34]. Subsequently, the increased level of CNBP protein expression results in the increased translation of ODC1 and associated activation of polyamine metabolism, which is essential for the SHH-dependent proliferation of medulloblastoma cells [34]. In their study, D'amico et al. also showed that targeting CNBP or ODC1 impairs SHH-driven medulloblastoma and thus indirectly inferred that AMPK may be required for tumor formation in this context [34]. Our observation that *AMPKα2* KO impairs tumor progression in the *SmoA1* mouse model of medulloblastoma directly verifies this hypothesis (Figures 1 and 2). To further assess the relevance of this SHH/AMPK/CNBP axis in SHH medulloblastoma, we investigated the level of protein expression of CNBP in both normal and tumorigenic cerebellums in the presence or in the absence of *AMPKα2*. We observed that, in both physiological (*SmoA1−*) and pathophysiological (*SmoA1+*) contexts, the loss of *AMPKα2* resulted in lower levels of the CNBP protein (Figure 3a). In conclusion, notwithstanding that other AMPK substrates [11,12] may also contribute to the pro-tumorigenic role of *AMPKα2* in medulloblastoma, these results further support the hypothesis of a critical role for the SHH/AMPK/CNBP axis in SHH-driven medulloblastoma.

Loss of *AMPKα2* can be associated with a compensatory increase in AMPKα1 protein, e.g., in the muscle [39]. We investigated this effect in the cerebellum. While the level of expression of AMPKα1 is higher in the *SmoA1+* mice than in the *SmoA1−* mice, it remains the same in the control and the *AMPKα2* KO mice in both the *SmoA1− and SmoA1+* contexts (Figure 3b). Thus, in both normal and tumorigenic cerebellums, the deficiency in AMPKα2 was not compensated for by AMPKα1 overexpression.

Similarly, we investigated the level of expression of phosphorylated acetyl-CoA carboxylase alpha (pACC1) in both normal and tumorigenic cerebellums in the presence or in the absence of *AMPKα2*. ACC1 is a key enzyme in the fatty acid synthesis pathway that mediates the conversion of acetyl-CoA to malonyl-CoA. ACC1 phosphorylation by AMPK results in its inactivation, thus inhibiting lipogenesis [40,41]. We observed that, in the *SmoA1+* context, the loss of *AMPKα2* resulted in lower levels of the pACC1 protein (Figure 3c). This analysis of pACC1 suggests a potential impact associated with *AMPKα2* KO on lipogenesis. Interestingly, while a pro-tumorigenic role of ACC1 has been previously described in various contexts [42], a tumor-suppressive role has also been suggested. Indeed, ACC1 inhibition by AMPK is required to maintain NADPH levels and in vivo growth of lung and breast tumor xenografts [23]. Similarly, impaired inhibition of ACC1 upon *AMPKα2* KO may result in low NADPH levels in medulloblastoma cells and thus may contribute to the pro-tumorigenic role of AMPKα2 in medulloblastoma.

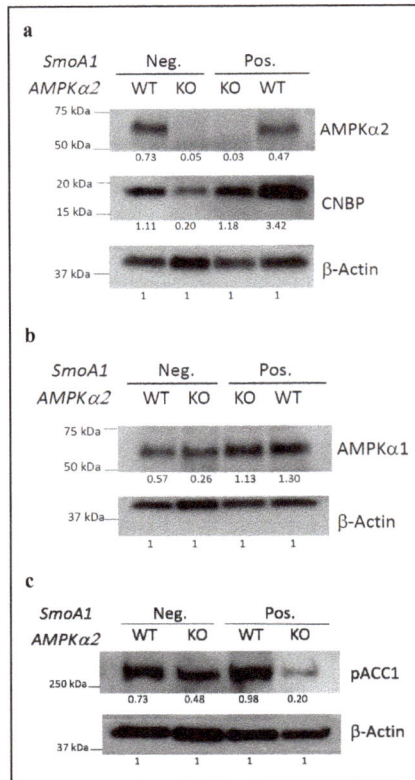

Figure 3. Loss of *AMPKα2* results in lower level of expression of the CNBP and pACC1 proteins in both physiological (*SmoA1*−) and pathophysiological (*SmoA1*+) contexts. Western blot analyses of the AMPKα2 (**a**), AMPKα1 (**b**), CNBP (**a**), pACC1 (**c**), and β-Actin proteins in whole cerebellum protein extracts obtained from WT, *AMPKα2*$^{-/-}$, [*GFAP-tTA;TRE-SmoA1; AMPKα2*$^{-/-}$], and [*GFAP-tTA;TRE-SmoA1*] Postnatal Day 16 (P16) littermate mice. Below each Western blot band, we show the ratio of the band intensity of the targeted protein over the band intensity of β-Actin.

2.3. Frequent Copy Number Gains for the AMPK, CNBP, and ODC1 Genes in SHH and Group #3 Medulloblastoma Patients

We analyzed the publicly available copy number data for 345 medulloblastoma samples [5] (Figure 4). Notably, *AMPK* genes do not have frequent copy number losses in medulloblastoma, the sole exception being the loss of *AMPKγ3* in 6% of Group #4 tumors. In contrast, *AMPK* genes tend to have frequent copy number gains in medulloblastoma in general, and particularly in Group #3 tumors, which have frequent gains for the *AMPKα1* (30%), *AMPKβ2* (38%), and *AMPKγ2* (43%) genes. *AMPKα2* also has frequent copy number gains in Group #3 tumors (11%) (Figure 4a,b). These observations resonate with the previous report that Group #3 medulloblastoma has frequent chromosome 1 gains [43]. In SHH medulloblastoma patients, we observed frequent copy number gains for *AMPKβ2* (6%), *AMPKγ2* (5%) and *AMPKγ3* (10%). Interestingly, strong copy number gains are also observed for *CNBP* and *ODC1* in SHH medulloblastoma (19% and 11%, respectively), as well as in Group #3 medulloblastoma (10% and 22%, respectively). These observations draw yet another interesting parallel with the D'amico et al. study in that their immunohistochemistry analysis of CNBP and ODC1 in a cohort of 42 medulloblastoma patients revealed high levels of expression of both proteins in both SHH and Group

#3 medulloblastoma [34]. The extent to which genetic events co-occur or not in patients can reveal functional relationships between genes, e.g., dependence or redundancy [44]. We found that the copy numbers for the nine genes under investigation are often positively correlated, with a remarkable 21 out of 33 non-syntenic pairs of genes having significant ($p < 0.01$) positive correlations (Figure 4c). In other words, not only are *AMPK*, *CNBP*, and *ODC1* genes frequently gained in medulloblastoma, but these gains tend to co-occur more often than expected at random, suggesting that the co-occurrence of these gains may provide a synergistic advantage. Together, these observations are in agreement with a pro-tumorigenic role for the SHH/AMPK/CNBP axis in medulloblastoma.

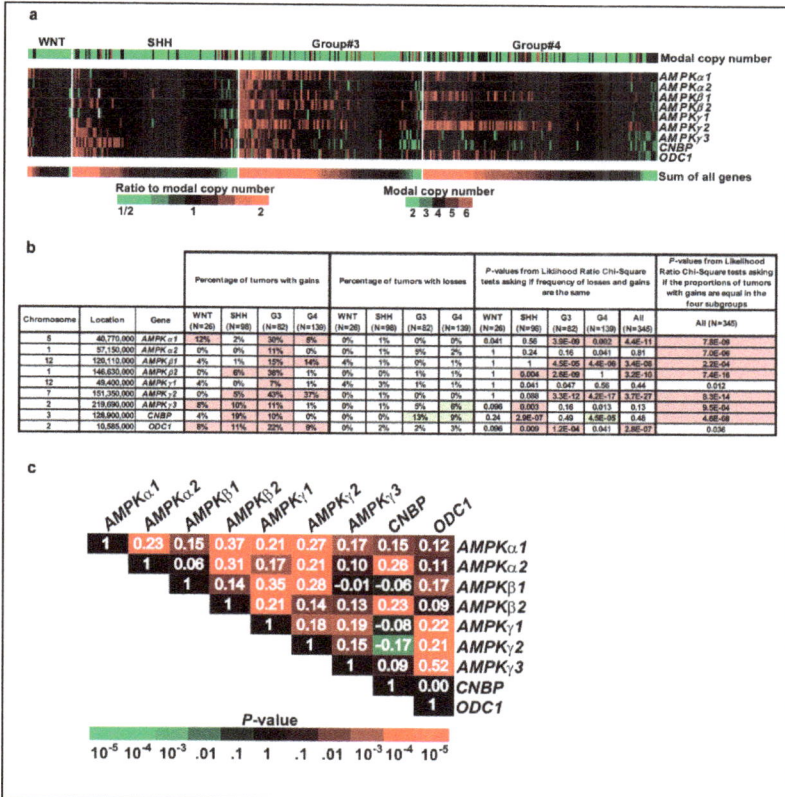

Figure 4. Frequent copy number gains for the AMPK, CNBP, and ODC1 genes in SHH and Group #3 medulloblastoma patients. (**a**) Estimated ratio of copy number to modal copy number for each tumor for 345 medulloblastomas, including 26 WNT, 98 SHH, 82 Group #3, and 139 Group #4 samples. Modal copy number estimates are shown in the first row and the rows for individual genes show copy number estimates normalized to the modal copy number. The "sum of all genes" row was computed from the sum of the log-transformed ratios for the nine genes and was used to determine the sorting of samples in each group. (**b**) The percentage of tumors within each of the four subgroups with copy number gains [$\log2(\frac{copy\ number}{modal\ copy\ number}) > 0.4$] or losses [$\log2(\frac{copy\ number}{modal\ copy\ number}) < -0.4$]. Gains or losses in greater than 5% of the tumors in a given group are highlighted with red or green background, respectively. Similarly, *p* values < 0.01 are highlighted. (**c**) Analysis of the correlations of log-transformed normalized copy number estimates shows that positive correlations are frequent. Pearson correlations are shown as numbers in white, while their significance is indicated by the color of the background in each cell.

3. Discussion

In this study, we demonstrate that loss of AMPKα2 impairs SHH-driven medulloblastoma tumorigenesis. This is a surprising observation given the previous reports supporting a tumor-suppressive role for AMPK in medulloblastoma tumorigenesis [28–33] and the fact that the subunit AMPKα2 alone can suppress tumorigenesis in other contexts [21,45,46]. In fact, together with a previous analysis of AMPKα2 in mammary carcinoma [25], our analysis represents the first in vivo evidence that the loss of AMPKα2 alone can impair oncogenesis.

Several reasons may underlie this apparent AMPK paradox, both in cancer in general and in medulloblastoma in particular. As previously reviewed, AMPK modulates numerous targets both to inhibit anabolism and to promote catabolism. As such, AMPK has a dual role in cancer, tumor-suppressive or pro-tumorigenic, depending on the cellular or tissue context [13–15]. For example, on the one hand, as an inhibitor of the mTOR complex 1 and the cell cycle, AMPK can be tumor-suppressive [19,20]. On the other hand, maintenance of the metabolic balance by AMPK may be critical for survival during metabolic stress that can occur in the tumor microenvironment—hence the requirement for AMPK activity in some cancer cells [22–27]. Similarly, in medulloblastoma, the role of AMPK as both an inhibitor of GLI1 and SHH signaling [28–30,33] (tumor-suppressive effect) as well as a promoter of CNBP and polyamine metabolism (pro-tumorigenic effect) supposes a dual role for AMPK in this disease context. In agreement with the observation that CNBP protein levels are decreased in the absence of AMPKα2, we observed that the pro-tumorigenic function of AMPKα2 prevails in the context of the highly penetrant [GFAP-tTA;TRE-SMOA1] mouse model of medulloblastoma. It would be interesting, though, to assess the loss of AMPKα2 in other mouse models of medulloblastoma with lower incidence/longer latency [47,48], in order to investigate which of AMPKα2's roles, tumor-suppressive or pro-tumorigenic, prevails in these contexts.

The apparent discrepancy between our results observed in mouse and the previously described tumor-suppressive role of AMPK in human medulloblastoma cell lines [28–30,33] could be due to the species-specific role of AMPK in the cerebellum. While all studies in which the role of AMPK has been directly tested genetically in human medulloblastoma cells agree, i.e., AMPK suppresses SHH-driven cell growth, the role of AMPK in mouse cell lines is debated [28–30,33]. For example, several studies by the Yang laboratory showed that AMPK phosphorylates GLI1 at multiple sites and targets it for proteasomal degradation in murine cells [28,29,33]. Yet, Di Magno et al. argue that AMPK phosphorylates GLI1 at the unique residue Ser408, which is conserved only in primates but not in other species [30]. If the regulation of GLI1 by AMPK does not occur in mouse, as suggested by Di Magno et al., this species-specific difference could explain why our observations obtained in mouse contradict the previous reports that AMPK suppresses growth in human SHH medulloblastoma [28–33]. Further investigation of the species-specific effect of AMPK on GLI1 and medulloblastoma is warranted.

In some contexts, AMPKα1 and AMPKα2 mediate specific, non-redundant functions, e.g., AMPKα2$^{-/-}$ but not AMPKα1$^{-/-}$, mice are resistant to hypoglycemic AICAR effects [37,49]. In other contexts, AMPKα1 and AMPKα2 are genetically redundant, e.g., while both AMPKα1 and AMPKα2 are dispensable for development, the combined loss of both genes is embryonic lethal at E10.5 [50]. While we cannot exclude the possibility that a complete KO of AMPKα may have an even more dramatic effect on medulloblastoma tumorigenesis and survival, our results suggest that AMPKα2 has a specific function in this context. This lack of genetic redundancy may underlie various mechanisms, as previously reviewed [51,52]. Could this be due to different biochemical function for AMPKα1- and AMPKα2-containing heterotrimeric complexes? AMPKα2 may be the target of different regulators and/or AMPKα2 may target different substrates. In that regard, it would be interesting to assess whether CNBP is specifically targeted by AMPKα2- but not by AMPKα1-containing heterotrimeric complexes. Could this be due to variations in the spatio-temporal expression for AMPKα1 and AMPKα2? Cell types may express one gene, but not the other. In agreement with an AMPKα2-specific function in SHH medulloblastoma and the fact that the cerebellar granular neuron progenitor cell

(CGNP) is the "cell of origin" in SHH-driven medulloblastoma [53], AMPKα2 is expressed at higher level than AMPKα1 in cerebellar granule neurons [54].

Genomics applied to medulloblastoma defined four medulloblastoma subgroups (WNT, SHH, and Groups #3 and #4) [3,4] and, more recently, different subtypes within them [5–7]. D'amico et al. previously showed that targeting CNBP or ODC1 impairs SHH-driven medulloblastoma and inferred that the SHH/AMPK/CNBP axis may be pro-tumorigenic in this context [34]. Notwithstanding the debate on the nature of the effect (pro- versus anti-tumorigenic) [28–34], our observation that AMPKα2 impairs tumor progression in the SmoA1 mouse model of medulloblastoma validates this hypothesis. Is AMPK functionally relevant in other medulloblastoma subgroups? We observed high copy numbers for the AMPK, CNBP, and ODC1 genes not only in SHH but also in Group #3 medulloblastoma. Taken together with the complementary report that both SHH and Group #3 medulloblastomas show a high level of expression of CNBP and ODC1 proteins [34], the SHH/AMPK/CNBP axis could then have a critical role in both the SHH and Group #3 subgroups. Group #3 medulloblastomas account for ~25% of all medulloblastomas, are characterized by a transcriptional signature associated with photoreceptors and gamma aminobutyric acid–secreting (GABAergic) neurons, and are associated with a poor prognosis [3]. Though differences in the metabolic profiles between medulloblastoma subgroups have not yet been comprehensively examined, the higher degree of aggressiveness associated with Group #3 medulloblastomas may translate to a higher need for energy and thus a higher sensibility to the proper maintenance of the metabolic balance by AMPK under metabolic stress conditions. Investigation of AMPK KO in mouse models of Group #3 medulloblastoma, when they become available, is warranted.

Our study reveals a pro-tumorigenic role for AMPKα2 in SHH medulloblastoma. In light of the previous reports on the tumor-suppressive role of AMPK in SHH medulloblastoma, this observation further underscores the multifaceted role of AMPK in cancer. It also warrants caution to the previously proposed use of AMPK agonists for the treatment of cancer patients in general, and in medulloblastoma patients in particular.

4. Materials and Methods

4.1. [GFAP-tTA;TRE-SmoA1] Mouse Model

We studied medulloblastoma tumorigenesis in the absence or presence of AMPKα2 in the previously published bitransgenic [GFAP-tTA;TRE-SmoA1] model [35,36], where the expression of the tetracycline-regulated transactivator (tTA) is driven by a GFAP promoter and the expression of oncogenic SmoA1 is under the control of the tetracycline responsive element (TRE). The experimental breeders used in this study were an AMPKα2 KO mouse [37], a TRE-SmoA1 mouse [35], and a GFAP-tTA mouse [55], all of which were maintained on a C57/BL6 background for at least five generations prior to initiating experiments. TRE-SmoA1, [GFAP-tTA;TRE-SmoA1], and [GFAP-tTA;TRE-SmoA1;AMPKα2$^{-/-}$] mouse littermates were generated by crossing [GFAP-tTA;AMPKα2$^{+/-}$] mice with [TRE-SmoA1;AMPKα2$^{+/-}$] mice. Animals that meet the guidelines for end-stage illness and/or found to be at protocol endpoint (evidence of large tumor formation, enlarged dome head, and/or severe neurological dysfunction) were humanely euthanized in accordance with the institutional guidelines for the welfare of experimental animals. Maintenance of mouse colonies and experimental procedures were approved by the University of Michigan Committee on the Use and Care of Animals.

4.2. Mouse Genotyping

The mouse genotyping experiments were performed as previously described [36], with minor modifications. Genotyping experiments were performed by PCR analysis using tail genomic DNA obtained from pups at Postnatal Days 10 (P10) and 14 (P14). Genotyping of *AMPKα2* was determined by PCR using the following oligonucleotides: forward: 5′-gcttagcacgttaccctggat-3′ and WT reverse: 5′-gtcttcactgaaatacatagca-3′ or mutant reverse: 5′-gcattgaaccacagtccttcctc-3′.

Genotyping of the *GFAP-tTA* transgene was determined by PCR using the following oligonucleotides: forward: 5′-ctcgcccagaagctaggtgt-3′ and reverse: 5′-ccatcgcgatgacttagt-3′. Genotyping of the *TRE-SmoA1* transgene was determined by PCR using the following oligonucleotides: forward: 5′-ggaactgatgaatgggagca-3′ and reverse: 5′-gggaggtgtgggaggttt-3′. For internal control genotyping, we used the following primers (forward: 5′-caaatgttgctgtctggtg-3′ and reverse: 5′-gtcagtcgagtgcacagttt-3′).

4.3. Immunohistochemistry

Immunohistochemistry (IHC) experiments were performed as previously described [36].

4.4. Western Blot Analyses

Whole cerebellum protein samples were prepared using RIPA buffer, sonicated three times on ice, and supernatants were collected after centrifugation. The resulting cerebellum protein extracts were separated on acrylamide gels, transferred to PVDF membranes, and proteins were detected using standard immunoblotting techniques. The following antibodies were used: AMPKα1 (Bethyl Laboratory®, Montgomery, TX, USA; Cat# A300-507A), AMPKα2 (Bethyl Laboratory®, Cat# A300-508A), pACC1 (Cell Signaling®, Danvers, MA, USA; Cat# 11818), β-actin (Cell Signaling®, Cat# 5125), and goat α-rabbit IgG (Jackson Immunoresearch Laboratory®, West Grove, PA, USA; Cat# 111-035-045). The CNBP antibody [34] was a gift from Dr. Gianluca Canettieri, Sapienza University of Rome, Italy. The intensities of Western blot signal bands were quantified using Gel Analysis in ImageJ.

4.5. Copy Number Variations in Medulloblastoma Patients

We analyzed the publicly available copy number data for medulloblastoma samples estimated from short-read DNA sequencing [5]. We downloaded a single segment file holding estimated $\log_2(\frac{copy\ number}{2})$ data from PedcBioportal (http://pedcbioportal.org/). Following the methods of the original paper [5], we excluded 16 samples and analyzed the remaining 345. Many tumors had chromosome doublings, so we normalized our data by first computing the average of the $\log_2(\frac{copy\ number}{2})$ estimates for each chromosome and then counted how many chromosomes gave estimates within 0.1 unit of each possible log-transformed integer copy number. We then selected the most common integer chromosomal copy number for each tumor, after excluding those chromosomes with the most frequent copy number changes (Chr. 6, 7, 8, and 17, X and Y), and referred to this as the modal copy number. Among the 345 tumors, we observed 212 tumors with a 2-modal copy number, 19 tumors with a 3-modal copy number, 91 tumors with a 4-modal copy number, 2 tumors with a 5-modal copy number, and 21 tumors with a 6-modal copy number. Our final estimates of the relative gene copy numbers were of $\log_2(\frac{copy\ number}{modal\ copy\ number})$.

Supplementary Materials: Supplementary Materials can be found at http://www.mdpi.com/1422-0067/19/11/3287/s1.

Author Contributions: J.-F.R. conceived and directed the project. H.Z., C.P., J.Y., E.C.F., B.V., and J.-F.R. designed and performed the in vivo mouse studies. H.Z., S.-S.P., J.W., D.T., K.I., and S.C.-P. performed and analyzed the molecular studies and the IHC experiments. R.K. performed the copy number variation and statistical analyses. J.-F.R. wrote the manuscript, with contributions from other co-authors.

funding: This work was supported by the Padnos Fund for Innovative Cancer Research of the University of Michigan Rogel Cancer Center, awarded to J.-F.R., the Bench to Bedside Translation Award of the Michigan Institute for Clinical and Health Research (MICHR), the National Center for Advancing Translational Sciences (NCAT) of the National Institutes of Health (NIH) under Award Number UL1TR000433 awarded to J.-F.R., the M-Cubed Grant, awarded to J.-F.R., K.I., and S.C.-P., NIH grants (RO1 DK083491, RO1 GM110019) awarded to K.I., and funds from the University of Michigan Department of Pathology provided to J.-F.R. R.K. was supported by an NIH support grant (P30 CA046592) awarded to the University of Michigan Rogel Cancer Center. E.C.F. was supported by the St. Baldrick's Staff Giving Campaign Summer Fellowship awarded by the St. Baldrick's Foundation.

Acknowledgments: We thank the University of Michigan Rogel Cancer Center Tissue Core for technical support with the immunohistochemistry experiments; Gianluca Canettieri, Sapienza University of Rome, Rome, Italy, for the CNBP antibody; Tao Xu for his comments and suggestions in the early development of the project.

Conflicts of Interest: The authors declare no conflict of interest. The funders had no role in the design of the study; in the collection, analyses, or interpretation of data; in the writing of the manuscript; or in the decision to publish the results.

Abbreviations

AMPK	AMP-activated protein kinase
CGNPs	Cerebellar granular neuron progenitor cells
Chr.	Chromosome
DAB	Diaminobenzidine
GABAergic	Gamma aminobutyric acid–secreting
H&E	Hematoxylin and eosin
HRP	Horseradish peroxidase
IHC	Immunohistochemistry
IGL	Internal granular cell
ML	Molecular
P10	Postnatal day 10
pACC1	phosphorylated acetyl-CoA carboxylase alpha
PC	Purkinje cell
SHH	Sonic Hedgehog
TRE	Tetracycline responsive element
WT	Wild type

References

1. Ostrom, Q.T.; Gittleman, H.; Liao, P.; Vecchione-Koval, T.; Wolinsky, Y.; Kruchko, C.; Barnholtz-Sloan, J.S. CBTRUS Statistical Report: Primary brain and other central nervous system tumors diagnosed in the United States in 2010–2014. *Neuro Oncol.* **2017**, *19*, v1–v88. [CrossRef] [PubMed]

2. Liu, K.W.; Pajtler, K.W.; Worst, B.C.; Pfister, S.M.; Wechsler-Reya, R.J. Molecular mechanisms and therapeutic targets in pediatric brain tumors. *Sci. Signal.* **2017**, *10*. [CrossRef] [PubMed]

3. Taylor, M.D.; Northcott, P.A.; Korshunov, A.; Remke, M.; Cho, Y.J.; Clifford, S.C.; Eberhart, C.G.; Parsons, D.W.; Rutkowski, S.; Gajjar, A.; et al. Molecular subgroups of medulloblastoma: The current consensus. *Acta Neuropathol.* **2012**, *123*, 465–472. [CrossRef] [PubMed]

4. Northcott, P.A.; Jones, D.T.; Kool, M.; Robinson, G.W.; Gilbertson, R.J.; Cho, Y.J.; Pomeroy, S.L.; Korshunov, A.; Lichter, P.; Taylor, M.D.; et al. Medulloblastomics: The end of the beginning. *Nat. Rev. Cancer* **2012**, *12*, 818–834. [CrossRef] [PubMed]

5. Northcott, P.A.; Buchhalter, I.; Morrissy, A.S.; Hovestadt, V.; Weischenfeldt, J.; Ehrenberger, T.; Grobner, S.; Segura-Wang, M.; Zichner, T.; Rudneva, V.A.; et al. The whole-genome landscape of medulloblastoma subtypes. *Nature* **2017**, *547*, 311–317. [CrossRef] [PubMed]

6. Lin, C.Y.; Erkek, S.; Tong, Y.; Yin, L.; Federation, A.J.; Zapatka, M.; Haldipur, P.; Kawauchi, D.; Risch, T.; Warnatz, H.J.; et al. Active medulloblastoma enhancers reveal subgroup-specific cellular origins. *Nature* **2016**, *530*, 57–62. [CrossRef] [PubMed]

7. Cavalli, F.M.G.; Remke, M.; Rampasek, L.; Peacock, J.; Shih, D.J.H.; Luu, B.; Garzia, L.; Torchia, J.; Nor, C.; Morrissy, A.S.; et al. Intertumoral Heterogeneity within Medulloblastoma Subgroups. *Cancer Cell* **2017**, *31*, 737–754. [CrossRef] [PubMed]

8. Mulhern, R.K.; Merchant, T.E.; Gajjar, A.; Reddick, W.E.; Kun, L.E. Late neurocognitive sequelae in survivors of brain tumours in childhood. *Lancet Oncol.* **2004**, *5*, 399–408. [CrossRef]

9. Laughton, S.J.; Merchant, T.E.; Sklar, C.A.; Kun, L.E.; Fouladi, M.; Broniscer, A.; Morris, E.B.; Sanders, R.P.; Krasin, M.J.; Shelso, J.; et al. Endocrine outcomes for children with embryonal brain tumors after risk-adapted craniospinal and conformal primary-site irradiation and high-dose chemotherapy with stem-cell rescue on the SJMB-96 trial. *J. Clin. Oncol.* **2008**, *26*, 1112–1118. [CrossRef] [PubMed]

10. Hardie, D.G.; Ross, F.A.; Hawley, S.A. AMPK: A nutrient and energy sensor that maintains energy homeostasis. *Nat. Rev. Mol. Cell Biol.* **2012**, *13*, 251–262. [CrossRef] [PubMed]

11. Hardie, D.G.; Schaffer, B.E.; Brunet, A. AMPK: An Energy-Sensing Pathway with Multiple Inputs and Outputs. *Trends Cell Biol.* **2016**, *26*, 190–201. [CrossRef] [PubMed]

12. Cheng, J.; Zhang, T.; Ji, H.; Tao, K.; Guo, J.; Wei, W. Functional characterization of AMP-activated protein kinase signaling in tumorigenesis. *Biochim. Biophys. Acta* **2016**, *1866*, 232–251. [CrossRef] [PubMed]

13. Liang, J.; Mills, G.B. AMPK: A contextual oncogene or tumor suppressor? *Cancer Res.* **2013**, *73*, 2929–2935. [CrossRef] [PubMed]

14. Zadra, G.; Batista, J.L.; Loda, M. Dissecting the Dual Role of AMPK in Cancer: From Experimental to Human Studies. *Mol. Cancer Res.* **2015**, *13*, 1059–1072. [CrossRef] [PubMed]

15. Dasgupta, B.; Chhipa, R.R. Evolving Lessons on the Complex Role of AMPK in Normal Physiology and Cancer. *Trends Pharmacol. Sci.* **2016**, *37*, 192–206. [CrossRef] [PubMed]

16. Hawley, S.A.; Boudeau, J.; Reid, J.L.; Mustard, K.J.; Udd, L.; Makela, T.P.; Alessi, D.R.; Hardie, D.G. Complexes between the LKB1 tumor suppressor, STRAD alpha/beta and MO25 alpha/beta are upstream kinases in the AMP-activated protein kinase cascade. *J. Biol.* **2003**, *2*, 28. [CrossRef] [PubMed]

17. Woods, A.; Johnstone, S.R.; Dickerson, K.; Leiper, F.C.; Fryer, L.G.; Neumann, D.; Schlattner, U.; Wallimann, T.; Carlson, M.; Carling, D. LKB1 is the upstream kinase in the AMP-activated protein kinase cascade. *Curr. Biol.* **2003**, *13*, 2004–2008. [CrossRef] [PubMed]

18. Shaw, R.J.; Kosmatka, M.; Bardeesy, N.; Hurley, R.L.; Witters, L.A.; DePinho, R.A.; Cantley, L.C. The tumor suppressor LKB1 kinase directly activates AMP-activated kinase and regulates apoptosis in response to energy stress. *Proc. Natl. Acad. Sci. USA* **2004**, *101*, 3329–3335. [CrossRef] [PubMed]

19. Shackelford, D.B.; Shaw, R.J. The LKB1-AMPK pathway: Metabolism and growth control in tumour suppression. *Nat. Rev. Cancer* **2009**, *9*, 563–575. [CrossRef] [PubMed]

20. Faubert, B.; Boily, G.; Izreig, S.; Griss, T.; Samborska, B.; Dong, Z.; Dupuy, F.; Chambers, C.; Fuerth, B.J.; Viollet, B.; et al. AMPK is a negative regulator of the Warburg effect and suppresses tumor growth in vivo. *Cell Metab.* **2013**, *17*, 113–124. [CrossRef] [PubMed]

21. Phoenix, K.N.; Devarakonda, C.V.; Fox, M.M.; Stevens, L.E.; Claffey, K.P. AMPKalpha2 Suppresses Murine Embryonic Fibroblast Transformation and Tumorigenesis. *Genes Cancer* **2012**, *3*, 51–62. [CrossRef] [PubMed]

22. Laderoute, K.R.; Amin, K.; Calaoagan, J.M.; Knapp, M.; Le, T.; Orduna, J.; Foretz, M.; Viollet, B. 5'-AMP-activated protein kinase (AMPK) is induced by low-oxygen and glucose deprivation conditions found in solid-tumor microenvironments. *Mol. Cell. Biol.* **2006**, *26*, 5336–5347. [CrossRef] [PubMed]

23. Jeon, S.M.; Chandel, N.S.; Hay, N. AMPK regulates NADPH homeostasis to promote tumour cell survival during energy stress. *Nature* **2012**, *485*, 661–665. [CrossRef] [PubMed]

24. Rios, M.; Foretz, M.; Viollet, B.; Prieto, A.; Fraga, M.; Costoya, J.A.; Senaris, R. AMPK activation by oncogenesis is required to maintain cancer cell proliferation in astrocytic tumors. *Cancer Res.* **2013**, *73*, 2628–2638. [CrossRef] [PubMed]

25. Hindupur, S.K.; Balaji, S.A.; Saxena, M.; Pandey, S.; Sravan, G.S.; Heda, N.; Kumar, M.V.; Mukherjee, G.; Dey, D.; Rangarajan, A. Identification of a novel AMPK-PEA15 axis in the anoikis-resistant growth of mammary cells. *Breast Cancer Res.* **2014**, *16*, 420. [CrossRef] [PubMed]

26. Saito, Y.; Chapple, R.H.; Lin, A.; Kitano, A.; Nakada, D. AMPK Protects Leukemia-Initiating Cells in Myeloid Leukemias from Metabolic Stress in the Bone Marrow. *Cell Stem Cell* **2015**, *17*, 585–596. [CrossRef] [PubMed]

27. Chhipa, R.R.; Fan, Q.; Anderson, J.; Muraleedharan, R.; Huang, Y.; Ciraolo, G.; Chen, X.; Waclaw, R.; Chow, L.M.; Khuchua, Z.; et al. AMP kinase promotes glioblastoma bioenergetics and tumour growth. *Nat. Cell Biol.* **2018**, *20*, 823–835. [CrossRef] [PubMed]

28. Li, Y.H.; Luo, J.; Mosley, Y.Y.; Hedrick, V.E.; Paul, L.N.; Chang, J.; Zhang, G.; Wang, Y.K.; Banko, M.R.; Brunet, A.; et al. AMP-Activated Protein Kinase Directly Phosphorylates and Destabilizes Hedgehog Pathway Transcription Factor GLI1 in Medulloblastoma. *Cell Rep.* **2015**, *12*, 599–609. [CrossRef] [PubMed]

29. Zhang, R.; Huang, S.Y.; Ka-Wai Li, K.; Li, Y.H.; Hsu, W.H.; Zhang, G.J.; Chang, C.J.; Yang, J.Y. Dual degradation signals destruct GLI1: AMPK inhibits GLI1 through beta-TrCP-mediated proteasome degradation. *Oncotarget* **2017**, *8*, 49869–49881. [CrossRef] [PubMed]

30. Di Magno, L.; Basile, A.; Coni, S.; Manni, S.; Sdruscia, G.; D'Amico, D.; Antonucci, L.; Infante, P.; De Smaele, E.; Cucchi, D.; et al. The energy sensor AMPK regulates Hedgehog signaling in human cells through a unique Gli1 metabolic checkpoint. *Oncotarget* **2016**, *7*, 9538–9549. [CrossRef] [PubMed]

31. Gershon, T.R.; Crowther, A.J.; Tikunov, A.; Garcia, I.; Annis, R.; Yuan, H.; Miller, C.R.; Macdonald, J.; Olson, J.; Deshmukh, M. Hexokinase-2-mediated aerobic glycolysis is integral to cerebellar neurogenesis and pathogenesis of medulloblastoma. *Cancer Metab.* **2013**, *1*, 2. [CrossRef] [PubMed]

32. Tech, K.; Gershon, T.R. Energy metabolism in neurodevelopment and medulloblastoma. *Transl. Pediatr* **2015**, *4*, 12–19. [CrossRef] [PubMed]

33. Huang, S.Y.; Chen, S.K.; Yang, J.Y. Activation of AMPK inhibits medulloblastoma cell growth and Gli1 activity. *Cancer Rep. Rev.* **2017**. [CrossRef]

34. D'Amico, D.; Antonucci, L.; Di Magno, L.; Coni, S.; Sdruscia, G.; Macone, A.; Miele, E.; Infante, P.; Di Marcotullio, L.; De Smaele, E.; et al. Non-canonical Hedgehog/AMPK-Mediated Control of Polyamine Metabolism Supports Neuronal and Medulloblastoma Cell Growth. *Dev. Cell* **2015**, *35*, 21–35. [CrossRef] [PubMed]

35. Michael, L.E.; Westerman, B.A.; Ermilov, A.N.; Wang, A.; Ferris, J.; Liu, J.; Blom, M.; Ellison, D.W.; van Lohuizen, M.; Dlugosz, A.A. Bmi1 is required for Hedgehog pathway-driven medulloblastoma expansion. *Neoplasia* **2008**, *10*, 1343–1349. [CrossRef] [PubMed]

36. Xu, T.; Zhang, H.; Park, S.S.; Venneti, S.; Kuick, R.; Ha, K.; Michael, L.E.; Santi, M.; Uchida, C.; Uchida, T.; et al. Loss of Pin1 Suppresses Hedgehog-Driven Medulloblastoma Tumorigenesis. *Neoplasia* **2017**, *19*, 216–225. [CrossRef] [PubMed]

37. Viollet, B.; Andreelli, F.; Jorgensen, S.B.; Perrin, C.; Geloen, A.; Flamez, D.; Mu, J.; Lenzner, C.; Baud, O.; Bennoun, M.; et al. The AMP-activated protein kinase alpha2 catalytic subunit controls whole-body insulin sensitivity. *J. Clin. Investig.* **2003**, *111*, 91–98. [CrossRef] [PubMed]

38. De Luca, A.; Cerrato, V.; Fuca, E.; Parmigiani, E.; Buffo, A.; Leto, K. Sonic hedgehog patterning during cerebellar development. *Cell. Mol. Life Sci.* **2016**, *73*, 291–303. [CrossRef] [PubMed]

39. Jorgensen, S.B.; Treebak, J.T.; Viollet, B.; Schjerling, P.; Vaulont, S.; Wojtaszewski, J.F.; Richter, E.A. Role of AMPKalpha2 in basal, training-, and AICAR-induced GLUT4, hexokinase II, and mitochondrial protein expression in mouse muscle. *Am. J. Physiol. Endocrinol. Metab.* **2007**, *292*, E331–E339. [CrossRef] [PubMed]

40. Carling, D.; Zammit, V.A.; Hardie, D.G. A common bicyclic protein kinase cascade inactivates the regulatory enzymes of fatty acid and cholesterol biosynthesis. *FEBS Lett.* **1987**, *223*, 217–222. [CrossRef]

41. Fullerton, M.D.; Galic, S.; Marcinko, K.; Sikkema, S.; Pulinilkunnil, T.; Chen, Z.P.; O'Neill, H.M.; Ford, R.J.; Palanivel, R.; O'Brien, M.; et al. Single phosphorylation sites in Acc1 and Acc2 regulate lipid homeostasis and the insulin-sensitizing effects of metformin. *Nat. Med.* **2013**, *19*, 1649–1654. [CrossRef] [PubMed]

42. Wang, C.; Ma, J.; Zhang, N.; Yang, Q.; Jin, Y.; Wang, Y. The acetyl-CoA carboxylase enzyme: A target for cancer therapy? *Expert Rev. Anticancer Ther.* **2015**, *15*, 667–676. [CrossRef] [PubMed]

43. Northcott, P.A.; Shih, D.J.; Peacock, J.; Garzia, L.; Morrissy, A.S.; Zichner, T.; Stutz, A.M.; Korshunov, A.; Reimand, J.; Schumacher, S.E.; et al. Subgroup-specific structural variation across 1000 medulloblastoma genomes. *Nature* **2012**, *488*, 49–56. [CrossRef] [PubMed]

44. Ashworth, A.; Lord, C.J.; Reis-Filho, J.S. Genetic interactions in cancer progression and treatment. *Cell* **2011**, *145*, 30–38. [CrossRef] [PubMed]

45. Fox, M.M.; Phoenix, K.N.; Kopsiaftis, S.G.; Claffey, K.P. AMP-Activated Protein Kinase alpha 2 Isoform Suppression in Primary Breast Cancer Alters AMPK Growth Control and Apoptotic Signaling. *Genes Cancer* **2013**, *4*, 3–14. [CrossRef] [PubMed]

46. Vila, I.K.; Yao, Y.; Kim, G.; Xia, W.; Kim, H.; Kim, S.J.; Park, M.K.; Hwang, J.P.; Billalabeitia, E.G.; Hung, M.C.; et al. A UBE2O-AMPKα2 Axis That Promotes Tumor Initiation and Progression Offers Opportunities for Therapy. *Cancer Cell* **2017**, *31*, 208–224. [CrossRef] [PubMed]

47. Goodrich, L.V.; Milenkovic, L.; Higgins, K.M.; Scott, M.P. Altered neural cell fates and medulloblastoma in mouse patched mutants. *Science* **1997**, *277*, 1109–1113. [CrossRef] [PubMed]

48. Hallahan, A.R.; Pritchard, J.I.; Hansen, S.; Benson, M.; Stoeck, J.; Hatton, B.A.; Russell, T.L.; Ellenbogen, R.G.; Bernstein, I.D.; Beachy, P.A.; et al. The SmoA1 mouse model reveals that notch signaling is critical for the growth and survival of sonic hedgehog-induced medulloblastomas. *Cancer Res.* **2004**, *64*, 7794–7800. [CrossRef] [PubMed]

49. Jorgensen, S.B.; Viollet, B.; Andreelli, F.; Frosig, C.; Birk, J.B.; Schjerling, P.; Vaulont, S.; Richter, E.A.; Wojtaszewski, J.F. Knockout of the alpha2 but not alpha1 5′-AMP-activated protein kinase isoform abolishes 5-aminoimidazole-4-carboxamide-1-beta-4-ribofuranosidebut not contraction-induced glucose uptake in skeletal muscle. *J. Biol. Chem.* **2004**, *279*, 1070–1079. [CrossRef] [PubMed]

50. Viollet, B.; Athea, Y.; Mounier, R.; Guigas, B.; Zarrinpashneh, E.; Horman, S.; Lantier, L.; Hebrard, S.; Devin-Leclerc, J.; Beauloye, C.; et al. AMPK: Lessons from transgenic and knockout animals. *Front. Biosci.* **2009**, *14*, 19–44. [CrossRef]

51. Ross, F.A.; MacKintosh, C.; Hardie, D.G. AMP-activated protein kinase: A cellular energy sensor that comes in 12 flavours. *FEBS J.* **2016**, *283*, 2987–3001. [CrossRef] [PubMed]

52. Herzig, S.; Shaw, R.J. AMPK: Guardian of metabolism and mitochondrial homeostasis. *Nat. Rev. Mol. Cell Biol.* **2018**, *19*, 121–135. [CrossRef] [PubMed]

53. Yang, Z.J.; Ellis, T.; Markant, S.L.; Read, T.A.; Kessler, J.D.; Bourboulas, M.; Schuller, U.; Machold, R.; Fishell, G.; Rowitch, D.H.; et al. Medulloblastoma can be initiated by deletion of Patched in lineage-restricted progenitors or stem cells. *Cancer Cell* **2008**, *14*, 135–145. [CrossRef] [PubMed]

54. Turnley, A.M.; Stapleton, D.; Mann, R.J.; Witters, L.A.; Kemp, B.E.; Bartlett, P.F. Cellular distribution and developmental expression of AMP-activated protein kinase isoforms in mouse central nervous system. *J. Neurochem.* **1999**, *72*, 1707–1716. [CrossRef] [PubMed]

55. Lin, W.; Kemper, A.; McCarthy, K.D.; Pytel, P.; Wang, J.P.; Campbell, I.L.; Utset, M.F.; Popko, B. Interferon-gamma induced medulloblastoma in the developing cerebellum. *J. Neurosci.* **2004**, *24*, 10074–10083. [CrossRef] [PubMed]

International Journal of
Molecular Sciences

MDPI

Review

AMP-Activated Protein Kinase and Host Defense against Infection

Prashanta Silwal [1,2], Jin Kyung Kim [1,2,3], Jae-Min Yuk [4] and Eun-Kyeong Jo [1,2,3,*]

1   Department of Microbiology, Chungnam National University School of Medicine, Daejeon 35015, Korea;
     pst.ktz@gmail.com (P.S.); pcjlovesh6@naver.com (J.K.K.)
2   Infection Control Convergence Research Center, Chungnam National University School of Medicine,
     Daejeon 35015, Korea
3   Department of Medical Science, Chungnam National University School of Medicine, Daejeon 35015, Korea
4   Department of Infection Biology, Chungnam National University School of Medicine, Daejeon 35015, Korea;
     yjaemin0@cnu.ac.kr
*   Correspondence: hayoungj@cnu.ac.kr; Tel.: +82-42-580-8243; Fax: +82-42-585-3686

Received: 27 September 2018; Accepted: 5 November 2018; Published: 6 November 2018

Abstract: 5′-AMP-activated protein kinase (AMPK) plays diverse roles in various physiological and pathological conditions. AMPK is involved in energy metabolism, which is perturbed by infectious stimuli. Indeed, various pathogens modulate AMPK activity, which affects host defenses against infection. In some viral infections, including hepatitis B and C viral infections, AMPK activation is beneficial, but in others such as dengue virus, Ebola virus, and human cytomegaloviral infections, AMPK plays a detrimental role. AMPK-targeting agents or small molecules enhance the antiviral response and contribute to the control of microbial and parasitic infections. In addition, this review focuses on the double-edged role of AMPK in innate and adaptive immune responses to infection. Understanding how AMPK regulates host defenses will enable development of more effective host-directed therapeutic strategies against infectious diseases.

Keywords: AMPK; infection; mycobacteria; host defense

1. Introduction

5′-AMP-activated protein kinase (AMPK) is an intracellular serine/threonine kinase and a key energy sensor that is activated under conditions of metabolic stress [1–3]. It governs a variety of biological processes for the maintenance of energy homeostasis in response to metabolic stresses such as adenosine triphosphate (ATP) depletion [2]. Due to its critical function in metabolic homeostasis, much research has focused on the roles of AMPK in metabolic diseases and cancers [4–6]. However, much less is known about the function of AMPK in infection [7]. Due to the energetic demands of infected cells, most infections by intracellular pathogens are associated with activation of host AMPK, presumably to promote microbial proliferation [8]. AMPK functions as a modulator of host defenses against intracellular bacterial, viral, and parasitic infections [9–12]. Indeed, numerous viruses have the ability to trigger metabolic changes, thereby modulating AMPK activity and substrate selection [13], and AMPK signaling could facilitate or inhibit intracellular viral replication depending on the virus infection [14].

This review focuses on the double-edged role of AMPK in the regulation of host antimicrobial defenses in infections of viruses, bacteria, and parasites. In this review, we describe the existing evidence for the defensive and inhibitory roles of AMPK and the mechanisms underlying its regulation of innate and inflammatory responses. Finally, we describe AMPK-targeting agents that enhance host defenses against infection or control harmful inflammation.

2. Overview of AMPK

5′-AMP-activated protein kinase (AMPK), a serine/threonine kinase, is a key player in bioenergetic homeostasis to preserve cellular ATP [1]. AMPK is activated in response to an increased cellular adenosine monophosphate (AMP)/ATP or adenosine diphosphate (ADP)/ATP ratio, thus promoting catabolic pathways and suppressing biosynthetic pathways [1–3]. Mammalian AMPK exists as a heterotrimeric complex comprising a catalytic subunit α (α1 and α2), a scaffolding β subunit (β1 and β2), and a regulatory γ subunit (γ1, γ2, and γ3) (Figure 1A) [15]. Multiple isoforms of AMPK are encoded by distinct genes of the subunit isotypes, depending on the cell/tissue or species [16]. The AMPK subunit composition and ligand-induced activities of each AMPK isoform complex can differ among cell types, although the α1, β1, and γ1 isoforms are ubiquitously expressed [16,17].

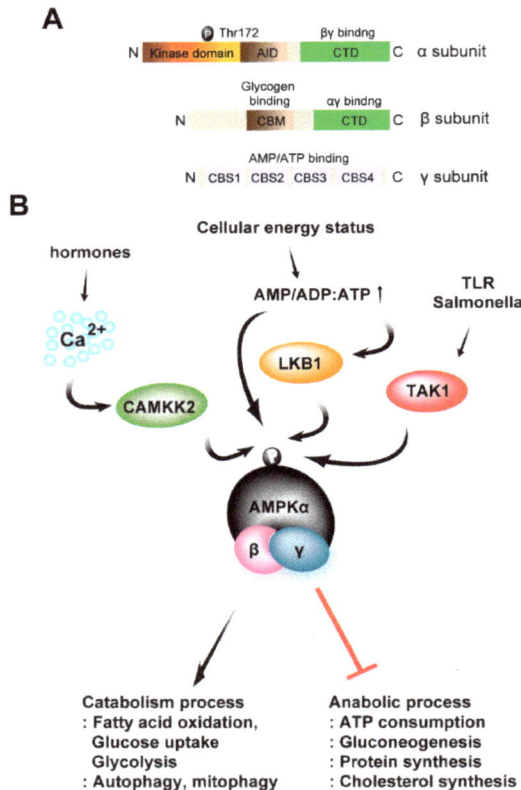

Figure 1. Domain structures of the 5′ AMP-activated protein kinase (AMPK) subunits and the mechanisms that regulate activation of AMPK signaling pathways. (**A**) Conserved domain structure of AMPK subunits consisting of a catalytic α subunit, scaffolding β subunit, and regulatory γ subunit. AID, autoinhibitory domain; CBM, carbohydrate-binding module; CBS, cystathionine-beta-synthase; CTD, C-terminal domain. (**B**) AMPK is activated by the upstream kinases LKB1, CAMKK2 and TAK1 associated with the canonical pathway (triggered by an increased cellular AMP/ATP ratio) or the non-canonical pathway (triggered by an increased intracellular Ca^{2+} concentration or infection/TLR activation). Activated AMPK modulates cellular homeostasis, such as energy metabolism and autophagy, and mitochondrial homeostasis. (black arrow indicate activation/increase; bar-headed red arrow indicates inhibition/decrease). CAMKK2, calcium/calmodulin-dependent kinase kinase 2; LKB1, liver kinase B1; TAK1, Transforming growth factor-β-activated kinase 1; TLR, Toll-like receptor.

The AMPK α subunit contains a kinase domain at the N terminus, which is activated by phosphorylation of Thr-172 by the major upstream liver kinase B1 (LKB1) [18,19]. In contrast to LKB1, the upstream Ca^{2+}-calmodulin-dependent kinase kinase (CaMKK) activates AMPK in response to an increased intracellular Ca^{2+} concentration in the absence of significant changes in ATP/ADP/AMP levels [20]. The regulatory β subunit of AMPK contains a glycogen-binding domain that can sense the structural state of glycogen [21]. Four consecutive cystathionine-β-synthase domains in the regulatory γ subunit are essential for binding to adenosine nucleotides to form an active αβγ complex (Figure 1B) [22,23].

Different AMPK isoforms may have distinct biological functions in different physiological and pathological systems. AMPK governs the cellular energy status by acting as a crucial regulator of energy homeostasis in response to various metabolic stresses, including starvation, hypoxia, and muscle contraction. AMPK activity can be altered by numerous factors, including hormones, cytokines, and nutrients, as well as diverse pathological changes such as metabolic disturbances [24,25]. Because AMPK is important in the adaptation to energy stress, dysregulation of or decreased AMPK activation is implicated in the development of metabolic disorders associated with insulin resistance [6]. In addition to its primary role in the regulation of energy metabolism, AMPK signaling plays a critical role in host–microbial interactions [7]. Furthermore, infections by several viruses result in dysregulation or stimulation of AMPK activity [13]. In mycobacterial infections, AMPK activation promotes activation of host defenses in macrophages and in vivo [12,26]. However, much less is known about the function of AMPK in innate host defenses compared with that in the regulation of metabolism and its mitochondrial function.

3. Multifaceted Role of AMPK in Antimicrobial Responses

Viruses have evolved strategies to manipulate the AMPK signaling pathway to escape host defenses. Indeed, several pathogens can modulate the activity of AMPK/mTOR to obtain sufficient energy for their growth and proliferation [8]. In this review, we discuss microbial manipulation of AMPK activity to affect host defenses against infections. Figure 2 summarizes the multiple roles of AMPK in the viral and bacterial infections addressed in this review. The detailed mechanisms and outcomes of host–pathogen interactions in terms of AMPK modulation are described in Tables 1–4.

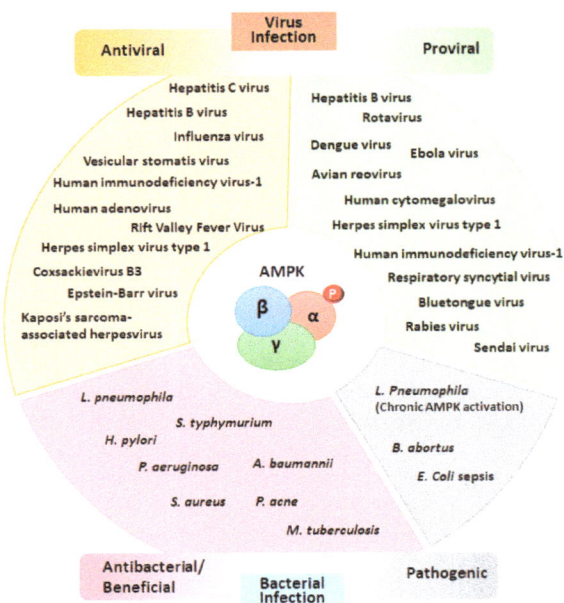

Figure 2. Multifaceted roles of AMPK in viral and bacterial infections. A variety of viruses and bacteria modulate host AMPK activity to promote their growth in host cells. Activation of the AMPK signaling pathway has been implicated in both beneficial antiviral (**left upper**) and detrimental proviral (**right upper**) responses. In addition, AMPK activation promotes the host response to infections by various bacteria (**left lower**) but, in some cases, promotes a detrimental response (**right lower**). The detailed mechanisms by which AMPK activation/inhibition affects infection outcomes are listed in Tables 1–4.

3.1. Roles of AMPK in Viral Infections

3.1.1. Beneficial Effects of AMPK on Virus Infections

Hepatitis C virus (HCV) is a major etiologic agent of chronic liver disease worldwide. HCV infection inhibits AMPKα phosphorylation and signaling [27], and the AMPK agonist metformin suppresses HCV replication in an autophagy-independent manner [28]. Moreover, HCV core protein increases the levels of reactive oxygen species (ROS) and alters the NAD/NADH ratio to decrease the activity and expression of sirtuin 1 (SIRT1) and AMPK, thereby altering the metabolic profile of hepatocytes. This mechanism is implicated in the pathogenesis of hepatic metabolic diseases [29]. As AMPK is a crucial regulator of lipid and glucose metabolism, pharmacological restoration of AMPK activity inhibits lipid accumulation and viral replication in HCV-infected cells [27]. In addition, metformin enhances type I interferon (IFN) signaling by activating AMPK, resulting in inhibition of HCV replication [30]. AMPK inhibition resulted in the downregulation of type I IFN signaling and rescue of the metformin-mediated decrease in the HCV core protein level [30]. Moreover, the AMPK activator 5-aminoimidazole-4-carboxamide 1-β-D-ribofuranoside (AICAR) inhibits HCV replication by activating AMPK signaling, although the anti-HCV effect of metformin is independent of AMPK activation [31]. In chronic HCV infection, the expression of Sucrose-non-fermenting protein kinase 1/AMP-activated protein kinase-related protein kinase (SNARK), an AMPK-related kinase, is increased to promote transforming growth factor β signaling, which is critical for hepatic fibrogenesis [32]. A more recent study showed that HCV-mediated ROS production triggers AMPK activation to attenuate lipid synthesis and promote fatty acid β-oxidation in HCV-infected cells [33]. These data suggest that HCV inhibits AMPK activation to promote its replication, and that the restoration of AMPK activity may be an effective therapeutic modality for HCV infection that acts by metabolic reprogramming or modulation of type I IFN production in host cells [27,28,30,33].

In hepatitis B viral (HBV) infection, AMPK can promote or inhibit viral replication. The detrimental effects of AMPK are described in the following section. Xie et al. reported that AMPK, which is activated by HBV-induced ROS accumulation, suppresses HBV replication [9]. Mechanistically, AMPK activation leads to HBV-mediated autophagic activation, which enhances autolysosome-dependent degradation to restrict viral proliferation [9]. AMPK activity is also involved in the defense against vesicular stomatitis virus, the causal agent of an influenza-like illness, by activating stimulator of IFN genes (STING) [10]. Treatment of mouse macrophages or fibroblasts with an AMPK inhibitor suppressed the production of type I IFN and TNF-α in response to a STING-dependent ligand or agonist, suggesting a role for AMPK in STING signaling [10]. AMPK plays a role in the excessive inflammatory cytokine/chemokine levels in Mint3/Apba3 depletion models of severe pneumonia due to influenza virus [34]. Indeed, food-derived polyphenols, such as epigallocatechin gallate and curcumin, are useful for controlling viral and bacterial infections [35]. Although a review of AMPK-modulating polyphenols is beyond our scope, we highlight the therapeutic promise of polyphenols against infection. For example, curcumin from *Curcuma longa* inhibits influenza A viral infection in vitro and in vivo, at least in part by activating AMPK [36]. The polyphenol epigallocatechin gallate attenuates Tat-induced human immunodeficiency virus (HIV)-1 transactivation by activating AMPK [37]. Further studies should examine the ability of food-derived polyphenols to activate AMPK signaling to control viral replication in host cells.

Human adenovirus type 36, which is associated with obesity, inhibits fatty acid oxidation and AMPK activity and increases accumulation of lipid droplets in infected cells [38]. The AMPK signaling pathway and its upstream regulator LKB1 repress replication of the bunyavirus Rift Valley Fever virus (RVFV), a re-emerging human pathogen [39]. The mechanisms of the antiviral effects of AMPK on RVFV and other viruses are mediated by AMPK inhibition of fatty acid synthesis [39]. Pharmacologic activation of AMPK suppresses RVFV infection and reduces lipid levels by inhibiting fatty acid biosynthesis [39]. In addition, the AMPK/Sirt1 activators resveratrol and quercetin significantly reduce the viral titer and gene expression, as well as increase the viability of infected neurons, in herpes simplex virus type 1 (HSV-1) infection [40]. Moreover, coxsackievirus B3 (CVB3) infection triggers AMPK activation, which suppresses viral replication in HeLa and primary myocardial cells [41]. The AMPK agonists AICAR and metformin suppress CVB3 replication and attenuate lipid accumulation by inhibiting lipid biosynthesis [41]. Thus, regulation of fatty acid metabolism by AMPK signaling is an essential component of cell autonomous immune responses [39].

Latent membrane protein 1 (LMP1) of Epstein-Barr virus (EBV) inactivates LKB1/AMPK, whereas AMPK activation by AICAR abrogated LMP1-mediated proliferation and transformation of nasopharyngeal epithelial cells, suggesting therapeutic potential for EBV-associated nasopharyngeal carcinoma [42]. Moreover, constitutive activation of AMPK inhibited lytic replication of Kaposi's sarcoma-associated herpesvirus in primary human umbilical vein endothelial cells [43]. These data suggest that AMPK suppresses cell transformation and infection-related tumorigenesis in a context-dependent manner. The roles of AMPK in viral infection are listed in Table 1.

3.1.2. Detrimental Effects of AMPK on Virus Infections

Several viruses manipulate AMPK signaling to promote their replication. Genome-scale RNA interference screening of host factors in rotaviral infection identified AMPK as a critical factor in the initiation of a rotavirus-favorable environment [44]. In dengue viral infections, the 3-hydroxy-3-methylglutaryl-CoA reductase (HMGGR) activity elevated by AMPK inactivation resulted in generation of a cholesterol-rich environment in the endoplasmic reticulum, which promoted formation of viral replication complexes [45]. Also, dengue viral infection stimulates AMPK activation to induce proviral lipophagy, thereby enhancing fatty acid β-oxidation and viral replication [46].

Table 1. Beneficial Effects of AMPK in viral infection.

Pathogen	Small Molecules/Chemicals	Agonist/Antagonist	Involvement of AMPK	Outcome (In Vitro/In Vivo)	Ref.
Hepatitis C virus (HCV)	HCV	-	HCV infection inhibit AMPKα phosphorylation and Akt-TSC-mTORC1 pathway	AMPK inhibition is required for HCV replication (in vitro)	[27,28]
	AICAR, Metformin, A769662	Agonist	Restoration of AMPKα activity	Antiviral effects (in vitro)	[27,28]
	Metformin	Agonist	Type I interferon signaling through AMPK pathway activation	Inhibits HCV replication (in vitro)	[30]
	AICAR	Agonist	AMPK activation (Indirect effects counteracted by compound C)	Suppression of HCV replication (in vitro)	[31]
Hepatitis B virus (HBV)	HBV	-	ROS-dependent AMPK activation in HBV-producing cells	Negatively regulates HBV production	[9]
	AICAR constitutive active AMPKα	Agonist	AMPK activation, autophagic flux activation	Inhibits HBV production (in vitro)	[9]
	Compound C dominant-negative AMPKα	Antagonist	AMPK inhibition	Enhances HBV production (in vitro and in vivo)	[9]
Vesicular stomatitis virus (VSV)	AICAR	Agonist	STING-dependent signaling activation	Type I IFN production and antiviral responses (in vitro)	[10]
	Compound C	Antagonist	Inhibition of STING-dependent signaling	Suppression of IFN-β production (in vitro)	[10]
Influenza virus	Mint3 depletion	-	AMPK activation	Attenuates severe pneumonia by influenza infection (in vivo)	[34]
	AICAR	Agonist	AMPK activation in Mint3 depletion model	Decreases inflammatory cytokine production in Mint3-deficient macrophages (in vitro)	[34]
	Curcumin	Activator	AMPK activation	Inhibits influenza A virus infection (in vitro and in vivo)	[36]
Human immunodeficiency virus-1	Epigallocatechin gallate	Activator	AMPK activation	Attenuation of Tat-induced human immunodeficiency virus 1 (HIV-1) transactivation	[37]
Human adenovirus	Adenovirus	-	Inhibit AMPK activity/signaling	Virus induces lipid droplets, presumably associated with obesity (in vitro)	[38]
Rift Valley Fever Virus (RVFV)	A769662, 2-deoxy-D-glucose (2-DG)	Agonist	LKB1/AMPK signaling activation; Inhibition of fatty acid synthesis	Restriction of viral infection (in vitro)	[39]
Herpes simplex virus type 1 (HSV-1)	AICAR, Resveratrol, Quercetin	Activator/agonist	AMPK/Sirt1 activation	Reduces viral titer and the expression of viral genes (in vitro)	[40]

Table 1. *Cont.*

Pathogen	Small Molecules/Chemicals	Agonist/Antagonist	Involvement of AMPK	Outcome (In Vitro/In Vivo)	Ref.
Coxsackievirus B3 (CVB3)	-	-	AMPK activation by CVB3	Restriction of viral replication; reversed by siRNA against AMPK	[41]
	AICAR, A769662, Metformin	Agonist	AMPK activation	Restriction of viral replication; improve the survival rate of infected mice (in vitro and in vivo)	[41]
Epstein-Barr virus (EBV)	LMP1 of EBV	-	LKB1-AMPK inactivation	AMPK inactivation leads to proliferation and transformation of epithelial cells associated with EBV infection (in vitro)	[42]
	AICAR	Agonist	AMPK activation	Inhibition of proliferation of nasopharyngeal epithelial cells (in vitro)	[42]
Kaposi's sarcoma-associated herpesvirus (KSHV)	AICAR, Metformin, Constitutive active AMPK	Agonist	AMPK as a KSHV restriction factor	Inhibits the expression of viral lytic genes and virion production (in vitro)	[43]
	Compound C, Knockdown of AMPKα1	Antagonist	AMPK inhibition	Enhances viral lytic gene expression and virion production (in vitro)	[43]

Int. J. Mol. Sci. **2018**, *19*, 3495

In HBV infection, the HBV X protein activates AMPK, and inhibition of AMPK reduces HBV replication in rat primary hepatocytes [47]. Inhibition of AMPK led to activation of mTORC1, which is required for inhibition of HBV replication in the presence of low AMPK activity [47]. The crosstalk between AMPK and mTORC1 may enable development of therapeutics that suppress HBV replication and thus also hepatocellular carcinoma (HCC) development [47]. However, as described in Section 3.1.1, AMPK activation by AICAR inhibits extracellular HBV production in HepG2 cells [9]. The discrepancy may be attributed to the use of different cell lines in the two studies [9,47]. Further work should address the role of AMPK in HBV infection in vitro and in vivo. In infection by Zaire Ebolavirus (EBOV), the expression levels of the γ2 subunit of AMPK are correlated with EBOV transduction in host cells. In mouse embryonic fibroblasts treated with a small-molecule inhibitor of AMPK (compound C), it was shown that AMPK activity is required for EBOV replication in host cells and EBOV glycoprotein-mediated entry/uptake [48]. In addition, Avian reoviral infection upregulates AMPK phosphorylation, which leads to activation of mitogen-activated protein kinase (MAPK) p38 in Vero cells, which enhances viral replication [49]. The nonstructural protein p17 of avian reovirus positively regulates AMPK activity, inducing autophagy and increasing viral replication [50].

In HSV-1 infection, the activated AMPK/Sirt1 axis inhibits host-cell apoptosis during early-stage infection, which promotes viral latency and protects neurons [51]. However, during the later stages of infection, HSV-1 induces apoptosis of host cells concomitantly with Sirt1 activation [51]. In HIV-infected cocaine abusers, AMPK signaling plays a role in energy deficit and neuronal dysfunction, which are associated with the development of neuroAIDS [52]. These data suggest that differential regulation of AMPK signaling is a determinant of the viral infection course.

Using a kinome-profiling approach, AMPK and related kinases were found to be effectors of human cytomegalovirus (HCMV) replication [53,54]. HCMV infection induces AMPK and CaMKK2 (upstream activator of AMPK)-dependent remodeling of core metabolism, both of which are required for optimal yield and replication of HCMV [53,54]. Notably, inhibition of AMPK activity by short-interfering RNA-mediated AMPK knockdown or an AMPK antagonist (compound C) prevents viral gene expression, providing valuable insight into the mechanisms of HCMV infection [53]. In addition, the AMPK activation-dependent modulation by HCMV of host-cell metabolism is associated with HCMV replication [55]. HCMV-mediated AMPK activation is dependent on CaMKK, and inhibition of AMPK activity abrogated HCMV replication and DNA synthesis [55]. Furthermore, the cardiac glycoside digitoxin induces phosphorylation of AMPK/ULK1, whereas it suppresses mTOR activity to increase autophagic flux and inhibit HCMV replication [56]. Moreover, HCMV induces production of the host protein viperin [57], which is required for AMPK activation, transcriptional activation of GLUT4 and lipogenic enzymes, and lipid synthesis [58]. The enhanced lipid synthesis promotes formation of the viral envelope and production of HCMV virions [58].

In infection, host-cell autophagy plays an important role in host defense and virus survival. Several viruses can manipulate or subvert the autophagic machinery to favor viral replication. For example, respiratory syncytial virus activates autophagy via the AMPK/mTOR signaling pathway to enhance its replication by inhibiting host-cell apoptosis [59]. Bluetongue virus, a double-stranded segmented RNA virus, also induces host-cell autophagy by activating AMPK [60]. Moreover, AMPK is an upstream regulator of rabies virus-induced incomplete autophagy to provide the scaffolds for viral replication [61]. In Sendai viral infection, AMPK activity is required for autophagic initiation to promote viral replication [62]. In oncogenic EBV infection, the increased cell survival caused by AMPK-mediated autophagic activation maintains early hyperproliferation of infected cells [63]. These data suggest AMPK activity to be a therapeutic target for the development of novel antiviral agents.

Importantly, type I IFN, a critical effector in the antiviral response, attenuates AMPK phosphorylation and increases the intracellular ATP level [64]. In addition, IFN-β-mediated glycolytic metabolism is important for the acute phase of the antiviral response to CVB3 [64]. The antiviral cytokine IFN-β regulates host-cell metabolism to enhance glucose uptake and ATP generation, which promote the antiviral response [64]. In infection by snakehead vesiculovirus, miR-214 targeting AMPK suppressed viral replication and upregulated IFN-α expression [65]. Thus, regulation of AMPK activity by the host–pathogen interaction mediates diverse metabolic effects, which modulate viral replication and the host defense response. The beneficial and detrimental effects of AMPK on viral infections are summarized in Tables 1 and 2, respectively.

3.2. Bacterial Infections and AMPK Activation

Intracellular pathogens manipulate the AMPK signaling pathway to alter their metabolic environment to favor bacterial survival or pathogenesis. Mitochondrial dysfunction triggers AMPK signaling, thus enhancing the proliferation of *Legionella pneumophila*, a respiratory pathogen, in *Dictyostelium* cells [66]. Inhibition of AMPK activation reversed the increased *Legionella* proliferation in host cells with mitochondrial disease [66]. However, the AMPK activator metformin triggers mitochondrial ROS generation and activates the AMPK signaling pathway to enhance the host response to *L. pneumophila* in macrophages and promote survival in a murine model of *L. pneumophila* pneumonia [67]. Thus, the role of AMPK activation in bacterial infections differs depending on the host species.

Salmonella typhimurium degrades SIRT1/AMPK to evade host xenophagy [68]. In addition, cytosolic *Salmonella* is ubiquitinated and targeted for xenophagy by AMPK activation [69,70]. AMPK activation by AICAR induces autophagy and colocalization of *Salmonella*-containing vacuoles with LC3 autophagosomes [68], whereas inhibition of AMPK by compound C increases bacterial replication by suppressing autophagy [69]. In *Salmonella*-infected cells, AMPK activation is mediated through toll-like receptor-activated TGF-β-activated kinase 1 (TAK1) [69] which is a direct upstream kinase of AMPK in addition to LKB1 and CaMKK2 [71]. In *Brucella abortus* infection, AMPK activation enhances intracellular growth of *B. abortus* by inhibiting nicotinamide adenine dinucleotide phosphate (NADPH) oxidase-mediated ROS generation [72]. In models of *Escherichia coli* sepsis, ATP-induced pyroptosis is blocked by piperine, a phytochemical present in black pepper (*Piper nigrum* Linn) [73]. The inhibitory effects of piperine on pyroptosis and systemic inflammation are mediated by regulation of the AMPK signaling pathway, as shown by suppression of ATP-mediated AMPK activation by piperine treatment in vitro and in vivo [73]. These data suggest that AMPK plays multiple roles in bacterial infections.

AMPK activation promotes host defenses against infections by several microbes. Transcriptomic and proteomic analyses of a *Caenorhabditis elegans* model indicated that AMPKs function as regulators and mediators of the immune response to infection by, for example, *Bacillus thuringiensis* [74]. Several small-molecule AMPK activators exert protective effects against *Helicobacter pylori*-induced apoptosis of gastric epithelial cells [70,75]. The AMPK agonists A-769662 and resveratrol, as well as AMPKα overexpression, inhibit apoptosis in *H. pylori*-infected gastric epithelial cells [76]. The AMPK activator compound 13 ameliorates *H. pylori*-induced apoptosis of gastric epithelial cells by modulating ROS levels via the AMPK-heme oxygenase-1 axis [76]. Blockade of AMPK signaling significantly abrogates the protective effect of compound 13 against *H. pylori* within gastric epithelial cells [76].

Metformin and AICAR repress infection by *Pseudomonas aeruginosa*, an important opportunistic pathogen, of airway epithelial cells by inhibiting bacterial growth and increasing transepithelial electrical resistance [77]. AMPKα1 depletion increased the susceptibility to *Staphylococcus aureus* endophthalmitis in mice [78], suggesting a protective role for AMPK in bacterial retinal inflammation. Moreover, AMPK activation by AICAR enhances its anti-inflammatory effects, phagocytosis, and bactericidal activity against *S. aureus* infection of various phagocytic cells including microglia, macrophages, and neutrophils [78]. Epigallocatechin gallate, a polyphenol in green tea, inhibits the viability of *Propionibacterium acnes*, a pathogen associated with acne, and exerted an antilipogenic effect in SEB-1 sebocytes by activating the AMPK/sterol regulatory element-binding protein pathway [35]. Moreover, *Acinetobacter baumannii*, an emerging opportunistic pathogen, activates autophagy via the AMPK/ERK/mTOR pathway to promote an antimicrobial response to intracellular *A. baumannii* [79]. The roles of AMPK in bacterial infection are summarized in Table 3.

Table 2. Detrimental Effects of AMPK in viral infection.

Pathogen	Small Molecules/Chemicals	Agonist/Antagonist	Involvement of AMPK	Outcome (In Vitro/In Vivo)	Ref.
	RNAi	-	AMPK-mediated glycolysis, fatty acid oxidation and autophagy	Development of a rotavirus replication-permissive environment (in vitro)	[44]
Rotavirus	AICAR, Metformin	Agonist	AMPK activation (AICAR, directly; Metformin, indirectly)	Upregulation of the proportion of viral infected cells (in vitro)	[44]
	Dorsomorphin	Inhibitor	Inhibition of AMPK activity	Reduces the number of infected cells (in vitro)	[44]
	Virus infection	-	Elevates 3-hydroxy-3-methylglutaryl-CoA reductase activity through AMPK inactivation	Promotes the formation of viral replicative complexes (in vitro)	[45]
	Metformin, A769662	Agonist	AMPK activation	Antiviral effects (in vitro)	[45]
Dengue virus	Compound C	Antagonist	AMPK inhibition	Augments the viral genome copies (in vitro)	[45]
	Virus infection	-	AMPK activation; induction of lipophagy	Increases viral replication (in vitro)	[46]
	Compound C siRNA against AMPKα1	Antagonist	Inhibition of proviral lipophagy	Decreases viral replication (in vitro)	[46]
Hepatitis B virus (HBV)	HBx protein	-	Decreased ATP, activates AMPK in rat primary hepatocytes	AMPK inhibition decreases HBV replication (in vitro)	[47]
	Compound C	Antagonist	Activates mTORC1	Reduces HBV replication (in vitro)	[47]
Ebola virus	Compound C	Antagonist	Less permissive to Ebola virus infection (Similar effects in AMPKα1- or AMPKα2-deleted mouse embryonic fibroblasts)	Inhibits EBOV replication in Vero cells (in vitro)	[48]
	Virus infection	-	Upregulates AMPK phosphorylation leading to p38 MAPK activation	Increases virus replication (in vitro)	[49]
	P17 protein	-	P17 protein activates AMPK to induce autophagy	Increases virus replication (in vitro)	[50]
Avian reovirus	AICAR	Agonist	AMPK activation (Indirect effects through p38 MAPK)	Increases virus replication (in vitro)	[49]
	Compound C	Antagonist	AMPK inhibition	Decreases virus replication (in vitro)	[49]
Herpes simplex virus type 1 (HSV-1)	HSV-1	-	In early infection, AMPK is down-regulated, and then recovered gradually	AMPK/Sirt1 axis inhibits host apoptosis in early infection (in vitro)	[51]
Human immunodeficiency virus-1 (HIV1)	Cocaine	-	Induces AMPK upregulation; AMPK plays a role in energy deficit and metabolic dysfunction	Cocaine exposure during HIV infection accelerates neuronal dysfunction (in vitro)	[52]

Table 2. *Cont.*

Pathogen	Small Molecules/Chemicals	Agonist/Antagonist	Involvement of AMPK	Outcome (In Vitro/In Vivo)	Ref.
	RNAi	-	AMPK may activate numerous metabolic pathways during HCMV infection	siRNA to AMPKα reduces HCMV replication (in vitro)	[53,54]
	HCMV	-	Upregulation of host AMPK	Favors viral replication (in vitro)	[53,54]
	Compound C	Antagonist	Interferes with normal accumulation of viral proteins and alters the core metabolism	Compound C inhibits the viral production of HCMV (in vitro); blocks the immediate early phase of viral replication (in vitro)	[53,54]
Human cytomegalovirus (HCMV)	RNAi to AMPK	-	Blocks glycolytic activation in HCMV-infected cells	RNA-based inhibition of AMPK attenuates HCMV replication (in vitro)	[55]
	Digitoxin	Activator	Digitoxin modulates AMPK-ULK1 and mTOR activity to increase autophagic flux	Viral inhibition (in vitro)	[56]
	Digitoxin + AICAR	-	Combination reduces autophagy	Viral replication (in vitro)	[56]
	HCMV	-	Induces targeting host protein viperin to mitochondria; viperin is required for AMPK activation and regulate lipid metabolism	Viperin-dependent lipogenesis promotes viral replication and production by infected host cells (in vitro)	[57]
Respiratory syncytial virus (RSV)	RSV	-	RSV induces autophagy through ROS and AMPK activation	RSV-induced autophagy favors viral replication (in vitro)	[59]
	Compound C	Antagonist	Inhibition of AMPK and autophagy	Compound C reduces viral gene and protein expression, and total viral titers (in vitro)	
Bluetongue virus	Bluetongue virus	-	Induces autophagy through activation of AMPK	Favors viral replication (in vitro)	[60]
	Compound C siRNA to AMPK	Antagonist	Inhibits BTV1-induced autophagy	AMPK inhibition decreases viral titers (in vitro)	
Rabies virus	Rabies virus		Incomplete autophagy induction via CASP2-AMPK-MAPK1/3/11-AKT1-mTOR pathways	Enhances viral replication (in vitro)	[61]
Sendai virus	Sendai virus	-	Induces host protein TDRD7, an inhibitor of autophagy-inducing AMPK	Host autophagy and viral replication is inhibited by TDRD7 (in vitro)	[62]
	Compound C, shRNA to AMPK	Antagonist	Inhibition of AMPK activity; inhibits viral protein	AMPK activity is required for viral replication (in vitro)	
Snakehead vesiculo-virus	Snakehead vesiculo-virus	-	Downregulates miR-214, which targets AMPK	AMPK upregulation promotes viral replication through reduction of IFN-α expression (in vitro)	[65]

Table 3. The roles of AMPK in bacterial infection.

Pathogen	Small Molecules/Chemicals	Agonist/Antagonist	Involvement of AMPK	Outcome (In Vitro/In Vivo)	Ref.
L. pneumophila	-	-	Chronic AMPK activation involved in host susceptibility to infection (Direct effects by AMPKα antisense)	Bacterial multiplication in host cells with mitochondrial dysfunction	[66]
	Metformin	Agonist	Bactericidal effects are mediated by mitochondrial ROS production (Indirect)	Antimicrobial responses (in vitro and in vivo)	[67]
S. typhimurium	-	-	S. typhimurium exhibits virulence through lysosomal degradation of SIRT1 and AMPK to impair autophagy	Bacterial evasion from autophagic clearance (in vitro)	[68]
	AICAR	Agonist	Upregulation of autophagy	Increased colocalization of salmonella containing vacuole with LC3 (in vitro)	[68]
	-	-	AMPK activation via TAK1; autophagy initiation by ULK1 phosphorylation	Autophagy activation (in vitro)	[69]
	Compound C	Antagonist	AMPK inhibition	Increased bacterial replication by suppression of autophagy (in vitro)	[69]
B. abortus	-	-	AMPK activation via inositol-requiring enzyme 1 (IRE1)	Promote intracellular growth of B. abortus (in vitro)	[72]
	Compound C	Antagonist	AMPK inhibition; activation of NADPH oxidase-mediated ROS production	Suppression of intracellular growth (in vitro)	[72]
E. coli	Piperine	Antagonist	Inhibits ATP-induced pyroptosis by suppressing AMPK activation	Inhibition of pyroptosis; Attenuation of systemic inflammation (in vitro and in vivo)	[73]
	ATP Metformin	Agonist	AMPK activation; increases pyroptosis by inflammasome activation	Activation of pyroptosis (in vitro)	[73]
B. thuringiensis	-	-	AMPK identified by transcriptome and proteome data analysis in vivo (Indirect)	Potentially related to regulation of immune defense (Not determined)	[74]
	-	-	TAK1-mediated AMPK activation	Protects gastric epithelial cells from H. pylori-induced apoptosis (in vitro)	[75]
H. pylori	A-769662 Resveratrol	Agonist	Inhibits H. pylori-induced apoptosis (Direct effects by overexpression of AMPKα)	Alleviates H. pylori-induced gastric epithelial cell apoptosis (in vitro)	[76]
	Compound 13	Agonist	Inhibits H. pylori-induced apoptosis through AMPK-heme oxygenase-1 signaling	Alleviates H. pylori-induced gastric epithelial cell apoptosis (in vitro)	[76]
	Compound C	Antagonist	Inhibitory effects upon compound 13-mediated anti-H. pylori activities (Direct effects by AMPKα1 shRNAs)	Aggravates H. pylori-induced gastric epithelial cell apoptosis (in vitro)	[76]
P. aeruginosa	AICAR Metformin	Agonist	Counteracts the bacterial effects on the reduction of transepithelial electrical resistance (Indirect effects)	Inhibits hyperglycemia-induced bacterial growth; Improve airway epithelial barrier function (in vitro)	[77]

Table 3. *Cont.*

Pathogen	Small Molecules/Chemicals	Agonist/Antagonist	Involvement of AMPK	Outcome (In Vitro/In Vivo)	Ref.
	AICAR	Agonist	AMPK activation	Reduces bacterial burden and intraocular inflammation; Increases bacterial killing in macrophages (in vitro and in vivo)	[78]
S. aureus	Compound C	Antagonist	Downregulates AMPK activity (Direct effects by AMPKα1 knockout mice)	Counteracts AICAR-mediated anti-inflammatory effects (in vivo); Increases susceptibility towards *S. aureus* endophthalmitis (in vivo)	[78]
P. acne	Epigallocatechin gallate	-	Activates AMPK-sterol regulatory element-binding proteins pathway activation	Antilipogenic effects in SEB-1 sebocytes (in vitro)	[35]
A. baumannii	-	-	Activates autophagy through Beclin-1-dependent AMPK/ERK/mTOR pathway (Indirect effects by different *A. baumannii* strains)	Autophagy may promote antimicrobial responses (in vivo)	[79]
	Metformin	Agonist	AMPK activation; Increased mtROS production; Increases phago-lysosomal fusion (Direct effects upon bacterial growth in vitro)	Inhibition of intracellular growth of *M. tuberculosis* (drug-resistant strain; in vitro); Increases the efficacy of conventional TB drugs in vivo	[80]
	AICAR	Agonist	AMPK-PPARGC1A signaling-mediated autophagy activation; Enhancement of phagosomal maturation (Direct effects by shRNA against AMPKα)	Upregulation of antimicrobial responses (in vitro and in vivo)	[12]
	Compound C	Antagonist	Counteracts the effects by AICAR upon intracellular inhibition of *M. tuberculosis* growth	Downregulation of antimicrobial responses (in vitro)	[12]
Mycobacterium tuberculosis	Vitamin D (1,25-D3)	-	Induces autophagy through LL-37 and AMPK activation (Indirect effects upon LL-37 function)	Promotes autophagy and antimicrobial response in human monocytes/macrophages (in vitro)	[81]
	Phenylbutyrate Vitamin D	-	Induces LL-37-mediated autophagy (Indirect effects; AMPK is involved in LL-37-mediated autophagy)	Improves intracellular killing of *M. tuberculosis* (in vitro)	[82]
	Gamma-aminobutyric acid (GABA)	Agonist	Induces autophagy (Direct effects by shRNA against AMPK)	Promotes antimicrobial effects against *M. tuberculosis* (in vitro and in vivo)	[83]
	Ohmyungsamycins	-	Activates AMPK and autophagy; Intracellular inhibition of bacterial growth; Amelioration of inflammation (Indirect effects upon host autophagy)	Promotes antimicrobial effects against *M. tuberculosis* (in vitro and in vivo)	[26]
	Compound C	Antagonist	Blocks the secretion of neutrophil Matrix metalloproteinase-8 (MMP-8)	Neutrophil MMP-8 secretion is related to matrix destruction in human pulmonary TB (in vitro and in human TB lung specimens)	[84]

3.3. Roles of AMPK in Mycobacterial Infection

The seminal study by Singhal et al. addressed the effect of metformin as an adjunctive therapy for tuberculosis. Importantly, metformin suppressed the intracellular growth of *Mycobacterium tuberculosis* (Mtb) in vitro, including drug-resistant strains, by activating the AMPK signaling pathway [80]. In vivo, metformin attenuated the immunopathology and enhanced the immune response and showed a synergistic effect with conventional anti-TB drugs in Mtb-infected mice [80]. The microbicidal effect of metformin in macrophages is due, at least in part, to mitochondrial ROS generation, which is associated with AMPK signaling [80]. Type 2 diabetes mellitus (DM) is re-emerging as a risk factor for human tuberculosis, thus candidate host directed therapeutic targets for tuberculosis combined with DM should be identified [85]. Human cohort studies showed that metformin treatment for DM is associated with a decreased prevalence of latent tuberculosis compared with alternative DM treatments, suggesting metformin to be a candidate HDT for tuberculosis patients with type 2 DM [80,85]. Indeed, Mtb infection inhibits AMPK phosphorylation but increases mTOR kinase activation in macrophages [12]. The AMPK activator AICAR via autophagic activation enhances phagosomal maturation and antimicrobial responses in macrophages in Mtb infection [12]. In human monocytes/macrophages, vitamin D-mediated antimicrobial responses are mediated by the antimicrobial peptide LL-37 via AMPK activation [81]. In addition, LL-37-induced autophagy by phenylbutyrate, alone or in combination with vitamin D, promotes intracellular killing of Mtb in human macrophages via AMPK- and PtdIns3K-dependent pathways [82]. Recent findings revealed the role of gamma-aminobutyric acid (GABA) in AMPK activation to enhance the autophagy and the antimicrobial responses [83]. Silencing of AMPK by a lentiviral short hairpin RNA (shRNA) specific to AMPK reduces GABA-induced autophagic activation as well as phagosomal maturation during Mtb infection [83].

MicroRNAs are small non-coding RNAs involved in the regulation of diverse physiological and pathological processes, including Mtb infection. Mycobacterial infection of macrophages upregulates miR-33 and miR-33*, which target and suppress AMPKα [86]. Interestingly, miR-33/miR-33* regulates autophagy by suppressing AMPK-dependent activation of the transcription of autophagy- and lysosome-related genes and promoting accumulation of lipid bodies in Mtb infection [86]. Mtb infection increases the expression of MIR144*/has-miR-144-5p, which targets DNA damage regulated autophagy modulator 2 (DRAM2), to inhibit the antimicrobial responses to Mtb infection in human monocytes/macrophages. In contrast, autophagic activators enhance production of the autophagy-related protein DRAM2 by activating the AMPK signaling pathway; this contributes to host defenses against Mtb in human macrophages [87].

Although AMPK may play a protective role in tuberculosis, it has also been reported to exert an immunopathological effect by driving the secretion of neutrophil matrix metalloproteinase-8 (MMP-8), resulting in matrix destruction and cavitation, which enhance the spread of Mtb [84]. Neutrophil-derived MMP-8 secretion is upregulated in Mtb infection and neutrophils from AMPK-deficient patients express lower levels of MMP-8, suggesting a key role for MMP-8 in tuberculosis immunopathology [84]. Because the pathogenesis of tuberculosis is complex, further information on the function of AMPK in the immune response to Mtb infection is needed for development of improved therapeutic strategies [88]. The roles of AMPK in mycobacterial infection are listed in Table 3.

3.4. Roles of AMPK in Parasite Infections

The immune response to parasitic helminths involves M2-type cells, CD4(+) Th2 cells, and group 2 innate lymphoid cells. AMPK activation regulates type 2 immune responses and ameliorates lung injury in response to hookworm infections [89]. Mice deficient in AMPK α1 subunit exhibited impaired type 2 responses, an increased intestinal worm burden, and exacerbated lung injury [89]. In *Leishmania*-infected macrophages, *Leishmania infantum* causes a metabolic switch to enhance oxidative phosphorylation by activating LKB1/AMPK and SIRT1 [90]. Impairment of metabolic reprogramming by SIRT1 or AMPK suppresses intracellular growth of the parasite, suggesting a role for AMPK/SIRT1 in intracellular proliferation of *L. infantum* [90]. In *Schistosoma japonicum* egg antigen (SEA)-mediated autophagy, which is modulated by IL-7 and the AMPK signaling pathway, ameliorate liver pathology, suggesting AMPK to be a therapeutic target factor for schistosomiasis [91].

Notably, host AMPK activity is decreased by hepatic *Plasmodium* infection. Activation of the AMPK signaling pathway by AMPK agonists, including salicylate, suppresses the intracellular replication of malaria parasites, including that of the human pathogen *Plasmodium falciparum* [92]. These data suggest that host AMPK signaling is a therapeutic target for hepatic *Plasmodium* infection [92]. In addition, resveratrol protects cardiac function and reduces lipid peroxidation and trypanosomal burden in the heart by activating AMPK, suggesting a role for AMPK in Chagas heart disease [93]. The roles of AMPK in parasitic infections are listed in Table 4.

Table 4. The role of AMPK in parasitic infection.

Pathogen	Small Molecules/Chemicals	Agonist/Antagonist	Involvement of AMPK (Direct/Indirect)	Outcome (In Vitro/In Vivo)	Ref.
Hookworm *Nippostrongylus brasiliensis*	-		AMPKα1 deficiency inhibit IL-13 and CCL17, and defective type 2 immune resistance (Direct effects using by AMPKα1 knockout mice)	AMPKα1 suppresses lung injury and drives M2 polarization during infection	[89]
L. infantum	-	-	Infection leads to a metabolic switch to activate AMPK through the SIRT1-LKB1 axis (Direct effects using by AMPKα1 knockout mice)	Ablation of AMPK promotes parasite clearance in vitro and in vivo	[90]
S. japonicum			Infection-driven IL-7-IL-7R signaling inhibits autophagy; IL-7 inhibits macrophage autophagy via AMPK	Anti-autophagic IL-7 increases liver pathology (in vivo)	[91]
	Metformin	Agonist	Decreases the autophagosome formation in macrophages	in vitro	
	Compound C siAMPKα	Antagonist	Increases autophagosome formation in macrophages (Direct effects by siAMPKα)	in vitro	
P. falciparum	-	-	AMPK activity is suppressed upon infection	Decreases *Plasmodium* hepatic growth	[92]
	Salicylate Metformin A769662	Agonist	AMPK activation impairs the intracellular replication of malaria	Antimalarial interventions (in vitro and in vivo)	
Trypanosoma cruzi	Resveratrol, Metformin	Agonist	AMPK activation reduces heart oxidative stress (Indirect effects)	Reduces heart parasite burden; Protects heart function in Chagas heart disease (in vivo)	[93]

3.5. AMPK in Fungal Infection

A recent phosphoproteomic analysis of *Cryptococcus neoformans* (Cn) infection showed that AMPK activation is triggered by fungal phagocytosis and is required for autophagic induction. Interestingly, AMPK depletion in monocytes promoted host resistance to fungal infection in mouse models, suggesting that AMPK represses the immune response to *Cryptococcus* infection [94].

4. Roles of AMPK in Innate and Adaptive Immune Responses

The roles of AMPK in modulation of the mitochondrial network and energy metabolism, which are associated with the immune response, have been investigated [95]. Here, we briefly review recent data on AMPK regulation of innate and adaptive immune responses in infection and inflammation. Figure 3 summarizes the regulatory roles and mechanisms of a variety of small-molecule AMPK activators in terms of innate immune and inflammatory responses.

The data must be interpreted cautiously, as the small-molecule activators (e.g., AICAR, metformin, and compound C) function via off-target mechanisms, such as AMPK-independent pathways or inhibition of protein kinases other than AMPK [31,96,97]. The beneficial effects of these compounds remain to be fully determined. Thus, selective compounds such as MK-8722 [98] and SC4 [99] and the selective inhibitor SBI-0206965 [100] should be considered for future AMPK-targeted treatment strategies.

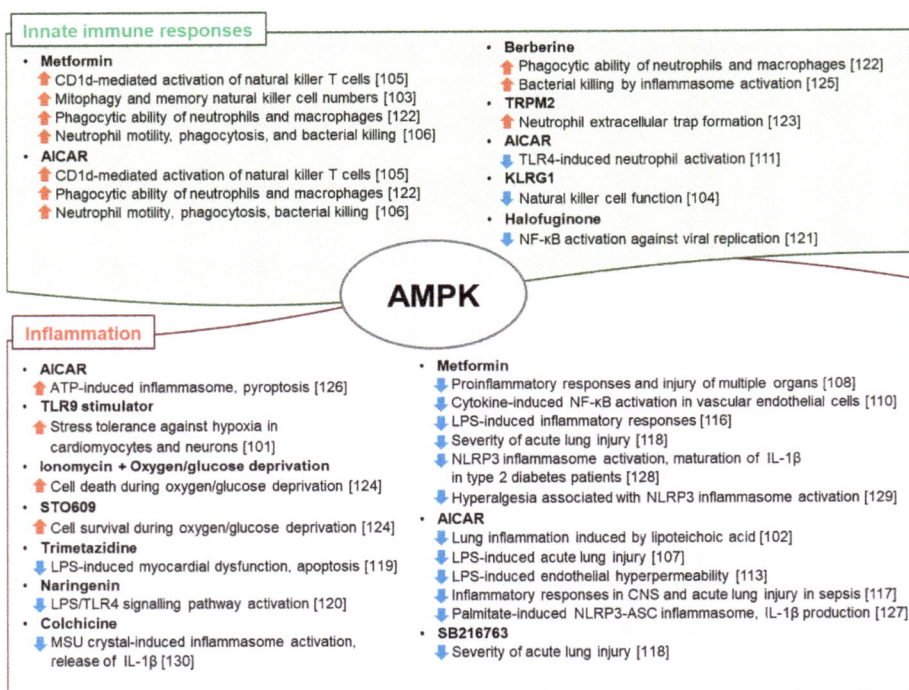

Figure 3. Regulatory effects and underlying mechanisms of small-molecule AMPK activators on the innate immune and inflammatory responses. (red upward arrows indicate activation/increase and blue downward arrows indicate inhibition/decrease)

4.1. Role of AMPK in Regulation of the Innate Immune Response

AMPK is involved in regulation of the innate immune response. For example, the innate-immune stimulator toll-like receptor (TLR) 9 inhibits energy substrates (intracellular ATP levels) and activates AMPK, which enhances stress tolerance in cardiomyocytes and neurons, while stimulation by the TLR9 ligand induces inflammation [101]. The AMPK activator AICAR suppresses the lung inflammation induced by lipoteichoic acid, a major component of the cell wall of Gram-positive bacteria [102]. Natural killer (NK) cells are crucial in the innate immune response to viral infections and transformed cells. Activation of the AMPK signaling pathway or inhibition of mTOR is associated with enhanced mitophagy and an increased number of memory NK cells in antiviral responses [103]. In contrast, increased expression of the inhibitory killer cell lectin-like receptor G1 in aged humans is related to AMPK activation, which has been implicated in disruption of NK cell function [104]. In addition, AMPK activation contributes to CD1d-mediated activation of NK T cells, an important cell type in the innate immune response [105]. The findings above suggest that AMPK plays a pleiotropic role in the regulation of the innate immune response depending on the stimulus and cell type in question.

4.2. AMPK Regulation of Local and Systemic Inflammation

AMPK activators enhance neutrophil chemotaxis, phagocytosis, and bacterial killing to protect against peritonitis-induced sepsis [106]. Indeed, AMPK activators including metformin inhibit injurious inflammatory responses, including neutrophil proinflammatory responses and injury to multiple organs such as the lung, liver, and kidney [107–109]. Pharmacologic activation of AMPK by metformin, berberine, or AICAR dampens excessive TLR4/NF-κB signaling, M2-type macrophage polarization, and the production of proinflammatory mediators in vitro and in models of sepsis [110–115]. The anti-inflammatory effect of metformin in mice with lipopolysachharide (LPS)-induced septic shock and in ob/ob mice is mediated at least in part by AMPK activation [116]. In septic mice, AMPK activation by AICAR or metformin reduces the severity of sepsis-induced lung injury, enhances AMPK phosphorylation in the brain, and attenuates the inflammatory response [117,118].

Treatment with trimetazidine protects against LPS-induced myocardial dysfunction, exerts an anti-apoptotic effect, and attenuates the inflammatory response due to its effect on the SIRT1/AMPK pathway [119]. Moreover, the flavonoid naringenin dampens inflammation in vitro and protects against murine endotoxemia in vivo; these effects are mediated by AMPK/ATF3-dependent inhibition of the TLR4 signaling pathway [120]. In severe acute HBV infection, halofuginone, a plant alkaloid, inhibits viral replication by activating AMPK-mediated anti-inflammatory responses [121]. AMPK activation also enhances the phagocytic capacities of neutrophils and macrophages [122]. Transient receptor potential melastatin 2, an oxidant sensor cation channel, promotes extracellular trap formation by neutrophils via the AMPK/p38 MAPK pathway, enhancing their antimicrobial activity [123]. AMPK activation not only modulates the acute inflammatory response but also promotes neutrophil-dependent bacterial uptake and killing [106].

In perinatal hypoxic–ischemic encephalopathy, prolonged activation of AMPK signaling suppresses the response to oxygen/glucose deprivation and promotes neonatal hypoxic–ischemic injury [124]. Although AMPK inhibition increases neuronal survival, blockade of AMPK prior to oxygen/glucose deprivation increases cell damage and death [124]. Therefore, the clinical implications of AMPK activation are complex, and further preclinical and clinical data are needed to enable therapeutic use of AMPK activators in patients with acute or chronic inflammation.

4.3. Role of AMPK in Inflammasome Activation

AMPK is implicated in modulation of NLRP3 inflammasome activation. The bactericidal activity of the isoquinoline alkaloid berberine exerts a bactericidal effect by augmenting inflammasome activation via AMPK signaling [125]. However, metformin increases mortality of mice with bacteremia, likely via an AMPK-mediated increase in ATP-induced inflammasome activation and pyroptosis [126].

AMPK is implicated in the inhibition of palmitate-induced inflammasome activation [127]. The AMPK activator AICAR inhibits palmitate-induced activation of the NLRP3 inflammasome and IL-1β secretion by suppressing ROS generation [127]. In addition, NLRP3 inflammasome activation and production of IL-1β are upregulated in the peripheral mononuclear cells of drug-naïve type-2 diabetic patients, suggesting a role of the inflammasome in the pathogenesis of type-2 diabetes [128]. Interestingly, AMPK activation is responsible for the significantly reduced mature IL-1β level in peripheral myeloid cells from type-2 diabetic patients after two months of metformin therapy [128]. In a model of hyperalgesia, which is associated with NLRP3 inflammasome activation, metformin attenuated the clinical symptoms and improved the biochemical parameters, whereas blockade of AMPK activation by compound C provoked hyperalgesia and increased the levels of IL-1β and IL-18 [129]. Furthermore, pharmacological activation of AMPK inhibits the monosodium urate (MSU) crystal-induced inflammatory response, suggesting a role for AMPK in gouty inflammation. Moreover, colchicine, an inhibitor of microtubule assembly used to treat gouty arthritis, enhances AMPKα-mediated phosphorylation, thereby inhibiting inflammasome activation and IL-1β release [130]. Further studies on the efficacy of AMPK activators against inflammasome-associated diseases are thus warranted.

4.4. Role of AMPK in the Regulation of the Adaptive Immune Responses

AMPKα1 is a key regulator of the adaptive immune response, particularly T helper (Th1) 1 and Th17 cell differentiation and the T-cell responses to viral and bacterial infections [131]. In addition, in models of simian immunodeficiency viral infection, AMPK activation is associated with the virus-specific CD8(+) cytotoxic T-lymphocyte population and control of Simian Immunodeficiency Virus (SIV) [132]. The mechanism(s) by which AMPK signaling activates innate and adaptive immune responses and controls excessive inflammation must be determined if the potential of AMPK-targeted therapy is to be realized.

5. Conclusions

Although much research has focused on the role of AMPK in the regulation of mitochondrial and metabolic homeostasis, several issues remain to be addressed. Further work should focus on the mechanism(s) by which AMPK modulates host defenses against infections in vivo. Several pathogens modulate the host metabolic environment to promote their survival and replication. Because of its role in regulating mitochondrial metabolism, dynamics, and biogenesis, AMPK signaling can provide energy to the pathogen and/or host, benefitting either. Stimulation of AMPK activity enhances host defenses against diverse viruses, bacteria, and parasites, notably Mtb. Moreover, AMPK links the innate and adaptive immune responses to infection. However, the molecular mechanisms underlying AMPK regulation of innate and adaptive immunity are unclear. AMPK-targeted small molecules have potential as antimicrobial agents as well as metabolic drugs. Further work is needed to enable development of therapeutics that target AMPK to control inflammation and promote host defenses against infection. This work should focus on elucidating the mechanisms by which AMPK and/or AMPK-targeting compounds modulate host defenses against infection.

funding: We are indebted to current and past members of our laboratory for discussions and investigations that contributed to this article. We apologize to colleagues whose work and publications could not be referenced owing to space constraints. This work was supported by the National Research Foundation of Korea (NRF) Grant funded by the Korean Government (MSIP) (No. 2017R1A5A2015385) and by the National Research Foundation of Korea (NRF) grant funded by the Korea government (MSIP) (No. NRF-2015M3C9A2054326).

Conflicts of Interest: The authors declare no conflict of interest.

Abbreviations

ADP	Adenosine diphosphate
AICAR	5-aminoimidazole-4-carboxamide 1-β-D-ribofuanoside
AMP	Adenosine monophosphate
AMPK	5′-AMP-activated protein kinase
ATF3	Activating transcription factor 3
ATP	Adenosine triphosphate
CaMKK	Ca2+/calmodulin-dependent protein kinase kinases
CCL17	Chemokine ligand 17
CVB3	Coxsackie virus B3

DM	Diabetes mellitus
EBOV	Zaire Ebolavirus
EBV	Epstein-Barr Virus
GABA	Gamma-aminobutyric acid
HBV	Hepatitis B virus
HCC	Hepatocellular carcinoma
HCMV	Human cytomegalovirus
HCV	Hepatitis C virus
HIV	Human immunodeficiency virus
HMGCR	3-hydroxy-3-methylglutaryl-CoA reductase
HSV-1	Herpex Simplex Virus Type 1
IFN	Interferons
IL-17	Interleukin-17
IL-1β	Interleukin-1β
IL-37	Interleukin-37
IL-7	Interleukin-7
KSHV	Kaposi's sarcoma-associated herpesvirus
LKB1	Liver kinase B1
LMP1	Latent membrane protein 1
MAPK	Mitogen-activated protein kinase
MMP-8	Matrix metalloproteinase-8
mTOR	Mammalian target of rapamycin
mTORC1	Mammalian target of rapamycin complex 1
NADPH	Nicotinamide adenine dinucleotide phosphate
NF-κB	Nuclear factor kappa-light-chain-enhancer of activated B cells
PPARGC1A	Peroxisome proliferator-activated receptor-gamma, coactivator 1α
ROS	Reactive oxygen species
RVFV	Rift Valley Fever Virus
SIRT1	Sirtuin 1
SIV	Simian immunodeficiency virus
SNARK	Sucrose-non-fermenting protein kinase 1/AMP-activated protein kinase-related protein kinase
STING	Stimulator of IFN genes
TAK1	Transforming growth factor (TGF)-β-activated kinase 1
TLR	Toll-like receptor
TSC	Tuberous sclerosis complex

References

1. Hardie, D.G.; Ross, F.A.; Hawley, S.A. AMPK: A nutrient and energy sensor that maintains energy homeostasis. *Nat. Rev. Mol. Cell Biol.* **2012**, *13*, 251–262. [CrossRef] [PubMed]
2. Gowans, G.J.; Hardie, D.G. AMPK: A cellular energy sensor primarily regulated by AMP. *Biochem. Soc. Trans.* **2014**, *42*, 71–75. [CrossRef] [PubMed]
3. Gowans, G.J.; Hawley, S.A.; Ross, F.A.; Hardie, D.G. AMP is a true physiological regulator of AMP-activated protein kinase by both allosteric activation and enhancing net phosphorylation. *Cell Metab.* **2013**, *18*, 556–566. [CrossRef] [PubMed]
4. Shackelford, D.B.; Shaw, R.J. The LKB1-AMPK pathway: Metabolism and growth control in tumour suppression. *Nat. Rev. Cancer* **2009**, *9*, 563–575. [CrossRef] [PubMed]
5. Faubert, B.; Boily, G.; Izreig, S.; Griss, T.; Samborska, B.; Dong, Z.; Dupuy, F.; Chambers, C.; Fuerth, B.J.; Viollet, B.; et al. AMPK is a negative regulator of the Warburg effect and suppresses tumor growth in vivo. *Cell Metab.* **2013**, *17*, 113–124. [CrossRef] [PubMed]
6. Hardie, D.G. AMP-activated protein kinase: A cellular energy sensor with a key role in metabolic disorders and in cancer. *Biochem. Soc. Trans.* **2011**, *39*, 1–13. [CrossRef] [PubMed]
7. Moreira, D.; Silvestre, R.; Cordeiro-da-Silva, A.; Estaquier, J.; Foretz, M.; Viollet, B. AMP-activated Protein Kinase as a Target for Pathogens: Friends or Foes? *Curr. Drug Targets* **2016**, *17*, 942–953. [CrossRef] [PubMed]

8. Brunton, J.; Steele, S.; Ziehr, B.; Moorman, N.; Kawula, T. Feeding uninvited guests: MTOR and AMPK set the table for intracellular pathogens. *PLoS Pathog.* **2013**, *9*, e1003552. [CrossRef] [PubMed]

9. Xie, N.; Yuan, K.; Zhou, L.; Wang, K.; Chen, H.N.; Lei, Y.; Lan, J.; Pu, Q.; Gao, W.; Zhang, L.; et al. PRKAA/AMPK restricts HBV replication through promotion of autophagic degradation. *Autophagy* **2016**, *12*, 1507–1520. [CrossRef] [PubMed]

10. Prantner, D.; Perkins, D.J.; Vogel, S.N. AMP-activated Kinase (AMPK) Promotes Innate Immunity and Antiviral Defense through Modulation of Stimulator of Interferon Genes (STING) Signaling. *J. Biol. Chem.* **2017**, *292*, 292–304. [CrossRef] [PubMed]

11. Shin, D.M.; Yang, C.S.; Lee, J.Y.; Lee, S.J.; Choi, H.H.; Lee, H.M.; Yuk, J.M.; Harding, C.V.; Jo, E.K. *Mycobacterium tuberculosis* lipoprotein-induced association of TLR2 with protein kinase C zeta in lipid rafts contributes to reactive oxygen species-dependent inflammatory signalling in macrophages. *Cell Microbiol.* **2008**, *10*, 1893–1905. [CrossRef] [PubMed]

12. Yang, C.S.; Kim, J.J.; Lee, H.M.; Jin, H.S.; Lee, S.H.; Park, J.H.; Kim, S.J.; Kim, J.M.; Han, Y.M.; Lee, M.S.; et al. The AMPK-PPARGC1A pathway is required for antimicrobial host defense through activation of autophagy. *Autophagy* **2014**, *10*, 785–802. [CrossRef] [PubMed]

13. Mankouri, J.; Harris, M. Viruses and the fuel sensor: The emerging link between AMPK and virus replication. *Rev. Med. Virol.* **2011**, *21*, 205–212. [CrossRef] [PubMed]

14. Mesquita, I.; Moreira, D.; Sampaio-Marques, B.; Laforge, M.; Cordeiro-da-Silva, A.; Ludovico, P.; Estaquier, J.; Silvestre, R. AMPK in Pathogens. *EXS* **2016**, *107*, 287–323. [CrossRef] [PubMed]

15. Kim, J.; Yang, G.; Kim, Y.; Kim, J.; Ha, J. AMPK activators: Mechanisms of action and physiological activities. *Exp. Mol. Med.* **2016**, *48*, e224. [CrossRef] [PubMed]

16. Wu, J.; Puppala, D.; Feng, X.; Monetti, M.; Lapworth, A.L.; Geoghegan, K.F. Chemoproteomic analysis of intertissue and interspecies isoform diversity of AMP-activated protein kinase (AMPK). *J. Biol. Chem.* **2013**, *288*, 35904–35912. [CrossRef] [PubMed]

17. Rajamohan, F.; Reyes, A.R.; Frisbie, R.K.; Hoth, L.R.; Sahasrabudhe, P.; Magyar, R.; Landro, J.A.; Withka, J.M.; Caspers, N.L.; Calabrese, M.F.; et al. Probing the enzyme kinetics, allosteric modulation and activation of alpha1- and alpha2-subunit-containing AMP-activated protein kinase (AMPK) heterotrimeric complexes by pharmacological and physiological activators. *Biochem. J.* **2016**, *473*, 581–592. [CrossRef] [PubMed]

18. Woods, A.; Johnstone, S.R.; Dickerson, K.; Leiper, F.C.; Fryer, L.G.; Neumann, D.; Schlattner, U.; Wallimann, T.; Carlson, M.; Carling, D. LKB1 is the upstream kinase in the AMP-activated protein kinase cascade. *Curr. Biol.* **2003**, *13*, 2004–2008. [CrossRef] [PubMed]

19. Carling, D.; Aguan, K.; Woods, A.; Verhoeven, A.J.; Beri, R.K.; Brennan, C.H.; Sidebottom, C.; Davison, M.D.; Scott, J. Mammalian AMP-activated protein kinase is homologous to yeast and plant protein kinases involved in the regulation of carbon metabolism. *J. Biol. Chem.* **1994**, *269*, 11442–11448. [PubMed]

20. Hawley, S.A.; Pan, D.A.; Mustard, K.J.; Ross, L.; Bain, J.; Edelman, A.M.; Frenguelli, B.G.; Hardie, D.G. Calmodulin-dependent protein kinase kinase-beta is an alternative upstream kinase for AMP-activated protein kinase. *Cell. Metab.* **2005**, *2*, 9–19. [CrossRef] [PubMed]

21. McBride, A.; Ghilagaber, S.; Nikolaev, A.; Hardie, D.G. The glycogen-binding domain on the AMPK beta subunit allows the kinase to act as a glycogen sensor. *Cell. Metab.* **2009**, *9*, 23–34. [CrossRef] [PubMed]

22. Viana, R.; Towler, M.C.; Pan, D.A.; Carling, D.; Viollet, B.; Hardie, D.G.; Sanz, P. A conserved sequence immediately N-terminal to the Bateman domains in AMP-activated protein kinase gamma subunits is required for the interaction with the beta subunits. *J. Biol. Chem.* **2007**, *282*, 16117–16125. [CrossRef] [PubMed]

23. Xiao, B.; Heath, R.; Saiu, P.; Leiper, F.C.; Leone, P.; Jing, C.; Walker, P.A.; Haire, L.; Eccleston, J.F.; Davis, C.T.; et al. Structural basis for AMP binding to mammalian AMP-activated protein kinase. *Nature* **2007**, *449*, 496–500. [CrossRef] [PubMed]

24. Steinberg, G.R.; Kemp, B.E. AMPK in Health and Disease. *Physiol. Rev.* **2009**, *89*, 1025–1078. [CrossRef] [PubMed]

25. Viollet, B.; Horman, S.; Leclerc, J.; Lantier, L.; Foretz, M.; Billaud, M.; Giri, S.; Andreelli, F. AMPK inhibition in health and disease. *Crit. Rev. Biochem. Mol. Biol.* **2010**, *45*, 276–295. [CrossRef] [PubMed]

26. Kim, T.S.; Shin, Y.H.; Lee, H.M.; Kim, J.K.; Choe, J.H.; Jang, J.C.; Um, S.; Jin, H.S.; Komatsu, M.; Cha, G.H.; et al. Ohmyungsamycins promote antimicrobial responses through autophagy activation via AMP-activated protein kinase pathway. *Sci. Rep.* **2017**, *7*, 3431. [CrossRef] [PubMed]

27. Mankouri, J.; Tedbury, P.R.; Gretton, S.; Hughes, M.E.; Griffin, S.D.; Dallas, M.L.; Green, K.A.; Hardie, D.G.; Peers, C.; Harris, M. Enhanced hepatitis C virus genome replication and lipid accumulation mediated by inhibition of AMP-activated protein kinase. *Proc. Natl. Acad. Sci. USA* **2010**, *107*, 11549–11554. [CrossRef] [PubMed]

28. Huang, H.; Kang, R.; Wang, J.; Luo, G.; Yang, W.; Zhao, Z. Hepatitis C virus inhibits AKT-tuberous sclerosis complex (TSC), the mechanistic target of rapamycin (MTOR) pathway, through endoplasmic reticulum stress to induce autophagy. *Autophagy* **2013**, *9*, 175–195. [CrossRef] [PubMed]

29. Yu, J.W.; Sun, L.J.; Liu, W.; Zhao, Y.H.; Kang, P.; Yan, B.Z. Hepatitis C virus core protein induces hepatic metabolism disorders through down-regulation of the SIRT1-AMPK signaling pathway. *Int. J. Infect. Dis.* **2013**, *17*, e539–e545. [CrossRef] [PubMed]

30. Tsai, W.L.; Chang, T.H.; Sun, W.C.; Chan, H.H.; Wu, C.C.; Hsu, P.I.; Cheng, J.S.; Yu, M.L. Metformin activates type I interferon signaling against HCV via activation of adenosine monophosphate-activated protein kinase. *Oncotarget* **2017**, *8*, 91928–91937. [CrossRef] [PubMed]

31. Nakashima, K.; Takeuchi, K.; Chihara, K.; Hotta, H.; Sada, K. Inhibition of hepatitis C virus replication through adenosine monophosphate-activated protein kinase-dependent and -independent pathways. *Microbiol. Immunol.* **2011**, *55*, 774–782. [CrossRef] [PubMed]

32. Goto, K.; Lin, W.; Zhang, L.; Jilg, N.; Shao, R.X.; Schaefer, E.A.; Zhao, H.; Fusco, D.N.; Peng, L.F.; Kato, N.; et al. The AMPK-related kinase SNARK regulates hepatitis C virus replication and pathogenesis through enhancement of TGF-beta signaling. *J. Hepatol.* **2013**, *59*, 942–948. [CrossRef] [PubMed]

33. Douglas, D.N.; Pu, C.H.; Lewis, J.T.; Bhat, R.; Anwar-Mohamed, A.; Logan, M.; Lund, G.; Addison, W.R.; Lehner, R.; Kneteman, N.M. Oxidative Stress Attenuates Lipid Synthesis and Increases Mitochondrial Fatty Acid Oxidation in Hepatoma Cells Infected with Hepatitis C Virus. *J. Biol. Chem.* **2016**, *291*, 1974–1990. [CrossRef] [PubMed]

34. Uematsu, T.; Fujita, T.; Nakaoka, H.J.; Hara, T.; Kobayashi, N.; Murakami, Y.; Seiki, M.; Sakamoto, T. Mint3/Apba3 depletion ameliorates severe murine influenza pneumonia and macrophage cytokine production in response to the influenza virus. *Sci. Rep.* **2016**, *6*, 37815. [CrossRef] [PubMed]

35. Yoon, J.Y.; Kwon, H.H.; Min, S.U.; Thiboutot, D.M.; Suh, D.H. Epigallocatechin-3-gallate improves acne in humans by modulating intracellular molecular targets and inhibiting *P. acnes*. *J. Investig. Dermatol.* **2013**, *133*, 429–440. [CrossRef] [PubMed]

36. Han, S.; Xu, J.; Guo, X.; Huang, M. Curcumin ameliorates severe influenza pneumonia via attenuating lung injury and regulating macrophage cytokines production. *Clin. Exp. Pharmacol. Physiol.* **2018**, *45*, 84–93. [CrossRef] [PubMed]

37. Zhang, H.S.; Wu, T.C.; Sang, W.W.; Ruan, Z. EGCG inhibits Tat-induced LTR transactivation: Role of Nrf2, AKT, AMPK signaling pathway. *Life Sci.* **2012**, *90*, 747–754. [CrossRef] [PubMed]

38. Wang, Z.Q.; Yu, Y.; Zhang, X.H.; Floyd, E.Z.; Cefalu, W.T. Human adenovirus 36 decreases fatty acid oxidation and increases de novo lipogenesis in primary cultured human skeletal muscle cells by promoting Cidec/FSP27 expression. *Int. J. Obes.* **2010**, *34*, 1355–1364. [CrossRef] [PubMed]

39. Moser, T.S.; Schieffer, D.; Cherry, S. AMP-activated kinase restricts Rift Valley fever virus infection by inhibiting fatty acid synthesis. *PLoS Pathog.* **2012**, *8*, e1002661. [CrossRef] [PubMed]

40. Leyton, L.; Hott, M.; Acuna, F.; Caroca, J.; Nunez, M.; Martin, C.; Zambrano, A.; Concha, M.I.; Otth, C. Nutraceutical activators of AMPK/Sirt1 axis inhibit viral production and protect neurons from neurodegenerative events triggered during HSV-1 infection. *Virus Res.* **2015**, *205*, 63–72. [CrossRef] [PubMed]

41. Xie, W.; Wang, L.; Dai, Q.; Yu, H.; He, X.; Xiong, J.; Sheng, H.; Zhang, D.; Xin, R.; Qi, Y.; et al. Activation of AMPK restricts coxsackievirus B3 replication by inhibiting lipid accumulation. *J. Mol. Cell Cardiol.* **2015**, *85*, 155–167. [CrossRef] [PubMed]

42. Lo, A.K.; Lo, K.W.; Ko, C.W.; Young, L.S.; Dawson, C.W. Inhibition of the LKB1-AMPK pathway by the Epstein-Barr virus-encoded LMP1 promotes proliferation and transformation of human nasopharyngeal epithelial cells. *J. Pathol.* **2013**, *230*, 336–346. [CrossRef] [PubMed]

43. Cheng, F.; He, M.; Jung, J.U.; Lu, C.; Gao, S.J. Suppression of Kaposi's Sarcoma-Associated Herpesvirus Infection and Replication by 5′-AMP-Activated Protein Kinase. *J. Virol.* **2016**, *90*, 6515–6525. [CrossRef] [PubMed]

44. Green, V.A.; Pelkmans, L. A Systems Survey of Progressive Host-Cell Reorganization during Rotavirus Infection. *Cell Host Microbe* **2016**, *20*, 107–120. [CrossRef] [PubMed]

45. Soto-Acosta, R.; Bautista-Carbajal, P.; Cervantes-Salazar, M.; Angel-Ambrocio, A.H.; Del Angel, R.M. DENV up-regulates the HMG-CoA reductase activity through the impairment of AMPK phosphorylation: A potential antiviral target. *PLoS Pathog.* **2017**, *13*, e1006257. [CrossRef] [PubMed]

46. Jordan, T.X.; Randall, G. Dengue Virus Activates the AMP Kinase-mTOR Axis to Stimulate a Proviral Lipophagy. *J. Virol.* **2017**, *91*. [CrossRef] [PubMed]

47. Bagga, S.; Rawat, S.; Ajenjo, M.; Bouchard, M.J. Hepatitis B virus (HBV) X protein-mediated regulation of hepatocyte metabolic pathways affects viral replication. *Virology* **2016**, *498*, 9–22. [CrossRef] [PubMed]

48. Kondratowicz, A.S.; Hunt, C.L.; Davey, R.A.; Cherry, S.; Maury, W.J. AMP-activated protein kinase is required for the macropinocytic internalization of ebolavirus. *J. Virol.* **2013**, *87*, 746–755. [CrossRef] [PubMed]

49. Ji, W.T.; Lee, L.H.; Lin, F.L.; Wang, L.; Liu, H.J. AMP-activated protein kinase facilitates avian reovirus to induce mitogen-activated protein kinase (MAPK) p38 and MAPK kinase 3/6 signalling that is beneficial for virus replication. *J. Gen. Virol.* **2009**, *90*, 3002–3009. [CrossRef] [PubMed]

50. Chi, P.I.; Huang, W.R.; Lai, I.H.; Cheng, C.Y.; Liu, H.J. The p17 nonstructural protein of avian reovirus triggers autophagy enhancing virus replication via activation of phosphatase and tensin deleted on chromosome 10 (PTEN) and AMP-activated protein kinase (AMPK), as well as dsRNA-dependent protein kinase (PKR)/eIF2alpha signaling pathways. *J. Biol. Chem.* **2013**, *288*, 3571–3584. [CrossRef] [PubMed]

51. Martin, C.; Leyton, L.; Arancibia, Y.; Cuevas, A.; Zambrano, A.; Concha, M.I.; Otth, C. Modulation of the AMPK/Sirt1 axis during neuronal infection by herpes simplex virus type 1. *J. Alzheimers Dis.* **2014**, *42*, 301–312. [CrossRef] [PubMed]

52. Samikkannu, T.; Atluri, V.S.; Nair, M.P. HIV and Cocaine Impact Glial Metabolism: Energy Sensor AMP-activated protein kinase Role in Mitochondrial Biogenesis and Epigenetic Remodeling. *Sci. Rep.* **2016**, *6*, 31784. [CrossRef] [PubMed]

53. Terry, L.J.; Vastag, L.; Rabinowitz, J.D.; Shenk, T. Human kinome profiling identifies a requirement for AMP-activated protein kinase during human cytomegalovirus infection. *Proc. Natl. Acad. Sci. USA* **2012**, *109*, 3071–3076. [CrossRef] [PubMed]

54. Hutterer, C.; Wandinger, S.K.; Wagner, S.; Muller, R.; Stamminger, T.; Zeittrager, I.; Godl, K.; Baumgartner, R.; Strobl, S.; Marschall, M. Profiling of the kinome of cytomegalovirus-infected cells reveals the functional importance of host kinases Aurora A, ABL and AMPK. *Antivir. Res.* **2013**, *99*, 139–148. [CrossRef] [PubMed]

55. McArdle, J.; Moorman, N.J.; Munger, J. HCMV targets the metabolic stress response through activation of AMPK whose activity is important for viral replication. *PLoS Pathog.* **2012**, *8*, e1002502. [CrossRef] [PubMed]

56. Mukhopadhyay, R.; Venkatadri, R.; Katnelson, J.; Arav-Boger, R. Digitoxin Suppresses Human Cytomegalovirus Replication via Na$^+$, K$^+$/ATPase alpha1 Subunit-Dependent AMP-Activated Protein Kinase and Autophagy Activation. *J. Virol.* **2018**, *92*. [CrossRef] [PubMed]

57. Seo, J.Y.; Yaneva, R.; Hinson, E.R.; Cresswell, P. Human cytomegalovirus directly induces the antiviral protein viperin to enhance infectivity. *Science* **2011**, *332*, 1093–1097. [CrossRef] [PubMed]

58. Seo, J.Y.; Cresswell, P. Viperin regulates cellular lipid metabolism during human cytomegalovirus infection. *PLoS Pathog.* **2013**, *9*, e1003497. [CrossRef] [PubMed]

59. Li, M.; Li, J.; Zeng, R.; Yang, J.; Liu, J.; Zhang, Z.; Song, X.; Yao, Z.; Ma, C.; Li, W.; et al. Respiratory Syncytial Virus Replication Is Promoted by Autophagy-Mediated Inhibition of Apoptosis. *J. Virol.* **2018**, *92*. [CrossRef] [PubMed]

60. Lv, S.; Xu, Q.Y.; Sun, E.C.; Zhang, J.K.; Wu, D.L. Dissection and integration of the autophagy signaling network initiated by bluetongue virus infection: Crucial candidates ERK1/2, Akt and AMPK. *Sci. Rep.* **2016**, *6*, 23130. [CrossRef] [PubMed]

61. Liu, J.; Wang, H.; Gu, J.; Deng, T.; Yuan, Z.; Hu, B.; Xu, Y.; Yan, Y.; Zan, J.; Liao, M.; et al. BECN1-dependent CASP2 incomplete autophagy induction by binding to rabies virus phosphoprotein. *Autophagy* **2017**, *13*, 739–753. [CrossRef] [PubMed]

62. Subramanian, G.; Kuzmanovic, T.; Zhang, Y.; Peter, C.B.; Veleeparambil, M.; Chakravarti, R.; Sen, G.C.; Chattopadhyay, S. A new mechanism of interferon's antiviral action: Induction of autophagy, essential for paramyxovirus replication, is inhibited by the interferon stimulated gene, TDRD7. *PLoS Pathog.* **2018**, *14*, e1006877. [CrossRef] [PubMed]

63. McFadden, K.; Hafez, A.Y.; Kishton, R.; Messinger, J.E.; Nikitin, P.A.; Rathmell, J.C.; Luftig, M.A. Metabolic stress is a barrier to Epstein-Barr virus-mediated B-cell immortalization. *Proc. Natl. Acad. Sci. USA* **2016**, *113*, E782–E790. [CrossRef] [PubMed]

64. Burke, J.D.; Platanias, L.C.; Fish, E.N. Beta interferon regulation of glucose metabolism is PI3K/Akt dependent and important for antiviral activity against coxsackievirus B3. *J. Virol.* **2014**, *88*, 3485–3495. [CrossRef] [PubMed]

65. Zhang, C.; Feng, S.; Zhang, W.; Chen, N.; Hegazy, A.M.; Chen, W.; Liu, X.; Zhao, L.; Li, J.; Lin, L.; et al. MicroRNA miR-214 Inhibits Snakehead Vesiculovirus Replication by Promoting IFN-alpha Expression via Targeting Host Adenosine 5'-Monophosphate-Activated Protein Kinase. *Front. Immunol.* **2017**, *8*, 1775. [CrossRef] [PubMed]

66. Francione, L.; Smith, P.K.; Accari, S.L.; Taylor, P.E.; Bokko, P.B.; Bozzaro, S.; Beech, P.L.; Fisher, P.R. *Legionella pneumophila* multiplication is enhanced by chronic AMPK signalling in mitochondrially diseased Dictyostelium cells. *Dis. Model Mech.* **2009**, *2*, 479–489. [CrossRef] [PubMed]

67. Kajiwara, C.; Kusaka, Y.; Kimura, S.; Yamaguchi, T.; Nanjo, Y.; Ishii, Y.; Udono, H.; Standiford, T.J.; Tateda, K. Metformin Mediates Protection against *Legionella* Pneumonia through Activation of AMPK and Mitochondrial Reactive Oxygen Species. *J. Immunol.* **2018**, *200*, 623–631. [CrossRef] [PubMed]

68. Ganesan, R.; Hos, N.J.; Gutierrez, S.; Fischer, J.; Stepek, J.M.; Daglidu, E.; Kronke, M.; Robinson, N. *Salmonella* Typhimurium disrupts Sirt1/AMPK checkpoint control of mTOR to impair autophagy. *PLoS Pathog.* **2017**, *13*, e1006227. [CrossRef] [PubMed]

69. Liu, W.; Jiang, Y.; Sun, J.; Geng, S.; Pan, Z.; Prinz, R.A.; Wang, C.; Sun, J.; Jiao, X.; Xu, X. Activation of TGF-beta-activated kinase 1 (TAK1) restricts *Salmonella* Typhimurium growth by inducing AMPK activation and autophagy. *Cell Death Dis.* **2018**, *9*, 570. [CrossRef] [PubMed]

70. Tattoli, I.; Sorbara, M.T.; Philpott, D.J.; Girardin, S.E. Bacterial autophagy: The trigger, the target and the timing. *Autophagy* **2012**, *8*, 1848–1850. [CrossRef] [PubMed]

71. Neumann, D. Is TAK1 a Direct Upstream Kinase of AMPK? *Int. J. Mol. Sci.* **2018**, *19*, 2412. [CrossRef] [PubMed]

72. Liu, N.; Li, Y.; Dong, C.; Xu, X.; Wei, P.; Sun, W.; Peng, Q. Inositol-Requiring Enzyme 1-Dependent Activation of AMPK Promotes *Brucella abortus* Intracellular Growth. *J. Bacteriol.* **2016**, *198*, 986–993. [CrossRef] [PubMed]

73. Liang, Y.D.; Bai, W.J.; Li, C.G.; Xu, L.H.; Wei, H.X.; Pan, H.; He, X.H.; Ouyang, D.Y. Piperine Suppresses Pyroptosis and Interleukin-1beta Release upon ATP Triggering and Bacterial Infection. *Front. Pharmacol.* **2016**, *7*, 390. [CrossRef] [PubMed]

74. Yang, W.; Dierking, K.; Esser, D.; Tholey, A.; Leippe, M.; Rosenstiel, P.; Schulenburg, H. Overlapping and unique signatures in the proteomic and transcriptomic responses of the nematode *Caenorhabditis elegans* toward pathogenic *Bacillus thuringiensis*. *Dev. Comp. Immunol.* **2015**, *51*, 1–9. [CrossRef] [PubMed]

75. Lv, G.; Zhu, H.; Zhou, F.; Lin, Z.; Lin, G.; Li, C. AMP-activated protein kinase activation protects gastric epithelial cells from *Helicobacter pylori*-induced apoptosis. *Biochem. Biophys. Res. Commun.* **2014**, *453*, 13–18. [CrossRef] [PubMed]

76. Zhao, H.; Zhu, H.; Lin, Z.; Lin, G.; Lv, G. Compound 13, an alpha1-selective small molecule activator of AMPK, inhibits *Helicobacter pylori*-induced oxidative stresses and gastric epithelial cell apoptosis. *Biochem. Biophys. Res. Commun.* **2015**, *463*, 510–517. [CrossRef] [PubMed]

77. Patkee, W.R.; Carr, G.; Baker, E.H.; Baines, D.L.; Garnett, J.P. Metformin prevents the effects of *Pseudomonas aeruginosa* on airway epithelial tight junctions and restricts hyperglycaemia-induced bacterial growth. *J. Cell Mol. Med.* **2016**, *20*, 758–764. [CrossRef] [PubMed]

78. Kumar, A.; Giri, S.; Kumar, A. 5-Aminoimidazole-4-carboxamide ribonucleoside-mediated adenosine monophosphate-activated protein kinase activation induces protective innate responses in bacterial endophthalmitis. *Cell Microbiol.* **2016**, *18*, 1815–1830. [CrossRef] [PubMed]

79. Wang, Y.; Zhang, K.; Shi, X.; Wang, C.; Wang, F.; Fan, J.; Shen, F.; Xu, J.; Bao, W.; Liu, M.; et al. Critical role of bacterial isochorismatase in the autophagic process induced by *Acinetobacter baumannii* in mammalian cells. *FASEB J.* **2016**, *30*, 3563–3577. [CrossRef] [PubMed]

80. Singhal, A.; Jie, L.; Kumar, P.; Hong, G.S.; Leow, M.K.; Paleja, B.; Tsenova, L.; Kurepina, N.; Chen, J.; Zolezzi, F.; et al. Metformin as adjunct antituberculosis therapy. *Sci. Transl. Med.* **2014**, *6*, 263ra159. [CrossRef] [PubMed]

81. Yuk, J.M.; Shin, D.M.; Lee, H.M.; Yang, C.S.; Jin, H.S.; Kim, K.K.; Lee, Z.W.; Lee, S.H.; Kim, J.M.; Jo, E.K. Vitamin D3 induces autophagy in human monocytes/macrophages via cathelicidin. *Cell Host Microbe* **2009**, *6*, 231–243. [CrossRef] [PubMed]

82. Rekha, R.S.; Rao Muvva, S.S.; Wan, M.; Raqib, R.; Bergman, P.; Brighenti, S.; Gudmundsson, G.H.; Agerberth, B. Phenylbutyrate induces LL-37-dependent autophagy and intracellular killing of *Mycobacterium tuberculosis* in human macrophages. *Autophagy* **2015**, *11*, 1688–1699. [CrossRef] [PubMed]

83. Kim, J.K.; Kim, Y.S.; Lee, H.M.; Jin, H.S.; Neupane, C.; Kim, S.; Lee, S.H.; Min, J.J.; Sasai, M.; Jeong, J.H.; et al. GABAergic signaling linked to autophagy enhances host protection against intracellular bacterial infections. *Nat. Commun.* **2018**, *9*, 4184. [CrossRef] [PubMed]

84. Ong, C.W.; Elkington, P.T.; Brilha, S.; Ugarte-Gil, C.; Tome-Esteban, M.T.; Tezera, L.B.; Pabisiak, P.J.; Moores, R.C.; Sathyamoorthy, T.; Patel, V.; et al. Neutrophil-Derived MMP-8 Drives AMPK-Dependent Matrix Destruction in Human Pulmonary Tuberculosis. *PLoS Pathog.* **2015**, *11*, e1004917. [CrossRef] [PubMed]

85. Restrepo, B.I. Metformin: Candidate host-directed therapy for tuberculosis in diabetes and non-diabetes patients. *Tuberculosis (Edinb)* **2016**, *101S*, S69–S72. [CrossRef] [PubMed]

86. Ouimet, M.; Koster, S.; Sakowski, E.; Ramkhelawon, B.; van Solingen, C.; Oldebeken, S.; Karunakaran, D.; Portal-Celhay, C.; Sheedy, F.J.; Ray, T.D.; et al. *Mycobacterium tuberculosis* induces the miR-33 locus to reprogram autophagy and host lipid metabolism. *Nat. Immunol.* **2016**, *17*, 677–686. [CrossRef] [PubMed]

87. Kim, J.K.; Lee, H.M.; Park, K.S.; Shin, D.M.; Kim, T.S.; Kim, Y.S.; Suh, H.W.; Kim, S.Y.; Kim, I.S.; Kim, J.M.; et al. MIR144* inhibits antimicrobial responses against *Mycobacterium tuberculosis* in human monocytes and macrophages by targeting the autophagy protein DRAM2. *Autophagy* **2017**, *13*, 423–441. [CrossRef] [PubMed]

88. Dorhoi, A.; Kaufmann, S.H. Pathology and immune reactivity: Understanding multidimensionality in pulmonary tuberculosis. *Semin. Immunopathol.* **2016**, *38*, 153–166. [CrossRef] [PubMed]

89. Nieves, W.; Hung, L.Y.; Oniskey, T.K.; Boon, L.; Foretz, M.; Viollet, B.; Herbert, D.R. Myeloid-Restricted AMPKalpha1 Promotes Host Immunity and Protects against IL-12/23p40-Dependent Lung Injury during Hookworm Infection. *J. Immunol.* **2016**, *196*, 4632–4640. [CrossRef] [PubMed]

90. Moreira, D.; Rodrigues, V.; Abengozar, M.; Rivas, L.; Rial, E.; Laforge, M.; Li, X.; Foretz, M.; Viollet, B.; Estaquier, J.; et al. *Leishmania infantum* modulates host macrophage mitochondrial metabolism by hijacking the SIRT1-AMPK axis. *PLoS Pathog.* **2015**, *11*, e1004684. [CrossRef] [PubMed]

91. Zhu, J.; Zhang, W.; Zhang, L.; Xu, L.; Chen, X.; Zhou, S.; Xu, Z.; Xiao, M.; Bai, H.; Liu, F.; et al. IL-7 suppresses macrophage autophagy and promotes liver pathology in *Schistosoma japonicum*-infected mice. *J. Cell. Mol. Med.* **2018**, *22*, 3353–3363. [CrossRef] [PubMed]

92. Ruivo, M.T.G.; Vera, I.M.; Sales-Dias, J.; Meireles, P.; Gural, N.; Bhatia, S.N.; Mota, M.M.; Mancio-Silva, L. Host AMPK Is a Modulator of Plasmodium Liver Infection. *Cell. Rep.* **2016**, *16*, 2539–2545. [CrossRef] [PubMed]

93. Vilar-Pereira, G.; Carneiro, V.C.; Mata-Santos, H.; Vicentino, A.R.; Ramos, I.P.; Giarola, N.L.; Feijo, D.F.; Meyer-Fernandes, J.R.; Paula-Neto, H.A.; Medei, E.; et al. Resveratrol Reverses Functional Chagas Heart Disease in Mice. *PLoS Pathog.* **2016**, *12*, e1005947. [CrossRef] [PubMed]

94. Pandey, A.; Ding, S.L.; Qin, Q.M.; Gupta, R.; Gomez, G.; Lin, F.; Feng, X.; Fachini da Costa, L.; Chaki, S.P.; Katepalli, M.; et al. Global Reprogramming of Host Kinase Signaling in Response to Fungal Infection. *Cell Host Microbe* **2017**, *21*, 637–649. [CrossRef] [PubMed]

95. Andris, F.; Leo, O. AMPK in lymphocyte metabolism and function. *Int. Rev. Immunol.* **2015**, *34*, 67–81. [CrossRef] [PubMed]

96. Vincent, E.E.; Coelho, P.P.; Blagih, J.; Griss, T.; Viollet, B.; Jones, R.G. Differential effects of AMPK agonists on cell growth and metabolism. *Oncogene* **2015**, *34*, 3627–3639. [CrossRef] [PubMed]

97. Bain, J.; Plater, L.; Elliott, M.; Shpiro, N.; Hastie, C.J.; McLauchlan, H.; Klevernic, I.; Arthur, J.S.; Alessi, D.R.; Cohen, P. The selectivity of protein kinase inhibitors: A further update. *Biochem. J.* **2007**, *408*, 297–315. [CrossRef] [PubMed]

98. Myers, R.W.; Guan, H.P.; Ehrhart, J.; Petrov, A.; Prahalada, S.; Tozzo, E.; Yang, X.; Kurtz, M.M.; Trujillo, M.; Gonzalez Trotter, D.; et al. Systemic pan-AMPK activator MK-8722 improves glucose homeostasis but induces cardiac hypertrophy. *Science* **2017**, *357*, 507–511. [CrossRef] [PubMed]

99. Ngoei, K.R.W.; Langendorf, C.G.; Ling, N.X.Y.; Hoque, A.; Varghese, S.; Camerino, M.A.; Walker, S.R.; Bozikis, Y.E.; Dite, T.A.; Ovens, A.J.; et al. Structural Determinants for Small-Molecule Activation of Skeletal Muscle AMPK alpha2beta2gamma1 by the Glucose Importagog SC4. *Cell Chem. Biol.* **2018**, *25*, 728–737. [CrossRef] [PubMed]

100. Dite, T.A.; Langendorf, C.G.; Hoque, A.; Galic, S.; Rebello, R.J.; Ovens, A.J.; Lindqvist, L.M.; Ngoei, K.R.W.; Ling, N.X.Y.; Furic, L.; et al. AMP-activated protein kinase selectively inhibited by the type II inhibitor SBI-0206965. *J. Biol. Chem.* **2018**, *293*, 8874–8885. [CrossRef] [PubMed]

101. Shintani, Y.; Kapoor, A.; Kaneko, M.; Smolenski, R.T.; D'Acquisto, F.; Coppen, S.R.; Harada-Shoji, N.; Lee, H.J.; Thiemermann, C.; Takashima, S.; et al. TLR9 mediates cellular protection by modulating energy metabolism in cardiomyocytes and neurons. *Proc. Natl. Acad. Sci. USA* **2013**, *110*, 5109–5114. [CrossRef] [PubMed]

102. Hoogendijk, A.J.; Pinhancos, S.S.; van der Poll, T.; Wieland, C.W. AMP-activated protein kinase activation by 5-aminoimidazole-4-carbox-amide-1-beta-D-ribofuranoside (AICAR) reduces lipoteichoic acid-induced lung inflammation. *J. Biol. Chem.* **2013**, *288*, 7047–7052. [CrossRef] [PubMed]

103. O'Sullivan, T.E.; Johnson, L.R.; Kang, H.H.; Sun, J.C. BNIP3- and BNIP3L-Mediated Mitophagy Promotes the Generation of Natural Killer Cell Memory. *Immunity* **2015**, *43*, 331–342. [CrossRef] [PubMed]

104. Muller-Durovic, B.; Lanna, A.; Covre, L.P.; Mills, R.S.; Henson, S.M.; Akbar, A.N. Killer Cell Lectin-like Receptor G1 Inhibits NK Cell Function through Activation of Adenosine 5′-Monophosphate-Activated Protein Kinase. *J. Immunol.* **2016**, *197*, 2891–2899. [CrossRef] [PubMed]

105. Webb, T.J.; Carey, G.B.; East, J.E.; Sun, W.; Bollino, D.R.; Kimball, A.S.; Brutkiewicz, R.R. Alterations in cellular metabolism modulate CD1d-mediated NKT-cell responses. *Pathog. Dis.* **2016**, *74*. [CrossRef] [PubMed]

106. Park, D.W.; Jiang, S.; Tadie, J.M.; Stigler, W.S.; Gao, Y.; Deshane, J.; Abraham, E.; Zmijewski, J.W. Activation of AMPK enhances neutrophil chemotaxis and bacterial killing. *Mol. Med.* **2013**, *19*, 387–398. [CrossRef] [PubMed]

107. Zhao, X.; Zmijewski, J.W.; Lorne, E.; Liu, G.; Park, Y.J.; Tsuruta, Y.; Abraham, E. Activation of AMPK attenuates neutrophil proinflammatory activity and decreases the severity of acute lung injury. *Am. J. Physiol. Lung Cell Mol. Physiol.* **2008**, *295*, L497–L504. [CrossRef] [PubMed]

108. Bergheim, I.; Luyendyk, J.P.; Steele, C.; Russell, G.K.; Guo, L.; Roth, R.A.; Arteel, G.E. Metformin prevents endotoxin-induced liver injury after partial hepatectomy. *J. Pharmacol. Exp. Ther.* **2006**, *316*, 1053–1061. [CrossRef] [PubMed]

109. Escobar, D.A.; Botero-Quintero, A.M.; Kautza, B.C.; Luciano, J.; Loughran, P.; Darwiche, S.; Rosengart, M.R.; Zuckerbraun, B.S.; Gomez, H. Adenosine monophosphate-activated protein kinase activation protects against sepsis-induced organ injury and inflammation. *J. Surg. Res.* **2015**, *194*, 262–272. [CrossRef] [PubMed]

110. Hattori, Y.; Suzuki, K.; Hattori, S.; Kasai, K. Metformin inhibits cytokine-induced nuclear factor κB activation via AMP-activated protein kinase activation in vascular endothelial cells. *Hypertension* **2006**, *47*, 1183–1188. [CrossRef] [PubMed]

111. Zmijewski, J.W.; Lorne, E.; Zhao, X.; Tsuruta, Y.; Sha, Y.; Liu, G.; Siegal, G.P.; Abraham, E. Mitochondrial respiratory complex I regulates neutrophil activation and severity of lung injury. *Am. J. Respir. Crit. Care Med.* **2008**, *178*, 168–179. [CrossRef] [PubMed]

112. Sag, D.; Carling, D.; Stout, R.D.; Suttles, J. Adenosine 5′-monophosphate-activated protein kinase promotes macrophage polarization to an anti-inflammatory functional phenotype. *J. Immunol.* **2008**, *181*, 8633–8641. [CrossRef] [PubMed]

113. Xing, J.; Wang, Q.; Coughlan, K.; Viollet, B.; Moriasi, C.; Zou, M.H. Inhibition of AMP-activated protein kinase accentuates lipopolysaccharide-induced lung endothelial barrier dysfunction and lung injury in vivo. *Am. J. Pathol.* **2013**, *182*, 1021–1030. [CrossRef] [PubMed]

114. Jeong, H.W.; Hsu, K.C.; Lee, J.W.; Ham, M.; Huh, J.Y.; Shin, H.J.; Kim, W.S.; Kim, J.B. Berberine suppresses proinflammatory responses through AMPK activation in macrophages. *Am. J. Physiol. Endocrinol. Metab.* **2009**, *296*, E955–E964. [CrossRef] [PubMed]

115. Vaez, H.; Rameshrad, M.; Najafi, M.; Barar, J.; Barzegari, A.; Garjani, A. Cardioprotective effect of metformin in lipopolysaccharide-induced sepsis via suppression of toll-like receptor 4 (TLR4) in heart. *Eur. J. Pharmacol.* **2016**, *772*, 115–123. [CrossRef] [PubMed]

116. Kim, J.; Kwak, H.J.; Cha, J.Y.; Jeong, Y.S.; Rhee, S.D.; Kim, K.R.; Cheon, H.G. Metformin suppresses lipopolysaccharide (LPS)-induced inflammatory response in murine macrophages via activating transcription factor-3 (ATF-3) induction. *J. Biol. Chem.* **2014**, *289*, 23246–23255. [CrossRef] [PubMed]

117. Mulchandani, N.; Yang, W.L.; Khan, M.M.; Zhang, F.; Marambaud, P.; Nicastro, J.; Coppa, G.F.; Wang, P. Stimulation of Brain AMP-Activated Protein Kinase Attenuates Inflammation and Acute Lung Injury in Sepsis. *Mol. Med.* **2015**, *21*, 637–644. [CrossRef] [PubMed]

118. Liu, Z.; Bone, N.; Jiang, S.; Park, D.W.; Tadie, J.M.; Deshane, J.; Rodriguez, C.A.; Pittet, J.F.; Abraham, E.; Zmijewski, J.W. AMP-Activated Protein Kinase and Glycogen Synthase Kinase 3beta Modulate the Severity of Sepsis-Induced Lung Injury. *Mol. Med.* **2016**, *21*, 937–950. [CrossRef] [PubMed]

119. Chen, J.; Lai, J.; Yang, L.; Ruan, G.; Chaugai, S.; Ning, Q.; Chen, C.; Wang, D.W. Trimetazidine prevents macrophage-mediated septic myocardial dysfunction via activation of the histone deacetylase sirtuin 1. *Br. J. Pharmacol.* **2016**, *173*, 545–561. [CrossRef] [PubMed]

120. Liu, X.; Wang, N.; Fan, S.; Zheng, X.; Yang, Y.; Zhu, Y.; Lu, Y.; Chen, Q.; Zhou, H.; Zheng, J. The citrus flavonoid naringenin confers protection in a murine endotoxaemia model through AMPK-ATF3-dependent negative regulation of the TLR4 signalling pathway. *Sci. Rep.* **2016**, *6*, 39735. [CrossRef] [PubMed]

121. Zhan, W.; Kang, Y.; Chen, N.; Mao, C.; Kang, Y.; Shang, J. Halofuginone ameliorates inflammation in severe acute hepatitis B virus (HBV)-infected SD rats through AMPK activation. *Drug Des. Dev. Ther.* **2017**, *11*, 2947–2955. [CrossRef] [PubMed]

122. Bae, H.B.; Zmijewski, J.W.; Deshane, J.S.; Tadie, J.M.; Chaplin, D.D.; Takashima, S.; Abraham, E. AMP-activated protein kinase enhances the phagocytic ability of macrophages and neutrophils. *FASEB J.* **2011**, *25*, 4358–4368. [CrossRef] [PubMed]

123. Li, L.; Xu, S.; Guo, T.; Gong, S.; Zhang, C. Effect of dapagliflozin on intestinal flora in MafA-deficient mice. *Curr. Pharm. Des.* **2018**. [CrossRef] [PubMed]

124. Rousset, C.I.; Leiper, F.C.; Kichev, A.; Gressens, P.; Carling, D.; Hagberg, H.; Thornton, C. A dual role for AMP-activated protein kinase (AMPK) during neonatal hypoxic-ischaemic brain injury in mice. *J. Neurochem.* **2015**, *133*, 242–252. [CrossRef] [PubMed]

125. Li, C.G.; Yan, L.; Jing, Y.Y.; Xu, L.H.; Liang, Y.D.; Wei, H.X.; Hu, B.; Pan, H.; Zha, Q.B.; Ouyang, D.Y.; et al. Berberine augments ATP-induced inflammasome activation in macrophages by enhancing AMPK signaling. *Oncotarget* **2017**, *8*, 95–109. [CrossRef] [PubMed]

126. Zha, Q.B.; Wei, H.X.; Li, C.G.; Liang, Y.D.; Xu, L.H.; Bai, W.J.; Pan, H.; He, X.H.; Ouyang, D.Y. ATP-Induced Inflammasome Activation and Pyroptosis Is Regulated by AMP-Activated Protein Kinase in Macrophages. *Front. Immunol.* **2016**, *7*, 597. [CrossRef] [PubMed]

127. Wen, H.; Gris, D.; Lei, Y.; Jha, S.; Zhang, L.; Huang, M.T.; Brickey, W.J.; Ting, J.P. Fatty acid-induced NLRP3-ASC inflammasome activation interferes with insulin signaling. *Nat. Immunol.* **2011**, *12*, 408–415. [CrossRef] [PubMed]

128. Lee, H.M.; Kim, J.J.; Kim, H.J.; Shong, M.; Ku, B.J.; Jo, E.K. Upregulated NLRP3 inflammasome activation in patients with type 2 diabetes. *Diabetes* **2013**, *62*, 194–204. [CrossRef] [PubMed]

129. Bullon, P.; Alcocer-Gomez, E.; Carrion, A.M.; Marin-Aguilar, F.; Garrido-Maraver, J.; Roman-Malo, L.; Ruiz-Cabello, J.; Culic, O.; Ryffel, B.; Apetoh, L.; et al. AMPK Phosphorylation Modulates Pain by Activation of NLRP3 Inflammasome. *Antioxid. Redox. Signal.* **2016**, *24*, 157–170. [CrossRef] [PubMed]

130. Wang, Y.; Viollet, B.; Terkeltaub, R.; Liu-Bryan, R. AMP-activated protein kinase suppresses urate crystal-induced inflammation and transduces colchicine effects in macrophages. *Ann. Rheum. Dis.* **2016**, *75*, 286–294. [CrossRef] [PubMed]

131. Blagih, J.; Coulombe, F.; Vincent, E.E.; Dupuy, F.; Galicia-Vazquez, G.; Yurchenko, E.; Raissi, T.C.; van der Windt, G.J.; Viollet, B.; Pearce, E.L.; et al. The energy sensor AMPK regulates T cell metabolic adaptation and effector responses in vivo. *Immunity* **2015**, *42*, 41–54. [CrossRef] [PubMed]

132. Iseda, S.; Takahashi, N.; Poplimont, H.; Nomura, T.; Seki, S.; Nakane, T.; Nakamura, M.; Shi, S.; Ishii, H.; Furukawa, S.; et al. Biphasic CD8+ T-Cell Defense in Simian Immunodeficiency Virus Control by Acute-Phase Passive Neutralizing Antibody Immunization. *J. Virol.* **2016**, *90*, 6276–6290. [CrossRef] [PubMed]

International Journal of
Molecular Sciences

MDPI

Review

AMPK Function in Mammalian Spermatozoa

**David Martin-Hidalgo [1,2,†], Ana Hurtado de Llera [1,3,†], Violeta Calle-Guisado [1],
Lauro Gonzalez-Fernandez [1], Luis Garcia-Marin [1] and M. Julia Bragado [1,*]**

[1]   Research Group of Intracellular Signaling and Technology of Reproduction (SINTREP),
      Institute of Biotechnology in Agriculture and Livestock (INBIO G+C), University of Extremadura,
      10003 Cáceres, Spain; davidmh@unex.es (D.M.-H.); anahl@unex.es (A.H.d.L.); violetacg@unex.es (V.C.-G.);
      lgonfer@unex.es (L.G.-F.); ljgarcia@unex.es (L.G.-M.)
[2]   Unit for Multidisciplinary Research in Biomedicine (UMIB), Laboratory of Cell Biology,
      Department of Microscopy, Institute of Biomedical Sciences Abel Salazar (ICBAS), University of Porto,
      40050-313 Porto, Portugal
[3]   Hormones and Metabolism Research Group, Faculty of Health Sciences, University of Beira Interior,
      6200-506 Covilhã, Portugal
[*]   Correspondence: jbragado@unex.es; Tel.: +34-927-257-160
[†]   These authors contributed equally to this work.

Received: 2 October 2018; Accepted: 20 October 2018; Published: 23 October 2018

Abstract: AMP-activated protein kinase AMPK regulates cellular energy by controlling metabolism through the inhibition of anabolic pathways and the simultaneous stimulation of catabolic pathways. Given its central regulator role in cell metabolism, AMPK activity and its regulation have been the focus of relevant investigations, although only a few studies have focused on the AMPK function in the control of spermatozoa's ability to fertilize. This review summarizes the known cellular roles of AMPK that have been identified in mammalian spermatozoa. The involvement of AMPK activity is described in terms of the main physiological functions of mature spermatozoa, particularly in the regulation of suitable sperm motility adapted to the fluctuating extracellular medium, maintenance of the integrity of sperm membranes, and the mitochondrial membrane potential. In addition, the intracellular signaling pathways leading to AMPK activation in mammalian spermatozoa are reviewed. We also discuss the role of AMPK in assisted reproduction techniques, particularly during semen cryopreservation and preservation (at 17 °C). Finally, we reinforce the idea of AMPK as a key signaling kinase in spermatozoa that acts as an essential linker/bridge between metabolism energy and sperm's ability to fertilize.

Keywords: AMP-activated protein kinase (AMPK); spermatozoa; motility; mitochondria; membranes; signaling; stress; assisted reproduction techniques

1. Introduction

The AMP-activated protein kinase (AMPK) acts as a sensor molecule of cellular energy charge that maintains energy homeostasis at both the whole-body and cellular levels [1,2] by controlling key anabolic and catabolic pathways under different energy or stressful conditions. Thus, AMPK can be activated by physiological conditions, mediated by metabolic hormones, by different cell stress conditions, and also by pharmacological molecules [1,2]. This kinase is highly conserved in eukaryotic species and is ubiquitously expressed. Structurally, AMPK is a heterotrimeric protein including an alpha (α) catalytic subunit, a scaffolding beta (β), and regulatory gamma (γ) subunits, which are encoded by seven genes, allowing the formation of different heterotrimer combinations of α, β, and γ isoforms of AMPK (Figure 1).

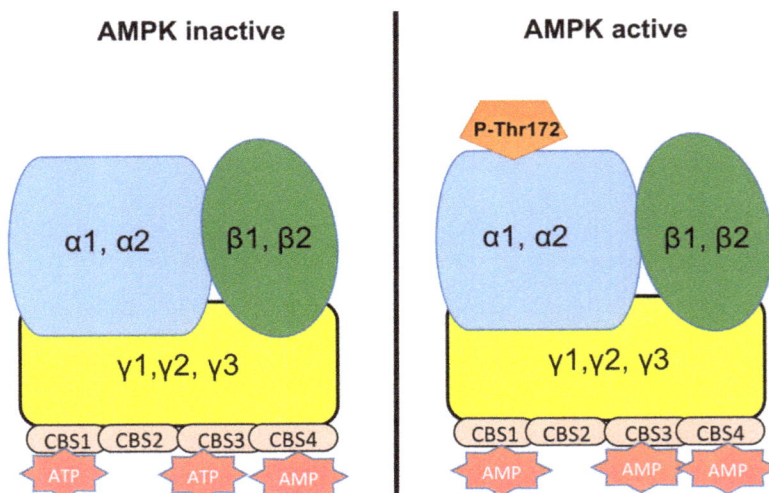

Figure 1. Structure of AMP-activated kinase (AMPK). AMPK is a heterotrimeric protein with 3 differents subunits: catalytic α (blue), scaffolding β (green) and regulatory γ (yelow), each one can be expressed as different isoforms, enabling distinct heterotrimeric combinations of AMPK. Regulatory subunit γ controls the activity of the α subunit through 4 tandem repeats of the motif CBS (cystathionine β-synthase) that bind adenosine nucleotides (AMP, ADP, ATP) with high affinity. AMPK remains inactive in presence of high levels of ATP, whereas during starvation or when ATP level decreases, CBS domains bind AMP, which favors the phosphorylation at Thr172 (catalytic α subunit) and subsequently leading to AMPK activation.

Given its central regulator role in metabolism and its pleiotropic action in many cellular processes, AMPK has received enormous interest as a therapeutic target, and therefore, there has been a great effort to develop pharmacological activators. In addition, AMPK has been the focus of relevant works, although the majority of them have been conducted in somatic cells. However, the ability to preserve energy homeostasis under fluctuating extracellular nutrients or conditions that physiologically occur through the female reproductive tract is a necessary characteristic of spermatozoa. As mentioned before, the direct molecular connector between the substrate's supply and energy requirement is AMPK. In 2012, two different research groups identified for the first time the presence of AMPK in mammalian spermatozoa [3,4]. Since then, interesting scientific efforts have been performed to elucidate the functions of AMPK in these male gametes in several species, including human, using different experimental approaches, including the use of α1AMPK knockout mice [3], and the pharmacological inhibitor and activators. To date, availability of AMPK inhibitors is very restricted and the most widely-used inhibitor of AMPK is the pyrazolopyrimidine derivate compound C or dorsomorphin (CC) that binds to the active site of AMPK and acts as an ATP-competitive inhibitor [5]. Very recently, a new pyrimidine derivative, SBI-0206965, has been discovered as a direct AMPK inhibitor with greater potency (40 fold) and lower kinase promiscuity than CC [6]. Regarding AMPK activators, great effort has been performed for the last few years in the development of direct small molecules and also in their mechanisms of action. The classical activator is the 5-aminoimidazol-4-carboxamide-1-B-D-ribofuranoside (AICAR), although its therapeutic use is not warranted in humans because of its elevated threshold for AMPK activation [7]. The first direct AMPK activator available to scientists was A769662, which activates AMPK both allosterically and by inhibiting AMPKα Thr172 dephosphorylation through binding at the ADaM site, and displays high specificity for complexes containing AMPK-β1 isoforms compared to AMPK-β2 [7]. An exhaustive review about other AMPK activators has been recently published [7] and includes cyclic benzimidazole derivatives (compound 991, PF-739, and MK-8722), the compound C2, the iminothiazolidione PT-1 and its optimized C24, the bi-quinoline JJO-1 and dihydroxyquinoline MT47-100, and recently indole- and indazole-acid-based AMPK synthetic activators, as well as two novel small activators, PXL770 and O304, which successfully completed phase I of clinical trials [7]. To date, the most clinically-used activator of AMPK is metformin and although its action was originally thought to be mediated by AMPK [8], likely through inhibition of AMP desaminase [9], it is currently established that many of metformin effects are AMPK-independent [10–16]. Thus, besides effects on glucose metabolism, metformin also directly inhibits complex I (NADH:ubiquinone oxidoreductase) of the mitochondrial electron transport chain [17,18] with obvious consequences in cell energy production. Overall, we believe that the modulation of AMPK activity would be an interesting tool for the future knowledge about exact roles of AMPK in mammalian spermatozoa, as well as for development of novel protocols for male fertility and/or sperm preservation in vitro.

In this review, we provide the current knowledge about AMPK function in spermatozoa under physiological and different extracellular conditions. AMPK activity is involved in the main physiological functions of mature spermatozoa, in particular in the maintenance of proper sperm motility, the integrity of sperm membranes, and the mitochondrial membrane potential, all of them adapted to the fluctuating medium of the female reproductive tract. Moreover, the intracellular signaling pathways leading to AMPK activation in mammalian spermatozoa are reviewed. We also discuss the role of AMPK in assisted reproduction techniques, particularly during semen cryopreservation and preservation at low temperature. Finally, we reinforce the idea of AMPK as a key signaling kinase in spermatozoa that acts as an essential connector between metabolism energy and sperm ability to fertilize.

2. Role of AMPK in Male Gonads and Spermatogenesis

Fertility can be affected by nutrition and energy metabolism; in this sense, AMPK plays an important role in the reproductive function, connecting, among other physiological regulators, the hypothalamus–pituitary–gonadal axis with energy balance [19]. Germ cells production in the testis occurs through spermatogenesis and terminates when the mature spermatozoa are released into the lumen of the seminiferous tubules. A proper energy balance is necessary for spermatozoa production in the testis, as well as for sperm function and quality after ejaculation. Nonetheless, only a few studies have focused on the role of AMPK in the control of male fertility.

In 2008, Towler et al. [20] indicated that kinases related to AMPK exert physiological functions in mammalian spermatozoa. Thereby, a shorter isoform of the tumor suppressor LKB1 (LKB1s) was shown to be mainly expressed in haploid spermatids [20] where it plays an essential role in spermiogenesis and in its fertilizing ability. *LKB1s* knockout mice have a high reduction in the number of mature spermatozoa in the epididymis, and the few spermatozoa produced are not motile and have abnormal head morphology, which causes sterile mice [20]. These data suggested that this variant of the LKB1 has a crucial role in spermiogenesis and fertility in mice. A new recent study has confirmed these results, where Kong et al. have shown that although LKB1s is the dominant isoform expressed in mice testis and plays a crucial role in spermiogenesis, the long isoform of the tumor suppressor $LKB1_L$ was also required [21]. This group suggests an indispensable role of $LKB1_L$ in spermatogonial stem cell maintenance and the cooperative regulations of both LKB1 isoforms in spermatids differentiation [21]. On the other hand, deletions of *TSSK1* and *TSSK2*, two members of testis specific TSSK family or Ser/Thr kinases, which belong to AMPK branch in the human kinome tree, cause infertility in chimera mice due to haploinsufficiency [22]. The above-mentioned studies point out that some AMPK-related kinases might play a key role in the spermatozoa production and function.

Male gonadal somatic cells include Sertoli (SCs) and Leydig cells (LCs). Proliferation and survival of somatic cells of the testis are crucial for fertility. In fact, testis size and sperm production are directly correlated to the total number of adult Sertoli cells. These cells play a central role in spermatogenesis by providing support and nutrition for spermatogenic cells; however, each SC nourishes a limited number of differentiating germ cells [23]. SCs maintain the blood–testis barrier, an essential feature of seminiferous tubules, which creates the proper environment for the spermatogenic process to be held [24]. It was demonstrated that maintenance of SCs cytoskeletal dynamics, polarity, and junctional communications are essential for fertility [25], and that disorders on their apical extensions, which direct migration of differentiating germ cells towards the lumen, leads to premature germ cell loss [26]. In 2012, Tanwar et al. showed the importance of the upstream AMPK kinase LKB1 signaling in rats SCs biology and spermatogenesis. They demonstrated that aberrant AMPK-mTOR signaling causes disruption of SCs polarity and spermatogenesis [27]. In the same year, Riera et al. demonstrated that the follicle-stimulating hormone (FSH) regulates SCs proliferation with the participation of the PI3K/Akt/mTORC1 pathway and that AMPK activation may be involved in the detention of proliferation by, at least in part, a decrease in mTORC1 signaling and an increase in cyclin-dependent kinase inhibitor (CDKI) expression [28]. Thus, well-coordinated AMPK-mTOR signaling is essential for SCs functions and survival in rats [27,28]. According to a recent study, heat treatment could reversibly perturb the expression of tight junction proteins in immature porcine SCs by inhibiting the AMPK signaling pathway [29]. The use of the AMPK activator AICAR can affect adhesion molecule expression and influences junction complex integrity in rat SCs, as it has been described in 2010 by Galardo et al. [30]. This group also suggested an important role of AMPK in modulating the nutritional function of SCs. They demonstrated that AMPK stimulation with AICAR increased lactate production and glucose transport [31]. On the other hand, the inactivation of α1AMPK in mice SCs reduces the expression of mitochondrial markers (cytochrome c and PGC1-α) and the ATP content, and conversely, increases the lactate production and lipid droplets [32]. Specific deletion of the α1AMPK gene in mice SCs resulted in a 25% reduction in male fertility associated with abnormal spermatozoa with a thin head. Testes showed no clear alterations of morphology or modification in the number of SCs in vivo, but a deregulation in energy metabolism in SCs was observed [32]. These authors suggested that deregulation of the energy-sensing machinery through disruption of α1AMPK in SCs leads to a reduction in the quality of germ cells and fertility [32]. These results supported those obtained by Tartarin et al. performed in mice, where the indirect AMPK activator metformin decreased testosterone production in vitro and reduced the testicular size and the population of SCs in vivo [3]. The involvement of AMPK activity has also been studied in SCs under different stimuli/conditions and from other species such as boar. Thus, Jiao et al., reported that 17β-estradiol inhibited immature boar SCs viability and cell cycle progression by activating the AMPK signaling pathway. Furthermore, they showed that AMPK activation by AICAR inhibited SCs viability, while the AMPK inhibitor compound C attenuated the effects of 17β-estradiol on SCs [33]. Besides these findings in mammalian testes, it has been reported that AMPK activity also exerts regulatory functions in avian gonads, as metformin treatment also reduced SCs proliferation and increased lactate secretion in chickens [34].

The precise composition of the extracellular environment of germinal cells in the seminiferous tubules remains unknown, but it is well established that SCs secrete molecules that are crucial for the spermatogenic process [35]. In this sense, SCs secrete proteins and cytokines, which participate in the control of spermatozoa movement from the testis to the epididymis, and in the control of the pH of the seminiferous fluid [36]. During spermatogenesis, lactate is the preferred substrate of spermatocytes and spermatids. Lactate production from glucose in SCs is an important point for the control of spermatogenesis [37]. Recently, it has been demonstrated that AMPK reduces heat-induced lactate secretion by decreasing the expression levels of the glucose transporter GLUT3, LDHA, and MCT1 [38]. This work also suggested that AMPK is a negative regulator of heat treatment-induced lactate secretion in cultured boar SCs [38].

It has been demonstrated that AMPK is involved in the regulation of spermatozoa quality through its action, not only on the proliferation of SCs, but also in another type of testicular somatic cells, Leydig cells (LCs), (Figure 2) which are responsible for producing testicular androgens, such as testosterone, for the paracrine regulation of spermatogenesis within the testis in adult life [39].

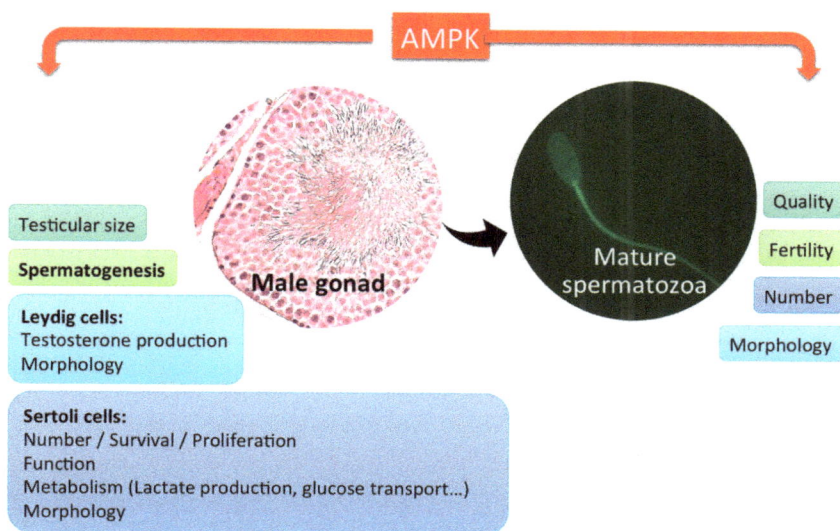

Figure 2. AMPK functions in male gonads and spermatozoa production. AMPK activity is involved in the control of the male gonad function, spermatozoa production and quality by its action also on the male gonadal somatic cells (Sertoli and Leydig cells).

Steroid hormones regulate essential physiological processes, thus, testosterone is also responsible for the differentiation and development of male reproductive organs, as well as for maintaining secondary sex characteristics and sexual function. Within the testis, the functions of the LCs are mainly controlled by the hypothalamus–pituitary axis. An *α1AMPK* knockout mice model has been studied by Tartarin et al. [3]. The male $\alpha1AMPK^{-/-}$ mice have high levels of testosterone due to hyperactive LCs [3]. Indeed, the LCs of these animals have an increased volume, an altered endoplasmic reticulum area, a high intratesticular cholesterol concentration, and a greater expression of proteins involved in steroid production [3]. In agreement with this data, Abdou et al. identified AMPK as a molecular rheostat that actively represses steroid hormone biosynthesis to preserve cellular energy homeostasis and prevent excess steroid production in LCs [40]. Recently, the role of AMPK in steroidogenesis has also been investigated in vivo in ovine testis by Taibi et al. These authors report that active AMPK is expressed in this gonad, and more importantly, they demonstrated that AMPK activity is regulated by nutritional status [41]. However, to date, the expression patterns of the different AMPK isoforms in spermatozoa are unknown. One of the most recent studies revealed that resveratrol enhanced AMPK phosphorylation and may exert its cytoprotective role against oxidative injury by the activation of autophagy via AMPK/mTOR pathway in LCs [42]. Beside mammals, the effect of metformin has also been studied in avian species; thus in 6-week-old chickens, metformin induces a decrease in testosterone levels and reduces testis size [34].

To date, AMPK has also been studied in male gonads from other mammal species such as rats [28,43–48], mice [49–55], lambs [56], monkeys [48], and humans [57]. Taking into account all the aforementioned articles, it can be concluded that the AMPK pathway is involved in the control of the male gonad function, spermatozoa production, and quality by its action also on the male gonadal somatic cells and point out AMPK as a novel gatekeeper of steroidogenesis and a target for modulating steroid hormone production. Therefore, AMPK appears as a key signaling protein for the spermatozoa and male fertility control.

3. Localization of AMPK in Mammalian Spermatozoa

The presence of the AMPK protein in mature spermatozoa was demonstrated independently in 2012 in two mammalian species, boar [4] and mouse [3]. Later, AMPK protein localization has been confirmed in mouse cauda epididymal spermatozoa [58] and also in spermatozoa from stallions [59,60], rats [61], and very recently, goats [62] and humans [63,64]. Immunolocalization techniques reveal that AMPK localization in spermatozoa slightly varies between species. AMPK is localized at the entire acrosome and in the midpiece of flagellum in boar spermatozoa [65]. Interestingly, in this species, when AMPK becomes phosphorylated at Thr172 (active) is specifically restricted to the most apical part of the acrosome and to the subequatorial segment, remaining in the midpiece of the flagellum [65]. In human spermatozoa, AMPK is localized at the entire acrosome, the midpiece and along the tail of the flagellum [63,64], whereas its active form is found only at the most apical part of the acrosome and along the entire tail, with very slight amounts in the post-acrosomal region and in the midpiece of the flagellum [63]. In stallion spermatozoa, active AMPK is mainly detected in the subequatorial region in the head and in the principal piece of the tail [59]. AMPK localization has also been studied in spermatozoa from avian spermatozoa [66]. Thus, in chicken spermatozoa, AMPK is present in the acrosome, the intermediate part, and the whole flagellum, whereas active AMPK is mainly localized in the flagellum and the acrosome [66].

4. Functions of AMPK in Mammalian Spermatozoa

4.1. Role of AMPK in the Regulation of Spermatozoa Motility

Sperm metabolic plasticity, and particularly the capacity to switch substrates and also to respond to stimuli, renders a transcendent adaptation to sperm extracellular medium fluctuations, which might include exhausted energy, high-energy demand, and low oxygen availability. AMPK plays a central role in the regulation of energy metabolism homeostasis [1,2], and therefore, this kinase has recently emerged as a new signaling pathway that exerts a necessary control of sperm function. The specific and primordial function of spermatozoa fertilization is highly dependent on the proper sperm motility. Motility is a characteristic sperm functional process that is tightly regulated in response to the spermatozoa environment and to the subsequent changes in energy demands during its transit trough the female reproductive tract. AMPK involvement on sperm motility has been demonstrated by different experimental approaches that were independently performed in two mammalian species, boar [4] and mice [3]. Thus, transgenic mice lacking the catalytic subunit α1 gene (*α1AMPK* knockout) have a dramatic reduction in sperm motility and curvilinear velocity [3]. Similarly, a pharmacological approach in boar spermatozoa demonstrates that AMPK inhibition by CC significantly decreases the percentage of motile spermatozoa, the sperm curvilinear velocity and subsequently diminishes the percentage of rapid spermatozoa (with average velocity >80 μm/s), also affecting other motility parameters and coefficients evaluated specifically using a computer-assisted sperm analysis system [4]. Later, the important regulatory role of AMPK in spermatozoa motility was demonstrated for other mammalian species such as human [63,64], rat [61], stallion [60], and goat [62]. This clear role of AMPK in sperm motility suggests that this kinase might phosphorylate downstream substrates including proteins of the axoneme or in other related structures that are indispensable for sperm flagellar motility, as it has been previously demonstrated for the AMPK related kinase TSSK2 [67]. Thus, sperm TSKK2 phosphorylates the axoneme protein SPAG16L (sperm associated antigen 16) in vitro, a protein necessary for flagellar motility in mouse spermatozoa [68]. Surprisingly, by using a specific pharmacological AMPK activator, A769662 [69], it has been shown that an increase in AMPK activity above physiological levels also adversely influences sperm motility in two mammalian species, boar [70] and human [71]. Therefore, it has been proposed [70,72] that either up or down fluctuations of sperm AMPK activity (away from energy charge-regulated physiological levels) cause a negative role in mammalian sperm motility. A range of physiological levels of AMPK activity is essential to accomplish optimal sperm motility (Figure 3) that must adapt to the fluctuating extracellular conditions that spermatozoa can be physiologically exposed to [19,70,72].

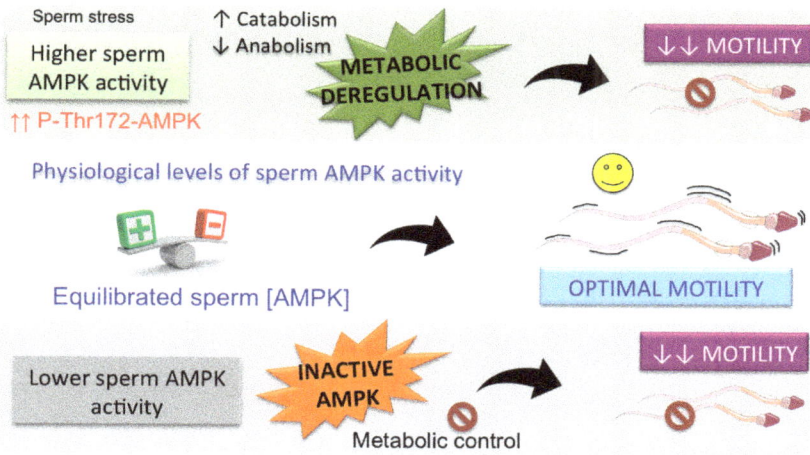

Figure 3. Different AMPK activity levels control spermatozoa motility. *In vivo*, during the spermatozoa transit trough the female reproductive tract, fluctuating levels of AMPK activity might occur in spermatozoa depending on the sperm energy demands. A physiological range of energy-charge regulated levels of AMPK activity is essential to accomplish optimal sperm motility that is adapted to the fluctuating extracellular conditions within the female reproductive tract. Up or down fluctuations of sperm AMPK activity cause a negative role in mammalian sperm motility.

In vivo, during the sperm transit trough the female reproductive tract, fluctuating levels of AMPK activity might occur in spermatozoa depending on the sperm energy demands. Whenever environment-depending demands of energy lead to a lower sperm AMPK activity that falls below the so-called physiological range, inactive AMPK is unable to adjust sperm metabolic pathways to control and maintain energy levels required for spermatozoa motility, which is negatively affected, as demonstrated in several species [3,4,62,63]. The proposal that a physiological range of AMPK activity is necessary for appropriate sperm motility is firmly supported by a study performed in transgenic mice lacking the catalytic subunit α1 gene [α1AMPK knockout (KO)] which presents a great reduction in sperm motility and curvilinear velocity [3]. Whenever environment-depending demands of energy lead to a sustained sperm AMPK activity raised above the physiological range (i.e., sperm stresses), the sustained AMPK activity would lead to a deregulation of sperm metabolism caused by a prolonged stimulation of ATP-generating catabolic pathways and by a sustained inhibition of ATP-consuming anabolic pathways (Figure 3). Thus, the deregulation of sperm metabolic pathways is not suitable to accomplish proper sperm motility under any extracellular conditions, as demonstrated in A769662-treated spermatozoa from boars [70] and humans [71]. In accordance with these findings, it has been demonstrated that spermatozoa exposure to stress stimuli, such as pollutants like hydrogen sulfide and/or ammonia, which lead to AMPK activation, causes a decline in boar spermatozoa motility [73]. In addition, this proposal is also supported by recent results obtained using a more physiological method of arresting the motility activation of rat caudal spermatozoa by isolating and lowering temperature (0–2 °C) without freezing [61]. AMPK activity in rat quiescent spermatozoa is higher than in motile spermatozoa [61], which supports the idea that AMPK is effectively regulating sperm motility by allowing these male gametes to adapt to extracellular-driven changes in sperm energy charge. Thus, the potent activation of AMPK in rat quiescent spermatozoa guarantees the most efficient pathways of energy metabolism and promotes their survival under a low energy-starved status with very limited energy resources [61]. Under these circumstances of an elevated AMPK activation, rat quiescent spermatozoa are not motile. When sperm substrates/stimuli become available during the initiation of sperm motility after ejaculation, AMPK activity levels diminish compared with quiescent spermatozoa [61], although they are still detectable and likely remain at the so-called physiological range of AMPK activity in rat spermatozoa. This female tract environment-driven shift in AMPK activity to a physiological level is necessary to adjust the proper sperm motility under these new requirements in energy demand [61].

However, very recently, one study found that other AMPK activators, such as AICAR and metformin, improve progressive motility in goat spermatozoa [62]. It is important to mention several arguments that might account for the different results found in this species: (a) The increase in phospho-AMPK levels obtained by AICAR and metformin in goat spermatozoa is very slight [62] compared with the high AMPK activity levels observed in boar [70] or rat [61] that lead to inhibition of sperm motility. Moreover, these two activators, AICAR and metformin, fail to activate AMPK in spermatozoa from other species such boar [70], where only A769662 was found to be effective activating AMPK. In addition, metformin also fails to effectively increase AMPK phosphorylation in human spermatozoa [71]. (b) It is reasonable to assume that besides great differences in the intensity of AMPK activation and the differences attributable to spermatozoa from distinct species, the specificity and effectiveness of each pharmacological activator should be kept in mind, also for further studies. This review tries to summarize the current knowledge about the role of AMPK activity in the regulation of sperm motility, and it seems clear that future research in this area is needed to elucidate the exact molecular mechanism by which AMPK exerts this regulation in sperm cells and also to discover AMPK protein substrates that belong to the intracellular signaling pathways controlling sperm motility.

4.2. Role of AMPK in the Regulation of Spermatozoa Mitochondrial Activity

As an energy regulator protein, AMPK has additionally been pointed out as important pathway that contributes to the maintenance of the mitochondrial membrane potential ($\Delta\Psi$m) in spermatozoa [3,65]. Different experimental approaches designed to inactivate sperm AMPK cause a decrease in sperm $\Delta\Psi$m in different species: goat [62], boar [65], stallion [60], and α1AMPK knockout mouse [3]. Besides a fall (50%) in sperm $\Delta\Psi$m, α1AMPK knockout (KO) mice exhibited a lower number of mitochondria and had a significant lower consumption (60%) of basal oxygen [3]. Furthermore, a short up-activation of sperm AMPK using A769662 prevents the decrease in the number of Ca^{2+} and bicarbonate-stimulated spermatozoa presenting high $\Delta\Psi$m in boar [70], whereas it has no significant effect in $\Delta\Psi$m of human spermatozoa [71]. However, the indirect activator of AMPK resveratrol significantly increased $\Delta\Psi$m in frozen–thawed human spermatozoa [64]. In mouse fresh spermatozoa, another indirect activator of AMPK metformin causes a concentration-dependent decrease of $\Delta\Psi$m [74]. Thus, AMPK activity seems to be involved in the regulation of sperm $\Delta\Psi$m, although there are differences between species, the (direct or indirect) pharmacological compound used to activate AMPK and the cell status of spermatozoa (fresh or frozen–thawed). In any case, the effect of AMPK in sperm $\Delta\Psi$m, when demonstrated, is clearly dependent on sperm extracellular stimuli [70,72], pointing out AMPK as a metabolic checkpoint by integrating stimuli-induced signaling with sperm metabolism [70]. Therefore, it has been proposed that AMPK activity (at specific levels) is essential to maintain sperm mitochondrial membrane potential [70,72]. Thus, AMPK contributes to modulate the mitochondrial membrane potential according to extracellular stimuli or conditions-derived sperm energy, and subsequently promotes the proper sperm motility, at least in some species. The involvement of AMPK in the modulation of sperm $\Delta\Psi$m is further supported by the localization of important amounts of active AMPK (Thr172 phosphorylated) at the midpiece of mammalian spermatozoa where mitochondria are located, as it has been demonstrated in boar [65], goats [62], stallions [60], and humans [63].

4.3. Role of AMPK in the Regulation of Spermatozoa Membranes

Mammalian spermatozoa need a tight modulation of energy also to maintain cellular structure, stability, and physiology of membranes during fluctuating extracellular conditions within the female reproductive tract that ultimately allow them oocyte fertilization [75,76]. An important factor that contributes to the correct function of spermatozoa is the degree of lipid organization of their plasma membrane. In fact, the energy charge-sensor kinase AMPK plays a role in the maintenance of sperm plasma membrane fluidity and lipid organization as it has been demonstrated in some species such as boar [65] and goat [62]. Thus, experiments using either an AMPK inhibitor (CC) in boar [65] and goat spermatozoa [62], or the activator A769662, demonstrated a significant increase in plasma membrane lipid disorganization [70]. Importantly, this effect of AMPK is clearly dependent of the stimuli present in the extracellular medium, as sperm plasma membrane disorganization is ≈3-fold higher when boar spermatozoa are incubated in the presence of capacitating stimuli (Ca^{2+}, bicarbonate, and serum albumin) than in a stimuli-free medium [65]. All together, these findings suggest that specific physiological levels of AMPK play a regulatory role in the preservation of the proper lipid organization and fluidity in sperm plasma membrane according to extracellular stimuli [65,77]. Furthermore, physiological levels of bicarbonate cause a quick collapse of the asymmetry of the sperm plasma membrane attributable to the scramblases activation, which translocate phospholipids such as phosphatidylethanolamine and phosphatidylserine (PS) outward of plasma membrane [78]. The PS externalization evidences plasma membrane scrambling which physiologically takes place in important sperm processes. Interestingly, AMPK inhibition in spermatozoa incubated in the presence of bicarbonate result in a significant inhibition of the outward exposure of PS in plasma membrane, suggesting that inhibition of AMPK, at least over a short time (4 h), could be causing a downstream inhibition of scramblases activity [65]. Moreover, an enhanced AMPK activity (for 24 h) induces a significant PS externalization in boar sperm plasma membrane [70]. These functional consequences of AMPK in boar sperm plasma membrane, which are lipid disorganization and the PS translocation, are presumably taking place at the plasma membrane surrounding the most apical acrosome region, where most of active phospho-Thr172 AMPK is restricted in this specie [65]. The involvement of AMPK in boar sperm membranes organization is supported by the finding that AMPK is downstream of the cAMP/PKA pathway [79], which is an essential regulatory pathway for the lipid architecture of the sperm plasma membrane [80,81].

Besides the regulatory role at the sperm plasma membrane, AMPK also regulates acrosome membrane integrity under capacitating conditions [65], which is additionally supported by the fact that high levels of active phospho-Thr172-AMPK are situated at this part of acrosome under physiological conditions [65]. By using different experimental approaches it has been demonstrated that any oscillation either up (A769662) or down (CC) of the AMPK activity outside of physiological levels lead to a loss of the outer acrosome membrane integrity [65,70]. In summary, physiological sperm AMPK activity is also essential to keep the integrity of acrosome membrane at a level suitable to the fluctuating extracellular conditions at which spermatozoa are physiologically exposed.

4.4. Role of AMPK in the Regulation of Spermatozoa Acrosome Reaction

Mammalian spermatozoa need to acquire the ability to reach the oocyte, penetrate the cumulus oophorus, and to bind to the zone pellucida of the oocyte, which subsequently triggers the acrosome reaction that leads to egg fertilization [76]. The acquisition of these spermatozoa functional competences occurs through important physiological and biochemical modifications that are collectively named sperm capacitation. Spermatozoa processes that necessarily must occur during capacitation are dependent on the energy levels of this gamete [82,83]. Therefore, it might be reasonable to assume that the energy charge-sensitive AMPK should be somehow involved in the acquisition of these sperm processes, although to date, few works have considered this issue. In fact, the activity of the energy charge-sensitive AMPK is regulated under capacitating conditions (including Ca^{2+}, bicarbonate, and serum albumin) in boar spermatozoa [65]. The inhibition of AMPK activity significantly reduces the integrity of the acrosomal membrane in a boar sperm-capacitating medium, whereas it has no effect in a stimuli-free medium [65]. These results indicate that AMPK is involved in the regulation of sperm events that take place at acrosome membrane during capacitation, at least in boar spermatozoa. However, although AMPK contributes to the maintenance of acrosome membrane integrity, is not likely involved in the sperm process of acrosome reaction, as the inhibition of AMPK in capacitated boar spermatozoa did not affect the acrosome reaction triggered by the calcium ionophore A23187 [65]. This lack of effect of AMPK activity in the acrosome reaction has been demonstrated also in fresh mouse spermatozoa using the indirect activator of AMPK metformin [74]. Moreover, AMPK up-activation does not affect the integrity of the sperm acrosome membrane. Thus, it has been demonstrated that the activator A769662 has no measurable effect in the integrity of the acrosome membrane in human [71] or boar spermatozoa, independently of stimuli present in the medium [79]. However, the effect of A769662 in boar seems to be dependent on the stimulation time as a sustained AMPK activity over 24 h causes a significant loss of outer acrosome membrane integrity [79]. In contrast to mammals, few works performed in avian sperm show that AMPK activity plays a role in the acrosome reaction [66,84]. These differences between mammals and avian spermatozoa highlight the importance of further research about the functional role of AMPK in the acrosome reaction. To date, it can be concluded that in parallel to modulate physiological changes (lipids organization) that occurs at the mammalian plasma membrane during the sperm capacitation, AMPK activity contributes to the maintenance of the outer acrosome membrane integrity, where a majority of phospho-Thr172-AMPK active is localized at physiological conditions [4].

5. Signaling Pathways Leading to AMPK Activation in Spermatozoa

The function of AMPK as a key regulatory molecule of the essential processes that contribute to the spermatozoa function of fertilization has been reviewed above. Therefore, an important issue in spermatozoa physiology is to elucidate the signaling pathways leading to AMPK activity (Figure 4). Several works using different experimental approaches demonstrate that soluble adenylate cyclase (sAC), cAMP [4], and cAMP-dependent protein kinase (PKA)-mediated pathway lie upstream of AMPK activity in boar spermatozoa [65,79].

The sperm cAMP/PKA pathway might promote AMPK activation through its upstream kinase LKB1, as it occurs in somatic cells [85], where it is reported that LKB1 can be directly phosphorylated at Ser431 by PKA in response to activation of adenylate cyclase by forskolin [86,87] or IBMX [86]. It has been demonstrated that the short splice variant of LKB1$_S$ is highly expressed in haploid spermatids in mice testis [20], where it is critically involved in spermiogenesis and fertility [88]. The knocking out LKB1$_S$ in male mice causes important reproductive consequences: sterility, a marked decrease in the number of mature spermatozoa, spermatozoa showing an abnormal head morphology, and are non-motile. An increase in cAMP concentration might also lead to AMPK activation trough cAMP degradation to AMP by phosphodiesterases, as occurs in somatic cells [89]. Thus, any sperm stimulus that increases intracellular cAMP levels could promote AMPK activation either (i) by direct (allosteric) activation of PKA, or (ii) by indirect activity of phosphodiesterases, which increase AMP levels that allosterically activate AMPK, or (iii) by both pathways.

The intracellular messenger Ca^{2+} is also an essential regulator of spermatozoa functional processes. Intracellular Ca^{2+} activates the specific sperm sAC and its downstream signaling through PKA, therefore potently activates AMPK in boar spermatozoa under physiological conditions [65,79]. However, spermatozoa from the same species under conditions considered high extracellular Ca^{2+} concentrations (>3 mM), sperm AMPK phosphorylation decreases [90]. The involvement of Ca^{2+} as an upstream regulator of AMPK activity has also been demonstrated in avian spermatozoa, as Ca^{2+} entry via store-operated Ca^{2+} channel (SOC) activates AMPK [84]. Moreover, Ca^{2+} can also lead to the activation of boar sperm AMPK through the activation of Ca^{2+}-calmodulin dependent kinase kinases II and CaMKKα/β as demonstrated in spermatozoa from boar [79] or chicken [91], which lie upstream of AMPK in somatic cells [92,93].

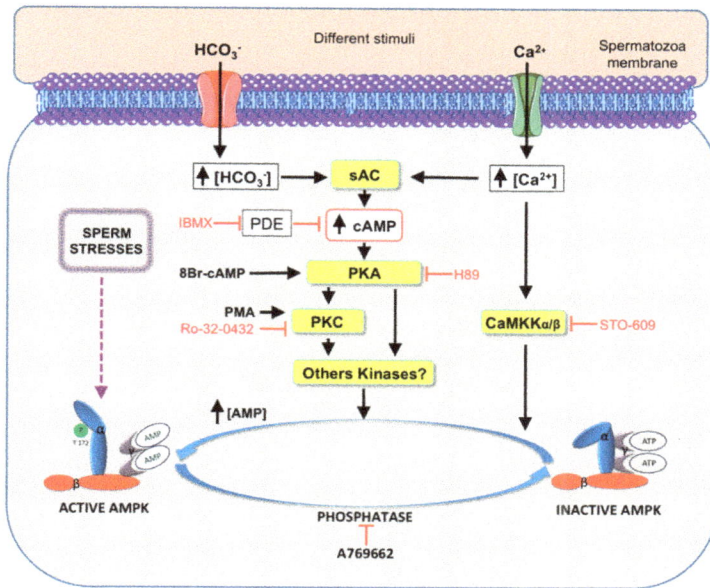

Figure 4. Signaling pathways underlying the regulation of AMPK activity in mammalian spermatozoa. This Figure summarizes the intracellular kinases and mechanisms that have been demonstrated to be involved in AMPK activation by Thr172 phosphorylation: soluble adenylyl cyclase (sAC), cAMP, protein kinase A (PKA), protein kinase C (PKC), intracellular Ca^{2+}, calcium and calmodulin kinase kinases α/β (CaMKK α/β) and different types of cell stresses detailed in the text (absence of Ca^{2+} by BAPTA-AM, hyperosmotic stress, inhibition of mitochondrial activity). Pharmacological inhibitors of different AMPK upstream kinases (H89, IBMX, STO-609 and Ro-0432) are also indicated with red lines.

Sperm AMPK phosphorylation is also stimulated by direct activation of PKC with phorbol 12-myristate 13-acetate (PMA), whereas the PKC inhibitor Ro-32-0432 inhibits HCO_3^- and Ca^{2+}-induced AMPK activation in boar spermatozoa [79]. These findings indicate that at least one isoform of PKC is upstream of sperm AMPK. Several isoforms of PKC have been identified in mammalian spermatozoa: PKCα and PKCβI in bovines [94], PKC-zeta in hamsters [95], and mouse spermatozoa [96]. It is therefore plausible that some of these PKC isoforms might exert in spermatozoa a similar function than in somatic cells by phosphorylating LKB1 and subsequently leading to AMPK activation. Thus, it has been demonstrated that PKC-zeta phosphorylates LKB1 at Ser 307 [97] and PKCζ phosphorylates LKB1 at Ser 399 [98]. An alternative explanation describing the pathway by which PKC is upstream of sperm AMPK activity is based on the fact that PKC activity lies downstream of PKA in the control pathway of boar sperm motility [99]. Previously, Harayama and Miyake (2006) demonstrated that cAMP/PKA signaling can induce the activation of calcium-sensitive PKCs, which are responsible for boar sperm hyperactivation [100]. Thus, it is proposed that another PKC isoform(s) besides PKC-zeta, which is not calcium sensitive, could likely be involved in AMPK activation, at least in response to an elevation of cAMP levels in boar spermatozoa [79].

In addition to the mentioned physiological mimicking conditions, AMPK becomes markedly activated in boar spermatozoa under different stimuli considered to be cell stress (Figure 1), such as inhibition of spermatozoa mitochondrial activity by blocking electron transport chain and sorbitol-induced hyperosmotic stress [79]. In somatic cells, cell stress-induced AMPK activation can be mediated by (i) an increase in AMP levels, and/or (ii) reactive oxygen species ROS generation that act as signaling molecules to activate AMPK [101] through LKB1 and CaMKKs pathways. Surprisingly, the incubation with the intracellular calcium chelator BAPTA-AM in a Ca^{2+}-free medium leads to a strong increase in AMPK activity [79], which may be mediated through an increase in nitric oxide NO· production. In this sense, de Lamirande et al. (2009) demonstrated in human sperm that BAPTA-AM promotes the production of a reactive oxygen specie, the nitric oxide NO· [102]. Accordingly, AMPK activation is also directly influenced by cellular redox status in somatic cells, as H_2O_2 activates AMPK through oxidative modification of cysteine residues in the AMPKα subunit [103]. An alternative or simultaneous explanation is that NO· produced by BAPTA-AM in boar spermatozoa might interacts with the cAMP pathway, as it occurs in human [102], leading to AMPK activity.

Moreover, other different types of stimuli that activate AMPK in mammalian spermatozoa have recently been described. Thus, pharmacological compounds, such as the anti-diabetic drug rosiglitazone, increase AMPK phosphorylation in stallions [60]. Also, some toxicant compounds that cause a marked reduction in sperm motility, such as the heavy metal cadmium, a major environmental toxicant, affected AMPK activity in mouse spermatozoa [104], or the air pollutants hydrogen sulfide and ammonia that activated AMPK activity in boar spermatozoa [73]. Additionally, the common natural mycotoxin ochratoxin A triggers AMPK activation to cause a clear decrease in boar sperm motility [105]. Based on all above-mentioned works, the exact molecular mechanisms involved in the signaling pathway(s) triggered by a variety of stimuli that lead to AMPK activity as well as the identity of AMPK downstream targets that ultimately control sperm function, undoubtedly deserve future investigations.

6. Role of AMPK during Assisted Reproduction Techniques: Semen Preservation

Assisted reproduction techniques (ART) include a wide range of technologies used to improve the chances to achieve pregnancy after the collection and handling of oocytes, sperm, and embryos in in vitro conditions. Nowadays, around 1.5 million human ART cycles are performed each year worldwide, with a reported 333,000 babies born. Data from human spermiograms show a decrease of sperm quality in the last decades [106–108], and as a consequence, an increase of the use of ART, such as intracytoplasmatic sperm injection (ICSI), sperm cryopreservation, in vitro fertilization (IVF), or the use of spermatozoa from testicle biopsies are the most commonly used ART. In the animal field, the use of ART is mostly aimed to quickly spread genetic material of selected animal to preserve gametes from endangered species, also to reduce disease transmission risk or to increase animal breading. This current and future reality leads to an increasing ART use and develop of new technical strategies to overcome fertility problems.

6.1. Why AMPK Protein Is Important in ART?

As mentioned before, sperm capacitation is a mandatory process to fertilize oocytes and the events related with sperm capacitation (i.e.,: hypermotility, acrosome reaction, protein phosphorylation, etc.) consume large amounts of energy (ATP). It was also stated that AMPK acts as an energy sensor and it is activated under physiological and/or stressful sperm conditions. Therefore, there is a growing interest in this kinase to improve ART procedures where gametes are handled and subjected to different types of stress such as mechanical (centrifugation, pipetting), changes in temperature, levels of CO_2 and O_2, medium composition, extracellular matrix (solids, plastics), or to light exposure [109] (Figure 5).

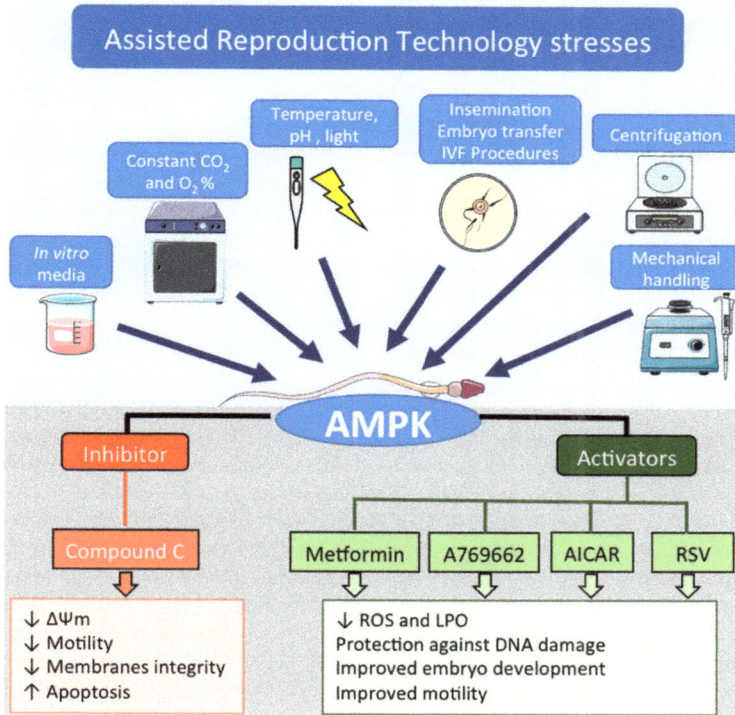

Figure 5. Effects of the modulation of AMPK activity (up or down) in spermatozoa during Assisted Reproduction Techniques (ART), including In Vitro Fertilization (IVF). During ART, spermatozoa are subjected to different stress conditions (non-physiological temperature and media, light exposure, mechanical handling, centrifugation or constant levels of CO_2 and O_2) that might modify AMPK activity. This figure summarizes the reported (beneficial or detrimental) actions in spermatozoa when extender media used in ART were supplemented with AMPK inhibitor (CC), AMPK activators (A769662 and AICAR) or indirect AMPK activators such as metformin and resveratrol (RSV).

6.2. AMPK as a Tool to Improve ART

The sector of animal farm is very interested in ART since livestock genetic background can be easily and successfully improved by the use of these techniques. Artificial Insemination (AI) was the first widely accepted ART for livestock. This technology allows genetically superior males to produce more offspring than it would be possible through conventional mating. More than 90% of western Europe's pig industry performs AI using boar seminal doses preserved at 17 °C as a routine technique [110]. The temperature decrease during this type of preservation is aimed to reduce sperm metabolism, which subsequently leads to lower rates of both acidification and reactive oxygen species (ROS) production in the storage extender, both derived from accumulation of CO_2 and lactic acid from oxidative phosphorylation and glycolysis respectively, obtaining the so called "quiescent spermatozoa". Based on the principle that AMPK acts as a stress and energy regulator, it was reported that AMPK becomes activated when quiescent rat spermatozoa are subjected to decreasing temperatures down to 4 °C [61]. Several works have studied AMPK with the aim to improve ART when spermatozoa are stored under non-physiological temperature during semen preservation techniques. A study about boar seminal doses conservation at a non-physiological temperature (17 °C), demonstrated that before semen preservation (day 0), sperm AMPK phosphorylation was undetectable but sperm AMPK activity increased during the following days of preservation, reaching a maximum phosphorylation at day 7 [77]. Therefore, during semen preservation at 17 °C, at least in boar spermatozoa, the activity of AMPK fluctuates according to extracellular conditions that include stress, and is tightly regulated during the conservation period. This AMPK activation could be associated with a decrease of intracellular ATP levels described during boar sperm conservation [111,112]. Effects of different pharmacological drugs that influence AMPK activity (anti-diabetic compounds, such as metformin and rosiglitazone, or the inhibitor CC) during semen preservation, have been studied in different species [60,77,113]. In boar spermatozoa, only the percentage of motile spermatozoa was improved after a short preservation period (<4 days) when compound C was added to the extender [77]. However, other beneficial effects in semen quality were not found with the addition of CC [77] or metformin [113] to the boar semen preservation media. Both pharmacological approaches (metformin and CC) used during boar sperm conservation for 10 days cause similar negative effects: a reduction of sperm motility and lower mitochondria membrane potential [77,113]. Therefore, as mentioned before, any alteration of AMPK activity away from physiological levels during semen preservation leads to a detrimental effect on the main sperm functions [72,77]. These results in boar have been supported by a recent work in equine spermatozoa where compound C had a detrimental effect by decreasing sperm motility [60]. However, incubation of equine spermatozoa at physiological temperature with rosiglitazone, another antidiabetic compound that activates AMPK [114], increases the percentages of motile and rapid spermatozoa [60]. These authors suggest that rosiglitazone either protects and enhances mitochondria metabolism or induces a shift in stallion spermatozoa to a glycolytic metabolism (glucose uptake is increased), subsequently decreasing ROS and increasing intracellular ATP levels [60]. These positive effects in sperm motility led to authors to test rosiglitazone effect on stallion seminal doses preserved at 24 °C and results showed an improvement of equine sperm motility, suggesting its addition to stallion extender during preservation [60].

Sperm cryopreservation is other ART widely used that allows preserved indefinitely spermatozoa. Nevertheless, this procedure is harmful to the cells causing mainly structural damages to sperm membrane, DNA fragmentation and metabolic changes that restrict its fertilization chances [115]. These deleterious effects during cryopreservation are caused by cold shock, intracellular ice formation, oxidative stress, hypertonic damage, and combinations of these and others stressful conditions [115]. Many studies have tried to improve male gamete cryopreservation by adding antioxidants to the semen extender to reduce oxidative stress [116]. In this regard, the natural compound resveratrol that activates AMPK in some somatic cell types, when added to cryopreservation media, had positive effects on human spermatozoa [64,117]. Thus, AMPK was activated and ROS levels were reduced in human spermatozoa cryopreserved in the presence of resveratrol [64]. Moreover, DNA damage on human cryopreserved spermatozoa was ameliorated when resveratrol was added before cryopreservation, likely mediated by AMPK [117]. However, when resveratrol was added to boar seminal doses preserved at 17 °C, no improvement was detected in any quality sperm parameter analyzed, and negative effects were reported in mitochondria membrane potential, intracellular ATP content and motility [118]. Thus, the possible beneficial effects of resveratrol during semen preservation seem to be dependent on the species studied or the preservation protocol, where different actions of resveratrol might preponderate depending of the temperature, i.e., at −196 °C, the scavenging capacity preponderates, while at 17 °C, resveratrol negative actions might be due to Ca^{2+} mobilization over time. Besides mammalian spermatozoa, the effects of the classical AMPK activator AICAR or the indirect activator metformin were studied in avian spermatozoa. Thus, chicken spermatozoa cryopreserved in the presence of AICAR or metformin had stimulated sperm anti-oxidative defenses by partially restoring superoxide dismutase, glutathione peroxidase, and glutathione reductase activities, which subsequently decreased ROS levels and lipid peroxidation, leading to an improved sperm cryopreserved quality [119]. Also, in order to decrease oxidative stress, metformin was added (5 to 5000 μM) before mouse spermatozoa cryopreservation [74]. Although only the higher concentration of metformin activated AMPK, any concentration tested doubled sperm survival, motility, and percentage of motile spermatozoa, as well as the fertilization rate and embryo development without any reduction in membrane lipid peroxidation [74]. These results observed in mice are surprising since it is assumed that mitochondria is the major cellular target of metformin that leads to an inhibition of mitochondrial respiration [120]. In fact, boar spermatozoa preserved at 17 °C in the presence of metformin effectively had lower mitochondria membrane potential and also inhibited sperm motility [113]. Interestingly, the quality of cryopreserved stallion spermatozoa was not improved when AMPK activity was modified by any drugs tested: CC, AICAR, or metformin [59]. No significant improvement was observed in stallion sperm survival, total motility, progressive motility or live sperm with a structurally intact acrosome [59].

In general, the effects of the manipulation (up or down) of AMPK activity during ART protocols seem contradictory and are basically dependent on the pharmacological agent, species, and/or the ART protocol studied. As an example, whereas resveratrol did not have any positive effect during boar spermatozoa preservation at 17 °C [118], positive effects were described in cryopreserved human spermatozoa [64,117]. While metformin did not have any positive effect during stallion spermatozoa cryopreservation [59], very positive outcomes were described for cryopreserved mouse spermatozoa [74]. Moreover, it should be kept in mind that the effects of the different AMPK activators in spermatozoa from the same species vary according to the ART used. For instance, AICAR or metformin do not exert any beneficial effect in stallion spermatozoa cryopreserved for a long period of time at −196 °C [59], whereas the indirect activator rosiglitazone has some positive actions when spermatozoa are preserved a room temperature [60]. Therefore, we cautiously suggest that experiments designed with pharmacological agents that act through modification of sperm AMPK activity should be performed in each particular animal species and also under each specific conditions of ART used in order to address stress-associated problems originated by ART protocols and that ultimately lead to a decrease in sperm quality.

Effects of the AMPK inhibitor CC supplementation to sperm conservation media during ART protocols are described in Table 1.

Table 1. Effects of the use of AMPK inhibitor C (compound C) in spermatozoa functions of several species during Assisted Reproduction Techniques (ART).

Specie	ART/Stress	Sperm Effects	Reference
Boar	Preserved (17 °C)	1. At short period (<4 days): Improves the % of motile sperm 2. At long period (≥4 days): Decreases sperm MMP Decreases acrosome membrane integrity Increases plasma membrane disorganization	[77]
Stallion	Cryopreserved (−196 °C)	No effect described	[59]
Chicken	Cryopreserved (−196 °C)	Decreases motility and antioxidant capacity	[119]
Stallion	Preserved (RT)	Reduces sperm motility	[60]
Human	Cryopreserved (−196 °C)	Deleterious effect in sperm motility and mitochondria Increases apoptotic-like spermatozoa	[64]

As expected, negative effects were found in spermatozoa when the AMPK pathway is inhibited by CC such as reduced mitochondrial membrane potential (by 50%) and basal oxygen consumption (by 60%) in *AMPKα1* knock out mice [3], and detrimental effects on important sperm functions in refrigerated boar spermatozoa [77], human cryopreserved spermatozoa [64], chicken cryopreserved spermatozoa [119], or room temperature preserved stallion spermatozoa [60]. These negative effects can be summarized as: (i) decrease of mitochondrial membrane potential; (ii) sperm motility inhibition; (iii) loss of acrosomal membrane integrity; (iv) increase of plasma membrane disorganization; and (v) increase of apoptotic-like spermatozoa (Table 1).

Nevertheless, although differences were observed between species and depending on the ART used, in general, beneficial effects were found when AMPK was shown to be over-activated by metformin in cryopreserved mouse spermatozoa [74], rosiglitazone in stallion spermatozoa preserved at room temperature [60], AICAR in cryopreserved chicken spermatozoa [119] and resveratrol in cryopreserved human spermatozoa [64,117]. The positive effects observed were: (i) improvement of sperm survival, motility, and velocity; (ii) improvement of sperm antioxidant defense and decrease of lipid peroxidation (LPO); and (iii) improvement of embryo quality (Table 2).

Table 2. Demonstrated effects of direct (AICAR) or indirect AMPK activators (resveratrol, metformin and rosiglitazone) in the spermatozoa functions of different species during Assisted Reproduction Techniques (ART). Cryopreservation was always at −196 °C. * It has not been showed that the indicated compound effectively activates AMPK in spermatozoa from the indicated specie.

Compound	Specie	ART/Stress	Sperm Effects	Ref.
Resveratrol *	Boar	Preserved (17 °C)	Decreases sperm motility, MMP and ATP	[118]
Metformin *	Stallion	Cryopreserved	No effect described	[59]
AICAR *	Stallion	Cryopreserved	No effect described	[59]
Metformin	Mouse	Cryopreserved	Increases sperm motility and viability Increases fertility rate and quality of embryos	[74]
AICAR	Chicken	Cryopreserved	Protects against ROS and lipid peroxidation Improves motility and % rapid spermatozoa	[119]
Metformin	Chicken	Cryopreserved	Protects against ROS and lipid peroxidation Improves motility and % rapid spermatozoa	[119]
Rosiglitazone	Stallion	Preserved (RT)	Improves % motile and % rapid spermatozoa Shifts to glycolytic metabolism, increases glucose uptake and reduces ROS	[60]
Resveratrol	Human	Cryopreserved	Reduces ROS and apoptosis-like spermatozoa	[64]
Resveratrol	Human	Cryopreserved	Protects against apoptotic-like spermatozoa and MMP damage	[117]
Metformin *	Boar	Preserved (17 °C)	Reduces sperm motility and MMP	[113]

However, this issue seems controversial, as negative effects have also been demonstrated, at least regarding metformin supplementation. Thus, sperm motility and mitochondrial membrane potential are negatively affected in boar spermatozoa preserved at 17 °C in the presence of metformin [113]. Above all, AMPK protein should be considered a suitable target to effectively improve assisted reproduction technologies' outcomes where spermatozoa are subjected to different types of stress.

7. AMPK in Spermatozoa: A Physiological Link between Fertility and Energy Metabolism

The central role of AMPK in energy metabolism is well known, although its involvement in fertility control has only recently been pointed out. AMPK is present in female and male gonads and seems to contribute to the different stages of maturation of germ cells and spermatozoa by modulating hormone production (steroidogenesis) and their interaction with nourishing gonadal somatic cells. In mature mammalian spermatozoa, the ability to preserve energy homeostasis under fluctuating extracellular nutrients, physiological stimuli, or even stress conditions that occur through the female reproductive tract is a necessary characteristic of successful spermatozoa. Moreover, the energy charge-sensitive kinase, AMPK, performs an important and essential regulator of the main physiological processes of spermatozoa and therefore represents a proper target to improve assisted reproduction technologies. The increasing scientific evidence about AMPK in the male fertility area points out AMPK as the physiological link between reproduction and energy metabolism.

Moreover, considerable progress has been performed recently in the molecular development of direct AMPK activators/inhibitors. A future scientific challenge will be to integrate some of these promising therapeutic compounds derived from pre-clinical animal studies into practical improvements in the field of assisted reproduction in animals and human.

Author Contributions: Conceptualization, D.M.-H., A.H.d.L., and M.J.B.; Writing-Original Draft Preparation, D.M.-H., A.H.d.L., and M.J.B.; Writing-Review & Editing, D.M.-H., A.H.d.L., V.C.-G., L.G.-F., and L.J.G.-M.; Supervision, M.J.B.; Project Administration, L.G.-M.; Funding Acquisition, L.G.-M., L.G.-F., and M.J.B.

funding: Part of this research was funded by Junta de Extremadura, Spain, and Fondo Europeo de Desarrollo Regional (FEDER) (grants IB13121, IB16184, and GR15118), and by Agencia Estatal de Investigación (AEI) of the Ministry of Economy, Industry and Competitiveness, Spain (grant number AGL2015-73249-JIN).

Acknowledgments: D. Martin-Hidalgo and A. Hurtado de Llera are recipients of post-doctoral fellowships of Junta de Extremadura (Spain) and the Fondo Social Europeo (Refs. PO17020 and PO17022, respectively). V. Calle-Guisado is recipient of a PhD Fellowship (Ref. FPU14/03449) from the Ministry of Education, Culture and Sport, Spain. L. Gonzalez-Fernandez is supported by the Grant AGL2015-73249-JIN of Agencia Estatal de Investigación (AEI) of the Ministry of Economy, Industry and Competitiveness, Spain.

Conflicts of Interest: The authors declare no conflict of interest. The funders had no role in the design of the study; in the collection, analyses, or interpretation of data; in the writing of the manuscript, or in the decision to publish the results.

Abbreviations

AMPK	AMP-activated protein kinase
ART	Assisted reproduction techniques
ICSI	Intracytoplasmatic sperm injection
IVF	In Vitro Fertilization
SC	Sertoli cells
LC	Leydig cells
sAC	Soluble adenylate cyclase
cAMP	Cyclic adenosine mono phosphate
PKA	cAMP-dependent protein kinase
PKC	Protein kinase C
LKB1	Liver kinase B1
CaMKKα/β	Ca^{2+}-calmodulin dependent kinase kinases II
TSSK1/2	Testis specific serine/threonine kinases 1/2
AICAR	5-aminoimidazol-4-carboxamide-1-B-D-ribofuranoside
CC	Compound C or dorsomorphin
$\Delta\Psi$m	Mitochondrial membrane potential
PMA	Phorbol 12-myristate 13-acetate
ROS	Reactive oxygen species
RSV	Resveratrol

References

1. Carling, D. AMPK signalling in health and disease. *Curr. Opin. Cell Biol.* **2017**, *45*, 31–37. [CrossRef] [PubMed]
2. Hardie, D.G.; Schaffer, B.E.; Brunet, A. AMPK: An Energy-Sensing Pathway with Multiple Inputs and Outputs. *Trends Cell Biol.* **2016**, *26*, 190–201. [CrossRef] [PubMed]
3. Tartarin, P.; Guibert, E.; Toure, A.; Ouiste, C.; Leclerc, J.; Sanz, N.; Brière, S.; Dacheux, J.L.; Delaleu, B.; McNeilly, J.R.; et al. Inactivation of AMPKalpha1 induces asthenozoospermia and alters spermatozoa morphology. *Endocrinology* **2012**, *153*, 3468–3481. [CrossRef] [PubMed]
4. Hurtado de Llera, A.; Martin-Hidalgo, D.; Gil, M.C.; Garcia-Marin, L.J.; Bragado, M.J. AMP-Activated Kinase AMPK Is Expressed in Boar Spermatozoa and Regulates Motility. *PLoS ONE* **2012**, *7*, e38840. [CrossRef] [PubMed]
5. Handa, N.; Takagi, T.; Saijo, S.; Kishishita, S.; Takaya, D.; Toyama, M.; Terada, T.; Shirouzu, M.; Suzuki, A.; Lee, S.; et al. Structural basis for compound C inhibition of the human AMP-activated protein kinase alpha2 subunit kinase domain. *Acta Crystallogr. D Biol. Crystallogr.* **2011**, *67*, 480–487. [CrossRef] [PubMed]

6. Dite, T.A.; Langendorf, C.G.; Hoque, A.; Galic, S.; Rebello, R.J.; Ovens, A.J.; Lindqvist, L.M.; Ngoei, K.R.W.; Ling, N.X.Y.; Furic, L.; et al. AMP-activated protein kinase selectively inhibited by the type II inhibitor SBI-0206965. *J. Biol. Chem.* **2018**, *293*, 8874–8885. [CrossRef] [PubMed]

7. Olivier, S.; Foretz, M.; Viollet, B. Promise and challenges for direct small molecule AMPK activators. *Biochem. Pharmacol.* **2018**, *153*, 147–158. [CrossRef] [PubMed]

8. Zhou, G.; Myers, R.; Li, Y.; Chen, Y.; Shen, X.; Fenyk-Melody, J.; Wu, M.; Ventre, J.; Doebber, T.; Fujii, N.; et al. Role of AMP-activated protein kinase in mechanism of metformin action. *J. Clin. Investig.* **2001**, *108*, 1167–1174. [CrossRef] [PubMed]

9. Ouyang, J.; Parakhia, R.A.; Ochs, R.S. Metformin activates AMP kinase through inhibition of AMP deaminase. *J. Biol. Chem.* **2011**, *286*, 1–11. [CrossRef] [PubMed]

10. Ben-Sahra, I.; Regazzetti, C.; Robert, G.; Laurent, K.; Le Marchand-Brustel, Y.; Auberger, P.; Tanti, J.F.; Giorgetti-Peraldi, S.; Bost, F. Metformin, independent of AMPK, induces mTOR inhibition and cell-cycle arrest through REDD1. *Cancer Res.* **2011**, *71*, 4366–4372. [CrossRef] [PubMed]

11. Corominas-Faja, B.; Quirantes-Pine, R.; Oliveras-Ferraros, C.; Vazquez-Martin, A.; Cufí, S.; Martin-Castillo, B.; Micol, V.; Joven, J.; Segura-Carretero, A.; Menendez, J.A. Metabolomic fingerprint reveals that metformin impairs one-carbon metabolism in a manner similar to the antifolate class of chemotherapy drugs. *Aging* **2012**, *4*, 480–498. [CrossRef] [PubMed]

12. Kelly, B.; Tannahill, G.M.; Murphy, M.P.; O'Neill, L.A. Metformin Inhibits the Production of Reactive Oxygen Species from NADH: Ubiquinone Oxidoreductase to Limit Induction of Interleukin-1beta (IL-1beta) and Boosts Interleukin-10 (IL-10) in Lipopolysaccharide (LPS)-activated Macrophages. *J. Biol. Chem.* **2015**, *290*, 20348–20359. [CrossRef] [PubMed]

13. Miller, R.A.; Birnbaum, M.J. An energetic tale of AMPK-independent effects of metformin. *J. Clin. Investig.* **2010**, *120*, 2267–2270. [CrossRef] [PubMed]

14. Rosilio, C.; Ben-Sahra, I.; Bost, F.; Peyron, J.F. Metformin: A metabolic disruptor and anti-diabetic drug to target human leukemia. *Cancer Lett.* **2014**, *346*, 188–196. [CrossRef] [PubMed]

15. Saeedi, R.; Parsons, H.L.; Wambolt, R.B.; Paulson, K.; Sharma, V.; Dyck, J.R.; Brownsey, R.W.; Allard, M.F. Metabolic actions of metformin in the heart can occur by AMPK-independent mechanisms. *Am. J. Physiol. Heart Circ. Physiol.* **2008**, *294*, H2497–H2506. [CrossRef] [PubMed]

16. Scotland, S.; Saland, E.; Skuli, N.; de Toni, F.; Boutzen, H.; Micklow, E.; Sénégas, I.; Peyraud, R.; Peyriga, L.; Théodoro, F.; et al. Mitochondrial energetic and AKT status mediate metabolic effects and apoptosis of metformin in human leukemic cells. *Leukemia* **2013**, *27*, 2129–2138. [CrossRef] [PubMed]

17. Owen, M.R.; Doran, E.; Halestrap, A.P. Evidence that metformin exerts its anti-diabetic effects through inhibition of complex 1 of the mitochondrial respiratory chain. *Biochem. J.* **2000**, *348*, 607–614. [CrossRef] [PubMed]

18. El-Mir, M.Y.; Nogueira, V.; Fontaine, E.; Averet, N.; Rigoulet, M.; Leverve, X. Dimethylbiguanide inhibits cell respiration via an indirect effect targeted on the respiratory chain complex I. *J. Biol. Chem.* **2000**, *275*, 223–228. [CrossRef] [PubMed]

19. Bertoldo, M.J.; Faure, M.; Dupont, J.; Froment, P. AMPK: A master energy regulator for gonadal function. *Front. Neurosci.* **2015**, *9*, 235. [CrossRef] [PubMed]

20. Towler, M.C.; Fogarty, S.; Hawley, S.A.; Pan, D.A.; Martin, D.M.; Morrice, N.A.; McCarthy, A.; Galardo, M.N.; Meroni, S.B.; Cigorraga, S.B.; et al. A novel short splice variant of the tumour suppressor LKB1 is required for spermiogenesis. *Biochem. J.* **2008**, *416*, 1–14. [CrossRef] [PubMed]

21. Kong, F.; Wang, M.; Huang, X.; Yue, Q.; Wei, X.; Dou, X.; Peng, X.; Jia, Y.; Zheng, K.; Wu, T.; et al. Differential regulation of spermatogenic process by Lkb1 isoforms in mouse testis. *Cell Death Dis.* **2017**, *8*, e3121. [CrossRef] [PubMed]

22. Xu, B.; Hao, Z.; Jha, K.N.; Zhang, Z.; Urekar, C.; Digilio, L.; Pulido, S.; Strauss, J.F., 3rd; Flickinger, C.J.; Herr, J.C. Targeted deletion of Tssk1 and 2 causes male infertility due to haploinsufficiency. *Dev. Biol.* **2008**, *319*, 211–222. [CrossRef] [PubMed]

23. Russell, L.D.; Tallon-Doran, M.; Weber, J.E.; Wong, V.; Peterson, R.N. Three-dimensional reconstruction of a rat stage V Sertoli cell: III. A study of specific cellular relationships. *Am. J. Anat.* **1983**, *167*, 181–192. [CrossRef] [PubMed]

24. Crisostomo, L.; Alves, M.G.; Gorga, A.; Sousa, M.; Riera, M.F.; Galardo, M.N.; Meroni, S.B.; Oliveira, P.F. Molecular Mechanisms and Signaling Pathways Involved in the Nutritional Support of Spermatogenesis by Sertoli Cells. *Methods Mol. Biol.* **2018**, *1748*, 129–155. [CrossRef] [PubMed]

25. Dong, H.; Chen, Z.; Wang, C.; Xiong, Z.; Zhao, W.; Jia, C.; Lin, J.; Lin, Y.; Yuan, W.; Zhao, A.Z.; et al. Rictor Regulates Spermatogenesis by Controlling Sertoli Cell Cytoskeletal Organization and Cell Polarity in the Mouse Testis. *Endocrinology* **2015**, *156*, 4244–4256. [CrossRef] [PubMed]

26. Tanwar, P.S.; Zhang, L.; Teixeira, J.M. Adenomatous polyposis coli (APC) is essential for maintaining the integrity of the seminiferous epithelium. *Mol. Endocrinol.* **2011**, *25*, 1725–1739. [CrossRef] [PubMed]

27. Tanwar, P.S.; Kaneko-Tarui, T.; Zhang, L.; Teixeira, J.M. Altered LKB1/AMPK/TSC1/TSC2/mTOR signaling causes disruption of Sertoli cell polarity and spermatogenesis. *Hum. Mol. Genet.* **2012**, *21*, 4394–4405. [CrossRef] [PubMed]

28. Riera, M.F.; Regueira, M.; Galardo, M.N.; Pellizzari, E.H.; Meroni, S.B.; Cigorraga, S.B. Signal transduction pathways in FSH regulation of rat Sertoli cell proliferation. *Am. J. Physiol. Endocrinol. Metab.* **2012**, *302*, E914–E923. [CrossRef] [PubMed]

29. Yang, W.R.; Liao, T.T.; Bao, Z.Q.; Zhou, C.Q.; Luo, H.Y.; Lu, C.; Pan, M.H.; Wang, X.Z. Role of AMPK in the expression of tight junction proteins in heat-treated porcine Sertoli cells. *Theriogenology* **2018**, *121*, 42–52. [CrossRef] [PubMed]

30. Galardo, M.N.; Riera, M.F.; Pellizzari, E.H.; Sobarzo, C.; Scarcelli, R.; Denduchis, B.; Lustig, L.; Cigorraga, S.B.; Meroni, S.B. Adenosine regulates Sertoli cell function by activating AMPK. *Mol. Cell. Endocrinol.* **2010**, *330*, 49–58. [CrossRef] [PubMed]

31. Galardo, M.N.; Riera, M.F.; Pellizzari, E.H.; Cigorraga, S.B.; Meroni, S.B. The AMP-activated protein kinase activator, 5-aminoimidazole-4-carboxamide-1-b-D-ribonucleoside, regulates lactate production in rat Sertoli cells. *J. Mol. Endocrinol.* **2007**, *39*, 279–288. [CrossRef] [PubMed]

32. Bertoldo, M.J.; Guibert, E.; Faure, M.; Guillou, F.; Ramé, C.; Nadal-Desbarats, L.; Foretz, M.; Viollet, B.; Dupont, J.; Froment, P. Specific deletion of AMP-activated protein kinase (alpha1AMPK) in mouse Sertoli cells modifies germ cell quality. *Mol. Cell. Endocrinol.* **2016**, *423*, 96–112. [CrossRef] [PubMed]

33. Jiao, Z.J.; Yi, W.; Rong, Y.W.; Kee, J.D.; Zhong, W.X. MicroRNA-1285 Regulates 17beta-Estradiol-Inhibited Immature Boar Sertoli Cell Proliferation via Adenosine Monophosphate-Activated Protein Kinase Activation. *Endocrinology* **2015**, *156*, 4059–4070. [CrossRef] [PubMed]

34. Faure, M.; Guibert, E.; Alves, S.; Pain, B.; Ramé, C.; Dupont, J.; Brillard, J.P.; Froment, P. The insulin sensitiser metformin regulates chicken Sertoli and germ cell populations. *Reproduction* **2016**, *151*, 527–538. [CrossRef] [PubMed]

35. Paillamanque, J.; Sanchez-Tusie, A.; Carmona, E.M.; Trevino, C.L.; Sandoval, C.; Nualart, F.; Osses, N.; Reyes, J.G. Arachidonic acid triggers [Ca^{2+}]i increases in rat round spermatids by a likely GPR activation, ERK signalling and ER/acidic compartments Ca^{2+} release. *PLoS ONE* **2017**, *12*, e0172128. [CrossRef] [PubMed]

36. Oliveira, P.F.; Sousa, M.; Barros, A.; Moura, T.; Rebelo da Costa, A. Intracellular pH regulation in human Sertoli cells: Role of membrane transporters. *Reproduction* **2009**, *137*, 353–359. [CrossRef] [PubMed]

37. Rato, L.; Alves, M.G.; Socorro, S.; Duarte, A.I.; Cavaco, J.E.; Oliveira, P.F. Metabolic regulation is important for spermatogenesis. *Nat. Rev. Urol.* **2012**, *9*, 330–338. [CrossRef] [PubMed]

38. Yu, C.L.; Guan, J.Y.; Ding, J.; Huang, S.; Lian, Y.; Luo, H.Y.; Wang, X.Z. AMP-activated protein kinase negatively regulates heat treatment-induced lactate secretion in cultured boar sertoli cells. *Theriogenology* **2018**, *121*, 35–41. [CrossRef] [PubMed]

39. Teerds, K.J.; Huhtaniemi, I.T. Morphological and functional maturation of Leydig cells: From rodent models to primates. *Hum. Reprod. Update* **2015**, *21*, 310–328. [CrossRef] [PubMed]

40. Abdou, H.S.; Bergeron, F.; Tremblay, J.J. A cell-autonomous molecular cascade initiated by AMP-activated protein kinase represses steroidogenesis. *Mol. Cell. Biol.* **2014**, *34*, 4257–4271. [CrossRef] [PubMed]

41. Taibi, N.; Dupont, J.; Bouguermouh, Z.; Froment, P.; Ramé, C.; Anane, A.; Amirat, Z.; Khammar, F. Expression of adenosine 5′-monophosphate-Activated protein kinase (AMPK) in ovine testis (*Ovis aries*): In vivo regulation by nutritional state. *Anim. Reprod. Sci.* **2017**, *178*, 9–22. [CrossRef] [PubMed]

42. Liu, S.; Sun, Y.; Li, Z. Resveratrol protects Leydig cells from nicotine-induced oxidative damage through enhanced autophagy. *Clin. Exp. Pharmacol. Physiol.* **2018**, *45*, 573–580. [CrossRef] [PubMed]

43. Chen, X.; Dong, Y.; Tian, E.; Xie, L.; Wang, G.; Li, X.; Chen, X.; Chen, Y.; Lv, Y.; Ni, C.; et al. 4-Bromodiphenyl ether delays pubertal Leydig cell development in rats. *Chemosphere* **2018**, *211*, 986–997. [CrossRef] [PubMed]

44. Cheung, P.C.; Salt, I.P.; Davies, S.P.; Hardie, D.G.; Carling, D. Characterization of AMP-activated protein kinase gamma-subunit isoforms and their role in AMP binding. *Biochem. J.* **2000**, *346*, 659–669. [CrossRef] [PubMed]

45. Duan, P.; Hu, C.; Quan, C.; Yu, T.; Zhou, W.; Yuan, M.; Shi, Y.; Yang, K. 4-Nonylphenol induces apoptosis, autophagy and necrosis in Sertoli cells: Involvement of ROS-mediated AMPK/AKT-mTOR and JNK pathways. *Toxicology* **2016**, *341–343*, 28–40. [CrossRef] [PubMed]

46. Duan, P.; Hu, C.; Quan, C.; Yu, T.; Huang, W.; Chen, W.; Tang, S.; Shi, Y.; Martin, F.L.; Yang, K. 4-Nonylphenol induces autophagy and attenuates mTOR-p70S6K/4EBP1 signaling by modulating AMPK activation in Sertoli cells. *Toxicol. Lett.* **2017**, *267*, 21–31. [CrossRef] [PubMed]

47. Guo, Z.; Yan, X.; Wang, L.; Wu, J.; Jing, X.; Liu, J. Effect of Telmisartan or Insulin on the Expression of Adiponectin and its Receptors in the Testis of Streptozotocin-Induced Diabetic Rats. *Horm. Metab. Res.* **2016**, *48*, 404–412. [CrossRef] [PubMed]

48. Hallows, K.R.; Alzamora, R.; Li, H.; Gong, F.; Smolak, C.; Neumann, D.; Pastor-Soler, N.M. AMP-activated protein kinase inhibits alkaline pH- and PKA-induced apical vacuolar H+-ATPase accumulation in epididymal clear cells. *Am. J. Physiol. Cell Physiol.* **2009**, *296*, C672–C681. [CrossRef] [PubMed]

49. Ahn, S.W.; Nedumaran, B.; Xie, Y.; Kim, D.K.; Kim, Y.D.; Choi, H.S. Bisphenol A bis(2,3-dihydroxypropyl) ether (BADGE.2H2O) induces orphan nuclear receptor Nur77 gene expression and increases steroidogenesis in mouse testicular Leydig cells. *Mol. Cells* **2008**, *26*, 74–80. [PubMed]

50. Cheng, Y.; Chen, G.; Wang, L.; Kong, J.; Pan, J.; Xi, Y.; Shen, F.; Huang, Z. Triptolide-induced mitochondrial damage dysregulates fatty acid metabolism in mouse sertoli cells. *Toxicol. Lett.* **2018**, *292*, 136–150. [CrossRef] [PubMed]

51. Li, R.; Luo, X.; Zhu, Y.; Zhao, L.; Li, L.; Peng, Q.; Ma, M.; Gao, Y. ATM signals to AMPK to promote autophagy and positively regulate DNA damage in response to cadmium-induced ROS in mouse spermatocytes. *Environ. Pollut.* **2017**, *231*, 1560–1568. [CrossRef] [PubMed]

52. Li, W.; Fu, J.; Zhang, S.; Zhao, J.; Xie, N.; Cai, G. The proteasome inhibitor bortezomib induces testicular toxicity by upregulation of oxidative stress, AMP-activated protein kinase (AMPK) activation and deregulation of germ cell development in adult murine testis. *Toxicol. Appl. Pharmacol.* **2015**, *285*, 98–109. [CrossRef] [PubMed]

53. Mancilla, H.; Maldonado, R.; Cereceda, K.; Villarroel-Espindola, F.; Montes de Oca, M.; Angulo, C.; Castro, M.A.; Slebe, J.C.; Vera, J.C.; Lavandero, S.; et al. Glutathione Depletion Induces Spermatogonial Cell Autophagy. *J. Cell. Biochem.* **2015**, *116*, 2283–2292. [CrossRef] [PubMed]

54. Wu, L.; Xu, B.; Fan, W.; Zhu, X.; Wang, G.; Zhang, A. Adiponectin protects Leydig cells against proinflammatory cytokines by suppressing the nuclear factor-kappaB signaling pathway. *FEBS J.* **2013**, *280*, 3920–3927. [CrossRef] [PubMed]

55. Zhang, J.; Zhu, Y.; Shi, Y.; Han, Y.; Liang, C.; Feng, Z.; Zheng, H.; Eng, M.; Wang, J. Fluoride-Induced Autophagy via the Regulation of Phosphorylation of Mammalian Targets of Rapamycin in Mice Leydig Cells. *J. Agric. Food Chem.* **2017**, *65*, 8966–8976. [CrossRef] [PubMed]

56. Pang, J.; Li, F.; Feng, X.; Yang, H.; Han, L.; Fan, Y.; Nie, H.; Wang, Z.; Wang, F.; Zhang, Y. Influences of different dietary energy level on sheep testicular development associated with AMPK/ULK1/autophagy pathway. *Theriogenology* **2018**, *108*, 362–370. [CrossRef] [PubMed]

57. Ham, S.; Brown, K.A.; Simpson, E.R.; Meachem, S.J. Immunolocalisation of aromatase regulators liver kinase B1, phosphorylated AMP-activated protein kinase and cAMP response element-binding protein-regulated transcription co-activators in the human testis. *Reprod. Fertil. Dev.* **2016**. [CrossRef] [PubMed]

58. Vadnais, M.L.; Kirkwood, R.N.; Tempelman, R.J.; Sprecher, D.J.; Chou, K. Effect of cooling and seminal plasma on the capacitation status of fresh boar sperm as determined using chlortetracycline assay. *Anim. Reprod. Sci.* **2005**, *87*, 121–132. [CrossRef] [PubMed]

59. Cordova, A.; Strobel, P.; Vallejo, A.; Valenzuela, P.; Ulloa, O.; Burgos, R.A.; Menarim, B.; Rodríguez-Gil, J.E.; Ratto, M.; Ramírez-Reveco, A. Use of hypometabolic TRIS extenders and high cooling rate refrigeration for cryopreservation of stallion sperm: Presence and sensitivity of 5′ AMP-activated protein kinase (AMPK). *Cryobiology* **2014**, *69*, 473–481. [CrossRef] [PubMed]

60. Swegen, A.; Lambourne, S.R.; Aitken, R.J.; Gibb, Z. Rosiglitazone Improves Stallion Sperm Motility, ATP Content, and Mitochondrial Function. *Biol. Reprod.* **2016**, *95*, 107. [CrossRef] [PubMed]

61. Kumar, L.; Yadav, S.K.; Kushwaha, B.; Pandey, A.; Sharma, V.; Verma, V.; Maikhuri, J.P.; Rajender, S.; Sharma, V.L.; Gupta, G. Energy Utilization for Survival and Fertilization-Parsimonious Quiescent Sperm Turn Extravagant on Motility Activation in Rat. *Biol. Reprod.* **2016**, *94*, 96. [CrossRef] [PubMed]

62. Zhu, Z.; Li, R.; Ma, G.; Bai, W.; Fan, X.; Lv, Y.; Luo, J.; Zeng, W. 5′-AMP-Activated Protein Kinase Regulates Goat Sperm Functions via Energy Metabolism in Vitro. *Cell. Physiol. Biochem.* **2018**, *47*, 2420–2431. [CrossRef] [PubMed]

63. Calle-Guisado, V.; Hurtado de Llera, A.; Martin-Hidalgo, D.; Mijares, J.; Gil, M.C.; Alvarez, I.S.; Bragado, M.J.; Garcia-Marin, L.J. AMP-activated kinase in human spermatozoa: Identification, intracellular localization, and key function in the regulation of sperm motility. *Asian J. Androl.* **2016**, *185*, 848.

64. Shabani, N.M.; Amidi, F.; Sedighi Gilani, M.A.; Aleyasin, A.; Bakhshalizadeh, S.; Naji, M.; Nekoonam, S. Protective features of resveratrol on human spermatozoa cryopreservation may be mediated through 5′ AMP-activated protein kinase activation. *Andrology* **2017**, *5*, 313–326. [CrossRef] [PubMed]

65. Hurtado de Llera, A.; Martin-Hidalgo, D.; Rodriguez-Gil, J.E.; Gil, M.C.; Garcia-Marin, L.J.; Bragado, M.J. AMP-activated kinase, AMPK, is involved in the maintenance of plasma membrane organization in boar spermatozoa. *Biochim. Biophys. Acta* **2013**, *1828*, 2143–2151. [CrossRef] [PubMed]

66. Nguyen, T.M.; Alves, S.; Grasseau, I.; Metayer-Coustard, S.; Praud, C.; Froment, P.; Blesbois, E. Central role of 5′-AMP-activated protein kinase in chicken sperm functions. *Biol. Reprod.* **2014**, *91*, 121. [CrossRef] [PubMed]

67. Xu, B.; Hao, Z.; Jha, K.N.; Digilio, L.; Urekar, C.; Kim, Y.H.; Pulido, S.; Flickinger, C.J.; Herr, J.C. Validation of a testis specific serine/threonine kinase [TSSK] family and the substrate of TSSK1 & 2, TSKS, as contraceptive targets. *Soc. Reprod. Fertil. Suppl.* **2007**, *63*, 87–101. [PubMed]

68. Zhang, Z.; Kostetskii, I.; Tang, W.; Haig-Ladewig, L.; Sapiro, R.; Wei, Z.; Patel, A.M.; Bennett, J.; Gerton, G.L.; Moss, S.B.; et al. Deficiency of SPAG16L causes male infertility associated with impaired sperm motility. *Biol. Reprod.* **2006**, *74*, 751–759. [CrossRef] [PubMed]

69. Cool, B.; Zinker, B.; Chiou, W.; Kifle, L.; Cao, N.; Perham, M.; Dickinson, R.; Adler, A.; Gagne, G.; Iyengar, R.; et al. Identification and characterization of a small molecule AMPK activator that treats key components of type 2 diabetes and the metabolic syndrome. *Cell Metab.* **2006**, *3*, 403–416. [CrossRef] [PubMed]

70. Hurtado de Llera, A.; Martin-Hidalgo, D.; Gil, M.C.; Garcia-Marin, L.J.; Bragado, M.J. AMPK up-activation reduces motility and regulates other functions of boar spermatozoa. *Mol. Hum. Reprod.* **2015**, *21*, 31–45. [CrossRef] [PubMed]

71. Calle-Guisado, V.; de Hurtado, L.A.; Gonzalez-Fernandez, L.; Bragado, M.J.; Garcia-Marin, L.J. Human sperm motility is downregulated by the AMPK activator A769662. *Andrology* **2017**, *5*, 1131–1140. [CrossRef] [PubMed]

72. De Hurtado, L.A.; Martin-Hidalgo, D.; Gil, M.C.; Garcia-Marin, L.J.; Bragado, M.J. New insights into transduction pathways that regulate boar sperm function. *Theriogenology* **2016**, *85*, 12–20. [CrossRef] [PubMed]

73. Zhao, Y.; Zhang, W.D.; Liu, X.Q.; Zhang, P.F.; Hao, Y.N.; Li, L.; Chen, L.; Shen, W.; Tang, X.F.; Min, L.J.; et al. Hydrogen Sulfide and/or Ammonia Reduces Spermatozoa Motility through AMPK/AKT Related Pathways. *Sci. Rep.* **2016**, *6*, 37884. [CrossRef] [PubMed]

74. Bertoldo, M.J.; Guibert, E.; Tartarin, P.; Guillory, V.; Ramé, C.; Nadal-Desbarats, L.; Foretz, M.; Viollet, B.; Dupont, J.; Froment, P. Effect of metformin on the fertilizing ability of mouse spermatozoa. *Cryobiology* **2014**, *68*, 262–268. [CrossRef] [PubMed]

75. Harayama, H. Roles of intracellular cyclic AMP signal transduction in the capacitation and subsequent hyperactivation of mouse and boar spermatozoa. *J. Reprod. Dev.* **2013**, *59*, 421–430. [CrossRef] [PubMed]

76. Yanagimachi, R. Mammalian fertilization. In *The Physiology of Reproduction*; Knobil, E., Neil, J.D., Eds.; Raven Press: New York, NY, USA, 1994; pp. 189–317.

77. Martin-Hidalgo, D.; Hurtado de Llera, A.; Yeste, M.; Gil, M.C.; Bragado, M.J.; Garcia-Marin, L.J. Adenosine monophosphate-activated kinase, AMPK, is involved in the maintenance of the quality of extended boar semen during long-term storage. *Theriogenology* **2013**. [CrossRef] [PubMed]

78. Gadella, B.M.; Harrison, R.A. The capacitating agent bicarbonate induces protein kinase A-dependent changes in phospholipid transbilayer behavior in the sperm plasma membrane. *Development* **2000**, *127*, 2407–2420. [PubMed]

79. Hurtado de Llera, A.; Martin-Hidalgo, D.; Gil, M.C.; Garcia-Marin, L.J.; Bragado, M.J. The calcium/CaMKKalpha/beta and the cAMP/PKA pathways are essential upstream regulators of AMPK activity in boar spermatozoa. *Biol. Reprod.* **2014**, *90*, 29. [CrossRef] [PubMed]

80. Gadella, B.M.; van Gestel, R.A. Bicarbonate and its role in mammalian sperm function. *Anim. Reprod. Sci.* **2004**, *82–83*, 307–319. [CrossRef] [PubMed]

81. Harrison, R.A. Rapid PKA-catalysed phosphorylation of boar sperm proteins induced by the capacitating agent bicarbonate. *Mol. Reprod. Dev.* **2004**, *67*, 337–352. [CrossRef] [PubMed]

82. Garrett, L.J.; Revell, S.G.; Leese, H.J. Adenosine triphosphate production by bovine spermatozoa and its relationship to semen fertilizing ability. *J. Androl.* **2008**, *29*, 449–458. [CrossRef] [PubMed]

83. Miki, K. Energy metabolism and sperm function. *Soc. Reprod. Fertil. Suppl.* **2007**, *65*, 309–325. [PubMed]

84. Nguyen, T.M.; Duittoz, A.; Praud, C.; Combarnous, Y.; Blesbois, E. Calcium channels in chicken sperm regulate motility and the acrosome reaction. *FEBS J.* **2016**, *283*, 1902–1920. [CrossRef] [PubMed]

85. Woods, A.; Johnstone, S.R.; Dickerson, K.; Leiper, F.C.; Fryer, L.G.; Neumann, D.; Schlattner, U.; Wallimann, T.; Carlson, M.; Carling, D. LKB1 is the upstream kinase in the AMP-activated protein kinase cascade. *Curr. Biol.* **2003**, *13*, 2004–2008. [CrossRef] [PubMed]

86. Collins, S.P.; Reoma, J.L.; Gamm, D.M.; Uhler, M.D. LKB1, a novel serine/threonine protein kinase and potential tumour suppressor, is phosphorylated by cAMP-dependent protein kinase (PKA) and prenylated in vivo. *Biochem. J.* **2000**, *345*, 673–680. [CrossRef] [PubMed]

87. Sapkota, G.P.; Kieloch, A.; Lizcano, J.M.; Lain, S.; Arthur, J.S.; Williams, M.R.; Morrice, N.; Deak, M.; Alessi, D.R. Phosphorylation of the protein kinase mutated in Peutz-Jeghers cancer syndrome, LKB1/STK11, at Ser431 by p90(RSK) and cAMP-dependent protein kinase, but not its farnesylation at Cys(433), is essential for LKB1 to suppress cell vrowth. *J. Biol. Chem.* **2001**, *276*, 19469–19482. [CrossRef] [PubMed]

88. Denison, F.C.; Smith, L.B.; Muckett, P.J.; O'Hara, L.; Carling, D.; Woods, A. LKB1 is an essential regulator of spermatozoa release during spermiation in the mammalian testis. *PLoS ONE* **2011**, *6*, e28306. [CrossRef] [PubMed]

89. Omar, B.; Zmuda-Trzebiatowska, E.; Manganiello, V.; Goransson, O.; Degerman, E. Regulation of AMP-activated protein kinase by cAMP in adipocytes: Roles for phosphodiesterases, protein kinase B, protein kinase A, Epac and lipolysis. *Cell Signal.* **2009**, *21*, 760–766. [CrossRef] [PubMed]

90. Li, X.; Wang, L.; Li, Y.; Zhao, N.; Zhen, L.; Fu, J.; Yang, Q. Calcium regulates motility and protein phosphorylation by changing cAMP and ATP concentrations in boar sperm in vitro. *Anim. Reprod. Sci.* **2016**, *172*, 39–51. [CrossRef] [PubMed]

91. Nguyen, T.M.; Combarnous, Y.; Praud, C.; Duittoz, A.; Blesbois, E. Ca^{2+}/Calmodulin-Dependent Protein Kinase Kinases (CaMKKs) Effects on AMP-Activated Protein Kinase (AMPK) Regulation of Chicken Sperm Functions. *PLoS ONE* **2016**, *11*, e0147559. [CrossRef] [PubMed]

92. Hawley, S.A.; Pan, D.A.; Mustard, K.J.; Ross, L.; Bain, J.; Edelman, A.M.; Frenguelli, B.G.; Hardie, D.G. Calmodulin-dependent protein kinase kinase-beta is an alternative upstream kinase for AMP-activated protein kinase. *Cell Metab.* **2005**, *2*, 9–19. [CrossRef] [PubMed]

93. Woods, A.; Dickerson, K.; Heath, R.; Hong, S.P.; Momcilovic, M.; Johnstone, S.R.; Carlson, M.; Carling, D. Ca^{2+}/calmodulin-dependent protein kinase kinase-beta acts upstream of AMP-activated protein kinase in mammalian cells. *Cell Metab.* **2005**, *2*, 21–33. [CrossRef] [PubMed]

94. Breitbart, H.; Naor, Z. Protein kinases in mammalian sperm capacitation and the acrosome reaction. *Rev. Reprod.* **1999**, *4*, 151–159. [CrossRef] [PubMed]

95. NagDas, S.K.; Winfrey, V.P.; Olson, G.E. Identification of ras and its downstream signaling elements and their potential role in hamster sperm motility. *Biol. Reprod.* **2002**, *67*, 1058–1066. [CrossRef] [PubMed]

96. Jungnickel, M.K.; Sutton, K.A.; Wang, Y.; Florman, H.M. Phosphoinositide-dependent pathways in mouse sperm are regulated by egg ZP3 and drive the acrosome reaction. *Dev. Biol.* **2007**, *304*, 116–126. [CrossRef] [PubMed]

97. Xie, Z.; Dong, Y.; Zhang, J.; Scholz, R.; Neumann, D.; Zou, M.H. Identification of the serine 307 of LKB1 as a novel phosphorylation site essential for its nucleocytoplasmic transport and endothelial cell angiogenesis. *Mol. Cell. Biol.* **2009**, *29*, 3582–3596. [CrossRef] [PubMed]

98. Zhu, H.; Moriasi, C.M.; Zhang, M.; Zhao, Y.; Zou, M.H. Phosphorylation of Serine 399 in LKB1 Protein Short Form by Protein Kinase Czeta Is Required for Its Nucleocytoplasmic Transport and Consequent AMP-activated Protein Kinase (AMPK) Activation. *J. Biol. Chem.* **2013**, *288*, 16495–16505. [CrossRef] [PubMed]

99. Bragado, M.J.; Aparicio, I.M.; Gil, M.C.; Garcia-Marin, L.J. Protein kinases A and C and phosphatidylinositol 3 kinase regulate glycogen synthase kinase-3A serine 21 phosphorylation in boar spermatozoa. *J. Cell. Biochem.* **2010**, *109*, 65–73. [CrossRef] [PubMed]

100. Harayama, H.; Miyake, M. A cyclic adenosine 3′,5′-monophosphate-dependent protein kinase C activation is involved in the hyperactivation of boar spermatozoa. *Mol. Reprod. Dev.* **2006**, *73*, 1169–1178. [CrossRef] [PubMed]

101. Emerling, B.M.; Weinberg, F.; Snyder, C.; Burgess, Z.; Mutlu, G.M.; Viollet, B.; Budinger, G.R.; Chandel, N.S. Hypoxic activation of AMPK is dependent on mitochondrial ROS but independent of an increase in AMP/ATP ratio. *Free Radic. Biol. Med.* **2009**, *46*, 1386–1391. [CrossRef] [PubMed]

102. De Lamirande, E.; Lamothe, G.; Villemure, M. Control of superoxide and nitric oxide formation during human sperm capacitation. *Free Radic. Biol. Med.* **2009**, *46*, 1420–1427. [CrossRef] [PubMed]

103. Zmijewski, J.W.; Banerjee, S.; Bae, H.; Friggeri, A.; Lazarowski, E.R.; Abraham, E. Exposure to hydrogen peroxide induces oxidation and activation of AMP-activated protein kinase. *J. Biol. Chem.* **2010**, *285*, 33154–33164. [CrossRef] [PubMed]

104. Wang, L.; Li, Y.; Fu, J.; Zhen, L.; Zhao, N.; Yang, Q.; Li, S.; Li, X. Cadmium inhibits mouse sperm motility through inducing tyrosine phosphorylation in a specific subset of proteins. *Reprod. Toxicol.* **2016**, *63*, 96–106. [CrossRef] [PubMed]

105. Zhang, T.Y.; Wu, R.Y.; Zhao, Y.; Xu, C.S.; Zhang, W.D.; Ge, W.; Liu, J.; Sun, Z.Y.; Zou, S.H.; Shen, W. Ochratoxin A exposure decreased sperm motility via the AMPK and PTEN signaling pathways. *Toxicol. Appl. Pharmacol.* **2018**, *340*, 49–57. [CrossRef] [PubMed]

106. Carlsen, E.; Giwercman, A.; Keiding, N.; Skakkebaek, N.E. Evidence for decreasing quality of semen during past 50 years. *BMJ* **1992**, *305*, 609–613. [CrossRef] [PubMed]

107. Levine, H.; Jorgensen, N.; Martino-Andrade, A.; Mendiola, J.; Weksler-Derri, D.; Mindlis, I.; Pinotti, R.; Swan, S.H. Temporal trends in sperm count: A systematic review and meta-regression analysis. *Hum. Reprod. Update* **2017**, *23*, 646–659. [CrossRef] [PubMed]

108. Swan, S.H.; Elkin, E.P.; Fenster, L. The question of declining sperm density revisited: An analysis of 101 studies published 1934-1996. *Environ. Health Perspect.* **2000**, *108*, 961–966. [CrossRef] [PubMed]

109. Agarwal, A.; Durairajanayagam, D.; du Plessis, S.S. Utility of antioxidants during assisted reproductive techniques: An evidence based review. *Reprod. Biol. Endocrinol.* **2014**, *12*, 112. [CrossRef] [PubMed]

110. Knox, R.V. Artificial insemination in pigs today. *Theriogenology* **2016**, *85*, 83–93. [CrossRef] [PubMed]

111. Gogol, P.; Szczesniak-Fabianczyk, B.; Wierzchos-Hilczer, A. The photon emission, ATP level and motility of boar spermatozoa during liquid storage. *Reprod. Biol.* **2009**, *9*, 39–49. [CrossRef]

112. Nguyen, Q.T.; Wallner, U.; Schmicke, M.; Waberski, D.; Henning, H. Energy metabolic state in hypothermically stored boar spermatozoa using a revised protocol for efficient ATP extraction. *Biol. Open* **2016**, *5*, 1743–1751. [CrossRef] [PubMed]

113. Hurtado de Llera, A.; Martin-Hidalgo, D.; Garcia-Marin, L.J.; Bragado, M.J. Metformin blocks mitochondrial membrane potential and inhibits sperm motility in fresh and refrigerated boar spermatozoa. *Reprod. Domest. Anim.* **2018**, *53*, 733–741. [CrossRef] [PubMed]

114. Fryer, L.G.; Parbu-Patel, A.; Carling, D. The Anti-diabetic drugs rosiglitazone and metformin stimulate AMP-activated protein kinase through distinct signaling pathways. *J. Biol. Chem.* **2002**, *277*, 25226–25232. [CrossRef] [PubMed]

115. Yeste, M. Sperm cryopreservation update: Cryodamage, markers, and factors affecting the sperm freezability in pigs. *Theriogenology* **2016**, *85*, 47–64. [CrossRef] [PubMed]

116. Amidi, F.; Pazhohan, A.; Shabani, N.M.; Khodarahmian, M.; Nekoonam, S. The role of antioxidants in sperm freezing: A review. *Cell Tissue Bank.* **2016**, *17*, 745–756. [CrossRef] [PubMed]

117. Shabani, N.M.; Nekoonam, S.; Naji, M.; Bakhshalizadeh, S.; Amidi, F. Cryoprotective effect of resveratrol on DNA damage and crucial human sperm messenger RNAs, possibly through 5′ AMP-activated protein kinase activation. *Cell Tissue Bank.* **2018**, *19*, 87–95. [CrossRef] [PubMed]

118. Martin-Hidalgo, D.; Hurtado de Llera, A.; Henning, H.; Wallner, U.; Waberski, D.; Bragado, M.J.; Gil, M.C.; Garcia-Marin, L.J. The Effect of Resveratrol on the Quality of Extended Boar Semen during Storage at 17 °C. *J. Agric. Sci.* **2013**, *5*, 231–242.

119. Nguyen, T.M.; Seigneurin, F.; Froment, P.; Combarnous, Y.; Blesbois, E. The 5′-AMP-Activated Protein Kinase (AMPK) Is Involved in the Augmentation of Antioxidant Defenses in Cryopreserved Chicken Sperm. *PLoS ONE* **2015**, *10*, e0134420. [CrossRef] [PubMed]

120. Bridges, H.R.; Jones, A.J.; Pollak, M.N.; Hirst, J. Effects of metformin and other biguanides on oxidative phosphorylation in mitochondria. *Biochem. J.* **2014**, *462*, 475–487. [CrossRef] [PubMed]

International Journal of
Molecular Sciences

MDPI

Review

AMP-Activated Protein (AMPK) in Pathophysiology of Pregnancy Complications

Asako Kumagai [1,2], Atsuo Itakura [2], Daisuke Koya [1,3] and Keizo Kanasaki [1,3,*]

[1] Department of Diabetology and Endocrinology, Kanazawa Medical University, Uchinada, Ishikawa 920-0293, Japan; a-kumagai@juntendo.ac.jp (A.K.); koya0516@kanazawa-med.ac.jp (D.K.)
[2] Department of Obstetrics and Gynecology, Juntendo University, Bunkyo-ku, Tokyo 113-0033, Japan; a-itakur@juntendo.ac.jp
[3] Division of Anticipatory Molecular Food Science and Technology, Medical Research Institute, Kanazawa Medical University, Uchinada, Ishikawa 920-0293, Japan
* Correspondence: kkanasak@kanazawa-med.ac.jp; Tel.: +81-76-286-2211 (ext. 3305); Fax: +81-76-286-6927

Received: 11 September 2018; Accepted: 1 October 2018; Published: 9 October 2018

Abstract: Although the global maternal mortality ratio has been consistently reduced over time, in 2015, there were still 303,000 maternal deaths throughout the world, of which 99% occurred in developing countries. Understanding pathophysiology of pregnancy complications contributes to the proper prenatal care for the reduction of prenatal, perinatal and neonatal mortality and morbidity ratio. In this review, we focus on AMP-activated protein kinase (AMPK) as a regulator of pregnancy complications. AMPK is a serine/threonine kinase that is conserved within eukaryotes. It regulates the cellular and whole-body energy homeostasis under stress condition. The functions of AMPK are diverse, and the dysregulation of AMPK is known to correlate with many disorders such as cardiovascular disease, diabetes, inflammatory disease, and cancer. During pregnancy, AMPK is necessary for the proper placental differentiation, nutrient transportation, maternal and fetal energy homeostasis, and protection of the fetal membrane. Activators of AMPK such as 5-Aminoimidazole-4-carboxamide ribonucleotide (AICAR), resveratrol, and metformin restores pregnancy complications such as gestational diabetes mellitus (GDM), preeclampsia, intrauterine growth restriction, and preterm birth preclinically. We also discuss on the relationship between catechol-*O*-methyltransferase (COMT), an enzyme that metabolizes catechol, and AMPK during pregnancy. It is known that metformin cannot activate AMPK in COMT deficient mice, and that 2-methoxyestradiol (2-ME), a metabolite of COMT, recovers the AMPK activity, suggesting that COMT is a regulator of AMPK. These reports suggest the therapeutic use of AMPK activators for various pregnancy complications, however, careful analysis is required for the safe use of AMPK activators since AMPK activation could cause fetal malformation.

Keywords: pregnancy; catechol-*O*-methyltransferase; 2-methoxyestradiol; preeclampsia; gestational diabetes mellitus

1. Introduction

Although the global maternal mortality ratio has been consistently reduced, in 2015 there were still 303,000 maternal deaths worldwide, 99% of which occurred in developing countries {Alkema, 2016 #3}. Among those countries, accelerated reduction of the maternal mortality rate was observed in countries that improved their transportation systems, health facilities (including the number of free-standing health facilities), training of health-care providers as well as education [1]. The proper prenatal care has a significant impact on educating the expectant mothers and their family members as well as the prevention of pregnancy complications. Pregnancy complications have two aspects, one is maternal complications such as gestational diabetes mellitus (GDM), and preeclampsia, the other is fetal complications such as intrauterine growth restriction (IUGR), and the risk of preterm birth (PTB). Many of those complications have previously reported to be associated with AMP-activated protein kinase (AMPK), a stress-induced enzyme.

AMPK is a serine/threonine kinase that is conserved within eukaryotes. It is formed of a heterotrimeric complex, which consists of a catalytic α subunit and regulatory β, γ subunits. Two isomers exist for the α and β subunits, (α1, α2), (β1, β2), and three isomers for the γ subunit (γ1, γ2, γ3), and the combinations are highly tissue specific [2–5]. Each subunit isoform is encoded by specific gene such as PRKAA1 for AMPK α1, and PRKAB1 for AMPK β1. AMPK regulates the cellular and whole-body energy homeostasis under stress condition. When cells are stressed, consumption of ATP increases, subsequently resulting in elevation of the AMP/ATP ratio [2]. This elevation of the AMP/ATP ratio activates AMPK via phosphorylation of αThr^{172} [3]. Activated AMPK shifts the cell metabolism from anabolism to catabolism to increase the cellular ATP concentration [2]. AMPK is also activated by intracellular calcium and oxidant signaling as well as extracellular signaling by hormones and cytokines [4]. The functions of AMPK are diverse and include glucose and lipid metabolism, protein synthesis, mitochondrial biogenesis, redox reaction, anti-inflammation, anti-oxidative stress, anti-apoptosis, and nitric oxide synthesis [5–7]. Due to its many functions as a regulator of energy balance, many disorders have been known to correlate with AMPK such as cardiovascular disease, diabetes, inflammatory disease, and cancer [5,8,9].

Activated AMPK exists in placental tissue and the uterine artery of humans and mice, contributing to placental differentiation and fetal growth [10]. On the other hand, AMPK activity is decreased during gestation in the hypothalamus, the brain area that maintains whole body energy balance. In non-pregnant rats, when energy balance was negative, AMPK activation increased, as well as fatty acid synthase (FAS) and the anorectic signal, malonyl-CoA, decreased in the hypothalamus to induce food intake. However, in pregnant rats, hypothalamic AMPK activation and FAS expression decreased and malonyl-CoA increased although food intake is enhanced, suggesting resistance to anorectic signals during gestation [11] (Table 1). Pregnancy changes the metabolic balance to maintain the energy demand for the embryonic growth. In this review, we focus on the relationship between AMPK activity and pregnancy complications.

Int. J. Mol. Sci. **2018**, *19*, 3076

Table 1. Phosphorylated AMP-activated protein kinase (p-AMPK) levels in different organs during pregnancies with complications.

| | | p-AMPK Levels | | | | | |
| | | Maternal | | | | | |
		Hypothalamus	Liver	Vessel	Placenta	Serum	Fetal Membrane
No complication	human						
	animal model	→ Ref. [11]					
IUGR	human						
	animal model				→ Refs. [12,13]		
GDM	human		→ Ref. [14]		→ Ref. [15]		
	animal model		→ Ref. [16]				
Preeclampsia	human			↓ (indirect) Ref. [17]	→ Ref. [17]	↑ Ref. [18]	
	animal model			↓ (indirect) Refs. [13,19]			
PTB	human						→ Ref. [20]
	animal model						→ Ref. [21]
Offspring of complicated pregnancy	human						
	animal model	↑ Ref. [22]	→ Ref. [23–25]				

(Fetal)

↑: Increased p-AMPK. ↓: Decreased p-AMPK. Ref: Reference number.

2. Intrauterine Growth Restriction (IUGR)

IUGR is one of the leading causes of perinatal mortality [26,27]. Except for genetic factors, fetal growth depends on maternal and utero-placental factors. Maternal factors are represented by maternal nutrition and hypoxia. Caloric restriction decreases maternal circulating insulin-like growth factor 1 (IGF-1), leptin, insulin, and increases cortisol in humans and animals. These maternal hormones are known as regulators for placental nutrient transport. In human IUGR, the activity of amino acid transporter system A (system A) is low in the microvillous membrane of the syncytiotrophoblast [28,29]. In vivo experiments using rodent and sheep suggested that IGF-1, leptin and insulin stimulated the activity of system A, whereas cortisol reduced placental nutrient transport [30,31]. This activation of system A by maternal hormones is known to be mediated by activated mammalian target of rapamycin (mTOR) signaling. mTOR induces cell growth and proliferation in a nutrition rich environment by sensing nutrient availability [32]. In human trophoblast cells, mTOR positively correlated with system A activation [28,32]. Human, baboon, and rodent studies also supported that calorie restriction reduced placental mTOR activity, resulting in decreasing nutrient transport and increasing the risk of IUGR. AMPK is an inhibitor of mTOR, however, baboon placental AMPK activity was unchanged in caloric restriction studies [28,32,33].

Sufficient oxygen supply is also required for nutrient transport [34]. Women living at higher altitudes have a decreased ability for placental nutrition exchange compared to women living at lower altitudes [35]. Under maternal hypoxia, utero-placental oxygen supply decreases regardless of defensive adaptation against low oxygen such as fetal polycythemia, resulting in a higher rate of IUGR [36,37]. Among utero-placental factors, uterine factors such as uterine malformation, uterine myoma, and adenomyosis directly obstruct uterine expansion and placental factors obstruct nutrient transport due to improper placental differentiation or reduced blood supply. Blood supply is essential for the fetus to acquire nutrients through the placenta. In vivo experiment of impaired utero-placental blood flow by ligating the uterine artery decreased the glucose and amino acid transportation in rats [12] (Table 1). The cellular mTOR signaling was inhibited under hypoxic condition via activation of AMPK in HEK293 cells and mouse embryonic fibroblasts (MEF) cells [38]. Indeed, placental mTOR expression is inhibited in pregnant women at high altitude who are known to have less ability for placental nutrient transport [39]. In vivo study revealed that activated AMPK increased uterine artery blood flow velocity either by inhibition of vasoconstriction prostanoids or by increasing nitric oxide production [12]. Resveratrol, the natural polyphenol that is found in grapes, cranberries, and red wine, is known to activate AMPK. Administration of resveratrol to the pregnant mice under severe hypoxic condition improved fetal survival and fetal growth [13] (Table 1). AMPK knockdown SM10 cells (mouse trophoblast progenitor cells), created by infecting lentivirus containing AMPK α1/2 shRNA, were shown to have less ability of cell growth (<50%). In addition, TGF-β induced SM10 cell differentiation was inhibited by AMPK knockdown. In terms of nutrient transport, AMPK knockdown reduced glucose transport by inhibiting expression of glucose transporter 3 (GLUT3) in SM cells. Immunohistochemistry revealed that the normal cellular localization of GLUT3 was mainly on the cell surface, indicating the proper glucose transportation, whereas AMPK knockdown cells exhibited GLUT3 localized near the nucleus [10].

These reports clearly demonstrate that AMPK activation is important for nutrient transport both by increasing uterine blood flow and increasing glucose receptor on cell surface.

3. Gestational Diabetes Mellitus (GDM)

GDM is defined as glucose intolerance that is first diagnosed after conception. The frequency of GDM differs depending on the ethnicity, but it is reported as 2–5% worldwide [15]. Obesity and family history of diabetes are the risk factors of GDM. GDM increases the risk of hypertensive disorders of pregnancy (HDP), large for gestational age, shoulder dystocia, nerve palsies, neonatal hypoglycemia, hyperbilirubinemia, and polycythemia. More than 90% of GDM resolves soon after the delivery, however, the long-term morbidity of type-2 diabetes is seven-times more frequent than women without GDM [40]. Prevention and treatment of GDM is important both for mothers and their offspring.

During pregnancy, the placenta produces hormones such as human placental lactogen (hPL), which decreases maternal insulin sensitivity and maternal glucose utilization to transport sufficient glucose through the placenta to the fetus. GDM develops when maternal insulin production by pancreatic β cells does not match with the insulin sensitivity of the organs [41,42].

In the first trimester, the development of the placenta induces drastic alternations in the trophoblast environment. Placenta and trophoblast are grown under a hypoxic environment in the beginning of gestation, however, once the spiral artery develops, the oxygen level increases, which results in an increase in the oxidative stress [43]. If GDM develops in the first trimester, the embryo is damaged both by oxidative and hyperglycemic stress, resulting in growth restriction of the placental embryonic unit [44,45]. In the mid and late trimester, fetal demand of oxygen is higher in GDM due to the enhanced metabolism by fetal high insulin levels. To meet the demand from the fetus, the placenta increases its volume and promotes angiogenesis to supply more oxygen, and erythropoiesis along with fetus growth [46,47]. Similar to in the first trimester, high oxygen supply induces oxidative stress as well as hyperglycemic stress in the fetal environment during the mid-late trimester. The maternal hyperglycemic environment enriches cellular ATP which inactivates AMPK and activates mTOR in the liver of humans, and mice [14,16] (Table 1). In the placenta, gene expression of AMPK is suppressed, and m-TOR activation is enhanced in GDM women [48] (Table 1).

There were several reports that AMPK activation by chemical compounds, such as resveratrol, 5-Aminoimidazole-4-carboxamide ribonucleotide (AICAR), and metformin, cured GDM preclinically [16,49,50]. Resveratrol was known to reduce high-glucose induced oxidative stress by activation of AMPK in type-2 diabetic animal models [51]. In GDM model mice, resveratrol increased phosphorylated AMPK (p-AMPK) in the maternal liver and lowered the maternal insulin resistance and the fetal body weight while increasing the fetal survival rate via increasing the activity of glucose-6-phosphatase in both mother and offspring [16]. Metformin is known to decrease hepatic glucose production, to increase glucose uptake in peripheral tissues, and to lower plasma triglyceride and free fatty acids. Metformin activates AMPK directly by phosphorylating Th172 on the α subunit, or indirectly by inhibiting mitochondrial complex I which increases the AMP/ATP ratio [52,53]. Active AMPK in skeletal muscle enhanced insulin-stimulated GLUT4 expression to increase glucose uptake, while in the hepatic tissue, gluconeogenic genes were inhibited by active AMPK [54,55]. Active AMPK also stimulated glucose uptake in skeletal muscle independent of insulin [56,57]. One concern about taking medication by pregnant women is placental transportability. It is known that embryonic AMPK activation stimulated by hyperglycemic and oxidative stress in GDM patients causes neural tube defect (NTD) through inhibiting the expression of pax3, an essential gene for neural tube closure [58]. Ex vivo study showed that metformin-treated mouse embryonic stem cell-derived neural progenitor cells increased activated AMPK and reduced pax3 expression [59]. However, in in vivo studies, metformin did not increase active AMPK in the embryos of pregnant mice [59]. It was reported that metformin did not pass through to the embryo due to insufficient expression of the metformin transporter, *Oct3/Slc22*, on the embryo during the period of organogenesis [59,60]. On the other hand, the placenta had sufficient expression of *Oct3/Slc22*, and some clinical studies reported that there were no significant differences in incidence of LGA, mean birthweight, and neonatal morbidity with metformin treatment and insulin treatment of GDM women [61,62]. Further study is expected for the safety of metformin in GDM.

In summary, activation of maternal AMPK ameliorates maternal diabetic features and normalizes fetal growth as a result. The safety of AMPK activators should be further studied.

4. Preeclampsia

HDP affects about 10% of all pregnancies in the world [63]. HDP is a group of four diseases, gestational hypertension, chronic hypertension, preeclampsia, and eclampsia. Among them, preeclampsia is the leading cause of maternal and perinatal mortality and morbidity. Preeclampsia is defined as development of new-onset hypertension with proteinuria after 20 weeks of gestation. The progression of preeclampsia results in placental insufficiency, which induces IUGR, and maternal organ dysfunction such as HELLP syndrome, a complication of pregnancy characterized by hemolysis, elevated liver enzymes, and low platelet counts, and eclampsia [63]. Although pathogenesis of preeclampsia is only partially understood, failure in placentation during the early stage of pregnancy has been thought to be a crucial factor that exposes the placenta and embryo to oxidative and inflammatory stress [64,65]. The mal-placentation causes a hypoxic environment, which induces angiogenic imbalance (vascular endothelial growth factor; VEGF< soluble fms-like tyrosine kinase-1; sFlt-1) and hypertension [64,66]. As mentioned in the IUGR section, AMPK activation is required for placental differentiation and vasodilation of uterine artery blood flow. Treatment of hypertension with AICAR restored blood pressure (BP) and angiogenic balance (VEGF > sFlt-1) in rats [19] (Table 1). Metformin exerted the reduction of sFlt-1 secretion on endothelial cells, villous cytotrophoblast cells, and preterm preeclamptic placental villous explants in primary human tissues [17] (Table 1). Metformin also improved vasorelaxation of human omental blood vessels which were cultured in placental villous explants obtained from patients with severe early onset preeclampsia, to the level of vessels cultured in normal media. Metformin also restored the outgrowth of omental vessel rings which was reduced when it was treated solely with sFlt-1 [17] (Table 1). Another study indicated that in preeclamptic maternal serum, p-AMPK was positively correlated with the severity of preeclampsia and BP, while it was negatively correlated with gestational week at delivery and birth weight [18] (Table 1). In summary, lack of AMPK induces mal-placentation, which results in angiogenic imbalance. The increase in serum AMPK in severely preeclamptic women suggests a compensatory mechanism for the angiogenic imbalance. AMPK activators ameliorate the preeclamptic symptoms, which indicates AMPK as a potential therapeutic target of preeclampsia

In the immunological view, imbalance between regulatory T (Treg) cells and Th17 cells is reported in preeclamptic women. Treg cells exhibit immunological tolerance during pregnancy. Th17 cells, on the other hand, induce inflammation. In normal pregnancies, increase in Treg cells and decrease in Th17 cells are found in peripheral blood compared to non-pregnant women [67]. However, in preeclampsia, Treg cells decrease and Th17 cells increase compared to non-pregnant levels [68,69]. AMPK activation restored the normal balance between Treg and Th17 cells and cured such an imbalance. Active AMPK is reported to induce Treg cells development and reduce Th17 cells differentiation, as a result, systemic inflammation improves, and immunological homeostasis is maintained [70].

5. Preterm Birth (PTB)

Complications of PTB are the major cause of neonatal deaths, and second leading cause of death among children under five years old. Many of the survived children suffer from lifelong disabilities [71]. Lowering the rate of PTB is in great demand in the world.

The common risk factors of PTB are inflammation and oxidative stress. By generating uterine specific depletion of p53 mice, the model mice of PTB, several studies have found that the activation of mTOR signaling induced decidual senescence during early pregnancy and phosphorylated mTOR increased COX2-derived prostaglandins, which resulted in spontaneous PTB in 50–60% of p53 depleted mice [21] (Table 1). In addition, AMPK activators, metformin and resveratrol, improved the decidual health and the rate of PTB was reduced in PTB model mice [21] (Table 1). In human fetal membranes, AMPK and p-AMPK exist in amnion epithelium, chorionic trophoblasts and decidua. A study reported that p-AMPK was significantly lower in fetal membrane of spontaneous labor at term compared to caesarean delivery at term and that p-AMPK levels in fetal membranes with pre-labor rupture of membrane was significantly lower compared to intact membranes. Preincubation of AMPK activators, AICAR, phenformin, A769662, decreased inflammatory cytokines such as TNF-α, IL-6, IL1-β, IL-8 levels when human fetal membranes were treated with LPS. These data suggested the anti-inflammatory effect of p-AMPK on fetal membranes [20] (Table 1).

6. Reprogramming

For decades, many reports have shown that adverse uteroplacental environments have strong associations with metabolic diseases, cardiovascular diseases, skeletal muscle deformity, and cognitive impairments in adult offspring; this concept is known as the developmental origin of health and disease (DOHaD) [72,73]. The DOHaD concept offers a reprogramming strategy that shifts the therapeutic intervention from adulthood to early-life [74]. IUGR rats had higher p-AMPK in hypothalamus regardless of feeding compared to appropriate for gestational age (AGA) rats. It has also been shown that IUGR rats expressed increased orexigenic and decreased anorexigenic mRNA expression in the hypothalamus, resulting in enhanced appetite drive, which contributes to adult obesity [22] (Table 1). On the other hand, there were reports in rats that offspring who were grown under a mal-uteroplacental environment such as mal-nutrition, hyperglycemia, and oxidative stress had reduced hepatic p-AMPK after several weeks from birth [23,24] (Table 1). Many studies have reported that administration of AMPK activators such as resveratrol, metformin and natural polyphenol containing foods (azuki bean, green tea etc.) to the pregnant mice or mice under lactation improved the offspring's outcome [24,25] (Table 1). Moreover, in rodents, resveratrol applied directly to the offspring also improved their adverse effects of growing under a mal-uteroplacental environment [8,75]. Growth hormone also reversed the dyslipidemia in small for gestational age (SGA) rat offspring grown under mal-nutrition. Hepatic p-AMPK, which was a regulator of lipid and glucose metabolism in the liver, showed no significant difference between SGA rats and AGA rats on neonatal day 1, however, it was significantly lower in SGA rats compared to AGA rats after three weeks from birth. The level of serum triglyceride was also identical between AGA rats and SGA rats at birth, but it significantly increased in the SGA rats at 10 weeks of age. Administration of growth hormone restored the level of hepatic p-AMPK as well as serum triglyceride level and body weight [23] (Table 1). Apparently, maternal environment gives both positive and negative impacts on intra-uterine environment through the placenta, and AMPK activity is key to reversing metabolic imbalance in offspring. Although earlier modification is better, too much activation of AMPK could result in the development of a fetal developmental anomaly such as NTD. Further study is needed for the timing and dose of AMPK activation treatment for both mothers and their offspring.

7. Perspective: Catechol-*O*-Methyltransferase and Pregnancy

Catechol-*O*-methyltransferase (COMT) is an enzyme that metabolizes catechol such as catecholamines and 2-hydroxyestradiol (2HE), one of the catechol estrogens. 2HE is converted into 2-methoxyestradiol (2-ME) by COMT [76]. In humans, the COMT single-nucleotide polymorphism (SNP) rs4680 (COMT$^{158Val-Met}$) exhibits reduction of enzymatic activity and stability in the Met allele carriers. COMT$^{158Val-Met}$ is associated with many diseases including diabetes, obesity and hypertension [77,78]. Indeed, in preeclamptic women, COMT protein levels and activity have shown to be lower and the COMT-mediated metabolite 2-ME was suppressed in the plasma. As a topic in the AMPK regulation, we would like to introduce our recent findings about the AMPK activation from the view of COMT/2-ME axis (Figure 1). We have recently shown that COMT is an essential enzyme to liver AMPK activity. COMT deficiency either created by a high-fat diet (HFD), COMT inhibitor or siRNA mediated knockdown, induced glucose tolerance defects associated with liver AMPK suppression in mice [79]. Such COMT deficient-associated metabolic defects and suppression of AMPK activation were ameliorated by 2-ME. Metformin recovered the activity of liver AMPK in HFD-treated mice, however, co-administration of the COMT inhibitor suppressed liver AMPK activation [79]. In addition, metformin increased COMT protein expression, suggesting that COMT could be involved in metformin-induced AMPK activation [79]. As well, 2-ME activated AMPK and induced insulin secretion in the cultured insulinoma cell line; MIN-6. The biological significance of such 2-ME-induced AMPK in the insulin secretion is a debatable issue; however in our analysis, AMPK suppression by siRNA in MIN-6 cells abolished 2-ME-induced insulin secretion [79]. A conundrum of this study is that while AMPK is needed for 2-ME-induced insulin secretion; AMPK activator

AICAR did not induce insulin secretion in MIN-6 [79]. In the pregnant mice, COMT deficiency is also associated with elevated BP, higher rate of preterm-birth, larger number of fetal wastages, and smaller placentae/decidua [76], all of which are related with lack of p-AMPK. The histological study of COMT deficient placenta showed vascular damage. The elevated BP in COMT deficient mice was explained by increase in angiotensin II receptor type 1 (ATR1) expression, which leads to the hypersensitivity of vascular smooth muscle cells to angiotensin II (AgII). 2-ME suppresses ATR1 expression, and normalizes BP [80]. AgII treatment in mice increased systolic BP and reduced urinary sodium excretion and p-AMPK level in kidney. The AgII antagonist, losartan, and metformin lowered systolic BP and increased urinary sodium excretion and p-AMPK level in kidney in mice [81]. Ex vivo experiments with embryonic rat cardiomyocytes also showed that metformin inhibited AgII-induced upregulation of AgII receptor [82]. Resveratrol was reported to regulate vascular smooth muscle contraction and BP by inhibiting AgII activity [75].

AgII-induced hypertension was ameliorated both by 2-ME and AMPK activators, indicating the close relationship between COMT and AMPK. Further study is required to identify the precise molecular mechanisms of COMT/2-ME axis-associated AMPK activation.

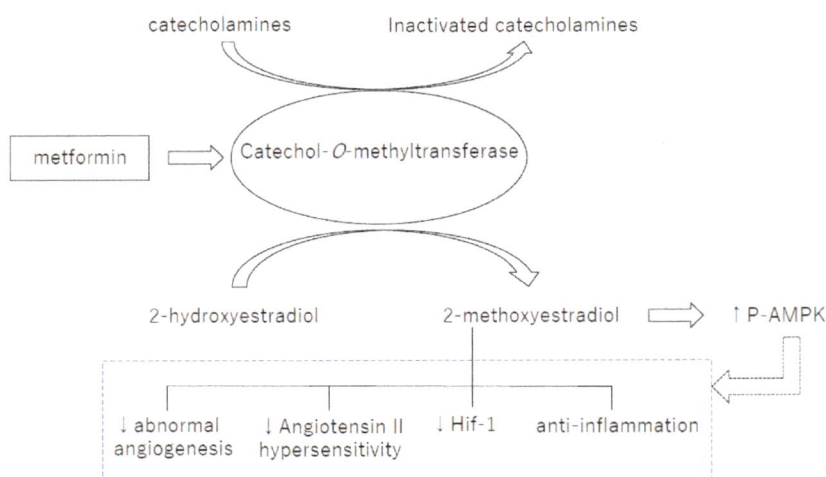

Figure 1. Catechol-*O*-methyltransferase (COMT) is an essential enzyme for the production of 2-methoxyestradiol (2-ME). 2-ME, a metabolite of COMT, induces activation of various molecular pathways, including activation of AMPK. ↑: Increase in activity. ↓: Decrease in activity.

8. Conclusions

AMPK maintains the maternal metabolic balance and protects fetal growth from diverse types of stress throughout the pregnancy. Pregnancy complications that are listed in this review have large impacts on maternal or fetal morbidity and mortality. Moreover, they also affect maternal health problems after labor, as well as their offspring's health problems in adolescence. Supplementation of AMPK activators seems effective for improving both maternal symptoms and fetal growth by restoring the metabolic balance. However, inappropriately activated AMPK in fetus could cause congenital developmental disorders. Thus the timing and the types of AMPK activators should be further studied for safe use.

funding: This work was partially supported by grants from the Japan Society for the Promotion of Science for KK (23790381, 26460403).

Conflicts of Interest: The authors declare no conflict of interest.

References

1. Alkema, L.; Chou, D.; Hogan, D.; Zhang, S.; Moller, A.B.; Gemmill, A.; Fat, D.M.; Boerma, T.; Temmerman, M.; Mathers, C.; et al. Global, regional, and national levels and trends in maternal mortality between 1990 and 2015, with scenario-based projections to 2030: A systematic analysis by the UN maternal mortality estimation inter-agency group. *Lancet* **2016**, *387*, 462–474. [CrossRef]

2. Hardie, D.G.; Ross, F.A.; Hawley, S.A. AMPK: A nutrient and energy sensor that maintains energy homeostasis. *Nat. Rev. Mol. Cell Biol.* **2012**, *13*, 251–262. [CrossRef] [PubMed]

3. Hawley, S.A.; Davison, M.; Woods, A.; Davies, S.P.; Beri, R.K.; Carling, D.; Hardie, D.G. Characterization of the AMP-activated protein kinase kinase from rat liver and identification of threonine 172 as the major site at which it phosphorylates AMP-activated protein kinase. *J. Biol. Chem.* **1996**, *271*, 27879–27887. [CrossRef] [PubMed]

4. Fogarty, S.; Hawley, S.A.; Green, K.A.; Saner, N.; Mustard, K.J.; Hardie, D.G. Calmodulin-dependent protein kinase kinase-beta activates AMPK without forming a stable complex: Synergistic effects of Ca^{2+} and AMP. *Biochem. J.* **2010**, *426*, 109–118. [CrossRef] [PubMed]

5. Steinberg, G.R.; Kemp, B.E. AMPK in health and disease. *Physiol. Rev.* **2009**, *89*, 1025–1078. [CrossRef] [PubMed]

6. Trewin, A.J.; Berry, B.J.; Wojtovich, A.P. Exercise and mitochondrial dynamics: Keeping in Shape with ROS and AMPK. *Antioxidants* **2018**, *7*. [CrossRef] [PubMed]

7. Moreira, D.; Silvestre, R.; Cordeiro-da-Silva, A.; Estaquier, J.; Foretz, M.; Viollet, B. AMP-activated Protein Kinase as a target for pathogens: Friends or foes? *Curr. Drug Targets* **2016**, *17*, 942–953. [CrossRef] [PubMed]

8. Tain, Y.L.; Hsu, C.N. AMP-activated protein kinase as a reprogramming strategy for hypertension and kidney disease of developmental origin. *Int. J. Mol. Sci.* **2018**, *19*. [CrossRef] [PubMed]

9. Li, W.; Saud, S.M.; Young, M.R.; Chen, G.; Hua, B. Targeting AMPK for cancer prevention and treatment. *Oncotarget* **2015**, *6*, 7365–7378. [CrossRef] [PubMed]

10. Carey, E.A.; Albers, R.E.; Doliboa, S.R.; Hughes, M.; Wyatt, C.N.; Natale, D.R.; Brown, T.L. AMPK knockdown in placental trophoblast cells results in altered morphology and function. *Stem Cells Dev.* **2014**, *23*, 2921–2930. [CrossRef] [PubMed]

11. Martinez de Morentin, P.B.; Lage, R.; Gonzalez-Garcia, I.; Ruiz-Pino, F.; Martins, L.; Fernandez-Mallo, D.; Gallego, R.; Ferno, J.; Senaris, R.; Saha, A.K.; et al. Pregnancy induces resistance to the anorectic effect of hypothalamic malonyl-CoA and the thermogenic effect of hypothalamic AMPK inhibition in female rats. *Endocrinology* **2015**, *156*, 947–960. [CrossRef] [PubMed]

12. Skeffington, K.L.; Higgins, J.S.; Mahmoud, A.D.; Evans, A.M.; Sferruzzi-Perri, A.N.; Fowden, A.L.; Yung, H.W.; Burton, G.J.; Giussani, D.A.; Moore, L.G. Hypoxia, AMPK activation and uterine artery vasoreactivity. *J. Physiol.* **2016**, *594*, 1357–1369. [CrossRef] [PubMed]

13. Poudel, R.; Stanley, J.L.; Rueda-Clausen, C.F.; Andersson, I.J.; Sibley, C.P.; Davidge, S.T.; Baker, P.N. Effects of resveratrol in pregnancy using murine models with reduced blood supply to the uterus. *PLoS ONE* **2013**, *8*, e64401. [CrossRef] [PubMed]

14. Perez-Perez, A.; Maymo, J.L.; Gambino, Y.P.; Guadix, P.; Duenas, J.L.; Varone, C.L.; Sanchez-Margalet, V. Activated translation signaling in placenta from pregnant women with gestational diabetes mellitus: Possible role of leptin. *Horm. Metab. Res.* **2013**, *45*, 436–442. [CrossRef] [PubMed]

15. American Diabetes Association. Gestational diabetes mellitus. *Diabetes Care* **2004**, *27*, S88–S90.

16. Yao, L.; Wan, J.; Li, H.; Ding, J.; Wang, Y.; Wang, X.; Li, M. Resveratrol relieves gestational diabetes mellitus in mice through activating AMPK. *Reprod. Biol. Endocrinol.* **2015**, *13*, 118. [CrossRef] [PubMed]

17. Brownfoot, F.C.; Hastie, R.; Hannan, N.J.; Cannon, P.; Tuohey, L.; Parry, L.J.; Senadheera, S.; Illanes, S.E.; Kaitu'u-Lino, T.J.; Tong, S. Metformin as a prevention and treatment for preeclampsia: Effects on soluble fms-like tyrosine kinase 1 and soluble endoglin secretion and endothelial dysfunction. *Am. J. Obstet. Gynecol.* **2016**, *214*, 356.e1–356.e15. [CrossRef] [PubMed]

18. Koroglu, N.; Tola, E.; Temel Yuksel, I.; Aslan Cetin, B.; Turhan, U.; Topcu, G.; Dag, I. Maternal serum AMP-activated protein kinase levels in mild and severe preeclampsia. *J. Matern. Fetal. Neonatal. Med.* **2018**, 1–6. [CrossRef] [PubMed]

19. Banek, C.T.; Bauer, A.J.; Needham, K.M.; Dreyer, H.C.; Gilbert, J.S. AICAR administration ameliorates hypertension and angiogenic imbalance in a model of preeclampsia in the rat. *Am. J. Physiol. Heart. Circ. Physiol.* **2013**, *304*, H1159–H1165. [CrossRef] [PubMed]

20. Lim, R.; Barker, G.; Lappas, M. Activation of AMPK in human fetal membranes alleviates infection-induced expression of pro-inflammatory and pro-labour mediators. *Placenta* **2015**, *36*, 454–462. [CrossRef] [PubMed]

21. Deng, W.; Cha, J.; Yuan, J.; Haraguchi, H.; Bartos, A.; Leishman, E.; Viollet, B.; Bradshaw, H.B.; Hirota, Y.; Dey, S.K. p53 coordinates decidual sestrin 2/AMPK/mTORC1 signaling to govern parturition timing. *J. Clin. Investig.* **2016**, *126*, 2941–2954. [CrossRef] [PubMed]

22. Fukami, T.; Sun, X.; Li, T.; Desai, M.; Ross, M.G. Mechanism of programmed obesity in intrauterine fetal growth restricted offspring: Paradoxically enhanced appetite stimulation in fed and fasting states. *Reprod. Sci.* **2012**, *19*, 423–430. [CrossRef] [PubMed]

23. Zhu, W.F.; Tang, S.J.; Shen, Z.; Wang, Y.M.; Liang, L. Growth hormone reverses dyslipidemia in adult offspring after maternal undernutrition. *Sci. Rep.* **2017**, *7*, 6038. [CrossRef] [PubMed]

24. Crescenti, A.; del Bas, J.M.; Arola-Arnal, A.; Oms-Oliu, G.; Arola, L.; Caimari, A. Grape seed procyanidins administered at physiological doses to rats during pregnancy and lactation promote lipid oxidation and up-regulate AMPK in the muscle of male offspring in adulthood. *J. Nutr. Biochem.* **2015**, *26*, 912–920. [CrossRef] [PubMed]

25. Mukai, Y.; Sun, Y.; Sato, S. Azuki bean polyphenols intake during lactation upregulate AMPK in male rat offspring exposed to fetal malnutrition. *Nutrition* **2013**, *29*, 291–297. [CrossRef] [PubMed]

26. Unterscheider, J.; O'Donoghue, K.; Daly, S.; Geary, M.P.; Kennelly, M.M.; McAuliffe, F.M.; Hunter, A.; Morrison, J.J.; Burke, G.; Dicker, P.; et al. Fetal growth restriction and the risk of perinatal mortality-case studies from the multicentre PORTO study. *BMC Pregnancy Childbirth* **2014**, *14*, 63. [CrossRef] [PubMed]

27. Longo, S.; Bollani, L.; Decembrino, L.; di Comite, A.; Angelini, M.; Stronati, M. Short-term and long-term sequelae in intrauterine growth retardation (IUGR). *J. Matern. Fetal. Neonatal. Med.* **2013**, *26*, 222–225. [CrossRef] [PubMed]

28. Kavitha, J.V.; Rosario, F.J.; Nijland, M.J.; McDonald, T.J.; Wu, G.; Kanai, Y.; Powell, T.L.; Nathanielsz, P.W.; Jansson, T. Down-regulation of placental mTOR, insulin/IGF-I signaling, and nutrient transporters in response to maternal nutrient restriction in the baboon. *FASEB J.* **2014**, *28*, 1294–1305. [CrossRef] [PubMed]

29. Jansson, N.; Greenwood, S.L.; Johansson, B.R.; Powell, T.L.; Jansson, T. Leptin stimulates the activity of the system A amino acid transporter in human placental villous fragments. *J. Clin. Endocrinol. Metab.* **2003**, *88*, 1205–1211. [CrossRef] [PubMed]

30. Yiallourides, M.; Sebert, S.P.; Wilson, V.; Sharkey, D.; Rhind, S.M.; Symonds, M.E.; Budge, H. The differential effects of the timing of maternal nutrient restriction in the ovine placenta on glucocorticoid sensitivity, uncoupling protein 2, peroxisome proliferator-activated receptor-gamma and cell proliferation. *Reproduction* **2009**, *138*, 601–608. [CrossRef] [PubMed]

31. Jansson, N.; Pettersson, J.; Haafiz, A.; Ericsson, A.; Palmberg, I.; Tranberg, M.; Ganapathy, V.; Powell, T.L.; Jansson, T. Down-regulation of placental transport of amino acids precedes the development of intrauterine growth restriction in rats fed a low protein diet. *J. Physiol.* **2006**, *576*, 935–946. [CrossRef] [PubMed]

32. Jansson, T.; Powell, T.L. Role of placental nutrient sensing in developmental programming. *Clin. Obstet. Gynecol.* **2013**, *56*, 591–601. [CrossRef] [PubMed]

33. Rosario, F.J.; Jansson, N.; Kanai, Y.; Prasad, P.D.; Powell, T.L.; Jansson, T. Maternal protein restriction in the rat inhibits placental insulin, mTOR, and STAT3 signaling and down-regulates placental amino acid transporters. *Endocrinology* **2011**, *152*, 1119–1129. [CrossRef] [PubMed]

34. Dimasuay, K.G.; Boeuf, P.; Powell, T.L.; Jansson, T. Placental responses to changes in the maternal environment determine fetal growth. *Front. Physiol.* **2016**, *7*, 12. [CrossRef] [PubMed]

35. Zamudio, S.; Moore, L.G. Altitude and fetal growth: Current knowledge and future directions. *Ultrasound Obstet. Gynecol.* **2000**, *16*, 6–8. [CrossRef] [PubMed]

36. Giussani, D.A.; Salinas, C.E.; Villena, M.; Blanco, C.E. The role of oxygen in prenatal growth: Studies in the chick embryo. *J. Physiol.* **2007**, *585*, 911–917. [CrossRef] [PubMed]

37. Nelson, D.M.; Smith, S.D.; Furesz, T.C.; Sadovsky, Y.; Ganapathy, V.; Parvin, C.A.; Smith, C.H. Hypoxia reduces expression and function of system A amino acid transporters in cultured term human trophoblasts. *Am. J. Physiol. Cell Physiol.* **2003**, *284*, C310–C315. [CrossRef] [PubMed]

38. Inoki, K.; Zhu, T.; Guan, K.L. TSC2 mediates cellular energy response to control cell growth and survival. *Cell* **2003**, *115*, 577–590. [CrossRef]

39. Yung, H.W.; Calabrese, S.; Hynx, D.; Hemmings, B.A.; Cetin, I.; Charnock-Jones, D.S.; Burton, G.J. Evidence of placental translation inhibition and endoplasmic reticulum stress in the etiology of human intrauterine growth restriction. *Am. J. Pathol.* **2008**, *173*, 451–462. [CrossRef] [PubMed]

40. Ashwal, E.; Hod, M. Gestational diabetes mellitus: Where are we now? *Clin. Chim. Acta* **2015**, *451*, 14–20. [CrossRef] [PubMed]

41. Huang, C.; Snider, F.; Cross, J.C. Prolactin receptor is required for normal glucose homeostasis and modulation of beta-cell mass during pregnancy. *Endocrinology* **2009**, *150*, 1618–1626. [CrossRef] [PubMed]

42. Le, T.N.; Elsea, S.H.; Romero, R.; Chaiworapongsa, T.; Francis, G.L. Prolactin receptor gene polymorphisms are associated with gestational diabetes. *Genet. Test. Mol. Biomark.* **2013**, *17*, 567–571. [CrossRef] [PubMed]

43. Jauniaux, E.; Watson, A.L.; Hempstock, J.; Bao, Y.P.; Skepper, J.N.; Burton, G.J. Onset of maternal arterial blood flow and placental oxidative stress. A possible factor in human early pregnancy failure. *Am. J. Pathol.* **2000**, *157*, 2111–2122. [CrossRef]

44. Bjork, O.; Persson, B.; Stangenberg, M.; Vaclavinkova, V. Spiral artery lesions in relation to metabolic control in diabetes mellitus. *Acta Obstet. Gynecol. Scand.* **1984**, *63*, 123–127. [CrossRef] [PubMed]

45. Desoye, G. The human placenta in diabetes and obesity: Friend or foe? The 2017 norbert freinkel award lecture. *Diabetes Care* **2018**, *41*, 1362–1369. [CrossRef] [PubMed]

46. Salvesen, D.R.; Brudenell, J.M.; Snijders, R.J.; Ireland, R.M.; Nicolaides, K.H. Fetal plasma erythropoietin in pregnancies complicated by maternal diabetes mellitus. *Am. J. Obstet. Gynecol.* **1993**, *168*, 88–94. [CrossRef]

47. Desoye, G.; Shafrir, E. Placental metabolism and its regulation in health and diabetes. *Mol. Aspects Med.* **1994**, *15*, 505–682. [CrossRef]

48. Martino, J.; Sebert, S.; Segura, M.T.; Garcia-Valdes, L.; Florido, J.; Padilla, M.C.; Marcos, A.; Rueda, R.; McArdle, H.J.; Budge, H.; et al. Maternal body weight and gestational diabetes differentially influence placental and pregnancy outcomes. *J. Clin. Endocrinol. Metab.* **2016**, *101*, 59–68. [CrossRef] [PubMed]

49. Tan, R.R.; Zhang, S.J.; Tsoi, B.; Huang, W.S.; Zhuang, X.J.; Chen, X.Y.; Yao, N.; Mao, Z.F.; Tang, L.P.; Wang, Q.; et al. A natural product, resveratrol, protects against high-glucose-induced developmental damage in chicken embryo. *J. Asian Nat. Prod. Res.* **2015**, *17*, 586–594. [CrossRef] [PubMed]

50. Liong, S.; Lappas, M. Activation of AMPK improves inflammation and insulin resistance in adipose tissue and skeletal muscle from pregnant women. *J. Physiol. Biochem.* **2015**, *71*, 703–717. [CrossRef] [PubMed]

51. Gonzalez-Rodriguez, A.; Santamaria, B.; Mas-Gutierrez, J.A.; Rada, P.; Fernandez-Millan, E.; Pardo, V.; Alvarez, C.; Cuadrado, A.; Ros, M.; Serrano, M.; et al. Resveratrol treatment restores peripheral insulin sensitivity in diabetic mice in a sirt1-independent manner. *Mol. Nutr. Food Res.* **2015**, *59*, 1431–1442. [CrossRef] [PubMed]

52. Kim, Y.D.; Park, K.G.; Lee, Y.S.; Park, Y.Y.; Kim, D.K.; Nedumaran, B.; Jang, W.G.; Cho, W.J.; Ha, J.; Lee, I.K.; et al. Metformin inhibits hepatic gluconeogenesis through AMP-activated protein kinase-dependent regulation of the orphan nuclear receptor SHP. *Diabetes* **2008**, *57*, 306–314. [CrossRef] [PubMed]

53. Zhou, G.; Myers, R.; Li, Y.; Chen, Y.; Shen, X.; Fenyk-Melody, J.; Wu, M.; Ventre, J.; Doebber, T.; Fujii, N.; et al. Role of AMP-activated protein kinase in mechanism of metformin action. *J. Clin. Investig.* **2001**, *108*, 1167–1174. [CrossRef] [PubMed]

54. Egawa, T.; Hamada, T.; Kameda, N.; Karaike, K.; Ma, X.; Masuda, S.; Iwanaka, N.; Hayashi, T. Caffeine acutely activates 5'adenosine monophosphate-activated protein kinase and increases insulin-independent glucose transport in rat skeletal muscles. *Metabolism* **2009**, *58*, 1609–1617. [CrossRef] [PubMed]

55. Lee, J.M.; Seo, W.Y.; Song, K.H.; Chanda, D.; Kim, Y.D.; Kim, D.K.; Lee, M.W.; Ryu, D.; Kim, Y.H.; Noh, J.R.; et al. AMPK-dependent repression of hepatic gluconeogenesis via disruption of CREB.CRTC2 complex by orphan nuclear receptor small heterodimer partner. *J. Biol. Chem.* **2010**, *285*, 32182–32191. [CrossRef] [PubMed]

56. Koh, H.J. Regulation of exercise-stimulated glucose uptake in skeletal muscle. *Ann. Pediatr. Endocrinol. Metab.* **2016**, *21*, 61–65. [CrossRef] [PubMed]

57. Witczak, C.A.; Sharoff, C.G.; Goodyear, L.J. AMP-activated protein kinase in skeletal muscle: From structure and localization to its role as a master regulator of cellular metabolism. *Cell Mol. Life Sci.* **2008**, *65*, 3737–3755. [CrossRef] [PubMed]

58. Wu, Y.; Viana, M.; Thirumangalathu, S.; Loeken, M.R. AMP-activated protein kinase mediates effects of oxidative stress on embryo gene expression in a mouse model of diabetic embryopathy. *Diabetologia* **2012**, *55*, 245–254. [CrossRef] [PubMed]

59. Lee, H.Y.; Wei, D.; Loeken, M.R. Lack of metformin effect on mouse embryo AMPK activity: Implications for metformin treatment during pregnancy. *Diabetes Metab. Res. Rev.* **2014**, *30*, 23–30. [CrossRef] [PubMed]

60. Ahmadimoghaddam, D.; Zemankova, L.; Nachtigal, P.; Dolezelova, E.; Neumanova, Z.; Cerveny, L.; Ceckova, M.; Kacerovsky, M.; Micuda, S.; Staud, F. Organic cation transporter 3 (OCT3/SLC22A3) and multidrug and toxin extrusion 1 (MATE1/SLC47A1) transporter in the placenta and fetal tissues: Expression profile and fetus protective role at different stages of gestation. *Biol. Reprod.* **2013**, *88*, 55. [CrossRef] [PubMed]

61. Ijas, H.; Vaarasmaki, M.; Morin-Papunen, L.; Keravuo, R.; Ebeling, T.; Saarela, T.; Raudaskoski, T. Metformin should be considered in the treatment of gestational diabetes: A prospective randomised study. *BJOG* **2011**, *118*, 880–885. [CrossRef] [PubMed]

62. Nanovskaya, T.N.; Nekhayeva, I.A.; Patrikeeva, S.L.; Hankins, G.D.; Ahmed, M.S. Transfer of metformin across the dually perfused human placental lobule. *Am. J. Obstet. Gynecol.* **2006**, *195*, 1081–1085. [CrossRef] [PubMed]

63. World Health Organization. *WHO Recommendations for Prevention and Treatment of Pre-Eclampsia and Eclampsia*; WHO Press: Geneva, Switzerland, 2011; p. 38, ISBN 9789241548335.

64. Gilbert, J.S.; Ryan, M.J.; LaMarca, B.B.; Sedeek, M.; Murphy, S.R.; Granger, J.P. Pathophysiology of hypertension during preeclampsia: Linking placental ischemia with endothelial dysfunction. *Am. J. Physiol. Heart Circ. Physiol.* **2008**, *294*, H541–H550. [CrossRef] [PubMed]

65. Conrad, K.P.; Benyo, D.F. Placental cytokines and the pathogenesis of preeclampsia. *Am. J. Reprod. Immunol.* **1997**, *37*, 240–249. [CrossRef] [PubMed]

66. Gilbert, J.S.; Babcock, S.A.; Granger, J.P. Hypertension produced by reduced uterine perfusion in pregnant rats is associated with increased soluble fms-like tyrosine kinase-1 expression. *Hypertension* **2007**, *50*, 1142–1147. [CrossRef] [PubMed]

67. Santner-Nanan, B.; Peek, M.J.; Khanam, R.; Richarts, L.; Zhu, E.; Fazekas de St Groth, B.; Nanan, R. Systemic increase in the ratio between Foxp3+ and IL-17-producing CD4+ T cells in healthy pregnancy but not in preeclampsia. *J. Immunol.* **2009**, *183*, 7023–7030. [CrossRef] [PubMed]

68. Norris, W.; Nevers, T.; Sharma, S.; Kalkunte, S. Review: hCG, preeclampsia and regulatory T cells. *Placenta* **2011**, *32*, S182–S185. [CrossRef] [PubMed]

69. Saito, S.; Nakashima, A.; Shima, T.; Ito, M. Th1/Th2/Th17 and regulatory T-cell paradigm in pregnancy. *Am. J. Reprod. Immunol.* **2010**, *63*, 601–610. [CrossRef] [PubMed]

70. Gualdoni, G.A.; Mayer, K.A.; Goschl, L.; Boucheron, N.; Ellmeier, W.; Zlabinger, G.J. The AMP analog AICAR modulates the Treg/Th17 axis through enhancement of fatty acid oxidation. *FASEB J.* **2016**, *30*, 3800–3809. [CrossRef] [PubMed]

71. World Health Organization. *WHO Recommendations on Interventions to Improve Preterm Birth Outcomes*; WHO Press: Geneva, Switzerland, 2015; ISBN 9789241508988.

72. Godfrey, K.M. Maternal regulation of fetal development and health in adult life. *Eur. J. Obstet. Gynecol. Reprod. Biol.* **1998**, *78*, 141–150. [CrossRef]

73. Cao, K.; Zheng, A.; Xu, J.; Li, H.; Liu, J.; Peng, Y.; Long, J.; Zou, X.; Li, Y.; Chen, C.; et al. AMPK activation prevents prenatal stress-induced cognitive impairment: Modulation of mitochondrial content and oxidative stress. *Free Radic. Biol. Med.* **2014**, *75*, 156–166. [CrossRef] [PubMed]

74. Tain, Y.L.; Hsu, C.N. Developmental programming of the metabolic syndrome: Can we reprogram with resveratrol? *Int. J. Mol. Sci.* **2018**, *19*. [CrossRef] [PubMed]

75. Tain, Y.L.; Lin, Y.J.; Sheen, J.M.; Lin, I.C.; Yu, H.R.; Huang, L.T.; Hsu, C.N. Resveratrol prevents the combined maternal plus postweaning high-fat-diets-induced hypertension in male offspring. *J. Nutr. Biochem.* **2017**, *48*, 120–127. [CrossRef] [PubMed]

76. Kanasaki, K.; Palmsten, K.; Sugimoto, H.; Ahmad, S.; Hamano, Y.; Xie, L.; Parry, S.; Augustin, H.G.; Gattone, V.H.; Folkman, J.; et al. Deficiency in Catechol-*O*-methyltransferase and 2-methoxyoestradiol is associated with pre-eclampsia. *Nature* **2008**, *453*, 1117–1121. [CrossRef] [PubMed]

77. Htun, N.C.; Miyaki, K.; Song, Y.; Ikeda, S.; Shimbo, T.; Muramatsu, M. Association of the Catechol-*O*-methyl transferase gene Val158Met polymorphism with blood pressure and prevalence of hypertension: Interaction with dietary energy intake. *Am. J. Hypertens.* **2011**, *24*, 1022–1026. [CrossRef] [PubMed]

78. Xiu, L.; Lin, M.; Liu, W.; Kong, D.; Liu, Z.; Zhang, Y.; Ouyang, P.; Liang, Y.; Zhong, S.; Chen, C.; et al. Association of DRD3, COMT, and SLC6A4 gene polymorphisms with type 2 diabetes in southern chinese: A hospital-based case-control study. *Diabetes Technol. Ther.* **2015**, *17*, 580–586. [CrossRef] [PubMed]

79. Kanasaki, M.; Srivastava, S.P.; Yang, F.; Xu, L.; Kudoh, S.; Kitada, M.; Ueki, N.; Kim, H.; Li, J.; Takeda, S.; et al. Deficiency in catechol-o-methyltransferase is linked to a disruption of glucose homeostasis in mice. *Sci. Rep.* **2017**, *7*, 7927. [CrossRef] [PubMed]

80. Ueki, N.; Kanasaki, K.; Kanasaki, M.; Takeda, S.; Koya, D. Catechol-*O*-Methyltransferase Deficiency leads to hypersensitivity of the pressor response against angiotensin II. *Hypertension* **2017**, *69*, 1156–1164. [CrossRef] [PubMed]

81. Deji, N.; Kume, S.; Araki, S.; Isshiki, K.; Araki, H.; Chin-Kanasaki, M.; Tanaka, Y.; Nishiyama, A.; Koya, D.; Haneda, M.; et al. Role of angiotensin II-mediated AMPK inactivation on obesity-related salt-sensitive hypertension. *Biochem. Biophys. Res. Commun.* **2012**, *418*, 559–564. [CrossRef] [PubMed]

82. Hernandez, J.S.; Barreto-Torres, G.; Kuznetsov, A.V.; Khuchua, Z.; Javadov, S. Crosstalk between AMPK activation and angiotensin II-induced hypertrophy in cardiomyocytes: The role of mitochondria. *J. Cell Mol. Med.* **2014**, *18*, 709–720. [CrossRef] [PubMed]

MDPI

St. Alban-Anlage 66

4052 Basel

Switzerland

Tel. +41 61 683 77 34

Fax +41 61 302 89 18

www.mdpi.com

International Journal of Molecular Sciences Editorial Office

E-mail: ijms@mdpi.com

www.mdpi.com/journal/ijms

www.ingramcontent.com/pod-product-compliance
Lightning Source LLC
Chambersburg PA
CBHW051703210326
41597CB00032B/5359